T0178376

Technical Thermodynamics for Engineers

Achim Schmidt

Technical Thermodynamics for Engineers

Basics and Applications

 Springer

Prof. Dr.-Ing. Achim Schmidt
Fakultät Technik und Informatik
HAW Hamburg
Hamburg, Germany

ISBN 978-3-030-20399-3   ISBN 978-3-030-20397-9   (eBook)
https://doi.org/10.1007/978-3-030-20397-9

© Springer Nature Switzerland AG 2019
This work is subject to copyright. All rights are reserved by the Publisher, whether the whole or part
of the material is concerned, specifically the rights of translation, reprinting, reuse of illustrations,
recitation, broadcasting, reproduction on microfilms or in any other physical way, and transmission
or information storage and retrieval, electronic adaptation, computer software, or by similar or dissimilar
methodology now known or hereafter developed.
The use of general descriptive names, registered names, trademarks, service marks, etc. in this
publication does not imply, even in the absence of a specific statement, that such names are exempt from
the relevant protective laws and regulations and therefore free for general use.
The publisher, the authors and the editors are safe to assume that the advice and information in this
book are believed to be true and accurate at the date of publication. Neither the publisher nor the
authors or the editors give a warranty, expressed or implied, with respect to the material contained
herein or for any errors or omissions that may have been made. The publisher remains neutral with regard
to jurisdictional claims in published maps and institutional affiliations.

This Springer imprint is published by the registered company Springer Nature Switzerland AG
The registered company address is: Gewerbestrasse 11, 6330 Cham, Switzerland

Dedicated to my parents.

Preface

Arnold Sommerfeld, the famous physicist, said:

> *Thermodynamic is a funny subject. The first time you go
> through it, you do not understand it at all. The second time
> you go through it, you think you understand it, except for one
> or two small points. The third time you go through it, you
> know you do not understand it, but by that time you are so
> used to it, it does not bother you anymore.*

In fact, thermodynamics is probably one of the most difficult and challenging subjects of mechanical engineering studies. Many students claim that it would be a great subject—if only it weren't for the written exam.

Now that I have given the lectures in *Technical Thermodynamics I/II* for several semesters at the University of Applied Sciences in Hamburg, I have decided to write an accompanying textbook. I believe that my personal gaps of understanding have become smaller. According to Arnold Sommerfeld, I'm one of those who've dealt with thermodynamics for at least the third time. Interestingly enough, my fascination for the subject is still growing. In addition, each semester there are a couple of questions from the students that cause me to constantly question the theory and that give me a new perspective on the subject. I have collected these and my questions during my studies and tried to summarise the answers as well as possible in this textbook.

Why another textbook—there are already so many available! During the preparation of my lectures I had the impression that there are many great textbooks on the subject, each with very specific merits. Nevertheless, my approach is to combine what I find to be best understandable and to go in the depths where it is required, always in order not to lose the red, thermodynamic thread. Numerous questions of the students have been helpful, especially in the context of exam preparations. So this book is a reference guide that has the same structure as my lecture. Although English is not my mother tongue, it was very important to me to write the book in English: First, I reach a larger readership than if the book were in German. Second, linguistic limitations force me to explain even complicated things in a simple way.

This book is categorised into three parts: Part I introduces the fundamentals of technical thermodynamics. First and second law of thermodynamics will be derived that enable us to understand the principle of energy conversion. Fluids are simply treated as ideal gases or incompressible liquids. The physical description of these fluids obey equations of state. Thermodynamic cycles will be discussed that convert thermal energy into mechanical energy, e.g. an internal combustion engine. Furthermore, thermodynamic cycles can be utilised to shift thermal energy from a low temperature level to a larger temperature level, e.g. a fridge. In Part II, real fluids will be investigated, i.e. fluids that can change their aggregate state for example. These fluids cannot be treated as the fluids from Part I. Furthermore, mixtures of fluids are introduced, e.g. humid air as a mixture of dry air and water. Changes of state of these mixtures will be treated as well. However, these mixtures will not be chemically reactive. Finally, Part III includes chemically reactive fluids. This is required to calculate combustion processes for instance. Combustion processes are part of many technical applications.

I have tried to find a good mixture of theory, examples and tasks. Since technical drawings are the language of the engineer, this book contains a large number of detailed illustrations which are intended to clarify even the most difficult aspects of theory.

Finally, any of your feedback is highly appreciated!

Hamburg, Germany   Prof. Dr.-Ing. Achim Schmidt
Winter 2018/19

Contents

Nomenclature

Roman Symbols

A	Area
a	Cohesion pressure, see equation 18.38
a	Mass fraction ashes
a	Velocity of sound
a	Acceleration
b	Co-volume, see equation 18.38
$\dot{B}_x$	Flux of anergy
$B(T)$	Virial coefficient
B_x	Anergy
b_x	Specific anergy
C	Capacity
C	Constant
C	Molar heat capacity $C = c_M$
C	Number of components
c	Mass fraction carbon
c	Specific heat capacity
$\bar{c}$	Specific, averaged heat capacity
a	Velocity of sound
c	Velocity
$C(T)$	Virial coefficient
c_i	Molarity of a component i
$D(T)$	Virial coefficient
D, d	Diameter
E	Energy
$\Delta \dot{E}_{x,V}$	Flux of loss of exergy
$\Delta E_{x,V}$	Loss of exergy
$\Delta e_{x,V}$	Specific loss of exergy
$\dot{E}_x$	Flux of exergy

$E_{xm,F}$	Molar specific absolute exergy of the fuel
E_{xm}	Molar specific absolute exergy
E_x	Exergy
e_x	Specific exergy
F	Degree of freedom
F	Force
f	Specific Helmholtz energy
G	Gibbs enthalpy
g	Acceleration of gravity, $g = 9.81\,\frac{m}{s^2}$
g	Specific Gibbs enthalpy
G_m	Molar specific Gibbs enthalpy
H	Enthalpy
h	Mass fraction hydrogen
h	Specific enthalpy
Δh_v	Specific enthalpy of vaporisation
$\Delta_B^0 H_m$	Molar specific enthalpy of formation at standard conditions
$\Delta_R^0 H_m$	Molar specific enthalpy of reaction at standard conditions
H_m	Molar specific absolute enthalpy
H, h	Height in a gravity field
H_{0M}	Molar specific upper heating value
H_{0v}	Volume specific upper heating value
H_0	Mass specific upper heating value
H_{UM}	Molar specific lower heating value
H_{Uv}	Volume specific lower heating value
H_U	Mass specific lower heating value
k	Heat transition coefficient
k_B	Boltzmann constant, $k_B = 1.3806 \times 10^{-23}\,\frac{J}{K}$
k_F	Spring constant
L_{min}	Minimum molar-specific air need
l_{min}	Minimum mass-specific air need
M	Molar mass
M	Torque
m	Mass
$\dot{m}$	Mass flux
$\dot{m}''$	Mass flux density
Ma	Mach-number
n	Mass fraction nitrogen
n	Molar quantity
n	Polytropic exponent
n	Speed
$\dot{n}$	Molar flux
$\vec{n}$	Normal
N_A	Avogadro constant, $N_A = 6.022\,045 \times 10^{23}\,\frac{1}{mol}$

o	Mass fraction oxygen
O_{min}	Minimum molar-specific oxygen need
o_{min}	Minimum mass-specific oxygen need
P	Number of phases
P	Power
p	Pressure
p_i	Partial pressure of a component i
Q	Electric charge
Q	Heat or thermal energy
$\dot{Q}$	Heat flux
q	Specific heat
R	Individual gas constant
R, r	Radius
R_M	General gas constant, $R_M = 8.3143 \frac{kJ}{kmol\,K}$
S	Entropy $\left(\frac{J}{K}\right)$
s	Mass fraction sulphur
$\Delta_R S_m$	Molar specific entropy of a reaction
$\dot{S}_a$	Flux of entropy carried with heat
$\dot{S}_i$	Flux of entropy generation
s	Distance
s	Specific entropy
s_a	Specific entropy carried with heat
s_i	Specific entropy generation
S_m	Molar specific absolute entropy
T	Absolute thermodynamic temperature
t	Time (s)
T_r	Reduced thermodynamic temperature according to equation 18.63
U	Internal energy
v	Specific internal energy
U_m	Molar specific absolute internal energy
V	Voltage
V	Volume
v	Specific volume
$\dot{V}$	Volume flux
W	Work
w	Mass fraction water
w	Specific work
x	Coordinate
x	Molar concentration of a component i
x	Molar fraction
x	Vapour ratio
x	Water content
y	Coordinate

y	Specific pressure work
Z	Compressibility factor
Z	Extensive state value
z	Coordinate
z	Distance
z	Specific state value, $z = \frac{Z}{m}$

Greek Symbols

α	Abbreviation
α	Angle
α	Heat transfer coefficient
β	Isobaric volumetric thermal expansion coefficient
δ_h	Isenthalpic throttle coefficient
δ_T	Isothermal throttle coefficient
ϵ	Compression ratio
η	Abbreviation
η	Efficiency
γ_i	Stoichiometric factor of a component i
κ	Isentropic coefficient
λ	Air-fuel equivalence ratio
μ	Chemical potential
μ_i	Mass-specific exhaust gas composition of component i
v_i	Molar-specific exhaust gas composition of component i
Ω	Flow function of a nozzle
Ω	Statistical weight, measure of the probability
ω	Acentric factor
Ψ	Dissipation
ψ	Relative saturation
$\dot{\psi}$	Flux of dissipation
ψ	Specific dissipation
ψ'	Specific dissipation per length
ψ_i	Volume ratio of a component i
ρ	Density
ρ_i	Partial density of a component i
σ_i	Volume concentration of a component i
ε	Coefficient of performance
φ	Relative humidity
ϑ	Celsius-temperature
ξ	Abbreviation
ξ_i	Mass concentration of a component i

Acronyms

$\prime$	Saturated liquid state, saturated humid air
$\prime\prime$	Saturated vapour state
δ	Process value
d	State value
1P	Single-phase
2P	Two-phase
C	Carnot
CM	Cold machine/fridge
cp	Critical point
HP	Heat pump
HP	High pressure
HT	Heat transfer
HVAC	Heating, ventilation and airconditioning technology
LP	Low pressure
Pr	Product
TE	Thermal engine
TP	Triple point

Subscripts

A	Air (wet)
a	Air (dry)
a	Outer
C	Cylinder
comp	Compression
cond.	Condenser
eff	Effective
EG	Exhaust gas
el	Electric
env	Environment
F	Fuel
fric.	Friction
G	Gas
gas	Gas
HP	High pressure
ice	Ice
in	Inlet
irrev	Irreversible
K	Control volume
kin	Kinetic

L	Liquid
liq	Liquid
LP	Low pressure
m	Mean
m	Melting
M,m	Molar
max	Maximum
Mech	Mechanic
min	Minimum
MP	Medium pressure
n	Narrowest cross section
out	Outlet
P	Piston
p	$p = $ const.
pot	Potential
R	Reservoir
ref	Reference state
rev	Reversible
S	Steel
s	Saturated
shift	Shifting
Source	Source term
spr	Spring
swing	Stroke
Sys	System
t	Technical
T	$T = $ const.
th	Thermal
total	Total
V	Volume
v	$v = $ const.
V,v	Vapour
W	Water

List of Figures

List of Tables

Chapter 1
Introduction

Thermodynamics is one of the most difficult and probably even one of the most challenging disciplines in mechanical engineering. However, thermodynamics is necessary to understand the principles of energy conversion, e.g. when designing thermal machines such as gas turbines, internal combustion engines or even fuel cells in modern applications. The fundamental principles of energy conversion are actually what thermodynamics is about: It is well known from physics lessons, that energy is conserved, i.e. energy can not be generated or destroyed. This is what the first law of thermodynamics summarises. Furthermore, energy conversion, e.g. from thermal energy to mechanical energy, is limited. A conventional combustion engine converts chemical bonded energy into mechanical energy, but as we know, part of this converted energy has to be released to the environment, e.g. by a cooler. This limitation, however, obeys the second law of thermodynamics.

Although the fundamentals of thermodynamics date back to the 18th century, their principles still apply today. The word *thermodynamics* is composed of the Greek words *therme* (heat) and *dynamis* (force): It covers the theory of of *energy* and its convertibility. Even though the word *dynamics* suggests that systems are in motion, this book deals with so-called equilibrium states.

As already mentioned, the conversion of energy plays an important role for mechanical engineers today. Examples from everyday life are:

- Vehicles with internal combustion engines convert chemical bonded energy of the fuel to mechanical energy for the powertrain. This is what mobility looks like at the beginning of the 21st century. For various reasons, it makes sense to strengthen renewable energies and make electric driving possible. New technologies, in particular electric storage systems, are needed to supply the electric drives with energy. It is also possible to recover energy, e.g. when braking the vehicle. The conversion of energy thus also plays a decisive role in new modern drive systems.
- Thermal power plants fire fossil fuels in order to generate electricity. Hence, chemically bound energy, i.e. fossil fuels, is converted into thermal energy in the com-

© Springer Nature Switzerland AG 2019
A. Schmidt, *Technical Thermodynamics for Engineers*,
https://doi.org/10.1007/978-3-030-20397-9_1

bustion chamber of a power plant. This thermal energy can then be converted into electrical energy in the turbine generator unit.

- Solar thermal power plants utilise solar energy to generate heat, which is converted into electricity via turbines.

All these examples are intended to show, that thermodynamics, as an important basic subject, is indispensable for the energy turnaround that has already been initiated. The sustainable energy supply of a constantly growing global population is one of the future megatrends. The aim of thermodynamics is to provide the laws of energy conversion. In addition, engineers are to be enabled to develop and design technical applications for future energy conversion machines.

1.1 How Is This Book Structured?

The structure of the book is derived from the module descriptions that can be found at German universities in the field of thermodynamics. Due to the complexity of the subject, the lecture is divided into two parts, namely *Technical Thermodynamics I* and *Technical Thermodynamics II*. However, I have decided to divide the book into three parts.

Part I deals with the basics of the topic: A conceptual clarification of thermodynamic systems and their states is given. Simple systems are investigated in thermodynamic equilibrium. Heat and work as process variables are introduced, which can drive a system into a new state of equilibrium. The first law of thermodynamics, the energy conservation principle that every system must follow, is applied. This principle makes it possible to understand energy transformations, e.g. from thermal energy to mechanical energy in a so-called thermal engine. In order to optimise the conversion rate, the second law of thermodynamics and thus entropy as a new state variable is indispensable. Thermodynamic cycles are also dealt with. Liquids in Part I are ideal, i.e. they behave either like an ideal gas or like an incompressible liquid. The thermodynamics of transient processes, e.g. the cooling of a system to a new state of equilibrium, is also covered with in Part I. The didactics of the first part follow Erich Hahne, see [1], whose textbook is excellently suited and designed for access to the subject of technical thermodynamics. I tried to modify the structure of the first chapters until I realised that his structure was already perfect - so I decided to leave it unchanged. It was also Erich Hahne who inspired me through personal conversations during my first semester at Hamburg University of Applied Sciences. I am therefore very grateful to him.

While liquids in Part I are regarded as ideal gases or incompressible liquids, Part II deals with fluids that can change their state of aggregation. For example, water can change its state of aggregation by supplying or releasing thermal energy, e.g. it freezes by cooling so that liquid water becomes a solid. The physical properties, i.e. the state values, vary with each change in the aggregate state. These so-called real fluids do no longer obey the thermal/caloric equations of state, that have been applied in Part I for ideal gases. The task of thermodynamics is to quantify the state of the system and to provide means for calculating the change of state. Part II also includes gas mixtures, i.e. the presence of several gaseous components without being chemically reactive. An example of a gas mixture is humid air as a mixture of dry air and vaporous water. The amount of steam in humid air depends on its temperature: as the temperature drops, liquid water or even ice can form. Clockwise or counterclockwise thermodynamic cycles, which are operated with real fluids, are introduced and calculated. Supersonic flows, usually part of the field of *Gas Dynamics*, are briefly discussed.

Part III finally implies chemically reactive systems and will complete this textbook. Focus will be on fossil fuels: Any combustion of fossil fuels, i.e. fuels containing hydrogen and carbon, with air usually leads to carbon dioxide, water, nitrogen and oxygen. Thus, the resulting flue gas is a mixture of various components, has it has been treated in Part II. For the design of a combustion process, it is important to determine the state of aggregation of the water in the exhaust gas. Condensed water for instance can easily lead to corrosive liquids, which cause several problems in technical systems. Furthermore, the specific lower/upper heating value is utilised to predict exhaust gas temperature as well as released heat. An alternative approach, following the principles of absolute enthalpy/entropy, is investigated in the last section of Part III. This approach allows to describe any chemical reaction energetically and to apply the second law of thermodynamics in order to specify the irreversibility of chemical reactions. However, the application of this approach allows the investigation of modern energy conversion systems, such as fuel cell systems.

Finally, Fig. 1.1 shows the classical fields of conventional technical thermodynamics. All three fields are part of this book and are introduced briefly below.

This books contains several examples and exercises. However, from time to time, especially in Part I, some of these examples require the knowledge of upcoming chapters. Don't let that demotivate you. Access to the questions is obtained by a second study of the book after the basic philosophy of thermodynamics is understood.

State of a System/Phase Change

It turns out that it is imperative to define a *system* and its *state* before energy conversion can be studied. A system can be open or closed with regard to the exchange of mass across system boundaries. In addition, energy in the form of work or heat can exceed the system boundary. Anyhow, the state of a system can be changed by external influences.

For example, the state of a mass in a gravitational field can be described by its temperature and its position in a gravitational field. By lifting the mass, i.e. by an external intervention, the vertical coordinate changes. As a result, the potential energy has increased. If this body falls to the ground, it is subjected to a (constant) acceleration and potential energy is converted into kinetic energy. When the mass hits the ground, its temperature rises.

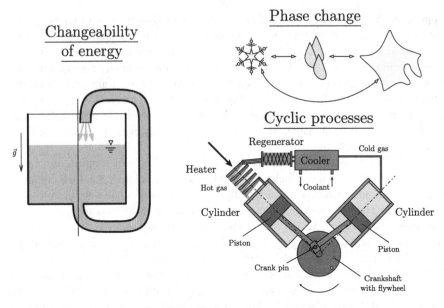

Fig. 1.1 What is thermodynamics all about? (Tasks of thermodynamics)

For more complex systems, it is not only the temperature or vertical position that describes the state, but other additional information such as pressure that is needed to accurately describe the system.

It is known from everyday experience that gases can be compressed as the pressure increases. At low temperatures and high pressures, gases can also be liquefied. This phase change cannot be treated with the theory of ideal gases and is discussed in Part II.

Changeability of Energy

The physical principle of energy conservation, which reads as the first law of thermodynamics, is introduced and applied to calculate the change of energy from one form to another. For example, an idealised pendulum converts potential energy into kinetic energy and vice versa.

In the first part it is derived which restrictions the energetic conversion is subject to. It is shown that it is impossible to design a machine that, for example, completely converts heat into work: It is well known that a coal-fired power plant cannot run without cooling towers, which obviously release thermal energy into the environment. The necessary cooling reduces the benefit of such a power plant, as the chemical energy in the fuel is not completely converted into electricity.

For an engineer, the question inevitably arises as to the maximum efficiency at which an ideal power plant can be operated. The derivation of this maximum efficiency leads to a new state value called entropy. Like all state values, the entropy within a system must be constant if the system is in a so-called steady state, i.e. if there is no temporal change in any state value. Entropy can be transported convectively through the flow, it can be transported by heat exceeding the system boundary, and it can be generated internally by dissipation, i.e. internal friction. Internal friction or dissipation is a measure of the irreversibility of a process. The second law of thermodynamics finally answers the question why cooling is necessary in a thermal power plant. A thermal engine converts heat into a mechanical/electrical energy, while another part is released as heat into the environment. Once this mechanism is understood, the exergy is introduced as that part of the total energy that can be converted into any other form of energy, e.g. electricity. The temperature of the combustion process on the one hand and the ambient temperature on the other determine the maximum efficiency.

Cyclic Processes

However, the examples of a coal-fired power plant or a combustion engine show an important field of work for the thermodynamic engineer: Heat is converted into mechanical/electrical work to operate a machine. These processes are called clockwise cycles or thermal engines. Such a process is shown in Fig. 1.1, which shows a Stirling machine. Unlike clockwise cycles, a counterclockwise cycle is a refrigerator or heat pump that uses ambient heat at low temperatures to heat a building.

1.2 Classification of Thermodynamics

Common thermodynamic textbooks are usually divided into three categories, which are briefly presented. Mechanical engineering students learn the basics of *technical thermodynamics* and, depending on their specialisation, *chemical thermodynamics*. For both phenomenological categories there are a number of textbooks, both in German and in English language. Students of natural sciences, on the other hand, deal with *statistical thermodynamics or statistical mechanics*, which is not thematised intensively in this book.

1.2.1 Technical Thermodynamics

In this area of thermodynamics, engineering tasks, i.e. the conversion and use of energy in thermal machines, are focused. Thermal machines, such as combustion engines and gas turbines, are presented in this book and are characterised by the conversion of thermal energy into mechanical energy. These machines operate as clockwise thermodynamic cycles. Refrigerators and heat pumps stand for so-called counterclockwise cycles.

1.2.2 Statistical Thermodynamics

The statistical thermodynamics examines the properties of systems consisting of many atoms or molecules. An attempt is made to calculate macroscopic state values (pressure, temperature, entropy, etc.) by means of statistics. Many findings of this statistical approach form the basis for the technical thermodynamics. In Part I the focus is on the so-called ideal gas and the corresponding ideal gas equation, that can be derived primarily from the statistical thermodynamics. According to Boltzmann the entropy S has a historical significance: Entropy can be interpreted as a statistical energy distribution. Hence, it is proportional to the logarithm of the probability of a state:

$$S = k_B \ln (\Omega) \qquad (1.1)$$

The statistical thermodynamics is not part of this book. For further details reference is made to Bošnjaković and Knoche, see [8].

1.2.3 Chemical Thermodynamics

The chemical thermodynamics describes chemical and physical/chemical processes. Degree respectively limits of conversion, i.e. dynamic chemical equilibrium, and the speed of reaction are, for example, of interest. A chemical reaction is a material conversion of reactants into reaction products, that have different material properties compared to the educts. Thus, a chemical reaction is characterised by heat and mass transfer. The chemical thermodynamics distinguishes between homogeneous and heterogeneous chemical equilibrium reactions. A homogeneous equilibrium occurs when the reaction products are present only in a single phase state, e.g. gaseous, liquid or solid. A heterogeneous equilibrium is reached when the reaction products do not only occur in one phase, but appear, e.g in liquid and gaseous state. Typical combustion engines, which are designed among others by a thermodynamic engineer, are internal combustion engines, jet engines, boilers and rocket engines. Methods of the chemical thermodynamics are applied in Chaps. 23 and 24. Hence, chemical thermodynamics is partially introduced in this book, see Part III.

1.3 Distinction Thermodynamics/Heat Transfer

In addition to thermodynamics, heat and mass transfer is also part of the course in *thermal engineering*. This raises the question of how these two subjects differ from each other. The main aspects are shown in Fig. 1.2 and are explained in the following.

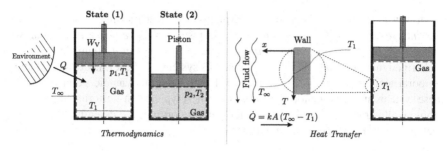

Fig. 1.2 Distinction thermodynamics (left)/heat transfer (right)

1.3.1 Thermodynamics

Thermodynamics defines systems and their state, that can be indicated by temperature, pressure, internal energy and several more. It investigates how a system's state can be modified by process values, i.e. external impacts respectively influences, that can act across the system boundary. These process values are work and heat, i.e. both forms of energy. So, after a long period of time, a system, that has been initially in equilibrium state, reaches a new state of equilibrium. However, in thermodynamics, the system itself is treated as homogeneous, i.e. there are no gradients: The temperature for instance is uniform, see Fig. 1.2.

Unfortunately, thermodynamics does not tell, how a system can release heat or how to supply heat to a system. The physical transport mechanisms are not explained in detail. Especially, to initiate heat transfer a temperature gradient is required. Thus, strictly speaking, a homogeneous system is not able to transfer heat. Though, energy conversion can be explained by means of thermodynamics, it is not possible to design a heat exchange in detail for example. Thermodynamics teaches what impact heat has on a system, but it does not explain, how heat can be supplied or released.

1.3.2 Heat Transfer

Thermodynamic explains the correlation between a thermal potential, i.e. temperature difference, and a resulting heat flux. Hence, its a physical principle of cause and effect. Any temperature difference means imbalance. Intensive state values, i.e. pressure and temperature, have in common to cause a transient balancing process.[1] Thermodynamics teaches, that this balancing process comes along with generation of entropy—the larger the imbalance is, the more entropy is generated.

---

[1]Everyone can observe this process, when a hot cup of coffee is placed in the ambient. There is an imbalance and the system *coffee* drives in a new equilibrium state with its *environment*. After a while coffee and environment are thermally balanced - indicated by a cold cup of coffee having ambient temperature.

However, the field of heat transfer explains how heat transfer works in detail: *Conductive heat transfer* in systems being in rest, *convective heat transfer* that is linked with fluid dynamics and *radiation*, that even works in vacuum. In contrast to thermodynamics, that handles homogeneous systems, in heat transfer spatial and temporal temperature profiles are treated. From a thermodynamic point of view, these systems are heterogenous, i.e. non-uniform. A typical non-uniform temperature profile is illustrated in Fig. 1.2. These temperature profiles are essential to transfer heat and are not covered by thermodynamics.

1.4 History of Thermodynamics

1.4.1 The Caloric Theory Around 1780

The french chemist *Lavoisier*[2] postulated around the year 1783 the so-called *Caloric theory*. Hence, he assumed that heat is an invisible and weightless substance, called "caloric", which is located in the material between the molecules/atoms. According to Lavoisier's approach, caloric can be transferred from one body to a second body, which expands and warms up. The transfer is initiated due to an internal repulsive force of caloric. Lavoisier assumed that caloric, like the elements, can not be destroyed or generated. Thus, the amount of caloric remains constant. Lavoisier was able to explain the effects of phase change with his theory: Supply of caloric causes a phase change from solid to liquid. The reversal change of state, i.e. from liquid to solid state, can be achieved by release of caloric.

The theory is, from today's perspective, outdated as clarified by the following simple example. Rubbing two bodies against each other, "generates heat" at the contact point of the bodies. This frictional process can be sustained over a long period of time without exhausting a finite reservoir of caloric. Thus, this is a contradiction to Lavoisier's theory. The process can rather be described with modern thermodynamics in the form of the first law of thermodynamics: According to the first law, frictional work causes an increase of the bodies' internal energy, that can be measured by a rising temperature.[3] However, heat is a process value,[4] which only occurs when two bodies being in contact have different temperatures. An exchange of energy, i.e. heat, between the two bodies can be observed until a thermal equilibrium between the bodies arrives. At this thermal equilibrium both bodies have the same temperature. However, the phenomena, that heat supply respectively release can cause a phase state, still corresponds to the caloric theory.

---

[2]Antoine Laurent de Lavoisier (∗26 August 1743 in Paris, †8 May 1794 in Paris).

[3]Mechanical energy is converted into dissipation/frictional energy, that increases the body's internal energy.

[4]In thermodynamics it is important to distinguish between state values and process values.

1.4.2 Thermodynamics as From the 18th Century

To show the history of modern thermodynamics, hereinafter important scientists on the field of thermodynamics are introduced. Their contribution to thermodynamics is first given briefly. Nevertheless, their theories are explained in detail in this book later on.

The french engineer officer *N.L.S. Carnot*[5] is considered to be one of the founders of modern thermodynamics.

- He postulated the first modern theory of how work can be gained from heat. These days, this principle is realised in every modern thermal engine. A thermal engine converts energy in form of heat, e.g. released by firing fossil fuels, into mechanical work.
- Thermodynamic cycles are important in modern technology. They are characterised by a working medium that runs through several changes of state until it finally reaches its initial state once again. Cycles can, as already shown, be designed for different purposes. Thus, in thermal engines, see above, heat is converted into work.[6] Later on, the thermal engine is called a clockwise cycle. Carnot raised the question how this cycle can be optimised in order to achieve the maximum work. However, Carnot postulated that the thermal engine has to avoid any internal friction—in thermodynamics the word dissipation is used to quantify internal friction. Additionally, he found how heat, i.e. the driver of a clockwise cycle, needs be supplied. Carnot derived a correlation for the maximum efficiency of reversible cycles, i.e. cycles free of dissipation. This phenomena is discussed in detail when the second law of thermodynamics is introduced.

The British private scholar *J.P. Joule*[7] from Manchester and the German physician *J.R. Mayer*[8] from Heilbronn refined the theory of heat.

- According to the principle of the equivalence of heat and work, heat is no longer an indestructible substance, as was assumed for the *caloric*, see Sect. 1.4.1. Rather, it is possible to convert work into heat and vice versa. This is an important finding in thermodynamics. Heat and work influence the state of a system in the same matter, they are convertible into each other. In the 19th century the principle of equivalence replaced the caloric theory. Finally this principle led to the law of the conservation of energy, also known as first law of thermodynamics.
- Joule proved the so-call "mechanical equivalence of heat" with a simple experiment. This experiment is known as Joule's Paddle Wheel and is explained in Chap. 2: by supply of mechanical work in an adiabatic, i.e. perfect insulated, vessel, filled with water, an increase of temperature can be observed.

---

[5]Nicolas Léonard Sadi Carnot (∗1 June 1796 in Paris, †24 August 1832 in Paris).

[6]In this case heat is the effort, since we do have to pay for the fuel, that is needed to generate the heat. Work is the benefit, as the thermal machine is designed to gain mechanical or electrical work.

[7]James Prescott Joule (∗24 December 1818 in Salford near Manchester, †11 October 1889 in Sale (Greater Manchester)).

[8]Julius Robert von Mayer (∗25 November 1814 in Heilbronn, †20 March 1878 in Heilbronn).

- Moreover, Joule showed, that electrical energy can be converted into heat - called Joule heating. The amount of heat is proportional to the electrical resistance and proportional to the square of the current.

The German professor *R.J.E. Clausius*,[9] Professor of Physics (Germany and Switzerland) has an important share in modern thermodynamics and will also guide through this book.

- He provided the mathematical formulation of the two major laws of thermodynamics. The first law of thermodynamics predicts how process values, i.e. heat and work passing the system boundary, influence the internal state of a system.
- Thus, Clausius introduced the state value *internal energy*, which can vary due to external influences.
- With the first law of thermodynamics the theory of conversion of energy can be explained, but it does not consider the limits of energy conversion. Consequently, according to the first law, it is possible to design a machine, that permanently removes heat from a hot reservoir and converts it completely into mechanical work, e.g. a coal fired power plant without any cooling tower. Since such a power plant does not exist, a cooling as part of the energy conversion is required: However, any heat release to the environment decreases the electrical output. This universal principle, that for steady state operation any thermal machine, i.e. heat is converted into work, needs a cooling, is introduced as second law of thermodynamics. The physical explanation is given by Clausius and is strongly linked with the term entropy.
- For a quantitative formulation of the changeability of energy, known as second law of thermodynamics, Clausius introduced a new state value "equivalent value of transformation". This equivalent value was named entropy, that has a fundamental importance in thermodynamics. For the state value entropy there is no conservation law like there is for energy, mass and momentum. However, the entropy can be balanced: It can be stored in systems, generated internally by dissipation as well as transferred by a convective flux or by an exchange of heat. Chapter 13 will show how the balancing is done in detail. By the entropy the changeability of energy gets a mathematical basis. Furthermore, it explains why e.g. heat always follows a temperature gradient.[10]

Another important thermodynamicist of modern science is the Scotsman *W. Thomson*,[11] who was a professor of natural philosophy and theoretical physics in Glasgow.

- Thomson gave an alternative formulation of the second law of thermodynamics, also known as a theorem of dissipation of mechanical energy. He realised, that any irreversibilities inside a system, i.e mechanical friction or natural balancing

---

[9]Rudolf Julius Emanuel Clausius (∗2 January 1822 in Koeslin, †24 August 1888 in Bonn).

[10]Image a hot plate that is touched by your hand. From the first law's point of view it would be possible, that your hand cools down while the hot plate's temperature further rises. It is the second law of thermodynamics, that fixes the direction of heat transfer. Do not touch a hot plate!

[11]William Thomson, 1. Baron Kelvin, mostly as Lord Kelvin or Kelvin of Largs (∗26 June 1824 in Belfast, Northern Ireland, †17 December 1907 in Netherhall near Largs, Scotland).

processes, are related with the generation of entropy. It is explained, that any generation of entropy inside a system decreases its working capability. Thus, a new term *exergy*, i.e. working capability, is introduced, that is linked strongly with the entropy. Hence, the loss of exergy is a function of entropy generation, see Chap. 16.

- Entropy can be, as already mentioned, stored, transferred by heat respectively transported by mass flux and generated due to irreversibilities. Since entropy can not be destroyed, but merely transferred from the system to the ambient, e.g. by heat release to the environmemt, the global amount of entropy is constantly increasing.
- Entropy S, absolute temperature T and heat Q are fundamental parameters in thermodynamics and correlate as follows:

$$\delta Q_{\text{rev}} = T \, dS \tag{1.2}$$

This is one of the essential equations that is investigated in this book. However, several authors claim, there is another physical more accessible explanation. In comparison with other physical definitions of work one realises the common principle structure: Energy is the product from an "amount" and a potential. In case of the potential energy, the amount is equal to the mass and the potential is given by the product of h and acceleration of gravity g. If this principle is transferred to Eq. (1.2), the entropy S should actually be treated as "amount of heat", while the thermodynamical temperature T is the potential. Finally, both form the thermal energy Q. If, according to the suggested nomenclature, a system is supplied by heat dS, this occurs against its inner thermal potential T. Thus, for the *thermal work* we get:

$$\delta W = T \, dS \tag{1.3}$$

Analogously, if a system's volume is increased by dV against its inner pressure p: The released *mechanical work* is given by

$$\delta W = p \, dV \tag{1.4}$$

The same with the *electrical work*, while Q being the electric charge and U being the voltage potential in this case:

$$\delta W = U \, dQ \tag{1.5}$$

It can be seen, that with today's definition of heat Q an interpretation of entropy S is difficult. If one had decided to take S as an equivalent for an amount of heat and Q as thermal energy, an interpretation would have been much easier and in accordance with definitions from the mechanics. However, in this book we keep the common definitions, i.e. S is the entropy, Q is the amount of heat.

In addition to the fundamental basics in the 18./19. century the first technical applications have been designed at this time. First of all the British inventor *T. Newcomen*[12] has to be mentioned.

---

[12]Thomas Newcomen (∗26 February 1663 in Dartmouth, †5 August 1729 in London).

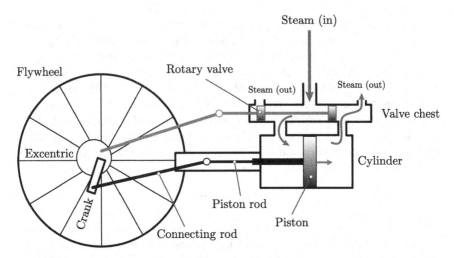

Fig. 1.3 Working principle of a steam machine

- Newcomen has already designed in 1712 the first atmospheric steam machine, that was applied in the mining industry in order to pump out the intruding water in the tunnels. His invention and the application of the steam machine is strongly related with the industrial revolution. The industrial revolution accelerated the technical progress as well as the social development and was originated in England.
- The steam machine can be regarded as the first thermal engine, that converts supplied heat into work. Figure 1.3 shows the working principle of a steam machine. The supplied heat evaporates the working fluid in a boiler, that is not shown in Fig. 1.3, first. However, the vapour is lead into the valve chest, that can can guide the steam at two positions inside a cylinder. By this a piston as well as a piston rod are set into motion so that the flywheel starts turning. By switching the steam inlet, the piston can be set into an axial oscillation, so keeps the flywheel turning. The switching of the steam inlet is achieved by a rotary valve, that is controlled by the flywheel. Finally, the thermal energy is converted into mechanical, rotational work. The efficiency of Newcomen's machine was less than 0.5 %, i.e. less than 0.5 % of the supplied thermal energy of the fuel is converted into mechanical work.
- J. Watt[13] performed several technical improvements at the machine in 1769, that lead to the English patent Nr. 913. The thermal efficiency of Watt's machine was up to 3 %.

1.4.3 Thermodynamics in the 21st Century

Even though one could get the impression, that thermodynamics is old-fashioned, its philosophy and its theoretical basis are still relevant and applicable, even in the 21st century. Thermodynamics these days still is an important subject, that is needed

---

[13] James Watt (*30 January 1736 in Greenock, †25 August 1819 in Heathfield, Staffordshire).

to design modern and innovative machines. Thus, this is the reason why mechanical engineer students still need a profound knowledge on this field. In the following section current topics, that have been designed based on thermodynamic principles, are listed. Most of the applications have been optimised over the years.

Combustion Engine

The combustion engine is a thermal engine with internal combustion, i.e. a fuel-air mixture is ignited inside a cylinder. Due to the chemical reaction heat is released. However, it differs from gas or steam turbines, since combustion engines actually work discontinuously. Strictly speaking, even the Stirling engine, which will be introduced in Chap. 17, is not an internal combustion engines, since the combustion is done externally. Over the years the internal combustion engine has been constantly developed further, as evidenced by the declining fuel consumption with rising vehicle mass. Another challenge are the ever stricter emission limits. According to the second law of thermodynamics, the engine needs to release heat to a cold reservoir in order to be operated continuously. As will be shown later, heat release should occur at ambient temperature in order to minimise the loss of exergy. Doing so maximises the temperature potential, that is the driver for the process. Nevertheless, the exhaust gas still contains energy, e.g. thermal energy since the exhaust gas has a larger temperature than the environment and kinetic energy due to the velocity of the exhaust gas. While in the past this energy leaving the vehicle was not recovered, current development try to gain exergy out of the exhaust gas. Two approaches will be presented in Sect. 1.4.4.

Gas and Steam Turbines

A combined gas/steam turbine plant is another example for a thermal engine. The exhaust heat of the gas turbine is utilised to operate a steam turbine process. Generally speaking, in a turbine a fluid expands from a high pressure to a lower pressure. Thus, thermal energy is converted into mechanical energy. Similar to other thermal engines, the thermal efficiency increases with the temperature at which heat is supplied. Obviously, while optimising a thermal engine other disciplines, e.g. material science, are required as well in order to design materials, that can withstand the increasing temperatures (Fig. 1.4).

Fossil and Nuclear Power Plants

Basically, what has been mentioned for gas and steam turbines applies for power plants as well: Continuously working thermal engine, working in-between two temperature levels, have a theoretically maximum thermal efficiency, that depends solely on the maximum (hot reservoir) and minimum (cold reservoir) temperature - called *Carnot efficiency*. In the 2000s approximately 44 % of the electricity produced in Germany comes from firing hard coal and lignite. The averaged thermal efficiency

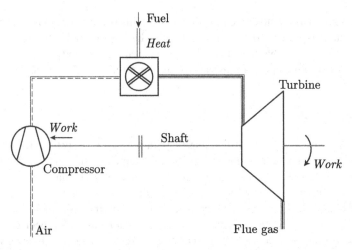

Fig. 1.4 Gas turbine (Schematic working principle)

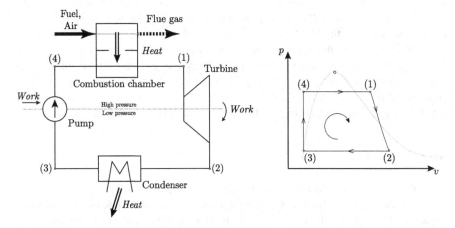

Fig. 1.5 Fossil power plant (left: Schematic working principle, right: Visualisation in a p, v-diagram, liquid water is supposed approximately to be incompressible)

of all installed power plants is approximately 38 %, compared with Watt's steam machine a considerable increase. Figure 1.5 shows the schematic working principle of a fossil power plant as well as a visualisation in a p, v-diagram. The p, v-diagram is an important tool in thermodynamics. In this case, it indicates why a thermal engine is supposed to be a clockwise cycle.

Compressors, Pumps, Blowers

These components are important components, that are required for designing thermodynamic cycles. Task of an engineer is to design most efficient components, that are, theoretically idealised, free of any irreversibilites. If these theoretical frictionless components are also adiabatic, i.e. perfectly insulated, the thermodynamicist calls this machine to be isentropic.

Propulsion Systems for Aircrafts and Rockets

Aircraft turbines follow the same principle as gas turbines. In a first step, air is compressed and lead into a combustion chamber. After the ignition of the air/fuel mixture, thermal energy is released at high pressure. Within the turbine the exhaust gas expands and thermal energy is converted in mechanical energy. The released energy can be partially used to operate the compressor. In a finally step, the exhaust gas leaves the nozzle and provides the aircraft's propulsion. For supersonic flights the nozzle needs to have a specific shape, called Laval nozzle, that will be introduced in Part II.

Combustion Systems

The fields of thermodynamics, that have already been introduced, show clearly, that many of them require a combustion process. This is the reason, why there is an entire chapter in this book teaching the theory of combustion. When designing a combustion process one needs to understand in what ratio fuel (e.g. fossil fuels like coal, gas or oil) and air have to be mixed in order to achieve the desired combustion temperature and in order to guarantee emission limits. Though this technique might be regarded as old-fashioned, since nowadays e.g. combustion engines are slightly replaced by new, emission-free systems, even new systems, i.e. fuel cells, are based on combustion effects.

Cryogenic Systems

In this area of thermodynamics processes at very low temperature are treated. These processes are essential for gas separation as well as for liquefaction of gases. A well-known technical application is the Linde process,[14] that utilises the Joule-Thomson effect. While the adiabatic throttling[15] of an ideal gas the temperature remains constant, real gases underlie the Joule-Thomson effect, that leads to a temperature decrease. A detailed explanation of this effect is given in Sect. 18.5.

---

[14]Named after the German engineer Carl von Linde (*26 June 1824 in Berndorf, †16 November 1934 in Munich).

[15]An adiabatic throttling is treated as isenthalpic - in case potential and kinetic energies are ignored!

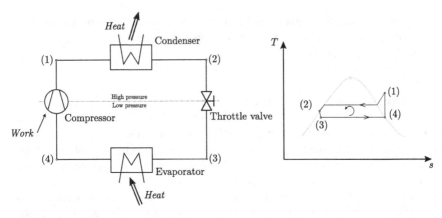

Fig. 1.6 Cooling machine (left: Schematic working principle, right: Illustration in a T, s-diagram, the compressor is supposed to be isentropic)

Heating, Ventilation and Air Conditioning (HVAC)

The air conditioning of buildings requires the knowledge of thermodynamics as well. In addition to the temperature, which can be affected by air conditioning, the humidity is another comfort criterion. For the energetic dimensioning of HVAC systems thermodynamic principles are applied. Especially in Chap. 20 we will focus on gas mixtures and humid air.

Compressions-/ and Sorption Cooling Machines

Cooling machines work according to a counterclockwise cycle, that needs to be supplied with mechanical/electrical energy. Heat is taken from a cold reservoir and finally released at a larger temperature into a hot reservoir. Figure 1.6 illustrates the principle working scheme of a cooling machine and the required components. In contrast to clockwise cycles, that are quantified with a thermal efficiency, a so-called coefficient of performance (COP) is defined for counterclockwise cycles. It is the ratio of effort and benefit. For a cooling machine the benefit is the supplied heat at low temperature, whereas the effort is the supplied mechanical/electrical energy of the required compressor. Cooling machines following the absorption principle, the mechanical operated compressor is replaced by a so-called thermal compressor.

Heat Pumps

Heat pumps do have the same technical layout as cooling machines, including the required components. However, the benefit differs: Heat pumps are commonly used for heating purpose, thus the benefit is the heat released at high temperature. Similar to the cooling machine, heat is taken from a cold reservoir, that might be ambient air, soil or even water. Since it is a contradiction to the second law of thermodynamics, i.e. heat follows the temperature gradient and does not move from cold to hot, mechanical work is required to "pump" heat from a cold reservoir into a hot reservoir. Once again the coefficient of performance characterises the efficiency this machine. Typically, the COP is greater than one. If the COP is equal to one, the machine is an instantaneous water heater. The entire heating purpose is covered electrically and thus is not the preferable solution. A COP e.g. of six means, that for a specific heating requirement, i.e. the benefit, of 6 kW a mechanical/electrical power of 1 kW needs to be applied. The rest of 5 kW is taken from the environment and does not constitute an effort.

Fuel Cell Systems

Conventional energy conversion systems are based, as described above, on combustion processes. The heat being released at the combustion can be converted in a thermal machine in order to generate mechanical/electrical work. However, in a fuel cell a so-called cold combustion, i.e. a chemical reaction of hydrogen and oxygen, occurs. Chemical bonded energy is directly converted into electrical energy. As the only reaction product is water, this energy conversion is emission-free. However, the required hydrogen needs to be generated, e.g. with renewable energies. In times, where politicians debate of banning vehicles with internal combustion engines, we currently live in an era of major changes. Hence, fuel cell technologies are very likely applied in the near future.

Solar Systems, Geothermal Systems, Wind Turbines

The thermodynamic principle apply for renewable energy systems as well. A detailed introduction to these systems is not given in this book but can be found for example in [9].

1.4.4 Modern Automotive Applications

In the late 2000s years the electrification of vehicles has started in Europe. This is a major step for the automotive industry, since the common internal combustion engines have to be replaced by new propulsion systems step by step. One option for electrical driving is the usage of Lithium ion batteries for energy storage. Energy recuperation with a generator while braking is possible with this technology. However, several technical options from full-electric driving up to hybrid drives, i.e. combing electric engine and internal combustion engine, are currently developed.

Nevertheless, at that time the optimisation of the internal combustion engine did not stop. Obviously, major steps of several percentage point for the thermal efficiency

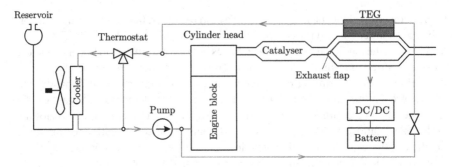

Fig. 1.7 Thermoelectric generator

can not be expected, but even small steps are acceptable. According to the second law of thermodynamics a thermal engine needs to release heat. As the temperature level of the released heat is commonly larger than the ambient temperature, it contains exergy. From a thermodynamic point of view it makes sense to focus on this exergy in order to utilise it. Two possible systems to recover the waste heat are introduced briefly.

Thermoelectric Generator

This technique is based on the thermoelectric effect, also known as Seebeck effect. Thermoelectric materials, e.g. Bi_2Te_3, generate an electrical potential once a temperature gradient is applied. The same effect is utilised for temperature measurement by a thermocouple. Thus, one side of this ceramic material needs to be hot while the other side is cool. From a thermodynamical point of view, due to the temperature gradient a heat flux is initiated, that is partially converted into electrical energy. An electric efficiency can be defined for the conversion rate of the thermoelectric element. Currently, the efficiency is, depending on the type of ceramic, less than 10%.

For a technical application in a vehicle, a hot as well as a cold reservoir are required. Figure 1.7 illustrates a possible automotive application. A thermoelectric element is mounted on the exhaust pipe. At this position the temperature is above ambient, thus a heat flux can be potentially measured. Depending on the vehicle's state the exhaust pipe can be bypassed with a flap. In order to provide a cold reservoir, there are several alternatives: Generally speaking, having the maximum temperature spread in mind, the best variant would be cooling with ambient air. Unfortunately, the heat transfer characteristic in this case is rather poor and would lead to an enlarged heat transfer area. This is a contradiction to a light and compact design. The preferred option is to use the cooling circuit of the combustion engine. The temperature is larger than ambient temperature, but still sufficient to provide a temperature gradient over the element. Using cooling water as cold reservoir guarantees a compact design, since the heat transfer is more efficient than e.g. with air.

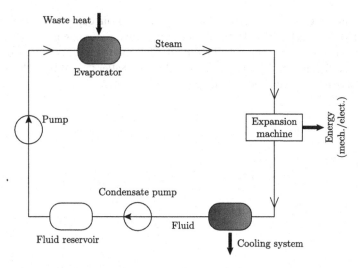

Fig. 1.8 Waste heat recovery

The recovered waste heat is available as electrical energy that can be used to operate electric auxiliaries or to relieve the alternator. This example shows, that even today technical improvements are possible. A detailed knowledge of energy conversion is required in order to optimise such technical systems.

Turbosteamer

The basic idea of the so-called turbosteamer, see [10], is the same as it is for the thermoelectric generator: Waste heat shall be recovered in order to improve the overall energy balance of the vehicle.

In this application, waste heat is utilised to heat up fluid in a secondary circuit as shown in Fig. 1.8. In fact, this system is a combination of internal combustion engine and a steam-based thermal engine. The recovered waste heat evaporates a fluid, that has been pressurised first.

The steam being on a high pressure level, when leaving the heat exchanger, can then be expanded, e.g. in a turbine. In this step thermal energy is converted into mechanical energy. A part of this mechanical energy can be applied to cover the energy consumption of the pump, that is required for pressurising the fluid. Once mechanical energy is available, it can be used for boosting or it can further be converted into electrical energy in a generator.

As indicated in Fig. 1.8 the thermodynamic process is designed as cycle. This is why another component is needed in order to close the loop. After leaving the expansion devise, the fluid needs to be condensed to reach the initial, liquid state. This is realised by another heat exchanger. At this point the cycle is closed and it can start again.

The entire process can be regarded as clockwise thermal engine. Thermal energy is converted into mechanical/electrical energy, whereby, according to the second law of thermodynamics, a cooling is still required. This cooling need is the reason, why the thermal efficiency of the turbosteamer is always smaller than 100%. This is what all thermal engines have in common!

Part I
Basics & Ideal Fluids

Chapter 2
Energy and Work

It is well known from physics lessons that energy is conserved. The principle of energy conservation reads as:

Theorem 2.1 *In a fully-closed system, with any internal mechanical, thermal, electrical or chemical processes, the total amount of energy remains constant.*

However, before clarifying in Chap. 3 what a fully-closed system is, we first want to focus on the term *energy*. Energy can be transferred to other objects or it can even be converted into other forms as we have already discussed in the previous introducing chapter. Any work performed on a mass increases its energy and enables it to perform its own work. Hence, energy is defined as the ability of a mass to provide work, see e.g. [11].

2.1 Mechanical Energy

First, let us focus on what is known from mechanics: Work is based on an interaction of a force F and a mass m. This force performs work W, when under its influence the mass is moved by a distance s. It is the scalar product of force and distance, see Fig. 2.1:

$$W = \vec{F} \cdot \vec{s} \tag{2.1}$$

The scalar product obeys

$$W = F \cdot s \cdot \cos \alpha \tag{2.2}$$

with α being the angle between force and distance. Hence, in case force and distance are perpendicular to each other, no work is performed. As we know from physics, power is the time derivative of the work, i.e. the work that is transferred per unit time

$$P = \frac{\delta W}{dt} \tag{2.3}$$

© Springer Nature Switzerland AG 2019
A. Schmidt, *Technical Thermodynamics for Engineers*,
https://doi.org/10.1007/978-3-030-20397-9_2

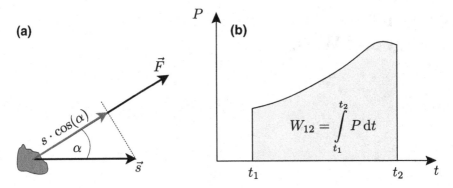

Fig. 2.1 a Definition of work and **b** power

According to this, work can be calculated by integration over a period of time:

$$W = \int_1^2 P\,(t)\,\mathrm{d}t. \tag{2.4}$$

Following this definition of work, see Eq. 2.1, correlations for any kind of mechanical energy can be derived, see below.

2.1.1 Kinetic Energy

The kinetic energy of a mass correlates with the amount of work, that is required to accelerate a mass from a state of rest to a certain velocity. A vehicle in rest, indicated by state (1), is accelerated to a velocity of c_2 in state (2). As illustrated in Fig. 2.2, the function $c\,(t)$ might have a non-linear characteristic. However, the question is, how much work is required to change the vehicle's state from (1) to (2). During the time period $\mathrm{d}t$ a distance of $\mathrm{d}s$ is covered. Thereby, according to Eq. 2.1, work is performed:

$$\delta W = F \cdot \mathrm{d}s. \tag{2.5}$$

The force obeys

$$F = m \cdot a = m \cdot \frac{\mathrm{d}c}{\mathrm{d}t}, \tag{2.6}$$

the differential distance reads as

$$\mathrm{d}s = c\,\mathrm{d}t. \tag{2.7}$$

Hence we get

$$\delta W = m \cdot \frac{\mathrm{d}c}{\mathrm{d}t} c\,\mathrm{d}t = m \cdot c\,\mathrm{d}c. \tag{2.8}$$

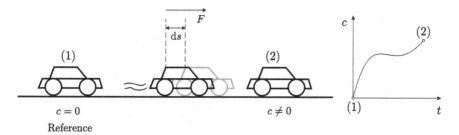

Fig. 2.2 Kinetic energy

As indicated in Fig. 2.2 the change of velocity dc is positive when accelerating or negative when decelerating. Hence, from the vehicle's perspective the related work can be positive (acceleration) or negative (deceleration). From integrating[1] we finally find the total amount of work that has to be supplied for changing the velocity:

$$W = \int_1^2 \delta W = m \int_1^2 c\,dc = \frac{1}{2}m\left(c_2^2 - c_1^2\right)\tag{2.9}$$

This work is stored in position (2) as kinetic energy:

$$\boxed{E_{\text{kin},12} = W_{12} = \frac{1}{2}m\left(c_2^2 - c_1^2\right)}\tag{2.10}$$

Hence, the function $c(t)$ is not required to calculated the work, but the initial state (1) and the final state (2). In order to assign kinetic energy to a state, the reference level, from where the counting starts, needs to be determined. This might be the state of rest for instance.

2.1.2 Potential Energy

A mass can be lifted in a gravitational field. This requires mechanical work, that is stored as energy within the mass, that comes to rest at an upper position depending how much work has been supplied. While moving a mass by a distance of dz within a gravitational field, see Fig. 2.3, work is supplied:

$$\delta W = F \cdot dz.\tag{2.11}$$

---

[1] Assuming, that the vehicle mass stays constant. If the car is equipped with an internal combustion engine, it would be required not the release the exhaust gas. Or, alternatively, it is an electric vehicle.

Fig. 2.3 Potential energy

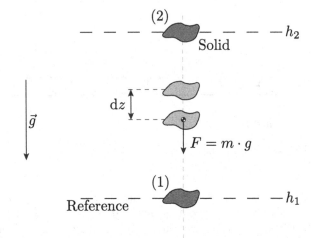

The force is given by

$$F = m \cdot g \tag{2.12}$$

so that we get

$$\delta W = m \cdot g \, dz. \tag{2.13}$$

By integrating, we finally find the total amount of work, that has to be supplied in order to lift the mass by a distance of Δh:

$$W = \int_1^2 \delta W = mg \int_1^2 dz = mg \, (h_2 - h_1) = mg \Delta h \tag{2.14}$$

This work is stored in position (2) as potential energy within the mass

$$\boxed{E_{\text{pot},12} = W_{12} = mg \Delta h} \tag{2.15}$$

As illustrated in Fig. 2.3 and given by Eq. 2.15 it is required to define and fix a reference level, from where the potential energy shall be counted.

2.1.3 Spring Energy

Another example for mechanical energy is related with the work it requires to compress a spring, see Fig. 2.4. The force ist a function of the distance, the spring is compressed, and follows

Fig. 2.4 Spring energy

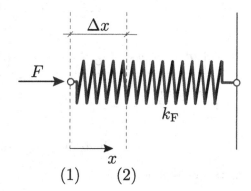

$$F(x) = k_F x \tag{2.16}$$

Mind, that if the spring expands work is released, i.e. depending on the coordinate x the force can be positive or negative. Let x start at reference position where there is no tension within the spring. In differential notation the work is

$$\delta W = F\,dx = k_F x\,dx. \tag{2.17}$$

In order to calculate the entire work for a compression by Δx, an integration is done

$$W_{12} = \int_1^2 \delta W = \int_0^{\Delta x} k_F x\,dx = \frac{1}{2}k_F\,\Delta x^2. \tag{2.18}$$

This work is stored in position (2) as spring energy:

$$\boxed{E_{\text{spr}} = W_{12} = \frac{1}{2}k_F\,\Delta x^2} \tag{2.19}$$

2.2 Thermal Energy—Heat

The previous examples have shown the correlation between work and energy from a mechanical point of view. However, energy is more than just the ability of providing *work*.

Heat for example is another form of energy: Systems at different temperatures transfer heat Q, i.e. *thermal energy* at their interface, once they get in contact. A switched-on hot plate transfers heat to your hand while touching it. Obviously, the amount of heat depends on the temperature difference between hand and hot plate and further on the time the contact lasts. The larger the temperature potential between

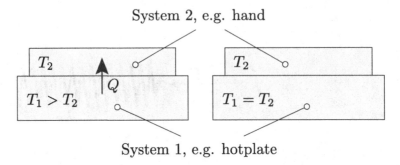

Fig. 2.5 Heat as thermal energy

hand and hot plate is, the larger the transferred heat is. If there is no temperature difference any more, the heat transfer stops and the two systems reach a thermal equilibrium. Obviously, heat is another form of energy as it has the power to change the systems' state, in this case for instance the hand's structure. What we know from experience is, that the transfer of thermal energy follows a temperature gradient. Thus, it has never been observed, that due to the contact, the hot plate gets even hotter while the hand is cooling down. In case no gradient occurs, i.e. the systems having identical temperatures, no thermal energy is transferred, see Fig. 2.5. This phenomena is explained later on with the second law of thermodynamics. Consequently, the definition of energy needs to be extended:

Theorem 2.2 *Energy is not only the ability to perform work, but also the ability of heating up a mass.*

In thermodynamics it is essential to distinguish between mechanical work W and thermal energy Q.

2.3 Chemical Energy

Fuels as well carry chemical bonded energy, that can be released by a chemical reaction. To do so it requires oxygen and fuel as educts to release energy and an exhaust gas as reaction product. This chemical bonded energy is commonly defined by a so-called specific lower/upper heating value and is discussed in detail in part III. During the combustion this chemical energy is converted into thermal energy. Anyhow, a fuel cell releases electrical energy as well.

2.4 Changeability of Energy

It is known from experience that it is possible to convert the form of energy. A mass fixed at a height h above a reference level carries potential energy, that is converted into kinetic energy as soon as the fixing is released. It is even possible to convert kinetic energy into thermal energy, e.g. when a vehicle is braking from a velocity c_1 to standstill. By friction at the braking disks energy is dissipated and finally released as heat to the environment. This leads to an extension of Theorem 2.1:

Theorem 2.3 *Energy can exist in a variety of forms, such as electrical, mechanical, chemical and thermal energy. It can be converted from one form to another. However, it can not be destroyed.*

This approach is further investigated in the next section with the Paddle Wheel experiment according to Joule.

2.4.1 Joule's Paddle Wheel

In this section the focus is on the changeability of energy and it is shown, that it is required to introduce a new form of energy: Internal energy describes the energetic state of a system. As we have seen in the previous section, energy may take several forms and can be transferred, though it can not be destroyed or generated from nothing—known as the principle of energy conservation. Joule suggested a simple experiment, that is briefly sketched in Fig. 2.6.

Initially, a thermal-insulated vessel is filled with a fluid, e.g. water. An impeller is immersed into the water and is connected with two masses by a spindle. The two masses are fixed outside at a well-defined height above ground. The entire system is supposed to be in an equilibrium state (1). Due to gravity the masses start falling down as soon as the fixing is released. Obviously, while the masses descend, the system is in imbalance. However, the potential energy of the two masses is converted into kinetic energy and rotational energy of the impeller. The impeller starts rotating and the fluid being in contact with the impeller is accelerated due to friction. Distant fluid elements are still in rest. As a consequence turbulent eddies within the water are generated. After a while the energy carried by the eddies are dissipated, the entire system comes to rest. A new equilibrium state (2) is reached. Imagine, an observer was not present during the change of state from (1) to (2), all he realises are the

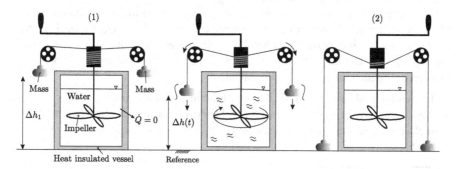

Fig. 2.6 Joule's paddle wheel experiment

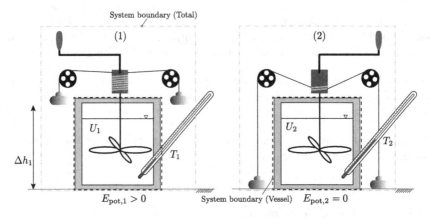

Fig. 2.7 Joule's paddle wheel experiment, state (1) and (2)

two balanced states (1) and (2). The question he will most likely ask is, where has the potential energy of the two masses gone, as he knows the principle of energy conservation.

Assuming the walls being massless, the entire energy of the masses needs to be transferred to the water. Thus, the observer suggests to estimate the water's temperature, i.e. a measure of the energy, initially in state (1) and finally in state (2), see Fig. 2.7.[2] Suppose it would be possible to measure the temperature accurately, a temperature rise from state (1) to state (2) could be observed.

2.4.2 Internal Energy

Going back the Joule's paddle wheel experiment in the previous Sect. 2.4.1, we just found out, that potential energy, that has been converted to kinetic/rotational energy, is finally dissipated and leads to an increase of temperature within the water. Since the vessel is perfectly insulated, the entire former potential energy needs to be within the water, see Theorem 2.3. The fluid's energy is called *internal energy U*. Thus, in state (1) the water has an amount U_1 of internal energy. In the new balanced state (2) the system has a larger amount of internal energy, i.e. $U_2 > U_1$. Obviously, the energetic state is indirectly characterised by temperature T, that is a measure of internal energy. The larger the temperature is, the more internal energy is present. Generally, the internal energy U consists of

- thermal internal energy (kinetic energy of the molecules/atoms)
- chemical internal energy (molecular bond energy)—will be covered in part III
- nuclear internal energy (forces within the atomic cores)—will not be treated in this book.

---

[2]His idea would be, the larger the temperature of the fluid is, the larger its energy content is.

Commonly, only the thermal internal energy is considered at least, as long as there are no chemical/nuclear reactions. According to the kinetic theory, the thermal motion on molecule/atom level correlates with the temperature. In Chap. 24, however, we will introduce an approach, that takes the chemical bond energy into account as well. Colloquially, the heat content of a body is characterised as internal energy.[3] Internal energy describes the state of a system, hence it is a state value. As we will learn, the internal energy never can be measured directly as e.g. pressure or temperature. Consequently, a state equation is required to *calculate* the internal energy of a system, see Chap. 12. This state function is rather simple for ideal gases, since its internal energy solely is a function of temperature. The internal energy for ideal gases is proportional to its temperature. However, for real fluids, as treated in part II, the internal energy depends on two state values, e.g. pressure and temperature. In analogy to the kinetic or potential energy, a reference level is required, to determine from where the counting of internal energy shall be started. Generally speaking, the reference level is arbitrary though it must not be changed when calculating a change of state as we will do later on. A reference level might be defined as:

$$U_0 \left(\vartheta = 0\,°C, \, p = 1\,bar \right) = 0 \tag{2.20}$$

Respectively, for ideal gases:

$$U_0 \left(\vartheta = 0\,°C \right) = 0 \tag{2.21}$$

Problem 2.4 Two equal, vertical standing pipes with a diameter of $D = 0.1$m are connected with each other through a thin pipe and a valve, see Fig. 2.8. One pipe is filled up to a height of $H = 10$ m with water. The water has a density of $\rho = 1000\,\frac{kg}{m^3}$, at first the other pipe is empty. Now the valve is opened. After a while a balance of the water volume will develop. By what absolute value does the potential energy of the water change?

Solution (Alternative 1)

Starting with state (1), each slice dz of the column of water possesses potential energy dE_{pot}, see sketch 2.9 and Eq. 2.15.

$$dE_{pot} = dm\, gz \tag{2.22}$$

Its mass follows

$$dm = \rho\, dV = \rho \frac{\pi}{4} D^2\, dz \tag{2.23}$$

---

[3]Later on we will learn that heat is not a state value. Thus, a body never contains heat, but it contains internal energy. Any body will have a temperature, a parameter that characterises the state. Thermodynamically, in contrast to this, *heat* is a process value, that passes a body due to a temperature difference.

Fig. 2.8 Sketch to
Problem 2.4

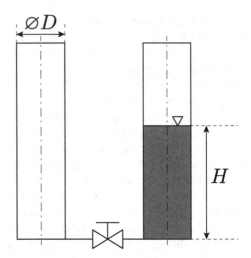

So we get

$$dE_{\text{pot}} = gz\rho\frac{\pi}{4}D^2\,dz \tag{2.24}$$

To calculate the entire potential energy in state (1) Eq. 2.24 needs to be integrated

$$E_{\text{pot},1} = \int\limits_0^H dE_{\text{pot}} = \int\limits_0^H gz\rho\frac{\pi}{4}D^2\,dz = \rho\frac{\pi}{4}D^2Hg\frac{H}{2} \tag{2.25}$$

By analogy we find for state (2)

$$E_{\text{pot},2} = 2\int\limits_0^{H/2} dE_{\text{pot}} = 2\int\limits_0^{H/2} gz\rho\frac{\pi}{4}D^2\,dz = \rho\frac{\pi}{4}D^2Hg\frac{H}{4} \tag{2.26}$$

We finally get the change of potential energy by

$$\boxed{\Delta E_{\text{pot},12} = E_{\text{pot},2} - E_{\text{pot},1} = -\rho\frac{\pi}{4}D^2Hg\frac{H}{4} = -1926.2\,\text{J}} \tag{2.27}$$

Solution (Alternative 2)

Instead of using a differential approach as in alternative 1, the entire mass can be regarded as concentrated in the center of gravity, illustrated in Fig. 2.9. Thus, we find for state (1)

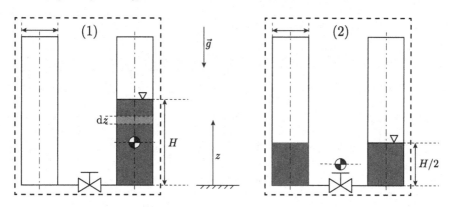

Fig. 2.9 Solution Problem 2.4

$$E_{\text{pot},1} = mg\frac{H}{2} \tag{2.28}$$

Respectively, for state (2)

$$E_{\text{pot},2} = mg\frac{H}{4} \tag{2.29}$$

We finally get the change of potential energy by

$$\Delta E_{\text{pot},12} = E_{\text{pot},2} - E_{\text{pot},1} = -mg\frac{H}{4} \tag{2.30}$$

Replacing the mass, see Eq. 2.23, leads to the same result as before in alternative 1

$$\boxed{\Delta E_{\text{pot},12} = -\underbrace{\rho\frac{\pi}{4}D^2 H}_{m}\, g\frac{H}{4} = -1926.2\,\text{J}} \tag{2.31}$$

The negative sign indicates, that the potential energy from state (1) to state (2) decreases.

Chapter 3
System and State

In this chapter it is clarified what a thermodynamic system is and how its state can be described. Any system is separated from an environment by a system boundary as it is done in other technical disciplines, e.g. technical mechanics, as well. First, the permeability of the system boundary is categorised. However, after classifying the system and the system boundary, it is the next step to identify the internal state. This leads to so-called state values, e.g. pressure and temperature, that fix the state of a system. Our everyday experience shows, that the state can be varied by external impacts across the boundary.[1]

3.1 System

What Is a System?

The definition of a thermodynamic system (brief: system) is prerequisite for performing thermodynamic balances. Figure 3.1 shows the principle of a thermodynamic system. It can consist of a single compound, a body or an arrangement of several bodies, apparatuses or machines. A system can be in rest, compared to its environment, or it even might be in motion. As indicated in Fig. 3.1 it is characterised by a system boundary, that separates system and environment. The system boundary can be an actual boundary, e.g. a cylinder wall, or it might be even virtual. Environment and system can potentially influence each other, since mass, momentum and energy can be transferred across the system boundary. The major difficulty, when solving thermodynamic problems, is to find an appropriate system boundary for the specific problem. Though a system boundary itself can never be wrong, an inaptly chosen

---

[1]The content of a bottle can be heated up, e.g. by a lighter. By doing so, its internal state, given by the temperature for instance, changes. Thus, the lighter is the external impact acting at the boundary.

© Springer Nature Switzerland AG 2019
A. Schmidt, *Technical Thermodynamics for Engineers*,
https://doi.org/10.1007/978-3-030-20397-9_3

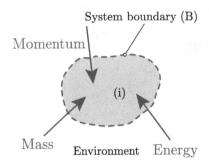

Fig. 3.1 Definition of a system, (i) indicates the internal state, (B) the system boundary, that might be permeable for mass, momentum and energy

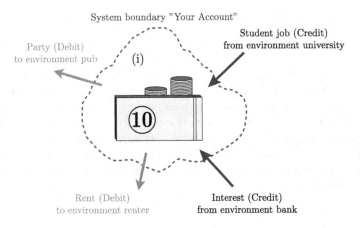

(i) Money supply M

Fig. 3.2 Example of a bank account

boundary might lead to many interfaces and causes difficulties to count for all relevant influences. Thus, in this book several examples are discussed, that clearly show how important the definition of a system boundary is. We will address and answer the following questions:

- What is transferred across the system boundary? Specifying the permeability leads to an important classification of systems.
- How can the state of a system be described? State values, that characterise the internal state, are required.
- How can we selectively influence the system's internal state by external impact? Answering this question will bring us to the first law of thermodynamics, see Chap. 11.

Example 3.1 Figure 3.2 shows a simple example of a system, everyone knows from everyday experience. Your bank account can be treated as a system. In this case the amount of money M is the internal state (i) of this system. The system boundary (B) separates your account from any financial environment and makes a balancing for your bank possible. As you know, your account underlies several influences across the boundary. Some of them increase the internal state, e.g. your salary or income from bank interest. However, there are some fluxes having the opposite direction, i.e. they decrease the amount of money M. This is what your monthly rent as well as your leisure activities do. Obviously, it does not cause any difficulties to manage a bank account: what comes in increases the inventory, what leaves decreases the inventory. Starting with the initially available money M_0 and the fluxes across the system boundary, the amount of money can be easily calculated:

$$M = M_0 + \sum_i M_{i,\text{in}} - \sum_j M_{j,\text{out}} \tag{3.1}$$

This sort of balancing can be done with any extensive state value[2] Z. Anyhow, this sort of balancing is what we do in thermodynamics. If there is no temporal change of any state value Z, e.g. the amount of money M for instance, the system has reached an equilibrium or steady state:

$$\frac{dZ}{dt} = 0 \tag{3.2}$$

Referring back to the previous example it needs to be

$$\sum_i M_{i,\text{in}} = \sum_j M_{j,\text{out}}. \tag{3.3}$$

On its way to steady state the system might follow several transient effects, indicated by unsteady states.

3.1.1 Classification of Systems

Heterogeneous Systems

A heterogenous system consists of several phases. A phase is supposed to be a region throughout any physical property is uniform. As shown in Fig. 3.3 the system consists of cylinder, piston, gas and environment. Each component forms a phase from a thermodynamic point of view. Wall and piston are in solid state, whereas gas and environment are supposed to be gaseous. Obviously, if the density is measured throughout the system for instance, there is no uniform distribution, but at each

---

[2]The expression *extensive state value* will follow in the next chapters. However, extensive state values can be counted.

Fig. 3.3 Heterogeneous
versus homogeneous systems

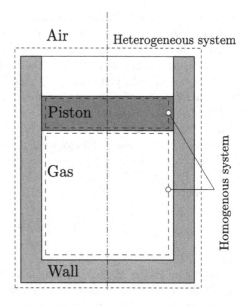

interface a step, i.e. a non-uniformity, can be measured. Thus, it is obvious, that it is impossible to assign a single density[3] for a heterogeneous system. However, in the given example a heterogeneous system can be split into several homogeneous sub-systems. Figure 3.4 shows an example of heterogenous systems. In many chemical engineering processes granular media needs a drying. This can be achieved by fluidising the media with air. However, the more air is used, the more turbulent the two-phase flow gets. Thus, it appears, that heat as well as mass transfer improve. The same technique can be applied for the combustion of coal and sludge. The required air is fluidising the two-phase flow and leads to a good heat transfer characteristic, see [12]. From a thermodynamic point of view this two-phase flow is a heterogeneous system. According to [13], Fig. 3.5 shows a diesel injection spray at high pressure. This is another example of a two-phase flow consisting of air as gaseous phase and diesel as liquid phase. Thus, this is a heterogeneous system as well. The fuel jet profile produced by the injection nozzle is crucial during direct diesel injection. In best case, the injection initialises small droplets in the outer region and coarser droplets in the internal region of the spray. The small droplets spark first when the combustion process starts. As a consequence, the fuel burns comparable slowly and uniformly. Hence, the pressure increase within the combustion chamber is rather smooth. As mentioned before, a system is heterogeneous for instance, if multiple aggregate states are present. A simple model, to explain aggregate states is illustrated in Fig. 3.6. Three aggregate states are possible:

---

[3]The density of a system is a state value like pressure, temperature and many more, see Sect. 3.2 Assigning state values to a heterogenous system can cause difficulties as they might be equivocal.

Bubbling fluidised bed

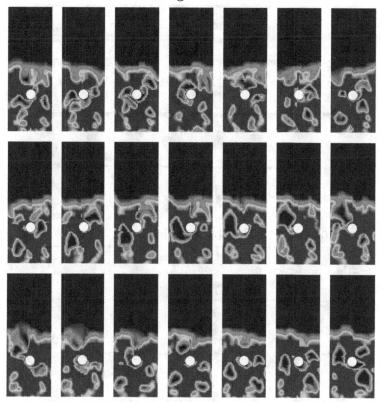

Fig. 3.4 Heterogeneous systems—fluidised bed, see [12]

- <u>Solid</u>: stable in shape and volume, large intermolecular forces, strong lattice among neighbouring molecules
- <u>Liquid</u>: stable in volume, interaction among neighbouring molecules, distance between molecules constant at first approach
- <u>Gaseous</u>: No bonding between the molecules, small intermolecular forces, elastic impact between the molecules

From now on, when talking about fluids, we have gases or liquids in mind. In part I we focus on idealised fluids: Gases are supposed to behave as an ideal gas, i.e. the molecules respectively atoms of the gas are treated punctiformly, interactions among neighbouring particles are neglected. This leads to rather simple equations of state. Unfortunately, not all phenomena can be explained with this simplified approach: The so-called Joule–Thomson effect, many gases cool down when being throttled adiabatically, is a well-known deviation from ideal gas behaviour. However, this phenomena will be explained in part II: Real gases are introduced in order to

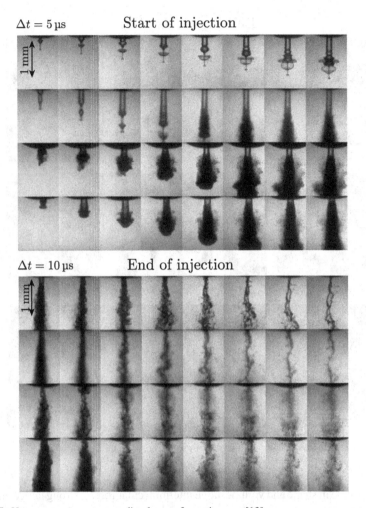

Fig. 3.5 Heterogeneous systems—diesel spray formation, see [13]

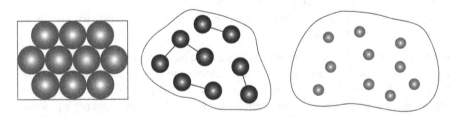

Fig. 3.6 Aggregate states from left to right: solid/liquid/gas

calculate the Joule–Thomson effect. The physical description of the gaseous state is getting more complex at that point than it is for ideal gases.

Liquid fluids in part I also follow a simplified idea: They are treated as incompressible, i.e. they keep their volume even if pressure or temperature vary. This characteristic will be refined in part II: Even though the compressibility is rather small, this effect is taken into consideration. Furthermore, changes in aggregate state are not considered in part I, but this phase change is investigated thermodynamically in the second part.

Homogeneous Systems

Any heterogenous system can be further divided as long as homogeneous systems occur. In contrast to heterogeneous systems, the properties within such a system are uniform, see Fig. 3.3. Taking the density as an example, it is homogeneously distributed within the selected sub-systems. Consequently, the number of interfaces, respectively boundaries, increases while sub-dividing a heterogeneous system. The homogeneous system *gas* has two environments, namely *piston* and *wall*, that both influence the system *gas*. However, when solving thermodynamic problems, in most cases it is a good idea to start with a homogeneous system. Any system is clearly defined by a system boundary, that in fact can have any shape. Most difficulties, when balancing a system, occur, if the boundary is not clearly chosen. In order to avoid these difficulties, appropriate sketches as well as experience are required. Later in this book several examples will be presented.

3.1.2 Permeability of Systems—Open Versus Closed Systems

Probably the most important distinction of systems is made in terms of the permeability of the system boundary with respect to the mass. Closed systems are characterised by no mass transfer across the system boundary, i.e. the mass inside the system remains constant. Open systems let mass pass across their system boundary. However, the total mass inside such an opne system can nevertheless be constant, e.g. in steady state, so that the incoming mass needs to be balanced by the outgoing mass. An overview regarding the different systems is presented in Fig. 3.7:

Insulated or Fully-Closed System

An insulated or fully-closed system does not allow

- mass transfer Δm
- transfer of work W
- transfer of heat Q

across the system boundary. Space for instance is supposed to be a fully-closed system, with no fluxes at its boundaries.[4] In terms of technical applications a closed,

---

[4]Though nobody has ever visited the boundary of space...

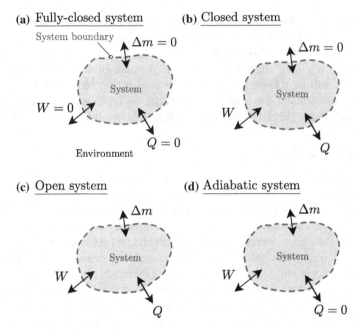

Fig. 3.7 Permeability of systems

perfectly insulated thermos can be treated as a fully-closed system as long as the
thermos is in rest: No mass crosses the system boundary, the insulation prevents heat
leaving the system[5] and no mechanical work is acting on the thermos as long as it
is in standstill. However, most technical applications are somehow ideally treated as
fully-closed systems, since a *perfect* thermal insulation does not occur in reality.

Closed System

A closed system does not allow mass transfer Δm across the system boundary.
However, transfer of work W and heat Q are possible. A cylinder filled with an ideal
gas and sealed with a movable piston is treated as a closed system, as long as the
sealing is perfect gas tight. Work can be supplied by moving the piston and the gas
can additionally be heated up or cooled down.

Open System

In contrast to a closed system an open system allows mass Δm to cross the system
boundary. Transfer of work W and heat Q are also possible. Many technical applications are treated as open systems: A fluid flow through a channel can be treated
as open system, since mass enters and leaves the channel and heat might be trans-

---

[5]Every student knows, that a perfectly insulated thermos does not exist. After a three-hour lecture
the coffee inside a thermos is inedible. However, in thermodynamics ideal systems are assumed to
exist.

ferred depending on the environment as well. Turbines and compressors are also open systems, but in contrast to the flow channel, additionally work W is transferred.

Heat Insulated or Adiabatic System

Both, open and closed systems, can be adiabatic, i.e. no heat passes the system boundary. Technically, there are two options for adiabatic systems: First, the system has a perfect insulation. Second, the system and the environment do have the same temperature, so there is no driver for a transfer of heat, see Fig. 2.5.

3.1.3 Examples for Thermodynamic Systems

Cooling Machine

Figure 3.8 shows several possibilities how to define a system. The sketch on the left of Fig. 3.8 shows a so-called counterclockwise thermodynamic cycle, i.e. a cooling machine or heat pump. Though cycles like this will be discussed later, see Chap. 15, the thermodynamic details of such cycles are not required in order to define systems at that time:

A system boundary can be chosen in this way that all components are part of the system, see Fig. 3.8a. This is called *overall* or *integral balance*. According to our recently made definitions, a system like this is closed: No mass enters or leaves the system. Regarding the permeability of the boundary the sketch shows, that a heat flux $\dot{Q}_0$ is entering, while a heat flux $\dot{Q}$ is leaving the system. Furthermore technical power P_t is supplied to the system. In Sect. 15.2.1, we will learn that a cooling machine, i.e. fridge, is supplied with a heat flux $\dot{Q}_0$ of a rather low temperature.[6] However, the component taking the heat flux is called *evaporator*. Since energy can not be destroyed, the heat, taken at the low temperature, needs to be released. This is done in the *condenser*, that releases heat at a higher temperature level.[7] Obviously, heat is shifted from a low temperature to a higher temperature. This is in contrast to our everyday experience: heat always follows the temperature gradient from hot to cold.[8]

In terms of a cooling machine a heat flux needs to be lifted from cold to hot, i.e. against the natural gradient. This requires technical power P_t.[9] In Sect. 15.2.1 the machine will be explained with the state value entropy: In steady state operation the

---

[6] Just think of your fridge at home: A bottle of beer, that needs to be cooled down, is placed inside the fridge. In this compartment the temperature is low. However, at this low temperature, a heat flux needs to leave the bottle of beer in order to decrease its internal energy, i.e. cooling it down. This heat is supplied to the refrigerant cycle of the fridge.

[7] You might have already realised that the backside of your fridge has a larger temperature than the environment and thus releases heat.

[8] Also known as second law of thermodynamics. Placing your hand on a hotplate proves the second law of thermodynamics impressively.

[9] This is the reason why the fridge needs to be connected with a plug.

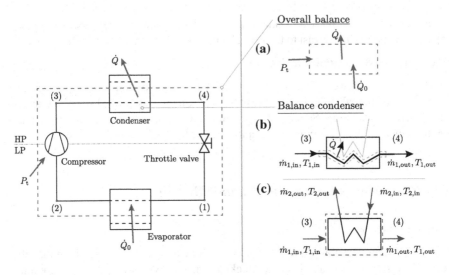

Fig. 3.8 Cooling machine on the left: thermodynamic cycle on the right: possible system boundaries

amount of entropy inside the machine, as all other state values, must be constant with respect to time. Heat entering the machine at low temperature carries much entropy with it. In order to keep the entropy inside the machine constant, on a higher temperature level a larger heat flux needs to be released. Following the first law of thermodynamics, increasing the released heat can be achieved, if additional power is supplied to the system.

Theorem 3.2 *For most thermodynamic problems it is wise to start with an integral balance first. If this is not sufficient in order to solve the problem, one should think of gathering additional information, e.g. by balancing further systems.*

Applying this theorem, one can get additional information by selecting a system boundary as shown in Fig. 3.8b: The boundary separates the fluid flow inside the component *condenser*. The condenser is an open system, since the mass flux $\dot{m}_{1,in}$ enters the system and $\dot{m}_{1,out}$ leaves the system. If the system boundary is just including the mass flux of the refrigerant, as it is in case (b), the heat flux $\dot{Q}$ becomes visible at the boundary. The second mass flow, that takes the released heat, is not part of the system boundary.

Alternatively, the boundary can include the entire condenser, shown in Fig. 3.8c. Within the condenser the working fluid needs to release heat, that is transferred to a second mass flow.[10] With this system boundary there are two mass flows entering, namely $\dot{m}_{1,in}$ and $\dot{m}_{2,in}$. Both mass flows need to leave the component in steady state. Consequently, since the heat flux $\dot{Q}$ occurs internally, i.e. not being visible at the

---

[10]In case of a fridge the working fluid releases heat to the surrounding air, that can be regarded as second mass flow.

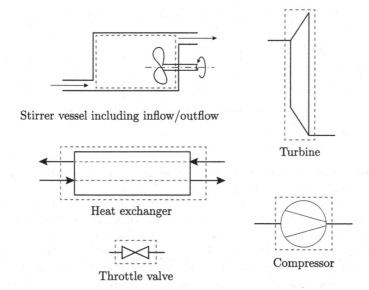

Stirrer vessel including inflow/outflow

Turbine

Heat exchanger

Throttle valve

Compressor

Fig. 3.9 Examples for thermodynamic open systems

system boundary, it is not included in (c) and it does not affect the thermodynamic balances directly.

Theorem 3.3 *It is essential to clearly identify the system boundary and the physical quantities passing it.*

Passive Versus Active Systems

Several components, that are needed in the course of this book, are introduced briefly, see Fig. 3.9. All of them are open systems, i.e. mass is passing the system boundary. However, it makes sense to distinguish between active and passive systems. Passive systems are actually work-insulated systems, so there is no work/power passing the system boundary. Examples for *passive systems* are:

- Heat exchangers that are required to transfer heat from one mass flow to another. By doing so, one flow cools down, while the other is heated up.[11]
- A throttle valve reduces the pressure of a mass flow by internal friction, i.e. dissipation.
- Tubes or channels that conducts a fluid flow.

In contrast to passive systems, an active system is characterised by work/power passing the system boundary. Examples for *active systems* are:

---

[11] At least as long as we talk about sensible heat, i.e. the fluids do not underlie a phase change. Part II covers fluids, that also can change their aggregate states. If this is done isobarically for instance, supplied heat is not utilized to vary the temperature but to perform the phase change. Heat then is called latent heat and the phase change runs isothermally.

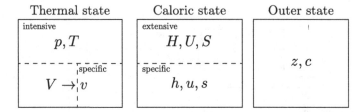

Fig. 3.10 Overview state values

- Turbines as thermal machines, that release work/power across the system boundary while the fluid's pressure decreases.
- Compressors are thermal machines as well. Instead of decreasing the fluid's pressure, the pressure rises when passing the machine. Thus, work needs to be supplied across the system boundary in order to increase the pressure.
- The stirrer vessel in Fig. 3.9 is an active system as long as the stirrer is transferring work across the boundary. Two options are possible: if the stirrer is supplied with electrical or mechanical work from the outside, work is entering the system. The other option is, that the flow inside the system causes the stirrer rotation. This work might be transferred mechanically to the environment, so work is released by the system. Shaft work, that is the cause for the stirrer to transfer work, is further investigated in Sect. 9.2.7.

3.2 State of a System

Now that we know, how to classify systems and how to define system boundaries, the next step is to describe the internal state of a system. The thermodynamic state of a system is quantified unequivocally by so-called *state values*, see also Sect. 5.1. Consequently, changes within a system can easily be detected and quantified by state values. A generic classification of state values is illustrated in Fig. 3.10. A more detailed explanation follows in the upcoming sections. Figure 3.11 shows exemplary state values for a closed system. The following sections summarise each category.

3.2.1 Thermal State Values

Thermal state values have in common, that they can be measured easily. Hence, the following variables can be utilised to identify the state of a system:

- Pressure p, [bar, Pa][12]

---

[12]Just keep in mind, that Pa is the SI-unit. Nevertheless, for many applications bar is the more common unit. The conversion follows $1 \times 10^5 \mathrm{Pa} \equiv 1\mathrm{bar}$.

Fig. 3.11 Overview state values for a closed system

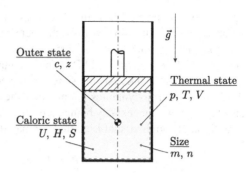

Table 3.1 Temperature measurement

Method	Measured quantity
Liquid thermometer (mercury, alcohol)	Thermal expansion of a fluid
Gas thermometer (He, Ar, N_2)	Thermal expansion of a gas
Thermocouple (e.g. Ni/Cr-Ni)	Electrical voltage (Seebeck-effect[a]) of a metal couple
Resistance thermometer (e.g. Pt 100)	El. resistance of a metal wire
Radiation thermometer (pyrometer)	Impact of the detected heat radiation (e.g. el. voltage of a photo diode)

[a]The Seebeck effect is a thermoelectric effect. Any thermoelectric element produces a voltage potential when there is a temperature difference between its two ends, i.e. a temperature potential

- Temperature ϑ, T, [°C, K]
- Volume V, $[\mathrm{m}^3]$, specific volume $v = \frac{V}{m} = \frac{1}{\rho}$, $\left[\frac{\mathrm{m}^3}{\mathrm{kg}}\right]$

The volume of a system can be quantified by geometrical means. If the volume of a system is divided by its mass, one gets the so-called specific volume, i.e. the reciprocal of density.

Sensors are available in order to measure the pressure as well as the temperature of a system. The temperature can be given on a Celsius scale, i.e. $[\vartheta] = 1\,°C$, for instance or on a Kelvin scale, i.e. as absolute temperature $[T] = 1\,K$. Most thermodynamic equations require the absolute temperature T, which is essential in Chap. 6.

How Can Temperature be Measured?

There are several possibilities how to determine the temperature of a system, see Table 3.1, that further includes the underlying physical principle.

Temperature Scales

As mentioned above, there are several temperature scales possible. However, in thermodynamics the absolute temperature T is most relevant. It is a base unit as length, mass and time for instance and is measured in

$$[T] = 1\,K. \tag{3.4}$$

1 K is defined to be the 273.16th part of the thermodynamic temperature of the triple point of water.[13] The triple point is the state, where a fluid can be liquid, solid and vaporous at the same time in thermodynamic equilibrium. For water this occurs at $T = 273.16\,\text{K}$ and $p \approx 611\,\text{Pa}$. The absolute temperature T starts counting at absolute zero $T = 0\,\text{K}$. At this state the molecular motion freezes. However, the third law of thermodynamics claims, that the absolute temperature can *approach* absolute zero without exactly being zero. Nevertheless, in this book the Celsius scale is additionally applied, that is shifted compared with the absolute temperature scale by

$$\vartheta = T - 273.15\,\text{K} \tag{3.5}$$

You should keep in mind, that temperature differences are always given in K, i.e.

$$\Delta T = \vartheta_2 - \vartheta_1 \tag{3.6}$$

with

$$[\Delta T] = 1\,\text{K}. \tag{3.7}$$

3.2.2 Caloric State Values

In contrast to thermal state values caloric state values can not be measured, i.e. there is no sensor available for quantifying a caloric state value. The internal energy U has already been introduced as a measure of the kinetic energy of fluctuating molecules respectively atoms. It has been required to explain Joule's paddle wheel experiment, see Sect. 2.4.1. This section has shown, that the internal energy correlates with the temperature for ideal gases respectively ideal liquids. However, there is no sensor available to determine the internal energy *directly*. Consequently, a state equation is necessary to *calulate* the internal energy. Though other caloric state values[14] have not been introduced yet, there are two more in addition to the internal energy, that have particular relevance in thermodynamics:

- Internal energy U, [J]
- Enthalpy H, [J]
- Entropy S, $\left[\frac{\text{J}}{\text{K}}\right]$

As mentioned before, it is not possible to measure these caloric state values. Thus, in Chap. 12 equations of state are derived in order to calculate caloric state values. A generic approach is followed, that simplifies for ideal gases and incompressible liquids. Ideal gases and incompressible liquids are covered in part I. Real fluids, that even can underlie a phase change, are treated in part II.

---

[13] The thermodynamic explanation of the triple point will be given in part II!

[14] Enthalpy and entropy are introduced at that point though the physical explanation is going to follow at a later point of time. Currently we just assume, they exist.

3.2.3 Outer State Values

In addition to the internal state, given by thermal respectively caloric state values, the so-called *outer state* quantifies a system as well: The outer state is related to the centre of gravity of a system, which might be in rest or in motion. Compared to a reference level, a system posseses potential energy, given by a coordinate z, perpendicular in the gravity field, as well as kinetic energy, that correlates with its velocity c. Kinetic as well as potential energy have already been introduced in Sects. 2.1.1 and 2.1.2.

3.2.4 Size of a System

So far the internal state of a system has been introduced. However, pressure and temperature for instance do not contain any information regarding the dimension of a system. In order to specify the size of a system, *extensive state values* are required. Extensive state values will be part of Sect. 3.2.5. Anyhow, the size of a thermodynamic system can easily be determined by its mass m, its volume V or even by its molar quantity n.

3.2.5 Extensive, Intensive and Specific State Values

Another important distinction in thermodynamics is made between intensive and extensive state values. The basic idea is illustrated in Fig. 3.12.

Intensive State Values

With respect to Fig. 3.12, intensive state values do not change when a system is divided into sub-systems. They are independent of the mass of the system:

- Pressure p
- Temperature T

Fig. 3.12 Extensive and intensive state values

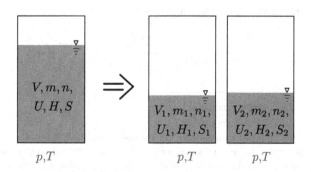

Imagine a cup of coffee that is decanted in two smaller cups. This has no impact on the pressure, that is dominated by the environment. Neglecting the process of decanting, the temperature of the coffee does also not vary. This behaviour is characteristic of intensive state values.

Extensive State Values

In contrast to intensive state values, extensive state values Z modify their amount by dividing the system or by merging sub-systems proportional to the mass. In case of the example documented in Fig. 3.12, the entire volume V can be calculated by adding the volumes of the two sub-systems

$$V = V_1 + V_2 \tag{3.8}$$

Extensive state values Z are

- Volume V
- Internal energy U
- Enthalpy H
- Entropy S

Extensive state values Z can be balanced, even in case they might not follow the quality of a conservation value, i.e. even if a source[15] or a sink[16] exist:

$$\frac{dZ}{dt} = \sum \dot{Z}_{in} - \sum \dot{Z}_{out} + \dot{Z}_{Source} \tag{3.9}$$

Following Fig. 3.13 the amount of any extensive state value Z inside a system varies by time due to an incoming flux, that enlarges the state value within the system, and an outgoing flux, that reduces the amount inside. Sources as well as sinks also have an impact on the temporal variation of Z.[17] Mind, that $\dot{Z}$ represents the flux of Z, i.e. the quantity of Z per second. A system in steady state is characterised by no temporal change of any state value inside the system:

$$\frac{dZ}{dt} = 0 = \sum \dot{Z}_{in} - \sum \dot{Z}_{out} + \dot{Z}_{Source} \tag{3.10}$$

This can be given in another notation

$$\sum \dot{Z}_{in} + \dot{Z}_{Source} = \sum \dot{Z}_{out} \tag{3.11}$$

---

[15]E.g. entropy generation.

[16]E.g. loss of exergy.

[17]Just think of your bathtub: The amount of water as a function of time in the tub is influenced by the flux of incoming water as well as by the flux at the sink. This example visualises Eq. 3.9 though this example does not include sinks or sources. Another example was given in Fig. 3.2.

Fig. 3.13 Extensive state value

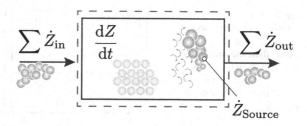

In other words: The inflow of an extensive state value equals the fluxes out of the system!

In case the temporal development of a state variable is of no interest but only the overall change counts, the differential Eq. 3.9 simplifies to a difference equation

$$\Delta Z = \sum Z_{\text{in}} - \sum Z_{\text{out}} + Z_{\text{Source}} \qquad (3.12)$$

Specific State Values

The dependency of extensive state values Z on the mass of the system can easily be eliminated by referring Z to the mass of the system m, i.e.:

$$z = \frac{Z}{m} \qquad (3.13)$$

respectively

$$z = \frac{\dot{Z}}{\dot{m}} \qquad (3.14)$$

z then is called *specific state value*. Examples for specific state values are:

- Specific volume $v = \frac{V}{m} = \frac{1}{\rho}$, $\left[\frac{\text{m}^3}{\text{kg}}\right]$
- Specific internal energy $u = \frac{U}{m}$, $\left[\frac{\text{kJ}}{\text{kg}}\right]$
- Specific enthalpy $h - \frac{H}{m}$, $\left[\frac{\text{kJ}}{\text{kg}}\right]$
- Specific entropy $s = \frac{S}{m}$, $\left[\frac{\text{kJ}}{\text{kg K}}\right]$

Specific state values stay constant when dividing a system[18] or merging several subsystems. Thus specific state values behave as intensive state values. In contrast to intensive state values, specific state values are no driving forces for changes of state, see Fig. 3.14. The left picture in Fig. 3.14 shows a liquid and a vapour both having the same pressure p and temperature T, i.e. they are in thermodynamic equilibrium. Systems like this will be investigated in part II: Real fluids can undergo a change of aggregate state. Just think of water that is heated up and finally evaporates. During

---

[18]Going back to the example with the cup of coffee: When the coffee is decanted in two cups, the mass m as well as the volume V sure vary, but its density $\rho = \frac{1}{v} = \frac{m}{V}$ remains constant.

Fig. 3.14 Intensive versus specific state value according to Hahne [1] left: thermodynamic equilibrium possible, though $\rho_1 < \rho_2$ right: thermodynamic equilibrium impossible, if $p_1 \neq p_2$

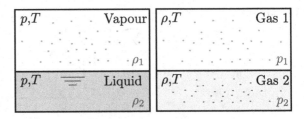

vaporisation liquid and vapour can occur coincident in thermodynamic equilibrium. In part II the term *wet steam* will be introduced to describe this phenomena. Referring to Fig. 3.14, initially both sub-systems are separated by a dividing wall. Once the wall is removed, the system does not drive itself into a new thermodynamic equilibrium, though the densities $\rho_1 = \frac{1}{v_1}$ and ρ_2 are different. Obviously, the specific state value v is not able to initiate a new thermodynamic state.

Now, let us have a closer look to the right picture in Fig. 3.14. In this case, two different, but ideal gases are separated by a dividing wall. Initially, both of them have the same temperature T and identical densities $\rho = \rho_1 = \rho_2$. However, the pressures are different $p_1 \neq p_2$. In Chap. 6 it will be clarified, how pressure, density and temperature correlate and how it is possible to have an arrangement as it is illustrated in Fig. 3.14.[19] According to the thermal equation of state it follows, for constant ρ and T:

$$\frac{p_1}{R_1} = \rho T \tag{3.15}$$

$$\frac{p_2}{R_2} = \rho T \tag{3.16}$$

$$\frac{p_1}{p_2} = \frac{R_1}{R_2} \tag{3.17}$$

Obviously, it is the difference of the gas constants that makes the state in the right picture of Fig. 3.14 possible. However, if the wall is removed, our everyday experience tells us, that a change of state arrives after a short while. This is due to the potential of the intensive state value p that forces the system in an equilibrium state.

Theorem 3.4 *Every change of a thermodynamic equilibrium state is based on a change of an intensive state value. A change of extensive state values does not cause a change of the thermodynamic equilibrium. Specific state values behave as intensive state values, though they do not have the power to drive a system in a new thermodynamic equilibrium!*

Molar State Values

Instead of using the mass of a system m to calculate specific state values z, one can use the molar quantity n alternatively. By doing so, we get *molar state values* z_M:

---

[19]However, the equation is given at this time, though the explanation follows later on.

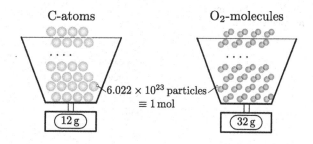

Fig. 3.15 Definition of molar mass

$$z_M = \frac{Z}{n} \tag{3.18}$$

The definition of molar quantity n is rather simple and related to the following question: How many atoms of carbon isotope[20] 12C are required in order to get 12 g of 12C?[21] As known from chemistry, $(6.022045 \pm 0.000031) \cdot 10^{23}$ atoms are required. This huge number of particles is summarised as 1 mol. In science the so-called Avogadro-constant N_A is defined as

$$N_A = (6.022045 \pm 0.000031) \times 10^{23} \frac{1}{\text{mol}}. \tag{3.19}$$

Thus, the molar quantity n has its base unit:

$$[n] = 1\,\text{mol} \tag{3.20}$$

Hence, 1 mol of carbon atoms[22] have a mass of 12 g. However, if one takes the same amount of particles, i.e. 1 mol, of a different element, the mass varies. Such as 1 mol of oxygen molecules O_2 have a mass of 32 g for instance,[23] see Fig. 3.15. This leads to the molar mass M:

$$M = \frac{m}{n} \tag{3.21}$$

In case of carbon atoms and oxygen molecules it follows:

$$M_C = 12\frac{\text{g}}{\text{mol}} = 12\frac{\text{kg}}{\text{kmol}} \tag{3.22}$$

respectively

---

[20]Isotopes are variants of a chemical element which differ in the neutron number but having the same number of protons in its core.

[21]However, this definition is arbitrary.

[22]Which equals $(6.022045 \pm 0.000031) \times 10^{23}$ atoms.

[23]This can be compared with buying fruits in a super-market: ten apples have a larger mass than ten grapes!

Table 3.2 Molar masses of relevant elements

Element	Chemical symbol	Molar mass M in $\frac{kg}{kmol}$
Carbon	C	12
Hydrogen	H	1
Oxygen	O	16
Nitrogen	N	14
Sulfur	S	32

$$M_{O_2} = 32\frac{kg}{mol} = 32\frac{kg}{kmol}. \tag{3.23}$$

The molar masses of some relevant elements are listed in Table 3.2. Based on the principle of mass conservation the molar mass of chemical bonds can be treated as a modular system, e.g.

$$M_{SO_2} = 1 \cdot M_S + 2 \cdot M_O = 64\frac{kg}{mol}. \tag{3.24}$$

Now that we know what the molar quantity n is, we can easily define the following molar state values:

- Molar volume $v_M = \frac{V}{n}$, $\left[\frac{m^3}{mol}\right]$
- Molar internal energy $u_M = \frac{U}{n}$, $\left[\frac{J}{mol}\right]$
- Molar enthalpy $h_M = \frac{H}{n}$, $\left[\frac{J}{mol}\right]$
- Molar entropy $s_M = \frac{S}{n}$, $\left[\frac{J}{mol K}\right]$

Chapter 4
Thermodynamic Equilibrium

For the further understanding of thermodynamics the principle of thermodynamic equilibrium is essential. Systems in thermodynamic equilibrium do not change their state without any impact from outside, i.e. they are perfectly in rest. This requires a mechanical equilibrium, i.e. the pressure inside the system needs to be balanced, a thermal equilibrium, i.e. the temperature needs to be balanced, and a chemical equilibrium, i.e. any chemical reaction comes to a standstill.

Theorem 4.1 *Any fully-closed thermodynamic system drives itself in a state of thermal, mechanical and chemical equilibrium!*

Space is an example for a fully-closed system that has not reached thermodynamic equilibrium yet: Any changes of state inside will let space strive into thermodynamic equilibrium. This is related with generation of entropy as Sect. 14.4 will show.

After reaching thermodynamic equilibrium the system is characterised by, see also Fig. 4.1:

- the pressure potential, that causes forces, disappears, i.e. $dp = 0$
- the temperature potential, that causes a heat flux, disappears, i.e. $dT = 0$ and
- the chemical potential, that is responsible for mass transfer respectively material conversation, see Sect. 24.5, disappears as well, i.e. $\sum_{i=1}^{k} \mu_i \, dn_i = 0$.

Thermodynamic equilibrium is further investigated once the entropy has been introduced, see Chap. 14.

© Springer Nature Switzerland AG 2019
A. Schmidt, *Technical Thermodynamics for Engineers*,
https://doi.org/10.1007/978-3-030-20397-9_4

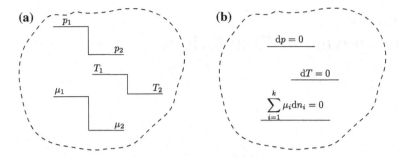

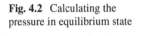

Fig. 4.1 a Imbalanced system, **b** system in thermodynamic equilibrium

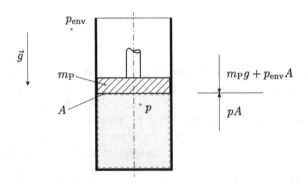

Fig. 4.2 Calculating the pressure in equilibrium state

4.1 Mechanical Equilibrium

As known from mechanics a body/system is accelerated as long as the sum of all forces is different to zero. A system is in rest, in case all forces are balanced, i.e.

$$\sum \vec{F} = 0. \tag{4.1}$$

A stone in a gravity field falls as long as it hits the ground for instance, i.e. the tracking force balances the weight force. Obviously, the stone strives for equilibrium state.

Mechanical Equilibrium in Closed Systems

In case of a closed thermodynamic system, see Fig. 4.2, the pressure inside can be gained from a balance of forces as well.

In equilibrium state there must be an equilibrium of forces, so that Eq. 4.1 is fulfilled at any position. Otherwise, the resulting force would accelerate the system and, consequently, the system would not be in equilibrium state. The system, indicated by the dashed line in Fig. 4.2, is assumed to be homogeneous.[1] Thus, the state value pressure p is uniform within the system. However, a balance of forces can be performed at any position, so e.g. at the interface gas/piston. The downward force is composed by the gravity force of the piston as well as by the environmental pressure acting on the piston's surface, i.e.

$$F_{\text{down}} = m_P g + \underbrace{p_{\text{env}} A}_{F_{\text{env}}} \tag{4.2}$$

The pressure p inside the system causes an upwards force on the lower surface of the piston, i.e.

$$F_{\text{up}} = pA. \tag{4.3}$$

From the balance of both forces

$$F_{\text{down}} = F_{\text{up}} \tag{4.4}$$

it follows, that

$$\boxed{p = p_{\text{env}} + \frac{m_P g}{A}} \tag{4.5}$$

4.2 Thermal Equilibrium

The state value temperature T is essential to quantify the internal state of a system. Before introducing the temperature itself, it is wise to start with the meaning of thermal equilibrium first. The idea of thermal equilibrium is sketched in Fig. 4.3. Just think of two separate systems: System A is supposed to be cold, while system B is supposed to be hot. Actually, both systems are closed systems, i.e. no exchange of mass is possible. Anyhow, the two systems can get in contact with each other—still separated by a dividing wall. From experience it s well known, that after a while both systems are thermally equalised, i.e. A and B are *warm*: The systems A and B have both changed their thermodynamic state. Obviously, energy was transferred among the two systems.[2] After a while both systems have the same *temperature*. During this transient process both systems transfer heat—system A is supplied with heat, system B releases heat. This process obviously is driven by a *temperature potential*. The state they both finally reach, and which they can not leave by themselves, is called thermal equilibrium—the temperature potential has gone and so has the driver for the process.

---

[1] This is an assumption. Actually, the pressure inside the system at the bottom is larger than at the interface gas/piston due to gravity. This effect is neglected though. However, this phenomena is investigated in Problem 4.5.

[2] This thermal energy has been introduced as heat, see Sect. 2.2.

Fig. 4.3 Thermal
equilibrium

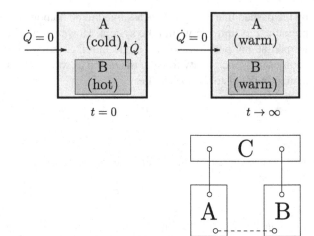

Fig. 4.4 Zeroth law of
thermodynamics

Temperature as Thermal State Value—Zeroth Law of Thermodynamics

The idea of a thermal equilibrium has such a significance in thermodyanmics, that it
has been postulated in the zeroth law of thermodynamics, illustrated in Fig. 4.4 and
explained in the following.

Theorem 4.2 *Two systems (A, B), that are respectively in a thermal equilibrium
with a third system (C), are also in a thermal equilibrium with each other.*

So, if there is a thermal equilibrium among A and C respectively B and C, the
conclusion is, that there needs to be an equilibrium among A and B as well.

What Does All This Have to Do with the State Value Temperature T?

Temperature is a measure of a system's thermal state. Different thermal states can
be determined by the temperature. Systems in thermal equilibrium have the same
temperature. In contrast, however, systems, not being in a thermal equilibrium with
each other, have different temperatures. Temperature measurement is a quantitative
measurement based on the thermal equilibrium, see Fig. 4.5. It requires a calibrated
instrument. However, the zeroth law of thermodynamics is the basis for any temperature measurement. With respect to Fig. 4.4, the thermometer is system A, that
is immersed into system B, i.e. iced water with a well known temperature. After a
while A and B are thermally balanced. If this thermometer is in contact with system
C, representing a cup of tea or coffee, afterwards, a transient balancing process takes
place. For any measurement it is required to reach a balanced state first[3]: If system

---

[3]Hence, this requires $t \to \infty$, see the very right sketch of Fig. 4.5.

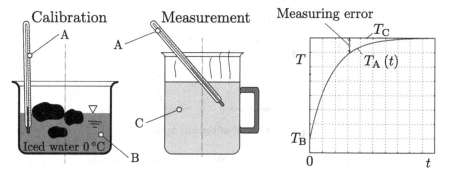

Fig. 4.5 Principle of temperature measurement

A and C are not balanced, the measurement is incorrect. Obviously, according to Fig. 4.5, system A and C do have different temperatures. A thermodynamic investigation of this example will follow once the first and second law of thermodynamics have been introduced, see Example 14.5.

4.3 Chemical Equilibrium

In case chemical reactions occur inside the system, a thermodynamic equilibrium additionally includes a chemical equilibrium. For a chemical reaction a chemical potential $\Delta \gamma \mu$ is required, analogue to a temperature potential ΔT for a heat flux or a pressure potential Δp for a mass flux. Details will be discussed in Sect. 24.5, see Example 24.7.

4.4 Local Thermodynamic Equilibrium

As mentioned above, each fully-closed system drives itself in a state of equilibrium. So question now is, what happens, if the system is not fully closed, see Fig. 4.6, that shows a heat conducting wall. The system is not fully closed, as a heat flux enters and leaves the system, but is supposed to be in steady state. Due to

$$\dot{Q}_1 = \dot{Q}_2 \tag{4.6}$$

incoming and leaving energy fluxes are equal—the prerequisite for steady state. Steady state actually means, that there is no *temporal* change of any state value inside the system. However, according to our previous considerations, the system is not in thermodynamic respectively thermal equilibrium, as the temperature distribution

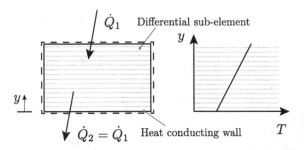

Fig. 4.6 Imbalanced systems in steady state—Example of a heat conducting wall

inside the system is not uniform, i.e. the system is heterogenous.[4] This is due to a temperature gradient inside the system, that is required in order to realise heat fluxes into and out of the system.[5] However, the thermal imbalance is maintained by external fluxes crossing the system boundary. Though the entire system *heat conducting wall* in Fig. 4.6 is not in thermodynamic equilibrium, it can be divided into several, differential elements. Each of these elements is in a local thermodynamic equilibrium, see [14].

Theorem 4.3 *The definition of a local thermodynamic equilibrium allows us to apply the laws of thermodynamics also to locally varying states.*

4.5 Assumptions in Technical Thermodynamics

Figure 4.7 shows the assumptions that are followed in technical thermodynamics:

- Due to gravity the gas pressure varies with the height of the system. However, a constant pressure with height is assumed,[6] that is derived by a balance of forces just beneath the piston, thus:

$$p = p_{\text{env}} + \frac{m_{\text{P}} g}{A} \tag{4.7}$$

This assumption is further investigated in Problem 4.5.
- In order to transfer heat Q into or out of a system a temperature gradient is required, see Fig. 4.7. This temperature gradient is ignored by common technical approaches: The system is supposed to be in thermal equilibrium, indicated by a uniform temperature T within the gas.

---

[4]The fluxes at the system's border indicate, that the system is permanently forced from outside to be imbalanced with the environment.

[5]This is part of the lecture Heat and Mass Transfer.

[6]Thus, the weight of the gas is neglected.

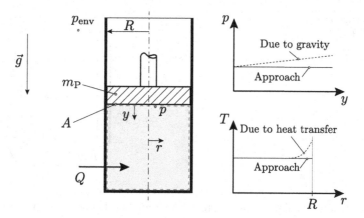

Fig. 4.7 Assumptions in thermodynamics

Fig. 4.8 Sketch to
Problem 4.4

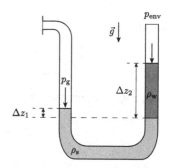

Problem 4.4 A U-pipe, that is attached to a low pressure gas conduit, is filled with silicone oil ($\rho_s = 1203 \frac{kg}{m^3}$) as sealing liquid. To enlarge the pressure display, water ($\rho_w = 998 \frac{kg}{m^3}$) is filled above the sealing liquid in the open flank (Fig. 4.8).

(a) What are the absolute pressure and the gauge pressure in the gas conduit, if $\Delta z_1 = 147$ mm, $\Delta z_2 = 336$ mm and $p_{env} = 0.953$ bar is applied?
(b) What level difference Δz would occur at the same pressure in the gas conduit, if no water was filled into the U-pipe?

Solution
Since the system is in equilibrium state, all forces need to balance at any position. If the forces are not balanced, there would be an acceleration, i.e. the system could not be in rest. So the position marked with the area A can be picked, see Fig. 4.9:

$$F_{down} = F_{up} \tag{4.8}$$

With

$$F_{down} = (p_{env} + \rho_w g \, \Delta z_2) A \tag{4.9}$$

Fig. 4.9 Solution to
Problem 4.4

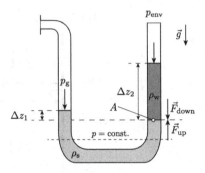

On the same horizontal position the pressure within the silicone oil is constant. This
leads to:

$$F_{up} = (p_g + \rho_s g\,\Delta z_1)\,A \tag{4.10}$$

Finally we get

$$p_g = p_{env} + \rho_w g\,\Delta z_2 - \rho_s g\,\Delta z_1 = 0.968\,55\,\text{bar} \tag{4.11}$$

The gauge pressure then is

$$\Delta p = p_g - p_{env} = 0.0155\,\text{bar} \tag{4.12}$$

If there was no water, the difference Δz would be

$$\Delta z = \frac{p_g - p_{env}}{\rho_s g} = 0.1317\,\text{m} \tag{4.13}$$

Problem 4.5 A basin has a height of $\Delta z_B = 15\,\text{m}$. The gauge pressure at the upper
end of the basin is measured by a U-pipe manometer and equals a water column
of $\Delta z_M = 1600\,\text{mm}$ against an atmosphere of $p_{env} = 1020\,\text{mbar}$ ($\rho_w = 1000\,\frac{kg}{m^3}$,
$g = 9.81\,\frac{m}{s^2}$).

(a) Determine the pressure p_1 at the upper end and the pressure p_2 at the bottom of
the tank. The tank is filled with nitrogen. For the dependency of the density of
nitrogen on the pressure please assume that

$$\rho_{N2} = C \cdot p \quad \text{with } C = 1.149\frac{kg}{\text{bar m}^3} \tag{4.14}$$

(b) What is the pressure p_2 if the tank is filled with oil instead of nitrogen? The
pressure p_1 shall be the same as before. Oil is considered to be an incompressible
liquid with $\rho_{oil} = 870\,\frac{kg}{m^3}$.

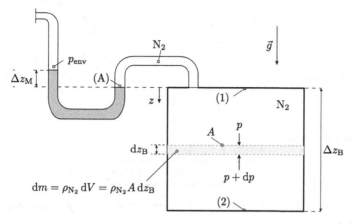

Fig. 4.10 Solution to Problem 4.5

Solution

The pressure p_1 can be calculated by a balance of forces at position (A)

$$\boxed{p_1 = p_{\text{env}} + \rho_w g\, \Delta z_M = 1.176\,96\,\text{bar}} \qquad (4.15)$$

Since the density is not constant, but a function of pressure, p_2 can not be calculated as easily as p_1. To get an overview, it is wise to start with a differential approach, see Fig. 4.10. Performing a balance of forces at the lower edge of the differential element leads to:

$$\underbrace{pA + \rho_{N_2} gA\, dz_B}_{F_{\text{down}}} = \underbrace{(p + dp)\, A}_{F_{\text{up}}} \qquad (4.16)$$

Thus, we find

$$dp = \rho_{N_2} g\, dz_B \qquad (4.17)$$

Replacing $\rho_{N2} = Cp$ brings

$$dp = Cpg\, dz_B \qquad (4.18)$$

This differential equation can be solved by separating the variables:

$$\frac{dp}{p} = Cg\, dz_B \qquad (4.19)$$

and integrating from (1) to (2)

$$\int_1^2 \frac{dp}{p} = \int_1^2 Cg\,dz_B. \tag{4.20}$$

Solving the integral

$$\ln\frac{p_2}{p_1} = Cg\,\Delta z_B \tag{4.21}$$

and rearranging finally leads to

$$\boxed{p_2 = p_1 e^{Cg\,\Delta z_B} = 1.179\,\text{bar}} \tag{4.22}$$

In case incompressible oil is used instead of nitrogen, the density would be constant. Thus, it follows, that

$$\boxed{p_2 = p_1 + \rho_{oil}g\,\Delta z_B = 2.457\,\text{bar}} \tag{4.23}$$

Chapter 5
Equations of State

Chapter 3 has shown, how to quantify the internal state of a system. Basically, a distinction has been made between thermal and caloric state values. Since not all state values can be measured directly,[1] equations to calculate them are required. The Paddle Wheel experiment for instance, see Sect. 2.4.1, has shown, that the internal energy describes the energetic state of a fluid. However, it is not possible to measure the internal energy of a system, so that its dependency on other state values is of interest. Furthermore, the question is, how many state values are mandatory to define a system unambiguously. Obviously, once a system is given unequivocally, it must be possible to estimate the other state values. Equations of state cover the physical dependency among state values. Preliminary considerations regarding equations of state are introduced in this chapter.

Theorem 5.1 *Characterisation of a system's state by state values is only unequivocal, if the system is in steady state respectively in a (local) thermodynamic equilibrium. Otherwise the system would be subject to a transient balancing process, i.e. the system's state is temporally changing.*

5.1 Gibbs' Phase Rule

First, it needs to be clarified, how many state values are required to fix the state of a system unambiguously. According to Gibbs' phase rule[2]

$$F = C - P + 2 \tag{5.1}$$

the degree of freedom F depends on the number of components C as well as the number of phases in thermodynamic equilibrium P. The degree of freedom F denotes

---

[1]Such as enthalpy, internal energy and entropy for instance.

[2]Josiah Willard Gibbs ($\star$11 February 1839 in New Haven, Connecticut, †28 April 1903 in New Haven, Connecticut)

© Springer Nature Switzerland AG 2019
A. Schmidt, *Technical Thermodynamics for Engineers*,
https://doi.org/10.1007/978-3-030-20397-9_5

the largest number of independent intensive variables, see Sect. 3.2.5, that can be varied simultaneously and arbitrarily without influencing one another.

Thermodynamic Proof for a Balanced System Without Chemical Reactions

The proof follows according to [15]. In case several components C exit in a single phase[3] $P = 1$, e.g. gaseous, one can specify its composition with the molar fraction[4] x_i of each component. Due to

$$\sum_{i=1}^{n} x_i = 1 \qquad (5.2)$$

only $(C - 1)$ molar fractions are required to specify the composition—the missing molar fraction follows from the mass balance. Now, let us assume, there are several phases P. Following our considerations, the total number of concentrations, i.e. intensive state values as well, to specify the thermodynamic system is

$$P(C - 1) \qquad (5.3)$$

In thermodynamic equilibrium the two intensive state values pressure p and temperature T are equal for all components as well as for all phases. So, the total number of intensive state values is

$$P(C - 1) + 2 \qquad (5.4)$$

Furthermore, the chemical potentials μ_i of each component in all phases are equal, since the system is in thermodynamic equilibrium. The temperature potential describes the potential for a heat flux, the pressure potential the potential for a mass flux and the chemical potential the potential for a chemical reaction respectively diffusion. In a thermodynamic balanced system, there is neither a temperature/pressure nor a chemical potential. Humid air is an excellent example, that shows the impact of the chemical potential: Let us assume humid air, i.e. an ideal mixture of dry air and vapour, is exposed to a reservoir of liquid water. As long as the humid air is *not saturated with vapour*, i.e. a chemical potential between the liquid water's surface and the vapour phase exists, water evaporates from the reservoir to the air. A thermodynamic equilibrium is reached as soon as the potential between surface and vapour has disappeared. The mass flux of liquid/vapour then stops, the air is finally in saturated state.

---

[3]E.g. air is an a gaseous state, i.e. $P = 1$, and it, simplified, contains $C = 2$ components, i.e. nitrogen and oxygen.
[4]The molar fraction will be introduced in part II and is defined as $x_i = \frac{n_i}{n_{total}}$. Thus, it has the meaning of a concentration of a component in a mixture.

A missing chemical potential reads for two phases and n components as:

$$\mu_{P=1,C=1} = \mu_{P=2,C=1} \tag{5.5}$$

$$\mu_{P=1,C=2} = \mu_{P=2,C=2}$$

$$\vdots$$

$$\mu_{P=1,C=n} = \mu_{P=2,C=n}$$

For m phases it is

$$\mu_{P=1,C=1} = \mu_{P=2,C=1} = \cdots = \mu_{P=m,C=1} \tag{5.6}$$

$$\mu_{P=1,C=2} = \mu_{P=2,C=2} = \cdots = \mu_{P=m,C=2}$$

$$\vdots$$

$$\mu_{P=1,C=n} = \mu_{P=2,C=n} = \cdots = \mu_{P=m,C=n}$$

The conclusion is, that there are

$$C(P-1) \tag{5.7}$$

of these mathematical equilibrium conditions. Finally, the degree of freedom F is the total number of variables reduced by the number of equilibrium correlations:

$$F = P(C-1) + 2 - C(P-1) \tag{5.8}$$

In other words

$$\boxed{F = C + 2 - P} \tag{5.9}$$

5.1.1 Single-component Systems Without Phase Change

In part I the focus is on single component systems, i.e. $C = 1$, that are not subject to a phase change, i.e. $P = 1$. This can be a gas, e.g. N_2, that stays gaseous throughout the entire change of state. According to Eq. 5.1 the degree of freedom is

$$F = C - P + 2 = 2 \tag{5.10}$$

Table 5.1 Exemplary combinations of state values for ideal gases in thermodynamic equilibrium

State value 1	State value 2	Statement
T	p	True[a]
T	$v(p, T)$	True
T	$u(T)$	False[b]
T	$h(T)$	False
T	$s(p, T)$	True
v	ρ	False[c]
$h(T)$	$u(T)$	False[d]
$h(T)$	$s(p, T)$	True
p	$s(p, T)$	True
p	$v(p, T)$	True
p	$u(T)$	True
p	$h(T)$	True
$s(p, T)$	$v(p, T)$	True

[a]True means state is unequivocal.
[b]False means state is equivocal.
[c]Due to $\rho = \frac{1}{v}$.
[d]Information regarding p is missing.

Thus, two independent intensive/specific state values, i.e. pressure p and temperature T, are selectable separately and independently from each other and fix the internal state of a system. According to this, combinations of independent state values for ideal gases are listed in Table 5.1. It also indicates what combinations of two state values are independent. However, the physical dependency among the state values will be derived in Chaps. 6 and 12. If additional information regarding the size of the system is required, one extensive state value must be stated additionally.

Gases in part I, however, are treated as ideal, Eq. 5.10 can be applied. Part I additionally focuses on so-called incompressible liquids.[5] These fluids keep their specific volume v no matter how much pressure they are exposed to. According to [4], the specific volume remains constant with respect to temperature variations as well, i.e.

$$\left(\frac{\partial v}{\partial T} \right)_p = \left(\frac{\partial v}{\partial p} \right)_T = 0 \tag{5.11}$$

Anyhow, Eq. 5.10 can be applied for incompressible liquids without phase change as well.

---

[5]Being incompressible is an idealised model. However, in part II real fluids are discussed, that do not follow this assumption any more.

5.1.2 Single-component Systems with Phase Change

In part II single-components ($C = 1$), that change their aggregate state, will be investigated, e.g. water being in liquid and vapour state, i.e. $P = 2$. Anyhow, these fluids
are treated as *real*. Figure 5.1, known as p, T-diagram shows the aggregate state of
water as a function of pressure p and temperature T.[6] Once water is heated up at
constant pressure, its temperature rises first, while it is still in liquid state.[7] At the
boiling point the fluid starts to evaporate. Hence, liquid as well as vaporous state, i.e.
$P = 2$, occur. During evaporation the temperature stays constant, the entire thermal
energy is used for the change of aggregate state.[8] According to Eq. 5.1, the degree
of freedom during evaporation is $F = 1$. This means, only one intensive state value
is free selectable, e.g. the pressure. It will be shown in part II, that in this two-phase
region the pressure is a function of temperature, known as vapour pressure curve.[9]
However, water can even exist in three phases (liquid, solid, vapour, i.e. $P = 3$)
coincidently. In this case the number of freedom is

$$F = 1 - 3 + 2 = 0 \tag{5.12}$$

Thus, none intensive state value is free selectable. This so-called triple point (TP)
exists in solely one specific point, e.g. for water at 0.01 °C and 611 Pa.

5.1.3 Multi-component Systems

Multi-component systems will be treated in parts II and III. For a binary mixture of
ideal gases,[10] for instance, the degree of freedom is

$$F = 2 - 1 + 2 = 3 \tag{5.13}$$

As a consequence, three independent intensive state values are free selectable. For
example, in a gaseous mixture of two components the pressure p and the temperature
T can be selected freely. However, the specific volume v of the mixture can be
selected independently as well by varying the mixture ratio. In this case, pressure p,

---

[6]This characteristic will be explained in Chap. 18.

[7]This is called sensible heat, since the supply of thermal energy can directly be measured by a
thermometre.

[8]Thus, this thermal energy is called latent heat, since it can not be measured by an increase of
temperature.

[9]You might already know, that the temperature, at what boiling starts, depends on the pressure.
Water on the Mount Everest starts boiling at lower temperature than on sea-level due to pressure
variation for instance.

[10]In such a case two components are available, i.e. $C = 2$. Both are ideal gases, i.e. only one
aggregate state occurs, so that $P = 1$.

Fig. 5.1 p, T-diagram for
water—Degrees of freedom

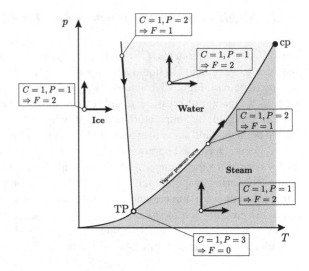

temperature T and molar fraction x_1 of one component are intensive state values,
that quantify the internal state of the system. The second molar fraction x_2 is not
independent any more, since the overall balance needs to be satisfied, i.e.

$$x_1 + x_2 = 1 \tag{5.14}$$

5.2 Explicit Versus Implicit Equations of State

It has been shown, that the state of a single, ideal gas ($C = 1$, $P = 1$) is unambiguously defined by two independent intensive/specific ($F = 2$) and one extensive state
value:

- The extensive state value quantifies the size of the system, for example by n, m,
  V, H, ...
- If just the intensive state and not the size of a system is of interest, the extensive
  state value can be ignored

Consequently, if a state is given unambiguously by two independent intensive/specific
state values, the other state values need to depend on these two independent state
values. Thus, a mathematical correlation, such as

$$z = f(x, y) \tag{5.15}$$

must exist. This correlation is called *equation of state*, that depends on the fluid's
properties. However, equations of state can be given implicitly as well as explicitly.

- Implicit:

$$f(x, y, z) = 0 \tag{5.16}$$

An example is the thermal equation of state, that is derived in Chap. 6 in implicit notation

$$f(p, v, T) = 0 \tag{5.17}$$

respectively

$$pv - RT = 0 \tag{5.18}$$

- Explicit:

$$z = f(x, y) \tag{5.19}$$

The thermal equation of state in an explicit notation is

$$T = f(p, v) \tag{5.20}$$

respectively

$$T = \frac{pv}{R} . \tag{5.21}$$

Chapter 6
Thermal Equation of State

In this chapter, the correlation between the thermal state variables pressure p, specific volume v and temperature T is investigated. As already discussed in the previous chapter, thermal state values can easily be measured. Furthermore, according to Gibbs' phase law for ideal gases two intensive state values, i.e. pressure and temperature, fix the state of a thermodynamic system unequivocally. Thus, a mathematical function to calculate the third thermal state value v must exist:

$$v = f(p, T) \tag{6.1}$$

6.1 Temperature Variations

To derive the thermal equation of state 6.1, first, the following experiment conducted by Gay-Lussac[1] is introduced: An ideal gas with a well-known mass m is filled into a cylinder, that is closed by a freely movable piston, see Fig. 6.1. The cylinder shall be insulated against the environment. The pressure inside the cylinder in thermodynamic equilibrium can be adjusted by a variation of the mass of the piston m_P. For the first part of the experiment the piston's mass shall be $m_{P,1} = $ const. With respect to the balance of forces, see Sect. 3.2.1, it is:

$$p_1 = p_{\text{env}} + \frac{m_{P,1} g}{A}. \tag{6.2}$$

However, the experiment starts with an initial temperature T_A, that can be determined with a thermometer. Due to supply of thermal energy the system is heated up, so that after a long period of time the system achieves a *new thermodynamic equilibrium*. Its state has changed, recognisable by a different temperature $T_B \neq T_A$ as well as a larger volume. Obviously, the gas inside the cylinder has expanded. The specific

---

[1] Joseph Louis Gay-Lussac (∗6 December 1778 in Saint-Léonard-de-Noblat, †9 May 1850 in Paris).

© Springer Nature Switzerland AG 2019
A. Schmidt, *Technical Thermodynamics for Engineers*,
https://doi.org/10.1007/978-3-030-20397-9_6

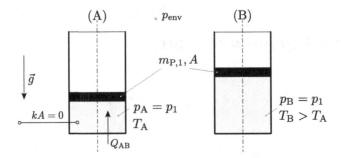

Fig. 6.1 Gay-Lussac law—Set-up

volume can be calculated according to

$$v_A = \frac{V_A}{m} \tag{6.3}$$

$$v_B = \frac{V_B}{m} \tag{6.4}$$

Since the piston's mass has not been changed while heating the system up, the pressure in state (B) is the same as in state (A).[2] The result of this first experiment is plotted in a v, T-diagram, see Fig. 6.2. In order to get sufficient resolution of the findings, the temperature can be varied in small steps, controllable by the amount of supplied heat. In a next step, the entire experiment is repeated with a modified pressure inside the cylinder. As mentioned before, this can easily be achieved by replacing the piston, i.e. variation of its mass. Once again, the state values can be plotted in the introduced v, T-diagram: If the pressure is decreased, the rise of volume while heating the system gets larger. This led Gay-Lussac to the following conclusion

Theorem 6.1 *The volume of ideal gases is at constant pressure and at constant mass directly proportional to its temperature. A gas expands when heated up and contracts when cooled down.*

According to Fig. 6.2 the mathematical function reads as:

$$v = v(\vartheta) = T \frac{v_0 \, (p)}{T_0} = T \, f(p) \tag{6.5}$$

$$\boxed{v = T \, f(p)} \tag{6.6}$$

Obviously, $\frac{v_0}{T_0}$ is the gradient of the linear function. This slope is a function of the pressure p. The larger the pressure p is, the smaller the gradient is in a v, T-diagram.

---

[2]Assuming, that the ambient pressure p_{env} is constant as well. This is an essential assumption for the environment: Its state is constant and homogeneous!

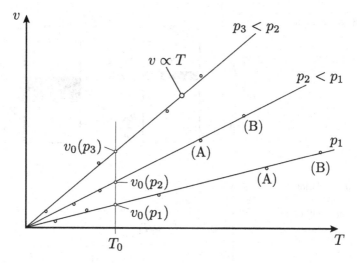

Fig. 6.2 Gay-Lussac law—Results for thermodynamic equilibrium

6.2 Pressure Variations

Boyle[3] and Mariotte[4] performed their investigations independently from each other. Similar to Gay-Lussac's experiment a cylinder is filled with a constant mass m of an ideal gas and closed by a freely movable piston. The cylinder shall have a perfect thermal contact to the environment: Once a *thermodynamic equilibrium* is achieved the temperature of the gas is ambient temperature $T_1 = T_A$, see Fig. 6.3. The pressure in state (A) depends on the mass of the piston $m_{P,A}$ and on ambient pressure, ie.

$$p_A = p_{env} + \frac{m_{P,A} g}{A}.$$ (6.7)

In the next step the mass of the piston is increase to $m_{P,B} > m_{P,A}$. As a consequence the gas pressure is

$$p_B = p_{env} + \frac{m_{P,B} g}{A} > p_A.$$ (6.8)

After a long period of time thermodynamic equilibrium is reached again. Thus, the temperature is $T_B = T_A = T_1$. However, the volume in state (B) is smaller than in state (A). Due to the constant mass of the gas it follows for the specific volume v:

$$v_B < v_A$$ (6.9)

---

[3]Robert Boyle (∗4 February 1627 in Lismore, Ireland, †10 January 1692 in London).
[4]Edme Mariotte (∗1620 in Dijon, †12 May 1684 in Paris).

Fig. 6.3 Boyle–Mariotte law—Set-up

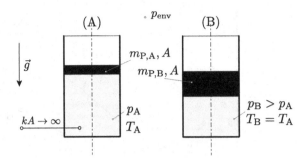

Fig. 6.4 Boyle–Mariotte law—Results for thermodynamic equilibrium

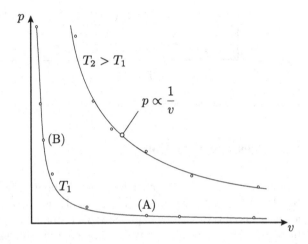

Specific volume v as well as pressure p can be visualised in a p, v-diagram, see Fig. 6.4. The experiment is repeated with a variation of the piston's mass in order to refine the plot for a constant ambient temperature T_1.

Finally, the entire experiment is repeated for another ambient temperature $T_2 > T_1$. The results for thermodynamic equilibrium are plotted in the p, v-diagram as well, see Fig. 6.4. Obviously, a second curve for T_2 exists. The larger the temperature of the gas is, the larger its specific volume is. Each curve, belonging to one specific temperature, has the form of a hyperbola. Thus, the conclusion reads as:

Theorem 6.2 *The pressure of ideal gases at constant temperature and at constant mass is inversely proportional to its volume. If the pressure is increased, the volume decreases. If the pressure decreases, the gas expands.*

For each $T = \text{const.}$, the experimental data shows, that each p, v-pair follows a hyperbola:

$$p = \frac{\text{const.}}{v} = \frac{g\,(T)}{v} \tag{6.10}$$

In another mathematical notation Boyle–Mariotte's law reads as:

$$pv = g(T)$$ (6.11)

6.3 Ideal Gas Law

From the Gay-Lussac law

$$v = Tf(p)$$ (6.12)

it follows by multiplying with p

$$pv = p\,f(p)\,T.$$ (6.13)

Comparing with the Boyle–Mariotte law

$$pv = g(T)$$ (6.14)

the right side of the Eq. 6.13 needs to be solely a function of temperature:

$$p\,f(p)\,T = g(T).$$ (6.15)

Thus, if the right side is a function of T, the left side of Eq. 6.15 as well needs to be solely a function of T. This can only be the case if

$$pf(p) = \text{const.} \equiv R.$$ (6.16)

Rearranging Eq. 6.13 leads to

$$pf(p) = \frac{pv}{T}.$$ (6.17)

Combining with Eq. 6.16, the thermal equation of state reads as:

$$pv = RT$$ (6.18)

In Eq. 6.16 a constant R has been defined, that is known as specific gas constant. For each gas R is individual. By this equation of state all thermal state values are linked with each other.

Equation 6.18 can be multiplied with the system's mass. Taking Eq. 3.13 into account as well, it is

$$pV = mRT$$ (6.19)

Replacing the mass m with the molar quantity n and the molar mass M:

$$pV = nMRT$$ (6.20)

Dividing by the molar quantity n:

$$pv_M = MRT \tag{6.21}$$

Avogadro[5] was an Italian scientist, most noted for his contribution to molecular theory known as Avogadro's law:

Theorem 6.3 *Equal volumes of gases under the same thermal conditions, i.e. temperature and pressure, contain an equal number of molecules, i.e. the same molar quantity.*

Let us assume two ideal gases A and B at the same pressure $p = p_A = p_B$ and the same temperature $T = T_A = T_B$. According to Avogadro, for the same molar quantity $n = n_A = n_B$, the volumes are equal as well, $V = V_A = V_B$. Applying Eq. 6.20 for each gas leads to:

$$pV = nM_A R_A T \tag{6.22}$$

$$pV = nM_B R_B T \tag{6.23}$$

Dividing Eqs. 6.22 and 6.23 shows that

$$M_A R_A = M_B R_B = \text{const.} \equiv R_M \tag{6.24}$$

Hence, the product of molar mass M and individual gas constant R is constant and the same for all ideal gases! This constant is called universal gas constant R_M. Its value is

$$\boxed{R_M = RM = 8.3143 \, \frac{\text{kJ}}{\text{kmol K}}} \tag{6.25}$$

Thus, the thermal equation of state might also be given in the following notation

$$\boxed{pv_M = R_M T} \tag{6.26}$$

Problem 6.4 Calculate the molar volume v_M of an ideal gas at standard conditions,[6] i.e. $\vartheta = 0\,°C$ and $p = 1.01325$ bar!

Solution

To answer this question, the thermal equation of state is applied:

$$v_M = \frac{R_M T}{p} \tag{6.27}$$

---

[5]Lorenzo Romano Amedeo Carlo Avogadro (∗9 August 1776 in Turin, †9 July 1856 in Turin).
[6]According to DIN 1343.

You should mind the units, when solving the equation. It is always a good idea to use SI-units:

$$v_M = \frac{8314.3 \, \frac{J}{kmol \, k} \, 273.15 \, K}{101325 \, \frac{N}{m^2}} = 22.414 \, \frac{m^3}{kmol} \tag{6.28}$$

1 kmol of any ideal gas cover a volume of $22.414 \, m^3$.

Problem 6.5 Calculate the mass of $V = 3.5 \, m^3$ of nitrogen (N_2) at a temperature of $\vartheta = 25 \, °C$ and a pressure of $p = 3 \, bar$!

Solution

In order to find the mass, the thermal equation of state is given in a rearranged notation

$$m = \frac{pV}{R_{N_2} T}. \tag{6.29}$$

However, the gas constant of nitrogen R_{N_2} is unknown. It can be determined by using the universal gas constant R_M. Thus, the molar mass of nitrogen ($M_{N_2} = 28 \, \frac{kg}{kmol}$) is required, so that

$$R_{N_2} = \frac{R_M}{M_{N_2}} = \frac{8314.3 \, \frac{J}{kmol \, K}}{28 \, \frac{kg}{mol}} = 296.94 \, \frac{J}{kg \, K} \tag{6.30}$$

$$m = \frac{pV}{RT} = \frac{3 \times 10^5 \, Pa \cdot 3.5 \, m^3}{296.94 \, \frac{J}{kg \, K} \cdot 298.15 \, K} = 11.86 \, kg \tag{6.31}$$

Chapter 7
Changes of State

So far thermodynamic systems have been discussed and state values to quantify their internal state have been categorised. In this chapter the focus will be on changes of state, i.e. bringing a system from an initial state to a new state. In order to do so the following distinction is required:

- In this first case, the initial state shall be an equilibrium state (1). External impacts, e.g. work and/or heat that act on the system, will bring the system into a new equilibrium state (2), see Fig. 7.1.
- The second case is about systems, that are *not* in equilibrium state. It has been discussed in the previous chapters, that state values have a different influence on systems: A heterogenous distribution, i.e. imperfection, of intensive state values will initiate a transient balancing process to force the system into thermodynamic equilibrium. In contrast to this, imperfections of non-intensive state vales do not have the power to cause a balancing process. Anyhow, imperfections with respect to temperature, pressure, concentration or even a position in a gravitational field force a system to drive into equilibrium state *without* additional external impacts, see also Chap. 14. Problem 2.4 is an example for a transient balancing process. Initially, the water-filled column is in equilibrium state, since the linking valve to the second column is closed. However, once the valve is opened, a mechanical imbalance between the two columns occurs, that makes the system drive into equilibrium state. Driver of this balancing process is gravity. Even space might be regarded as a closed system without any external impacts. However, since it has not reached an equilibrium, transient balancing processes run inside.

Initially, the system in Fig. 7.1 is in equilibrium state (1), i.e. all state values are constant with respect to time. Thus, the thermal equation of state

$$pv = RT \tag{7.1}$$

can be applied. Due to external impacts, e.g. by a supply of heat or work across the system boundary, the system changes its state. After a while the system is in an equilibrium state (2) as well. Equation 7.1 can be applied again.

© Springer Nature Switzerland AG 2019
A. Schmidt, *Technical Thermodynamics for Engineers*,
https://doi.org/10.1007/978-3-030-20397-9_7

Fig. 7.1 Change of state from one equilibrium state to another

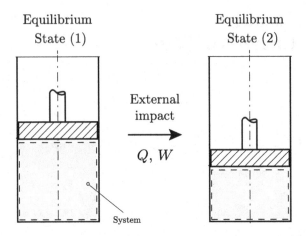

Within this chapter it is clarified how changes of state run. For the further understanding of thermodynamics this is essential, since process values, i.e. work, that influence a system, always depend on the path a change of state takes. So before solving a thermodynamic problem, you should figure out, which path the system takes.[1]

Theorem 7.1 *A change of state is the transition from one state of equilibrium into another by external impacts. However, imbalanced systems, i.e. systems that are not in thermodynamic equilibrium, are subject to transient balancing processes. Drivers for these changes of state are always potentials of intensive state values, e.g. pressure or temperature differences within the system. External impacts are not required.*

7.1   The p, v-Diagram

A p, v-diagram is commonly used in thermodynamics to visualise changes of state and to understand the efficiency of thermodynamic cycles for instance. As already shown, see Chap. 5, pressure p and temperature T clearly define a thermodynamic state. Since the specific volume v is given by

$$v = f(p, T) = \frac{RT}{p} \tag{7.2}$$

each pair of p, v unequivocally gives a thermodynamic state for ideal gases as well. The specific volume v is plotted as abscissa, the pressure p as ordinate. In Sect. 9.2.2

---

[1] Key to solve problems in thermodynamics is to understand the path a change of state takes!

the importance of a p, v-diagram even increases since work can be visualised in such a diagram. Anyhow, the p, v-diagram is explained within the next sections exemplary for typical thermodynamic changes of state.

7.1.1 Isothermal Change of State

An isothermal change of state is characterised by

$$T_1 = T_2 = T = \text{const.,} \tag{7.3}$$

so that the temperature stays constant for the entire change of state. What does it mean in terms of specific volume v and pressure p? According to the thermal equation of state

$$p = f(v) = \frac{RT}{v} = \frac{\text{const.}}{v} \tag{7.4}$$

an isothermal, i.e. a curve representing a constant temperature, has the form of a hyperbola in a p, v-diagram, see Fig. 7.2. Let us assume, an ideal gas is filled in a cylinder, that is closed by a freely movable piston. Obviously, since no gas can leave the cylinder, it is a closed system. On its way from state (1) to state (2), following a hyperbola, the gas is compressed by moving the piston down. Thus, the volume filled by the ideal gas decreases. Since its mass is constant, the specific volume decreases as well:

$$v_2 < v_1 \tag{7.5}$$

Consequently, according to the p, v-diagram the pressure increases. This is additionally indicated by the sketch on the left in Fig. 7.2. State (1) and state (2) are both in thermodynamic equilibrium, see Chap. 4 and Sect. 5.1:

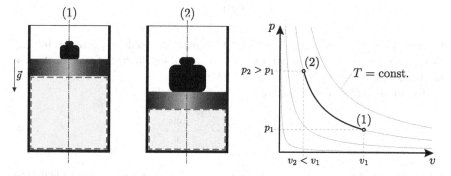

Fig. 7.2 Isothermal change of state

1. The system is thermally balanced, i.e. it is characterised by a uniform temperature distribution inside the system in states (1) and (2).
2. The system is mechanical in rest.

Hence, a thermal as well as a mechanical equilibrium is achieved. The pressure though results of a balance of forces, see Sect. 3.2.1. The larger pressure in state (2) compared to state (1) is compensated by a larger mass of the piston. However, the assumption is, that the weight of the gas can be ignored, so that the gas pressure inside the system is constant and does not vary by height, see also Sect. 4.5. As shown in Problem 4.5 the mechanical balance inside the gas actually, due to gravity, causes a larger pressure at the bottom of the cylinder than at the top. This effect is ignored from now on.

Theorem 7.2 *The pressure results instantaneously out of a local balance of forces at the edge to the piston and is assumed to be uniform within the system! Thus, the weight of the gas is ignored.*

7.1.2   Isobaric Change of State

In this second case, the ideal gas undergoes an isobaric change of state, so that

$$p_1 = p_2 = p = \text{const.} \tag{7.6}$$

Once again, the gas in the cylinder is compressed exemplary as illustrated in Fig. 7.3. In this example, the specific volume in state (2) is smaller than in state (1). Both states are at equilibrium, i.e. state (1) as well as state (2) do not vary temporally,[2] see also Sect. 7.1.1. Since the piston's mass is kept constant, the overlaying pressure is constant. What does is mean for the temperature? This can easily be answered by means of the p, v-diagram and the thermal equation of state

$$T = f(v) = \frac{pv}{R}. \tag{7.7}$$

Obviously, if $p = \text{const.}$ and $v_2 < v_1$, the temperature needs to decrease. In case there is a thermal equilibrium with the environment, the change of state of the gas has been initiated due to a cooling down environment.

The isothermal in a p, v-diagram follows a hyperbola

$$p = f(v) = \frac{RT}{v} = \frac{\text{const.}}{v}. \tag{7.8}$$

---

[2]The gas as part of the investigated system may also be at equilibrium though it is not in thermal equilibrium with the outer environment. In this case, the gas is adiabatically insulated and is not affected by the environment. In this case the temperature inside the cylinder is constant with respect to time and uniform!

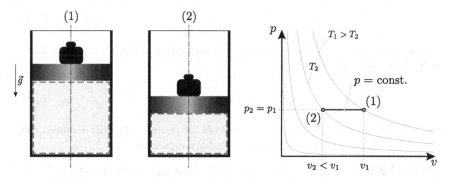

Fig. 7.3 Isobaric change of state

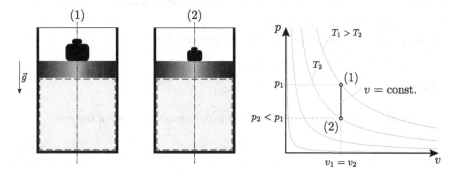

Fig. 7.4 Isochoric change of state

Figure 7.3 shows the isothermals as well. Obviously, according to Eq. 7.8, the larger the temperature is, the more the hyperbolas are shifted upwards. For the further understanding of thermodynamics this knowledge of the isothermals is essential.

7.1.3 Isochoric Change of State

Another change of state is possible, if the piston is fixed. Under this premise all changes of state occur at constant volume, see Fig. 7.4. Since the mass of the gas inside the cylinder is constant, the specific volume is constant as well, also known as isochoric change of state:

$$v_1 = v_2 = v = \text{const.} \tag{7.9}$$

In the example given in Fig. 7.4 an ideal gas in a closed cylinder is cooled down isochorically. As indicated in the p, v-diagram, state (2) lies on a lower isothermal than state (1), i.e.

$$T_2 < T_1. \tag{7.10}$$

The change in terms of pressure can also been visualised in the p, v-diagram. Obviously, the pressure decrease as well

$$p_2 < p_1. \tag{7.11}$$

This is documented in the sketch, as the mass of the piston in state (2) is smaller than in state (1). States (1) and (2) shall be at equilibrium, see Sect. 7.1.1.

Problem 7.3 An air mass with a volume of $540\,\text{cm}^3$, a temperature of $15\,°\text{C}$ and a pressure of 950 mbar is given. The volume of the gas is reduced to $400\,\text{cm}^3$. During this compression the gas is heated up to $45\,°\text{C}$. The gas constant is $R = 0.287\,\frac{\text{kJ}}{\text{kg K}}$.

(a) What mass does the air possess?
(b) What is the pressure of the gas after the compression?

Solution

(a) Applying the thermal equation of state to determine the mass of the gas:

$$pV = mRT \tag{7.12}$$

Since state (1) is well known, we can estimate the mass by

$$\boxed{m = \frac{p_1 V_1}{RT_1} = \frac{0.95 \times 10^5 \text{Pa} \cdot 5.4 \times 10^{-4} \text{m}^3}{287\,\frac{\text{N m}}{\text{kg K}} \cdot 288.15\,\text{K}} = 0.62\,\text{g}} \tag{7.13}$$

(b) Due to the closed system, the mass stays constant. This leads to

$$p = \frac{mRT_2}{V_2}. \tag{7.14}$$

Replacing the mass gives

$$\boxed{p_2 = \frac{p_1 V_1 RT_2}{RT_1 V_2} = p_1 \frac{V_1 T_2}{V_2 T_1} = 1.416\,\text{bar}} \tag{7.15}$$

Problem 7.4 A gas tank with a freely moveable upper limiting piston is filled with $500000\,\text{kg}$ of city gas. The gas takes up a volume of $760000\,\text{m}^3$. After extracting $300000\,\text{kg}$ with constant temperature the swimming piston subsides accordingly. Determine the new volume of the gas, as well as the specific volumes before and after the extraction. What are the densities?

Fig. 7.5 Solution to
Problem 7.4

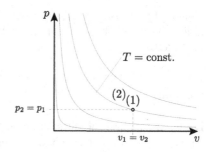

Solution

How does the change of state run?

- Constant temperature → isothermal change of state
- Freely moveable piston → isobaric change of state

This can be fixed in a p, v-diagram, see Fig. 7.5. In case the change of state is
isothermal and isobaric at the same time, it needs to be isochoric as well. This can
even be derived from the thermal equation of state:

$$v = \frac{RT}{p} = \text{const.} \tag{7.16}$$

$$v_1 = \frac{V_1}{m_1} = \frac{V_2}{m_2} = v_2 \tag{7.17}$$

$$V_2 = \frac{m_2}{m_1} V_1 = 304000 \, \text{m}^3 \tag{7.18}$$

$$v_1 = v_2 = \frac{V_1}{m_1} = 1.52 \frac{\text{m}^3}{\text{kg}} \tag{7.19}$$

Respectively

$$\rho_1 = \rho_2 = \frac{1}{v_1} = 0.657 \frac{\text{kg}}{\text{m}^3} \tag{7.20}$$

Problem 7.5 A pressure tank with a total volume of $V_t = 8 \, \text{m}^3$ contains carbon
dioxide CO_2 at a pressure of $p_t = 220 \, \text{bar}$ and a temperature of 293 K. How many
containers with a holding capacity of $V_c = 10 \, \text{dm}^3$ and a pressure of $p_c = 150 \, \text{bar}$
can be filled out of it, if the filling commences so slowly, that the temperature of the
gas does not change?

Remark Filling is only possible until the pressure in the tank equals the pressure in the containers. The remnant gas in the containers from before the filling can be neglected.

Solution

Let us first answer the question, how much gas is taken out of the pressure tank. Answering this question is possible, since state (1), i.e. the initial state, and state (2), i.e. when filling stops, are well known:

- State (1)

$$m_{t,1} = \frac{p_t V_t}{RT} \tag{7.21}$$

- State (2)

$$m_{t,2} = \frac{p_{t,2} V_t}{RT} \tag{7.22}$$

The refilling stops, when an equilibrium between tank and container has been reached, i.e.

$$p_{t,2} = p_c. \tag{7.23}$$

Thus, we know the mass, that needs to be refilled:

$$\Delta m = m_{t,1} - m_{t,2} = \frac{V_t}{RT} (p_t - p_c) \tag{7.24}$$

Each container can hold

$$m_c = \frac{p_c V_c}{RT} \tag{7.25}$$

The total number of required containers is

$$\boxed{n = \frac{\Delta m}{m_c} = \frac{V_t}{V_c} \cdot \frac{p_t - p_c}{p_c} = 373} \tag{7.26}$$

7.2 Equilibrium Thermodynamics

The previous discussion has shown the importance of the equilibrium principle for thermodynamic systems. Systems at equilibrium follow the thermal equation of state for ideal gases.

However, in contrast to idealised conditions, real changes of state, may run through imbalanced states inbetween initial and final state. From a thermodynamic point of view, these intermediate states are not defined unambiguously, as the thermal equation

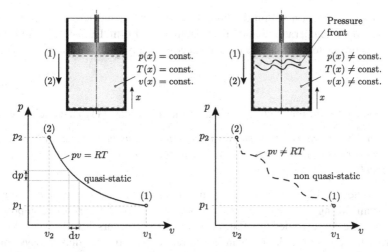

Fig. 7.6 Quasi-static versus non-quasi-static change of state: Illustration in a p, v–diagram for gas compression (The pressure is assumed to be homogeneous. Thus, the gravitational effect has been ignored, see Sect. 7.1.1!)

of state can not be applied.[3] If an intermediate state is not at equilibrium, the system would, if sufficient time was available, follow a transient balancing process[4] until equilibrium is reached, see Chaps. 4 and 14.

7.2.1 Quasi-Static Changes of State

In conventional thermodynamics the assumption is made, that changes of state of a thermodynamic system occur in infinitesimal sub-steps, e.g. dv respect. dp—in case the the change of state is illustrated in a p, v-diagram. By doing so, the system is supposed to be always *approximately* in an equilibrium state. Thus, the system is a homogeneous phase, see Sect. 3.1.1, and the pressure for instance can be estimated unequivocally. This quasi-static approach is shown on the left side of Fig. 7.6.

The major premise of this approach is, that all sub-steps are thermodynamically at equilibrium,[5] i.e. pressure as well as temperature are balanced. The change of state runs sufficiently slow in infinitesimal sub-steps. The summation of all of these sub-steps forms the entire change of state. In case the weight of the gas and the

---

[3]The experiments performed by Boyle-Mariotte as well as by Gay-Lussac, see Chap. 6, have been conducted under the premise of thermodynamic equilibrium.

[4]The system would drive itself into thermal, mechanical as well as chemical equilibrium!

[5]Think of taking pictures of a car that is accelerating. If your camera has a very short exposure time, each of the plenty of photos looks like static, all images are sharp. In this very short period of time the car movement is neglectable. If the shutter speed is too large, the resulting pictures do not look like static at all, since the movement results in blurred images.

temperature distribution within the boundary layer are neglected, see Fig. 4.7, the
pressure and the temperature within the system are uniform. Due to the equilibrium
state, pressure and temperature fix the state unequivocally. Thus, the thermal equation
of state

$$pv = RT \tag{7.27}$$

can be applied for the entire change of state from initial state (1) to the final state (2)!

Theorem 7.6 *In a quasi-static change of state the system behaves as a phase, i.e. a
homogeneous distribution of all state values occurs. No transient balance processes
take place!*

The right side of Fig. 7.6 shows what may happen, in case the system is not at
equilibrium during the change of state from (1) to (2). If the piston compresses
the gas inside the cylinder too fast, a pressure front occurs for instance. Within
this front the pressure is larger than in the remaining volume, thus the system is
not at mechanical equilibrium. As proven before, the system wants to strive for
equilibrium state, but if the piston's velocity is too high,[6] equilibrium can not be
reached. According to the pressure front, a temperature front may be observed as
well. Consequently, the pressure respectively temperature distribution within the
gas is uneven. The system is heterogenous. Obviously, it is impossible to specify a
pressure p as well as a temperature T for the system. Hence, as indicated in Fig. 7.6,
a well-defined function in a p, v-diagram can not be found. Furthermore, since the
system is not at equilibrium, the thermal equation of state can not be applied. The
system is not in unequivocal state. However, equilibrium states (1) and (2) can be
reached. If the system is in rest, when the external impact stops, the system strives
for equilibrium.

Theorem 7.7 *A quasi-static change of state passes several balanced sub-states. All
states between initial and final state can then be calculated with thermodynamic
equation of states for homogeneous systems, e.g. ideal gas law.*

7.2.2 Requirement for a Quasi-static Change of State

The criteria for a quasi-static change of state is, that the microscopic balancing must
be much faster than the macroscopic movement: In terms of the example given in
Fig. 7.6, the balancing within the system, i.e. the striving for equilibrium, needs to
be faster than the macroscopic velocity of the piston. In such a case, the system has
homogenised itself before the new pressure front occurs. In Chap. 21 it is derived,
that pressure fluctuations spread with the velocity of sound, i.e.

$$a = \sqrt{\kappa RT}. \tag{7.28}$$

---

[6]In the following sections, it is investigated what *too high* means!

Depending on the gas and its temperature, velocities of some 300 ... 1 000 m s^{-1} are not unusual. Compared to typical velocities of the piston in technical applications, the requirement for quasi-static behaviour is sufficiently fulfilled. Hence, a quasi-static change of state is often approximately applicable and the system can be treated as a phase without any inhomogeneities.

Theorem 7.8 *So even fast changes of state can be quasi-static, though the probability of internal dissipation rises with velocity, e.g. due to turbulences inside the system!*

In contrast to this, an inflow of a gas in a vacuum for instance is not quasi-static, since local and temporal gradients in pressure, density, temperature and velocity occur during the process. Microscopic as well as macroscopic velocity are the same, i.e. the speed of sound. The approximation of a quasi-static change of state is undue. But, once the inflow is completed, the system homogenises and strives for equilibrium. Once thermodynamic equilibrium is reached, the thermal equation of state, of course, can be applied.

Theorem 7.9 *Obviously, the equilibrium state of a system is independent of the path the change of state takes.*

Figure 7.7 shows this thesis. Starting from equilibrium state (1) one might follow path (A), characterised by quasi-static behaviour from (1) to (2). Instead of taking the direct path (A), one can also reach (2) by an isochoric plus isobaric change of state, both of them being quasi-static. This is indicated as path (B) in Fig. 7.7. However, equilibrium state (2) might also be reached by a non-static change of state, e.g. path (C), that actually can not be illustrated in a p, v-diagram unequivocally.

Theorem 7.10 *All non-static changes of state can be replaced by a quasi-static change of state. Based on the initial state and equations of state the final equilibrium state can be predicted.*

The deviation of the state values between initial and final state can be calculated, no matter which path has been taken:

Fig. 7.7 Path-independent equilibrium state

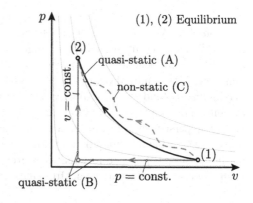

$$\Delta p = p_2 - p_1 \qquad (7.29)$$

$$\Delta v = v_2 - v_1 \qquad (7.30)$$

$$\Delta T = T_2 - T_1 \qquad (7.31)$$

7.3 Reversible Versus Irreversible Changes of State

Besides the newly introduced quasi-static approach, an essential distinction in thermodynamics is made between reversible respectively irreversible changes of state. This classification is important when introducing the entropy as new state value in Chap. 12. Reversible changes of state are the benchmark for all thermodynamic cycles as well as all thermodynamic machines. However, quasi-static and irreversible/reversible changes of state must be considered mutually since they are closely related, as elaborated in this section.

Theorem 7.11 *A reversible change of state can be reversed in any particular detail respectively sub-step. After a thermodynamic system has moved from state (1) to state (2), the process is reversible, if state (1) can be exactly reached again. This does not only include the system itself but the environment as well. Any reversible change of state is an idealised change of state. In contrast to a reversible change of state, an irreversible change of state means, that the initial state, i.e. regarding the system as well as the environment, can not be reached again. Sure, the system can still reach its initial state, even for an irreversible change of state, but the environment then is different than before.*

The following examples clarify the meaning of reversible/irreversible changes of state. A distinction is made between mechanical and thermal problems.

7.3.1 Mechanical

Example 7.12 With the help of simple mechanical applications the principle of reversibility can be illustrated easily. Figure 7.8 shows the physical behaviour of a so-called wire pendulum for instance, as probably known from mechanics. Initially, the mass is moved out of its rest position in a state (0), characterised by the maximum amplitude x_0. Due to gravity, the pendulum starts swinging. Let us plot the amplitude as a function of time, i.e. $x(t)$, indicated in Fig. 7.8. After a while the pendulum persists in its rest position again, since its energy has been converted into friction, e.g. at the mounting point of the wire and due to the air's drag. Of course the initial state (0) can be restored: However, the system cannot perform this by itself, but needs an external impact from the environment, e.g. mechanical work can

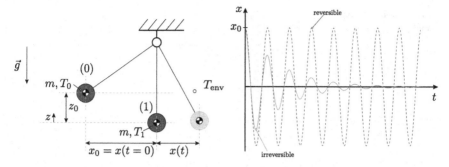

Fig. 7.8 Wire pendulum

Fig. 7.9 Dissipation in a closed systems

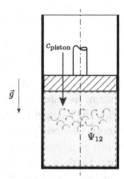

be supplied. Thus, the environment, that releases the required energy, has changed. Though state (0) is restored, the entire process is irreversible.

In case, there was no friction at all, the pendulum would periodically restore state (0), see Fig. 7.8, and no modifications in the environment can be measured.

Theorem 7.13 *It is common experience that reversible changes of state are always frictionless respectively free of dissipation.*

However, friction, like in the previously discussed wire pendulum, can occur in any thermodynamic system. Sketch 7.9 shows a gas being compressed in a cylinder. Due to the displacement of the piston the fluid inside moves. According to the no-slip condition the gas at the piston's surface has the same velocity as the piston, while the fluid at the bottom of the cylinder is still in rest. Hence, a relative velocity among the fluid particles occurs. The faster the piston moves, the more turbulence within the gas occurs and so-called turbulent eddies are formed. Thus, the fluid particles are rubbing against each other, i.e. internal friction arises.

This internal friction is named dissipation Ψ and will be deeply discussed in Sect. 9.2.4. Anyhow, the movement of these eddies always *consumes* and never releases energy, that is, similar to the wire pendulum, dissipated, i.e. converted from mechanical energy to frictional energy.

Fig. 7.10 Bouncing ball
driven by Δz

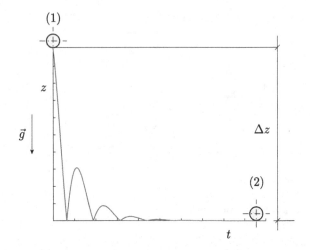

Example 7.14 Figure 7.10 shows another example for an irreversible mechanical system. In state (1) a ball is fixed at a height of Δz above ground in a gravitational field. The fixing is released so that the ball's potential energy is converted into kinetic energy. Obviously, the ball is bouncing several times but after a while it finally comes to rest at ground level. It is the friction, i.e. with the surrounding air and due to the impact at the ground, that lets the ball come to rest. Sure, it is possible to bring the ball back into state (1), but that would require an interference from outside, i.e. mechanical energy has to be supplied to the ball in order to lift it back to state (1). Thus, the process is irreversible. Anyhow, the system is driven by Δz in the gravitational field. Once, Δz to the ground has gone, there is no more driver to change the state of the system.

Example 7.15 A transient balancing also occurs, if there is a vessel, filled with an ideal gas at a pressure of $p_1 > p_{\text{env}}$, see Fig. 7.11. In case there is a small orifice in the tank, a pressure balancing takes place, see Chap. 4. The volume flow rate is driven by the pressure potential. Once, the system is at equilibrium, the volume flow disappears and the system can not move itself back into state (1). Hence, this balancing process is irreversible as well. Once again the environment is supposed to be huge in size, so that the ambient pressure is assumed to be constant.

7.3.2 Thermal

Example 7.16 Changes of state are even irreversible when they strive for a thermal equilibrium, see Fig. 7.12. Initially, the fluid has a temperature of $T_1 > T_{\text{env}}$. The temperature potential $T_1 - T_{\text{env}}$ causes a heat flux from fluid to ambient:

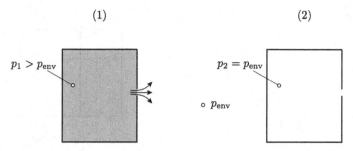

Fig. 7.11 Pressure balancing driven by Δp

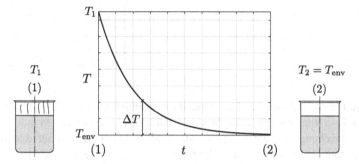

Fig. 7.12 Cooling down of a liquid driven by ΔT

$$\dot{Q} = kA\,(T_1 - T_{env}) \tag{7.32}$$

This reduces the internal energy of the fluid, indicated by a decreasing temperature. On the other side, the internal energy of the environment increase. Since in thermodynamics the environment is supposed to be huge, its temperature is assumed to be constant. As a consequent, the driving potential for this change of state reduces by time. After a while the temperature of the liquid approaches ambient temperature as long as the system is not adiabatic. This new equilibrium has no more temperature gap to the environment. Thus, the temperature potential reaches zero. As a consequence of this decreasing potential, the system finally can not move reversely! State (1) can only be reached if energy is supplied across the system boundary, e.g. by heating the fluid up externally. Thus, this process is irreversible.

7.3.3 Chemical

Example 7.17 This example shows the ideal mixing of two non-reactive components. According to Dalton, see Chap. 19, each component takes the entire available volume. Initially, in the left chamber the concentration of gas A is $\xi_A = 1$, while it is

Fig. 7.13 Mixing of gases driven by ξ_i

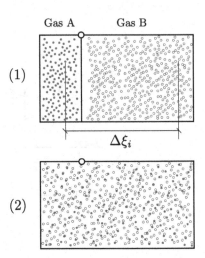

$\xi_A = 0$ in the right chamber.[7] This concentration imbalance causes—once the wall is removed—a transient balancing process until state (2) is reached. At that point the driver for the change of state comes to a standstill and the system is not able to move back into state (1) by itself. Hence, the process is irreversible (Fig. 7.13).

Theorem 7.18 *The shown examples have in common, that all transient balancing processes are irreversible: Any balancing process needs a potential as driver. This might be pressure, temperature or concentration. During the balancing the potential becomes smaller; once it is zero the balancing is done and due to the missing potential the system can not move back without external influences.*

7.4 Conventional Thermodynamics

An overview of possible scenarios is given in Fig. 7.14. This example shows two successive changes of state. Step 1 is an adiabatic compression followed by step 2, an adiabatic expansion until mechanical equilibrium with the environment is reached.

- Case (A) according to Fig. 7.14 ($c_{\text{piston}} \to 0$)
  Reversible changes of state are always quasi-static as well. If a sub-step was not quasi-static (i.e. the system is not at equilibrium), a transient balancing process would proceed. Balancing processes are always irreversible as indicated by Theorem 7.18.

Theorem 7.19 *For a reversible process a quasi-static change of state is required!*

---

[7]Mass fractions ξ_i will be introduced in part II though.

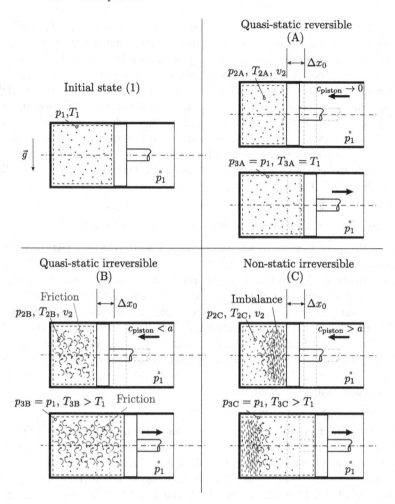

Fig. 7.14 Overview irreversible/reversible—Example of an adiabatic compression/expansion

Turbulence, that generates eddies and consequently leads to inhomogeneities, i.e. irreversibilities, can be avoided in case the velocity of the piston is very small. Strictly speaking, dissipation even occurs for very small velocities. Thus, case (A) is just a theoretical reference case, as reversible process turns out to be a technical benchmark, see Chap. 15!

- Case (B) according to Fig. 7.14 ($c_{\text{piston}} < a$)
  In this case internal friction occurs due to the gas being in motion, indicated by eddies inside the cylinder. In this case, the fluid swirls, i.e. internal friction among the fluid particles takes place. However, this internal friction is called dissipation Ψ. This phenomena is rather complex and leads in fact to inhomogeneities within the fluid. Anyhow, Theorem 7.13 has shown, that systems with friction can never be

reversible. Thus, strictly speaking, the system is not a phase any more and it is not quasi-static. As a consequence the approach of equilibrium thermodynamics, e.g. the thermal equation of state, could not be applied. This is a dilemma. Furthermore, these inhomogeneities initiate transient balancing processes as well. Hence, the process is irreversible. Actually, any change of state is based on a disturbance of the quasi-static state inside a system. In order to apply the means of thermodynamics, the approximation is made, that systems with internal, and thus irreversible friction, behave as phases and run through quasi-static changes of state. Consequently, in this quasi-homogeneous system all state values are in fact given by *mean values*— imperfections due to the dissipation are then part of these mean values. According to [1] this is only permitted in case the inhomogeneities are not too large. So the conclusion for case (B) is:

Theorem 7.20 *Not all quasi-static changes of state are reversible!*

Based on the approximation made, in case (A) as well as in case (B) the thermal equation of state can be applied, since the system is assumed to be in equilibrium[8] in each sub-step while the change of state runs. The appearance of dissipation, i.e. small turbulence eddies, correlates with the macroscopic velocity of the change of state. The larger the velocity, the higher the chance of dissipation. In contrast, the smaller the velocity, the larger the chance of avoiding dissipation. Dissipation is, what will be discussed further in Sect. 9.2.4, driven by the process and thus a process value. Since case (A) and case (B) are supposed to be quasi-static, the corresponding changes of state can be illustrated in a p, v-diagram. However, due to dissipation the curves run differently, see Problem 7.26.

- Case (C) according to Fig. 7.14 ($c_{piston} > a$)
  In case (C) the change of state runs so fast, that a thermodynamic equilibrium can not be reached in every single sub-step. This causes a pressure front as shown in the example by the uneven distribution of the dots that shall represent the gas. The change of state is non-static, since there is a heterogeneous distribution of the state values. As clarified before, this imbalance causes a transient balancing process. The system tries to achieve equilibrium, see Chap. 4. What the previous examples have shown is, that every balancing process is irreversible.

Theorem 7.21 *Non-static changes of state are always irreversible!*

The thermal equation of state can not be applied for Case (C). Furthermore, a non-static change of state also causes dissipation.

The previous examples (A), (B) and (C) and Theorem 7.18 have shown:

Theorem 7.22 *Irreversibility is either due to friction[9] or due to deviations from thermodynamic equilibrium. These deviations force the system to run a transient balancing process in order to achieve equilibrium. However, all balancing processes in nature are irreversible!*

---

[8]And thus a homogeneous phase!
[9]Later on, internal friction will be called dissipation!

Reversible process control

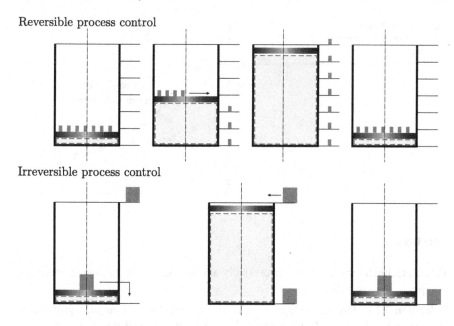

Irreversible process control

Fig. 7.15 Analogous model according to [16]

The need of a process to be quasi-static in order to be reversible as well, can be illustrated with a simplified example, see Fig. 7.15. The gas inside a cylinder/piston has to be expanded first. Once the expansion is finished, the original state shall be restored. The question is, what is required in order to let the process run reversibly.

First, the piston is covered with a huge number of weights. During the change of state weight by weight is moved to a storage on the right side of the cylinder. When the last weight is removed the maximum expansion is achieved. In the reversed change of state, weight by weight is moved back on the piston. When the last weight has been placed on the piston, the compression is finished and the system has been brought back in is initial state. Due to the huge number of weights, i.c. quasi-static approach, it is not only the gas that is in initial state, but the environment, i.e. the storage in this case, as well. Thus, the process is reversible! However, reversibility requires a rather huge number of weights and thus a huge number of shelves. For only a small number of weights, the weights still need to be lifted externally to store them on the shelves. This omits for an infinite number of weights.

Another way to handle this problem is illustrated in the lower part of Fig. 7.15. Instead of performing the change of state quasi-static, it is done by in one respectively two steps. The entire mass is removed from the piston and the gas expands at once. In order to compress the gas another weight, that has been stored at the upper position, now is placed on the piston. This causes the compression. Consequently, the gas is back in its initial state. However, the entire process is not reversible, though the gas looks like before, but the environment does not.

Fig. 7.16 Approach
technical thermodynamics
according to [1]

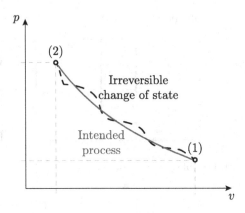

Conclusions

Though technical processes run irreversibly and non-static, the following approach
is applied, see Fig. 7.16.

Definition 7.23 If irreversibility is not due to a missing thermodynamic equilibrium,
shown previously as case (C), but due to the release of frictional energy, i.e. case (B),
a slow, quasi-static change of state, similar to a reversible change of state, see case
(A), can be assumed, see [1].

Such cases are given by a continuous curve in a p, v-diagram for instance, see
Fig. 7.16. At each point along the curve a thermodynamic equilibrium is reached,
though friction/dissipation occurs. Consequently, the thermal equation of state is
valid for the entire process and the methods of equilibrium thermodynamics can be
applied. However, if friction occurs, case (B), the process runs differently compared
to case (A). Thus, the illustration in a T, s- as well as in a p, v-diagram show these
differences, see Problem 7.26.

Example 7.24 In this example[10] a closed system, that undergoes an isobaric, slow
change of state is investigated. However, it is intended for the advance readers! The
closed system that is filled with an ideal gas is heated up, according to Fig. 7.17a.
The piston closes the system to an environment and it can move up and down. Thus,
during the very slow change of state, the pressure shall be constant all the time.[11]
However, this example will be investigated, in order to find out *the requirements for
an isobaric change of state*. This example will finally lead to Eq. 4.5. The first law
in differential notation[12] for a small movement dz of the piston, see Fig. 7.17b, is:

$$\sum \delta W + \sum \delta Q = dU_P + m_P g\, dz + m_{PCP}\, dc_P \qquad (7.33)$$

---

[10]To cope with this example a knowledge of first and second law of thermodynamics is required.

[11]The system will be further discussed later in Problem 11.8!

[12]Benefit of a differential notation is, that it shows what is ongoing *inbetween* initial and final
equilibrium state.

Fig. 7.17 Isobaric change of state—Differential notation

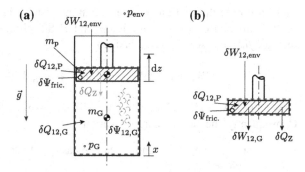

If this equation is integrated, the conventional, non-differential, notation is:

$$\sum W_{12} + \sum Q_{12} = \Delta U_P + m_P g \, \Delta z + \frac{1}{2} m_P \left(c_{P,2}^2 - c_{P,1}^2 \right) \qquad (7.34)$$

However, in this case, see Fig. 7.17b and considering all terms that exceed the system boundary, Eq. 7.33 is as follows:

$$\delta W_{12,\text{env}} - \delta W_{12,G} + \delta Q_{12,P} - \delta Q_Z = dU_P + m_P g \, dz + m_P c_P \, dc_P \qquad (7.35)$$

This equation does not only take into account the initial state (1) and final state (2), but gives a general description for the entire change of state inbetween states (1) and (2). The fundamental equation of thermodynamics, see Chap. 12, can be applied for the piston:

$$dS = \frac{dU_P + p \, dV}{T} = \delta S_i + \delta S_a \qquad (7.36)$$

For a solid body ($dV = 0$) it follows[13]:

$$\frac{dU_P}{T} = \delta S_i + \delta S_a = \frac{\delta \Psi_{\text{fric.}}}{T} + \sum \frac{\delta Q}{T} \qquad (7.37)$$

Applied to the discussed example:

$$\frac{dU_P}{T} = \frac{\delta \Psi_{\text{fric.}}}{T} + \frac{\delta Q_{12,P}}{T} - \frac{\delta Q_Z}{T} \qquad (7.38)$$

Hence, after multiplying with the temperature T:

$$dU_P = \delta \Psi_{\text{fric.}} + \delta Q_{12,P} - \delta Q_Z \qquad (7.39)$$

---

[13]The term *friction* $\Psi_{\text{fric.}}$ is explained in Sect. 9.2.5.

Thus, it is only the friction and the heat at the piston's boundary, that influence the piston's internal energy, i.e. the temperature! Substituting dU_P in the first law of thermodynamics, see Eq. 7.35, leads to

$$\delta W_{12,\text{env}} - \delta W_{12,\text{G}} = \delta \Psi_{\text{fric.}} + m_P g \, dz + m_P c_P \, dc_P \tag{7.40}$$

Now let us have a closer look at the work that crosses the piston's system boundary. Consequently, the partial energy equations are:

- The work at the upper boundary can be easily calculated from the environmental point of view[14]:

$$\delta W_{12,\text{env}} = -p_{\text{env}} \, dV \tag{7.41}$$

  The ambient is compressed, i.e. volume work is released to the ambient. This work is provided by the sketched system. As the ambient is supposed to be homogeneous there is no dissipation.
- The work at the lower boundary can be calculated from the gas point of view. It is composed of volume work at the gas, dissipation within the gas and mechanical work in order to move the centre of gravity of the gas, see Fig. 7.17a:

$$\delta W_{12,\text{G}} = -p_{\text{G}} \, dV + \delta \Psi_{12,\text{G}} + m_{\text{G}} g \, dx + m_{\text{G}} c_{\text{G}} \, dc_{\text{G}} \tag{7.42}$$

If the piston moves by dz the centre of gravity of the gas moves by $dx = \frac{dz}{2}$!

Hence, Eq. 7.40 reads as

$$-p_{\text{env}} \, dV + p_{\text{G}} \, dV - \delta \Psi_{12,\text{G}} - m_{\text{G}} g \, dx - m_{\text{G}} c_{\text{G}} \, dc_{\text{G}} = \\ \delta \Psi_{\text{fric.}} + m_P g \, dz + m_P c_P \, dc_P. \tag{7.43}$$

With the velocity of the piston

$$c_P = \frac{dz}{dt} \tag{7.44}$$

and the velocity of the gas

$$c_{\text{G}} = \frac{dx}{dt} \tag{7.45}$$

it is

$$-p_{\text{env}} \, dV + p_{\text{G}} \, dV - \delta \Psi_{12,\text{G}} - m_{\text{G}} g \, dx - m_{\text{G}} \frac{dx}{dt} \, dc_{\text{G}} = \\ \delta \Psi_{\text{fric.}} + m_P g \, dz + m_P \frac{dz}{dt} \, dc_P \tag{7.46}$$

---

[14]If the gas inside the cylinder expands, i.e. $dV > 0$, work is released by the system to the environment. Thus, $\delta W_{12,\text{env}}$, as assumed in the sketch, is negative! The related work can be calculated easily since the ambient pressure is constant.

With the acceleration of the piston

$$a_P = \frac{dc_P}{dt} \tag{7.47}$$

and the acceleration of the gas

$$a_G = \frac{dc_G}{dt} \tag{7.48}$$

it follows:

$$-p_{env}\, dV + p_G\, dV - \delta\Psi_{12,G} = \delta\Psi_{fric.} + m_P g\, dz + m_P a_P\, dz + m_G g\, dx + m_G a_G\, dx \tag{7.49}$$

Since

$$dx = \frac{1}{2} dz \tag{7.50}$$

it follows

$$c_G = \frac{dx}{dt} = \frac{1}{2}\frac{dz}{dt} = \frac{1}{2}c_P \tag{7.51}$$

and

$$a_G = \frac{dc_G}{dt} = \frac{1}{2}\frac{d^2z}{dt^2} = \frac{1}{2}a_P. \tag{7.52}$$

This leads to

$$-p_{env}\, dV + p_G\, dV - \delta\Psi_{12,G} = \delta\Psi_{fric.} + m_P g\, dz + m_P a_P\, dz + \frac{1}{2}m_G g\, dz + \frac{1}{4}m_G a_P\, dz \tag{7.53}$$

Dividing by dz:

$$-p_{env} A + p_G A - \frac{\delta\Psi_{12,G}}{dz} = \frac{\delta\Psi_{fric.}}{dz} + m_P g + m_P a_P + \frac{m_G}{2}\left(g + \frac{1}{2}a_P\right) \tag{7.54}$$

Respectively, after dividing by A

$$p_G = p_{env} + \frac{\delta\Psi_{12,G} + \delta\Psi_{fric.}}{A\, dz} + \frac{m_P}{A}(g + a_P) + \frac{1}{2A}m_G\left(g + \frac{1}{2}a_P\right) \tag{7.55}$$

The dissipation follows[15]

---

[15]For a technical application mathematical functions for $\psi_{12,G}$ and $\psi_{fric.}$ are required. These functions need to correlate the specific dissipation with the distance a system moves and its velocity. A detailed example is given with Problem 11.9.

$$\delta \Psi_{12,G} = \psi_{12,G} \, dm = \psi_{12,G} \rho A \, |dz| \tag{7.56}$$

and

$$\delta \Psi_{\text{fric.}} = \psi_{\text{fric.}} \, dm = \psi_{\text{fric.}} \rho A \, |dz| \tag{7.57}$$

Now let us additionally assume, that $m_G \ll m_P$, i.e. kinetic and potential energy of the gas are neglected, so we get

$$p_G = p_{\text{env}} + \rho \left(\psi_{12,G} + \psi_{\text{fric.}} \right) \frac{|dz|}{dz} + \frac{m_P}{A} (g + a_P) \tag{7.58}$$

Consequently, the larger the dissipation is, the larger the pressure inside is during the change of state. As mentioned in the problem description, the change of state shall run very slowly. Under such circumstances, the dissipation disappears in this mathematical limiting case.

Theorem 7.25 *Very slow is a synonym for a process that does not cause any turbulence inside the system and thus is free of dissipation.*

The process runs *isobarically* for an ideal gas under the following premises:

- The change of state runs very slowly. According to Theorem 7.25, the dissipation within the gas disappears, i.e.

$$\psi_{12,G} = 0. \tag{7.59}$$

- The piston moves frictionless, i.e.

$$\psi_{\text{fric.}} = 0. \tag{7.60}$$

- The system is non-accelerated, i.e.

$$a_P = 0. \tag{7.61}$$

The mass of the gas is small, so that kinetic and potential energy of the gas are ignored compared to the piston, i.e.

$$\Delta e_{a,12,G} = 0. \tag{7.62}$$

Under such circumstances the dissipation disappears and Eq. 7.58 is identical with Eq. 4.5 for a balance of forces at rest:

$$\Rightarrow p_G = p_{\text{env}} + \frac{m_P}{A} g = \text{const.} \tag{7.63}$$

Thus, the pressure is constant for the entire change of state from (1) to (2)! Furthermore, Eq. 7.54 can be rearranged,[16] so that

$$m_P a_P = -p_{env} A + p_G A - \frac{\delta \Psi_{12,G} + \delta \Psi_{fric.}}{dz} - m_P g \qquad (7.64)$$

Replacing the friction, as described above, leads to

$$m_P a = -p_{env} A + p_G A - \rho A \left(\psi_{12,G} + \psi_{fric.} \right) \frac{|dz|}{dz} - m_P g \qquad (7.65)$$

This equation might be known from mechanics and has the form

$$m_P a = \sum F. \qquad (7.66)$$

In case not all forces equalise themselves, the mass is accelerated.

Problem 7.26 This advanced problem[17] refers back to the classification of conventional thermodynamics, see Example 7.24. How do the changes of state (A) and (B) according to Fig. 7.14 look like in a p, v- respectively T, s-diagram?

Solution

Case (A) is rather simple. Since the entire process shall be adiabatic and reversible, i.e. free of dissipation, the process is isentropic. Starting from state (1) during the isentropic expansion the temperature rises until the final position has been reached (2A). The way back is also isentropic, so it is exactly the same path as for the compression. Both paths can easily been drawn in the p, v- and T, s-diagram, see Fig. 7.18.

Case (B) is somehow more difficult. From (1) to (2B) more work has to be supplied, since dissipation occurs, i.e. the turbulent eddies, that consume energy, need to be supplied with energy as well. Consequently at the final position (2B) the temperature is larger than in state (2A). However, the specific volume is the same, since the piston shall stop at the same position as in case (A):

$$v_{2A} = v_{2B} \qquad (7.67)$$

From (1) to (2B) the entropy increases. This is due to friction and not due to heat, since the process is assumed to be adiabatic. As indicated in the p, v-diagram the pressure is larger than in state (2). On its way back during expansion the entropy further increases, since friction still occurs. Mechanical equilibrium occurs, when

---

[16]With the mass of the gas still ignored.

[17]To cope with this problem a knowledge of first and second law of thermodynamics is required.

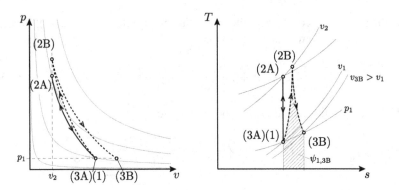

Fig. 7.18 p, v- and T, s-diagram according to Fig. 7.14

$$p_{3B} = p_1 \tag{7.68}$$

Both, p, v- and T, s-diagram, show, that the volume in the final position is larger than in state (1):

$$v_{3B} > v_1 \tag{7.69}$$

This is, since

$$T_{3B} > T_1. \tag{7.70}$$

Theorem 7.27 *This problem has shown, that internal friction, i.e. dissipation, is the cause for the change of state being irreversible—similar to a conventional mechanical problem as given with Example 7.12!*

Anyhow, case (C) can not be illustrated in p, v- and T, s-diagrams, since no equilibrium is reached!

Chapter 8
Thermodynamic Processes

So far it has been clarified what a thermodynamic system is and how its state can be determined. Within conventional thermodynamics the principle of thermodynamic equilibrium forms the basis for thermodynamic calculations. However, due to external impacts a system can be shifted from one equilibrium state to another. On its way a change of state takes place:

Theorem 8.1 *A thermodynamic process can consist of one or several successive changes of state.*

8.1 Equilibrium Process

In fact, equilibrium processes have been described in detail within Chap. 7. Most of the tasks and problems covered by this book are based on this principle. The required prerequisite is that the change of state follows a quasi-static process, i.e. in every sub-step the system is at equilibrium. Furthermore, at equilibrium state, there are no discontinuities within the system, which would initiate a balancing process, see Fig. 7.14 for further explanation.

An example of such a process is shown in Fig. 8.1. Obviously, on its way from state (1) to state (2) dissipation Ψ_{12}, i.e. internal friction, occurs. Within this book, friction is indicated by the small eddy symbols within the system. Though it might look as if these dissipation eddies are local spots,[1] they are distributed uniformly in order to achieve an equilibrium state. The quasi-static change of state can be visualised as a smooth curve in a p, v-diagram, see Fig. 8.1. At each point on this curve the thermal equation of state

$$pv = RT \tag{8.1}$$

can be applied.

---

[1] If they would be local spots, there would be discontinuities. In such a case the system would not be at equilibrium.

© Springer Nature Switzerland AG 2019
A. Schmidt, *Technical Thermodynamics for Engineers*,
https://doi.org/10.1007/978-3-030-20397-9_8

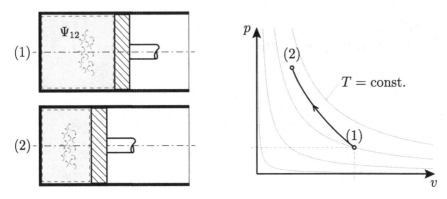

Fig. 8.1 Equilibrium process

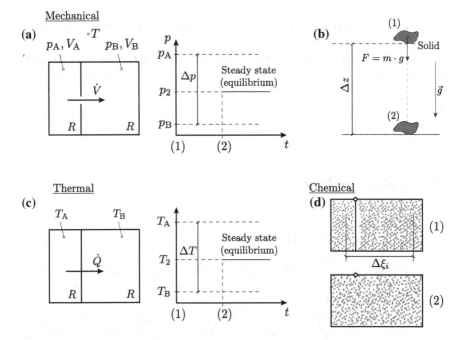

Fig. 8.2 Transient state

8.2 Transient State

A transient state occurs, if the thermodynamic equilibrium of a system is disturbed. Such processes are given in Fig. 8.2. Mind, that an imperfection regarding intensive state values cause a transient balancing process, see also Chap. 7: In state (1) the two tanks are separated from each other and do have different pressures inside, see Fig. 8.2a. Nevertheless, each of the tanks is at its own equilibrium.

At $t = 0$ the barrier between the two tanks is removed and a pressure balancing is initiated. As discussed previously this balancing runs with velocity of sound. Thus, the process is non-static and can consequently not been sketched in a p, v-diagram. While the balancing takes place, the thermal equation of state can not be applied. However, after a short period of time a new equilibrium state (2) arrives. For the equilibrium states (1) and (2) the principles of conventional thermodynamics can be applied.

Figure 8.2 additionally shows transient balancing processes due to (b) gravity, (c) temperature and (d) chemical potential.

Problem 8.2 Let us have a closer look at Fig. 8.2a. Initially, the pressure in both vessels shall be p_A respectively p_B. The volumes are V_A and V_B, both filled with the same ideal gas. Due to a diabatic[2] wall, the gas temperature inside both vessels is equal with ambient temperature T. What is the pressure p_2 after the system has changed into the new equilibrium state (2)? Again, in state (2) the system is also in thermal equilibrium with environment.

Solution

Since state (1) and state (2) are both at equilibrium, the thermal equation of state can be applied

- State (1)

$$p_A V_A = m_A R T \tag{8.2}$$

$$p_B V_B = m_B R T \tag{8.3}$$

- State (2)

$$p_2 (V_A + V_B) = (m_A + m_B) R T \tag{8.4}$$

The entire mass shares the total volume.

$$p_2 (V_A + V_B) = m_A R T + m_B R T \tag{8.5}$$

Combining Eqs. 8.2, 8.3 and 8.5 leads to

$$\boxed{p_2 = \frac{p_A V_A + p_B V_B}{V_A + V_B}} \tag{8.6}$$

---

[2]Diabatic is the opposite of adiabatic. Thus, heat can pass the wall.

Fig. 8.3 Thermodynamic cycle

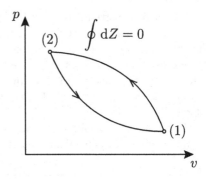

8.3   Thermodynamic Cycles

Figure 8.7 shows a thermodynamic cycle in a black box illustration. The cycle runs within the system boundary and it will be investigated later which changes of state the fluid inside underlies, see Chap. 17. The definition of a permanently running cycle, from the fluid perspective, is:

Theorem 8.3 *If a system, after several changes of state, finally reaches the initial state it is called a thermodynamic cycle. Cycles play an important role in technical applications (e.g. thermal engine, heat pump).*

Consequently, a thermodynamic cycle consists at least out of two successive changes of state, see Fig. 8.3.

Mathematically, all changes of a state value Z neutralise during one cycle, since once the initial state is reached again, all state values are as they used to be at the starting point:

$$\oint dZ = 0 \text{ resp. } \oint dz = 0 \tag{8.7}$$

Taking the specific volume for instance, see Fig. 8.3:

- Change of state $(1) \to (2)$

$$\Delta v_{12} = v_2 - v_1 \tag{8.8}$$

- Change of state $(2) \to (1)$

$$\Delta v_{21} = v_1 - v_2 \tag{8.9}$$

- In total:

$$\Delta v_{12} + \Delta v_{21} = 0 \tag{8.10}$$

  respectively

$$\Rightarrow \oint dv = 0 \tag{8.11}$$

Regarding the reversibility of thermodynamic cycles, the following rule is true:

Fig. 8.4 Stirling cycle

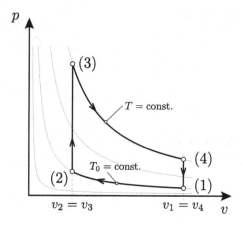

Theorem 8.4 *In case all changes of state are reversible, the entire cycle is reversible. If one single change of state is irreversible, the entire cycle is irreversible!*

As mentioned before, thermodynamic cycles play an important role in thermodynamics, so the focus is on cycles in Chap. 17. However, one example of a thermodynamic cycle is given in Fig. 8.4. It shows a Stirling cycle that consists of four changes of state, two isothermal and two isochoric steps.

In Chap. 15 it is distinguished between clockwise and counterclockwise cycles. These cycles help to understand the meaning of the state value *entropy*. Clockwise cycles, i.e. thermal engines, convert thermal energy into mechanical energy, while counterclockwise cycles are known as cooling machines respectively as heat pumps.

8.4 Steady State Process

For steady state processes any state value Z within the system is *constant in time* while it may differ locally.

8.4.1 Open Systems

Consequently, see Sect. 3.2.5, an open system in steady state is characterised by the following balance of any extensive state value:

$$\frac{dZ}{dt} = 0 = \sum \dot{Z}_{\text{in}} - \sum \dot{Z}_{\text{out}} + \dot{Z}_{\text{Source}} \tag{8.12}$$

As indicated in Fig. 8.5 a steady state process can occur in open systems. For any steady state flow it follows that the mass inflow is identical with the mass outflow.

Fig. 8.5 Steady state flow process

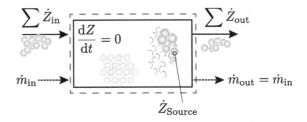

Fig. 8.6 Steady state closed system

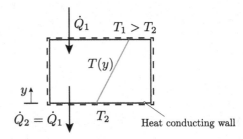

$$\dot{m}_{\text{in}} = \dot{m}_{\text{out}} \tag{8.13}$$

If this would not be the case, the mass inside the system would vary temporally. A system in steady state is not solely determined by constant extensive state values Z, but by temporally constant intensive state values, e.g. pressure p and temperature T, as well. If a balance for any extensive state value Z is performed, one has to distinguish between conservation values[3] and non-conservation values, e.g. entropy. In case Z is a non-conservation value, a source respectively sink term, i.e. $\dot{Z}_{\text{Source}}$, can occur in the system!

8.4.2 Closed Systems

However, even closed systems can be operated in steady state as well. Figure 8.6 shows exemplary a heat conducting wall. According to the previously made definitions, this is a closed system, since no mass crosses the system boundary.

In this case, steady state means, that state values are constant with respect to time though they may vary locally. $T(y)$ shows the temperature distribution within the wall, i.e. the temperature on top is larger compared to the temperature at the bottom. Anyhow, this profile is temporally fix and is the driver for heat crossing the wall on top as well as on the bottom. In order to keep this state fixed, in steady state the supplied energy needs to be balanced by the released energy, thus

$$\dot{Q}_2 = \dot{Q}_1. \tag{8.14}$$

---

[3]Conservation values are mass, momentum and energy.

Fig. 8.7 Steady state thermodynamic cycle (black box)

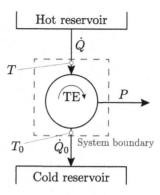

8.4.3 Cycles

Though thermodynamic cycles will be discussed later, see Sect. 8.3 and Chap. 17, Fig. 8.7 shows a thermodynamic cycle in a black box notation. A thermodynamic cycle can run permanently respectively cyclically. Thus, it is treated as to be in steady state. All state values within the system boundary are temporally constant. To achieve this, it is clarified—once the first law of thermodynamic is introduced—that the fluxes of energy need to be balanced[4] in steady state, i.e.

$$\dot{Q} = \dot{Q}_0 + P. \tag{8.15}$$

However, as indicated by the temperatures T and T_0, there is, since the system is operated between hot and cold reservoir, a local temperature spread in the system. The system has a cold and warm side as well!

---

[4]Incoming energy is equal to the outgoing energy!

Chapter 9
Process Values Heat and Work

Over the previous chapters the internal state of a system has been described. A thermodynamic system can change its state and several possible changes of state have been discussed. For the further understanding it is essential to be familiar with the terms reversibility/irreversibility as well as with the idea of a quasi-static change of state. These approaches have been clarified in the previous Chaps. 7 and 8: By external impacts a system's thermodynamic equilibrium can be disturbed and the system moves into a new balanced state.

In this chapter the focus is on so-called *process values* heat and work. They can be regarded as the external impacts, that force a system into a new equilibrium state. First, the thermal energy, known as heat, is introduced, though it has already been mentioned several times previously. Next, the thermodynamic *work* is investigated. In this section mechanical principles are applied. However, when talking about work, the type of system needs to be part of the considerations.

9.1 Thermal Energy—Heat

Figure 9.1 has already been discussed previously, but it shows what heat is about. Systems A and B, each having a different temperature, drive themselves into a thermal equilibrium in case they get in contact, i.e. the temperatures of A and B are equalised after a while. In order to achieve this equilibrium, thermal energy needs to be exchanged. This thermal energy is named *heat*. Obviously, heat appears *at the interface* of systems A and B: Heat crosses the interface of both systems as long as the systems have different temperatures. Once, the temperatures are balanced, the heat at the interface disappears.

Theorem 9.1 *Heat is an interaction between system and environment.*

© Springer Nature Switzerland AG 2019
A. Schmidt, *Technical Thermodynamics for Engineers*,
https://doi.org/10.1007/978-3-030-20397-9_9

Fig. 9.1 Thermal equilibrium is achieved by heat flux

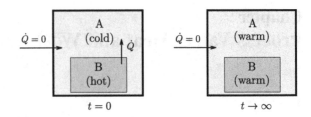

Hence, heat can not be observed within a system. Inside a system the state can be determined, e.g. by temperature or internal energy. Heat is an extensive value:

$$[Q] = 1J = 1Nm \tag{9.1}$$

For flow processes it makes sense to define a heat flux,[1] i.e. the temporal change of heat:

$$\dot{Q} = \frac{\delta Q}{dt} \tag{9.2}$$

The unit of a heat flux is

$$[\dot{Q}] = 1\frac{J}{s} = 1W. \tag{9.3}$$

Obviously, a heat flux represents a thermal power. As mentioned before, this is often required for open systems as well as for machines that operate permanently, e.g thermodynamic cycles. Thermodynamically, it makes sense to define a specific notation as well, i.e.:

- Specific heat by the mass of the system m:

$$q = \frac{Q}{m} \tag{9.4}$$

- Specific heat by the fluid's mass flow rate $\dot{m}$:

$$q = \frac{\dot{Q}}{\dot{m}} \tag{9.5}$$

In a next step a sign convention for the heat crossing a system boundary needs to be fixed. Figure 9.4 shows the convention followed in this book. The sign convention is always from the system's perspective. This means, heat supply has a positive sign, whereas heat release has a negative. Systems without any heat release/supply are called adiabatic. This includes two different scenarios:

1. The system is adiabatic due to a perfect insulation towards its environment.

---

[1] Just imagine a gas-turbine in steady state: It does not release a single portion of heat, but it releases heat permanently.

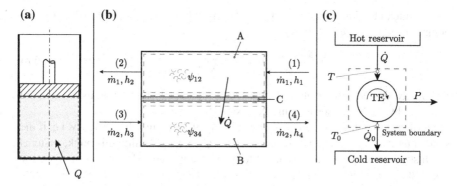

Fig. 9.2 Sign convention heat—Advice

2. The system is adiabatic since system and environment have the same temperature, i.e. they are in thermal equilibrium.

When solving thermodynamic problems, it is wise to have a solution strategy in mind. One approach may be to indicate heat always with an arrow leading *into* the system,[2] see Fig. 9.2a. This requires to prepare a sketch of the specific problem first. When the thermodynamic balancing is performed, the result for the heat, respectively work as well, needs an interpretation: If the heat is positive the assumption of the arrow direction was correct: The system is supplied with heat.[3] In contrast, if during the calculation the sign of the heat turns out to be negative, heat is released by the system.

However, this advice in some cases needs to be given up. Let's have a look at Fig. 9.2b for instance. This sketch shows a heat exchanger, i.e. an open system of two separated mass fluxes. Heat is transferred from one flow to the other. As indicated in this figure, several system boundaries can be applied, e.g. systems A and B. The recommendation given above can easily be applied for system B, as the arrow for the heat flux $\dot{Q}$ leads into system B. By this, the direction of the flux is fixed and thus the recommendation has to be violated for system A, as for this system the arrow leads *out*. Anyhow, in this case, the heat flux is the *coupling of the two systems A and B*. For one of the systems the recommendation can not be applied. This is typical for coupled systems and is further investigated when the first law of thermodynamics is introduced.

Figure 9.2c shows another application, when the advice is broken with. Thermodynamic cycles, e.g. the shown thermal engine, usually take the absolute values of heat/work respect. heat flux/power into account. However, the direction of the external impacts for cycles is considered as it is in reality. Thus, all fluxes shown in Fig. 9.2c are, as mentioned before, absolute values. The heat flux $\dot{Q}$ is supplied to the machine, whereas the mechanical power P and the heat flux $\dot{Q}_0$ leave the ther-

---

[2]In fact, this will be an applicable approach for the work later on as well!

[3]Remember that the sign convention is always from the system's perspective!

mal engine (TE). Later on, the first law of thermodynamic is applied to this thermal engine in steady state as follows

$$\dot{Q} = P + \dot{Q}_0. \tag{9.6}$$

Physically, this equation can be interpreted as that energy fluxes into the system are balanced by energy fluxes leaving the system.[4]

Furthermore, closed systems as shown in Fig. 9.2a are usually unsteady, i.e. if the system undergoes a change of state, its state values vary. Within this book, systems as illustrated in Fig. 9.2c are regarded to be in steady state. The heat exchanger, see Fig. 9.2b, i.e. a typical example of an open system, can be investigated in steady state or even in unsteady state. Anyhow, once the starting up of the system is finished and incoming and outgoing fluxes are constant the open system is in steady state. In such a case, the internal state does not vary temporally, though it might vary locally.

9.2 Work

Next to thermal energy it is external work, that has the power to change the state of a system. Just like heat, work can be supplied or released across the system boundary. However, though work has already been introduced in Chap. 2, a further clarification is given in this section. Subsequently, it is investigated, how work can be calculated for systems operated with compressible fluids. Anyhow, mechanical principles are applied.

Theorem 9.2 *Work is an interaction between system and environment. It has, as well as heat, the power to change the state of a thermodynamic system.*

9.2.1 Definition of Work

From a physical perspective, work is the scalar product of a force and a distance, see also Chap. 2:

$$W = \vec{F} \cdot \vec{s} \tag{9.7}$$

According to this mechanical principle, several forms of work are conceivable, see Fig. 9.3. The first example takes a closed system, consisting of cylinder and movable piston, into consideration. This system shall be filled with a gas, i.e. the fluid is compressible. Consequently, the fluid can expand or it can be compressed without varying its mass. If the gas is ideal, a quasi-static change of state follows the thermal equation of state

---

[4]In case of steady state!

Volume work

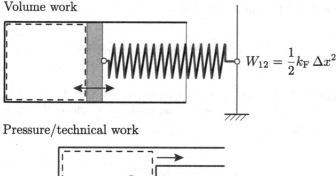

$$W_{12} = \frac{1}{2} k_F \, \Delta x^2$$

Pressure/technical work

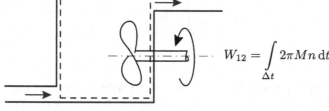

$$W_{12} = \int_{\Delta t} 2\pi M n \, dt$$

Electrical work

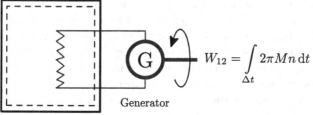

$$W_{12} = \int_{\Delta t} 2\pi M n \, dt$$

Generator

Fig. 9.3 Definition of work

$$pv = RT. \tag{9.8}$$

According to Fig. 9.3, while the gas expands in the cylinder, the spring is compressed. Obviously, following Eq. 9.7, work is released by the gas and stored as energy within the spring:

$$W_{12} = \frac{1}{2} k_F \, \Delta x^2 \tag{9.9}$$

Since the work is related to the volume change of the fluid, it is called *volume work*.

Another example is related with an open system as given in Fig. 9.3. A fluid flow enters at high pressure. On its way a stirrer, that is mounted into the system, starts to rotate and transfers its rotational energy to a shaft leading out of the system. When leaving the system at the outlet, the fluid's pressure has decreased compared to the inlet. Once again, mechanical principles, similar to the spring, can be applied in order to prove that *work* has been released by the system. Hence, the work of the rotating shaft is proportional to torque and speed, i.e.

$$W_{12} = \int_{\Delta t} 2\pi M n \, dt \text{ resp. } P = 2\pi M n \tag{9.10}$$

Work related with the flow energy of an open system is called *technical work*. Specific technical work $w_{t,12}$ is composed of pressure work y_{12}, mechanical work as well as of dissipated energy[5]:

$$w_{t,12} = \frac{P}{\dot{m}} = \underbrace{\int_1^2 v \, dp}_{y_{12}} + \psi_{12} + \Delta e_{a,12} \tag{9.11}$$

However, dissipation has an ambivalent character: On the one hand, dissipation finally leads to halting of all movements. Thus, dissipation should be avoided in technical applications. Dissipation even reduces the power output respectively increases the power consumption, see Chap. 16. Anyhow, on the other hand, it is obvious that many technical processes do not run without dissipation respectively friction. Without friction a car could not start moving and driving into a curve would be impossible for instance. Referring to the blades of the stirrer in Fig. 9.3, without the no-slip condition, according to Stokes,[6] there is no boundary layer, that is required in order to convert fluid energy to the stirrer and to make it move.

However, the stirrer in Fig. 9.3 can even *supply* technical work to the system, in case its shaft is driven from outside. Consequently, a pressure increase from inlet to outlet would be measured.

With respect to Fig. 9.3 even another form of work is possible. A shaft drives a generator, that starts producing electrical energy. Thus, mechanical work at the shaft is converted into electrical energy. An electrical resistance is connected, the current dissipates the electrical energy and converts it into heat.

Theorem 9.3 *Energy crossing the system boundary, that can be calculated according to mechanical principles, is called work!*

The SI-unit for work follows:

$$[W] = 1 \, J = 1 \, Nm \tag{9.12}$$

For flow processes it once again makes sense to define a power, i.e. the temporal change of work:

$$P = \frac{\delta W}{dt} \tag{9.13}$$

---

[5]A detailed explanation will follow in Sect. 11.3.4!
[6]Sir George Gabriel Stokes (*13 August 1819 in Skreen, County Sligo, †1 February 1903 in Cambridge).

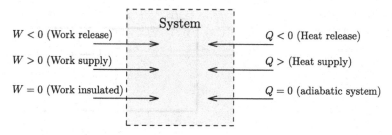

Fig. 9.4 Sign convention for the process values heat and work

with its unit

$$[P] = 1\frac{J}{s} = 1\,W. \tag{9.14}$$

Usually specific work is defined in thermodynamics as follows:

- Specific work by the mass of the system m:

$$w = \frac{W}{m} \tag{9.15}$$

- Specific technical work by the fluid's mass flow rate $\dot{m}$:

$$w_t = \frac{P_t}{\dot{m}} \tag{9.16}$$

Theorem 9.4 *Work as well as heat only occur, when crossing a system boundary! They both are process values and no state values.*

Regarding the sign of work, the same principles are followed as already introduced for the heat, i.e. work that is released by a system has a negative sign, whereas work that is supplied to a system has a positive sign, see Fig. 9.4.

9.2.2 Volume Work

Volume work has already been introduced in the previous section, see Fig. 9.3. Due to expansion inside a closed cylinder gas can release work. Reversely, if the gas is compressed, the system is supplied with work. Within this section an equation to calculate this volume work is derived.

Figure 9.5 illustrates the approach to calculate the volume work. In order to apply the definition

$$W = \vec{F} \cdot \vec{s} \tag{9.17}$$

Fig. 9.5 Deriving volume
work–Expansion (1) → (2)

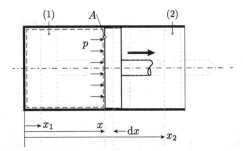

it first has to be clarified what force F is required for the change of volume. As
mentioned in Chap. 4 only systems in thermodynamic equilibrium are regarded.
Furthermore, for any change of state quasi-static behaviour is assumed. Thus, the
force can be easily predicted by the pressure inside the system by

$$F = p \cdot A \tag{9.18}$$

as the system on its way from state (1) to state (2) always is at mechanical equilibrium. According to everyday experience the pressure increases constantly during
compression. Thus, the force required for compression can not be constant, but a
function of the position's current position:

$$p = f(x). \tag{9.19}$$

Hence, it is wise to calculate the *differential* work first. If the piston is just moved by
a differential distance dx, see Fig. 9.5, the differential amount of work follows:

$$\delta W_v = F \cdot dx \tag{9.20}$$

Substituting the force brings:

$$\delta W_v = p(x) \, A \, dx \tag{9.21}$$

Due to the geometrical correlation

$$dV = A \, dx \tag{9.22}$$

it is

$$\delta W_v = p \, dV \tag{9.23}$$

Since the occupied volume is a function of the piston's position as well, the function
obeys

$$p = f(V). \tag{9.24}$$

This follows due to

$$x = \frac{V}{A}.$$
(9.25)

The sign of work is determined by the differential dV. In order to fulfil the sign convention, see Fig. 9.2, Eq. 9.23 finally needs to be adjusted:

$$\boxed{\delta W_{\mathrm{v}} = -p\,(V)\,dV}$$
(9.26)

Hence, three scenarios, all following the given sign convention, are conceivable:

- Compression ($dV < 0$):
  According to Eq. 9.26 work is positive, i.e. work needs to be supplied to the system.
- Expansion ($dV > 0$):
  According to Eq. 9.26 work is negative, i.e. work is released by the system.
- No change of volume[7] ($dV = 0$):
  According to Eq. 9.26 no work is exchanged.

Equation 9.26 is in differential notation, so if the entire *volume work* from state (1) to state (2) is requested, an integration needs to be done:

$$\boxed{W_{\mathrm{v},12} = \int_1^2 \delta W_{\mathrm{v}} = -\int_1^2 p\,dV}$$
(9.27)

For the *specific volume work* a division by the system's mass is needed, i.e.

$$\boxed{w_{\mathrm{v},12} = -\int_1^2 p\,dv}$$
(9.28)

For deriving Eqs. 9.27 and 9.28 no assumption has been made regarding the reversibility. Thus, these equations can be applied for reversible as well as for irreversible changes of state. The only premise is, that the system must be in thermodynamic equilibrium. Hence, if the change of state is irreversible, this is not due to imbalance but due to dissipation, see Sect. 7.4.

Visualisation in a p, v-Diagram

Let us have a closer look at the specific volume work in differential notation as given by

$$\boxed{\delta w_{\mathrm{v}} = -p\,dv}$$
(9.29)

---

[7]Isochoric change of state!

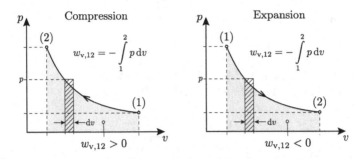

Fig. 9.6 Specific volume work $w_{v,12}$—illustration in a p, v-diagram

As discussed previously, any quasi-static change of state can be illustrated unequivocally in a p, v-diagram. Figure 9.6 shows a compression as well as an expansion of a closed system. The differential work δw_v for a small change of specific volume dv, according to Eq. 9.29, can be visualised by the hatched area. For the entire volume work the differential volume dv needs to be integrated from $(1) \rightarrow (2)$, i.e.

$$w_{v,12} = -\int_1^2 p \, dv. \tag{9.30}$$

This integral is the area beneath the curve, that describes the change of state from state (1) to state (2) in a p, v-diagram. Thus, specific volume work is represented by the grey coloured area in Fig. 9.6. Areas in a p, v-diagram, that are on the left hand side when moving from $(1) \rightarrow (2)$ indicate supplied work ($w_{v,12} > 0$), whereas areas on the right hand side indicate released work ($w_{v,12} < 0$). Obviously, the quantity of the specific volume work depends on the path the systems moves from (1) to (2). Hence, volume work is a so-called *process value*, since it is path-dependent. This is in contrast to state values, that never depend on the direction a system takes. It needs to be pointed out, that the volume work is valid for both: systems with and without dissipation. If dissipation within the system occurs, the path from (1) to (2) is different from as if it is when there is no dissipation.

Problem 9.5 Air with a temperature of $T_1 = 273$ K and a pressure of $p_1 = 2.8$ bar is enclosed in a cylinder $V_1 = 2.47$ m^3 with a frictionless moveable and gas-tight fitted piston. The inner diameter of the cylinder is $D = 920$ mm. Calculate the extensive and the specific volume work, if the piston movement of $\Delta s = 32$ cm leads to an isothermal compression. The gas constant of air is $R = 287.11 \frac{J}{kg\,K}$.

Solution

The volume of the gas in state (2) is

$$V_2 = V_1 - \frac{\pi}{4} D^2 \, \Delta s = 2.257 \, \text{m}^3 \tag{9.31}$$

The change of state is supposed to be quasi-static, so that the thermal equation of state can be applied for the entire change of state. It reads, since the change of state is isothermal, as:

$$pV = mRT = \text{const.} \tag{9.32}$$

This leads to

$$pV = p_1 V_1 = p_2 V_2 \tag{9.33}$$

This equation shows, that the pressure is a function of the volume. The smaller the volume is, the larger the pressure is.

$$\Rightarrow p = f(V) = p_1 \frac{V_1}{V} \tag{9.34}$$

The volume work follows

$$W_{v,12} = -\int_1^2 p\, dV = -\int_1^2 p_1 \frac{V_1}{V}\, dV = -p_1 V_1 \int_1^2 \frac{1}{V}\, dV \tag{9.35}$$

Solving the integral leads to

$$W_{v,12} = -p_1 V_1 \ln\frac{V_2}{V_1} = 62369.6 \text{ J} > 0 \tag{9.36}$$

Applying the thermal equation of state, finally gives:

$$W_{v,12} = -mRT \ln\frac{V_2}{V_1} \tag{9.37}$$

Dividing by the mass, leads to the specific volume work

$$w_{v,12} = RT \ln\frac{V_2}{V_1} = 7068.3 \ \frac{\text{J}}{\text{kg}} > 0 \tag{9.38}$$

9.2.3 Effective Work

Volume work describes the work a gas releases during expansion respectively the work that is supplied while a gas is compressed. Thus, volume work can be positive or negative. However, volume work is a process value from the gas point of view. For technical applications the *effective* work is of importance, that is illustrated in Fig. 9.7. While the volume of the gas inside the cylinder changes, work is transferred from the gas to the piston, that starts to move. The moving piston can be applied to operate a vehicle or a generator - hence volume work can be converted into mechanical

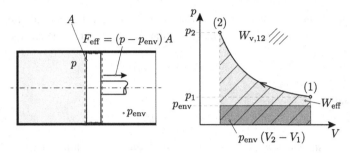

Fig. 9.7 Effective work W_{eff}—Illustration and visualisation in a p, V-diagram

respectively electrical energy. But as soon as an atmosphere is on the opposite side
of the piston, the effective, usable work varies. In case of an expansion, the piston's
energy can not be utilised completely in a technical application, since energy is
needed partially to *compress* the outer atmosphere. If the piston is moving into
the other direction, i.e. gas inside the cylinder is compressed, the outer atmosphere
supports this compression. In that case, the outer atmosphere expands.[8] As illustrated
in Fig. 9.7, the piston is faced by a force from the left, due to the internal gas pressure,
and an opposite pressure on the right from the outer atmosphere. Thus, the effective
force is:

$$F_{\text{eff}} = (p - p_{\text{env}}) A \tag{9.39}$$

In order to get the effective work, the effective force needs to be scalar multiplied with
the distance the piston is moving. Since the internal pressure can vary, a differential
approach is followed:

$$\delta W_{\text{eff}} = -F_{\text{eff}}\, dx \tag{9.40}$$

While the gas inside the cylinder is compressed, i.e. $dx < 0$, the outer atmosphere
supports. Vice versa, in case the gas expands, i.e. $dx > 0$, the outer atmosphere
impedes the expansion.

$$\delta W_{\text{eff}} = -(p - p_{\text{env}})\, dV \tag{9.41}$$

By integration the entire effective work is

$$\boxed{W_{\text{eff}} = -\int_{1}^{2} p\, dV + p_{\text{env}} (V_2 - V_1)} \tag{9.42}$$

---

[8]However, though the outer atmosphere might expand or might be compressed, the ambient pressure
is supposed to be constant due to its large size.

Hence, the effective work describes the compression respectively expansion from the piston's perspective as indicated by the system boundary in Fig. 9.7. The p, V-diagram shows the effective work in case of a compression. The environmental part of the work is indicated by

$$p_{\text{env}} (V_2 - V_1).$$ (9.43)

In this specific case, the effective work is smaller than the volume work, that would be required if there was no environment. This is due to the outer ambient support. In cyclic processes, e.g. an internal combustion engine, the work at respectively from the environment has to be neglected, since it neutralises while compression and expansion.

In some cases, when the ambient pressure is large, even for expansion the effective work can be positive. Thus mechanical work needs to be supplied at the piston, i.e. the piston needs to be pulled, though from the gas perspective work is released. This occurs, when

$$p_{\text{env}} > \frac{1}{(V_2 - V_1)} \int_1^2 p \, dV$$ (9.44)

Problem 9.6 A frictionless sliding piston compresses an ideal gas with a volume of $V_1 = 0.18 \, \text{m}^3$ and a pressure of $p_1 = 1340 \, \text{hPa}$ isothermally to $V_2 = 0.03 \, \text{m}^3$. Ambient pressure is $p_{\text{env}} = 980 \, \text{hPa}$. Determine the effective work performed by the piston rod force.

Solution

The effective, usable work can be calculated according to

$$W_{\text{eff}} = - \int_1^2 p \, dV + p_{\text{env}} (V_2 - V_1)$$ (9.45)

In order to solve the integral it is required to understand how the pressure varies with the volume. As the change of state is isothermal, see also Problem 9.5, we know that

$$pV = mRT = \text{const.}$$ (9.46)

This leads to

$$pV = p_1 V_1 = p_2 V_2.$$ (9.47)

This equation shows, that the pressure is a function of volume. The smaller the volume is, the larger the pressure is.

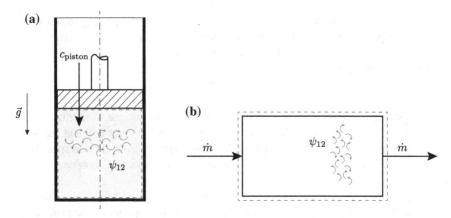

Fig. 9.8 Dissipation in closed (**a**) and open systems (**b**)

$$\Rightarrow p = f(V) = p_1 \frac{V_1}{V} \tag{9.48}$$

So, the effective work can be calculated according to

$$W_{\text{eff}} = -p_1 V_1 \ln\frac{V_2}{V_1} + p_{\text{env}} (V_2 - V_1) = 28.517\,\text{kJ} \tag{9.49}$$

9.2.4 Systems with Internal Friction—Dissipation

In thermodynamic systems, dissipation might occur: Due to a moving fluid turbulent eddies are generated, that cause internal friction. Figure 9.8 illustrates how these eddies can be generated in closed and in open systems. In case a piston moves in a closed system, as shown in Fig. 9.8a, the fluid velocity inside a system is not constant, so that, due to relative movement of the molecules, turbulence increases. Consequently, if no work is exchanged in an isochoric system for instance, no internal friction results due to the missing fluid movement. In open systems, see Fig. 9.8b, the fluid velocity also causes turbulence and thus internal friction. Experience shows, that this energy of dissipation leads to an increase of temperature inside the system. Anyhow, the internal friction, i.e. dissipation needs a driver: The required energy might be supplied

- from the outside across the system boundary, e.g. a piston that is moved, or an impeller that is immersed and rotates inside the system
- by changing the potential of kinetic energies,
- by pressure drop in an open system or
- in a fully-closed system by internal balancing processes.

Energy of dissipation is given by:

$$\Psi_{12} \text{ [J] respect. } \psi_{12} = \frac{\Psi_{12}}{m} \left[\frac{J}{kg} \right] \tag{9.50}$$

Since dissipation depends on the path the change of state takes, it is a process value similar to work or heat for instance. Thus, it does not describe the system's state, since it only occurs if the system change its state. However, dissipation can occur or it can not. Thus, negative dissipation does not exist, since turbulent eddies always *consume* energy, i.e.

$$\Psi \geq 0 \tag{9.51}$$

Theorem 9.7 *The wire pendulum experiment, see Sect. 7.3, has already proven, that processes with dissipation are always irreversible. Experience shows: If a process is irreversible, the work that has to be applied to achieve the same result as in a reversible process increases:*

$$W_{in,out} = W_{reversible,in,out} + \Psi \tag{9.52}$$

Actually, this is due to the energy consumption of the turbulent eddies. The consequences of Eq. 9.52 are as follows:

- Case 1: Compression (closed system)
  Let us assume, that in reversible operation the work for compression is

$$W_{reversible} = 10 \text{ kJ}. \tag{9.53}$$

If the compression is irreversible, the energy of dissipation might be

$$\Psi = 2 \text{ kJ}. \tag{9.54}$$

The entire work for the compression is

$$W = W_{reversible} + \Psi = 12 \text{ kJ}. \tag{9.55}$$

Thus, the energy supply is larger than in reversible mode, since the dissipated energy needs to be supplied across the system boundary as well.
- Case 2: Expansion (closed system)
  Let us assume, that in reversible operation the released work for an expansion is

$$W_{reversible} = -10 \text{ kJ}. \tag{9.56}$$

If the compression is irreversible, the energy of dissipation might be

$$\Psi = 2 \text{ kJ}. \tag{9.57}$$

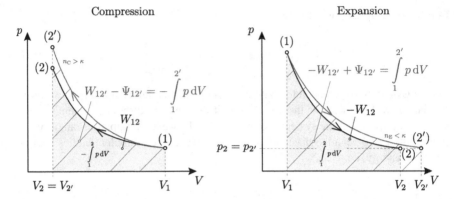

Fig. 9.9 Left: Compression to the same volume, right: Expansion to the same pressure (both adiabatic). $(1) \rightarrow (2)$ reversible, $(1) \rightarrow (2')$ irreversible

The total released work is

$$W = W_{\text{reversible}} + \Psi = -8 \text{ kJ}. \tag{9.58}$$

Thus, the released work is smaller than in reversible mode, since the dissipated energy is taken from the released work.

These examples show, that from an engineering point of view, dissipation should be avoided in any case: Dissipation enlarges the work required for compression and lowers the released work at expansion.

Process Value Dissipation—Example Compression/Expansion

Figure 9.9 shows an adiabatic compression as well as an adiabatic expansion of a closed system in a p, V-diagram.[9]

- Adiabatic compression
  Let us investigate what happens in case of an adiabatic and reversible, i.e. isentropic,[10] compression. In order to compress the gas inside the cylinder, work needs to be supplied from $(1) \rightarrow (2)$. The first law of thermodynamics in this case while ignoring the changes of outer energies[11] reads as:

$$W_{12} + \underbrace{Q_{12}}_{=0} = U_2 - U_1 = mc_v (T_2 - T_1) \tag{9.59}$$

with the partial energy equation

---

[9]For this section the knowledge of first law of thermodynamic is required!

[10]See Sect. 13.5.

[11]I.e. kinetic and potential energies!

$$W_{12} = -\int_1^2 p\,dV > 0. \tag{9.60}$$

The entire supplied work is volume work and can be visualised as grey-coloured area in Fig. 9.9. In this case total work and volume work are identical. Following the first law of thermodynamics, the temperature increases, i.e. $T_2 > T_1$. Now, from $(1) \rightarrow (2')$ dissipation shall occur, i.e. the process is irreversible but shall reach the same degree of compression as before, i.e. $V_{2'} = V_2$. The first law of thermodynamics in this case is

$$W_{12'} + \underbrace{Q_{12'}}_{=0} = U_{2'} - U_1 = mc_v\,(T_{2'} - T_1) \tag{9.61}$$

However, the partial energy equation follows

$$W_{12'} = -\int_1^{2'} p\,dV + \Psi_{12'} > 0 \tag{9.62}$$

The entire work $W_{12'}$ is composed of volume work, that is indicated by the hatched area in Fig. 9.9, and dissipation. According to Theorem 9.7 the supplied work, in case dissipation appears, is larger than if the process would be free of dissipation:

$$W_{12'} > W_{12} \tag{9.63}$$

Thus, following the first law of thermodynamics it is

$$T_{2'} > T_2 \tag{9.64}$$

Hence, the temperature rise with dissipation is larger than in case the process is free of dissipation. That extra amount of work, that is required to run the dissipative process, is obviously converted into internal energy. Equation 9.62 can be rearranged, so that

$$W_{12'} - \Psi_{12'} = -\int_1^{2'} p\,dV > 0 \tag{9.65}$$

Thus, the hatched area does not only indicate the supplied volume work but also $W_{12'} - \Psi_{12'}$.

- Adiabatic expansion
  Now let us investigate what happens in case of an adiabatic and reversible, i.e. isentropic, expansion. While the gas inside the cylinder expands, work is released

from $(1) \rightarrow (2)$. The first law of thermodynamics in this case while ignoring the change of outer energies reads as:

$$W_{12} + \underbrace{Q_{12}}_{=0} = U_2 - U_1 = mc_v (T_2 - T_1) \tag{9.66}$$

with the partial energy equation

$$W_{12} = - \int_1^2 p \, dV < 0. \tag{9.67}$$

The entire released work is volume work and can be visualised as grey-coloured area in Fig. 9.9. In this case total work and volume work are identical, since no dissipation needs to be driven. Following the first law of thermodynamics, the temperature decreases, i.e. $T_2 < T_1$.

Now, from $(1) \rightarrow (2')$ dissipation shall occur, i.e. the process is irreversible but shall reach equilibrium with the environment as before, i.e. $p_{2'} = p_2$. The first law of thermodynamics for this case is

$$W_{12'} + \underbrace{Q_{12'}}_{=0} = U_{2'} - U_1 = mc_v (T_{2'} - T_1) \tag{9.68}$$

However, the partial energy equation follows

$$W_{12'} = \underbrace{- \int_1^{2'} p \, dV}_{<0} + \underbrace{\Psi_{12'}}_{>0} < 0 \tag{9.69}$$

Mind, that the dissipation is driven partly by the released volume work! The entire released work $W_{12'} < 0$ is composed of volume work, that is indicated by the hatched area in Fig. 9.9, and dissipation. According to Theorem 9.7 the absolute value of the work, in case dissipation appears, is smaller than if the process would be free of dissipation:

$$|W_{12'}| < |W_{12}| \tag{9.70}$$

Rearranging this equation leads to

$$W_{12'} > W_{12} \tag{9.71}$$

Thus, following the first law of thermodynamics:

$$T_{2'} > T_2 \tag{9.72}$$

Hence, the temperature rise with dissipation is larger than in case the process is free of dissipation. The work required to run the dissipative process is obviously taken from the released volume work and is converted into internal energy. Equation 9.69 can be rearranged, so that

$$-W_{12'} + \Psi_{12'} = \int_1^{2'} p \, dV > 0 \qquad (9.73)$$

Thus, the hatched area does not only indicate the released volume work but also $-W_{12'} + \Psi_{12'}$.

Theorem 9.8 *The energy of dissipation decreases the released work* $-W_{out,rev}$ *and increases the supplied work* $W_{in,rev}$.

$$W_{12} = W_{12,rev} + \Psi_{12} \Rightarrow W_{12} = -\int_1^2 p \, dV + \Psi_{12} \qquad (9.74)$$

9.2.5 Dissipation Versus Outer Friction

Dissipation

Let us summarise what we have learned about dissipation so far by taking a closer look at a closed system as exemplary shown in Fig. 9.10a. Due to the piston movement turbulent eddies are induced within the fluid. The eddies' energy is transferred to the fluid and increases its internal energy once the eddies break up and release their previously consumed energy. Following the no-slip condition at the wall according to Stoke, the fluid movement stops and thus there are no eddies directly at the wall. However, the eddies occur within the boundary layer: According to the radial velocity distribution $c(r)$ relative motion of the fluid occurs and turbulence is generated. The larger the fluid's motion driven by the moving piston is, the more turbulent the flow gets.

Thus, the limiting case of a very slow ($\rightarrow$ no eddies!) change of state, see Fig. 7.14 and Theorem 7.25, reads as:

$$W = W_{rev} = -\int_1^2 p \, dV \qquad (9.75)$$

Thus

$$\Psi = 0. \qquad (9.76)$$

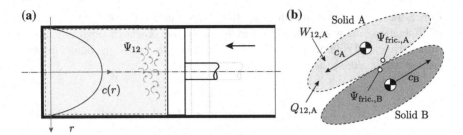

Fig. 9.10 Dissipation in closed systems due to fluid turbulence (left), outer friction (right)

Outer Friction

In contrast to this, outer friction might also occur, see Fig. 9.10b. Outer friction is related to *solids* that rub against each other, i.e. friction is among solid surfaces. Let us see how the first law of thermodynamics reads for solid A:

$$W_{12,\mathrm{A}} + Q_{12,\mathrm{A}} = \Delta U_\mathrm{A} + m_\mathrm{A}\left[\frac{1}{2}\left(c_2^2 - c_1^2\right)_\mathrm{A} + h\,(z_2 - z_1)_\mathrm{A}\right] \qquad (9.77)$$

The partial energy equation for $W_{12,\mathrm{A}}$ is

$$W_{12,\mathrm{A}} = W_{\mathrm{mech,A}} + \Psi_{\mathrm{fric.,A}} = m_\mathrm{A}\left[\frac{1}{2}\left(c_2^2 - c_1^2\right)_\mathrm{A} + h\,(z_2 - z_1)_\mathrm{A}\right] + \Psi_{\mathrm{fric.,A}}. \tag{9.78}$$

Finally, the first law of thermodynamics simplifies to:

$$\boxed{\Psi_{\mathrm{fric.,A}} + Q_{12,\mathrm{A}} = \Delta U_\mathrm{A}} \qquad (9.79)$$

Consequently, it is the heat and the outer friction that have an impact on the temperature change. The outer friction is, as the partial energy equation shows, supplied by the work $W_{12,\mathrm{A}}$. Thus, the supplied work $W_{12,\mathrm{A}}$ is not entirely mechanical work in order to move solid A, but obviously partly frictional work as well. This frictional work increases the internal energy of system A once it dissipates. Hence, outer friction has the same impact as heat, that might be supplied additionally.

   Though the second law of thermodynamics and the corresponding specific entropy s is introduced in Sect. 12.2, the second law of thermodynamics is now applied for that problem. Starting with the first law of thermodynamics according to Eq. 9.79 in differential and specific notation, i.e.

$$\delta\psi_{\mathrm{fric.}} + \delta q = \mathrm{d}u \qquad (9.80)$$

dividing by the temperature T leads to

$$\frac{\delta\psi_{\text{fric.}}}{T} + \frac{\delta q}{T} = \frac{\mathrm{d}u}{T}. \tag{9.81}$$

Applying the caloric equation of state[12] reads as

$$\frac{\mathrm{d}u}{T} = c\frac{\mathrm{d}T}{T}. \tag{9.82}$$

The integration from state (1) to state (2) results in

$$\int_{1}^{2} \frac{\mathrm{d}u}{T} = \int_{1}^{2} c\frac{\mathrm{d}T}{T} = c\ln\frac{T_2}{T_1}. \tag{9.83}$$

Obviously, a path information (1) $\rightarrow$ (2) was not required for solving the integral. Consequently, $\frac{\mathrm{d}u}{T}$ must be a state value. This state value is named specific entropy s:

$$\underbrace{\frac{\delta\psi_{\text{fric.}}}{T}}_{\delta s_i} + \underbrace{\frac{\delta q}{T}}_{\delta s_a} = \frac{\mathrm{d}u}{T} = \mathrm{d}s \tag{9.84}$$

Finally, the second law of thermodynamics reads as:

$$\boxed{\mathrm{d}s = \delta s_i + \delta s_a} \tag{9.85}$$

9.2.6 Mechanical Work

For changing kinetic as well as potential energy of a system work is required respectively released. The amount of work can be calculated by means of mechanics, see Chap. 2, i.e.

$$W_{\text{mech},12} = \frac{m}{2}\left(c_2^2 - c_1^2\right) + mg\left(z_2 - z_1\right) \tag{9.86}$$

Mechanical work occurs in systems when the centre of gravity is modified regarding the position within the gravitational field and its velocity. Lifting a crate of beer for instance, see Fig. 9.11 requires mechanical work:

$$W_{12} = W_{\text{mech},12} = mg\left(z_2 - z_1\right) \tag{9.87}$$

---

[12]A solid is supposed to be incompressible, so that there is no distinction between c_v and c_p, see Sect. 12.4.2.

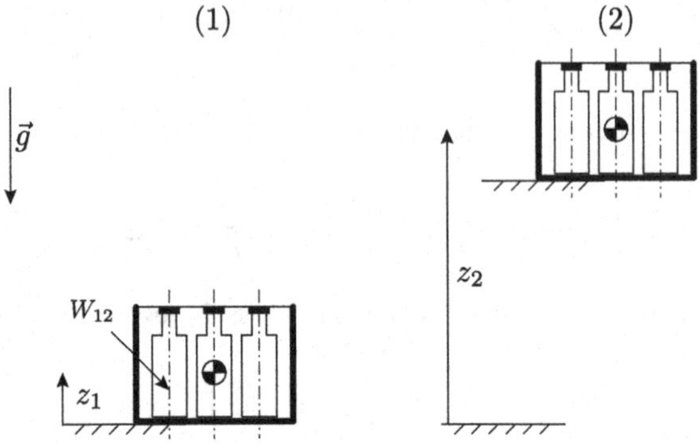

Fig. 9.11 Illustration of mechanical work

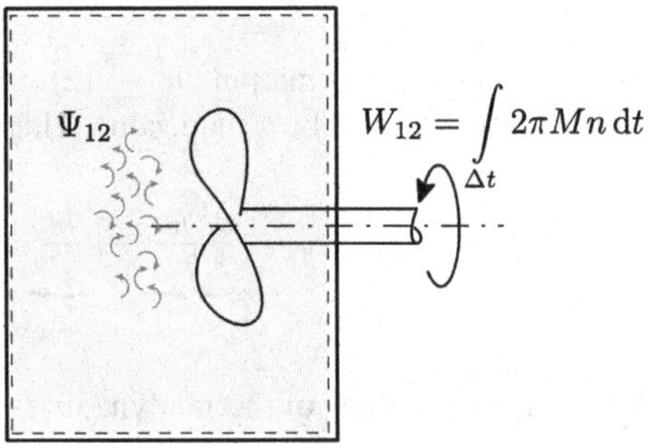

(1)   (2)

Fig. 9.12 Shaft work for a closed system

$$W_{12} = \int_{\Delta t} 2\pi M n \, \mathrm{d}t$$

9.2.7 Shaft Work

By a rotating shaft work is supplied to a closed system over its system boundary, see Fig. 9.12. Within the system the fluid is accelerated, so that the kinetic energy of the fluid rises. The fluid movement is rather complex and might be calculated by means of fluid dynamics, thus it is not part of thermodynamics.

Anyhow, the energy within the system is increased by the quantity of work which enters the system over a period of time. After the system has come to standstill a new equilibrium state is reached. The supplied work has finally dissipated, i.e. it has increased the fluid's internal energy and hence its temperature. In case of a closed system, see Fig. 9.12, the *first law of thermodynamics* from (1) $\rightarrow$ (2) reads as[13]

$$W_{12} + Q_{12} = U_2 - U_1 + \Delta E_{a,12} \tag{9.88}$$

The change of the outer energies follows

$$\Delta E_{a,12} = 0, \tag{9.89}$$

---

[13]The following equations will be introduced and explained in Chap. 11ff!

since the centre of gravity has not changed from (1) → (2) and it is in rest in both states. The *partial energy equation* for the work W_{12} reads as

$$W_{12} = W_{v,12} + W_{mech,12} + \Psi_{12} \tag{9.90}$$

- The system volume is constant, thus no volume work is supplied, i.e.

$$W_{v,12} = 0 \tag{9.91}$$

- No mechanical work is supplied either, since the centre of gravity is in the same rest state in (1) and (2), i.e.

$$W_{mech,12} = 0 \tag{9.92}$$

Consequently, the entire work W_{12} is converted into dissipation:

$$W_{12} = \Psi_{12} \tag{9.93}$$

Thus, it follows

$$\boxed{W_{shaft} = \Psi_{12}} \tag{9.94}$$

This equation shows, that shaft work in a closed system with a motionless fluid can only be positive. It has never been observed that the system drives a shaft, i.e. $W_{12} < 0$. Thus, according to the *second law of thermodynamics* the process is irreversible

$$S_{i,12} > 0. \tag{9.95}$$

However, the increased internal energy, measured by a temperature rise, can only be converted *partly* into mechanical work, e.g. by a thermal engine, since just the temperature potential towards *environment* can be utilised, see Sect. 16.1. Dissipation is always related with a loss of exergy, see Sect. 16.4:

$$\Delta E_{x,v,12} = S_{i,12} T_{env} > 0 \tag{9.96}$$

9.2.8 Shifting Work

In this section open systems as illustrated in Fig. 9.13 are investigated. An open system is characterised by mass passing the system boundary. Let the inlet be named (1) and the outlet named (2). Within the system the pressure might be distributed, so that the pressure at the inlet is p_1 and at the outlet it is p_2. The question now is, how much work is required to shift a piece of fluid into the system respectively how much energy leaves the system at the outlet?

Fig. 9.13 Shifting work in
open systems

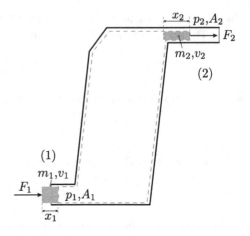

A piece of mass m_1 has to be moved across the system boundary at the inlet (1). In order to overcome the local pressure p_1 a force F_1 is required, i.e.

$$F_1 = p_1 A_1. \tag{9.97}$$

The piece of mass needs to be shifted by a distance x_1 to be fully moved into the system. Consequently, the work required for this step is

$$W_{\text{in}} = F_1 x_1. \tag{9.98}$$

Substituting the force leads to:

$$W_{\text{in}} = p_1 A_1 x_1 = p_1 V_1 \tag{9.99}$$

Analogue, the work at the outlet (2) can be calculated. At that point the mass m_2, and thus shifting work, leaves the system:

$$W_{\text{out}} = -p_2 A_2 x_2 = -p_2 V_2 \tag{9.100}$$

Summarising, the shifting work is given by

$$\boxed{W_{\text{shift}} = \pm p V} \tag{9.101}$$

respectively in specific notation:

$$\boxed{w_{\text{shift}} = \pm p v}. \tag{9.102}$$

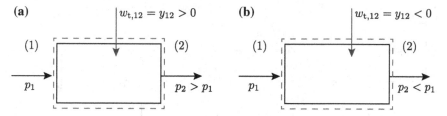

Fig. 9.14 Technical work—no dissipation, outer energies ignored

9.2.9 Technical Work Respectively Pressure Work

The term *technical work* is related to open systems, that are characterised by mass passing a system boundary. Actually, this name comes from fluid dynamics and represents the flow energy. In Sect. 21.1, it is shown, that technical work in fact follows the Bernoulli equation for incompressible fluids. In case dissipation as well as the outer energies are ignored, the specific technical work $w_{t,12}$ reduces to the so-called specific pressure work y_{12}, as illustrated in Fig. 9.14.

It is now investigated how much work is required to increase the pressure of a flow, see Fig. 9.14a, respectively how much work a system releases if the pressure decreases, see Fig. 9.14b. As mentioned above, dissipation and the change of outer energies shall be ignored at that point. Under these premises the specific technical work is

$$w_{t,12} = y_{12} = \int_1^2 v \cdot dp. \tag{9.103}$$

This equation is derived in Sect. 11.3.4. Anyhow, three cases can occur for the premises made:

- The pressure rises, i.e. $dp > 0$, so that the technical work is positive. Work is supplied to the system.
- The pressure decreases, i.e. $dp < 0$, so that the technical work is negative. The system releases work.
- The pressure stays constant, i.e. $dp = 0$, so that no technical work is transferred.

The technical power follows by multiplying with the mass flux:

$$P_t = \dot{m} w_{t,12} = \dot{m} y_{12} = \dot{m} \int_1^2 v \cdot dp \tag{9.104}$$

Integration over a period of time for a system in steady state leads to:

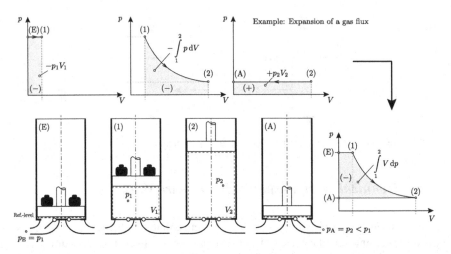

Fig. 9.15 Analogous model for technical work @ pressure decrease, related to Fig. 9.14b

$$W_{t,12} = \int_t P \, dt = m \int_1^2 v \cdot dp = \int_1^2 V \cdot dp \qquad (9.105)$$

Motivation of the Pressure Work

As mentioned above, a detailed physical proof will be given in Sect. 11.3.4. However, the idea of pressure work can even be motivated with principles that have been introduced so far. The approach for such an explanation is shown in Fig. 9.15.

An open system in steady state according to Fig. 9.14b is replaced by an analogous model, that is realised by a cylinder piston system: Such a system has two valves at its bottom, one for the inlet and one for the outlet. Anyhow, the change of outer energies of the gas are ignored in the following considerations. Furthermore, the entire change of state shall run reversibly and thus very slowly. In a first step the fluid enters the cylinder, so that its pressure is constant[14]: Hence, during the filling process the piston needs to moves upwards and needs to be in mechanical equilibrium with the gas for the entire process of filling the cylinder. The outlet valve is closed. From the gas point of view, see system boundary in Fig. 9.15, work is released to the piston, since the potential energy of the piston rises. As the pressure is constant, the work exchanged at the interface gas/piston for the filling process (E) → (1) obeys

$$W_{E1} = -\underbrace{p_1 A}_{=F_{gas}} z_1 = -p_1 V_1. \qquad (9.106)$$

---

[14] According to Example 7.24 the requirements for an isobaric change of state are fulfilled!

Since the goal is to lower the pressure, in a second step from (1) → (2), the inlet and outlet valves are both closed, while the piston's mass is reduced step by step. Hence, the volume increases—dependent on the process control. This step is performed quasi-statically with a small velocity, so that the principles of equilibrium thermodynamics can be applied and the process shall additionally be free of dissipation. The shown p, v-diagram visualises the work that is released from the system's point of view. Thus, since the piston moves upwards, work is released at the interface gas/piston and hence is negative, i.e.

$$W_{12} = - \int_1^2 p \, dV. \tag{9.107}$$

In the final step from (2) → (A) the outlet valve is opened, while the inlet valve is still closed, and the piston moves down slowly, so that the released fluid keeps its pressure constant. Once again the shifting work for releasing the gas can easily be calculated and has a positive sign, since the piston moves down,[15] i.e.

$$W_{2A} = \underbrace{p_2 A}_{=F_{\text{gas}}} z_2 = p_2 V_2. \tag{9.108}$$

Thus, the entire technical work at the interface gas/piston is the summation of all sub-steps

$$W_t = W_{E1} + W_{12} + W_{2A}. \tag{9.109}$$

Hence, it is

$$W_t = p_2 V_2 - p_1 V_1 - \int_1^2 p \, dV \tag{9.110}$$

Applying the product rule:

$$d(pV) = V \, dp + p \, dV \tag{9.111}$$

brings

$$\int_1^2 d(pV) = \int_1^2 V \, dp + \int_1^2 p \, dV. \tag{9.112}$$

Rearranging leads to

---

[15]The piston's potential energy decrease!

$$p_2 V_2 - p_1 V_1 = \int_1^2 V \, dp + \int_1^2 p \, dV \tag{9.113}$$

Thus, Eq. 9.110 simplifies to

$$W_t = \int_1^2 V \, dp \tag{9.114}$$

Calculating the time derivative since the real process, see Fig. 9.14b, runs continuously reads as

$$P_t = \int_1^2 \dot{V} \, dp. \tag{9.115}$$

Dividing by the mass flow rate $\dot{m}$ leads to

$$w_t = \int_1^2 v \, dp. \tag{9.116}$$

The integral obeys

$$\int_E^A v \, dp = \underbrace{\int_E^1 v \, dp}_{=0} + \int_1^2 v \, dp + \underbrace{\int_2^A v \, dp}_{=0}. \tag{9.117}$$

Hence, it finally is

$$\boxed{w_{t,EA} = \int_E^A v \, dp} \tag{9.118}$$

Chapter 10
State Value Versus Process Value

The previous chapters have shown, that it is essential to distinguish between state and process values. State values specify the state of a thermodynamic system. Furthermore, they can be visualised in a p, v-diagram for instance. The state of an ideal gas is fixed by two independent state values, so that each p, v-pair in a p, v-diagram defines the state of a system unambiguously, see Fig. 10.1. Consequently, the change of a specific state value Δz can be calculated by initial and final state. Hence, state values are always independent from the path the system takes:

$$\Delta z = \int_1^2 dz = z_2 - z_1 \tag{10.1}$$

In this notation d denotes the *differential of a state value*. No matter, which path, i.e. (a), (b) or (c) is followed, see Fig. 10.1, changes of the state values Δp and Δv are unaffected. According to Eq. 10.1, it is

$$\Delta p = p_2 - p_1 \tag{10.2}$$

$$\Delta v = v_2 - v_1 \tag{10.3}$$

However, mathematical combinations of state values lead to new state values, see [17], e.g. the specific enthalpy h, that will be introduced in Sect. 11.3.1 is

$$h = u + pv \tag{10.4}$$

Obviously, the specific enthalpy h is a mathematical combination of the state values specific internal energy u, pressure p and specific volume v. For calculating the specific enthalpy h no information regarding the path is required. Once the energy

© Springer Nature Switzerland AG 2019
A. Schmidt, *Technical Thermodynamics for Engineers*,
https://doi.org/10.1007/978-3-030-20397-9_10

Fig. 10.1 Process versus
state value

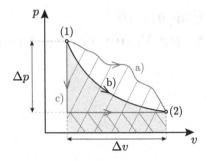

balance of an open system is discussed, the state value specific enthalpy h will get a physical meaning.[1] However, not every combination of state values leads to relevant state values with a vivid physical interpretation.

In contrast to state values, process values only occur when the state of a system is varying. Furthermore, they are even path-dependent, as illustrated in Fig. 10.1. According to Chap. 9, volume work is shown as area beneath the path of a change of state in a p, v-diagram for instance. Obviously, the amount of work depends on the process, i.e. on the mathematical function $p = f(v)$ that actually represents a path of a changing system. Specific heat q and specific work w are both process values, so is the specific dissipation. The *differential of a process value* is indicated by a δ to separate them clearly from state values. However, they follow:

$$\int_1^2 \delta w = w_{12} \tag{10.5}$$

By this, it is emphasised, that work is *not available* in state (1) or (2), but only occurs if a system moves from (1) to (2):

$$\int_1^2 \delta w \neq w_2 - w_1 \tag{10.6}$$

Theorem 10.1 *State values always provide total differentials.*

---

[1]In open systems the specific enthalpy h represents the sum of specific internal energy u and the *specific shifting work*, that is required to bring a piece of fluid into the system against the local pressure.

Fig. 10.2 Total differential

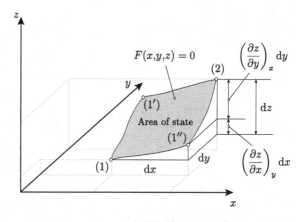

10.1 Total Differential

In order to prove Theorem 10.1, first the definition of a total differential is given. As shown in the previous chapters, the state of an ideal gas is unambiguously fixed with two state values, i.e. x and y for instance. Consequently, any other state value z must be a function of these two state values:

$$z = f(x, y),$$ (10.7)

respectively

$$F(x, y, z) = 0.$$ (10.8)

This function can be drawn in a graph, see Fig. 10.2. The question now is, how to calculate the total change of z, if the function depends on two independent variables x and y. To answer this, the variable y is first fixed, so that the partial change of z from (1) to (1″) follows:

$$\partial z_y = \left(\frac{\partial z}{\partial x}\right)_y dx$$ (10.9)

In a second step, from (1″) to (2), the variable x is kept constant, so that

$$\partial z_x = \left(\frac{\partial z}{\partial y}\right)_x dy$$ (10.10)

The entire change of z is then given by the total differential

$$\boxed{dz = \left(\frac{\partial z}{\partial x}\right)_y dx + \left(\frac{\partial z}{\partial y}\right)_x dy}$$ (10.11)

Fig. 10.3 Schwarz's
theorem

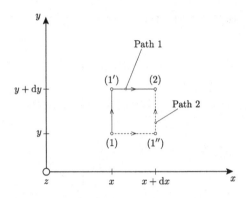

10.2 Schwarz's Theorem

Theorem 10.2 *The requirement for a total differential is, that point (2) according to Fig. 10.2 can be reached path-independently.*

In order to verify this theorem, Fig. 10.3 shows a top view. State (2) can be reached by path 1 and alternatively by path 2. The following abbreviations are introduced:

$$\xi(x, y) = \left(\frac{\partial z}{\partial x}\right)_y \tag{10.12}$$

$$\eta(x, y) = \left(\frac{\partial z}{\partial y}\right)_x \tag{10.13}$$

Now, starting at state (1) both paths to reach state (2) are investigated. Combining Theorems 10.1 and 10.2, for any state value its total differential must be path-independent:

- Path 1
  This path follows (1) → (1′) → (2)

$$dz = \eta(x, y)\, dy + \xi(x, y + dy)\, dx \tag{10.14}$$

- Path 2
  This path follows (1) → (1″) → (2)

$$dz = \xi(x, y)\, dx + \eta(x + dx, y)\, dy \tag{10.15}$$

For a state value it must be irrelevant if path 1 or path 2 is taken,[2] so that Eqs. 10.14 and 10.15 are equal:

---

[2]State values are never path-dependent!

$$\eta\,(x,\,y)\,\mathrm{d}y + \xi\,(x,\,y+\mathrm{d}y)\,\mathrm{d}x = \xi\,(x,\,y)\,\mathrm{d}x + \eta\,(x+\mathrm{d}x,\,y)\,\mathrm{d}y \qquad (10.16)$$

Rearranging leads to:

$$\frac{\xi\,(x,\,y+\mathrm{d}y) - \xi\,(x,\,y)}{\mathrm{d}y} = \frac{\eta\,(x+\mathrm{d}x,\,y) - \eta\,(x,\,y)}{\mathrm{d}x} \qquad (10.17)$$

The left hand side shows the derivate of ξ with respect to y, the right hand side the derivate of η with respect to x:

$$\left(\frac{\partial\xi}{\partial y}\right)_x = \left(\frac{\partial\eta}{\partial x}\right)_y \qquad (10.18)$$

Substitution of the previously made abbreviations leads to what is known as Schwarz's theorem:

$$\boxed{\frac{\partial^2 z}{\partial y\,\partial x} = \frac{\partial^2 z}{\partial x\,\partial y}} \qquad (10.19)$$

Theorem 10.3 *In case Eq. 10.19 is fulfilled, a total differential exists.*

Example 10.4 Ideal gas law

$$p = p\,(T,\,v) = \frac{RT}{v} \qquad (10.20)$$

Path 1

$$\frac{\partial p}{\partial T} = \frac{R}{v} \rightarrow \frac{\partial^2 p}{\partial T\,\partial v} = -\frac{R}{v^2} \qquad (10.21)$$

Path 2

$$\frac{\partial p}{\partial v} = -\frac{RT}{v^2} \rightarrow \frac{\partial^2 p}{\partial v\,\partial T} = -\frac{R}{v^2} \qquad (10.22)$$

$$\Rightarrow \frac{\partial^2 p}{\partial T\,\partial v} = \frac{\partial^2 p}{\partial v\,\partial T} \qquad (10.23)$$

Conclusion: Ideal gas law is an equation of state.

Example 10.5 Volume work

$$w = w\,(p,\,v) \qquad (10.24)$$

$$\delta w = \delta w\,(p,\,v) = -p\,\mathrm{d}v \qquad (10.25)$$

Thus, the total differential is

$$\delta w = \left(\frac{\partial w}{\partial v}\right)_p \mathrm{d}v + \left(\frac{\partial w}{\partial p}\right)_v \mathrm{d}p \qquad (10.26)$$

Comparison of coefficients leads to

$$\left(\frac{\partial w}{\partial v}\right)_p = -p \to \frac{\partial^2 w}{\partial v\,\partial p} = -1 \tag{10.27}$$

and

$$\left(\frac{\partial w}{\partial p}\right)_v = 0 \to \frac{\partial^2 w}{\partial p\,\partial v} = 0. \tag{10.28}$$

Consequently, it is

$$\frac{\partial^2 w}{\partial v\,\partial p} \neq \frac{\partial^2 w}{\partial p\,\partial v} \tag{10.29}$$

Conclusion: Volume work is path-dependent and thus no equation of state. A total differential does not exist.

Problem 10.6 Derive the so-called isobaric volumetric thermal expansion coefficient

$$\beta = \frac{1}{V}\left(\frac{\partial V}{\partial T}\right)_{p,m} \tag{10.30}$$

for an ideal gas!

Solution

According to the thermal equation of state the volume of an ideal gas follows

$$V = \frac{mRT}{p}. \tag{10.31}$$

For keeping pressure[3] and mass constant it is

$$\left(\frac{\partial V}{\partial T}\right)_{p,m} = \frac{mR}{p}. \tag{10.32}$$

This leads to

$$\beta = \frac{1}{V}\frac{mR}{p}. \tag{10.33}$$

Finally, one gets by applying the thermal equation of state:

$$\boxed{\beta = \frac{1}{T}} \tag{10.34}$$

---

[3]Since the isobaric thermal expansion coefficient is asked for!

Chapter 11
First Law of Thermodynamics

In this chapter the causality between process variables and the change of state variables of a thermodynamic system is derived. The first section introduces the principle of equivalence between work and heat—an essential prerequisite in order to formulate the energy conservation principle known as first law of thermodynamics. Finally, the first law of thermodynamics is applied for closed respectively open systems as well as for thermodynamic cycles. Anyhow, thermodynamic cycles play an important role in the following chapters.

11.1 Principle of Equivalence Between Work and Heat

In this section the impact of mechanical, electrical and thermal energy on a system will be analysed, see Fig. 11.1. Assuming, that the initial state (1) is at equilibrium for all three cases, the same amount of energy will be supplied to the system, though the form of energy differs. In order to clarify the initial state of the system its specific internal energy u_1 is estimated first. In case the system is filled with an ideal gas, the specific internal energy is purely a function of temperature, see Chap. 12. Thus, all three cases posses an identical temperature ϑ_1. By supply of energy the systems' internal state will strive for a new equilibrium state, i.e. the temperature ϑ_2 in state (2) will be increased compared to the initial state. Obviously, it does not make a difference if thermal energy by a hotplate, mechanical energy due to an impeller or electrical energy is supplied. In each case the temperature will rise. Furthermore, if the same quantity of energy is supplied, the temperature ϑ_2 will be identical in all cases.

© Springer Nature Switzerland AG 2019   149
A. Schmidt, *Technical Thermodynamics for Engineers*,
https://doi.org/10.1007/978-3-030-20397-9_11

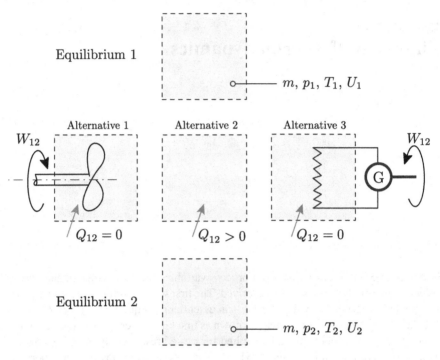

Fig. 11.1 Equivalence of heat and work

Theorem 11.1 *Heat and work can have the same effect, e.g. temperature increase,*[1] *on a system. Heat and work are equivalent. The same change of state can be achieved.*

This finding is important to derive the first law of thermodynamics, i.e. the law of energy conservation. Consequently, thermal, mechanical and electrical energy can be superimposed in order to investigate their interactions with the change of internal state.

Obviously, according to Fig. 11.2, heat can even be converted into work. In case the system is located in an environment with a larger temperature than the system itself, e.g. on a hotplate, the gas inside the system heats up and expands. Due to this expansion the gas releases volume work to the piston. Work on the other hand might be converted into heat as well, e.g. by converting it into electrical energy first. The electricity can then cause a current flow, that dissipates in an electrical resistance and releases heat finally.

Theorem 11.2 *Heat and work are equivalent. Heat can be generated by work and converted into work.*

---

[1] To be more precise, it should be *the same change of internal energy* that can be observed. If a real fluid changes its aggregate state, the temperature for instance can even be constant, though its internal energy changes.

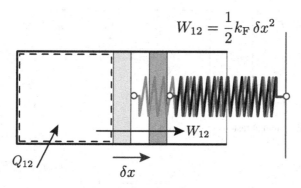

Fig. 11.2 Converting thermal to mechanical energy

Fig. 11.3 Closed system at rest—supply of heat

11.2 Closed Systems

In this section the first law of thermodynamics for closed systems is derived. However, a distinction is made between systems in rest and systems in motion.

11.2.1 Systems at Rest

Since thermal as well as mechanical energy are equivalent, the focus is initially on the impact of thermal energy on a system at rest, see Fig. 11.3.

By supply of heat the system in state (2) contains a larger amount of energy than in state (1). Mind, that the bubbles in sketch 11.3 shall represent energy. As energy can not be destroyed or generated out of nothing, the number of bubbles needs to be constant: Hence, according to the law of conservation of energy, the amount of supplied heat Q_{12} needs to be equal to the change of internal energy within the system, thus it is

$$Q_{12} = \Delta U = U_2 - U_1. \tag{11.1}$$

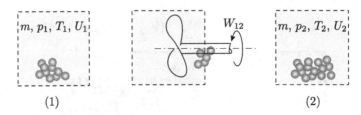

Fig. 11.4 Closed system at rest—supply of work

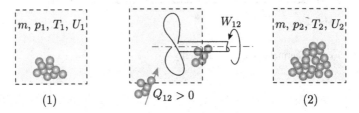

Fig. 11.5 Closed system at rest

In Chap. 12 the rise of internal energy is correlated with an increase of temperature for ideal gases respectively incompressible liquids.[2]

In a next step, mechanical energy is now supplied to the system, see Fig. 11.4. Again, according to the principle of energy conservation, supply of mechanical work influences the internal energy. Hence, the balance obeys

$$W_{12} = \Delta U = U_2 - U_1. \tag{11.2}$$

Obviously, the principle of energy conservation for a closed system at rest can be summarised according to Fig. 11.5. Heat and work can be superimposed, since they are equivalent regarding their influence on the system, see Theorem 11.2.

Thus, the first law of thermodynamics for closed system at rest reads as

$$\boxed{Q_{12} + W_{12} = U_2 - U_1} \tag{11.3}$$

It is only the internal energy that changes, since the system is supposed to be at rest, i.e. kinetic and potential energy related with the centre of gravity remain unchanged.

According to Eq. 11.3, internal energy in initial state (1) and final state (2) are relevant. Hence, intermediate information in-between (1) → (2) is not required. Consequently, the formulation of the first law of thermodynamics can be applied independently from its reversibility as well as for quasi-static/non-static changes of state. However, the path-dependent information is part of the process values Q_{12} and

---

[2]As mentioned before, real fluids do not necessarily heat up, since so-called latent heat causes a phase change!

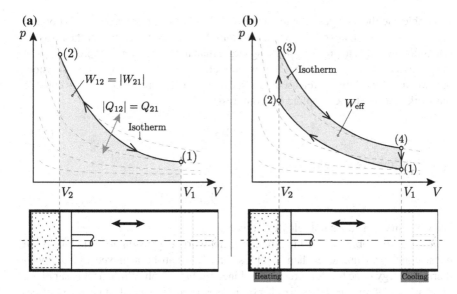

Fig. 11.6 Converting energy in a closed system

W_{12}. The first law of thermodynamics should always be applied in extensive notation[3] as shown in Eq. 11.3. However, since the mass of a closed system is constant, the division by the mass m leads to the following specific notation of the first law of thermodynamics:

$$q_{12} + w_{12} = u_2 - u_1 \qquad (11.4)$$

Form time to time it is advantageous to formulate the first law of thermodynamics in differential notation, especially when fundamental thermodynamic principles are derived. The differential notation obeys:

$$\delta q + \delta w = \mathrm{d}u \qquad (11.5)$$

Remember, that the differentials of process values are indicated by δ, whereas differentials of state values are denoted by d, see Chap. 10. Anyhow, the integration of Eq. 11.5 leads to Eq. 11.3.

Example 11.3 With the following example the operation of a closed system in order to gain net work is investigated. The proposed engine consists of a horizontal cylinder filled with an ideal gas closed by a frictionless movable piston, see Fig. 11.6.

In case (a) the gas is compressed from state (1) to state (2) as indicated in the p, V-diagram. While compressed, the gas' volume decreases, its pressure and its temperature rise. Furthermore, as derived in Chap. 9, the required work W_{12} is the grey-coloured area beneath the curve $(1) \rightarrow (2)$ as long as the change of state is

---

[3]It is the energy that remains constant, not the *specific* energy!

reversible and the changes of outer energies are ignored.[4] From the system's point of view this work is positive since work is required for the compression. If the piston is released now—due to pressure p_2 being larger than the initial, ambient pressure p_1—the gas expands. Let us assume it follows the same path as the previous compression, the amount of work being released during expansion W_{21} from (2) → (1) is equal to the work that had to be supplied for compression, i.e.

$$W_{12} = |W_{21}|. \tag{11.6}$$

Obviously, the overall work is

$$W_{\text{total}} = W_{12} + W_{21} = 0 \tag{11.7}$$

Thus, the machine does not release net work.

In order to enable the machine to release net work, expansion needs to take place on a higher pressure level than compression. That can be achieved by supply of thermal energy once the compression is done, see Fig. 11.6b. While heat is supplied, the position of the piston is fixed, i.e. an isochoric change of state takes place. Consequently state (3) is at the same volume as state (2) but the temperature as well as the pressure have increased. The following expansion to state (4) is now on a higher pressure level. Hence, the amount of released work is larger than the supplied work at compression:

$$|W_{34}| > W_{12} \tag{11.8}$$

If the gas in state (4) is compressed once again, state (3) is reached again. Let us assume heat can further be supplied isochorically,[5] the gas' temperature and pressure further rise. At this higher pressure the expansion can be initiated, so that work is released. If these steps are repeated frequently, temperature and pressure at bottom dead centre keep on rising until the machine overheats. Since expansion and compression in such a machine follow the same curve, the machine furthermore does not release work. The supplied heat purely leads to an increase of temperature, see Fig. 11.7.

In order to operate the machine continuously, thermal energy needs to be released at state (4), e.g. by an isochorically cooling, see Fig. 11.6b. By doing so, the temperature decreases and the initial state (1) can be re-reached. Once back in state (1) the entire process can run again, so that the machine can be operated permanently without overheating the machine. According to Fig. 11.6b work is released from (3) → (4), whereas work is supplied from (1) → (2). The net work can be visualised in a p, V-diagram by the grey enclosed area. Obviously, the thermodynamic cycle is

---

[4]Under these premises the entire work is purely composed of volume work! Areas in a p, V-diagram represent volume work respectively pressure work in open systems.

[5]This obviously depends on the temperature of the environment!

Fig. 11.7 Closed
system—no cooling

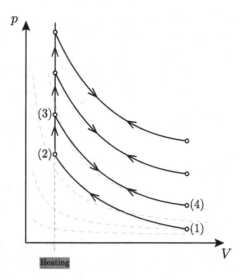

composed of four sequential changes of state. For each change of state the first law
of thermodynamics can be applied[6]:

- Change of state (1) to (2)

$$Q_{12} + W_{12} = U_2 - U_1 \tag{11.9}$$

- Change of state (2) to (3)

$$Q_{23} + \underbrace{W_{23}}_{=0} = U_3 - U_2 \tag{11.10}$$

- Change of state (3) to (4)

$$Q_{34} + W_{34} = U_4 - U_3 \tag{11.11}$$

- Change of state (4) to (1)

$$Q_{41} + \underbrace{W_{41}}_{=0} = U_1 - U_4 \tag{11.12}$$

A summation of Eqs. 11.9–11.12, leads to

---

[6]Changes of kinetic as well as potential energies are ignored, see Sect. 11.2.2. However, if these
effects are not neglected, the conclusion would be the same, since due to the addition of all sequential
changes of state, they disappear anyway!

Fig. 11.8 Closed system in motion

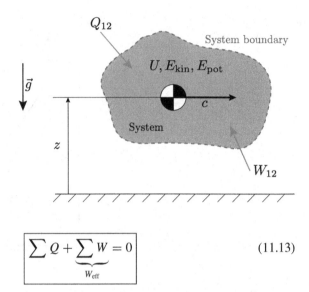

$$\boxed{\sum Q + \underbrace{\sum W}_{W_{\text{eff}}} = 0}\qquad(11.13)$$

The state values $U_1 \ldots U_4$ disappear, since after several changes of state the initial state (1) is reached again. Equation 11.13 can be applied to any thermodynamic cycle. Equation 11.13 proves, that the net work is equal to the sum of thermal energies. In case of the proposed machine, the released net work, i.e. the grey enclosed area[7] in Fig. 11.6b, is the difference of supplied and released heat, since the released heat is negative:

$$W_{\text{eff}} = -Q_{\text{Heating}} - Q_{\text{Cooling}} < 0\qquad(11.14)$$

Mind, that $Q_{\text{Heating}} > |Q_{\text{Cooling}}|$, so that the net work is negative, i.e. from the system's view work is released.

11.2.2   Systems in Motion

Now the focus is on closed systems that are in motion respectively that change their position within the gravity field. The entire energy of such a system is illustrated in Fig. 11.8.

Obviously, a system does not only posses internal energy U, but potential energy E_{pot} and kinetic energy E_{kin} as well. Both, potential and kinetic energy, are also summarised as *outer energies*[8]:

---

[7]Mind, that the p, V-diagram shows the volume work. Volume work is identical with the entire work, in case there is no dissipation and changes of outer energies are ignored!

[8]However, if the centre of gravity is fixed, neither potential nor kinetic energy *vary*. As a consequence they are ignored in Eq. 11.5 for systems at rest!

$$E = U + E_{\text{kin}} + E_{\text{pot}} = (U - U_0) + \frac{1}{2}m\left(c^2 - c_0^2\right) + mg(z - z_0) \quad (11.15)$$

Remember, that a reference level is required in order to specify energies. This is considered in Eq. 11.15 by reference levels for internal energy U_0, velocity c_0 and vertical position z_0. Following the law of energy conservation, a change of an internal energetic state correlates with an exchange of energy across the system boundary. Hence, it follows:

$$E_2 - E_1 = Q_{12} + W_{12} \quad (11.16)$$

Substituting the entire energy with Eq. 11.15 leads to a generic notation of the first law of thermodynamics for closed systems:

$$\boxed{U_2 - U_1 + \frac{1}{2}m \cdot \left(c_2^2 - c_1^2\right) + mg \cdot (z_2 - z_1) = Q_{12} + W_{12}} \quad (11.17)$$

Obviously, since two states are *compared*, the reference level disappears. Consequently, during a change of state from (1) → (2) the reference level *must not be varied*. By dividing with the system's mass, a specific notation of the first law of thermodynamics obeys:

$$u_2 - u_1 + \frac{1}{2}\left(c_2^2 - c_1^2\right) + g \cdot (z_2 - z_1) = q_{12} + w_{12} \quad (11.18)$$

Furthermore, in differential notation it is

$$\boxed{\underbrace{du + c\,dc + g\,dz}_{de_a} = \delta q + \delta w} \quad (11.19)$$

The first law of thermodynamics for closed systems can be motivated with the following Fig. 11.9. Energy shall be represented by bubbles, so that the system's entire energy in states (1) and (2) can be counted easily. Obviously, the change of the system's total energy, i.e. the difference of bubbles inside the system in states (2) and (1), is balanced by the energies W_{12} and Q_{12}, that both cross the system's boundary. If external energy is supplied, i.e. being positive, the number of bubbles increases. In case external energies are negative, indicating that the system releases energy, the number of bubbles in state (2) is smaller than in state (1). This correlation is given by Eqs. 11.16 respectively 11.17. On the left hand side of these equations the change of energy is quoted, whereas the right hand side designates the cause of this internal change, i.e. the external energetic influences.

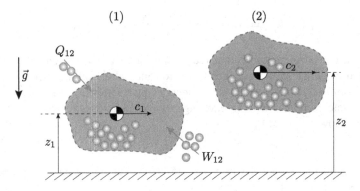

Fig. 11.9 First law of thermodynamics for closed systems, see Sect. 3.2.5

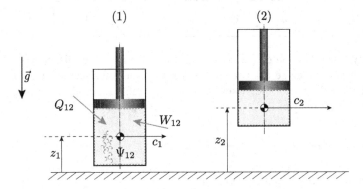

Fig. 11.10 Partial energy equation for closed systems

11.2.3   *Partial Energy Equation*

Figure 11.10 illustrates the procedure how to balance closed systems energetically. It shows an ideal gas inside a cylinder closed by a moving piston. However, it is wise and recommended to follow a clear, structured approach:

1. Formulation of the *first law of thermodynamics*:
   The energy balance, see Eq. 11.17, takes into account all external impacts, i.e. thermal energy Q_{12} as well as non-thermal energy W_{12}.

$$U_2 - U_1 + \frac{1}{2}m \cdot \left(c_2^2 - c_1^2\right) + mg \cdot (z_2 - z_1) = Q_{12} + W_{12} \qquad (11.20)$$

   Obviously, according to Fig. 11.10, the external energies initiate a change of the internal state, i.e. internal energy, as well as of the outer energies such as kinetic and potential energy that are related with the centre of gravity.

2. Formulation of the *partial energy equation* for the exchanged work:
   In this second step it should be precised how the non-thermal energy W_{12} is

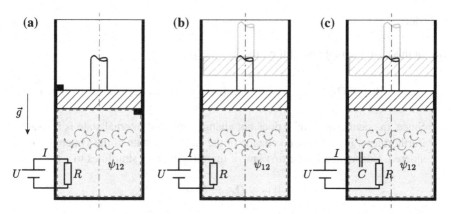

Fig. 11.11 **a** Supply of electrical energy in an isochoric system, **b** Supply of electrical energy in a non-isochoric system, **c** Supply of electrical energy with an electrical capacitor

composed. This is done by the so-called partial energy equation, that always needs to be adapted to the specific problem. The underlying question should be: What kind of work is part of W_{12}? A generic approach is given with Eq. 11.21:

$$W_{12} = W_{V,12} + W_{\text{Mech},12} + \Psi_{12} \tag{11.21}$$

In this case, i.e. Fig. 11.10, the entire total non-thermal energy W_{12} consists of volume work $W_{V,12}$, since the gas is compressed, and additional mechanical work $W_{\text{Mech},12}$ to lift and accelerate the centre of gravity. Furthermore, indicated by the sketched turbulent eddies, the dissipated energy Ψ_{12} needs to be supplied with energy: For sustaining the turbulence, respectively dissipation, energy is required. Specifically, this energy is delivered externally by W_{12}.[9] It is important to mention, that the partial energy equation is not an all-inclusive energy equation, as it only focuses on the *non-thermal energies*. Hence no thermal energies are covered by this equation.

Problem

Problem 11.4 Please state the first law of thermodynamics as well as the partial energy equation for the three cases given in Fig. 11.11!

---

[9]It is known from experience, e.g. with an air-pump, that it takes a larger effort to achieve the same result, for instance compressing to the same volume, in case there is friction than if the process runs reversibly.

Solution

(a) In this case the first law of thermodynamics reads as

$$W_{12} + Q_{12} = U_2 - U_1 + \frac{1}{2}m\left(c_2^2 - c_1^2\right) + mg\left(z_2 - z_1\right) \qquad (11.22)$$

Since, the centre of gravity does not change, only the internal state is relevant. Furthermore, in this case no change of the *electrical* state of the system is considered. Consequently, the first law of thermodynamics simplifies to

$$W_{12} + Q_{12} = U_2 - U_1. \qquad (11.23)$$

The partial energy equation for that case follows[10]

$$W_{12} = UI = \underbrace{W_{V,12}}_{=0} + \underbrace{W_{\text{Mech},12}}_{=0} + \Psi_{12} \qquad (11.24)$$

This equation expresses, that the entire work W_{12} is supplied electrically. No mechanical work is transferred, since the centre of gravity is fixed. Furthermore, the volume in that case is constant, so that no volume work is transferred either! According to

$$W_{12} = UI = \Psi_{12} \qquad (11.25)$$

the entire electrical energy is dissipated. This is the cause for an increase of temperature. However, this process can never be reversible, since the supplied energy can only be partially converted back, see Chap. 16.

(b) Starting with the first law of thermodynamics for the second case

$$W_{12} + Q_{12} = U_2 - U_1 + \frac{1}{2}m\left(c_2^2 - c_1^2\right) + mg\left(z_2 - z_1\right) \qquad (11.26)$$

Assuming, that the system in states (1) and (2) is at rest, it is the vertical position of the centre of gravity that has changed. Analogous to (a) the electrical state of the system is not regarded! Thus, the first law of thermodynamics obeys

$$W_{12} + Q_{12} = U_2 - U_1 + mg\left(z_2 - z_1\right) \qquad (11.27)$$

For this case, the partial energy equation follows:

$$W_{12} = UI = W_{V,12} + \underbrace{W_{\text{Mech},12}}_{=mg(z_2-z_1)} + \Psi_{12} \qquad (11.28)$$

---

[10] U indicates the voltage potential!

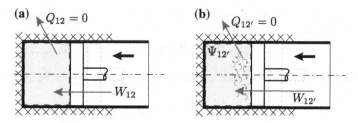

Fig. 11.12 **a** Reversible, adiabatic compression, **b** Irreversible, adiabatic compression

In this case, dissipation, mechanical as well as volume work occur, see Fig. 11.11b.

(c) This last case shows a very particular example, as the electrical state of the system shall be regarded as well. This is indicated by the symbol for a plate capacitor within the system boundary. An electrical capacitor is able to store electrical energy according to[11]

$$\Delta E_{el,12} = \frac{1}{2} C (U_2^2 - U_1^2) = W_{el,12} \tag{11.29}$$

Hence, the first law for that case, with assuming the states (1) and (2) to be in rest, is

$$W_{12} + Q_{12} = U_2 - U_1 + mg(z_2 - z_1) + \Delta E_{el,12} \tag{11.30}$$

However, for this case the partial energy equation reads as

$$W_{12} = W_{V,12} + W_{Mech,12} + W_{el,12} + \Psi_{12} \tag{11.31}$$

Example: Reversible Versus Irreversible

The first law of thermodynamics can be applied for any changes of a closed system, no matter if the change of state runs reversible or irreversible. However, let us investigate these two case, see Fig. 11.12. Figure 11.12a shows an adiabatic, reversible compression, while Fig. 11.12b represents an adiabatic, irreversible compression. The fluid inside the cylinder shall be an ideal gas.

- Reversible, adiabatic (1) → (2), Fig. 11.12a
  The first law of thermodynamics for this case reads as[12]:

$$\underbrace{Q_{12}}_{=0} + W_{12} = U_2 - U_1 \tag{11.32}$$

---

[11]Once again, in this equation U denotes the voltage potential and not the internal energy U!

[12]Change of outer energies can be ignored, since the system is at rest in states (1) and (2) and the vertical position of the centre of gravity does not vary.

Since, the work W_{12} is purely composed of volume work.[13] Thus, the partial energy equation follows:

$$W_{12} = W_{V,12} > 0 \tag{11.33}$$

Consequently, the supplied work increases the internal energy of the system.

$$W_{V,12} = U_2 - U_1 > 0 \tag{11.34}$$

With the caloric equation of state, that will be introduced in Chap. 12, it follows

$$U_2 - U_1 = mc_v\,(T_2 - T_1) > 0. \tag{11.35}$$

So, that the temperature from (1) to (2) increases:

$$T_2 > T_1 \tag{11.36}$$

- Irreversible, adiabatic (1) → (2′), Fig. 11.12b
  In this case the first law of thermodynamics has the same structure as in the reversible case. Thus, the dissipation does not occur directly:

$$\underbrace{Q_{12'}}_{=0} + W_{12'} = U_{2'} - U_1 \tag{11.37}$$

However, the partial energy equation includes the dissipation $\Psi_{12'}$:

$$W_{12'} = W_{V,12'} + \Psi_{12'} \tag{11.38}$$

The work that passes the system boundary $W_{12'}$ is utilised for the compression as well as for providing the dissipation.[14] This finally leads to

$$W_{V,12'} + \Psi_{12'} = U_{2'} - U_1 \tag{11.39}$$

Both, $W_{V,12'}$ and $\Psi_{12'}$, are positive, so that

$$U_{2'} - U_1 = mc_v\,(T_{2'} - T_1) > 0 \tag{11.40}$$

Again, from (1) → (2′) the temperature increases!

Experience shows, that for irreversible changes of state the amount of work is larger than at reversible changes of state to achieve the same effect, i.e.

$$W_{12'} > W_{12} \tag{11.41}$$

---

[13]No dissipation, no mechanical work due to horizontal cylinder.
[14]The dissipation always consumes energy. This energy has to be supplied!

Hence, it follows, that $U_{2'} > U_2$! Applying the caloric equation of state, see Chap. 12, leads to $T_{2'} > T_2$. Both changes of state can be illustrated in a p, V-diagram, see Fig. 11.13a. As indicated, reversible and irreversible compression reach the same final volume, i.e. both achieve the same effect. Since temperature $T_{2'}$ is larger than T_2, state $(2')$ lies vertically above (2). Remember, that the area beneath a change of state in a p, V-diagram stands for the *volume* work. Consequently, the larger temperature in case of an irreversible change of state leads to a larger pressure, so that the volume work from $(1) \rightarrow (2')$ is larger than the volume work $(1) \rightarrow (2)$:

$$W_{V,12'} > W_{V,12} \tag{11.42}$$

Mind, that the dissipation is not visible in the diagram. However, the *total* work is

$$\underbrace{W_{V,12'} + \overbrace{\Psi_{12'}}^{>0}}_{=W_{12'}} > W_{V,12} = W_{12} \tag{11.43}$$

so, that

$$W_{12'} > W_{12}. \tag{11.44}$$

With the second law of thermodynamics, see Chap. 13, the entire process can be explained in detail.[15] Furthermore, the changes of state can be stated in a T, s-diagram, see Fig. 11.13b. However, the T, s-diagram will be introduced in Sect. 13.4. For the entropy it follows:

$$ds = \frac{\delta q}{T} + \frac{\delta \psi}{T} \tag{11.45}$$

An adiabatic, reversible change of state is isentropic.[16] Obviously, the entropy remains constant from $(1) \rightarrow (2)$. Chapter 13 will show, that the area beneath a change of state in a T, s-diagram represents the sum of specific heat q and specific dissipation ψ. According to Eq. 11.45 the entropy rises from $(1) \rightarrow (2')$. In order to reach the same final volume as in the reversible case, i.e. the same specific volume, since the mass is constant, state $(2')$ lies to the right and above of state (2). Obviously, temperature $T_{2'}$ is larger than T_2. Hence p, V- and T, s-diagram lead to the same finding.

Example 11.5 Let us apply the principles of the first law of thermodynamics to the following example of a horizontal cylinder filled with an ideal gas, see Fig. 11.14.

---

[15]So we do not just have to rely on our experience, that work with friction requires a larger effort than work without friction.

[16]This will be covered discussed in Sect. 13.5!

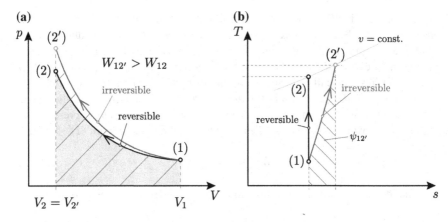

Fig. 11.13 Partial energy equation for closed systems

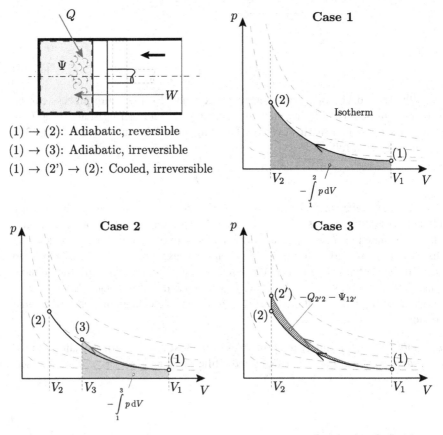

Fig. 11.14 Partial energy equation for closed systems

- Case 1 (frictionless, adiabatic) $(1) \rightarrow (2)$
  The first law of thermodynamics obeys

$$\underbrace{W_{12}}_{>0} + \underbrace{Q_{12}}_{=0} = U_2 - U_1 \tag{11.46}$$

Applying the caloric equation of state as well, so that temperature T_2 follows according to

$$T_2 = T_1 + \frac{W_{12}}{mc_v} \tag{11.47}$$

- Case 2 (friction, adiabatic, same W_{12}) $(1) \rightarrow (3)$
  The same temperature is reached, since the adiabatic system is supplied with the same amount of work as in case 1:

$$W_{13} = W_{12} = mc_v \, (T_3 - T_1) \tag{11.48}$$

However, the supplied work W_{13} is composed of volume work, i.e. the area beneath the curve in the p, V-diagram, and of dissipation. Hence, the partial energy equation reads as:

$$W_{13} = W_{12} = \underbrace{- \int_1^3 p \, dV}_{W_{V,13}} + \Psi_{13} \tag{11.49}$$

Obviously, the volume work is reduced by the dissipation Ψ_{13}:

$$W_{V,13} = W_{12} - \Psi_{13} \tag{11.50}$$

State (3) can be fixed in the p, V-diagram, see Fig. 11.14. States (2) and (3) need to lie on the same isothermal. However, the area beneath $(1) \rightarrow (2)$, i.e. the volume work $(1) \rightarrow (2)$, is larger than the area beneath $(1) \rightarrow (3)$, i.e. the volume work $(1) \rightarrow (3)$! Thus, the same amount of work in case of an irreversible change of state does not allow to achieve the same final volume as in case of reversible operation.

- Case 3 (Friction, cooled) $(1) \rightarrow (2') \rightarrow (2)$
  In order to reach the same final volume V_2 in case of irreversible operation, the amount of work exceeds[17] W_{12}:

$$W_{12'} > W_{12} \tag{11.51}$$

---

[17] As we have seen in case 2, the same amount of work in reversible/irreversible operation does not lead to the same compression rate. In irreversible operation the compression stops before reaching the final position. Obviously, in order to reach the same compression ratio further work needs to be supplied.

The first step $(1) \rightarrow (2')$ is done adiabatically. The first law of thermodynamics follows according to

$$W_{12'} = U_{2'} - U_1 > U_2 - U_1 \qquad (11.52)$$

Hence, temperature $T_{2'}$ is larger than temperature T_2! The partial energy equations obeys

$$W_{12'} = W_{V,12'} + \Psi_{12'} > W_{12} \qquad (11.53)$$

Now, that the compression is fully done to V_2, the temperature needs to be adjusted in order to reach state (2). Hence, the change of state $(2') \rightarrow (2)$ is performed isochorically, i.e. without any transfer of work. Heat needs to be released, so that the temperature decreases. According to the first law of thermodynamics it is

$$\underbrace{W_{2'2}}_{=0} + Q_{2'2} = U_2 - U_{2'} < 0 \qquad (11.54)$$

Obviously, it follows, that

$$Q_{2'2} = U_2 - U_{2'} = W_{12} - W_{12'}. \qquad (11.55)$$

Combining with the partial energy equation reads as

$$Q_{2'2} = W_{V,12} - W_{V,12'} + \Psi_{12'} - \Psi_{12'}. \qquad (11.56)$$

Thus, the hatched area in Fig. 11.14 represents

$$\boxed{W_{V,12'} - W_{V,12} = -Q_{2'2} - \Psi_{12'}} \qquad (11.57)$$

Problem 11.6 Two equal, vertical standing pipes with a diameter of $D = 0.1\,\mathrm{m}$ are connected with each other through a thin pipe and a valve, see Problem 2.4. One pipe is filled up to a height of $H = 10\,\mathrm{m}$ with water. The water has a density of $\rho = 1000\,\frac{\mathrm{kg}}{\mathrm{m}^3}$, at first the other pipe is empty. Now the valve is opened. After a while a balance of the water volume develops. By what absolute value does the internal energy of the water change, if the system does not exchange any energy with the environment?

Solution

Applying the first law of thermodynamics for closed systems reads as, see Fig. 11.15:

$$\underbrace{Q_{12} + W_{12}}_{\text{external impact}} = \underbrace{U_2 - U_1}_{\text{change of internal state}} + \underbrace{\Delta E_{\text{pot}} + \Delta E_{\text{kin}}}_{\text{change of outer state}} \qquad (11.58)$$

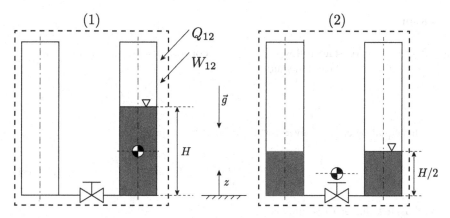

Fig. 11.15 Solution Problem 11.6

Since there is no energetic exchange with the environment, it follows that $Q_{12} = 0$ and $W_{12} = 0$. Furthermore the system is at rest in states (1) and (2), so that there is no change of the kinetic energy. Hence, Eq. 11.58 simplifies to:

$$0 = U_2 - U_1 + \Delta E_{\text{pot}} \tag{11.59}$$

The solution of Problem 2.4 has shown, that for calculating the change of potential energy the centre of gravity is relevant:

$$\Delta U = U_2 - U_1 = -\Delta E_{\text{pot}} = -mg\,(z_2 - z_1) \tag{11.60}$$

$$\boxed{\Delta U = -mg\left(\frac{H}{4} - \frac{H}{2}\right) = 1926.2\,\text{J}} \tag{11.61}$$

While the potential energy of the system decreases, its internal energy rises, so that the entire energy remains constant!

Problem 11.7 A horizontal cylinder, closed by a movable piston, contains $20\,\text{dm}^3$ of an ideal gas under a pressure of 1 bar. The gas is compressed by a piston to a pressure of 5 bar. The dissipated energy is 10% of the total energy feed. The change of state can be approximated by $p \cdot V^{1.25} = \text{const.}$ for the entire change of state.

(a) How far does the piston have to be pushed into the cylinder? The diameter of the cylinder is $d = 20\,\text{cm}$.
(b) How much work has to be supplied to the system?
(c) How much heat needs to be transferred across the cylinder walls, if the internal energy of the gas increases by $1\,\text{kJ}$ during the compression?

Solution

(a) The change of state follows $p \cdot V^{1.25} = \text{const.}$, so that this equation can be applied in state (1) and in state (2):

$$p_1 V_1^{1.25} = p_2 V_2^{1.25} \tag{11.62}$$

This equation can be solved for the volume V_2 in state (2):

$$V_2 = V_1 \cdot \sqrt[1.25]{\frac{p_1}{p_2}} = 5.52 \, \text{dm}^3 \tag{11.63}$$

Thus, the change of volume is

$$\Delta V = |V_2 - V_1| = 14.48 \, \text{dm}^3 \tag{11.64}$$

Following geometrical considerations:

$$\Delta V = \frac{\pi}{4} d^2 \, \Delta s \tag{11.65}$$

Thus, the distance, the piston has to be moved, is

$$\boxed{\Delta s = \frac{4 \, \Delta V}{\pi d^2} = 4.61 \, \text{dm}} \tag{11.66}$$

(b) The first law of thermodynamics applied to this problem is

$$W_{12} + Q_{12} = U_2 - U_1 \tag{11.67}$$

The dissipation Ψ_{12} does not occur directly, since energy is dissipated within the system. However, the partial energy equation reads as:

$$W_{12} = W_{12,\text{v}} + \Psi_{12} \tag{11.68}$$

What we know is, that $\Psi_{12} = 0.1 W_{12}$. Hence, we get

$$W_{12} = W_{12,\text{v}} + 0.1 W_{12} \tag{11.69}$$

Solving for W_{12}:

$$W_{12} = 1.11 W_{12,\text{v}} \tag{11.70}$$

The volume work can be calculated according to

$$W_{12,V} = - \int_1^2 p\, dV \tag{11.71}$$

Thus, it is

$$W_{12} = -1.11 \int_1^2 p\, dV \tag{11.72}$$

In order to solve the integral, the direction of the change of state is required. However, that description is given as $p \cdot V^{1.25} = $ const.. In Chap. 12 this change of state will be introduced as polytropic change of state. Applied to this problem it is

$$p = p\,(V) = p_1 \left(\frac{V_1}{V}\right)^{1.25} \tag{11.73}$$

Thus, Eq. 11.72 simplifies to

$$W_{12} = -1.11 \int_1^2 p_1 \left(\frac{V_1}{V}\right)^{1.25} dV = -1.11 p_1 V_1^{1.25} \int_1^2 V^{-1.25}\, dV \tag{11.74}$$

Its solution is

$$\boxed{W_{12} = 4.44\, p_1 V_1^{1.25}\left[V_2^{-0.25} - V_1^{-0.25}\right] = 3.3714 \times 10^3\, \text{J}} \tag{11.75}$$

(c) Applying the first law of thermodynamics leads to

$$Q_{12} + W_{12} = U_2 - U_1 \tag{11.76}$$

Thus,

$$\boxed{Q_{12} = U_2 - U_1 \quad W_{12} = -2.372 \times 10^3\, \text{J}} \tag{11.77}$$

Obviously, the system needs to be cooled!

Problem 11.8 An upright standing cylinder with an inner diameter of 250 mm is filled with a gaseous mixture, that is in thermodynamic equilibrium with the environment. The cylinder is closed against the environment by a gas-tight, frictionless moving piston. The weight of the piston as well as ambient pressure lead to a gas pressure of 4 bar in the cylinder. Now the mixture is ignited and a combustion takes place. Hence, the piston moves upwards and heat is exchanged with the environment. The entire change of state runs so slowly, that the mixture is under a constant pressure of 4 bar. Finally, thermodynamic equilibrium is re-reached. By what absolute value

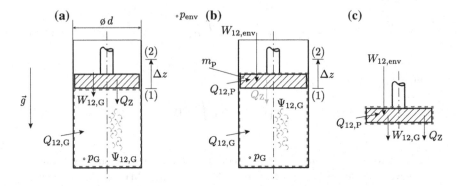

Fig. 11.16 Solution to Problem 11.8—isobaric change of state

does the internal energy of the gas change, if the gas releases 10 kJ of heat and the piston is lifted by 75 mm? Ambient pressure is $p_{env} = 1$ bar. The potential energy of the gas can be ignored. Derive the energy balances for the following systems:

(I) Cylinder content on its own
(II) Cylinder content and piston

Solution

The solution is generally derived, i.e. dissipation within the gas is taken into account in order to clarify how to handle it. However, as the change of state runs slowly and any acceleration of the piston can be ignored,[18] dissipation does not occur, see Example 7.24.

Part (I)

System boundary: see Fig. 11.16a

- First law of thermodynamics:

$$W_{12,G} + \underbrace{Q_{12,G} + Q_Z}_{Q_{total}=-10\,kJ} = \Delta U_G + \underbrace{\Delta E_{Pot,G}}_{=0} \qquad (11.78)$$

respectively

$$W_{12,G} + Q_{total} = \Delta U_G \qquad (11.79)$$

- With the partial energy equation according to Eq. 11.21:

$$W_{12,G} = W_{12,G,v} + \Psi_{12,G} + \underbrace{W_{12,G,mech}}_{=0} \qquad (11.80)$$

---

[18] A quasi-static change of state is assumed.

Since the change of state runs very slowly,[19] there is no dissipation, i.e. $\Psi_{12,G} = 0$, see Theorem 7.25:

$$W_{12,G} = W_{12,G,V} + \underbrace{\Psi_{12,G}}_{=0} = - \int_{1}^{2} p_G \, dV = -p_G A \Delta z \qquad (11.81)$$

Combining Eqs. 11.79 and 11.81 finally leads to

$$\boxed{\Delta U_G = -p_G A \Delta z + Q_{total} = -11.47 \, kJ} \qquad (11.82)$$

Part (II)
System boundary: see Fig. 11.16b

- First law of thermodynamics:

$$Q_{12,G} + Q_{12,P} + W_{12,env} = \Delta U_G + \Delta U_P + \Delta E_{Pot,P} + \underbrace{\Delta E_{Pot,G}}_{=0} \qquad (11.83)$$

respectively

$$Q_{12,G} + Q_{12,P} + W_{12,env} = \Delta U_G + \Delta U_P + m_P g \Delta z \qquad (11.84)$$

- The partial energy equation for $W_{12,env}$ can be given in two notations, since $W_{12,env}$ appears at the *interface* of system and environment:

  1. From the system's point of view, according to the partial energy Eq. 11.21, it is

$$W_{12,env} = W_{12,G,V} + \underbrace{\Psi_{12,G}}_{=0} + W_{12,P,mech} = -p_G A \Delta z + m_P g \Delta z \qquad (11.85)$$

Obviously, from this perspective work is required for the volume change of the gas and for lifting the piston.[20] These two mechanisms are part of the system boundary. As mentioned before, there is no dissipation within the gas. Combining Eqs. 11.84 and 11.85 leads to

$$\boxed{Q_{12,G} + Q_{12,P} - p_G A \Delta z = \Delta U_G + \Delta U_P} \qquad (11.86)$$

  2. From the environmental's point of view the work $W_{12,env}$ is purely required for compressing the environment. $\Psi_{12,env}$ does not appear since the environment

---

[19] As mentioned explicitly in the problem description.
[20] Work for lifting the centre of gravity of the gas shall be ignored according to the problem description!

is always supposed to be homogeneous,[21] $\Delta E_{a,env}$ does not appear since the environment is huge and at rest:

$$W_{12,env} = W_{12,env,V} = \int_1^2 p_{env}\, dV_{env} = p_{env}\Delta V_{env} \qquad (11.87)$$

There is no minus sign in front of the integral in this case, since according to Fig. 11.6b the arrow of $W_{12,env}$ points *out of* the environment, i.e. against the sign convention, see sketch 9.4. Anyhow, when the volume of the system increases, the environmental volume decrease and vice versa, so that

$$\Delta V_{env} = -\Delta V = -p_{env}A\Delta z \qquad (11.88)$$

Hence, it is

$$W_{12,env} = -p_{env}A\,\Delta z. \qquad (11.89)$$

Combining Eqs. 11.84 and 11.89 leads to

$$\boxed{\Delta U_G = Q_{12,G} + Q_{12,P} - p_{env}A\Delta z - m_P g \Delta z - \Delta U_P} \qquad (11.90)$$

How is the correlation between part (I) and (II)? The solution must be independent from the choice of system boundary. In order to clarify this, a third system boundary is investigated, see Fig. 11.16c.

- The first law of thermodynamics for the piston obeys

$$W_{12,env} - W_{12,G} + Q_{12,P} - Q_Z = \Delta U_P + \underbrace{\Delta E_{Pot,P}}_{m_P g \Delta z} \qquad (11.91)$$

Substitution of the transferred works by Eqs. 11.81 and 11.89 leads to

$$- p_{env}A\Delta z + p_G A\Delta z + Q_{12,P} - Q_Z = \Delta U_P + m_P g \Delta z \qquad (11.92)$$

respectively

$$0 = \Delta U_P + m_P g \Delta z + p_{env}A\Delta z - p_G A\Delta z - Q_{12,P} + Q_Z. \qquad (11.93)$$

- Summation of Eqs. 11.90 and 11.93 brings

$$\boxed{\Delta U_G = -p_G A\Delta z + Q_{total}} \qquad (11.94)$$

---

[21]Homogeneous means free of any imperfections. Hence, there is no dissipation, see Sect. 14.4 for details.

Fig. 11.17 Solution
Problem 11.9

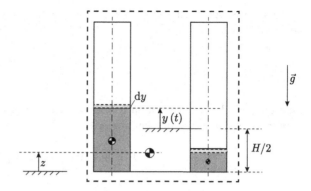

Hence, Eqs. 11.82 and 11.94 are identical. Consequently, it does not matter which system boundary is chosen: The result must always be boundary-independent.

Example 7.24 has shown that a balance of forces, see also Eq. 4.5, according to

$$p_G = p_{env} + \frac{m_P g}{A} \tag{11.95}$$

can be applied for slow changes of state, a non-accelerated and frictionless piston and for neglecting the kinetic and potential energy of the gas. Thus, Eq. 11.93 simplifies to

$$\Delta U_P = Q_{12,P} - Q_Z \tag{11.96}$$

The internal energy of the piston is determined solely[22] by the net heat! In this particular case, since the system is at thermal equilibrium with the environment in states (1) and (2), the change of the piston's internal energy is zero,[23] i.e.

$$\Delta U_P = 0 \tag{11.97}$$

The change of internal energy of the gas, however, though the temperatures T_1 and T_2 are identical, is not zero, since the chemical bonded energy of the gas has been released during the combustion process![24]

Problem 11.9 Let us refer back to Problem 11.6! The change of state starts at initial position (1) with one column filled with water, while the other being empty. Derive an equation, that shows the level of water as a function of time, see Fig. 11.17. Consider

---

[22] As long as the piston itself moves frictionless, see Example 7.24!

[23] Since the temperature of the piston does not change. It has ambient temperature in states (1) and (2).

[24] In fact, the gas from state (1) to state (2) has been replaced! Its chemical composition has changed.

a reversible as well as an irreversible change of state. The system still is supposed to be adiabatic.

Solution

First the partial energy equation for the work is applied. Since there is no work exchanged with the environment, it reads as:

$$W_{12} = 0 = \Psi + \underbrace{W_{V,12}}_{=0} + W_{\text{mech},12} \tag{11.98}$$

Since the volume is constant, there is no volume work, In differential notation the partial energy equation follows:

$$0 = \underbrace{\delta\Psi}_{\text{Dissipation}} + \underbrace{m_{\text{total}}c\,dc}_{\text{kinetic}} + \underbrace{m_{\text{total}}g\,dz}_{\text{potential}} \tag{11.99}$$

- Regarding the dissipation a model is required. What is known from fluid dynamics[25] is, that the dissipation for a laminar flow is proportional to the distance the fluid takes and its velocity. The distance is the way the fluid column is moved, i.e. $|dy|$

$$\delta\Psi \propto |\frac{dy}{dt}|\,|dy| \tag{11.100}$$

At this point it is important to keep the dissipation positive, as dy as well as $\frac{dy}{dt}$ can be positive and negative. Introducing a proportional constant ξ it follows that

$$\delta\Psi = \xi|\frac{dy}{dt}|\,|dy| \tag{11.101}$$

However, dy and $\frac{dy}{dt}$ always have the same sign, so that the product is always positive, i.e.

$$\delta\Psi = \xi\frac{dy}{dt}\,dy \tag{11.102}$$

- The kinetic energy obeys

$$m_{\text{total}}c\,dc = m_{\text{total}}\frac{dy}{dt}\,dc = m_{\text{total}}\frac{dc}{dt}\,dy = m_{\text{total}}\frac{d^2y}{dt^2}\,dy \tag{11.103}$$

---

[25]For a laminar flow the pressure drop is $\Delta p = \rho\frac{c^2}{2}\lambda\frac{l}{d}$ with $\lambda = \frac{64\eta}{\rho c d}$. The combination brings $\Delta p = \frac{32u\eta l}{d^2}$. The specific dissipation is $\psi = \frac{\Delta p}{\rho} = \frac{32u\eta l}{\rho d^2}$, the entire dissipation is $\Psi = \psi m_{\text{total}}$.

Since the velocity of the entire mass is equal to $\frac{dy}{dt}$, i.e. the mass has the same velocity as the water surface.

- Now, the position z of the entire centre of gravity as a function of y needs to be calculated. The mass of the left column is

$$m_{\text{left}} = \rho A \left(\frac{H}{2} + y \right) \tag{11.104}$$

The vertical position of its centre of gravity is

$$z_{\text{left}} = \frac{1}{2} \left(\frac{H}{2} + y \right) \tag{11.105}$$

The mass in the right column follows

$$m_{\text{right}} = \rho A \left(\frac{H}{2} - y \right) \tag{11.106}$$

The vertical position of its centre of gravity is

$$z_{\text{right}} = \frac{1}{2} \left(\frac{H}{2} - y \right) \tag{11.107}$$

In order to find the vertical position z of the entire centre of gravity, the following equation can be applied

$$z = \frac{z_{\text{left}} m_{\text{left}} + z_{\text{right}} m_{\text{right}}}{m_{\text{left}} + m_{\text{right}}} = \frac{H}{4} + \frac{y^2}{H} \tag{11.108}$$

Thus, its differential dz is

$$dz = d \left(\frac{y^2}{H} \right) = \frac{1}{H} d \left(y^2 \right) = \frac{2y}{H} dy \tag{11.109}$$

Hence, the partial energy Eq. 11.99 simplifies to:

$$0 = \xi \frac{dy}{dt} dy + m_{\text{total}} \frac{d^2 y}{dt^2} dy + m_{\text{total}} g \frac{2y}{H} dy \tag{11.110}$$

Dividing by the mass of the system m_{total} and dividing by the change of position of the water surface dy lead to:

$$\boxed{\frac{d^2 y}{dt^2} + \frac{\xi}{m_{\text{total}}} \frac{dy}{dt} + \frac{2g}{H} y = 0} \tag{11.111}$$

A numerical solution is given in Fig. 11.18. Obviously it is the dissipation, that is damping the system. A function for the temporal temperature development can be derived by a comparison of the first law of thermodynamics

$$\delta W + \delta Q = 0 = \underbrace{\mathrm{d}U}_{m_{\text{total}}c_{\mathrm{W}}\,\mathrm{d}T} + m_{\text{total}}g\,\mathrm{d}z + m_{\text{total}}c\,\mathrm{d}c \tag{11.112}$$

and the partial energy equation

$$\delta W = 0 = \underbrace{\delta W_{\mathrm{V}}}_{=0} + \delta\Psi + m_{\text{total}}g\,\mathrm{d}z + m_{\text{total}}c\,\mathrm{d}c. \tag{11.113}$$

Obviously, it is

$$\delta\Psi = mc_{\mathrm{W}}\,\mathrm{d}T \tag{11.114}$$

The dissipation can be substituted by Eq. 11.102, so that

$$\xi\frac{\mathrm{d}y}{\mathrm{d}t}\,\mathrm{d}y = m_{\text{total}}c_{\mathrm{W}}\,\mathrm{d}T \tag{11.115}$$

Rearranging leads to

$$\mathrm{d}T = \frac{\xi}{m_{\text{total}}c_{\mathrm{W}}}\frac{\mathrm{d}y}{\mathrm{d}t}\,\mathrm{d}y \tag{11.116}$$

Dividing by $\mathrm{d}t$ brings

$$\frac{\mathrm{d}T}{\mathrm{d}t} = \frac{\xi}{m_{\text{total}}c_{\mathrm{W}}}\frac{\mathrm{d}y}{\mathrm{d}t}\frac{\mathrm{d}y}{\mathrm{d}t} \tag{11.117}$$

Finally it is

$$\boxed{\frac{\mathrm{d}T}{\mathrm{d}t} = \frac{\xi}{m_{\text{total}}c_{\mathrm{W}}}\left(\frac{\mathrm{d}y}{\mathrm{d}t}\right)^2 > 0} \tag{11.118}$$

The numerical solution is illustrated in Fig. 11.18 as well.

11.3   Open Systems

An example for an open system is illustrated in Fig. 11.19. It shows two components in serial connection: A gas is compressed in a compressor and cooled down in a subsequent heat exchanger. However, open systems are characterised by an inflow respectively outflow, i.e. mass passes the system boundary. In many technical applications, e.g. a compressor plant, changes of state do not run as batch process, i.e.

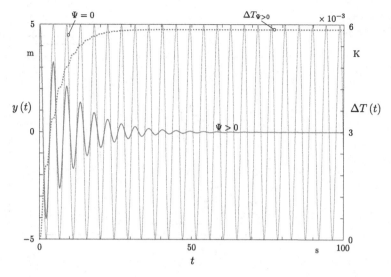

Fig. 11.18 Numerical solution to Problem 11.9

$$P_{t12} = \frac{\delta W_{t12}}{dt} \qquad \dot{Q}_{12} = \frac{\delta Q_{12}}{dt}$$

$$\dot{m}_{in} = \frac{dm_{in}}{dt} \qquad\qquad \dot{m}_{out} = \frac{dm_{out}}{dt}$$

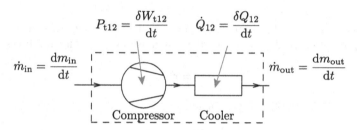

Fig. 11.19 Compression process

in discontinuous operation, but continuously.[26] In the illustrated example a gas flow enters the system indicated by the dashed boundary. The compressor increases the fluid's pressure level, while the fluid temperature rises as well. This compression requires technical work, that crosses the chosen system boundary. In the following heat exchanger, the fluid is cooled down, i.e. heat is released to the environment. Thus, heat crosses the system boundary as well. According to Fig. 11.19 the system boundary is open with respect to a heat flux in the cooler. Furthermore, there might be insulation losses in the compressor respectively in the connecting pipes. The system is further open to mechanical power, that is supplied to the compressor. Anyhow, the shown open system can be simplified with an analogous model, see Fig. 11.20. For inlet (1) a mass flux $\dot{m}_1$ and a corresponding velocity c_1 are given, i.e.

---

[26]Heat is not transferred once for instance but continuously. Hence, a heat flux is relevant instead of an amount of heat.

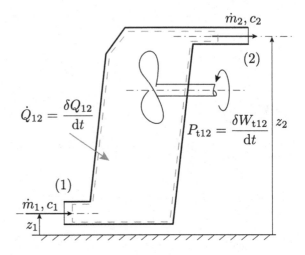

Fig. 11.20 Schematic diagram (analogous model)

$$m_1 = \rho_1 \dot{V}_1 = \frac{1}{v_1} A_1 c_1 \tag{11.119}$$

Additionally, the vertical position in the gravitational field is represented by z_1. The outlet (2) follows accordingly. In case the system is in steady state, it follows

$$\dot{m}_1 = \dot{m}_2 \tag{11.120}$$

Thus, the volume flows in steady state obey

$$\frac{\dot{V}_1}{\dot{V}_2} = \frac{\rho_2}{\rho_1} = \frac{v_1}{v_2} = \frac{p_2}{p_1} \frac{T_1}{T_2} \tag{11.121}$$

Next to the mass fluxes heat as well as technical work[27] pass the system boundary. It is important to emphasise, that the heat flux $\dot{Q}_{12}$ includes the entire heat, that passes the system boundary from (1) to (2), i.e. heat conduction within the fluid at inlet and outlet are part of $\dot{Q}_{12}$ as well.

11.3.1 Formulation of the First Law of Thermodynamics for Open Systems

The principles that have been derived for closed systems in the previous section are now applied to open systems. At $t = 0$ the mass inflow is achieved by a piston that supplies shifting work to a piece of mass Δm_1 at the inlet, see Fig. 11.21. Within a

---

[27] *Technical* work is always related to open systems!

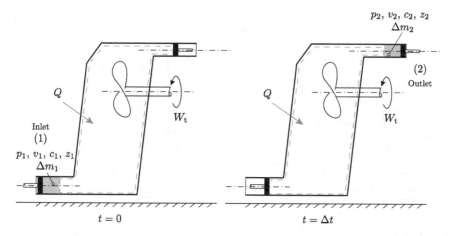

Fig. 11.21 From closed to open systems

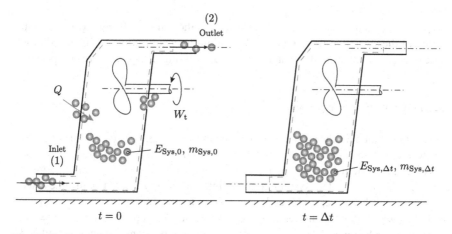

Fig. 11.22 Energy balance open system

time period Δt technical work W_t as well as heat Q are supplied. After the regarded time step Δt, the outflow in this analogous model is realised by a second piston. Thus, the system boundary is permeable. In Fig. 11.22 the extensive state value energy is illustrated as bubbles: Energy is within the system and it also passes the system boundary. As already known, energy is conserved, i.e. it can neither be generated nor destroyed. Consequently, the sum of the entire energy, that passes the system boundary, is equal to the change of energy E_{Sys}, that can be determined within the system:

$$E_{Sys,\Delta t} - E_{Sys,0} = \sum_i E_{in,i} - \sum_j E_{out,j} \qquad (11.122)$$

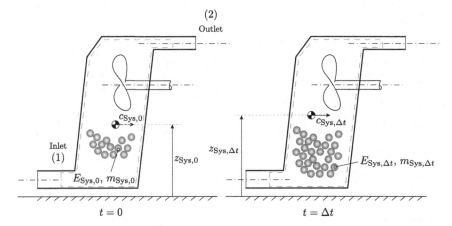

Fig. 11.23 Energy within an open system

$E_{\text{Sys},0}$ represents the total energy of the system at the time $t = 0$. $E_{\text{Sys},\Delta t}$ is the total energy of the system at the time $t = \Delta t$. Within the next section it is clarified how these energies, i.e. within the system as well as crossing the system, are composed.

Energy Content of the System

The energy content of the system is illustrated in Fig. 11.23. At this point it is important to emphasise, that any energy balance should be done extensively,[28] i.e. the mass[29] of the system must be known. Let the mass at $t = 0$ be $m_{\text{Sys},0}$ and the mass at $t = \Delta t$ be $m_{\text{Sys},\Delta t}$. The system contains internal energy U_{Sys} at both times, see Sect. 2.4.2. Furthermore, it posses kinetic and potential energy, due to c_{Sys} respectively z_{Sys} with respect to the centre of gravity. Obviously, at times $t = 0$ and $t = \Delta t$ the system can be regarded as closed system whose energy content can be calculated easily, i.e.

$$E_{\text{Sys},0} = m_{\text{Sys},0} \cdot \left[u_{\text{Sys},0} + \frac{1}{2} c_{\text{Sys},0}^2 + g \cdot z_{\text{Sys},0} \right] \tag{11.123}$$

and

$$E_{\text{Sys},\Delta t} = m_{\text{Sys},\Delta t} \cdot \left[u_{\text{Sys},\Delta t} + \frac{1}{2} c_{\text{Sys},\Delta t}^2 + g \cdot z_{\text{Sys},\Delta t} \right] \tag{11.124}$$

However, since mass is conserved, the overall mass balance obeys

$$m_{\text{Sys},\Delta t} = m_{\text{Sys},0} + \sum_i m_{\text{in},i} - \sum_j m_{\text{out},j} \tag{11.125}$$

---

[28]Energy is constant, not specific energy!

[29]Mass as an extensive state value represents the size of a system.

The mass at $t = \Delta t$ is the initial mass at $t = 0$ increased by the inflow mass and decreased by the outflow mass. In another notation it reads as

$$m_{\text{Sys},\Delta t} - m_{\text{Sys},0} = \sum_i m_{\text{in},i} - \sum_j m_{\text{out},j} \qquad (11.126)$$

Energy Supplied to the System

Heat, i.e. thermal energy Q, can pass the boundary regardless of whether the system is open or closed. This can happen due to an insufficient insulation or even due to an intended process. It is already known, that heat is initiated by a temperature potential. Furthermore, according to the second law of thermodynamics, see Sect. 14.2, heat is always transferred from larger to smaller temperature. Consequently, in the following approach Q includes the total thermal energy transferred due to temperature differences: This can be at the system boundary, i.e. housing for instance, as well as conductive heat transfer within the fluid at inlet and outlet. Next to the thermal energy, technical work W_t might also pass the system boundary. Both, heat Q as well as technical work W_t, can be positive or negative.[30] In case heat or technical work are negative, energy is released by the system. Energy is supplied to the system, while they are positive.

Now, energy transferred into the system at the inlet is investigated, i.e.

- Due to the velocity of the mass entering the system, kinetic energy is carried into the system. It follows

$$\frac{1}{2}\Delta m_1 c_1^2 \qquad (11.127)$$

- Let us assume, that the inlet has a vertical position of z_1 in the gravitational field. Consequently, the mass flow posses potential energy

$$\Delta m_1 g z_1 \qquad (11.128)$$

- Furthermore, the fluid contains internal energy, see Sect. 2.4.2, i.e.

$$\Delta m_1 u_1 \qquad (11.129)$$

- At the inlet the piece of mass Δm_1 is faced by a local pressure p_1 at the system boundary. In order to transfer the mass into the system, a force is required to overcome the local pressure. This force needs to move the piece of mass until the mass is entirely part of the system. The corresponding work has already been introduced in Sect. 9.2.8 and is called shifting work:

$$\Delta m_1 p_1 v_1 \qquad (11.130)$$

---

[30] Sure, both can also disappear!

Hence, the total energy, that is supplied to the system follows

$$E_{in} = E_1 = Q + W_t + \Delta m_1 \left[\frac{1}{2} c_1^2 + g z_1 + u_1 + p_1 v_1 \right] \tag{11.131}$$

Specific Enthalpy as New State Value

In Chap. 10 the specific enthalpy as a new state value has already been introduced, i.e.

$$h = u + pv. \tag{11.132}$$

Obviously, the specific enthalpy h is a state value, since it is purely composed of state values, i.e. no information regarding the path a system takes is required. Within this chapter the specific enthalpy can be explained descriptively, as it includes the specific internal energy u and the specific shifting work pv, that is needed in order to pass a piece of mass into a system against its local pressure p. All state values have in common, that they characterise the internal state a system has.[31] Though the specific enthalpy is comprehensible for open systems, any system, i.e. also a closed system, contains a state value enthalpy. The total differential for the specific enthalpy obeys

$$dh = \left(\frac{\partial h}{\partial u} \right)_{p,v} du + \left(\frac{\partial h}{\partial v} \right)_{p,u} dv + \left(\frac{\partial h}{\partial p} \right)_{u,v} dp \tag{11.133}$$

Thus, the total differential reads as

$$dh = du + p \, dv + v \, dp \tag{11.134}$$

However, for ideal gases Eq. 11.132 follows by applying the thermal equation of state

$$h = u + RT. \tag{11.135}$$

Hence, the entire energy entering the system is

$$\boxed{E_{in} = E_1 = Q + W_t + \Delta m_1 \left[\frac{1}{2} c_1^2 + g z_1 + h_1 \right].} \tag{11.136}$$

Energy Released by the System

According to Fig. 11.24 the entire energy leaving the system follows

---

[31] However, some state values have a vivid physical meaning, others are regarded as more or less artificial.

Fig. 11.24 Energy crossing the system boundary of an open system

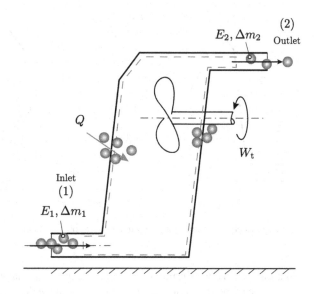

$$E_{\text{out}} = E_2 = \Delta m_2 \left[\frac{1}{2} c_2^2 + g z_2 + h_2 \right].\tag{11.137}$$

Similar to the *supplied* energy as given by Eq. 11.136, the energy *leaving the system* is composed out of kinetic, potential, internal and shifting work. As discussed before internal energy as well as shifting work are summarised as enthalpy. Nevertheless, the process values heat and work might *leave* the system as well. However, that case is taken into account by a negative sign for Q or W_t in Eq. 11.136.

Conclusion

Now that energy content of an open system at two different times and energy supply respectively energy release across the system boundary have been clarified, Eq. 11.122 can be applied. Due to the energy conservation principle, the temporal change of a system's energy content depends on released and supplied energy, see Fig. 11.25.[32]

$$m_{\text{Sys},\Delta t} \left(u_{\text{Sys},\Delta t} + \frac{c_{\text{Sys},\Delta t}^2}{2} + g z_{\text{Sys},\Delta t} \right) - m_{\text{Sys},0} \left(u_{\text{Sys},0} + \frac{c_{\text{Sys},0}^2}{2} + g z_{\text{Sys},0} \right) =$$
$$\Delta m_1 \left(h_1 + \frac{c_1^2}{2} + g z_1 \right) - \Delta m_2 \left(h_2 + \frac{c_2^2}{2} + g z_2 \right) + Q + W_t\tag{11.138}$$

---

[32]Once again, energy shall be represented by energy-bubbles, so that the energy conservation principle actually means counting bubbles.

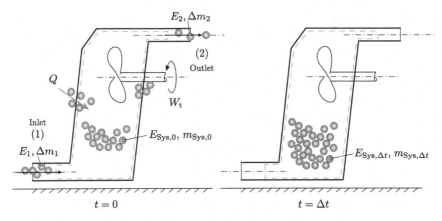

Fig. 11.25 Energy crossing system boundary of an open system

Equation 11.138 gives the mathematical correlation of what is illustrated in Fig. 11.25: The change of energy, represented by bubbles, inside the system within a time period of Δt is balanced by the flow of energy, i.e. bubbles, across the system boundary. The *change of energy* is given in the first line of Eq. 11.138, the *flows across the boundary* in the second line of that equation.

11.3.2   Non-steady State Flows

In this case the inflow and outflow over a period of time are variable. Hence, the mass balance reads as

$$m_{\text{Sys},\Delta t} - m_{\text{Sys},0} = \int\limits_0^{\Delta t} \dot m_1(t)\, \mathrm{d}t - \int\limits_0^{\Delta t} \dot m_2(t)\, \mathrm{d}t \qquad (11.139)$$

The change of mass within the system over a period of time correlates with the mass flows across the system boundary. An inflow increases the mass inside, an outflow decreases the system's content. According to Eq. 11.138 the energy balance can be adjusted for time-invariant flows:

$$m_{\text{Sys},\Delta t}\left(u_{\text{Sys},\Delta t} + \frac{c_{\text{Sys},\Delta t}^2}{2} + gz_{\text{Sys},\Delta t}\right) - m_{\text{Sys},0}\left(u_{\text{Sys},0} + \frac{c_{\text{Sys},0}^2}{2} + gz_{\text{Sys},0}\right) =$$

$$\int\limits_0^{\Delta t}\left(h_1 + \frac{c_1^2}{2} + gz_1\right)\dot m_1(t)\, \mathrm{d}t - \int\limits_0^{\Delta t}\left(h_2 + \frac{c_2^2}{2} + gz_2\right)\dot m_2(t)\, \mathrm{d}t + Q + W_t$$

$$(11.140)$$

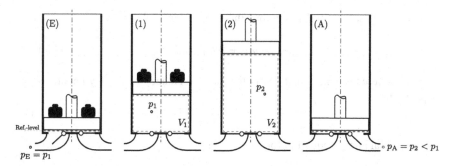

Fig. 11.26 Sketch to Problem 11.10, see also Fig. 9.15

Problem 11.10 In Sect. 9.2.9 an analogous model has been introduced, see Fig. 9.15, to derive the technical work. An open system with a constant mass flux is treated by means of a closed system: A flow enters with an inlet pressure of p_E and shall leave the system with an outlet pressure of $p_A < p_E$. Within the analogous model the cylinder is filled first with the fluid isobarically. Then the gas within the closed cylinder expands. Finally, the gas is released isobarically to reach the demanded state (A), see Fig. 11.26. Please derive the energy balance for the entire change of state, i.e. (E) → (A)!

Solution

The continuous flow of an open system is modeled by a so-called batch, i.e. discontinuous, process. Its changes of state are as follows:

(1) Isobaric filling of the reference volume: (E) → (1)

The system can be treated as an *open* system, since mass passes the system boundary. Hence, Eq. 11.138 can be applied. According to this problem it reads as

$$\underbrace{m_1 \left(u_1 + \frac{c_1^2}{2} + gz_1 \right) - 0}_{\text{System at } \Delta t} = \underbrace{\Delta m_E \left(h_E + \frac{c_E^2}{2} + gz_E \right) - 0}_{\text{Inflow}} + Q_{E1} + W_{E1}$$

(11.141)

Example 7.24 has shown that for an *isobaric* process, the change of state has to run so slowly,[33] that no dissipation/friction occurs. Furthermore, the potential energy of he gas needs to be neglected. Thus, it is

$$m_1 (u_1 + 0) = \Delta m_E (h_E + 0) + Q_{E1} + W_{E1}$$

(11.142)

---

[33] Thus, the kinetic energy of the inflow is ignored!

Since the system initially is empty, the mass balance obeys

$$m_1 = m_E + \Delta m_E = \Delta m_E \tag{11.143}$$

Δm_E denotes the mass that enters the system. Hence, the energy balance simplifies to

$$U_1 = H_E + Q_{E1} + W_{E1} \tag{11.144}$$

The partial energy equation for the work is

$$W_{E1} = W_{E1,V} + W_{E1,mech} + \Psi_{E1} \tag{11.145}$$

with the mechanical work[34]

$$W_{E1,mech} = 0 \tag{11.146}$$

and the volume work for an isobaric change of state[35]

$$W_{E1,V} = -p_1 V_1. \tag{11.147}$$

Hence, the work can be summarised as[36]

$$W_{E1} = -p_1 V_1 \tag{11.148}$$

Combining these equations brings

$$U_1 = H_E + Q_{E1} - p_1 V_1 \tag{11.149}$$

respectively

$$U_1 + p_1 V_1 = H_E + Q_{E1}. \tag{11.150}$$

In other words

$$\boxed{H_1 = H_E + Q_{E1}} \tag{11.151}$$

(2) Expansion of the enclosed mass: (1) → (2)
The system can be treated as a *closed* system, since the mass inside the system is constant.

- Exchanged work, i.e. partial energy equation

---

[34]Potential as well as kinetic energy of the gas are neglected, see Example 7.24!

[35]The volume change lifts the piston. Thus, work is released to increase the piston's potential energy.

[36]Mind, that there is no dissipation since the state of state needs to run very slowly, see Sect. 9.2.5!

$$W_{12} = -\int_1^2 p\,dV + \Delta E_{\text{pot},12} + \Psi_{12} \tag{11.152}$$

Hence, volume work is exchanged and the potential energy of the gas increases. In addition to Sect. 9.2.9 dissipation is regarded in this change of state!

- Energy balance (closed system)

$$Q_{12} + W_{12} = U_2 - U_1 + \Delta E_{\text{pot},12} \tag{11.153}$$

Combining both equations brings

$$\boxed{Q_{12} - \int_1^2 p\,dV + \Psi_{12} = U_2 - U_1} \tag{11.154}$$

(3) Isobaric releasing of the entire mass: (2) → (A)

The system can be treated as an *open* system, since mass passes the system boundary. Hence, Eq. 11.138 can be applied. According to this problem it reads as

$$\underbrace{0 - m_2\left(u_2 + \frac{c_2^2}{2} + gz_2\right)}_{\text{System at } t=0} = \underbrace{0 - \Delta m_A\left(h_A + \frac{c_A^2}{2} + gz_A\right)}_{\text{Outflow}} + Q_{2A} + W_{2A}$$

$$\tag{11.155}$$

Example 7.24 has shown that for an isobaric process, the change of state has to run so slowly,[37] that no dissipation/friction occurs. Furthermore, the potential energy of the gas needs to be neglected. Thus, it is

$$- m_2\left(u_2 + 0\right) = -\Delta m_A\left(h_A + 0\right) + Q_{2A} + W_{2A} \tag{11.156}$$

Since the system finally is empty, the mass balance obeys

$$m_A = m_2 - \Delta m_A \rightarrow m_2 = \Delta m_A \tag{11.157}$$

Δm_A denotes the mass that leaves the system. Hence, the energy balance simplifies to

$$- U_2 = -H_A + Q_{2A} + W_{2A} \tag{11.158}$$

The partial energy equation for the work is

---

[37] Thus, the kinetic energy of the inflow is ignored!

$$W_{2A} = W_{2A,V} + W_{2A,mech} + \Psi_{2A} \tag{11.159}$$

with the mechanical work[38]

$$W_{2A,mech} = 0 \tag{11.160}$$

and the volume work for an isobaric change of state[39]

$$W_{2A,V} = p_2 V_2. \tag{11.161}$$

Hence, the work can be summarised as[40]

$$W_{2A} = p_2 V_2 \tag{11.162}$$

Combining these equations brings

$$-U_2 = -H_A + Q_{2A} + p_2 V_2 \tag{11.163}$$

respectively

$$-U_2 - p_2 V_2 = -H_A + Q_{2A}. \tag{11.164}$$

In other words

$$\boxed{H_A = H_2 + Q_{2A}} \tag{11.165}$$

The summation of the energy balances of all sub-steps, i.e. Eqs. 11.151, 11.154 and 11.165, leads to the overall energy balance:

$$Q_{EA} + \Psi_{EA} - \int_1^2 p \, dV = H_1 - H_E + U_2 - U_1 + H_A - H_2 \tag{11.166}$$

With

$$Q_{EA} = Q_{E1} + Q_{12} + Q_{2A} \tag{11.167}$$

and

$$\Psi_{EA} = \Psi_{12}. \tag{11.168}$$

Substitution of $U_1 = H_1 - p_1 V_1$ and $U_2 = H_2 - p_2 V_2$ reads as

---

[38]Potential as well as kinetic energy of the gas are neglected, see Example 7.24!

[39]The potential energy of the piston is reduced and supplied to the gas.

[40]Mind, that there is no dissipation since the state of state needs to run very slowly, see Sect. 9.2.5!

$$Q_{EA} + \Psi_{EA} - \int_1^2 p \, dV = H_1 - H_E + H_2 - p_2 V_2 - H_1 + p_1 V_1 + H_A - H_2$$

$$(11.169)$$

Rearranging leads to

$$Q_{EA} + \Psi_{EA} - \int_1^2 p \, dV = H_A - H_E - (p_2 V_2 - p_1 V_1) \qquad (11.170)$$

Applying Eq. 9.113 brings the energy balance of the entire process (E) $\rightarrow$ (A)

$$Q_{EA} + \int_1^2 V \, dp + \Psi_{EA} = H_A - H_E \qquad (11.171)$$

Since first and last step are isobaric it further is

$$\int_1^2 V \, dp = \int_E^A V \, dp. \qquad (11.172)$$

Thus, the first law of thermodynamics finally obeys

$$\boxed{Q_{EA} + \int_E^A V \, dp + \Psi_{EA} = H_A - H_E} \qquad (11.173)$$

Problem 11.11 A hot air flow $\dot{m}$ ($u_1 = 250 \, \frac{kJ}{kg}$, $\rho_1 = 2.0 \, \frac{kg}{m^3}$, $p_1 = 2$ bar) flows through a throttle valve into a large tank. The velocity in the inlet pipe amounts to $c = 10 \, \frac{m}{s}$. The cross section of the inlet pipe is $A = 1 \times 10^2 \, m^2$ and its height above reference level z_0 is 10 m. To ensure the homogeneity in the tank an impeller is mounted with a power consumption of 10 kW. The tank is at rest, i.e. its velocity is zero, and the reference level, i.e. $z_0 = 0$, is at the tank's centre of gravity, see Fig. 11.27. Calculate the exchanged heat with the environment for an interval of $\Delta t = 10 \, s$, so that the specific internal energy of the tank content remains constant ($u_{Sys} = 214.2 \, \frac{kJ}{kg} = $ const.).

Solution

Obviously, the given problem is an open system since mass enters the system. However, this specific problem can not be in steady state, since there is no mass leaving the system. Hence, the mass inside increases. Figure 11.27 shows a sketch of the

Fig. 11.27 Sketch to
Problem 11.11

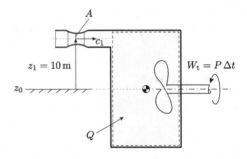

problem. Now, let us apply Eq. 11.138 with the assumption that $\Delta m_2 = 0$, since there is no mass flow leading out of the system:

$$m_{\mathrm{Sys},\Delta t}\left(u_{\mathrm{Sys},\Delta t} + \frac{c_{\mathrm{Sys},\Delta t}^2}{2} + gz_{\mathrm{Sys},\Delta t}\right) - m_{\mathrm{Sys},0}\left(u_{\mathrm{Sys},0} + \frac{c_{\mathrm{Sys},0}^2}{2} + gz_{\mathrm{Sys},0}\right) =$$
$$\Delta m_1\left(h_1 + \frac{c_1^2}{2} + gz_1\right) + Q + W_{\mathrm{t}} \tag{11.174}$$

Remember, that $m_{\mathrm{Sys},\Delta t}$ is the system's mass after $\Delta t = 10\,\mathrm{s}$ and $m_{\mathrm{Sys},0}$ is the initial mass, that is unknown in this case. The system's centre of gravity is in rest and does not vary vertically, since air, treated as an ideal gas, always fills out the entire volume. Thus, the first law of thermodynamics simplifies to

$$m_{\mathrm{Sys},\Delta t} u_{\mathrm{Sys},\Delta t} - m_{\mathrm{Sys},0} u_{\mathrm{Sys},0} = \Delta m_1\left(h_1 + \frac{c_1^2}{2} + gz_1\right) + Q + W_{\mathrm{t}} \tag{11.175}$$

According to Eq. 11.139 the mass balance follows

$$m_{\mathrm{Sys},\Delta t} - m_{\mathrm{Sys},0} = \int_0^{\Delta t} \dot{m}_1(t)\, \mathrm{d}t = \dot{m}_1 \Delta t = \Delta m_1 \tag{11.176}$$

Since the specific internal energy shall be constant, i.e. $u_{\mathrm{Sys},\Delta t} = u_{\mathrm{Sys},0} = u_{\mathrm{Sys}} = \mathrm{const.}$, the first law of thermodynamics obeys

$$u_{\mathrm{Sys}} \underbrace{\left(m_{\mathrm{Sys},\Delta t} - m_{\mathrm{Sys},0}\right)}_{=\Delta m_1} = \Delta m_1\left(h_1 + \frac{c_1^2}{2} + gz_1\right) + Q + W_{\mathrm{t}} \tag{11.177}$$

The following terms can be calculated:

- Technical work W_{t}

$$W_{\mathrm{t}} = P_{\mathrm{t}}\,\Delta t = 100\,\mathrm{kJ} \tag{11.178}$$

- Mass Δm_1

$$\Delta m_1 = \rho_1 \dot{V}_1 \, \Delta t = \rho_1 c_1 A \, \Delta t = 2 \, \text{kg} \tag{11.179}$$

- Specific outer energies

$$\frac{c_1^2}{2} + gz = 148 \, \frac{\text{J}}{\text{kg}} \tag{11.180}$$

- Specific enthalpy h_1

$$h_1 = u_1 + p_1 v_1 = u_1 + \frac{p_1}{\rho_1} = 350 \, \frac{\text{kJ}}{\text{kg}} \tag{11.181}$$

Thus, the heat obeys

$$\boxed{Q = u_{\text{Sys}} \Delta m_1 - \Delta m_1 \left(h_1 + \frac{c_1^2}{2} + gz_1 \right) - W_{\text{t}} = -371.9 \, \text{kJ}} \tag{11.182}$$

11.3.3 Steady State Flows

In this section open systems in steady state is investigated. The internal state of a system in steady state does not vary with time. Consequently, it is:

1. The mass of a system needs to be constant. With respect to Eq. 11.126 it implies that

$$\sum_i m_{\text{in},i} = \sum_j m_{\text{out},j}. \tag{11.183}$$

In case incoming and outgoing mass differ, the mass inside the system would either rise or decrease.
2. All state values need to be temporally constant as well, i.e.

$$z \neq f(t) \tag{11.184}$$

However, state values z can vary locally inside an open system. Pressure, as intensive state value, for instance can be different in an open system at inlet and outlet. Consequently, the pressure develops from inlet to outlet. Pressure sensors, located at different positions inside the system, would prove that. In case a steady state occurs, though the state values are locally different, there is no change with time, i.e. at any time the local sensor measures constant values. However, the system is supposed to be in local equilibrium, see Sect. 4.4. Hence, according to Eq. 11.122, an open system in steady state obeys[41]

---

[41] This is due to the *energetic* state of a system does not vary by time in steady state!

$$E_{K,\Delta t} - E_{K,0} = 0 = \sum_i E_{in,i} - \sum_j E_{out,j} \qquad (11.185)$$

Thus, the energy content of the system is constant with time, i.e.

$$E_{K,\Delta t} = E_{K,0} \qquad (11.186)$$

In case the open system consists of one inlet and one outlet the energy balance can be derived from Eq. 11.138 and reads as

$$0 = \Delta m_1 \left(h_1 + \frac{c_1^2}{2} + gz_1 \right) - \Delta m_2 \left(h_2 + \frac{c_2^2}{2} + gz_2 \right) + Q + W_t \qquad (11.187)$$

with

$$\Delta m_1 = \Delta m_2 = \Delta m. \qquad (11.188)$$

If mass continuously passes the system, the time derivative of Eq. 11.187 is

$$0 = \dot{m} \left(h_1 + \frac{c_1^2}{2} + gz_1 \right) - \dot{m} \left(h_2 + \frac{c_2^2}{2} + gz_2 \right) + \dot{Q} + P_t \qquad (11.189)$$

Finally, the first law of thermodynamics for open systems with one inlet/outlet in steady state obeys

$$\dot{Q} + P_t = \dot{m} \left[(h_2 - h_1) + \frac{1}{2} \left(c_2^2 - c_1^2 \right) + g\,(z_2 - z_1) \right] \qquad (11.190)$$

The specific notation follows by dividing by the constant mass flux $\dot{m}$:

$$q_{12} + w_{t,12} = (h_2 - h_1) + \frac{1}{2} \left(c_2^2 - c_1^2 \right) + g\,(z_2 - z_1) \qquad (11.191)$$

Mind, that $\dot{Q} = \dot{m} q_{12}$ and $P_t = \dot{m} w_{t,12}$!

Open Systems in Steady State with Multiple Inlets/Outlets

Now, the focus is on open systems in steady state with *multiple inlets/outlets*. Applying Eq. 11.122 leads to:

$$E_{K,\Delta t} - E_{K,0} = 0 = \sum_i E_{in,i} - \sum_j E_{out,j} \qquad (11.192)$$

respectively

$$\sum_i E_{in,i} = \sum_j E_{out,j}. \qquad (11.193)$$

The derivative with respect to time reads as

$$\boxed{\sum_i \dot{E}_{in,i} = \sum_j \dot{E}_{out,j}}$$

(11.194)

In other words:

$$\boxed{\text{Energy flux in} = \text{Energy flux out}}$$

This principle can be easily adapted to any open system in steady state. For the example illustrated in Fig. 11.28 it reads as[42]

$$\dot{Q}_{12} + P_{t12} + \dot{m}_1 \left(h_1 + \frac{c_1^2}{2} + gz_1 \right) + \dot{m}_2 \left(h_2 + \frac{c_2^2}{2} + gz_2 \right) =$$
$$\dot{m}_3 \left(h_3 + \frac{c_3^2}{2} + gz_3 \right)$$

(11.195)

In combination with the mass balance[43]

$$\boxed{\text{Mass flux in} = \text{Mass flux out}}$$

it is for the shown example, see Fig. 11.28:

$$\dot{m}_1 + \dot{m}_2 = \dot{m}_3.$$

(11.196)

Hence, the first law of thermodynamics obeys

$$\dot{Q}_{12} + P_{t12} = \dot{m}_1 \left[(h_3 - h_1) + \frac{c_3^2 - c_1^2}{2} + g(z_3 - z_1) \right] +$$
$$\dot{m}_2 \left[(h_3 - h_2) + \frac{c_3^2 - c_2^2}{2} + g(z_3 - z_2) \right]$$

(11.197)

This notation, i.e. *differences* of state values on the right hand side, is advantageous, see Chap. 12. In this case caloric equations of state can be applied easily.[44]

---

[42]Mind, that arrows pointing into the system are energy fluxes in, while arrows pointing out of the system represent energy fluxes out. Anyhow, a mass flux carries enthalpy, potential and kinetic energy.

[43]In steady state!

[44]Caloric equations of state indicate how the change of a caloric state value can be calculated. Remember, that caloric state values, such as specific enthalpy or specific internal energy, can not be measured, but must be calculated.

Fig. 11.28 Open systems
with multiple inlets/outlets

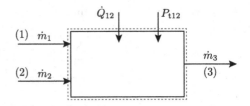

11.3.4   Partial Energy Equation

Similar to Sect. 11.2.3, in which the partial energy equation for the process value *work* has been derived for closed system, the *technical* work w_t, which is characteristic for open systems, see Eq. 11.191, is now further investigated. A partial energy equation is not an entire energy balance, but it clarifies how the work is utilised by a system. Mind, that caloric state values such as specific internal energy/enthalpy/entropy are not covered by the partial energy equation, but occur in the first law of thermodynamics. Anyhow, for an open system in steady state the first law of thermodynamics obeys

$$q_{12} + w_{t12} = h_2 - h_1 + \frac{c_2^2 - c_1^2}{2} + g \cdot (z_2 - z_1). \qquad (11.198)$$

Mind, that q_{12} as well as w_{t12} only take into account energies that cross the system boundary. Consequently, it is essential to identify the system boundary clearly. Nevertheless, technical work is only relevant, when mechanical or electrical work is obviously transferred: This can for instance be the case, when a mechanical shaft crosses the system boundary, see Fig. 11.29. Such a shaft can supply the system with energy,[45] or the flow inside the systems drives the shaft.[46] Furthermore, electrical energy can be supplied to a system, so that technical work is positive. If no such electrical wires or mechanical shafts are mounted, e.g. a section of a heating tube not including the water pump, then $P_t = \dot{m} w_{t12} = 0$.

In order to derive the partial energy equation for the specific technical work, a mass particle dm is regarded within an open system. Let this piece of mass be connected with a moving coordinate system, see Fig. 11.29. However, this mass particle can be treated as a closed system, since its mass shall be constant on its way from inlet (1) to outlet (2).

Thus, the first law of thermodynamics for closed systems, see Eq. 11.19, can be applied for the particle dm. In differential notation it obviously reads as:

$$\delta q + \delta w = du + de_a \qquad (11.199)$$

---

[45] The technical work then is positive.

[46] The technical work then is negative.

Fig. 11.29 Technical work

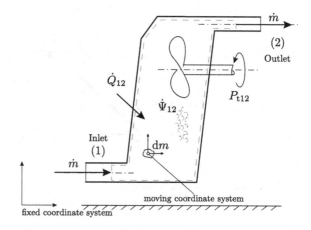

Applying the partial energy equation for the specific work of a closed system, according to Sect. 11.2.3, leads to

$$\delta w = \delta w_\mathrm{V} + \delta w_\mathrm{mech} + \delta \psi = -p\,\mathrm{d}v + \mathrm{d}e_\mathrm{a} + \delta \psi \qquad (11.200)$$

Obviously, the particle, treated as a closed system, is faced with volume work, since it might be compressed or it might expand while its mass remains constant. Furthermore, work is required for lifting and acceleration its centre of gravity. Mind, that in Eq. 11.200, the change of outer energies, that are related to the centre of gravity, require mechanical work.[47] Finally, energy might be dissipated due to internal friction.

The specific enthalpy of the mass particle follows, see Eq. 11.134:

$$\mathrm{d}h = \mathrm{d}u + p\,\mathrm{d}v + v\,\mathrm{d}p \qquad (11.201)$$

Rearranging for $\mathrm{d}u$ leads to:

$$\mathrm{d}u = \mathrm{d}h - p\,\mathrm{d}v - v\,\mathrm{d}p \qquad (11.202)$$

Hence, it is

$$\delta q + \delta \psi = \mathrm{d}h - v\,\mathrm{d}p \qquad (11.203)$$

Following the particle from (1) to (2) means integrating Eq. 11.203, so that

$$q_{12} + \psi_{12} = h_2 - h_1 - \int_1^2 v\,\mathrm{d}p \qquad (11.204)$$

---

[47] According to Eq. 11.21 lifting or accelerating the centre of gravity requires mechanical work!

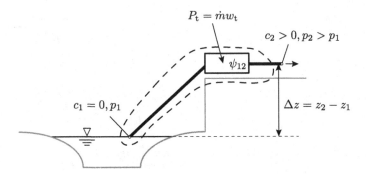

Fig. 11.30 What does technical work mean?

Solving for the specific heat leads to:

$$q_{12} = h_2 - h_1 - \int_1^2 v \, dp - \psi_{12} \tag{11.205}$$

Replacing the specific heat in Eq. 11.198 for an open system, i.e. joining the two coordinate systems,[48] leads to the desired equation for the technical work:

$$w_{t12} = \underbrace{\int_1^2 v \, dp}_{y_{12}} + \psi_{12} + \underbrace{\frac{c_2^2 - c_1^2}{2} + g \, (z_2 - z_1)}_{\Delta e_{a12}} \tag{11.206}$$

Fortunately, Eq. 11.206 can be visualised: think of a high-pressure pump that shall be operated with water from a lower lake. At the inlet the velocity is zero, since the lake is at rest. At the outlet, which is at a higher vertical position than the inlet, the fluid shall be accelerated to a velocity c_2. Furthermore, its pressure shall be increased as well, see sketch 11.30. The question now is, what power the pump needs to have. Anyhow, Eq. 11.206 explains how the required technical work is composed:

• Specific work to increase the pressure, i.e. specific pressure work y_{12}

$$y_{12} = \int_1^2 v \, dp. \tag{11.207}$$

---

[48]The moving coordinate system represents the single particle, while the fixed coordinate system represents the energy balance as given by Eq. 11.198.

Fig. 11.31 Specific pressure work y_{12}

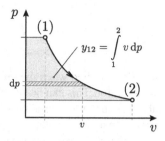

Pressure work only occurs in open systems![49]

- Due to dissipation ψ_{12} inside the system the required specific work is increased.[50]
- Specific work to increase the kinetic energy of the fluid $\frac{c_2^2 - c_1^2}{2}$. The velocity in Fig. 11.30 at the inlet is supposed to be zero. However, though the velocity is zero there still can be a mass flux, since the cross section of the inlet is supposed to be large.
- Specific work to increase the vertical position of the fluid $g\,(z_2 - z_1)$.

The next step is to investigate how the specific pressure work y_{12} is to be calculated. Hence, states (1) and (2) are illustrated in a p, v-diagram, see Fig. 11.31. The integral $\int_1^2 v\,dp$ is represented by the projected area to the p-axis as indicated by diagram 11.31. Thus, the p, v-diagram can not only be utilised to visualise the specific volume work w_V in a closed systems, but additionally the specific pressure work y_{12} for open systems. Just as for closed systems, the path (1) to (2) must also be known for open systems in order to calculate the specific pressure work. Hence, specific pressure work is a process value just like specific volume work for closed systems.

Theorem 11.12 *Pressure work is part of the technical work in an open system. It describes the work that needs to be supplied to an open system in order to increase the fluid's pressure—respectively, the work that is released by a system, when the pressure decreases!*

In Sect. 9.2.9 the pressure work has already been derived on an alternate way. This analogous approach was based on cylinder/piston system with inlet and outlet valves, that could be described with means of a closed system. However, both approaches lead to the same correlation for the pressure work.

Example—Why Do We Need the Partial Energy Equation?

The partial energy equation

---

[49]Volume work only occurs in closed systems! However, both, volume and pressure work, satisfy the physical definition of work, i.e. force times distance.

[50]This is well known from everyday experience: Systems with friction require a larger effort than frictionless systems!

$$w_{t12} = \int_1^2 v\,\mathrm{d}p + \psi_{12} + \frac{c_2^2 - c_1^2}{2} + g\,(z_2 - z_1) \tag{11.208}$$

should be applied when the pressure change in open systems and/or the dissipation are asked for. This can be shown with the following example. Think of a horizontal tube section that is passed by a fluid in steady state conditions. Let us further assume, that the changes of kinetic energies can be ignored. Under these conditions the first law of thermodynamics for open, steady state systems obeys

$$\dot{Q}_{12} + P_{t12} = \dot{m}\,(h_2 - h_1) \tag{11.209}$$

As can be seen, the first law of thermodynamics does not include any information regarding the pressure as well as the dissipation. Obviously, the partial energy equation contains such information and simplifies with the premises made to

$$w_{t12} = \int_1^2 v\,\mathrm{d}p + \psi_{12} \tag{11.210}$$

However, since the tube section is a work-insulated system, i.e. no work crosses the system boundary,[51] so that

$$w_{t12} = 0 = \int_1^2 v\,\mathrm{d}p + \psi_{12} \tag{11.211}$$

This leads to

$$\int_1^2 v\,\mathrm{d}p = -\psi_{12} \leq 0 \tag{11.212}$$

The right side of this equation is negative because the specific dissipation is $\psi_{12} \geq 0$! It is obvious that the specific dissipation causes a pressure drop under the given conditions. If there is no dissipation, the flow is isobaric. If dissipation occurs, a pressure drop is measured since

$$\int_1^2 v\,\mathrm{d}p < 0 \tag{11.213}$$

---

[51] Mind, as documented in Fig. 11.29, technical work only crosses the system boundary, if there is a mechanical shaft, electrical wires or any other possibility to carry energy in form of work into or out of the system. This is not the case—the presented example purely is a passive tube section!

This is only possible if

$$dp < 0. \tag{11.214}$$

The integration from (1) to (2) reads as

$$p_2 - p_1 < 0. \tag{11.215}$$

Chapter 12
Caloric Equations of State

In the previous chapters the law of energy conservation has been thoroughly discussed. By this, it was possible to evaluate thermodynamic systems energetically. A distinction has been made between closed and open systems. However, next to thermal state values, such as pressure p or temperature T, that can be measured easily, a new category of state values has been introduced: These state values, i.e. specific internal energy u and specific enthalpy h for instance, can not be determined by a sensor and thus need to be calculated. They are named caloric state values and occur in the first law of thermodynamics for instance:

- Closed systems

$$q_{12} + w_{12} = u_2 - u_1 + \Delta e_{a12} \tag{12.1}$$

- Open systems

$$q_{12} + w_{t12} = h_2 - h_1 + \Delta e_{a12} \tag{12.2}$$

What caloric state values have in common is, that they all include a dimension of energy.[1] Within this chapter caloric state values are further investigated and equations of state are derived in order to calculate and predict them. Furthermore, a new state value, i.e. the specific entropy s, is introduced. This state value quantifies the limitation of energy conversion, see Chap. 15. Anyhow, in Sect. 5.1 the Gibbs' phase rule has been introduced, which proved, that it requires two independent specific state values in order to define a state of a single ideal gas unequivocally. According to that, the following mathematical correlations can be stated.:

$$u = u(v, T) \tag{12.3}$$

$$h = h(p, T) \tag{12.4}$$

$$s = s(p, T) \tag{12.5}$$

However, the choice of which dependencies are formulated in Eqs. 12.3–12.5 is arbitrary. It would also be possible to vary, e.g. $u = u(p, T)$, but for the thermo-

---

[1] Such as $[u] = 1\frac{kJ}{kg}$ and $[h] = 1\frac{kJ}{kg}$.

© Springer Nature Switzerland AG 2019
A. Schmidt, *Technical Thermodynamics for Engineers*,
https://doi.org/10.1007/978-3-030-20397-9_12

dynamic derivation, i.e. introducing the specific heat capacities, it make sense to proceed with the given dependencies. Later on it will be possible to formulate other correlations, for instance such as

$$h = h\,(s, v)\,. \tag{12.6}$$

12.1 Specific Internal Energy u and Specific Enthalpy h for Ideal Gases

In fact the caloric equations of state for ideal gases are rather simple and are summarised briefly. For those of the readers, who are interested in a more detailed thermodynamic derivation, please refer to Sect. 12.3. The total differentials[2] for the specific internal energy and the specific enthalpy follow from Eqs. 12.3 to 12.4. According to Chap. 10, these total differentials exist, since u and h are state values, see Theorem 10.1:

- Specific internal energy u

$$du = \left(\frac{\partial u}{\partial T}\right)_v \cdot dT + \left(\frac{\partial u}{\partial v}\right)_T \cdot dv \tag{12.7}$$

The first partial derivative is defined as follows

$$\left(\frac{\partial u}{\partial T}\right)_v = c_v\,(v, T) \tag{12.8}$$

c_v is called specific heat capacity.[3] Since u is a function of two independent state values, i.e. in this case v and T, c_v also depends on these values.
- Specific enthalpy h

$$dh = \left(\frac{\partial h}{\partial T}\right)_p \cdot dT + \left(\frac{\partial h}{\partial p}\right)_T \cdot dp \tag{12.9}$$

The first partial derivative is defined as follows

$$\left(\frac{\partial h}{\partial T}\right)_p = c_p\,(p, T) \tag{12.10}$$

c_p is called specific heat capacity.[4] Since h is a function of two independent state values, i.e. in this case p and T, c_p also depends on these values.

---

[2]The indices at the brackets indicate, that this variable is kept constant!
[3]At constant specific volume v.
[4]At constant pressure p.

Fig. 12.1 Joule expansion

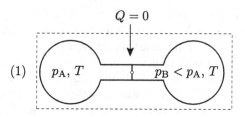

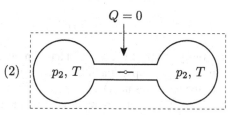

According to [18], experiments, e.g. performed by Joule, see also Example 12.1, have shown, that for ideal gases the specific internal energy and the specific enthalpy depend only on the temperature, i.e.[5]:

$$\left(\frac{\partial u}{\partial v}\right)_T = 0 \tag{12.11}$$

and

$$\left(\frac{\partial h}{\partial p}\right)_T = 0 \tag{12.12}$$

Since u and h are only functions of the temperature T, so are the specific heat capacities c_v and c_p. This leads to the following caloric equations of state for ideal gases:

$$\boxed{du = c_v(T)\,dT} \tag{12.13}$$

$$\boxed{dh = c_p(T)\,dT} \tag{12.14}$$

The fluid's specific heat capacities and the handling of the caloric equations of state will be further discussed in Sect. 12.4. In summary, for ideal gases the caloric equations of state, i.e. Eqs. 12.13 and 12.14, as well as the thermal equation of state, i.e.

$$pv = RT, \tag{12.15}$$

have been validated by experiments, see Chap. 6 and Example 12.1.

Example 12.1 In this example the Joule expansion is introduced, see Fig. 12.1. Two vessels containing an *ideal* gas are connected by a tube that, in state (1), is closed by

---

[5]The thermodynamic proof is given in Sect. 12.3.2.

a valve. The pressure in vessel B shall be smaller than the pressure in A. However, initially the two vessels are thermally balanced with the environment, i.e. each gas shall have the same temperature T. The entire system shall further be adiabatic. Now, the valve is opened, so that a mechanical balancing process runs. In state (2) the two vessels are in thermodynamic equilibrium, i.e. a new pressure p_2 occurs. Joule found, that the change of state is isothermal, i.e. the temperature remains constant for any chosen pressure difference between A and B. According to the first law it is

$$W_{12} + Q_{12} = 0 = U_2 - U_1 \qquad (12.16)$$

Obviously, the pressure does not have an influence on the internal energy. If pressure would have an impact the internal energy in (1) and (2) would have to be different, since the pressures in (1) and (2) are different. The first law of thermodynamics and the observation, that T remains constant, thus lead to

$$U = f(T). \qquad (12.17)$$

For an ideal gas it further is

$$H = U + pV = U(T) + RT. \qquad (12.18)$$

Thus, this leads to

$$H = f(T). \qquad (12.19)$$

However, in case the deviations from ideal gas behaviour increase, the change of state is not isothermal anymore, i.e. the internal energy is a function of temperature and pressure, see also Sect. 18.5.

12.2 Specific Entropy s as New State Value for Ideal Gases

In this section a new state value is introduced. The first part of this section, i.e. the mathematical derivation, is rather theoretical. In Chap. 13, however, this new state value is thermodynamically motivated with simple examples.

At this point, the question arises why a new state value is indispensable for understanding energy conversion, since the first law of thermodynamics, that shows the principle of energy conservation, has already been discussed. Unfortunately, the first law of thermodynamics does not restrict the *direction* of energy conversion: According to the first law of thermodynamics it is possible, when placing a hot cup of tea in a room at ambient conditions, the tea's temperature might even rise, while the room further cools down. In other words, the energy released by the tea is supplied to the environment, i.e. the amount of energy would be constant, and the first law of thermodynamics would be fulfilled. However, everyday experience rather shows that the opposite is true, i.e. the tea cools down. From this point of view it becomes clear that the first law of thermodynamics by itself is not sufficient for a comprehensive understanding. In order to understand the direction of energy conversion, a new state value, i.e. the entropy, is required.

Before starting with the theoretical derivation, it is important to distinguish between state and process value once again, see Chap. 10:

Theorem 12.2 *A state value describes the state of a system and does not take into account the path the system follows to reach this state. In contrast to a state value, a process value takes into account the path that the system follows from an initial state to a new state!*

Volume work in a closed system, respectively pressure work in an open system are process values, since they only occur, when the system underlies a change of state. A thermodynamic state, however, does not carry work or heat. Process values, i.e. heat, work and dissipation, depend on the path the systems follows. In contrast to that, state values characterise the system path-independently. Some state values have a comprehensible physical meaning, as they are part of our daily life, e.g. temperature T or pressure p. Others, anyhow, are more difficult to interpret, e.g. the specific internal energy. Some others even seem to be artificial: The specific enthalpy was introduced with the open systems, but as all state values, it also occurs in closed systems. It has been shown that the combination of state values lead to new state values, see Chap. 10. When deriving the caloric equations of state in Sect. 12.3 two new state values, that do not have a comprehensible physical meaning so far, the specific Helmholtz free energy f and the specific Gibbs free enthalpy g, occur. Regarding these two state values, both of them get a physical interpretation once the chemical reacting systems are introduced in Chap. 24.

Finally, a new state value is now derived, that accompanies us until the end of this book. Experience has shown that this new state value *entropy* s causes difficulties, since one tries from the beginning to find a physical motivation.[6] After all, it is much easier to accept s as a new state value first. In Chap. 13 the specific entropy is utilised and its meaning becomes more accessible. Finally, limitation of energy conversion, e.g. in a power plant, can be motivated with this new state value and its characteristics. By the end of part I, most readers will have become accustomed to this new state value.

The first law of thermodynamics for a closed system in differential, specific notation reads as

$$\delta q + \delta w = \mathrm{d}u + \mathrm{d}e_{\mathrm{a}}. \tag{12.20}$$

With the partial energy equation it is possible to specify the specific work δw, i.e.

$$\delta w = -p\,\mathrm{d}v + \delta\psi + \underbrace{\delta w_{\mathrm{mech}}}_{=\mathrm{d}e_{\mathrm{a}}}. \tag{12.21}$$

---

[6]To be honest, it is even difficult to find a physical meaning of the internal energy. Though you probably have accepted its existence by now.

Combining these two equations leads to

$$\delta q + \delta \psi = du + p \, dv. \tag{12.22}$$

By introducing the specific enthalpy h, see Eq. 11.134:

$$du = dh - p \, dv - v \, dp \tag{12.23}$$

the equation also reads as

$$\delta q + \delta \psi = dh - v \, dp. \tag{12.24}$$

Equation 12.22 can be divided[7] by the temperature T, so that

$$\frac{\delta q + \delta \psi}{T} = \frac{du}{T} + \frac{p \, dv}{T} \tag{12.25}$$

Equation 12.24 can alternatively be derived with the first law of thermodynamics for open systems, i.e.

$$\delta q + \delta w_t = dh + de_a. \tag{12.26}$$

The partial energy equation for the technical work δw_t is

$$\delta w_t = v \, dp + \delta \psi + de_a. \tag{12.27}$$

Combining these two equations leads to

$$\delta q + \delta \psi = dh - v \, dp. \tag{12.28}$$

Dividing by the temperature T (integrating factor) leads alternatively to

$$\frac{\delta q + \delta \psi}{T} = \frac{dh}{T} - \frac{v \, dp}{T}. \tag{12.29}$$

With the caloric and the thermal equations of state for ideal gases

$$du = c_v \, dT \tag{12.30}$$

respectively

$$\frac{p}{T} = \frac{R}{v} \tag{12.31}$$

Equation 12.25 obeys

---

[7]Known as integrating factor!

$$\frac{\delta q + \delta \psi}{T} = c_v \frac{dT}{T} + R \frac{dv}{v}.$$

(12.32)

The integration of the right side of Eq. 12.32 is path-independent, since

$$\int_1^2 c_v \frac{dT}{T} + \int_1^2 R \frac{dv}{v} = c_v \ln \frac{T_2}{T_1} + R \ln \frac{v_2}{v_1}$$

(12.33)

Obviously, the result of this integration only depends on initial and final state. No information regarding the path from state (1) to state (2) is required, i.e. it has the quality of a state value! Thus, the left side of Eq. 12.25 is a *state value* as well. This new state value is denoted as specific entropy s:

$$ds = \frac{\delta q + \delta \psi}{T} = \frac{du}{T} + \frac{p \, dv}{T}$$

(12.34)

Alternatively, by applying thermal and caloric equations of state Eq. 12.29 obeys:

$$\frac{\delta q + \delta \psi}{T} = c_p \frac{dT}{T} - R \frac{dp}{p}$$

(12.35)

Accordingly, the integration of the right side of Eq. 12.35 is path-independent as well, since

$$\int_1^2 c_p \frac{dT}{T} - \int_1^2 R \frac{dp}{p} = c_p \ln \frac{T_2}{T_1} - R \ln \frac{p_2}{p_1}$$

(12.36)

Obviously, the result of this integration only depends on initial and final state. No information regarding the path from state (1) to state (2) is required, i.e. it has the quality of a state value! Thus, the left side of Eq. 12.29 is a state value as well. This state value is denoted as specific entropy s:

$$ds = \frac{\delta q + \delta \psi}{T} = \frac{dh}{T} - \frac{v \, dp}{T}$$

(12.37)

Hence, Eqs. 12.34 and 12.37 are equal:

$$ds = \frac{\delta q + \delta \psi}{T} = \frac{dh}{T} - \frac{v \, dp}{T} = \frac{du}{T} + \frac{p \, dv}{T}$$

(12.38)

The following definition is made:

$$\boxed{ds = \frac{\delta q}{T} + \frac{\delta \psi}{T} \equiv \delta s_a + \delta s_i}$$

(12.39)

According to this equation, the state value specific entropy s is influenced by two mechanisms[8]:

1. The process value

$$\delta s_a = \frac{\delta q}{T}$$

(12.40)

that is related to *heat passing the system boundary*. In extensive notation, i.e. by multiplying with the mass of the system, it is

$$\delta S_a = \frac{\delta Q}{T}.$$

(12.41)

This equation can be divided by dt, so that it follows

$$\frac{\delta S_a}{dt} = \frac{1}{T}\frac{\delta Q}{dt}.$$

(12.42)

This equation represents derivatives with respect to time: A heat flux causes a flux of entropy:

$$\dot{S}_a = \frac{\dot{Q}}{T}$$

(12.43)

2. The process value

$$\delta s_i = \frac{\delta \psi}{T}$$

(12.44)

that is related to the *dissipation within the system*. In extensive notation, i.e. by multiplying with the mass of the system, it is

$$\delta S_i = \frac{\delta \Psi}{T}.$$

(12.45)

This equation can be divided by dt, so that it follows

$$\frac{\delta S_i}{dt} = \frac{1}{T}\frac{\delta \Psi}{dt}.$$

(12.46)

This equation represents derivatives with respect to time: A flux of dissipated energy causes a flux of entropy:

$$\dot{S}_i = \frac{\dot{\Psi}}{T}$$

(12.47)

---

[8] Similar to the first law of thermodynamics: The state value internal energy is influenced by process values work and heat!

However, Eqs. 12.43 and 12.47 are important, when open systems are treated in Sect. 13.7.2.

The change of entropy can be calculated by integration, see Eq. 12.33:

$$\Delta s = s_2 - s_1 = c_v \ln\frac{T_2}{T_1} + R \ln\frac{v_2}{v_1} \qquad (12.48)$$

Or, alternatively, according to Eq. 12.36:

$$\Delta s = s_2 - s_1 = c_p \ln\frac{T_2}{T_1} - R \ln\frac{p_2}{p_1} \qquad (12.49)$$

Both Eqs. 12.48 and 12.49, are equal and are part of the caloric equations of state for ideal gases! Anyhow, the conclusion of this theoretical derivation of the specific entropy s is:

Theorem 12.3 *The specific entropy s has been derived as new state value. It is influenced by heat crossing the system boundary and by dissipation within the system!*

In Chap. 13 this new state value will further be motivated and it will be shown how to utilise the specific entropy s.

Theorem 12.4 *Anyhow, the conclusion, that specific entropy s is a state value, does not depend on the type of fluid. Any state value can be determined for a system, independent of the fluid.*

However, caloric equations of state depend on the type of fluid. In this section such a caloric equation of state has been derived for an ideal gas. Nevertheless, entropy of real fluids will be investigated in the following section. As mentioned several times before, real fluids will be covered in part II.

12.3 Derivation of the Caloric Equations for Real Fluids

Mind, that this section is for the advanced readers. It is possible to skip this section and proceed with Sect. 12.4. More specifically, there are the correlations for the specific internal energy u, the specific enthalpy h and the specific entropy s, which are derived thermodynamically in this section.

12.3.1 Specific Internal Energy u

For the caloric state value specific internal energy

$$u = u(v, T) \qquad (12.50)$$

the total differential can be given as:

$$du = \underbrace{\left(\frac{\partial u}{\partial T}\right)_v}_{c_v} dT + \left(\frac{\partial u}{\partial v}\right)_T dv \tag{12.51}$$

As mentioned before, the specific heat capacity c_v is a function of v and T, since, according to Eq. 12.50, u depends on v and T. Goal is now to find a correlation for $\left(\frac{\partial u}{\partial v}\right)_T$ for real fluids.

In Sect. 12.2 Eq. 12.38 has been derived. This can be mathematically transformed and is called fundamental equation of thermodynamics:

$$\boxed{du = T\,ds - p\,dv} \tag{12.52}$$

respectively

$$\boxed{dh = T\,ds + v\,dp} \tag{12.53}$$

Any fluid obeys the fundamental equation of thermodynamics, since not restrictions have been made so far. At that point, it is important to emphasise, though the specific enthalpy h has been motivated with an open system, h is a state value, that can be determined for any system, i.e. also in a closed system!

Combining Eqs. 12.51 and 12.52 leads to

$$\left(\frac{\partial u}{\partial T}\right)_v dT + \left(\frac{\partial u}{\partial v}\right)_T dv = T\,ds - p\,dv \tag{12.54}$$

Mathematical reformulation results in

$$ds = \frac{1}{T}\underbrace{\left(\frac{\partial u}{\partial T}\right)_v}_{c_v} dT + \frac{1}{T}\left[p + \left(\frac{\partial u}{\partial v}\right)_T\right] dv \tag{12.55}$$

The specific entropy is a state value, see Sect. 12.2. Thus it depends on two independent state values for a single fluid without phase change, e.g.

$$s = s(v, T) \tag{12.56}$$

As for all state values a total differential exists, see Theorem 10.1, the following applies

$$ds = \left(\frac{\partial s}{\partial T}\right)_v dT + \left(\frac{\partial s}{\partial v}\right)_T dv. \tag{12.57}$$

Comparison of the coefficients for dT and dv in Eqs. 12.55 and 12.57 leads to:

$$\left(\frac{\partial s}{\partial v}\right)_T = \frac{1}{T}\left[p + \left(\frac{\partial u}{\partial v}\right)_T\right] \tag{12.58}$$

$$\left(\frac{\partial s}{\partial T}\right)_v = \frac{c_v}{T} = \frac{1}{T}\left(\frac{\partial u}{\partial T}\right)_v \tag{12.59}$$

Rearrange Eq. 12.58 gives

$$\left(\frac{\partial u}{\partial v}\right)_T = T\left(\frac{\partial s}{\partial v}\right)_T - p \tag{12.60}$$

In order to proceed, $\left(\frac{\partial s}{\partial v}\right)_T$ in Eq. 12.60 needs to be calculated. To do so, the so-called specific Helmholtz free energy f is defined as

$$f = u - Ts. \tag{12.61}$$

Obviously, f is a state value, since it is solely composed of state values. Hence, it is path-independent. Anyhow, the total differential obeys

$$df = du - T\,ds - s\,dT \tag{12.62}$$

It follows with Eq. 12.52:

$$df = -s\,dT - p\,dv \tag{12.63}$$

Since the specific Helmholtz free energy f is a state value, it is composed of two independent state values, e.g. as:

$$f = f(T, v) \tag{12.64}$$

According to that definition the total differential is

$$df = \left(\frac{\partial f}{\partial T}\right)_v dT + \left(\frac{\partial f}{\partial v}\right)_T dv \tag{12.65}$$

Comparison of the coefficients in Eqs. 12.63 and 12.65 leads to:

$$\left(\frac{\partial f}{\partial T}\right)_v = -s \tag{12.66}$$

$$\left(\frac{\partial f}{\partial v}\right)_T = -p \tag{12.67}$$

The second derivatives of the Eqs. 12.66 and 12.67 are as follows

$$\left(\frac{\partial \left(\frac{\partial f}{\partial T}\right)_v}{\partial v}\right)_T = \frac{\partial^2 f}{\partial v \, \partial T} = -\left(\frac{\partial s}{\partial v}\right)_T \qquad (12.68)$$

$$\left(\frac{\partial \left(\frac{\partial f}{\partial v}\right)_T}{\partial T}\right)_v = \frac{\partial^2 f}{\partial T \, \partial v} = -\left(\frac{\partial p}{\partial T}\right)_v \qquad (12.69)$$

According to Schwarz's Theorem 10.3, see Sect. 10.2, it follows, that Eqs. 12.68 and 12.69 must be identical, otherwise it would not be a state function. This leads to the derivative $\left(\frac{\partial s}{\partial v}\right)_T$ we have been looking for:

$$\left(\frac{\partial s}{\partial v}\right)_T = \left(\frac{\partial p}{\partial T}\right)_v \qquad (12.70)$$

Combination of Eqs. 12.60 and 12.70 gives:

$$\left(\frac{\partial u}{\partial v}\right)_T = T \left(\frac{\partial p}{\partial T}\right)_v - p \qquad (12.71)$$

In other words, the caloric Eq. 12.51 for the specific internal energy u is:

$$\boxed{du = c_v\,(v, T) \cdot dT + \left[T \left(\frac{\partial p}{\partial T}\right)_v - p\right] \cdot dv} \qquad (12.72)$$

Furthermore, for the specific entropy s it follows by combining Eqs. 12.57, 12.59 and 12.70

$$\boxed{ds = \frac{c_v}{T} \cdot dT + \left(\frac{\partial p}{\partial T}\right)_v \cdot dv} \qquad (12.73)$$

12.3.2   Specific Enthalpy h

For the caloric state value specific enthalpy

$$h = h(p, T) \qquad (12.74)$$

the total differential obeys

$$dh = \underbrace{\left(\frac{\partial h}{\partial T}\right)_p}_{c_p} dT + \left(\frac{\partial h}{\partial p}\right)_T dp \qquad (12.75)$$

As mentioned before, the specific heat capacity c_p is a function of p and T, since according to Eq. 12.74 h depends on p and T. Goal is now to find a correlation for $\left(\frac{\partial h}{\partial p}\right)_T$.

Combining Eq. 12.75 and the fundamental equation of thermodynamics Eq. 12.53 leads to

$$\left(\frac{\partial h}{\partial T}\right)_p dT + \left(\frac{\partial h}{\partial p}\right)_T dp = T\,ds + v\,dp \tag{12.76}$$

Mathematical reformulation results in

$$ds = \frac{1}{T}\left(\frac{\partial h}{\partial T}\right)_p dT + \frac{1}{T}\left[\left(\frac{\partial h}{\partial p}\right)_T - v\right]dp \tag{12.77}$$

The entropy as new state value also depends on two independent state values, e.g.

$$s = s(p, T) \tag{12.78}$$

Since s is a state value, its total differential exists:

$$ds = \left(\frac{\partial s}{\partial T}\right)_p dT + \left(\frac{\partial s}{\partial p}\right)_T dp \tag{12.79}$$

Comparison of the coefficients for dT and dp in Eqs. 12.77 and 12.79 leads to:

$$\left(\frac{\partial s}{\partial p}\right)_T = \frac{1}{T}\left[\left(\frac{\partial h}{\partial p}\right)_T - v\right] \tag{12.80}$$

$$\left(\frac{\partial s}{\partial T}\right)_p = \frac{c_p}{T} = \frac{1}{T}\left(\frac{\partial h}{\partial T}\right)_p \tag{12.81}$$

Rearranging Eq. 12.80 gives

$$\left(\frac{\partial h}{\partial p}\right)_T = T\left(\frac{\partial s}{\partial p}\right)_T + v \tag{12.82}$$

In order to proceed, $\left(\frac{\partial s}{\partial p}\right)_T$ in Eq. 12.82 needs to be calculated. To do so, the so-called Gibbs free enthalpy[9] g is defined as

$$g = h - Ts \tag{12.83}$$

---

[9]For the Gibbs free enthalpy there is a physical motivation for chemical reactive systems, e.g. fuel cells or Lithium Ion batteries. This will be handled in part III of this book, see Sect. 24.3!

Obviously, g is a state value, since it is solely composed of state values. Hence, it is path-independent. Hence, its total differential obeys

$$dg = dh - T\,ds - s\,dT \tag{12.84}$$

It follows with Eq. 12.53:

$$dg = v\,dp - s\,dT \tag{12.85}$$

Since the specific Gibbs free enthalpy g is a state value, it is composed of two independent state values, e.g. as:

$$g = g(T, p) \tag{12.86}$$

According to that definition the total differential is

$$dg = \left(\frac{\partial g}{\partial T}\right)_p dT + \left(\frac{\partial g}{\partial p}\right)_T dp \tag{12.87}$$

Comparison of the coefficients in Eqs. 12.85 and 12.87 leads to:

$$\left(\frac{\partial g}{\partial T}\right)_p = -s \tag{12.88}$$

$$\left(\frac{\partial g}{\partial p}\right)_T = v \tag{12.89}$$

The second derivatives of the Eqs. 12.88 and 12.89 are as follows:

$$\left(\frac{\partial \left(\frac{\partial g}{\partial T}\right)_p}{\partial p}\right)_T = \frac{\partial^2 g}{\partial p\,\partial T} = -\left(\frac{\partial s}{\partial p}\right)_T \tag{12.90}$$

$$\left(\frac{\partial \left(\frac{\partial g}{\partial p}\right)_T}{\partial T}\right)_p = \frac{\partial^2 g}{\partial T\,\partial p} = \left(\frac{\partial v}{\partial T}\right)_p \tag{12.91}$$

According to Schwarz's Theorem 10.3, see Sect. 10.2, it follows, that Eqs. 12.90 and 12.91 must be identical, otherwise it would not be a state function. This leads to the derivative $\left(\frac{\partial s}{\partial p}\right)_T$ we have been looking for:

$$-\left(\frac{\partial s}{\partial p}\right)_T = \left(\frac{\partial v}{\partial T}\right)_p \tag{12.92}$$

Combining Eqs. 12.82 and 12.92

$$\left(\frac{\partial h}{\partial p}\right)_T = -T\left(\frac{\partial v}{\partial T}\right)_p + v \tag{12.93}$$

In other words, the caloric Eq. 12.75 for the specific enthalpy h is:

$$\boxed{dh = c_p\,(p,T)\cdot dT + \left[-T\left(\frac{\partial v}{\partial T}\right)_p + v\right]\cdot dp} \tag{12.94}$$

Furthermore, for the specific entropy s it follows by combining Eqs. 12.79, 12.81 and 12.92

$$\boxed{ds = \frac{c_p}{T}\cdot dT - \left(\frac{\partial v}{\partial T}\right)_p\cdot dp} \tag{12.95}$$

Consequences for Ideal Gases

So far, no constraints regarding the fluid have been made. However, the focus now is on ideal gases. The thermal equation of state obeys

$$pv = RT. \tag{12.96}$$

Mathematical reformulation results in

$$p = T\frac{R}{v}. \tag{12.97}$$

The partial derivative follows accordingly:

$$\left(\frac{\partial p}{\partial T}\right)_v = \frac{R}{v} \tag{12.98}$$

Combination of Eqs. 12.72, 12.96 and 12.98 reads as:

$$du = c_v\,(v,T)\cdot dT + \left[T\frac{R}{v} - p\right]\cdot dv \tag{12.99}$$

respectively

$$\boxed{du = c_v\,(v;T)\cdot dT} \tag{12.100}$$

Obviously, the specific internal energy u for ideal gases is solely a function of temperature T. With reference to Eq. 12.8, the specific heat capacity c_v is only a function of temperature as well. Hence, the initial Eq. 12.51 simplifies for ideal gases to

$$du = c_v (T) \, dT \tag{12.101}$$

The thermal equation of state can also be given in the following notation

$$v = \frac{RT}{p} \tag{12.102}$$

Hence, the partial derivative follows accordingly:

$$\left(\frac{\partial v}{\partial T}\right)_p = \frac{R}{p} \tag{12.103}$$

Combining Eqs. 12.94, 12.102 and 12.103 reads as:

$$dh = c_p (p, T) \cdot dT + \left[-T\frac{R}{p} + v\right] \cdot dp \tag{12.104}$$

respectively

$$dh = c_p (p, T) \cdot dT \tag{12.105}$$

The specific enthalpy h for ideal gases is solely a function of temperature T. With reference to Eq. 12.10, the specific heat capacity c_p is only a function of temperature as well. Hence, the initial Eq. 12.75 simplifies for ideal gases to

$$dh = c_p (T) \, dT \tag{12.106}$$

In accordance with the Joule-Thomson experiment for ideal gases, see Example 12.1, the specific internal energy as well as the specific enthalpy are purely functions of temperature. For the specific entropy it follows, see Eqs. 12.73 and 12.95:

$$ds = \frac{\delta q + \delta \psi}{T} = \frac{c_v}{T} \cdot dT + \left(\frac{\partial p}{\partial T}\right)_v \cdot dv = \frac{c_v}{T} \cdot dT + \frac{R}{v} \cdot dv \tag{12.107}$$

$$ds = \frac{\delta q + \delta \psi}{T} = \frac{c_p}{T} \cdot dT - \left(\frac{\partial v}{\partial T}\right)_p \cdot dp = \frac{c_p}{T} \cdot dT - \frac{R}{p} \cdot dp \tag{12.108}$$

The integration leads to:

$$\Delta s = s_2 - s_1 = c_v \ln\frac{T_2}{T_1} + R \ln\frac{v_2}{v_1} \tag{12.109}$$

and

$$\Delta s = s_2 - s_1 = c_p \ln\frac{T_2}{T_1} - R \ln\frac{p_2}{p_1} \tag{12.110}$$

Consequences for Incompressible Liquids

Part I does not only cover ideal gases but incompressible liquids as well. According to [4] these idealised liquids follow the simple thermal state equation

$$v(p, T) = \text{const.} \tag{12.111}$$

respectively

$$dv = 0. \tag{12.112}$$

The simplification is that the specific volume does neither depend on the pressure nor on the temperature, i.e.

$$\left(\frac{\partial v}{\partial p}\right)_T = 0 \tag{12.113}$$

$$\left(\frac{\partial v}{\partial T}\right)_p = 0 \tag{12.114}$$

What are the consequences of such a simplification for the specific internal energy u and the specific enthalpy h?

- Specific internal energy u
  Following Eq. 12.72 it is

$$du = c_v(v, T) \cdot dT + \left[T\left(\frac{\partial p}{\partial T}\right)_v - p\right] \cdot dv = c_v(v, T) \cdot dT \tag{12.115}$$

The specific internal energy u is purely a function of T:

$$\boxed{du = c_v(T) \cdot dT} \tag{12.116}$$

Consequently, $c_v(T)$ also is just a function of T!
- Specific enthalpy h
  Following Eq. 12.94 it is

$$dh = c_p(p, T) \cdot dT + \left[-T\left(\frac{\partial v}{\partial T}\right)_p + v\right] \cdot dp = c_p(p, T) \cdot dT + v \cdot dp \tag{12.117}$$

Additionally, the specific enthalpy h is by definition, see Sect. 11.3.1:

$$h = u + pv \tag{12.118}$$

The total differential for an incompressible liquid, i.e. $dv = 0$, follows accordingly

$$dh = du + p\,dv + v\,dp = du + v\,dp \tag{12.119}$$

In combination with Eq. 12.116 it is

$$dh = c_v\,(T) \cdot dT + v\,dp \tag{12.120}$$

A comparison of Eqs. 12.117 and 12.120 results in

$$c_p\,(p, T) = c_v\,(T). \tag{12.121}$$

In other words, the specific heat capacity c_p is also only a function of T. Furthermore, for incompressible liquids it is:

$$\boxed{c\,(T) = c_p\,(T) = c_v\,(T)} \tag{12.122}$$

In summary, the following can be stated for incompressible liquids:

$$\boxed{du = c\,(T) \cdot dT} \tag{12.123}$$

$$\boxed{dh = c\,(T) \cdot dT + v \cdot dp} \tag{12.124}$$

The conclusion, that the specific entropy s is a state value, does not depend on the type of fluid. Thus, it must be possible to find a caloric equation of state in order to calculate the state value specific entropy s for an incompressible liquid as well. For incompressible liquids, i.e. $dv = 0$ as well as $c_v = c_p = c$, Eq. 12.25 simplifies to

$$\frac{\delta q + \delta \psi}{T} = \frac{du}{T} = c\frac{dT}{T} \tag{12.125}$$

The integration of the right hand side is path-independent, since

$$\int_1^2 c\frac{dT}{T} = c\ln\frac{T_2}{T_1}. \tag{12.126}$$

Consequently, the term needs to be a state and not a process value. This is the newly introduced specific entropy s:

$$ds = \frac{\delta q}{T} + \frac{\delta \psi}{T} = c\frac{dT}{T} \tag{12.127}$$

Its integration leads to the caloric equation for the specific entropy of incompressible liquids:

$$\boxed{\Delta s = s_2 - s_1 = c\ln\frac{T_2}{T_1}} \tag{12.128}$$

Mind, that solids, as long as they are incompressible, can be treated with the same means as the recently introduced incompressible liquids!

12.4 Handling of the Caloric State Equations

Initially, the goal was to find correlations for $\Delta u = u_2 - u_1$ and $\Delta h = h_2 - h_1$ in order to handle the first law of thermodynamics for closed systems

$$q_{12} + w_{12} = u_2 - u_1 + \Delta e_{a12} \tag{12.129}$$

as well as for open systems

$$q_{12} + w_{t12} = h_2 - h_1 + \Delta e_{a12}. \tag{12.130}$$

12.4.1 Ideal Gases

As mentioned, Δu and Δh can not be measured, so they need to be calculated. Integration of the caloric equations of state for ideal gases Eqs. 12.101 and 12.106 leads to[10]:

$$\int_1^2 du = u_2 - u_1 = \int_1^2 c_v \, dT = c_v \, (T_2 - T_1) \tag{12.131}$$

and

$$\int_1^2 dh = h_2 - h_1 = \int_1^2 c_p \, dT = c_p \, (T_2 - T_1) \tag{12.132}$$

Under these conditions the first law of thermodynamics simplifies to

- Closed system
$$q_{12} + w_{12} = c_v \, (T_2 - T_1) + \Delta e_{a12} \tag{12.133}$$

- Open system
$$q_{12} + w_{t12} = c_p \, (T_2 - T_1) + \Delta e_{a12}. \tag{12.134}$$

---

[10]With the assumptions, that $c_v = $ const. and $c_p = $ const.

12.4.2 Distinction Between c_v and c_p for Ideal Gases

Both, c_v and c_p are specific heat capacities. The specific heat capacity c_v indicates by what amount the specific internal energy u increases in case the temperature of 1 kg rises by 1 K. However, c_p is a measure of what amount the specific enthalpy increases under the same conditions. Thus, both heat capacities are thermal properties of the fluid. During the deduction of the caloric equations of state c_v was introduced as specific heat capacity with constant specific volume[11] and c_p as the specific heat capacity with constant pressure.[12]

Theorem 12.5 *No matter what change of state the system underlies, it is always c_p that is required to calculate Δh and always c_v that is required for Δu:*

$$\Delta h \rightarrow c_p$$

$$\Delta u \rightarrow c_v$$

The definition of the specific enthalpy h is

$$h = u + pv. \tag{12.135}$$

With the thermal equation for ideal gases it results in

$$h = u + pv = u + RT. \tag{12.136}$$

Hence, the derivative with respect to temperature follows

$$\frac{dh}{dT} = \frac{du}{dT} + R. \tag{12.137}$$

Following Eqs. 12.101 and 12.106 it is

$$du = c_v(T)\, dT \tag{12.138}$$

and

$$dh = c_p(T)\, dT. \tag{12.139}$$

Thus, Eq. 12.137 simplifies for ideal gases to

$$\boxed{c_p(T) = c_v(T) + R} \tag{12.140}$$

---

[11] Indicated by the v-subscript.
[12] Indicated by the p-subscript.

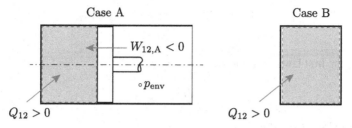

Fig. 12.2 Isobaric (Case A) versus isochoric (Case B) change of state

Example

Obviously, according to Eq. 12.140 the specific heat capacity with constant pressure c_p is always larger than the specific heat capacity with constant volume. This is verified with the following example, see Fig. 12.2. Both closed systems are filled with the same type and mass m of an ideal gas. System A is closed with a freely movable, horizontal piston, so that the fluid inside is always in mechanical equilibrium with the ambient, i.e. the fluid's pressure inside the cylinder is always equal to ambient pressure $p_A = p_{env}$. Hence, the change of state is isobaric. According to Example 7.24, only for very slow changes of state[13] ($\Psi_A = 0$) and a non-accelerated ($a = 0$),[14] frictionless ($\Psi_{fric.} = 0$) piston the process is isobaric. Thus, case A needs to be reversible!

System B is a rigid vessel, so that its volume is constant, i.e. the change of state is isochoric. Initially, the temperature in both systems is the same. Now, the same amount of heat, i.e. Q_{12}, is supplied to both systems. The question is, how the temperature rise differs in both systems:

- Case A: Isobaric change of state ($p =$ const.)

First law of thermodynamics in differential notation[15]:

$$\delta Q + \underbrace{\delta W_A}_{<0} = dU_A \tag{12.141}$$

Obviously, work δW_A is released: Due to the supply of heat the gas inside the cylinder heats up and, while the pressure remains constant, expands.[16] This expansion is the cause for the system to release of work.
Partial energy equation:

---

[13] Very slow is a synonym for no turbulence inside, see also Theorem 7.25.

[14] Gravity constant g is not relevant, since the piston is operated horizontally.

[15] The change of kinetic energy can be ignored, since the change of state is quasi-static! There is no change of potential energy, since the cylinder is horizontal!

[16] According to $V = \frac{mRT}{p}$.

$$\delta W_A = \delta W_{V,A} + \underbrace{\delta \Psi_A}_{=0} = -p_A \, dV \tag{12.142}$$

Hence, the first law of thermodynamics obeys

$$\delta Q = dU_A + p_A \, dV \tag{12.143}$$

The enthalpy H is defined as:

$$H = U + pV. \tag{12.144}$$

Its differential for an isobaric, i.e. $dp = 0$, change of state is:

$$dH = dU + p \, dV + V \, dp = dU + p \, dV \tag{12.145}$$

Thus, for case A it obviously is

$$dU_A = dH_A - p_A \, dV. \tag{12.146}$$

According to Eq. 12.143 this means:

$$\delta Q = dH_A \tag{12.147}$$

In combination with the caloric equation of state for the enthalpy[17] it results in

$$\delta Q = mc_p \, dT_A \tag{12.148}$$

Integration leads to

$$\boxed{Q_{12} = mc_p \, \Delta T_A}. \tag{12.149}$$

In other words:

$$c_p = \frac{Q_{12}}{m \, \Delta T_A} \tag{12.150}$$

- Case B: Isochoric change of state ($v = $ const.)

  First law of thermodynamics in differential notation

$$\delta Q + \delta W_B = dU_B \tag{12.151}$$

Partial energy equation

$$\delta W_B = 0 = -p_B \, dV + \delta \Psi_B \tag{12.152}$$

---

[17] Mind, that $H = mh$.

Obviously, according to Fig. 12.2, no work δW_B passes the system boundary. Furthermore, the volume does not vary, i.e. $dV = 0$. Consequently, there can not be any dissipation, i.e. $\delta \Psi_B = 0$. This makes sense, since there is no fluid motion inside the cylinder![18] Thus, the first law of thermodynamics simplifies to

$$\delta Q = dU_B \tag{12.153}$$

In combination with the caloric equation for the internal energy[19] it results in

$$\delta Q = mc_c \, dT_B. \tag{12.154}$$

Integration leads to

$$\boxed{Q_{12} = mc_v \, \Delta T_B}. \tag{12.155}$$

In other words:

$$c_v = \frac{Q_{12}}{m \, \Delta T_B} \tag{12.156}$$

Let us now compare Eqs. 12.150 and 12.156: Obviously, the temperature rise in case B is larger than in case A. This is a result of the first law of thermodynamics Eq. 12.141 respectively Eq. 12.153: In case B the entirely supplied heat is leading to an increase of internal energy, whereas in case A part of the supplied energy is released as work. Consequently, the remaining energy to increase the internal energy is less than in case B. The ratio of Eqs. 12.150 and 12.156 results in

$$\boxed{\frac{c_p}{c_v} = \frac{\Delta T_B}{\Delta T_A} > 1} \tag{12.157}$$

Due to $c_p > c_v$ the temperature rise of B must be larger than the temperature increase of A.

Problem 12.6 In this problem we refer to case A according to Fig. 12.2. The cylinder is filled with $m = 1$ kg of air (ideal gas, $R = 287 \frac{J}{kg\,K}$, $c_v = 717 \frac{J}{kg\,K}$). The initial temperature is $T_1 = 300$ K, the pressure is supposed to be $p = 1$ bar $=$ const. while the system is heated up by $Q = 10$ kJ. The entire change of state shall be reversible.

(a) Calculate the initial volume V_1.
(b) Estimate the final temperature T_2.
(c) What is volume V_2?
(d) How much work W_{12} is released?
(e) Sketch the change of state $(1) \rightarrow (2)$ in a p, V- and T, S-diagram.[20]

---

[18]Fluid motion would need to be initiated, e.g. by a moving piston.
[19]Mind, that $U = mu$.
[20]This part is for the advanced readers, who are already familiar with the T, s-diagram, see Sect. 13.4.

Solution

(a) In order to calculate the initial volume V_1, the thermal equation of state is applied:

$$V_1 = \frac{m R T_1}{p} = 0.861 \, \text{m}^3 \tag{12.158}$$

(b) The first law of thermodynamics obeys

$$Q_{12} + W_{12} = U_2 - U_1. \tag{12.159}$$

The partial energy equation for W_{12} results in

$$W_{12} = W_{\text{V},12} + \underbrace{\Psi_{12}}_{=0} = -\int_1^2 p \, dV. \tag{12.160}$$

There is no dissipation because the change of state shall be reversible. Since the change of state is isobaric as well, the volume work can easily be calculated:

$$W_{12} = -p \, \Delta V = -p \, (V_2 - V_1) \tag{12.161}$$

Hence, the first law of thermodynamics follows

$$Q_{12} = U_2 - U_1 + p \, (V_2 - V_1) = H_2 - H_1 = m \, (h_2 - h_1) \tag{12.162}$$

Applying the caloric equation of state Eq. 12.132 leads to[21]

$$Q_{12} = m c_p \, (T_2 - T_1) \tag{12.163}$$

with

$$c_p = c_v + R = 1004 \frac{\text{J}}{\text{kg K}}. \tag{12.164}$$

For T_2 it is

$$T_2 = T_1 + \frac{Q_{12}}{m c_p} = 309.96 \, \text{K} \tag{12.165}$$

(c) The thermal equation of state results in

$$V_2 = \frac{m R T_2}{p} = 0.8896 \, \text{m}^3 \tag{12.166}$$

(d) According to Eq. 12.161 it is

---

[21] Assuming, that $c_p = \text{const.}$!

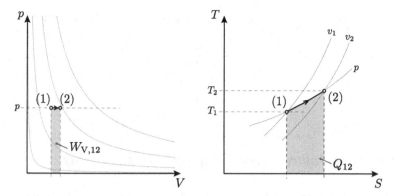

Fig. 12.3 Solution for Problem 12.6

$$W_{12} = -p\,\Delta V = -p\,(V_2 - V_1) = -2.8586\,\text{kJ} \tag{12.167}$$

Obviously, of the supplied $10\,\text{kJ}$ of thermal energy, $2.8586\,\text{kJ}$ are released as mechanical work. Thus, $7.1414\,\text{kJ}$ are utilised to increase the internal energy of the air inside the cylinder.

(e) The change of state $(1) \to (2)$ is visualised in a p, V- and a T, S-diagram, see Fig. 12.3.

12.4.3 Isentropic Exponent

The so-called isentropic exponent

$$\boxed{\kappa = \frac{c_p}{c_v}} \tag{12.168}$$

will be important when the isentropic change of state is introduced, see Sect. 13.5. However, it makes sense to define κ at this stage. For ideal gases, depending on the number of atoms they consist of, the isentropic exponent follows Table 12.1.

Applying Eq. 12.140 and the definition of the isentropic exponent leads to

$$c_p - c_v = R \tag{12.169}$$

and

$$c_p = \frac{\kappa}{\kappa - 1}R \tag{12.170}$$

and

$$c_v = \frac{1}{\kappa - 1}R. \tag{12.171}$$

Table 12.1 Isentropic exponent for ideal gases at standard conditions, i.e. temperature $\vartheta = 0\,°C$ according to DIN 1343

Number of atoms	κ
1	1.67
2	1.40
3	1.33

12.4.4   Temperature Dependent Specific Heat Capacity

For ideal gases the specific heat capacities c_v and c_p are purely functions of the temperature ϑ, see Sect. 12.4.2. The following correlation between c_p and c_v has been derived:

$$c_p(\vartheta) = R + c_v(\vartheta) \tag{12.172}$$

For small temperature variations the temperature dependency of the specific heat capacities can be ignored. However, in this section the temperature dependency is considered—this becomes relevant for significant temperature changes. The caloric equations of state obey

$$du = c_v \, dT \Rightarrow u_2 - u_1 = \int_1^2 c_v(\vartheta) \, d\vartheta \tag{12.173}$$

respectively

$$dh = c_p \, dT \Rightarrow h_2 - h_1 = \int_1^2 c_p(\vartheta) \, d\vartheta. \tag{12.174}$$

It does not make any difference which temperature scale (°C or K) is applied, since

$$\vartheta = T - 273.15\,K \Rightarrow d\vartheta = dT. \tag{12.175}$$

Instead of denoting the temperature dependency separately for c_v and c_p, c represents both in this section. Figure 12.4 illustrates how the specific heat capacity c correlates with the temperature ϑ. With rising temperature, the specific heat capacity increases as well. Goal now is to find a solution for the integrals in Eqs. 12.173 and 12.174. Thus, an averaged specific heat capacity $\bar{c}|_{\vartheta_1}^{\vartheta_2}$ within the temperature range from ϑ_1 and ϑ_2 is introduced:

$$\int_1^2 c(\vartheta) \, d\vartheta = \bar{c}|_{\vartheta_1}^{\vartheta_2}(\vartheta_2 - \vartheta_1) \tag{12.176}$$

Fig. 12.4 Temperature dependency of the specific heat capacity

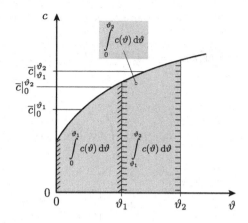

Thus, the averaged specific heat capacity is:

$$\overline{c}|_{\vartheta_1}^{\vartheta_2} = \frac{1}{(\vartheta_2 - \vartheta_1)} \cdot \int_1^2 c(\vartheta)\, d\vartheta \tag{12.177}$$

In tables, e.g. [19], averaged values are usually given. The reference level in these tables commonly is $0\,°C$ which simplifies the calculation:

$$\overline{c}|_0^{\vartheta} = \frac{1}{\vartheta} \cdot \int_0^{\vartheta} c(\vartheta)\, d\vartheta. \tag{12.178}$$

By this it follows

$$\vartheta_1 \cdot \overline{c}|_0^{\vartheta_1} = \int_0^{\vartheta_1} c(\vartheta)\, d\vartheta \tag{12.179}$$

respectively

$$\vartheta_2 \cdot \overline{c}|_0^{\vartheta_2} = \int_0^{\vartheta_2} c(\vartheta)\, d\vartheta. \tag{12.180}$$

For the integrals it is, see Fig. 12.4

$$\int_{\vartheta_1}^{\vartheta_2} c(\vartheta)\, d\vartheta = \int_0^{\vartheta_2} c(\vartheta)\, d\vartheta - \int_0^{\vartheta_1} c(\vartheta)\, d\vartheta \tag{12.181}$$

Thus, it finally results in

Table 12.2 Temperature dependency of an ideal gas according to Problem 12.7

ϑ (°C)	200	1200
$\overline{c_p}\vert_0^\vartheta \left(\frac{kJ}{kg\,K}\right)$	1.052	1.242

$$\int_{\vartheta_1}^{\vartheta_2} c(\vartheta)\, d\vartheta = \left[\overline{c}\vert_0^{\vartheta_2} \cdot \vartheta_2 - \overline{c}\vert_0^{\vartheta_1} \cdot \vartheta_1\right] \tag{12.182}$$

respectively

$$\boxed{\overline{c}\vert_{\vartheta_1}^{\vartheta_2} = \frac{\overline{c}\vert_0^{\vartheta_2} \cdot \vartheta_2 - \overline{c}\vert_0^{\vartheta_1} \cdot \vartheta_1}{\vartheta_2 - \vartheta_1}} \tag{12.183}$$

Problem 12.7 The specific heat capacity at constant pressure c_p of an ideal gas is temperature dependent and follows Table 12.2. Please calculate the averaged specific heat capacity at constant pressure within the temperatures $\vartheta_1 = 200\,°C$ and $\vartheta_2 = 1200\,°C$. What is $\Delta h = h_2 - h_1$?

Solution

The averaged heat capacity follows Eq. 12.183, so that

$$\overline{c_p}\vert_{\vartheta_1}^{\vartheta_2} = \frac{\overline{c_p}\vert_0^{\vartheta_2} \cdot \vartheta_2 - \overline{c_p}\vert_0^{\vartheta_1} \cdot \vartheta_1}{\vartheta_2 - \vartheta_1} \tag{12.184}$$

$$\Rightarrow \overline{c_p}\vert_{\vartheta_1}^{\vartheta_2} = \frac{1.242 \cdot 1200 - 1.052 \cdot 200}{1200 - 200}\,\frac{kJ}{kg\,K} = 1.28\frac{kJ}{kg\,K} \tag{12.185}$$

For the difference of specific enthalpy it follows

$$h_2 - h_1 = \int_1^2 c_p\,(\vartheta)\, d\vartheta = \overline{c_p}\vert_{\vartheta_1}^{\vartheta_2}\,(\vartheta_2 - \vartheta_1) \tag{12.186}$$

respectively

$$h_2 - h_1 = \overline{c_p}\vert_{\vartheta_1}^{\vartheta_2}\,(\vartheta_2 - \vartheta_1) = 1280\frac{kJ}{kg}. \tag{12.187}$$

12.4.5 Incompressible Fluids, Solids

Keep in mind, that $c = c_p = c_v$ for incompressible liquids, i.e. it is not required to distinguish between them. Solids are treated in part I as incompressible as well, so that $c = c_p = c_v$ also applies. Thus, the caloric equations of state follow, see Sect. 12.3.2:

- Specific internal energy u

$$du = c\,(\vartheta)\,d\vartheta \tag{12.188}$$

Integration leads to[22]

$$\boxed{u_2 - u_1 = \overline{c}\,|_{\vartheta_1}^{\vartheta_2}\,(\vartheta_2 - \vartheta_1)} \tag{12.189}$$

- Specific enthalpy h

$$dh = c\,(\vartheta)\,d\vartheta + v\,dp \tag{12.190}$$

Since $v = $ const., integration leads to

$$\boxed{h_2 - h_1 = \overline{c}\,|_{\vartheta_1}^{\vartheta_2}\,(\vartheta_2 - \vartheta_1) + v\,(p_2 - p_1)} \tag{12.191}$$

Thus, the term $v\,(p_2 - p_1)$ is *additionally* relevant in case the change of state from (1) to (2) is non-isobaric!

Problem 12.8 Two equal, vertical standing pipes with a diameter of $D = 0.1\,\text{m}$ are connected with each other by a thin pipe and a valve, see Problem 11.6. One pipe is filled up to a height of $H = 10\,\text{m}$ with water. Water has a density of $\rho = 1000\,\frac{\text{kg}}{\text{m}^3}$. First, the second pipe is empty. Now, the connecting valve is opened. After a while a balance of the water volume develops. What is the temperature rise of the water? Water shall be treated as an incompressible liquid with $c = 4.18\,\frac{\text{kJ}}{\text{kg K}}$.

Solution

The rise of internal energy ΔU has previously been calculated in Problem 11.6 by applying the first law of thermodynamics for closed systems:

$$\Delta U = mg\frac{H}{4} = 1926.2\,\text{J} \tag{12.192}$$

In order to calculate the temperature increase, the caloric equation of state has to be applied:

---

[22]For the calculation a temperature *difference* is required, so it does not make any difference, if you apply $\Delta\vartheta$ or ΔT, see Eq. 12.175!

$$\Delta U = m \, \Delta u = mc \, \Delta T \tag{12.193}$$

Hence, it is

$$\Delta T = \frac{\Delta U}{mc} = \frac{mg\frac{H}{4}}{mc} = \frac{gH}{4c} = 5.8672 \times 10^{-3} \text{ K}. \tag{12.194}$$

The problem has even been calculated dynamically, see Problem 11.9. Figure 11.18 shows the same temperature rise ΔT once the system comes to rest.

Problem 12.9 In a heat exchanger of a thermal power plant steam has to be condensed. Therefore, a thermal power of $\dot{Q} = 75 \times 10^3$ kW has to be transferred. For that purpose cooling water can be taken from a river, that flows by with a total water flow rate of $\dot{m}_{\text{total}} = 20 \times 10^3 \frac{\text{kg}}{\text{s}}$, see sketch 12.5. Outer energies can be ignored. The system is regarded in steady state.

(a) What mass flow rate $\dot{m}_{\text{cond.}}$ has to be taken from the river, if a temperature rise of $\Delta T_{\text{cond.}} = 3$ K of the coolant stream is tolerated?
(b) What temperature rise ΔT_{total} results for the river after complete admixing of the coolant flow, if this can be regarded as being adiabatic?

Hint: There is to be no pressure loss in the condenser. Change of outer energies can be ignored, the specific heat capacity of water is $c_W = 4.19 \frac{\text{kJ}}{\text{kg K}}$.

Solution

(a) In order to calculate the mass flow rate, the first law of thermodynamics for open systems according to system border A is applied, see Fig. 12.5. With reference to Sect. 11.3.3 the energy flux into the system needs to be balanced by the energy flux out of the system in steady state. Thus, the first law of thermodynamics results in

$$\dot{m}_{\text{cond.}} h_1 + \dot{Q} = \dot{m}_{\text{cond.}} h_2 \tag{12.195}$$

In order to apply the caloric equation of state a *difference* of enthalpies is required, so the first law of thermodynamics needs to be modified:

$$\dot{Q} = \dot{m}_{\text{cond.}} (h_2 - h_1) \tag{12.196}$$

Applying the caloric equation of state for an incompressible liquid without pressure loss

$$h_2 - h_1 = c_W (T_2 - T_1) + \underbrace{v (p_2 - p_1)}_{=0} \tag{12.197}$$

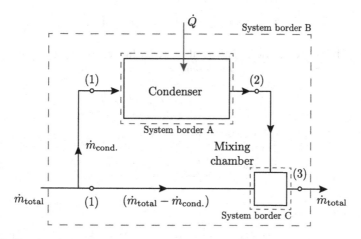

Fig. 12.5 Sketch to Problem 12.9

leads to

$$\dot{Q} = \dot{m}_{cond.} (h_2 - h_1) = \dot{m}_{cond.} c_W \underbrace{(T_2 - T_1)}_{=3\,K}. \tag{12.198}$$

Thus, the mass flow rate is

$$\dot{m}_{cond.} = \frac{\dot{Q}}{c_W (T_2 - T_1)} = 5.967 \times 10^3 \frac{kg}{s}. \tag{12.199}$$

(b) In order to calculate the temperature rise of the river $\Delta T_{total} = T_3 - T_1$, the first law of thermodynamics for system boundary B is applied:

$$\dot{m}_{total} h_1 + \dot{Q} = \dot{m}_{total} h_3 \tag{12.200}$$

Rearranging leads to

$$\dot{Q} = \dot{m}_{total} (h_3 - h_1). \tag{12.201}$$

Applying the caloric equation of state for isobaric, incompressible liquids results in

$$\dot{Q} = \dot{m}_{total} c_W (T_3 - T_1). \tag{12.202}$$

Thus, the temperature rise of the river follows

$$\Delta T_{total} = T_3 - T_1 = \frac{\dot{Q}}{\dot{m}_{total} c_W} = 0.895\,K. \tag{12.203}$$

Alternatively, the first law of thermodynamics can be applied for system boundary C. Steady state balancing of incoming and outgoing energy fluxes results in

$$(\dot{m}_{\text{total}} - \dot{m}_{\text{cond.}})\, h_1 + \dot{m}_{\text{cond.}} h_2 = \dot{m}_{\text{total}} h_3. \qquad (12.204)$$

For applying the caloric equations of state, the first law of thermodynamics has to be rearranged for enthalpy *differences*, i.e.

$$\dot{m}_{\text{total}}\,(h_3 - h_1) = \dot{m}_{\text{cond.}}\,(h_2 - h_1). \qquad (12.205)$$

By substituting the caloric equations of state it is

$$\dot{m}_{\text{total}} c_{\text{W}} \underbrace{(T_3 - T_1)}_{=\Delta T_{\text{total}}} = \dot{m}_{\text{cond.}} c_{\text{W}} \cdot \underbrace{(T_2 - T_1)}_{=\Delta T_{\text{cond.}}} \qquad (12.206)$$

Thus, it finally is

$$\Delta T_{\text{total}} = \frac{\Delta T_{\text{cond.}} \dot{m}_{\text{cond.}}}{\dot{m}_{\text{total}}} = 0.895\,\text{K}. \qquad (12.207)$$

Problem 12.10 Air ($T_1 = 300\,\text{K}$ and $p_1 = 1000\,\text{kPa}$) flows through an adiabatic throttle. Thus, its pressure changes to $p_2 = 700\,\text{kPa}$. The inlet velocity is $c_1 = 20\frac{\text{m}}{\text{s}}$. The cross sections of the tube at inlet A_1 and outlet A_2 shall be the same. Air can be treated as an ideal gas ($c_v = 0.717\frac{\text{J}}{\text{kg K}}$, $R = 287\frac{\text{J}}{\text{kg K}}$). What is the temperature T_2 after throttling? The change of kinetic energy shall not be ignored!

Solution

Let us start with the first law of thermodynamics for this steady state problem, see Fig. 12.6. The energy flux into the system needs to be balanced by the energy flux out of the system, otherwise the problem could not be steady state:

$$\dot{m}_1 h_1 + \dot{m}_1 \left[\frac{1}{2}c_1^2 + g z_1\right] + \underbrace{\dot{Q}_{12}}_{=0} + \underbrace{P_{t12}}_{=0} = \dot{m}_2 h_2 + \dot{m}_2 \left[\frac{1}{2}c_2^2 + g z_2\right] \qquad (12.208)$$

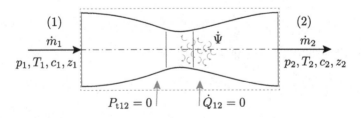

Fig. 12.6 Sketch to Problem 12.10

The mass balance in steady state means, that the mass flux into the system is equal to the mass flux out of the system, so that

$$\dot{m}_1 = \dot{m}_2 \tag{12.209}$$

Since no information regarding the change of potential energy is given, it is ignored. Hence, the first law of thermodynamics simplifies to

$$\frac{1}{2}\left(c_2^2 - c_1^2\right) + (h_2 - h_1) = 0 \tag{12.210}$$

Applying the caloric equation of state:

$$\frac{1}{2}\left(c_2^2 - c_1^2\right) + c_p\left(T_2 - T_1\right) = 0 \tag{12.211}$$

with

$$c_p = c_v + R = 1004\frac{J}{\text{kg K}}. \tag{12.212}$$

In order to gather information regarding the velocities, the equation of continuity is applied, i.e.

$$\dot{m}_1 = \dot{m}_2 \Rightarrow \rho_1 c_1 A_1 = \rho_2 c_2 A_2. \tag{12.213}$$

Hence, it is

$$c_2 = c_1\frac{\rho_1}{\rho_2} = c_1\frac{v_2}{v_1}. \tag{12.214}$$

The fluid is an ideal gas, so that the thermal equation of state can be applied

$$p_1 v_1 = RT_1 \Rightarrow v_1 = \frac{RT_1}{p_1} \tag{12.215}$$

and

$$p_2 v_2 = RT_2 \Rightarrow v_2 = \frac{RT_2}{p_2}. \tag{12.216}$$

Consequently, the outlet velocity is

$$c_2 = c_1\frac{p_1 T_2}{p_2 T_1}. \tag{12.217}$$

Substituting in the first law of thermodynamics 12.210 results in

$$T_2 - T_1 = -\frac{1}{2c_p}\left[c_1^2\left(\frac{p_1 T_2}{p_2 T_1}\right)^2 - c_1^2\right] \tag{12.218}$$

$$\Rightarrow c_1^2 \left(\frac{p_1 T_2}{p_2 T_1} \right)^2 T_2^2 + T_2 - \left[\frac{c_1^2}{2c_p} + T_1 \right] = 0. \tag{12.219}$$

This quadratic equation has two solutions:

$$T_{2,1} = 299.7932 \, \text{K} \tag{12.220}$$

respectively

$$T_{2,2} = -2.2168 \times 10^5 \, \text{K}. \tag{12.221}$$

Thus, only solution $T_{2,1} = 299.7932$ K makes sense. The corresponding velocity follows Eq. 12.217 and leads to $c_2 = 28.55 \frac{m}{s}$.

12.4.6 Adiabatic Throttle

As mentioned before, see Problem 12.10, a throttle is a simple work-insulated component, that is utilised for decreasing the pressure of a fluid. The principle sketch of an *adiabatic* throttle is illustrated in Fig. 12.6. As part of Problem 12.10 this component has already been calculated in detail, but now relevant premises are made:

In steady state the equation of continuity reads as:

$$\dot{m}_1 = \dot{m}_2 = \dot{m} \tag{12.222}$$

The first law of thermodynamics for an *adiabatic* throttle, i.e. an work-insulated component, in steady state obeys

$$\dot{m}h_1 + \dot{m} \left[\frac{1}{2}c_1^2 + gz_1 \right] + \underbrace{\dot{Q}_{12}}_{=0} + \underbrace{P_{t12}}_{=0} = \dot{m}h_2 + \dot{m} \left[\frac{1}{2}c_2^2 + gz_2 \right]. \tag{12.223}$$

In a different notation it results in

$$0 = \dot{m} \left[h_2 - h_1 + \frac{1}{2} \left(c_2^2 - c_1^2 \right) + g \left(z_2 - z_1 \right) \right]. \tag{12.224}$$

Since the dimensions of such a component are rather small, it is allowed to ignore the change of potential energy. Hence, the first law of thermodynamics further simplifies to

$$\left[h_2 - h_1 + \frac{1}{2} \left(c_2^2 - c_1^2 \right) \right] = 0. \tag{12.225}$$

In case the kinetic energy is disregarded as well, the first law of thermodynamics reduces to:

$$\boxed{h_2 - h_1 = 0} \tag{12.226}$$

Consequently, the change of state is called isenthalpic.[23] In case the fluid is an ideal gas, the specific enthalpy is purely a function of temperature, i.e.

$$c_p \left(T_2 - T_1 \right) = 0 \Rightarrow T_2 = T_1. \tag{12.227}$$

The assumption of neglecting the kinetic energy is sufficiently accurate as Problem 12.10 has shown. The conclusion from that problem has been, that the inlet and outlet temperature are *approximately* the same, i.e. the kinetic energy does not significantly contribute.

Theorem 12.11 *In case the variation of outer energies is ignored, the change of state in an adiabatic throttle is isenthalpic. Furthermore, in case an adiabatic throttle is operated with an ideal gas, the change of state is isothermal!*

However, if the fluid is a real fluid, see part II, i.e. non-ideal, the enthalpy is solely be a function of temperature. Thus, a change of temperature is measured while throttling the real fluid. This so-called Joule-Thomson effect will be treated in part II, see Sect. 18.5.

How is the pressure loss in a throttle technically achieved?
Let us have a look at the partial energy equation. Since the component is a work-insulated system, it obeys

$$w_t = 0 = \int_1^2 v \, dp + \psi_{12} + \frac{1}{2} \left(c_2^2 - c_1^2 \right) + g \left(z_2 - z_1 \right). \tag{12.228}$$

As discussed before, the outer energies are ignored, so that it simplifies to:

$$\boxed{\int_1^2 v \, dp = -\psi_{12} \le 0} \tag{12.229}$$

Hence, it finally is

$$dp \le 0. \tag{12.230}$$

Theorem 12.12 *Obviously, due to dissipation within the throttle the fluid's pressure decreases!*

---

[23] The specific enthalpy remains constant!

Chapter 13
Meaning and Handling of Entropy

In the previous Chap. 12 the caloric equations of state have been introduced and derived thermodynamically. A new state value, the specific entropy s has been introduced. However, at that point it was not clear how to utilise entropy in order to evaluate thermodynamic systems. This chapter focuses on clarifying why entropy is beneficial. First, a comparison is made with the first law of thermodynamics, so that entropy balancing is comprehensible: In contrast to energy, entropy is not a conservation value, since entropy can be generated within a system. Nevertheless, entropy can be balanced. A distinction is made between closed/open systems and thermodynamic cycles. Furthermore, a new state diagram, i.e. the T, s-diagram, is derived. Such a state diagram visualises the process values specific heat q and specific dissipation ψ. Obviously, together with the p, v-diagram, that illustrates the other process value, i.e. the specific work, it is an important diagram in thermodynamics. Finally, two new changes of state, namely the isentropic and polytropic change of state are treated in this chapter.

13.1 Entropy—Clarification

Let us take a closer look at what we have learned so far about entropy. According to Eq. 12.39 it is

$$\mathrm{d}s = \underbrace{\frac{\delta q}{T}}_{=\delta s_{\mathrm{a}}} + \underbrace{\frac{\delta \psi}{T}}_{=\delta s_{\mathrm{i}}} . \tag{13.1}$$

Hence, the state value specific entropy s of a system can be modified by two mechanisms, see Fig. 13.1:

© Springer Nature Switzerland AG 2019
A. Schmidt, *Technical Thermodynamics for Engineers*,
https://doi.org/10.1007/978-3-030-20397-9_13

1. Specific heat that passes the system boundary always carries specific entropy with it. Obviously, this part of the specific entropy is supplied from the environment, or released to the environment. This part follows

$$\delta s_a = \frac{\delta q}{T}, \tag{13.2}$$

integration leads to:

$$\int_1^2 \delta s_a = \int_1^2 \frac{\delta q}{T} = s_{a,12}. \tag{13.3}$$

In case, no specific heat crosses the system boundary, i.e. the system is adiabatic, there is no transport of specific entropy across the system boundary as well. In addition to the amount of specific heat, the temperature T, inside the system at the location where the specific heat transfer takes place, is relevant.

2. The second mechanism for influencing the specific entropy of a system takes place *inside* the system, i.e.

$$\delta s_i = \frac{\delta \psi}{T}, \tag{13.4}$$

integration leads to:

$$\int_1^2 \delta s_i = \int_1^2 \frac{\delta \psi}{T} = s_{i,12}. \tag{13.5}$$

In contrast to δs_a, that can have any sign, δs_i can only be positive, or in best case become zero, due to $\psi \geq 0$. The best case would be a reversible change of state, i.e. $\psi = 0$. Mind, that T in Eq. 13.4 is the temperature in the system at the position where the dissipation occurs. In case the system is homogeneous, see Sect. 3.1.1, the temperature T for δs_a is the same as for δs_i. Homogeneous systems have a uniform temperature.

The integration of Eq. 13.1 reads as

$$\boxed{\int_1^2 ds = s_2 - s_1 = s_{a,12} + s_{i,12}} \tag{13.6}$$

However, according to Eq. 13.6 the change of entropy can be positive, negative or it can become zero:

- In case specific heat is supplied and specific dissipation occurs, the specific entropy inside the system rises.
- In case the system is adiabatic, the specific entropy can be constant or it can rise. However, it can not sink.

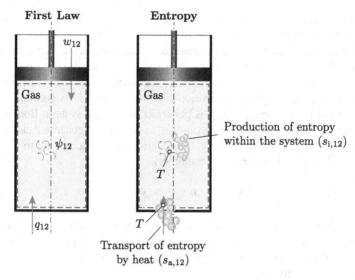

Fig. 13.1 Mechanism of entropy in a closed systems

- In case the system is cooled, the specific entropy can rise, it can decrease or it can be constant. It depends on ratio of released specific heat to specific dissipation.

The extensive entropy results in

$$S = ms \tag{13.7}$$

with

$$[S] = 1\frac{J}{K} \tag{13.8}$$

respectively

$$[s] = 1\frac{J}{K\,kg}. \tag{13.9}$$

13.2 Comparison Entropy Balance Versus First Law of Thermodynamics

In Fig. 13.2 the balance of entropy for a closed system is illustrated. Mind, that for thermodynamic balances extensive values should always be applied. In state (1) the system can be described by its state values, such as pressure p_1 and temperature T_1 for instance. Among these intensive state values there are additionally extensive state values, i.e. U_1 and H_1. Anyhow, as derived in Chap. 12, the extensive entropy $S_1 = ms_1$ is a state value as well. Thus, the system contains entropy S_1 in state (1), which is to be represented as bubbles in Fig. 13.2. However, due to external influences the system can be driven in a new equilibrium state (2), i.e. depending on the change of state some state values have changed. Hence, the amount of entropy might have changed as well. The mechanisms *how* entropy varies have been discussed already: Due to heat transfer Q_{12} entropy $S_{a,12}$ is carried across the system boundary. Additionally, there might be dissipation inside the system. Dissipation, however, is a source term for entropy, i.e. $S_{i,12}$ is generated inside the system. Both mechanisms, heat transfer from or to the environment, as well as the generated entropy, are quantified by *entropy bubbles*, see Fig. 13.2. Consequently, according to Fig. 13.2, an entropy balance is like counting of bubbles, i.e.

$$S_2 = S_1 + S_{a,12} + S_{i,12} \tag{13.10}$$

The entropy in state (2) is the initial amount of entropy in state (1) plus entropy crossing the system boundary and plus entropy, that has been generated internally. Mind, that both, $S_{a,12}$ and $S_{i,12}$, depend on the process, i.e. are process values. In other words the entropy balance obeys

$$\boxed{S_2 - S_1 = S_{a,12} + S_{i,12}} \tag{13.11}$$

However, the same result occurs when multiplying Eq. 13.6 with the mass of the system m. On the left hand side of Eq. 13.11 there is a *difference of state values*, that is equalised by *process values* on the right hand side. Entropy balancing is from now on denoted as second law of thermodynamics.[1]

However, the second law of thermodynamics Eq. 13.11 has the same structure as the first law of thermodynamics, that is illustrated in Fig. 13.3: In state (1) the system contains a state value extensive energy[2] E_1. Due to external influence of process values heat and work on the system, the amount of energy inside the system is varied. Consequently, in state (2) the system's energy is E_2. The first law of thermodynamics as derived in the previous chapters obeys

$$\boxed{E_2 - E_1 = W_{12} + Q_{12}} \tag{13.12}$$

Comparing Eqs. 13.11 and 13.12 shows the same structure: A difference of an extensive state value is balanced by process values. Even more: The difference of the internal energy in Eq. 13.12 can be calculated with a caloric equation of state, e.g. for an ideal gas:

$$U_2 - U_1 = m\,(u_2 - u_1) = mc_v\,(T_2 - T_1) \tag{13.13}$$

The same can be done with Eq. 13.11. In case the fluid in Fig. 13.2 is an ideal gas, the caloric equation of state reads as, see Sect. 12.3.2:

---

[1] Though, the actual second law of thermodynamics and the according principle will be introduced as late as in Chap. 15.

[2] This is internal energy as well as mechanical energy!

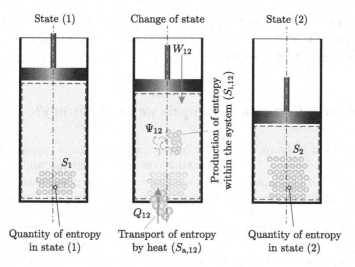

Fig. 13.2 Balance of entropy in a closed system

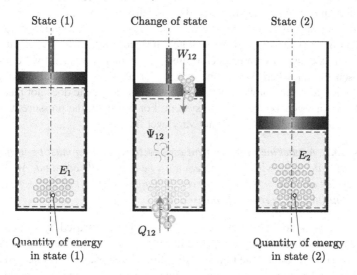

Fig. 13.3 Balance of energy in a closed system

$$S_2 - S_1 = m\,(s_2 - s_1) = m\left[c_v \ln\frac{T_2}{T_1} + R \ln\frac{v_2}{v_1}\right] = m\left[c_p \ln\frac{T_2}{T_1} - R \ln\frac{p_2}{p_1}\right]$$
(13.14)

The conclusion of this section is, that first as well as second law of thermodynamics have the same structure: The differences of energy respectively entropy can be calculated by a balance taking process values into account. Furthermore, the caloric equations of state can be applied to calculate the difference of internal energy as well as the difference of entropy from state (2) to state (1).

Theorem 13.1 *Entropy can be balanced, see Eq. 13.11. However, in contrast to energy, entropy is no conservation value, since entropy can be generated inside a system, i.e. a source term can occur!*

13.3   Energy Conversion—Why do we Need Entropy?

In this section it is shown, why the entropy as new state value is required. Thus, a so-called thermal engine, i.e. a machine that converts heat into mechanical energy, is investigated. An example for such a machine is a thermal power plant, that, in a first step, converts chemical bonded energy into thermal energy by firing the fuel. This thermal energy is, in a next step, converted into mechanical energy, for instance by a turbine. Finally, the turbine can run a generator that provides electrical energy. Machines like this play an important role in technical thermodynamics. Since not that much is known much about power plants so far, it is advantageous to apply a so-called *black-box notation* as given in Fig. 13.4. The power plant with its components is inside the system boundary. What can be noticed are the fluxes across the system boundary, without having any information regarding the thermodynamic cycle[3] inside. The power plant in this example shall be operated in steady state operation, i.e. the state of the system is constant with respect to time. With other words: If it were possible to place sensors anywhere within the system for all state values, they would all send a signal that is temporally constant. However, steady state does not mean, that the signal is homogeneous within the system. For example, the pressure would be different at each point, but it would not vary over time.

Theorem 13.2 *A machine in permanent and cyclic operation must be in a steady state, i.e. all state value within the system must be temporally constant. All thermodynamic cycles are characterised by several changes of state that run periodically, so that the initial state is reached consistently.*

Now, let us have a closer look at the power plant: Heat, released by a combustion process, enters the thermal engine on top of the black-box. At that location the process has the largest temperature due to the combustion. Anyhow, work is released by the thermal engine: From the management's point of view this amount of work should be maximised in order to make a profit. However, thermal power plants are usually built next to a river, since cooling is obviously required. Cooling takes place at the bottom side of the black box where temperature is rather small. Thus, the thermal engine operates between a so-called hot reservoir, i.e. the combustion chamber, and a cold reservoir which can be a river for instance. The question now is, how much work can be released in best case, i.e. the power plant should be free of any dissipation und thus be reversible. In order to answer this question, the first law of thermodynamics

---

[3]However, there need to be a cycle inside, since the power plant is supposed to be in around-the-clock operation.

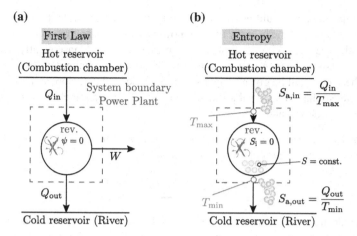

Fig. 13.4 Power plant as a block-box

for such a thermodynamic cycle is applied. The energy conservation principle for such a process has already been derived in Sect. 11.2.1, see Eq. 11.13:

Theorem 13.3 *The energy leading into the system needs to be balanced by the energy that is released:*

$$\sum E_{in} = \sum E_{out} \qquad (13.15)$$

If this is not the case, the energy inside might either increase by time causing a temperature rise, or the energy and thus the temperature inside decreases.[4] Hence, according to Fig. 13.4a the first law of thermodynamics obeys

$$Q_{in} = W + Q_{out}. \qquad (13.16)$$

When investigating thermodynamic cycles,[5] such as a power plant, the direction of the arrows pointing into respectively out of the system should be assumed as they are in reality,[6] i.e. in case heat is supplied to the system the arrow should point into the system, whereas it should point out in case the system is cooled. Solving the energy balance for the work, that is released by the power plant, leads to

$$W = Q_{in} - Q_{out}. \qquad (13.17)$$

The initial question was, how to maximise the work W that is released. Following Eq. 13.17 it is not wise to cool the machine, since cooling reduces the released work. From this point of view cooling should be avoided completely: In such a case the

---

[4]Similar to a bathtub, whose level of water is not be constant, if inflow and outflow differ.

[5]Thermodynamic cycles have already been discussed in Sects. 8.3 and 8.4.3.

[6]This is just a recommendation.

entirely supplied heat could be converted into work. Obviously, that does not work out, since cooling is done in every thermal power plant. This contradiction can not be solved with purely applying the first law of thermodynamics. Consequently, a second equation is required, since with Q_{out} and W there are two unknowns, whereas the thermal heat Q_{in} shall be given. Anyhow, the premise is, that the power plant runs in steady state, so that all state values inside need to be constant, including the entropy as newly introduced state value:

Theorem 13.4 *In steady state the entropy leading into the system needs to be balanced by the entropy that is leaving the system:*

$$\sum S_{in} = \sum S_{out} \qquad (13.18)$$

If that would not be the case, the entropy inside the system could not be constant by time. It would either rise or decrease. As a consequence, the state of the system could not be constant by time. This would contradict steady state operation. In our case, there is entropy being transferred[7] into the system by Q_{in} and entropy leaving the system with Q_{out}. In order to determine the entropies $S_{a,in}$ and $S_{a,out}$, the temperatures inside the system where the heat crosses the system boundary are required. As mentioned before, the machine is mounted in-between a hot and a cold reservoir. The heat Q_{in} enters the system at the hot side, where the temperature is T_{max}, Q_{out} leaves at the cold side, where T_{min} is measured, see Fig. 13.4. Since we focus on the best case, there is no dissipation and thus no generation of entropy S_i inside the system. Hence, the entropy balance in steady state obeys

$$S_{a,in} = \int \frac{\delta Q_{in}}{T_{max}} = \frac{Q_{in}}{T_{max}} = \frac{Q_{out}}{T_{min}} = \int \frac{\delta Q_{out}}{T_{min}} = S_{a,out}. \qquad (13.19)$$

Anyhow, this equation shows, that a cooling is required, in order to release entropy. If there was no cooling, the amount of entropy would rise permanently, since heat Q_{in} carries entropy into the system. In case there would be generation of entropy due to dissipation inside the machine as well, even more entropy would have to leave the machine, in order to keep the state value entropy inside constant. Thus, cooling must even be increased. Back to the best case, the entropy balance provides the required second equation. It indicates how much cooling is required for steady state operation, i.e.

$$Q_{out} = \frac{Q_{in}}{T_{max}} T_{min}. \qquad (13.20)$$

Now, having two equations, i.e. Eqs. 13.17 and 13.20, for two unknowns, i.e. W and Q_{out}, the cycle is fully described. However, the maximised work released by a thermal power plant in best case is

---

[7]Mind, that heat is a carrier for entropy!

$$W = Q_{in} - \frac{Q_{in}}{T_{max}} T_{min} = Q_{in} \left(1 - \frac{T_{min}}{T_{max}} \right). \tag{13.21}$$

Obviously, the larger the maximum temperature within the combustion chamber is, the more power can be gained out of the given thermal energy Q_{in}. The thermal efficiency of the thermal power plant η_{th} is defined by the ratio of benefit and effort, i.e.

$$\boxed{\eta_{th} = \frac{\text{Benefit}}{\text{Effort}} = \frac{W}{Q_{in}} = 1 - \frac{T_{min}}{T_{max}}} \tag{13.22}$$

This example shows, that the first law of thermodynamics can not state the limits of energy conversion. Only by applying an entropy balance, the thermodynamics of a thermal engine can be fully understood.

13.4 The T, s-Diagram

In this section the focus is on a new state diagram, that is required to analyse thermodynamic changes of state in its entirety. Some of the previous examples have already made the T, s-diagram a subject of discussion, though it is formally introduced in this chapter. It is shown, that with this new diagram it is possible to visualise all the process values learned so far: Volume/pressure work can be illustrated in a p, v-diagram, whereas heat and dissipation can be visualised in a T, s-diagram.

13.4.1 Benefit of a New State Diagram

At least for single component fluids without phase change, Gibb's phase rule has shown, that two independent state values fix a state, see Sect. 5.1. Since the specific entropy s is a state value, the state of a thermodynamic system can be given unambiguously by the system's temperature T and its specific entropy s. Hence, states as well as changes of state can be visualised in a T, s-diagram, that contains additional information compared to a p, v-diagram. In Fig. 13.5 the principle of a T, s-diagram is shown. States (1) and (2) represent thermodynamic states. These states are explicitly given and it is possible to derive all the other state values, e.g. pressure p, temperature T, specific volume v, specific enthalpy h and specific internal energy u, by equations of state. If the extensive state values are of interest, such as enthalpy H, additional information regarding the size of the system is required. Among the states of a system it is furthermore important to understand which path the system takes from one state to another. Figure 13.5 shows two possible directions the system might take. Let is now focus on path 1. Obviously, the grey coloured area is the integration of $\int T \, ds$ from state (1) to state (2). In the previous sections, the specific entropy s has been derived as

Fig. 13.5 T, s-diagram:
principle

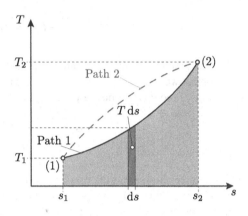

$$ds = \frac{\delta q}{T} + \frac{\delta \psi}{T}. \tag{13.23}$$

Multiplying with T:

$$T\,ds = \delta q + \delta \psi \tag{13.24}$$

Integration from state (1) to state (2) leads to the area beneath the change of state in a T, s-diagram

$$\int_1^2 T\,ds = \int_1^2 \delta q + \int_1^2 \delta \psi. \tag{13.25}$$

Obviously, the area beneath the change of state represents the sum of specific heat q_{12} and specific dissipation ψ_{12}, i.e.

$$\boxed{\int_1^2 T\,ds = q_{12} + \psi_{12}} \tag{13.26}$$

Since no restrictions regarding the type of fluid have been made, this conclusion counts for ideal gases and real fluids equally. However, the T, s-diagram proves, that specific heat as well as specific dissipation are *process* values, since they are both path-dependent. If path 2 would have been taken for instance, the amount of $q_{12} + \psi_{12}$ would differ from path 1 as Fig. 13.5 indicates. Nevertheless, state (2) can be reached as well.

Figure 13.6 shows the reversible and the irreversible case:

- The reversible case given in Fig. 13.6 (left) shows the heating of a fluid. The only mechanism for changing the entropy is due to heat transfer, since there is no entropy generation due to irreversibilites. Reversible changes of state are always free of

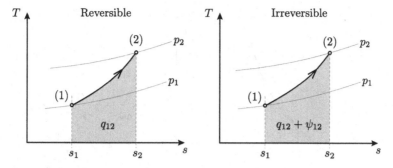

Fig. 13.6 T, s-diagram: reversible (left) and irreversible (right)

dissipation, see Theorem 7.13 and Example 14.1. Since the specific entropy rises, i.e. $s_2 > s_1$, entropy needs to be supplied to the system. This can only be achieved by supply of heat. Consequently, a reversible cooling of the system would lead to $s_2 < s_1$.

- Due to irreversibility, caused by dissipation, entropy inside the system is generated, so that the amount of entropy rises. Heat crossing the system boundary is a carrier for entropy as well. Depending on the sign of thermal energy, the system's entropy can rise or decrease. Regarding Fig. 13.6 (right) heating or cooling is possible. In case of cooling, the amount of released entropy needs to be smaller than the generated entropy, since $s_2 > s_1$.

13.4.2 Physical Laws in a T, s-Diagram for Ideal Gases

In order to illustrate changes of state it is required to understand how to navigate in a T, s-diagram. This is clarified in this section. To do so, the focus is on the isolines, such as isobar and isochor. Once the specific entropy has been derived in Sect. 12.2 a combination of first and second law of thermodynamics lead to Eq. 12.38, that is also known as fundamental equation of thermodynamics

$$T\,\mathrm{d}s = \mathrm{d}u + p\,\mathrm{d}v \tag{13.27}$$

respectively

$$T\,\mathrm{d}s = \mathrm{d}h - v\,\mathrm{d}p \tag{13.28}$$

These equations[8] can be applied to derive isobars as well as isochores in a T, s-diagram, see Fig. 13.7.

---

[8]The fundamental equations of thermodynamics apply for all fluids, so not just for ideal gases.

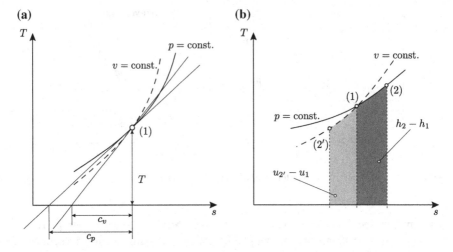

Fig. 13.7 T, s-diagram: isochore and isobar

Isobaric Change of State

For an isobaric change of state the pressure stays constant, so that

$$\mathrm{d}p = 0. \tag{13.29}$$

Equation 13.28 then reads as

$$T\,\mathrm{d}s = \mathrm{d}h. \tag{13.30}$$

For ideal gases the caloric equation of state, see Eq. 12.106, can be applied, i.e.

$$T\,\mathrm{d}s = c_p \mathrm{d}T. \tag{13.31}$$

Thus, the slope of the isobar in a T, s-diagram obeys

$$\boxed{\left(\frac{\mathrm{d}T}{\mathrm{d}s}\right)_p = \frac{T}{c_p} > 0} \tag{13.32}$$

Obviously, the slope is positive and rises with increasing temperature T, see Fig. 13.7a. In order to describe a change of state from (1) to (2), see Fig. 13.7b, Eq. 13.30 needs to be integrated, so that

$$\int_1^2 T\,\mathrm{d}s = h_2 - h_1 \tag{13.33}$$

With other words, for an isobaric change of state, the area beneath the curve from (1) to (2) represents the difference of the specific enthalpies $h_2 - h_1$, see Fig. 13.7b.

Isochoric Change of State

For an isochoric change of state the specific volume stays constant, so that

$$dv = 0. \tag{13.34}$$

Equation 13.27 then reads as

$$T \, ds = du. \tag{13.35}$$

For ideal gases the caloric equation of state, see Eq. 12.101, can be applied, i.e.

$$T \, ds = c_v dT. \tag{13.36}$$

Thus, the slope of the isochore in a T, s-diagram obeys

$$\boxed{\left(\frac{dT}{ds}\right)_v = \frac{T}{c_v} > 0} \tag{13.37}$$

Obviously, the slope is positive and rises with increasing temperature T, see Fig. 13.7a. Due to

$$c_p = c_v + R \tag{13.38}$$

c_p is larger than c_v, so that

$$\left(\frac{dT}{ds}\right)_v > \left(\frac{dT}{ds}\right)_p. \tag{13.39}$$

Theorem 13.5 *Thus, for ideal gases the isochore has a larger gradient than an isobar in a T, s-diagram, see Fig. 13.7a.*

The gradients $\frac{T}{c_p}$ respectively $\frac{T}{c_v}$ of an isobar and an isochore can even be constructed geometrically for state (1), see Fig. 13.7a. In order to describe a change of state from (1) to (2′), see Fig. 13.7b, Eq. 13.35 is integrated, so that

$$\int_1^{2'} T \, ds = u_{2'} - u_1. \tag{13.40}$$

With other words, for an isochoric change of state, the area beneath a curve from (1) to (2′) represents the difference of the specific internal energies $u_{2'} - u_1$, see Fig. 13.7b.

Fig. 13.8 Isobar and
isochore in a T, s-diagram
(ideal gas)

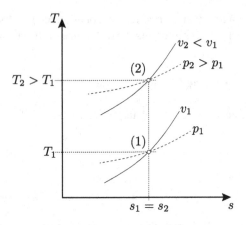

Figure 13.8 gives an overview of the isolines in a T, s-diagram. We already know, that the gradient of an isochore is larger than the gradient of an isobar. Furthermore, according to states (1) and (2), as given in Fig. 13.8, it is

$$s_2 - s_1 = 0 = c_p \ln \frac{T_2}{T_1} - R \ln \frac{p_2}{p_1} \tag{13.41}$$

respectively

$$c_p \ln \frac{T_2}{T_1} = R \ln \frac{p_2}{p_1}. \tag{13.42}$$

Consequently, the pressure in state (2) needs to be *larger* than in state (1). On the other hand it is

$$s_2 - s_1 = 0 = c_v \ln \frac{T_2}{T_1} + R \ln \frac{v_2}{v_1} \tag{13.43}$$

respectively

$$c_v \ln \frac{T_2}{T_1} = -R \ln \frac{v_2}{v_1} = R \ln \frac{v_1}{v_2}. \tag{13.44}$$

Consequently, the specific volume in state (2) needs to be *smaller* than in state (1). The conclusion is:

Theorem 13.6 *The higher the isobar in a T, s-diagram is positioned, the larger the pressure is. The higher the isochore in a T, s-diagram is located, the smaller the specific volume is.*

Fig. 13.9 Adiabatic,
reversible change of state

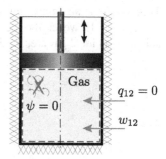

13.5 Adiabatic, Reversible Change of State

A new change of state is introduced at that point, see Fig. 13.9: The system shall be
adiabatic and free of any dissipation, i.e. reversible. Hence, it is

$$\delta q = 0 \Rightarrow \delta s_a = 0 \tag{13.45}$$

and

$$\delta \psi = 0 \Rightarrow \delta s_i = 0. \tag{13.46}$$

Following Eq. 12.39 it is

$$\boxed{ds = \frac{\delta q}{T} + \frac{\delta \psi}{T} = \delta s_a + \delta s_i = 0} \tag{13.47}$$

This equation shows, that for any adiabatic, reversible change of state the entropy
inside the system stays constant.[9] Such a change of state is named *isentropic*, i.e.
$ds = 0$. Under these conditions the fundamental equations, see Eqs. 13.27 and 13.28
simplify to

$$0 = du + p \, dv \tag{13.48}$$

and

$$0 = dh - v \, dp. \tag{13.49}$$

Rearranging and combining of these two equations results in

$$\boxed{\frac{du}{dh} = -\frac{p \, dv}{v \, dp}}. \tag{13.50}$$

---

[9]No entropy is carried across the system boundary by heat and there is no generation of entropy
inside the system! So, similar to a bathtub, the level of entropy inside the system must be constant.

This equation can be applied for any kind of fluid, since no restrictions have been made so far. However, the focus is now on ideal gases: The caloric equations for u and h for an ideal gas are, see Sect. 12.3:

$$du = c_v(v, T) \cdot dT + \left[T \left(\frac{\partial p}{\partial T} \right)_v - p \right] \cdot dv = c_v \, dT \tag{13.51}$$

and

$$dh = c_p(p, T) \cdot dT + \left[-T \left(\frac{\partial v}{\partial T} \right)_p + v \right] \cdot dp = c_p \, dT. \tag{13.52}$$

Hence, Eq. 13.50 simplifies for ideal gases with $\kappa = \frac{c_p}{c_v}$, i.e.

$$\frac{1}{\kappa} = -\frac{p \, dv}{v \, dp}. \tag{13.53}$$

Separating the variables leads to

$$-\frac{dp}{p} = \kappa \frac{dv}{v}. \tag{13.54}$$

An integration results in

$$- \ln p_2 + \ln p_1 = \kappa (\ln v_2 - \ln v_1) \tag{13.55}$$

respectively

$$\ln \frac{p_1}{p_2} = \kappa \ln \frac{v_2}{v_1}. \tag{13.56}$$

Rearranging reads as

$$\frac{p_1}{p_2} = \left(\frac{v_2}{v_1} \right)^\kappa \tag{13.57}$$

With other words, an adiabatic, reversible, i.e. isentropic, change of state for ideal gases obeys

$$\boxed{pv^\kappa = \text{const.}} \tag{13.58}$$

In a different notation it is

$$p_1 v_1^\kappa = p_2 v_2^\kappa. \tag{13.59}$$

Applying the thermal equation of state leads to

$$\boxed{\frac{T_2}{T_1} = \left(\frac{p_2}{p_1} \right)^{\frac{\kappa-1}{\kappa}}} \tag{13.60}$$

As defined before, an isentropic change of state is defined as $ds = 0$. However, there are two options to achieve an isentropic change of state:

- The change of state is adiabatic and reversible, e.g. according to Fig. 13.9:

$$ds = \underbrace{\frac{\delta q}{T}}_{=0} + \underbrace{\frac{\delta \psi}{T}}_{=0} = 0 \tag{13.61}$$

- In case dissipation occurs, i.e. the change of state is irreversible, the change of state can still be isentropic. According to Eq. 12.39 cooling is required:

$$ds = \underbrace{\frac{\delta q}{T}}_{<0} + \underbrace{\frac{\delta \psi}{T}}_{>0} = 0 \tag{13.62}$$

Hence, it follows

$$q_{12} = -\psi_{12}. \tag{13.63}$$

Under these conditions the released heat, i.e. cooling, needs to be equal to the dissipated energy. The entropy inside the system would then be constant.

Any heated system can never be isentropic, since under these conditions entropy inside the system would have to be destroyed. This is not possible, since $s_{i,12}$ can only be positive or zero.

Problem 13.7 An ideal gas is compressed reversibly in a horizontal cylinder. Two different options shall be investigated:

- $(1) \rightarrow (2)$: Isothermal compression
- $(1) \rightarrow (3)$: Adiabatic compression, so that $v_3 = v_2$

Sketch both changes of state in a p, v- and in a T, s-diagram!

Solution

Let us start with the first law of thermodynamics for the isothermal case:

$$Q_{12} + W_{12} = U_2 - U_1 = mc_v (T_2 - T_1) = 0. \tag{13.64}$$

Hence, it is

$$Q_{12} = -W_{12}. \tag{13.65}$$

The partial energy equation reads as

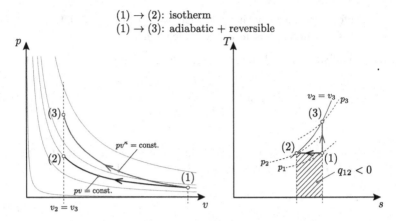

$(1) \rightarrow (2)$: isotherm
$(1) \rightarrow (3)$: adiabatic + reversible

Fig. 13.10 Reversible compression of an ideal gas: isothermal versus adiabatic according to Problem 13.8

$$W_{12} = W_{V,12} = - \int_1^2 p \, dV > 0. \qquad (13.66)$$

The work W_{12} is positive, since for a compression $dV < 0$. This leads to

$$Q_{12} = -W_{12} < 0. \qquad (13.67)$$

For an isothermal compression, heat needs to be released. In a p, v-diagram the change of state follows an isothermal, see Fig. 13.10. The volume decreases and accordingly, since the mass is constant in a closed system, the specific volume. Thus, the new state (2) lies left from state (1). The mathematical function $p = f(v)$ for an isothermal can be derived from the thermal equation of state

$$pv = RT = \text{const.} \qquad (13.68)$$

and reads as

$$p(v) = \frac{RT}{v} = \frac{p_1 v_1}{v} \propto \frac{1}{v} \qquad (13.69)$$

In a T, s-diagram an isothermal change of state needs to follow a horizontal line. Since the system must be cooled, entropy is released by the system, so that the specific entropy s_2 is smaller than the initial specific entropy s_1. Thus, state (2) is left from state (1) with the same temperature, see Fig. 13.10. A comparison of the isobars p_1 and p_2 in the T, s-diagram indicates that the position of state (2) makes sense, since the isobar p_2 lies above the isobar p_1. This correlates with the p, v-diagram as well. The specific released heat is represented by the hatched area beneath the curve (1) $\rightarrow$ (2) in the T, s-diagram.

For an adiabatic, reversible, i.e. isentropic, change of state $(1) \rightarrow (3)$ the first law of thermodynamics obeys

$$\underbrace{Q_{13}}_{=0} + W_{13} = U_3 - U_1 = mc_v \left(T_3 - T_1 \right). \tag{13.70}$$

The work follows the partial energy equation

$$W_{13} = W_{V,13} = -\int_1^3 p \, dV > 0. \tag{13.71}$$

Hence, the first law of thermodynamics states

$$W_{13} = U_3 - U_1 = mc_v \left(T_3 - T_1 \right) > 0. \tag{13.72}$$

Consequently, for an adiabatic, reversible change of state the temperature rises:

$$T_3 > T_1. \tag{13.73}$$

The function for an isentropic change of state has been recently derived and results in

$$pv^\kappa = \text{const.} \tag{13.74}$$

In other words, the function $p \, (v)$ is

$$p \, (v) = \frac{p_1 v_1^\kappa}{v^\kappa} \propto \frac{1}{v^\kappa} \tag{13.75}$$

Obviously, due to $\kappa > 1$ with decreasing specific volume, the pressure rises faster in case of an isentropic than in case of an isothermal change of state, see Fig. 13.10. This corresponds with the considerations regarding the first law of thermodynamics. Mind, that the thermal equation of state can be applied over the entire curve from $(1) \rightarrow (3)$, in case the change of state is quasi-static[10]:

$$pv = RT \neq \text{const.} \tag{13.76}$$

With Eq. 13.76 the system's temperature can be calculated at any point along the curve following Eq. 13.75. However, in a T, s-diagram state $(1) \rightarrow (3)$ follows a vertical line, since the entropy stays constant. Due to $T_3 > T_1$, state (3) is above state (1). This also correlates with the isobars p_1 and p_3 in the T, s-diagram.

---

[10] The prerequisite of a reversible change of state is that it is quasi-static as well!

Fig. 13.11 Polytropic
change of state for an ideal
gas in a p, v-diagram

$(1) \rightarrow (2)$: isotherm
$(1) \rightarrow (3)$: adiabatic + reversible
$(1) \rightarrow (4)$: polytropic

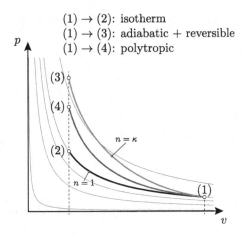

13.6 Polytropic Change of State

In Problem 13.8 two important changes of state, i.e. an isothermal and an isentropic change of state, have been compared. It has been shown, that an isentropic change of state can be realised with two alternatives: Any system with dissipation can only be isentropic in case heat is released. Additionally, an adiabatic system can be isentropic once it is free of any dissipation. Heated systems never can be isentropic. Figure 13.10 shows an isothermal as well as an isentropic change of state in a T, s-diagram and in a p, v-diagram accordingly. However, these two cases are special cases in technical applications: As everyday experience shows, perfectly insulated, i.e. adiabatic, systems do not exist, since heat transfer with an environment[11] can never be avoided completely.

Let us assume, that T_1 is identical with ambient temperature, see Fig. 13.11. In case $(1) \rightarrow (4)$ the system's thermal insulation is not perfect. However, the system's temperature rises during compression. Due to the thermal coupling with the environment, heat is released to the cooler environment. Consequently, the temperature in final state T_4 is lower than temperature T_3, that results from an adiabatic compression. In case the system is adiabatic but irreversible, the final temperature is even larger than T_3. This case is discussed as change of state $(1) \rightarrow (5)$ in Fig. 13.14. These changes of state are called polytropic.

As these examples show, varying temperatures can be achieved by the compression, it purely depends on the process control. In order to calculate the process values volume work respectively technical work, a mathematical function $p = f(v)$ is required. Two cases have already been discussed:

- Isothermal change of state (ideal gas)

$$pv = RT = \text{const.} \tag{13.77}$$

---

[11]In case the system has a different temperature than environment.

The explicit notation of the required function is

$$p(v) = \frac{\text{const.}}{v} \qquad (13.78)$$

Anyhow, in thermodynamics, usually an implicit notation is given, i.e.

$$\boxed{pv = \text{const.}} \qquad (13.79)$$

- Isentropic change of state (ideal gas)

This case has already been discussed in Sect. 13.5. Its implicit notation obeys

$$\boxed{pv^\kappa = \text{const.}} \qquad (13.80)$$

Isentropic changes of state are important in Chap. 21, in which flow processes, such as supersonic flows, will be investigated. In case the flow is very fast, there is almost no heat transfer to the environment, i.e. the flow is nearly adiabatic. Consequently, neglecting the dissipation, the flow is approximately isentropic.

- Polytropic change of state (ideal gas)

Obviously, a polytropic change of state (1) → (4) in a p, v-diagram has a different gradient than an isothermal/isentropic change of state, see Fig. 13.11. Comparing the correlations for an isothermal, see Eq. 13.79, and an isentrope, see Eq. 13.80, the gradient is obviously given by the exponent with respect to the specific volume v. Thus, a polytrope is defined by:

$$\boxed{pv^n = \text{const.}} \qquad (13.81)$$

Depending on the process n can be any number $-\infty < n < \infty$. However, some cases have a special technical meaning, see Table 13.1.

Table 13.1 Polytropic change of state—Technical meaning for ideal gases

Polytropic exponent n	Change of state
0	$pv^0 = \text{const.} \Rightarrow p = \text{const.}$
1	$pv^1 = \text{const.} \Rightarrow T = \text{const.}$
κ	$pv^\kappa = \text{const.} \Rightarrow$ isentrop
∞	$pv^\infty = \text{const.} \Rightarrow p^{1/\infty}v^1 = \text{const.} \Rightarrow v = \text{const.}$

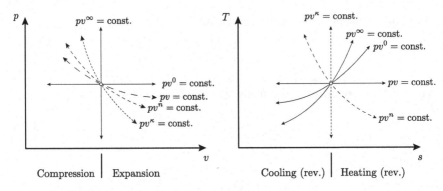

Fig. 13.12 Polytropic change of state for an ideal gas in a p, v- and in a T, s-diagram according to [1]

In Fig. 13.12 these cases are illustrated in a p, v- as well as in a T, s-diagram. Thus, e.g. for case (1) → (4), see Fig. 13.11, it is

$$1 < n < \kappa, \tag{13.82}$$

since the polytrope lies between the isothermal and the isentrope. However, additionally the thermal equation of state

$$pv = RT \tag{13.83}$$

can be applied for all cases, if the gas is ideal and in thermodynamic equilibrium. This leads to another notation of the polytropic change of state:

$$\boxed{\frac{T_1}{T_2} = \left[\frac{v_2}{v_1}\right]^{n-1} = \left[\frac{p_1}{p_2}\right]^{\frac{n-1}{n}}} \tag{13.84}$$

In technical processes it is assumed, that $n =$ const. for a change of state, see [8]. Depending on the technical problem, it makes sense, to calculate the polytropic exponent n in order to determine the process values work respectively technical work. If state (1) and state (2) are thermodynamical unambiguous, the exponent can be easily derived from Eq. 13.81, i.e.

$$n = \frac{\ln \frac{p_2}{p_1}}{\ln \frac{v_1}{v_2}}. \tag{13.85}$$

Once, the direction of the change of state is determined by the polytropic exponent n, the following process values can be calculated:

– Specific volume work (closed system)

$$w_{12,v} = -\int_1^2 p\,dv = -p_1 v_1^n \int_1^2 \frac{dv}{v^n} = \frac{p_1 v_1}{n-1}\left[\left(\frac{v_1}{v_2}\right)^{n-1} - 1\right] \quad (13.86)$$

– Specific pressure work (open system)

$$y_{12} = \int_1^2 v\,dp = v_1 p_1^{1/n}\int_1^2 p^{-1/n}\,dp = n\frac{p_1 v_1}{n-1}\left[\left(\frac{v_1}{v_2}\right)^{n-1} - 1\right] \quad (13.87)$$

• Polytropic change of state (real fluids)

Though real fluids will be treated as late as in part II, the definition of a polytropic change of state for real fluids follows according to Eq. 13.81, i.e.

$$\boxed{pv^n = \text{const.}} \quad (13.88)$$

However, for real fluids, the thermal equation of state $pv = RT$ is not fulfilled. Consequently, Eq. 13.84 can not be applied for real fluids. Furthermore, an isothermal change of state does not obey $n = 1$ and an isentropic change of state does not follow $n = \kappa$. Consequently, Table 13.1 only counts for ideal gases.

An overview of possible changes of state for ideal gases is given in Fig. 13.13.

Problem 13.8 An ideal gas is compressed in a horizontal cylinder. Several changes of state shall be investigated:

• (1) → (2): isothermal (reversible)
• (1) → (2'): isothermal (irreversible)
• (1) → (3): adiabatic (reversible)
• (1) → (4): polytropic (reversible), $1 < n < \kappa$
• (1) → (5): adiabatic (irreversible)

The final volume shall be the same in all cases. Please illustrate all changes of state in a p, v- and T, s-diagram. There shall be no friction for the piston. Irreversibility is due to dissipation within the gas.

Solution

The diagrams are shown in Fig. 13.14.

Problem 13.9 An ideal gas expands in a horizontal cylinder. Several changes of state shall be investigated:

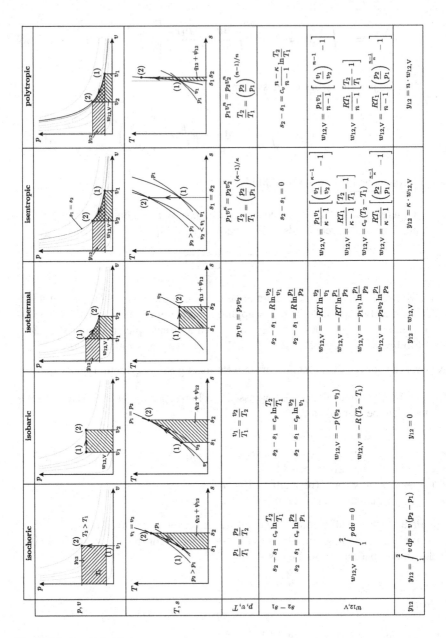

Fig. 13.13 Overview ideal gases

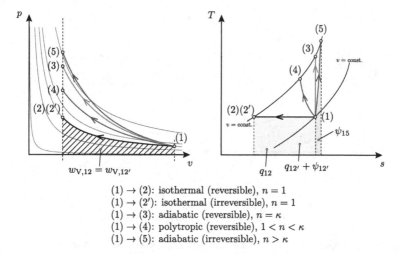

$(1) \rightarrow (2)$: isothermal (reversible), $n = 1$
$(1) \rightarrow (2')$: isothermal (irreversible), $n = 1$
$(1) \rightarrow (3)$: adiabatic (reversible), $n = \kappa$
$(1) \rightarrow (4)$: polytropic (reversible), $1 < n < \kappa$
$(1) \rightarrow (5)$: adiabatic (irreversible), $n > \kappa$

Fig. 13.14 Polytropic change of state for an ideal gas–Compression (closed system), see Problem 13.8

- $(1) \rightarrow (2)$: isothermal (reversible)
- $(1) \rightarrow (2')$: isothermal (irreversible)
- $(1) \rightarrow (3)$: adiabatic (reversible)
- $(1) \rightarrow (4)$: adiabatic (irreversible)

The final pressure shall be the same in all cases, since the system strives for mechanical equilibrium with the environment. Please illustrate all changes of state in a p, v- and T, s-diagram. There shall be no friction for the piston. Irreversibility is due to dissipation within the gas.

Solution

The diagrams are shown in Fig. 13.15. Anyhow, let us have a closer look at state (4), that is achieved by an adiabatic but irreversible change of state. Since, the gas inside the horizontal cylinder expands, the piston compresses the environment. Obviously, volume work is supplied to the environment. This work is supplied by the expanding gas, so that the work W_{14} needs to be negative.[12] Hence, the partial energy equation is

$$W_{14} = W_{14,\mathrm{V}} + \Psi_{14} = -\int_1^4 p \, \mathrm{d}V + \Psi_{14} < 0. \qquad (13.89)$$

---

[12] W_{14} is *identical* with the work supplied to the environment, since the piston moves horizontally, i.e. W_{14} is not utilised to lift a mass in a gravity field. Furthermore there is no friction between piston and cylinder wall.

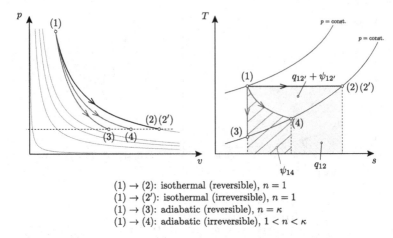

$(1) \rightarrow (2)$: isothermal (reversible), $n = 1$
$(1) \rightarrow (2')$: isothermal (irreversible), $n = 1$
$(1) \rightarrow (3)$: adiabatic (reversible), $n = \kappa$
$(1) \rightarrow (4)$: adiabatic (irreversible), $1 < n < \kappa$

Fig. 13.15 Polytropic change of state for an ideal gas–Expansion (closed system), see Problem 13.9

The first law of thermodynamics follows

$$\underbrace{W_{14}}_{<0} + \underbrace{Q_{14}}_{=0} = U_4 - U_1 = mc_v (T_4 - T_1) \tag{13.90}$$

so that $T_4 < T_1$. However, according to the T, s-diagram in Fig. 13.15, temperature T_4 must be larger than T_3, which is reached by an isentropic change of state. This is due to $p_3 = p_4$ and $s_4 > s_3$. The specific entropy rises because of the specific dissipation ψ_{14}.

Problem 13.10 A cooled compressor sucks in air out of the environment ($p_{env} = 1\,bar$, $T_{env} = 288\,K$) with a volume flow rate of $3500\frac{m^3}{h}$. While passing the compressor the air is compressed in a steady state process to a final pressure of 3.5 bar. As a consequence the temperature of the air in the outflow pipe ($d = 160\,mm$) reaches 393 K. During this process a heat flux of 49 kW is released.

(a) Calculate the power consumption of the compressor.
(b) Calculate the power consumption of a compressor, with the same ratio of pressure and the same mass flow rate of air, in case the compression is isentropic and frictionless.

The isentropic exponent is $\kappa = const. = 1.4$ and the air can be regarded as an ideal gas ($R = 287 \frac{J}{kg\,K}$).

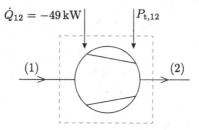

Fig. 13.16 Sketch of the compressor, see Problem 13.10

Solution

A compressor is a component, that is utilised for increasing the pressure of a gas. In order to do so, work needs to be supplied, see Fig. 13.16.

(a) In order to calculate the power consumption the first law of thermodynamics is applied. In accordance with the sketch it is

$$\dot{Q}_{12} + P_{t,12} + \dot{m}\left(h_1 + \frac{1}{2}c_1^2 + gz_1\right) = \dot{m}\left(h_2 + \frac{1}{2}c_2^2 + gz_2\right). \qquad (13.91)$$

The requested power consumption is

$$P_{t,12} = \dot{m}\left[h_2 - h_1 + \frac{1}{2}\left(c_2^2 - c_1^2\right) + g\left(z_2 - z_1\right)\right] - \dot{Q}_{12}. \qquad (13.92)$$

The mass flow rate can be calculated with the thermal equation of state for state (1), since the volume flow rate $\dot{V}_1$ is given:

$$m_1 R T_1 = p_1 V_1. \qquad (13.93)$$

The time derivative results in

$$\dot{m}_1 R T_1 = p_1 \dot{V}_1. \qquad (13.94)$$

Hence, the mass flow rate in steady state is

$$\dot{m} = \dot{m}_1 = \frac{p_1 \dot{V}_1}{R T_1} = 1.176 \, \frac{\text{kg}}{\text{s}}. \qquad (13.95)$$

Since state (1) is ambient state, its velocity[13] is $c_1 = 0$. The velocity c_2 can be calculated by applying the equation of continuity:

---

[13]However, even when the velocity is almost zero, according to $\dot{V}_1 = c_1 A_1$, there still can be a volume flow in case the cross section $A_1 \rightarrow \infty$, i.e. the environment is supposed to be huge.

$$\dot{m} = \rho_2 c_2 A_2 = \frac{c_2 A_2}{v_2} = \frac{c_2 \pi d_2^2}{4 v_2} = \frac{p_2 c_2 \pi d_2^2}{4 R T_2}. \tag{13.96}$$

The velocity c_2 is

$$c_2 = \frac{4 \dot{m} R T_2}{p_2 \pi d_2^2} = 18.84 \frac{m}{s} \tag{13.97}$$

The specific heat capacity is:

$$c_p = \frac{\kappa}{\kappa - 1} R = 1004.5 \frac{J}{kg\,K}. \tag{13.98}$$

Finally, the power consumption can be calculated by applying the caloric equation of state[14]:

$$\boxed{P_{t,12} = \dot{m} \left[c_p \left(T_2 - T_1 \right) + \frac{1}{2} \left(c_2^2 - c_1^2 \right) \right] - \dot{Q}_{12} = 173.2\,kW} \tag{13.99}$$

(b) The process shall be isentropic ($ds = 0$) and reversible ($\delta s_i = 0$). Thus, it follows:

$$ds = \delta s_i + \delta s_a \Rightarrow \delta s_a = ds - \delta s_i = 0. \tag{13.100}$$

Consequently, the change of state needs to be adiabatic, so that $\dot{Q}_{12'} = 0$. The first law of thermodynamics for the change of state $(1) \rightarrow (2')$ obeys

$$P_{t,12'} = \dot{m} \left[c_p \left(T_{2'} - T_1 \right) + \frac{1}{2} \left(c_{2'}^2 - c_1^2 \right) + g \underbrace{\left(z_{2'} - z_1 \right)}_{=0} \right]. \tag{13.101}$$

In order to determine the final temperature $T_{2'}$ for an isentropic change of state Eq. 13.60 can be applied. As the pressure ratio is the same as in (a), the pressure $p_{2'}$ is equal to p_2

$$T_{2'} = T_1 \left(\frac{p_2}{p_1} \right)^{\frac{\kappa-1}{\kappa}} = 412\,K. \tag{13.102}$$

Since the mass flow rate shall be the same as in (a), the velocity $c_{2'}$ is

$$c_{2'} = \frac{4 \dot{m} R T_{2'}}{p_2 \pi d_2^2} = 19.7 \frac{m}{s}. \tag{13.103}$$

Finally, the power consumption in case (b) is smaller than in case (a). It follows:

---

[14]Since there is no information regarding the vertical position of inlet and outlet the change of potential energy is ignored.

$$\boxed{P_{t,12'} = \dot{m}\left[c_p\left(T_{2'} - T_1\right) + \frac{1}{2}\left(c_{2'}^2 - c_1^2\right)\right] = 146.4\,\text{kW}} \tag{13.104}$$

13.7 Entropy Balancing

In order to evaluate thermodynamic processes it is essential not only to balance the energy, but to perform entropy balances as well. The entropy balance is the key to understand the direction energy conversion can take and its limitation. Consequently, the focus is on closed systems first, followed by open systems. Thermodynamic cycles, however, will be handled in Chap. 15 separately, as they are part of the classic formulation of the second law of thermodynamics.

13.7.1 Entropy Balance for Closed Systems

The analogy between energy balancing and entropy balancing has been worked out in Sect. 13.2. With Figs. 13.2 and 13.3, the principle of how to balance entropy in a closed system has been introduced in comparison with the first law of thermodynamics, see also Fig. 13.17. Nevertheless, in this section the major findings are summarised and motivated with a simple example.

It has been shown, that heat passing the system boundary, is a carrier for entropy, i.e.

$$\delta S_a = \frac{\delta Q}{T} \Rightarrow \int_1^2 \delta S_a = S_{a,12} = \int_1^2 \frac{\delta Q}{T}. \tag{13.105}$$

The generation of entropy within the system is due to dissipation and obeys

First law   **Second law**

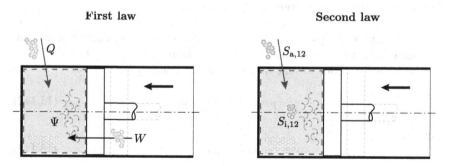

Fig. 13.17 Energy and entropy balance for a closed system

$$\delta S_{\mathrm{i}} = \frac{\delta \Psi}{T} \Rightarrow \int\limits_{1}^{2} \delta S_{\mathrm{i}} = S_{\mathrm{i},12} = \int\limits_{1}^{2} \frac{\delta \Psi}{T}. \tag{13.106}$$

In these equations T is the temperature within the system, where heat passes the boundary respectively where the dissipation occurs. In a homogeneous system, heat passing the boundary and dissipation are faced by the same uniform temperature. The change of the state value entropy of the system follows

$$\boxed{\mathrm{d}S = \delta S_{\mathrm{i}} + \delta S_{\mathrm{a}}} \tag{13.107}$$

The amount of entropy is influenced by the entropy crossing the system boundary, as well as by the entropy generated within the system. Thus, the change of entropy inside the system is achieved by integration:

$$\boxed{S_2 - S_1 = m(s_2 - s_1) = S_{\mathrm{i},12} + S_{\mathrm{a},12}} \tag{13.108}$$

Compare with the similar notation of the first law of thermodynamics[15]:

$$\boxed{U_2 - U_1 = m(u_2 - u_1) = Q_{12} + W_{12}} \tag{13.109}$$

For both equations, i.e. energy and entropy balance, the caloric equations of state can be applied. In case an ideal gas is regarded, the caloric equations are

$$s_2 - s_1 = c_v \cdot \ln \frac{T_2}{T_1} + R \cdot \ln \frac{v_2}{v_1} = c_p \cdot \ln \frac{T_2}{T_1} - R \cdot \ln \frac{p_2}{p_1} \tag{13.110}$$

respectively

$$u_2 - u_1 = c_v \cdot (T_2 - T_1). \tag{13.111}$$

Example—Isothermal Irreversible Compression

With this example the question is answered, by what amount the entropy of an ideal gas changes due to an isothermal, irreversible compression, see Fig. 13.18.
   The thermal equation of state for an isothermal change of state obeys

$$pv = RT = \mathrm{const.} \tag{13.112}$$

Thus, it is

$$p_1 v_1 = p_2 v_2. \tag{13.113}$$

---

[15]In case the change of the outer energies is ignored, e.g. due to a horizontal cylinder being in rest in state (1) and state (2).

Fig. 13.18 Isothermal irreversible compression of an ideal gas

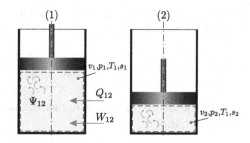

Since for a compression it is $v_2 < v_1$, it follows

$$p_2 > p_1. \tag{13.114}$$

The first law of thermodynamics for that case is

$$W_{12} + Q_{12} = U_2 - U_1 + \Delta E_{a,12} \tag{13.115}$$

- The partial energy equation reads as

$$W_{12} = W_{V,12} + \underbrace{W_{\text{mech},12}}_{=\Delta E_{a,12}} + \Psi_{12} \tag{13.116}$$

Mechanical work $W_{\text{mech},12}$ is required to change the outer energy by $\Delta E_{a,12}$.

- The caloric equation is

$$U_2 - U_1 = mc_v \left(T_2 - T_1 \right) = 0 \tag{13.117}$$

Hence, the first law of thermodynamics follows

$$\Psi_{12} + Q_{12} = -W_{V,12} \tag{13.118}$$

The volume work $W_{V,12}$ can be calculated according to Fig. 13.13, so that

$$\Psi_{12} + Q_{12} = -W_{V,12} = -mRT \ln\frac{p_2}{p_1}. \tag{13.119}$$

Now, let us focus on the change of entropy for that change of state. It follows

$$\underbrace{S_2 - S_1}_{\text{Alternative 1}} = \underbrace{S_{i,12} + S_{a,12}}_{\text{Alternative 2}} \tag{13.120}$$

There are two alternatives, to calculate the change of entropy:

- Alternative 1 follows the caloric equation of state, i.e. it investigates its *states* (1) and (2)

$$S_2 - S_1 = m \, (s_2 - s_1) = m \left[c_p \cdot \ln \frac{T_2}{T_1} - R \cdot \ln \frac{p_2}{p_1} \right] \qquad (13.121)$$

Since the change of state is isothermal, temperature T_2 and temperature T_2 are equal. Thus, it is

$$\boxed{S_2 - S_1 = -mR \cdot \ln \frac{p_2}{p_1}} \qquad (13.122)$$

- Alternative 2 investigates the *cause* for the change of entropy:

$$S_2 - S_1 = S_{i,12} + S_{a,12} = \int_1^2 \frac{\delta \Psi}{T} + \int_1^2 \frac{\delta Q}{T} \qquad (13.123)$$

Since the temperature T is constant, the equation simplifies to

$$S_2 - S_1 = S_{i,12} + S_{a,12} = \frac{1}{T} \int_1^2 \delta \Psi + \frac{1}{T} \int_1^2 \delta Q \qquad (13.124)$$

respectively

$$S_2 - S_1 = S_{i,12} + S_{a,12} = \frac{\Psi_{12}}{T} + \frac{Q_{12}}{T} \qquad (13.125)$$

Combining with Eq. 13.118 leads to

$$\boxed{S_2 - S_1 = S_{i,12} + S_{a,12} = -mR \cdot \ln \frac{p_2}{p_1}} \qquad (13.126)$$

However, both alternatives lead to the same result:

$$\boxed{S_2 - S_1 = -mR \ln \frac{p_2}{p_1} < 0} \qquad (13.127)$$

The change of state is illustrated as change of state (1) → (2') in Fig. 13.14.

13.7.2 Entropy Balance for Open Systems

The energy conservation principle for open systems has already been derived in Sect. 11.3.1—for unsteady as well as for steady state conditions. With the energy

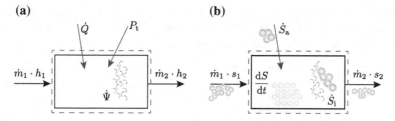

Fig. 13.19 Energy and entropy balance—open system

fluxes passing the system boundary it is possible to balance the amount of energy within the system.

This principle is illustrated in Fig. 13.19a. However, in this sketch the kinetic and the potential energies at the inlet and the outlet are ignored, solely the enthalpies are represented. Anyhow, energy fluxes leading into the system increase the amount of energy inside the system, whereas energy fluxes leaving the system decrease its energy content. Actually, this principle[16] can be applied for any extensive state value Z and has been discussed recently in Sect. 3.2.5.

Thus, the entropy follows accordingly, see Fig. 13.19b: Any entropy flux that crosses the system boundary into the system causes entropy within the system to increase over time. Entropy fluxes that leave the system make the entropy inside decrease over time. Furthermore, in contrast to the energy balance there can additionally be a *source term* for entropy, i.e. an entropy generation rate $\dot{S}_i$ due to a flux of dissipated energy. This source term also causes the amount of entropy in the system to increase over time. Dissipation Ψ leads to generation of entropy S_i; a flux of dissipation $\dot{\Psi}$ in open systems causes a flux of entropy generation $\dot{S}_i$, see Sect. 12.2, i.e.

$$\Psi \to S_i \text{ respect. } \dot{\Psi} \to \dot{S}_i. \tag{13.128}$$

Let us have a closer look at the entropy fluxes crossing the system boundary. Heat as a carrier for entropy causes entropy S_a stream into the system in case the system is heated, or out of the system, in case it is cooled. In case it is a flux of heat, e.g. in an open system that is operated permanently, a flux of entropy $\dot{S}_a$ is carried, see Sect. 12.2, i.e.

$$Q \to S_a \text{ respect. } \dot{Q} \to \dot{S}_a. \tag{13.129}$$

In contrast to closed systems, entropy is also carried *convectively* into the system by the mass flow at the inlet $\dot{S}_1$, respectively out of the system at the outlet $\dot{S}_2$. The mass flows $\dot{m}_1$ and $\dot{m}_2$ possess a thermodynamic state, e.g. measured by pressure p and temperature T. As already discussed, two independent state values fix the state unequivocally.[17] Any other state value for inlet and outlet can then be determined, e.g. by thermal or caloric equations of state. In addition to $\dot{S}_i$ and $\dot{S}_a$ the system's

---

[16]Bathtub-principle!

[17]At least for single component fluids without phase change, see Sect. 5.1.

entropy is increased by entropy with the inflow $\dot{S}_1 = \dot{m}_1 s_1$ and decreased by entropy with the outflow $\dot{S}_2 = \dot{m}_2 s_2$. Hence, according to Sect. 11.3.1 the conclusion for the temporal change of entropy within the system, see Fig. 13.19b, is[18]

$$\frac{\mathrm{d}S}{\mathrm{d}t} = \dot{m}_1 s_1 + \dot{S}_a + \dot{S}_i - \dot{m}_2 s_2 \qquad (13.130)$$

The term on the left hand side of Eq. 13.130 counts for the temporal change of entropy within the system. This change of the system's entropy is caused by the fluxes respectively the source term, that are summarised on the right hand side of Eq. 13.130.

Steady State Flow Systems

Let us assume the system given in Fig. 13.19 is in steady state. In this case the equation of continuity obeys

$$\dot{m}_1 = \dot{m}_2. \qquad (13.131)$$

With other words, the mass flux into the system needs to be equal to the mass flux leaving the system. Regarding the entropy balance Eq. 13.130, for steady state conditions $\frac{\mathrm{d}S}{\mathrm{d}t}$ needs to be zero, i.e.

$$0 = \dot{m}_1 s_1 + \dot{S}_a + \dot{S}_i - \dot{m}_2 s_2. \qquad (13.132)$$

This is due to the definition of steady state, that means, that all state values inside the system need to be *temporally constant*, though they might differ locally. However, that does not only affect the state values pressure p, temperature T but also the state value entropy S. Consequently, the entropy balance can be written in the following notation

$$\dot{m}_1 s_1 + \dot{S}_a + \dot{S}_i = \dot{m}_2 s_2. \qquad (13.133)$$

In other words:

$$\boxed{\text{Entropy flux in} = \text{Entropy flux out}}$$

Mind, that the source term $\dot{S}_i$ is treated as a flow into the system, since it increases the system's entropy.

Multiple Inlets/Outlets

Of course, open system are not limited to one outlet respectively inlet. In many technical applications mass flows are supplied at several inlets, or a mass flow is

---

[18]Bathtub-principle!

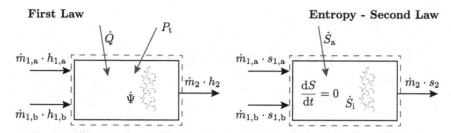

Fig. 13.20 Entropy balance—steady state flow system, multiple inlets/outlets

extracted, so that multiple inlets/outlets shall be covered as well. The general entropy balance for system like this follows, according to our considerations in Sect. 13.7.2:

$$\boxed{\frac{dS}{dt} = \sum_{in,i} \dot{m}_i s_i + \dot{S}_a + \dot{S}_i - \sum_{out,j} \dot{m}_j s_j}$$ (13.134)

In steady state this equation simplifies to

$$0 = \sum_{in,i} \dot{m}_i s_i + \dot{S}_a + \dot{S}_i - \sum_{out,j} \dot{m}_j s_j$$ (13.135)

respectively

$$\sum_{in,i} \dot{m}_i s_i + \dot{S}_a + \dot{S}_i = \sum_{out,j} \dot{m}_j s_j.$$ (13.136)

As before, the conclusion of multiple inlet/outlet systems in steady state is:

> Entropy flux in = Entropy flux out

Example 13.11 Let us apply this finding to the example of an open system in steady state given in Fig. 13.20.

- First, it is wise to start with the equation of continuity:

> Mass flux in = Mass flux out

In this case:

$$\dot{m}_{1,a} + \dot{m}_{1,b} = \dot{m}_2$$ (13.137)

- However, the next step is the first law of thermodynamics:

> Energy flux in = Energy flux out

Applied to Fig. 13.20:

$$\dot{Q} + P_t + \dot{m}_{1,a}\left(h_{1,a} + \frac{1}{2}c_{1,a}^2 + gz_{1,a}\right) + \dot{m}_{1,b}\left(h_{1,b} + \frac{1}{2}c_{1,b}^2 + gz_{1,b}\right) =$$
$$\dot{m}_2\left(h_2 + \frac{1}{2}c_2^2 + gz_2\right)$$

(13.138)

Let us ignore the kinetic and potential energy. Rearranging and combining with the equation of continuity, i.e. Eq. 13.137, results in

$$\dot{Q} + P_t = \dot{m}_{1,a}\left(h_2 - h_{1,a}\right) + \dot{m}_{1,b}\left(h_2 - h_{1,b}\right).$$

(13.139)

This notation is advantageous, since for the differences of the specific enthalpies the caloric equation of state can be applied.[19]

• Finally, the entropy is balanced, i.e. the <u>second law of thermodynamics</u> is applied:

Entropy flux in = Entropy flux out

In this example it is

$$\dot{S}_a + \dot{S}_i + \dot{m}_{1,a}s_{1,a} + \dot{m}_{1,b}s_{1,b} = \dot{m}_2 s_2.$$

(13.140)

Applying Eq. 13.137 and rearranging in order to get differences for the specific entropy s leads to

$$\dot{S}_a + \dot{S}_i = \dot{m}_{1,a}\left(s_2 - s_{1,a}\right) + \dot{m}_{1,b}\left(s_2 - s_{1,b}\right).$$

(13.141)

By doing so, the caloric equations of state, in case it is an ideal gas, see e.g. Eq. 12.49, can be applied, i.e.

$$s_2 - s_{1,a} = c_p \ln\frac{T_2}{T_{1,a}} - R\ln\frac{p_2}{p_{1,a}}$$

(13.142)

and

$$s_2 - s_{1,b} = c_p \ln\frac{T_2}{T_{1,b}} - R\ln\frac{p_2}{p_{1,b}}.$$

(13.143)

13.7.3   Thermodynamic Mean Temperature

It has been shown that heat crossing the system boundary leads to a transfer of entropy. In case a heat flux occurs, a flux of entropy passes the system boundary:

---

[19]Remember, that the caloric equations of state always indicate the *change* of a caloric state value!

$$S_{a,12} = \frac{Q_{12}}{T} \quad \text{respect.} \quad \dot{S}_a = \frac{\dot{Q}}{T} \tag{13.144}$$

Additionally, entropy is generated within the system, if energy dissipates. In case a flux of dissipated energy is present, a flux of entropy is released, i.e.

$$S_{i,12} = \frac{\Psi_{12}}{T} \quad \text{respect.} \quad \dot{S}_i = \frac{\dot{\Psi}}{T}. \tag{13.145}$$

In this section these two mechanisms are further investigated. It has been mentioned, that temperature T is the temperature within the system, where the heat crosses the system boundary respectively where the dissipation occurs. Obviously, T does not need to be locally constant. Consequently, temperature, heat and dissipation can depend on place and time. Accordingly, entropy carried by heat and entropy being generated inside a system can be functions of place and time. In order to clarify how to handle this phenomenon, the focus is on steady state open systems first and then on closed systems.

Steady State Flow Systems

In open systems the determination of $\dot{S}_a$ and $\dot{S}_i$ sometimes causes difficulties, since the temperature inside the system is non-uniform, e.g. the inlet temperature T_1 can be different from the outlet temperature T_2, see Fig. 13.21. However, the system is in local thermodynamic equilibrium, see Sect. 4.4. Since the system is supposed to be in steady state, the temperature profile is constant with time. Accordingly, if dissipation occurs inside the system, this is at a non-uniform, but temporally constant temperature.

In order to handle this problem the system is therefore discretised spatially, see Fig. 13.21. Transferred heat and dissipation now face a varying temperature T_i from element to element. Anyhow, the entire heat being transferred is

$$\dot{Q}_{12} = \int_1^2 \delta \dot{Q}_i \tag{13.146}$$

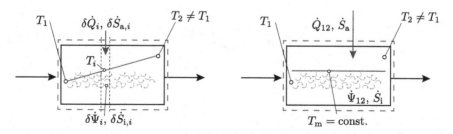

Fig. 13.21 Thermodynamic mean temperature—steady state flow system

and the entire dissipation is

$$\dot{\Psi}_{12} = \int_1^2 \delta \dot{\Psi}_i. \qquad (13.147)$$

Entropy is transferred differentially into[20] each element due to heat transfer:

$$\delta \dot{S}_a = \frac{\delta \dot{Q}_i}{T_i} \qquad (13.148)$$

The entire entropy being carried with the heat is

$$\dot{S}_a = \int_1^2 \delta \dot{S}_a = \int_1^2 \frac{\delta \dot{Q}_i}{T_i} \qquad (13.149)$$

In order to solve this integral a thermodynamic mean temperature T_m is defined, that is supposed to be constant within the entire system, see Fig. 13.21:

$$\dot{S}_a = \int_1^2 \frac{\delta \dot{Q}_i}{T_i} \equiv \frac{1}{T_m} \int_1^2 \delta \dot{Q}_i = \frac{\dot{Q}_{12}}{T_m} \qquad (13.150)$$

This representative temperature is

$$\boxed{T_m = \frac{\dot{Q}_{12}}{\dot{S}_a}} \qquad (13.151)$$

The dissipation can be treated accordingly, i.e.

$$\dot{S}_i = \int_1^2 \delta \dot{S}_i = \int_1^2 \frac{\delta \dot{\Psi}_i}{T_i} \qquad (13.152)$$

With the already introduced thermodynamic mean temperature T_m it results in

$$\dot{S}_i = \int_1^2 \frac{\delta \dot{\Psi}_i}{T_i} \equiv \frac{1}{T_m} \int_1^2 \delta \dot{\Psi}_i = \frac{\dot{\Psi}_{12}}{T_m}. \qquad (13.153)$$

---

[20] Or out of the element, in case the system releases heat!

This equation can be solved for T_m as well

$$\boxed{T_m = \frac{\dot{\Psi}_{12}}{\dot{S}_i}}$$
(13.154)

Combining Eqs. 13.151 and 13.154 gives

$$T_m \left(\dot{S}_a + \dot{S}_i \right) = \dot{\Psi}_{12} + \dot{Q}_{12}$$
(13.155)

Hence, the thermodynamic mean temperature also obeys

$$\boxed{T_m = \frac{\dot{\Psi}_{12} + \dot{Q}_{12}}{\dot{S}_a + \dot{S}_i}}$$
(13.156)

According to Eq. 13.136 for an open system with multiple inlets/outlets in steady state it is

$$\dot{S}_a + \dot{S}_i = \sum_{out,j} \dot{m}_j s_j - \sum_{in,i} \dot{m}_i s_i.$$
(13.157)

Thus, in such a case the thermodynamic mean temperature follows

$$\boxed{T_m = \frac{\dot{\Psi}_{12} + \dot{Q}_{12}}{\sum_{out,j} \dot{m}_j s_j - \sum_{in,i} \dot{m}_i}}$$
(13.158)

This equation can be simplified for systems with a single inlet and single outlet, i.e.

$$\boxed{T_m = \frac{\psi_{12} + q_{12}}{s_2 - s_1}}$$
(13.159)

Closed Systems

The change of state for a closed system $(1) \rightarrow (2)$ can be initiated by process values work and heat. On its way the system's temperature can vary from T_1 to T_2 while heat passes the system boundary. Thus, the heat faces a temporal changing temperature. Obviously, this varying temperature has an impact on the entropy that passes the system boundary. In order to calculate the entropy carried with the heat, the change of state is discretised into sub-steps, see Fig. 13.22.

The entire heat follows

$$Q_{12} = \int_1^2 \delta Q_i$$
(13.160)

Fig. 13.22 Thermodynamic mean temperature—closed system

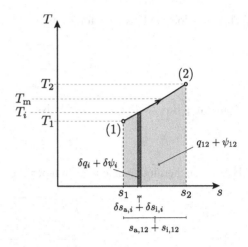

However, the entropy passing in each sub-step is

$$\delta S_{a,i} = \frac{\delta Q_i}{T_i} \tag{13.161}$$

In order to calculate the entire entropy, this equation needs to be integrated over all sub-steps

$$S_{a,12} = \int\limits_1^2 \delta S_{a,i} = \int\limits_1^2 \frac{\delta Q_i}{T_i} \tag{13.162}$$

With introducing a representative, constant thermodynamic mean temperature T_m it results in

$$S_{a,12} = \int\limits_1^2 \frac{\delta Q_i}{T_i} \equiv \frac{1}{T_m} \int\limits_1^2 \delta Q_i = \frac{Q_{12}}{T_m} \tag{13.163}$$

Hence, the thermodynamic mean temperature is

$$\boxed{T_m = \frac{Q_{12}}{S_{a,12}}} \tag{13.164}$$

Furthermore, during the change of state, dissipation[21] can occur accordingly at varying temperature. Hence, it affects the generated entropy. Analogous to the heat it is

$$\Psi_{12} = \int\limits_1^2 \delta \Psi_i \tag{13.165}$$

---

[21]Remember, that dissipation is a process value as well!

However, the entropy generated in each sub-step is

$$\delta S_{i,i} = \frac{\delta \Psi_i}{T_i} \tag{13.166}$$

In order to calculate the entire entropy this equation needs to be integrated over all sub-steps

$$S_{i,12} = \int_1^2 \delta S_{i,i} = \int_1^2 \frac{\delta \Psi_i}{T_i} \tag{13.167}$$

With introducing a representative, constant thermodynamic mean temperature T_m it results in

$$S_{i,12} = \int_1^2 \frac{\delta \Psi_i}{T_i} \equiv \frac{1}{T_m} \int_1^2 \delta \Psi_i = \frac{\Psi_{12}}{T_m} \tag{13.168}$$

Hence, the thermodynamic mean temperature is

$$\boxed{T_m = \frac{\Psi_{12}}{S_{i,12}}} \tag{13.169}$$

Combining Eqs. 13.164 and 13.169 leads to

$$T_m \left(S_{a,12} + S_{i,12} \right) = \Psi_{12} + Q_{12} \tag{13.170}$$

So that the thermodynamic mean temperature also obeys

$$\boxed{T_m = \frac{\Psi_{12} + Q_{12}}{S_{a,12} + S_{i,12}}} \tag{13.171}$$

Due to Eq. 13.108 it follows accordingly

$$\boxed{T_m = \frac{Q_{12} + \Psi_{12}}{S_2 - S_1} = \frac{q_{12} + \psi_{12}}{s_2 - s_1}} \tag{13.172}$$

13.7.4 Entropy and Process Evaluation

In this section the evaluation of thermodynamic processes is clarified. The previous sections have shown, that generation of entropy and entropy transport due to heat exchange can be characterised as follows:

- Process values S_a respect. $\dot{S}_a$

  Same sign as the heat that passes the system boundary:

  - $\delta S_a = 0 \Rightarrow S_{a,12} = 0$ respect. $\dot{S}_a = 0$ (adiabatic process)
  - $\delta S_a > 0 \Rightarrow S_{a,12} > 0$ respect. $\dot{S}_a > 0$ (system is heated)
  - $\delta S_a < 0 \Rightarrow S_{a,12} < 0$ respect. $\dot{S}_a < 0$ (system is cooled)

- Process value S_i respect. $\dot{S}_i$

  Quantitative measure for the level of irreversibility:

  - $\delta S_i = 0 \Rightarrow S_{i,12} = 0$ respect. $\dot{S}_i = 0$ (reversible process)
  - $\delta S_i > 0 \Rightarrow S_{i,12} > 0$ respect. $\dot{S}_i > 0$ (irreversible process)
  - $\delta S_i < 0 \Rightarrow S_{i,12} < 0$ respect. $\dot{S}_i < 0$ (impossible process)

Theorem 13.12 *Thus, from now on the generation of entropy is utilised for a process evaluation.*

Example 13.13 Let us have a closer look at the crate-of-beer-problem given in Fig. 13.23. We have learned so far, that it requires mechanical work to lift the beer from position z_1 to position z_2 in a gravity field. Everyday experience proves, that this change of state is rather exhausting.

An inventor claims to have a special device that adds heat to the beer without changing its internal state, i.e. temperature and pressure remain constant. His idea is, that the entire thermal energy is converted into potential energy. The task is to investigate if such a device can exist. According to the first law of thermodynamics for closed systems

$$\underbrace{W_{12}}_{=0} + Q_{12} = U_2 - U_1 + \underbrace{\frac{1}{2}m\left(c_2^2 - c_1^2\right)}_{=0} + mg\left(z_2 - z_1\right) \qquad (13.173)$$

in combination with the caloric equation of state

$$Q_{12} = \underbrace{mc\left(T_2 - T_1\right)}_{=0} + mg\left(z_2 - z_1\right) \qquad (13.174)$$

Fig. 13.23 Lifting a crate of beer

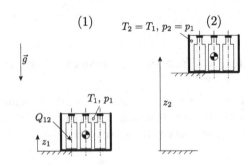

it is

$$Q_{12} = mg(z_2 - z_1). \tag{13.175}$$

Thus, according to the first law of thermodynamics it would be possible to convert the entire thermal energy into potential energy. Nevertheless, let us state the partial energy equation as well:

$$W_{12} = 0 = \underbrace{W_{V,12}}_{=0} + \underbrace{W_{mech,12}}_{mg(z_2-z_1)>0} + \Psi_{12} \tag{13.176}$$

Solving for the dissipation results in

$$\Psi_{12} = -mg(z_2 - z_1) < 0. \tag{13.177}$$

This is not possible, since dissipation can never be negative, see Sect. 9.2.4. However, let us finally analyse the entropy balance, i.e. the second law of thermodynamics. For a closed system it obeys

$$S_2 - S_1 = S_{a,12} + S_{i,12} \tag{13.178}$$

Since the crate of beer is supposed to be incompressible, the caloric equation of state for the entropy reads as:

$$S_2 - S_1 = mc \ln\frac{T_2}{T_1} = 0 \tag{13.179}$$

Hence, the generation of entropy is

$$S_{i,12} = -S_{a,12} = -\frac{Q_{12}}{T_1} < 0. \tag{13.180}$$

The entropy carried by heat can easily be calculated, since the temperature shall remain constant. As thermal energy is supplied, Q_{12} is positive. Due to

$$S_{i,12} < 0 \tag{13.181}$$

the process is impossible, see Sect. 13.7.4.

Problem 13.14 Two equal, vertical standing pipes with a diameter of $D = 0.1$ m are connected with each other by a thin pipe and a valve, see Problem 12.8. One vertical pipe is filled up to a height of $H = 10$ m with water. Water has a density of $\rho = 1000 \frac{kg}{m^3}$. The second vertical pipe is empty. Now the connecting valve is opened. After a while a balance of the water volume develops. Water is supposed to be an incompressible fluid with $c = 4.18 \frac{kJ}{kg\,K}$. Is the process reversible? Calculate the generated entropy. The temperature in state (1) is $T_1 = 293.15$ K. Calculate the thermodynamic mean temperature for the change of state (1) $\rightarrow$ (2). Mind, that no energy is exchanged with the environment.

Solution

The temperature rise ΔT has previously been calculated in Problem 12.8 by applying the first law of thermodynamics for closed systems and by applying the caloric equation of state. Hence, the temperature T_2 of the new equilibrium state can be calculated as

$$T_2 = \Delta T + T_1 = 293.1558 \text{ K}. \tag{13.182}$$

According to the entropy balance for closed systems it is

$$S_2 - S_1 = m (s_2 - s_1) = S_{i,12} + S_{i,12}. \tag{13.183}$$

Since no energy is exchanged with the environment, it follows that

$$S_{a,12} = 0. \tag{13.184}$$

Thus, the generation of entropy follows:

$$S_{i,12} = m (s_2 - s_1) \tag{13.185}$$

States (1) and (2) are given unequivocally. The caloric equation of state for the specific entropy can be applied. For an incompressible liquid it reads as

$$S_{i,12} = m (s_2 - s_1) = mc \ln \frac{T_2}{T_1}. \tag{13.186}$$

Finally, the generation of entropy results in

$$S_{i,12} = 6.5706 \, \frac{\text{J}}{\text{K}} > 0. \tag{13.187}$$

Since the generation of entropy is positive, the process is possible but irreversible. Thus, once state (2) is reached, the system can not move back into state (1) without external influences from the environment.

Equation 13.169 can be applied to calculate the thermodynamic mean temperature, i.e.

$$T_{\text{m}} = \frac{\Psi_{12}}{S_{i,12}} = \frac{mgH}{4mc \ln\frac{T_2}{T_1}} = \frac{gH}{4c \ln\frac{T_2}{T_1}} = 293.1529 \text{ K}. \tag{13.188}$$

Problem 13.15 An ideal gas flows isobarically through an adiabatic horizontal channel. In doing so its velocity decreases from $c_1 = 100 \, \frac{\text{m}}{\text{s}}$ to $c_2 = 20 \, \frac{\text{m}}{\text{s}}$.

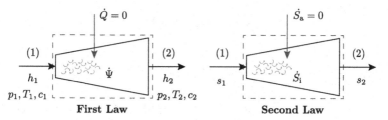

Fig. 13.24 Sketch to Problem 13.15

(a) Calculate the specific dissipated energy ψ_{12}.
(b) Does the temperature of the fluid increase, decrease or does it stay constant?
(c) Deduce a dependency to determine the thermodynamic mean temperature!

Solution

The problem is sketched in Fig. 13.24, that shows the first as well as the second law of thermodynamics.

(a) The specific dissipated energy ψ_{12} can be derived from the partial energy equation, i.e.

$$w_{t,12} = \int_1^2 v \, dp + \psi_{12} + \frac{1}{2}\left(c_2^2 - c_1^2\right) + g\left(z_2 - z_1\right). \tag{13.189}$$

Since no technical work is transferred and the channel is horizontal the equation simplifies to

$$0 = \underbrace{\int_1^2 v \, dp}_{y_{12}=0} + \psi_{12} + \frac{1}{2}\left(c_2^2 - c_1^2\right). \tag{13.190}$$

The specific pressure work y_{12} is zero, since the flow is supposed to be isobaric, i.e. $p_2 = p_1$. Thus, the specific dissipation results in

$$\psi_{12} = -\frac{1}{2}\left(c_2^2 - c_1^2\right) = 4800 \, \frac{J}{kg}. \tag{13.191}$$

(b) In order to gain information regarding the temperature, the first law of thermodynamics for open system in steady state is applied with ignoring the potential energy, i.e.

$$\underbrace{q_{12}}_{=0} + \underbrace{w_{t,12}}_{=0} = (h_2 - h_1) + \frac{1}{2}\left(c_2^2 - c_1^2\right). \tag{13.192}$$

With the caloric equation of state it follows

$$0 = c_p\,(T_2 - T_1) + \frac{1}{2}\left(c_2^2 - c_1^2\right). \tag{13.193}$$

Thus, the change of temperature is

$$\Delta T = T_2 - T_1 = -\frac{1}{2c_p}\left(c_2^2 - c_1^2\right) = \frac{\psi_{12}}{c_p} > 0. \tag{13.194}$$

The temperature rises.

Alternatively, the second law of thermodynamics can be applied, i.e.

$$\dot{S}_a + \dot{S}_i + \dot{m}s_1 = \dot{m}s_2. \tag{13.195}$$

$\dot{S}_a$ is zero, since the system is adiabatic. $\dot{S}_i$ must be positive, since dissipation occurs. Thus, it is

$$\dot{m}\,(s_2 - s_1) = \dot{S}_i > 0. \tag{13.196}$$

Applying the caloric equation of state leads to

$$c_p\,\ln\frac{T_2}{T_1} - \underbrace{R\,\ln\frac{p_2}{p_1}}_{=0} = \frac{\dot{S}_i}{\dot{m}} > 0 \tag{13.197}$$

Equation

$$c_p\,\ln\frac{T_2}{T_1} > 0 \tag{13.198}$$

can only be fulfilled in case $T_2 > T_1$.

(c) It is known from Sect. 13.7.3 that the thermodynamic mean temperature obeys

$$T_m = \frac{\dot{\psi}}{\dot{S}_i} = \frac{\psi_{12}}{s_{i,12}}. \tag{13.199}$$

With Eq. 13.196 and the caloric equation of state it finally is

$$T_m = \frac{\psi_{12}}{s_2 - s_1} = \frac{\psi_{12}}{c_p\,\ln\frac{T_2}{T_1}} = -\frac{c_2^2 - c_1^2}{2c_p\,\ln\frac{T_2}{T_1}}. \tag{13.200}$$

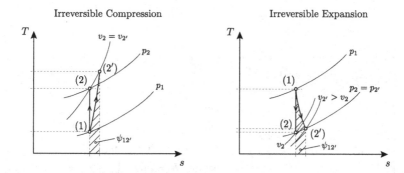

Fig. 13.25 Left: compression to the same volume, right: expansion to the same pressure (both adiabatic)

Problem 13.16 The adiabatic compression/expansion shown in Fig. 9.9 shall be plotted in a T, s-diagram!

Solution

See Fig. 13.25.

Problem 13.17 An ideal gas G with a mass m_G is trapped in an vertical cylinder C (diameter d). The cylinder is closed by a freely movable, frictionless piston P. Piston and cylinder are supposed to be incompressible and are made of the same steel. The masses of cylinder and piston are given, see sketch Fig. 13.26. In state (1) cylinder, piston and gas have the same temperature T_1 that is larger than ambient temperature T_{env}. During the change of state the entire systems cools slowly down until ambient temperature is reached, i.e. state (2). The change of state runs so slowly, that piston, cylinder and gas always have the same, homogeneous temperature. Changes of the potential energy of the gas as well as the acceleration of the piston can be ignored. Calculate for the change of state $(1) \rightarrow (2)$

(a) the gas pressure,[22]
(b) the change of the potential energy of the system (cylinder, piston and gas),
(c) the entire release heat to the ambient,
(d) the change of the entropy of the system (cylinder, piston and gas),
(e) the total generation of entropy (including the ambient!),
(f) the thermodynamic mean temperature!

---

[22]Assume, that the weight of the gas is small compared to the mass of the piston!

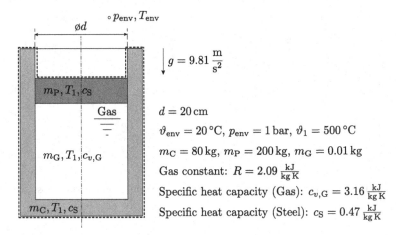

Fig. 13.26 Sketch to Problem 13.17

Solution

(a) This case has already been treated in Sect. 7.24. Since the process shall be very slow, i.e. free of dissipation, the weight of the gas is neglected, the piston moves frictionless and acceleration can be ignored, the change of state is isobaric. Thus, a balance of forces for the piston can be applied, i.e.

$$pA = m_P g + p_{env} A \qquad (13.201)$$

with

$$A = \frac{\pi}{4} d^2 = 0.0314 \, \text{m}^2. \qquad (13.202)$$

Hence, the pressure is

$$p = \frac{m_P g}{A} + p_{env} = 1.6245 \, \text{bar} = \text{const.} \qquad (13.203)$$

(b) According to the system boundary, see Fig. 13.26, the entire change of potential energy is

$$\Delta E_{pot,12} = \Delta E_{pot,12,P} + \underbrace{\Delta E_{pot,12,G}}_{=0} + \underbrace{\Delta E_{pot,12,C}}_{=0} \qquad (13.204)$$

respectively

$$\Delta E_{pot,12} = m_P g \, \Delta h \qquad (13.205)$$

with

$$\Delta h = h_2 - h_1. \tag{13.206}$$

Applying the thermal equation of state for states (1) and (2) leads to

$$p V_{G,1} = p A h_1 = m_G R T_1 \tag{13.207}$$

and

$$p V_{G,2} = p A h_2 = m_G R T_{\text{env}}. \tag{13.208}$$

Hence, the change of height is

$$\Delta h = h_2 - h_1 = \frac{m_G R}{p A} (T_{\text{env}} - T_1) = -1.9657 \, \text{m}. \tag{13.209}$$

Consequently, the total change of potential energy is

$$\Delta E_{\text{pot},12} = m_P g \, \Delta h = -3.8567 \times 10^3 \, \text{J}. \tag{13.210}$$

(c) Let us apply the first law of thermodynamics for system PGC as given by the system boundary in Fig. 13.27a, i.e.

$$W_{12} + Q_{12} = \Delta U_{\text{total},12} + \Delta E_{\text{pot},12}. \tag{13.211}$$

Hence, the exchanged heat follows

$$Q_{12} = \Delta U_{\text{total},12} + \Delta E_{\text{pot},12} - W_{12}. \tag{13.212}$$

Applying the caloric equation of state for the internal energy results in

$$\Delta U_{\text{total},12} = (m_P + m_C) c_S (T_{\text{env}} - T_1) + m_G c_{v,G} (T_{\text{env}} - T_1). \tag{13.213}$$

In order to estimate the work W_{12}, that is exchanged with the environment,[23] the partial energy equation obeys

$$W_{12} = W_{12,V} + \Psi_{\text{env}}. \tag{13.214}$$

With respect to the chosen system boundary, in this equation $W_{12,V}$ is at the interface to the environment, so that the work can be formulated from the environment's perspective: It includes *volume work at the environment* and *dissipation in the environment* Ψ_{env}. Both can easily be calculated in order to find W_{12}.

---

[23] Always be careful, when calculating the work, i.e. it must correlate with the chosen system boundary!

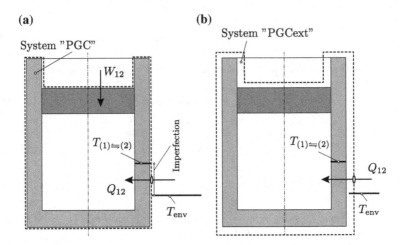

Fig. 13.27 System boundaries for Problem 13.17

Since the environment is supposed to be homogeneous,[24] it follows that

$$\Psi_{\mathrm{env}} = 0. \tag{13.215}$$

Thus, the partial energy equation simplifies to

$$W_{12} = W_{12,\mathrm{V}} = -\int_{1}^{2} p_{\mathrm{env}}\, \mathrm{d}V = -p_{\mathrm{env}}\,(V_2 - V_1) = -p_{\mathrm{env}}\,A\,\Delta h > 0 \tag{13.216}$$

This makes sense, since the piston moves down, i.e. the environment expands, so that work is supplied to the system! Thus, the transferred heat follows

$$Q_{12} = \left[(m_{\mathrm{P}} + m_{\mathrm{C}})\, c_{\mathrm{S}} + m_{\mathrm{G}} c_{v,\mathrm{G}}\right](T_{\mathrm{env}} - T_1) + m_{\mathrm{P}} g\,\Delta h + p_{\mathrm{env}} A\,\Delta h. \tag{13.217}$$

In a different notation it results in

$$Q_{12} = \left[(m_{\mathrm{P}} + m_{\mathrm{C}})\, c_{\mathrm{S}} + m_{\mathrm{G}} c_{v,\mathrm{G}}\right](T_{\mathrm{env}} - T_1) + \\ + (m_{\mathrm{P}} g + p_{\mathrm{env}} A)\,\frac{m_{\mathrm{G}} R}{pA}\,(T_{\mathrm{env}} - T_1) = -6.3193 \times 10^7\,\mathrm{J}. \tag{13.218}$$

(d) The entire change of entropy is composed of piston, cylinder and gas, see Fig. 13.27a. In contrast to the gas, steel is supposed to be incompressible. Consequently, the caloric equation of state leads to

---

[24]From now on this is always the case!

$$S_2 - S_1 = (m_P + m_C)\, c_S \ln\frac{T_{env}}{T_1} + m_G \left(c_{p,G} \ln\frac{T_{env}}{T_1} - R \ln\frac{p_{env}}{p_1} \right) \quad (13.219)$$

with

$$c_{p,G} = c_{v,G} + R = 5250\,\frac{J}{kg\,K}. \quad (13.220)$$

Hence, the entire change of entropy is

$$S_2 - S_1 = \Delta S_{PGC} = -1.2768 \times 10^5 \,\frac{J}{K}. \quad (13.221)$$

(e) When transferring heat, a temperature potential is required, i.e. an imperfection or discrepancy in homogeneity, see Fig. 13.27a. This imperfection is the cause for generation of entropy in this specific problem, see Sect. 14.2 for a detailed explanation. Consequently, when the total generation of entropy is asked for, this imperfection has to be part of the consideration, since the gas itself is compressed reversibly due to the slow change of state. To be more precise: The initial temperature spread between PGC and the environment is the driver for the change of state. It causes heat transfer, which decreases by time since the temperature potential gets smaller. Once the temperatures are equalised, the system can not move back into initial state. Thus, the change of state must be irreversible. Consequently, the system boundary has been slightly extended to system PGCext, so that the temperature imperfection, that is responsible for the generation of entropy, is *just* part of the system as well, see Fig. 13.27b. Anyhow, the additional mass of the surrounding ambient is so small, that it does not contribute to the extensive change of entropy. Nevertheless, goal is to catch the imperfection. Anyhow, the entire change of entropy then results in

$$\Delta S_{total} = \Delta S_{PGCext} = \Delta S_{PGC} + \underbrace{\Delta S_{env}}_{=0}$$
$$\equiv S_{i,12,total} + S_{a,12,total} = S_{i,12,total} + \frac{Q_{12}}{T_{env}} \quad (13.222)$$

Thus, the generation of entropy results in

$$S_{i,12,total} = \underbrace{\Delta S_{PGC}}_{\text{see d)}} - \frac{Q_{12}}{T_{env}} = 8.7891 \times 10^4 \,\frac{J}{K} > 0 \quad (13.223)$$

Hence, $S_{i,12,total} > 0$ proves that the process is irreversible.

Alternatively, the imperfection, i.e. the cause for generated entropy can be further investigated, see Fig. 13.28. Therefore, the boundary where the heat transfer takes place, is balanced. However, the boundary shall be so small, that it is almost massless, i.e. entropy going in needs to go out. The incoming entropy from the

Fig. 13.28 Massless
boundary, Problem 13.17e

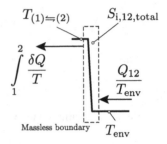

Massless boundary

environment[25] is at constant ambient temperature, whereas the outgoing entropy on the left is at a varying temperature. Furthermore, generation of entropy is assumed. Hence, the balance reads as[26]:

$$S_{i,12,\text{total}} + \frac{Q_{12}}{T_{\text{env}}} = \int_1^2 \frac{\delta Q}{T} \qquad (13.224)$$

The generated entropy results in

$$S_{i,12,\text{total}} = \int_1^2 \frac{\delta Q}{T} - \frac{Q_{12}}{T_{\text{env}}} \qquad (13.225)$$

However, the differential heat δQ is required. Thus, Eq. 13.218 leads to the differential notation, i.e.

$$\delta Q = \left[(m_P + m_C)\, c_S + m_G c_{v,G}\right] \mathrm{d}T + (m_P g + p_{\text{env}} A)\, \frac{m_G R}{pA}\, \mathrm{d}T \qquad (13.226)$$

Thus, for the generated entropy it follows

$$S_{i,12,\text{total}} = \left[(m_P + m_C)\, c_S + m_G c_{v,G}\right] \int_1^2 \frac{1}{T}\mathrm{d}T +$$

$$\qquad (13.227)$$

$$+ (m_P g + p_{\text{env}} A)\, \frac{m_G R}{pA} \int_1^2 \frac{1}{T}\mathrm{d}T - \frac{Q_{12}}{T_{\text{env}}}.$$

---

[25] Actually, it is outgoing, since $Q_{12} < 0$.
[26] Entropy in is equal to entropy out!

Finally, the generated entropy results in

$$S_{i,12,total} = \left[(m_P + m_C)\, c_S + m_G c_{v,G}\right] \ln\frac{T_{env}}{T_1} +$$

$$+ (m_P g + p_{env} A)\frac{m_G R}{pA} \ln\frac{T_{env}}{T_1} - \frac{Q_{12}}{T_{env}} = 8.7891 \times 10^4\,\frac{J}{K} > 0. \tag{13.228}$$

(f) For calculating the thermodynamic mean temperature of the system PGC, the following equation is applied

$$T_m = \frac{Q_{12}}{S_{a,12,PGC}}. \tag{13.229}$$

The heat Q_{12} has already been calculated. In order to determine the entropy carried by the heat it is

$$S_{a,12,PGC} = \int_1^2 \frac{\delta Q}{T}. \tag{13.230}$$

For solving the integral, Eq. 13.218 is applied in differential notation, i.e.

$$\delta Q = \left[(m_P + m_C)\, c_S + m_G c_{v,G}\right] dT + (m_P g + p_{env} A)\frac{m_G R}{pA}\, dT \tag{13.231}$$

Thus, for the transferred entropy the integral in Eq. 13.230 is solved

$$S_{a,12,PGC} = \left[(m_P + m_C)\, c_S + m_G c_{v,G}\right] \int_1^2 \frac{1}{T}dT +$$

$$+ (m_P g + p_{env} A)\frac{m_G R}{pA} \int_1^2 \frac{1}{T}dT \tag{13.232}$$

The solution results in

$$S_{a,12,PGC} = \left[(m_P + m_C)\, c_S + m_G c_{v,G}\right] \ln\frac{T_{env}}{T_1} +$$

$$+ (m_P g + p_{env} A)\frac{m_G R}{pA} \ln\frac{T_{env}}{T_1} = -1.2768 \times 10^5\,\frac{J}{K} \tag{13.233}$$

According to Eq. 13.229, the thermodynamic mean temperature finally obeys

$$T_m = \frac{Q_{12}}{S_{a,12,PGC}} = 494.95\ K. \tag{13.234}$$

Problem 13.18 Exhaust gas, to be treated as an ideal gas with $\kappa = 1.38$ and $R = 307.08 \frac{J}{kg\,K}$, expands adiabatically and reversibly in a turbine.[27] Its initial state (1) is given by $\vartheta_1 = 585\,°C$. State (2) is characterised by $\vartheta_2 = 390\,°C$ and $p_2 = 0.981$ bar. Potential as well as kinetic energy can be ignored.

(a) Calculate the exhaust gas pressure at the inlet p_1.
(b) What is the thermodynamic mean temperature from $(1) \rightarrow (2)$?

Solution

(a) The change of state runs adiabatically $(\delta s_a = 0)$ and reversibly $(\delta s_i = 0)$, i.e. it is isentropic $(ds = 0)$. Thus, it is

$$p_2 = p_1 \left(\frac{T_1}{T_2} \right)^{\frac{\kappa}{\kappa-1}} = 2.502\,\text{bar}. \tag{13.235}$$

(b) According to the second law of thermodynamics it follows

$$ds = \delta s_a + \delta s_i = \frac{\delta q}{T} + \frac{\delta \psi}{T}. \tag{13.236}$$

Rearranged, it is

$$T\,ds = \delta q + \delta \psi. \tag{13.237}$$

Introducing a thermodynamic mean temperature T_m the integration leads to

$$T_m \int_1^2 ds = \int_1^2 \delta q + \int_1^2 \delta \psi. \tag{13.238}$$

Finally, T_m obeys

$$T_m = \frac{q_{12} + \psi_{12}}{s_2 - s_1}. \tag{13.239}$$

Since the change of state is adiabatic, i.e. $q_{12} = 0$, and reversible, i.e. $\psi_{12} = 0$, the change of state is isentropic, i.e. $s_2 - s_1 = 0$. According to Eq. 13.239 the thermodynamic mean temperature can not be calculated directly, i.e.

$$T_m = \frac{0}{0}. \tag{13.240}$$

---

[27] The turbine will be treated in Chap. 17. However, it is a component, that releases work while a fluid expands from p_1 to $p_2 < p_1$.

In order to solve this problem, it is assumed that the change of state is adiabatic but irreversible. In this case, the process is polytropic with an polytropic exponent n. In the borderline case $n \rightarrow \kappa$ the change of state is identical with the given, isentropic conditions from (1) to (2), so that

$$T_m = \lim_{n \rightarrow \kappa} \frac{q_{12} + \psi_{12}}{s_2 - s_1} = \lim_{n \rightarrow \kappa} \frac{\psi_{12}}{s_{i,12}}. \tag{13.241}$$

To calculate T_m the following equations need to be solved sequentially for the polytropic change of state $n \rightarrow \kappa$

$$\boxed{\frac{T_2}{T_1} = \left(\frac{p_2}{p_1}\right)^{\frac{n-1}{n}}} \tag{13.242}$$

The partial energy equation is

$$w_{t,12} = \int_1^2 v \, dp + \psi_{12} = n \frac{RT_1}{n-1} \left(\frac{T_2}{T_1} - 1\right) + \psi_{12} \tag{13.243}$$

Applying the first law of thermodynamics for an adiabatic change of state, i.e. $q_{12} = 0$, leads to

$$w_{t,12} = h_2 - h_1 = c_p (T_2 - T_1) \tag{13.244}$$

Accordingly, the specific dissipation for $n \rightarrow \kappa$ results in

$$\boxed{\psi_{12} = c_p (T_2 - T_1) - n \frac{RT_1}{n-1} \left(\frac{T_2}{T_1} - 1\right)} \tag{13.245}$$

The change of specific entropy for $n \rightarrow \kappa$ is

$$\boxed{s_2 - s_1 = s_{i,12} = (c_p - R) \frac{n - \kappa}{n-1} \ln\left(\frac{T_2}{T_1}\right)} \tag{13.246}$$

Solving Eqs. 13.242, 13.245 and 13.246 numerically for $n \rightarrow \kappa$ leads to

$$T_m = \lim_{n \rightarrow \kappa} \frac{\psi_{12}}{s_{i,12}} = 756.47 \text{ K} \tag{13.247}$$

respectively

$$\vartheta_m = 483.32 \,°\text{C}. \tag{13.248}$$

Chapter 14
Transient Processes

Irreversibility plays an important role in evaluating thermodynamic processes. In case friction occurs, a system can not be operated reversibly. This has been treated in Sect. 7.3: A wire pendulum for instance does not reach its initial, starting position after a number of oscillations. The amplitude decreases by time until the pendulum finally stops in its rest position. This is due to dissipation, i.e. friction at the mounting point as well as due to interactions between environment and pendulum. Consequently, kinetic energy is dissipated and transferred to the environment from where it can not be gained back into the system. Consequently, the state value entropy in such a case rises due to dissipation. It has been shown, that the rate of generation of entropy is a quantitive measure for the degree of irreversibility.

14.1 Mechanical Driven Process

So far dissipation has been treated based on mechanical effects: A piston causes friction with the cylinder walls, or turbulent eddies being generated in a fluid for instance. What all these mechanical effects have in common is, that *motion* is required. This motion is induced by a *mechanical imbalance*. A homogeneous fluid in rest does not cause any irreversibilities. However, in this chapter other effects, that can cause irreversibility, are introduced.

Example 14.1 Anyhow, first and second law of thermodynamics have already been introduced. Thus, a wire pendulum, see Fig. 14.1, is further investigated.

- At the beginning at $t = 0$ the system possesses potential energy, since the mass m has been lifted in a field of gravity. In case the process is <u>reversible</u>, this potential energy is converted into kinetic energy, that has its maximum in state (1).

© Springer Nature Switzerland AG 2019
A. Schmidt, *Technical Thermodynamics for Engineers*,
https://doi.org/10.1007/978-3-030-20397-9_14

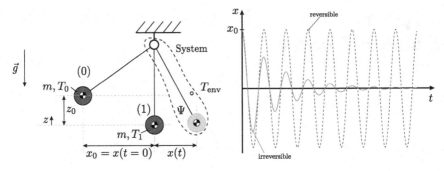

Fig. 14.1 Wire pendulum

First law of thermodynamics from $(0) \rightarrow (1)_{\text{rev}}$:

$$W_{01\text{rev}} + Q_{01\text{rev}} = \underbrace{mc\,(T_{1\text{rev}} - T_0)}_{=\Delta U} + \underbrace{mg\,(z_{1\text{rev}} - z_0)}_{\Delta E_{\text{pot}}} + \underbrace{\frac{1}{2}mg\,(c_{1\text{rev}}^2 - c_0^2)}_{\Delta E_{\text{kin}}} \quad (14.1)$$

Partial energy equation for $W_{01\text{rev}}$:

$$W_{01\text{rev}} = 0 = W_{01\text{rev,mech}} + \underbrace{\Psi_{01\text{rev}}}_{=0} \quad\quad\quad (14.2)$$

The work $W_{01\text{rev}}$ is zero, since no work crosses the system boundary! The mechanical work can be substituted, i.e.

$$W_{01\text{rev}} = 0 = mg\,(z_{1\text{rev}} - z_0) + \frac{1}{2}mg\,(c_{1\text{rev}}^2 - c_0^2). \quad\quad (14.3)$$

Thus, the velocity in state (1) results in

$$\boxed{c_{1\text{rev}} = \sqrt{2gz_0}} \quad\quad\quad\quad (14.4)$$

Combining Eqs. 14.3 and 14.1 leads to

$$Q_{01\text{rev}} = mc\,(T_{1\text{rev}} - T_0). \quad\quad\quad (14.5)$$

Let us assume, that pendulum and environment are in perfect thermal contact, i.e.

$$\boxed{T_{1\text{rev}} = T_0 = T_{\text{env}}} \quad\quad\quad (14.6)$$

Consequently, the process needs to be adiabatic, i.e.

$$Q_{01\text{rev}} = mc\,(T_{1\text{rev}} - T_0) = 0. \quad\quad\quad (14.7)$$

Second law of thermodynamics:

The pendulum is treated as an incompressible solid, so that

$$S_{1rev} - S_0 = mc \ln \frac{T_{1rev}}{T_0} = 0 = S_{i,01rev} + S_{a,01rev}. \tag{14.8}$$

The entropy stays constant, because $T_{1rev} = T_0$. Since the system further is adiabatic, i.e. $S_{a,01rev} = 0$, it results in

$$\boxed{S_{i,01rev} = 0} \tag{14.9}$$

This is the thermodynamic proof, that the process, if frictionless, is reversible!

• Now let us investigate what happens, in case the process is irreversible, i.e. friction occurs. Potential energy from state (1) is converted into kinetic energy and finally into frictional energy, measurable by a temperature increase. But let us take step by step:

First law of thermodynamics from $(0) \rightarrow (1)_{irrev}$:

$$W_{01irrev} + Q_{01irrev} = \underbrace{mc\,(T_{1irrev} - T_0)}_{= \Delta U} + \underbrace{mg\,(z_{1irrev} - z_0)}_{\Delta E_{pot}} + \underbrace{\frac{1}{2}mg\left(c_{1irrev}^2 - c_0^2\right)}_{\Delta E_{kin}}$$

$$\tag{14.10}$$

Partial energy equation for $W_{01irrev}$:

$$W_{01irrev} = 0 = W_{01irrev,mech} + \underbrace{\Psi_{01irrev}}_{>0} \tag{14.11}$$

The work $W_{01irrev}$ is zero, since no work crosses the system boundary! The mechanical work can be substituted, i.e.

$$W_{01irrev} = 0 = mg\,(z_{1irrev} - z_0) + \frac{1}{2}mg\left(c_{1irrev}^2 - c_0^2\right) + \Psi_{01irrev}. \tag{14.12}$$

Thus, the velocity in state (1) results in

$$\boxed{c_{1irrev} = \sqrt{2gz_0 - \frac{\Psi_{01irrev}}{m}} < c_{1rev}} \tag{14.13}$$

Combining Eqs. 14.12 and 14.10 leads to

$$Q_{01irrev} + \Psi_{01irrev} = mc\,(T_{1irrev} - T_0). \tag{14.14}$$

Let us assume, that pendulum and environment are in perfect thermal contact, i.e.

$$T_{1irrev} = T_{env} \tag{14.15}$$

Consequently, the system would have to release heat, according to

$$\boxed{Q_{01irrev} = -\Psi_{01irrev} < 0} \tag{14.16}$$

Second law of thermodynamics:

The pendulum is treated as an incompressible solid, so that

$$S_{1irrev} - S_0 = mc \ln\frac{T_{1irrev}}{T_0} = 0 = S_{i,01irrev} + S_{a,01irrev}. \tag{14.17}$$

It further is

$$S_{a,01irrev} = \frac{Q_{01irrev}}{T_{env}}. \tag{14.18}$$

Mind, that entropy carried by heat requires the temperature within the system where heat crosses the system boundary. In this case, see system boundary in Fig. 14.1, it is ambient temperature.[1]

$$\boxed{S_{i,01rirev} = -S_{a,01irrev} = -\frac{Q_{01irrev}}{T_{env}} > 0} \tag{14.19}$$

This is the thermodynamic proof, that the process, if friction occurs, is irreversible!

Problem 14.2 Two adiabatic tanks A and B contain nitrogen, to be treated as an ideal gas, in the following states:

- Tank A: $V = 1\,m^3$, $p = 22\,bar$, $T_1 = 300\,K$
- Tank B: $V = 1\,m^3$, $p = 2\,bar$, $T_1 = 300\,K$

The two tanks are connected, so that a pressure compensation between the two tanks can occur. No heat is exchanged with the environment during this process.

(a) What is the entropy change of the entire system and the entropy generation during this balancing process?
(b) What would be the change of entropy, if tank B was initially fully evacuated?

---

[1] Due to the perfect thermal contact, the pendulum itself always possesses ambient temperature as well.

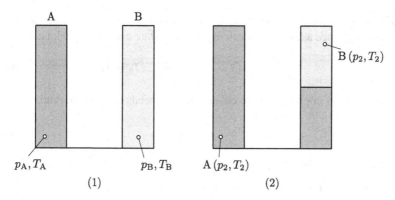

Fig. 14.2 Sketch to Problem 14.2

Solution

(a) The problem is sketched in Fig. 14.2. Though, gas A and gas B are identical, they are coloured differently in order to separate them for the following considerations. The change of entropy results in

$$\Delta S_{\text{total}} = S_2 - S_1. \tag{14.20}$$

The entropy of both masses A and B can be superimposed, since entropy is an extensive state value, i.e.

$$\Delta S_{\text{total}} = (S_2 - S_1)_A + (S_2 - S_1)_B \tag{14.21}$$

Since the gas is supposed to be ideal, the caloric equation of state can be applied as follows

$$\Delta S_{\text{total}} = m_A \left(c_p \ln \frac{T_2}{T_A} - R \ln \frac{p_2}{p_A} \right) + m_B \left(c_p \ln \frac{T_2}{T_B} - R \ln \frac{p_2}{p_D} \right) \tag{14.22}$$

with

$$T_A = T_B = T_1 = 300 \,\text{K}. \tag{14.23}$$

In a next step temperature T_2 and pressure p_2 in the new equilibrium state (2) needs to be calculated. Let us start with temperature T_2, that can be derived with the first law of thermodynamics[2]:

$$\underbrace{W_{12}}_{=0} + \underbrace{Q_{12}}_{=0} = U_2 - U_1 = (U_2 - U_1)_A + (U_2 - U_1)_B \tag{14.24}$$

---

[2]Mind, that the internal energy is an extensive state value, thus it can be superimposed as well!

The work W_{12} is zero, since there is no work being transferred across the system boundary. Once again, the caloric equation of state can be applied, i.e.

$$0 = m_A c_v (T_2 - T_1) + m_B c_v (T_2 - T_1).$$ (14.25)

This equation implies, that the change of state needs to be isothermal, i.e.

$$T_2 = T_1.$$ (14.26)

In order to calculate the pressure p_2 the thermal equation can be applied:

• Single systems A and B in state (1)

$$p_A V_A = m_A R T_1 \Rightarrow m_A = \frac{p_A V}{R T_1}$$ (14.27)

$$p_B V_B = m_B R T_1 \Rightarrow m_B = \frac{p_B V}{R T_1}$$ (14.28)

• Entire system in state (2)

$$p_2 2V = (m_A + m_B) R T_2$$ (14.29)

With Eqs. 14.27 and 14.28 it results in

$$p_2 = \frac{(m_A + m_B) R T_2}{2V} = \frac{p_A + p_B}{2} = 12 \, \text{bar}.$$ (14.30)

Now that we know T_2 and p_2, the change of entropy according to Eq. 14.22 can be calculated, i.e.

$$\Delta S_{total} = -m_A R \ln \frac{p_2}{p_A} - m_B R \ln \frac{p_2}{p_B}.$$ (14.31)

With Eqs. 14.27 and 14.28 it results in

$$\Delta S_{total} = -\frac{p_A V}{T_1} R \ln \frac{p_2}{p_A} - \frac{p_B V}{T_1} R \ln \frac{p_2}{p_B} = 3250 \, \frac{\text{J}}{\text{K}}.$$ (14.32)

In the following step, it is clarified how this change of state is achieved. Generally speaking, the change of state is initiated by entropy carried with the heat and by entropy being generated internally, i.e.

$$\Delta S_{total} = \underbrace{S_{a,12}}_{=0} + S_{i,12}$$ (14.33)

$S_{a,12}$ needs to be zero since the system is adiabatic, i.e.

$$S_{i,12} = \Delta S_{total} = 3250 \, \frac{J}{K} > 0. \tag{14.34}$$

Hence, the change of state is possible, but irreversible. Once, the thermodynamic equilibrium is reached, the system can not drive itself back into initial state without any external impact from outside.

(b) The same calculation as in (a) can be performed in case system B is fully evacuated, i.e. $m_B = 0$. The total change of entropy follows accordingly, i.e.

$$\Delta S_{total} = m_A \left(c_p \ln \frac{T_2}{T_A} - R \ln \frac{p_2}{p_A} \right). \tag{14.35}$$

With the same considerations as in (a) it follows, that

$$T_2 = T_1 \tag{14.36}$$

and

$$p_2 = \frac{1}{2} p_1 = 11 \, \text{bar}. \tag{14.37}$$

This leads to

$$\Delta S_{total} = -\frac{p_A V}{T_1} R \ln \frac{p_2}{p_A} = \underbrace{S_{a,12}}_{=0} + S_{i,12} = 5080 \, \frac{J}{K} > 0 \tag{14.38}$$

Obviously, the pressure difference between the two tanks A and B in case (b) is larger than in case (a). Consequently, the balancing process is more intense in case (b) than in case (a). Obviously, the generated entropy is a measure for the intensity of the balancing process.

14.2 Thermal Driven Process

As mentioned, the phenomena causing the generation of entropy have been based on mechanical effects so far. The driver for this mechanism is the motion of a fluid or a component. However, in systems driven by *thermal imbalances* for instance, there is another mechanism for entropy generation.

Conductive Wall

In order to understand these thermal effects, let us take a look at a simple example represented by Fig. 14.3. A solid wall separates the inside of a house from the environment. Within the house there shall be a larger temperature than ambient tem-

First Law   **Second Law**

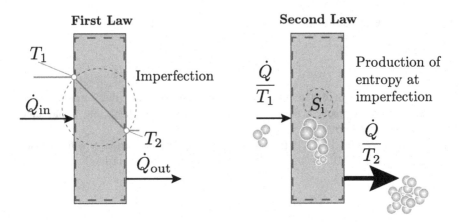

Fig. 14.3 Wall—Heat conduction

perature. Let us assume, that the wall's surface temperature on the inside is T_1 and on the outside it is T_2. From a thermodynamic point of view this is a thermal imbalance and the system has the tendency to drive into a new equilibrium state. Hence, a heat flux is initiated from the inside to the outside.[3] Obviously, the wall is mechanically in perfect rest, so there is no motion-based dissipation as known from systems, that have been treated before.

According to first law of thermodynamics in steady state the incoming energy fluxes equal outgoing energy fluxes, i.e.

$$\dot{Q}_{\mathrm{in}} = \dot{Q}_{\mathrm{out}} = \dot{Q}. \tag{14.39}$$

Otherwise the wall's temperature would vary temporally. A flux of entropy comes along with heat, i.e.

$$\dot{S}_{\mathrm{a}} = \frac{\dot{Q}}{T} \tag{14.40}$$

T is the temperature at the system boundary within the system, where the heat passes. Due to $T_1 > T_2$ less entropy enters than entropy leaves:

$$\frac{\dot{Q}}{T_1} < \frac{\dot{Q}}{T_2} \tag{14.41}$$

However, in steady state the incoming entropy needs to be equal to the outgoing entropy. If this is not the case, then entropy within the system would temporally vary. Since entropy is a state value, the state itself would temporally vary, i.e. a contradiction to steady state. In order to solve this dilemma, entropy must be *generated* within the system, i.e. $\dot{S}_{\mathrm{i}} > 0$:

---

[3]Everyday experience shows, that a house needs a heating system in order to compensate the heat being released to environment!

$$\frac{\dot{Q}}{T_1} + \dot{S}_i = \frac{\dot{Q}}{T_2} \tag{14.42}$$

Hence, the generated entropy follows

$$\dot{S}_i = \frac{\dot{Q}}{T_2} - \frac{\dot{Q}}{T_1}. \tag{14.43}$$

In a different mathematical notation it results in

$$\boxed{\dot{S}_i = \dot{Q}\frac{T_1 - T_2}{T_1 T_2} \geq 0} \text{ due to } T_1 \geq T_2 \tag{14.44}$$

Theorem 14.3 *Entropy is generated at the imperfection, i.e. the wall impedes the balancing process between inside and outside. In contrast, there is no generation of entropy in homogeneous systems, i.e. systems without an imperfection.*

It is known, that $\dot{S}_i$ can never be negative, i.e. entropy can not be destroyed. This leads to the conclusion, that the *direction* of heat transfer, as assumed in Fig. 14.3, can only occur, if $T_1 \geq T_2$:

Theorem 14.4 *Heat always follows the temperature gradient from warm to cold.*

According to Eq. 14.44 the generation of entropy rises with increasing temperature difference. Temperature differences lead to a balancing process. Once again,[4] the generated entropy is a measure for the intensity of a process.

Heat Transfer Between Two Systems

With this example, the view on balancing processes is extended. A distinction is made between homogeneous systems and imperfections, i.e. non-homogeneous systems, see Fig. 14.4.

Let us focus on the heat transfer between two systems A and B, that are both adiabatic to the environment. The problem shall be guided by the following premises:

- Both systems are separated from each other by a wall, i.e. there is no mass flux. They have different, but uniform temperatures with

$$T_B > T_A. \tag{14.45}$$

This temperature potential initiates a heat flux between system B and A, since the overall system strives for thermal equilibrium.

- The systems are supposed to be large in size, i.e. the temperatures remain constant temporally, though thermal energy leaves systems A and enters system B.

---

[4]Comparable to the pressure compensation in Problem 14.2.

Fig. 14.4 Heat transfer between two homogeneous systems

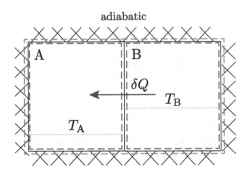

- Thus, each system is homogeneous. There are no imperfections within the systems. Thus, there is no generation of entropy within system A respectively system B.

However, the overall process is irreversible, since heat is transferred between two sub-systems of different temperatures. This is deduced within this section.

The change of entropy in system B follows[5]:

$$dS_B = -\frac{\delta Q}{T_B} + \delta S_{i,B} \text{ with } \delta S_{i,B} = 0 \text{ (homogeneous phase)} \qquad (14.46)$$

The change of entropy in system A follows:

$$dS_A = \frac{\delta Q}{T_A} + \delta S_{i,A} \text{ with } \delta S_{i,A} = 0 \text{ (homogeneous phase)} \qquad (14.47)$$

Hence, the overall change of entropy, as extensive state value, is composed of systems A and B:

$$dS_{total} = dS_A + dS_B = \underbrace{\delta S_{a,total}}_{=0,\,adiabatic} + \delta S_{i,total} = \delta S_{i,total} \qquad (14.48)$$

Obviously, the overall change of state is due to internal heat transfer[6] and due to generation of entropy. There are no influences respectively impacts from outside. Combining Eqs. 14.46 to 14.48 results in

$$\delta S_{i,total} = dS_A + dS_B = \frac{\delta Q}{T_A} - \frac{\delta Q}{T_B}. \qquad (14.49)$$

In a different notation it reads as

$$\boxed{\delta S_{i,total} = \delta Q \frac{T_B - T_A}{T_A T_B} > 0} \qquad (14.50)$$

---

[5]See Theorem 14.3.

[6]In this case $\delta S_{a,total}$ is zero, since the entire system shall be adiabatic to the environment.

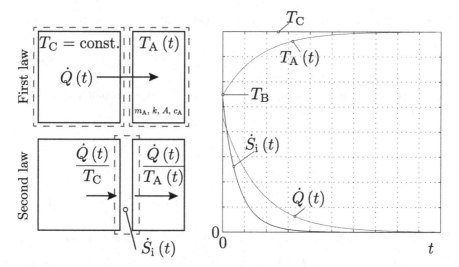

Fig. 14.5 Thermal balancing process

Thus, the conclusion is

- The demand of the second law of thermodynamics ($\delta S_{i,total} > 0$) is fulfilled, when $T_B > T_A$. In accordance with Theorem 14.4, it follows, that heat is always transferred from warm to cold!
- Irreversible generation of entropy occurs at the interface between the systems A and B. At that interface there is an imperfection and the temperature profile is inhomogeneous.
- The energy of the overall system remains constant, since no external influences cross the system boundary. However, entropy inside the fully-closed system rises due to the thermal balancing process.
- Generation of entropy increases with increasing temperature difference, i.e. the thermal potential.
- There is a limiting borderline case of reversible heat transfer in case the temperatures of system A and B are asymptotical equal. In order to still transfer heat, the effective surface needs to be huge under these circumstances. This can be shown with the following equation[7]:

$$\dot{Q} = kA\,\Delta T \tag{14.51}$$

Example 14.5 This example refers to the temperature measurement following the zeroth law of thermodynamics, see Fig. 4.5. The according balancing process is now described thermodynamically. Sketch 14.5 shows the first and second law of thermodynamics as well as a temperature profile. Let us assume, that the temperature of system C is constant even if the thermometer, i.e. system A, is in contact with C.

---

[7]This equation is derived in the lecture *Heat and Mass Transfer*.

However, a thermal balancing occurs and the temperature of A varies by time. The
temporal heat flux is known from *Heat and Mass Transfer*, i.e.

$$\dot{Q}(t) = kA[T_C - T_A(t)] \tag{14.52}$$

Under the premise, that the temperature of the thermometer A is uniform, the first
law of thermodynamics obeys

$$\frac{dU_A}{dt} = m_A c_A \frac{T_A}{dt} = +\dot{Q}(t). \tag{14.53}$$

Thus, it is

$$m_A c_A \frac{T_A}{dt} = kA[T_C - T_A(t)]. \tag{14.54}$$

This differential equation can be solved by separating the variables and integration,
i.e.

$$\int \frac{dT_A}{T_A - T_C} = -\int \frac{kA}{m_A c_A} dt. \tag{14.55}$$

By applying the initial condition, see Fig. 4.5, it follows

$$T_A(t = 0) = T_B. \tag{14.56}$$

Temperature $T_A(t)$ results in

$$\boxed{T_A(t) = T_C - (T_C - T_B)e^{-\frac{kA}{m_A c_A}t}} \tag{14.57}$$

The heat flux follows from Eq. 14.52, i.e.

$$\boxed{\dot{Q}(t) = kA(T_C - T_B)e^{-\frac{kA}{m_A c_A}t}} \tag{14.58}$$

These functions are plotted in Fig. 14.5. Obviously, after a long time the temperatures
are equalised. Initially, the temperature potential $T_C - T_A$ is maximal. Consequently,
the heat flux is maximal at the beginning and decreases with ongoing time t. Now let
us investigate the consequences for the entropy. A system boundary just at the interface of systems A and B is chosen. Thus, the system is massless, i.e. no accumulation
of entropy takes place. The incoming flux of entropy due to the heat flux is

$$\dot{S}_{a,in}(t) = \frac{\dot{Q}(t)}{T_C}. \tag{14.59}$$

Accordingly, the outgoing flux of entropy results in

$$\dot{S}_{a,out}(t) = \frac{\dot{Q}(t)}{T_A(t)}. \tag{14.60}$$

As long as $T_C > T_A$ the incoming flux of entropy is smaller than the outgoing. Since no accumulation of entropy is possible due to the massless system, i.e. interface between A and C, it needs to be

$$\sum \dot{S}_{in} = \sum \dot{S}_{out}. \tag{14.61}$$

Thus, in order to balance incoming and outgoing flux of entropy, a flux of generation of entropy $\dot{S}_i$ is required

$$\dot{S}_{a,in}(t) + \dot{S}_i(t) = \dot{S}_{a,out}(t). \tag{14.62}$$

For the flux of generation of entropy it is

$$\dot{S}_i(t) = \frac{\dot{Q}(t)}{T_A(t)} - \frac{\dot{Q}(t)}{T_C}. \tag{14.63}$$

This leads to:

$$\boxed{\dot{S}_i(t) = kA(T_C - T_B)e^{-\frac{kA}{m_A c_A}t}\left[\frac{1}{T_C - (T_C - T_B)e^{-\frac{kA}{m_A c_A}t}} - \frac{1}{T_C}\right]} \tag{14.64}$$

As soon as the temperatures A and C are balanced, the generation of entropy disappears! The larger the temperature spread between A and C is, the more entropy is generated. The generation of entropy occurs at the imperfection at the interface A/C, indicated by the temperature difference. When this imperfection disappears, the generation of entropy stops, see Fig. 14.5. This is at $t \to \infty$.

14.3 Chemical Driven Process

Chemical driven processes will be thermodynamically investigated in part II and III.

14.4 Conclusions

In Chap. 13 the generation of entropy has been investigated. Any mechanical system can cause internal friction, i.e. dissipation, due to motion of mechanical parts, e.g. a piston moving in a cylinder, or due to motion of a fluid.[8] In such processes energy is dissipated, given by $\psi_{12} > 0$. This leads to generation of entropy, i.e. $s_{i,12} > 0$. In Chap. 16 exergy will be introduced: it will be shown, that generation of entropy

---

[8]This can be in open as well as in closed systems.

reduces the working capability of a system. However, in thermodynamics there is a limiting borderline case of systems being free of any dissipation. This is a theoretical benchmark of systems working reversibly. Such systems possess the highest efficiency, see Chap. 15. Nevertheless, these ideal systems do not exist in reality since dissipation can not be prevented.

In addition to the mechanically caused generation of entropy, there is another mechanism of non-homogeneous systems characterised by imperfection: Within these systems entropy is generated at the imperfection, see Sect. 14.2.

Theorem 14.6 *Any disruption of thermodynamic equilibrium[9] initiates a balancing process in order to achieve a state of equilibrium, i.e. mechanical, thermal and chemical. This balancing process is irreversible and characterised by generation of entropy.*

Examples of transient balancing processes due to mechanically imbalanced systems are:

- pressure, see Problem 14.2
- or height, see Problem 13.14

Thermal imbalanced system are characterised by a heterogeneous temperature distribution. Chemical imperfections are due to a concentration imbalance. In Sect. 3.2 it has been shown, that thermodynamic systems always strive for mechanical, thermal and chemical equilibrium.

Theorem 14.7 *The larger the imperfection is, the more entropy is generated. Homogeneous systems, that do not have imperfections, i.e. gradients, do not produce entropy!*

Once the mechanical, thermal or chemical potential is equalised due to a balancing process, the system can not move back into its initial state without external impact.[10]

Problem 14.8 A heat flux of $\dot{Q} = 20 \, \text{kW}$ is released into the environment, with a temperature of $\vartheta_{\text{env}} = -5 \, °\text{C}$, through the wall of a house, with an inside temperature of $\vartheta = 22 \, °\text{C}$. What is the entropy flow $\dot{S}_{\text{i}}$, that is generated in this process?

Solution

The problem is sketched in Fig. 14.6. Obviously, in steady state, the heat flux entering the system is equal to the heat flux leaving the system. Furthermore, all state values inside the system need to be constant by time, so does the entropy inside the system, i.e.

$$\frac{\text{d}S}{\text{d}t} = 0. \tag{14.65}$$

---

[9]Pressure, temperature and chemical potential.

[10]A car does not re-climb a hill without external impact, once it has reached its rest position. A cup of tea does not heat up, once it has reached ambient temperature. Components in a gaseous mixture do not segregate once they are perfectly mixed.

Fig. 14.6 Sketch to
Problem 14.8

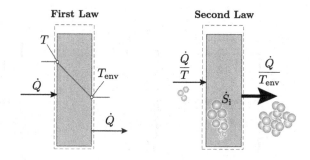

Thus, the entropy flux into the system must be balanced by the entropy flux out.
According to Fig. 14.6 it is:

$$\frac{\dot{Q}}{T} + \dot{S}_i = \frac{\dot{Q}}{T_{env}}. \tag{14.66}$$

Mind, that generation inside the system $\dot{S}_i$ is a source term. Finally, the generation
of entropy results in

$$\dot{S}_i = \dot{Q}\left(\frac{1}{T_{env}} - \frac{1}{T}\right) = 6.82\,\frac{W}{K} > 0. \tag{14.67}$$

Hence, the process is irreversible, i.e. once the heat flux has left the building it does
not return.

Problem 14.9 An ideal gas ($R = 0.280\,\frac{kJ}{kg\,K}$) with a mass of $m_G = 1\,kg$ and a temperature of $\vartheta_G = 20\,°C$ is to be compressed reversibly and isothermally from an
initial pressure $p_1 = 100\,kPa$ to a final pressure $p_2 = 600\,kPa$. The cylinder is in
thermal contact with a heat reservoir of constant temperature.[11]

Calculate the change of entropy for the gas in the cylinder and for the heat reservoir,
if

(a) the temperature of the reservoir is $\vartheta_R = \vartheta_G = 20\,°C$ (Case 1)?
(b) the temperature of the reservoir is $\vartheta_R = 15\,°C$ (Case 2)?
(c) What is the generated entropy $S_{i,12}$ in both cases?

Solution

(1) Let us start with the change of entropy for the gas, as illustrated by the system
     boundary "G" in Fig. 14.7. Independently from case 1 or case 2 there are two
     alternatives to estimate ΔS_G:

---

[11] The heat reservoir is huge!

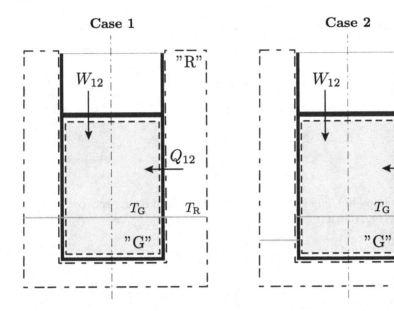

Fig. 14.7 Sketch to Problem 14.9

- Alternative 1—Caloric equation of state

$$\Delta S_G = (S_2 - S_1)_G = m_G (s_2 - s_1)_G \qquad (14.68)$$

Applying the caloric equation of state for an isothermal change of state results in

$$\Delta S_G = m_G \left(c_p \ln \frac{T_{G,2}}{T_{G,1}} - R \ln \frac{p_2}{p_1} \right) = -m_G R \ln \frac{p_2}{p_1} = -0.50169 \, \frac{\text{kJ}}{\text{K}} \qquad (14.69)$$

- Alternative 2—Entropy balance

The change of entropy for the gas is composed of entropy passing the system boundary due to heat transfer and of entropy generated inside the system:

$$\Delta S_G = S_{a,12} + S_{i,12} \qquad (14.70)$$

Since, the gas is compressed reversibly, there is no generation of entropy due to the compression, i.e. $S_{i,12} = 0$. Consequently, the change of entropy obeys

$$\Delta S_G = S_{a,12} = \int_1^2 \frac{\delta Q}{T} = \frac{Q_{12}}{T_G}. \qquad (14.71)$$

Mind, that in this equation T is the temperature inside the system, where the heat passes the boundary. In this case, since the system "G" is regarded, it is the isothermal temperature T_G. In a next step, the heat Q_{12} needs to be calculated. Thus, the first law of thermodynamics for system "G" is applied, i.e.

$$Q_{12} + W_{12} = (U_2 - U_1)_G = m_G c_v (T_2 - T_1)_G = 0. \tag{14.72}$$

Hence, it is

$$Q_{12} = -W_{12} \tag{14.73}$$

The partial energy equation for W_{12} in reversible operation reads as[12]

$$W_{12} = W_{V,12} + \underbrace{\Psi_{12}}_{=0} = W_{V,12} = m_G R T_G \ln\frac{p_2}{p_1}. \tag{14.74}$$

Consequently, the heat follows

$$Q_{12} = -m_G R T_G \ln\frac{p_2}{p_1} = -147.071\,\text{kJ}. \tag{14.75}$$

Finally, the change of entropy obeys

$$\Delta S_G = \frac{Q_{12}}{T_G} = -0.50169\,\frac{\text{kJ}}{\text{K}}. \tag{14.76}$$

Both alternatives lead to the same result, The result is independent from the heat reservoir, since only system "G" has been regarded!

(2) Now, let us focus on the heat reservoir "R". A caloric approach as for the gas (see alternative 1), i.e.

$$\Delta S_R = (S_2 - S_1)_R = \underbrace{m_R}_{\to\infty} \underbrace{(s_2 - s_1)_R}_{=0}, \tag{14.77}$$

is not expedient, since the mass of the heat reservoir is huge and its state constant. However, a balance of entropy leads to

$$\Delta S_R = S_{a,12} + \underbrace{S_{i,12}}_{=0} \tag{14.78}$$

---

[12] See Fig. 13.13!

Since the heat reservoir is supposed to be huge and homogeneous, there is no generation of entropy in the reservoir, i.e. $S_{i,12} = 0$. Since Q_{12} leaves system "R" at T_R, it is

$$\Delta S_R = -\frac{Q_{12}}{T_R}. \tag{14.79}$$

Thus, according to case 1 and 2 it finally is

$$\Delta S_R = \begin{cases} 0.50169\,\frac{kJ}{K} & @\ \vartheta_R = 20\,°C \\ 0.510398\,\frac{kJ}{K} & @\ \vartheta_R = 15\,°C \end{cases} \tag{14.80}$$

(3) Entire system (extended system boundary)

According to the extended system boundary as given in Fig. 14.8, the total change of state results in

$$\Delta S_{total} = \Delta S_G + \Delta S_R + \underbrace{\Delta S_{Cylinder}}_{=0} = S_{i,12} + \underbrace{S_{a,12}}_{=0}. \tag{14.81}$$

The cylinder is ignored, since it is regarded as incompressible, isothermal body.[13] The entire system is huge, so that it is adiabatic,[14] i.e. $S_{a,12} = 0$.
Thus, the equation simplifies to

$$\Delta S_{total} = S_{i,12} = \Delta S_G + \Delta S_R. \tag{14.82}$$

According to case 1 and 2 it finally is

$$\Delta S_{i,12} = \begin{cases} 0\,\frac{kJ}{K} & @\ \vartheta_R = 20\,°C \\ 0.0087\,\frac{kJ}{K} & @\ \vartheta_R = 15\,°C \end{cases} \tag{14.83}$$

In summary, there is no generation of entropy in case 1, since the gas is compressed reversibly, i.e. no dissipation occurs, and the heat transfer takes place in the borderline case of $\Delta T = 0$. This is a theoretical benchmark, as it would require a huge area for the heat transfer, see Eq. 14.51. However, the entire system is homogeneous, as indicated by the temperature profile. The overall change of state is reversible, the heat transfer could theoretically run in the opposite direction.[15]

---

[13] Hence, its state is constant!

[14] Due to its homogeneity, the *outer edge* of the environment has no temperature gradients, so that no heat is transferred. Furthermore, Q_{12} is *within* the system, i.e. it does not occur at the system boundary!

[15] Due to the large heat transfer area, that allows heat transfer with $\Delta T \to 0$.

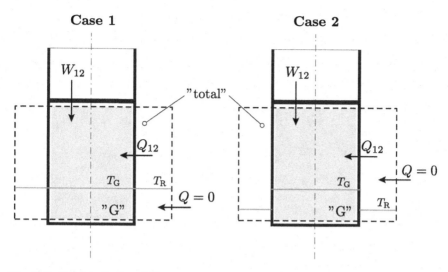

Fig. 14.8 Sketch to Problem 14.9

In contrast, case 2 is irreversible, though there is no dissipation within the gas as well. The cause for the irreversibility lies on the heat transfer, that takes place with $\Delta T = 5\,\text{K}$. Thus, the overall system is not homogeneous, but is characterised by an imperfection, see the overall temperature profile for case 2 in Fig. 14.8. This temperature step makes the entire change of state irreversible, since for the given system it is impossible to initiate a heat transfer from cold to warm.

Chapter 15
Second Law of Thermodynamics

Though the balance of entropy has been denoted as *Second Law of Thermodynamics* in the previous chapters, its classical formulation comes along with thermodynamic cycles. Thus, the focus now is on these cycles. However, there are two different types of thermodynamic cycles: Clockwise cycles on the one hand convert heat into mechanical energy and are named *thermal engines.* Counterclockwise cycles on the other hand are *fridges and heat pumps*, that lift thermal energy from a lower to a higher temperature level. However, at that stage cycles are represented in a so-called black-box notation, i.e. the fluxes at its border are balanced without focussing on the detailed physical processes that run *inside* the machine.

15.1 Formulation According to Planck—Clockwise Cycle Processes

According to Planck a permanently or cyclically running machine,[1] that takes heat out of a reservoir and converts it solely into work, is impossible, see Fig. 15.1. A fictive machine, that would be able to so, violates the second law of thermodynamics and is called *Perpetuum Mobile of the second kind.*[2] However, according to the first law of thermodynamics there is no contradiction: For closed systems in steady state, see Fig. 15.1a, the incoming energy is balanced by the energy leaving the machine, i.e.

$$Q = W. \tag{15.1}$$

Thus, with respect to the energy conservation principle such a machine would exist. Thermal energy would be solely converted into mechanical work. However, in steady

---

[1] Such a machine would be operated in steady state.

[2] A fictive machine, that violates the first law of thermodynamics, i.e. releases more energy than energy is supplied, is called a Perpetuum Mobile of the first kind.

© Springer Nature Switzerland AG 2019
A. Schmidt, *Technical Thermodynamics for Engineers*,
https://doi.org/10.1007/978-3-030-20397-9_15

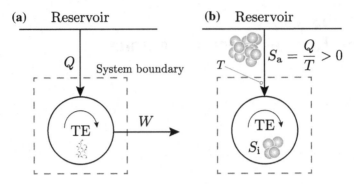

Fig. 15.1 Contradiction to the second law according to Planck

state each state value, e.g. pressure p, temperature T and internal energy U, needs to be constant over time, though it can differ locally. Since entropy S is a state value as well, it also needs to be constant over time. Hence, incoming entropy must be equal to outgoing entropy, see Fig. 15.1b:

$$\underbrace{\frac{Q}{T}}_{>0} + S_i = 0 \tag{15.2}$$

This results in

$$S_i = -\frac{Q}{T} < 0. \tag{15.3}$$

Thus, the process is impossible. The amount of entropy inside the system can not be constant over time, since entropy enters with the supplied heat and entropy might be generated inside, while no entropy leaves the system!

Theorem 15.1 *A machine, operated in steady state, is never able to convert thermal energy completely into mechanical energy.*

Example 15.2 The following example shows, how a thermal engine, that is designed for converting thermal energy into mechanical energy, might be realised. It has already been introduced in Sect. 11.2.1. Thus, Fig. 15.2 actually means opening the black box, as shown in Fig. 15.1, to become a look into the system. A gas can be compressed and it can expand. In order to release more work during expansion than work being required for compression, the expansion needs to be done on a larger pressure level. If this is not realised, the changes of state run from $(1)\rightarrow(2)\rightarrow(1)$ and expansion respectively compression work have the same amount.

Thus, the pressure needs to be increased once the minimum volume is reached by supplying thermal energy, see state (3). Now, during expansion of the gas until state (4) is reached a larger amount of work is released than work is supplied for compression. As the machine shall run permanently, i.e. in a steady state, the compression

Fig. 15.2 Thermal engine without cooling

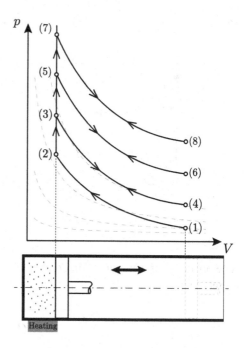

is done again and state (3) is re-reached. Now, while heating again, a larger pressure and temperature as in (3) is reached. If this cycle is repeated permanently, the temperature/pressure at the bottom dead centre rises continually without releasing effective work. The entire supplied thermal energy is taken for heating up the gas. Hence, a steady state can not be achieved and after a while the machine exceeds its temperature/pressure limits. In order to realise a steady state operation a cooling is required once the maximum volume is reached, see Fig. 11.6b. This is in accordance with Planck's formulation of the second law of thermodynamics, that says, that for converting heat into mechanical work continuously a cooling is required!

Theorem 15.3 *Thus, at least two reservoirs are required: A warm reservoir from which heat can be taken as well as a cold reservoir, where heat can be transferred to!*

15.1.1 The Thermal Engine

Figure 15.3 shows a so-called thermal engine in black box notation, i.e. a technical machine that converts thermal energy into mechanical energy: Theorem 15.3 claims, that two reservoirs are required and heat needs to be released. The thermodynamic proof is given in this section. Following the first law of thermodynamics in steady

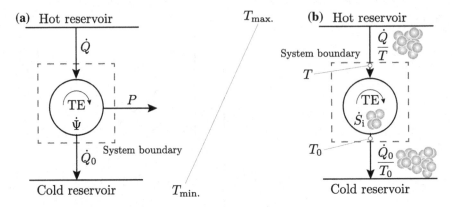

Fig. 15.3 Thermal engine in black box notation

state operation, the amount of internal energy inside the system must not vary with respect to time, so that $\frac{dU}{dt} = 0$, i.e.

$$\frac{dU}{dt} = 0 = \dot{Q} - \dot{Q}_0 - P \tag{15.4}$$

According to sketch Fig. 15.3a, $\dot{Q}$ increases the internal energy, and thus is positive, while $\dot{Q}_0$ and P leave the system and are negative. This notation is equivalent with the principle, that incoming and outgoing energy are identical:

$$\dot{Q} = \dot{Q}_0 + P \tag{15.5}$$

The question is, how much power P can be gained out of the thermal energy Q, i.e.

$$P = \dot{Q} - \dot{Q}_0 \tag{15.6}$$

Consequently, cooling $\dot{Q}_0$ reduces the maximum power output and thus it should be as small as possible. However, the first law of thermodynamics is limited at that stage, as it leads to one equation but two unknowns. In order to quantify the maximum power output, a balance of entropy is applied: Entropy is a state value, so that its amount inside the machine needs to be constant over time in steady state operation. Hence, the balance of entropy obeys

$$\frac{dS}{dt} = 0 = \frac{\dot{Q}}{T} - \frac{\dot{Q}_0}{T_0} + \dot{S}_i \tag{15.7}$$

However, in other words, the incoming entropy[3] balances the outgoing entropy, see also Fig. 15.3b:

$$\frac{\dot{Q}}{T} + \dot{S}_i = \frac{\dot{Q}_0}{T_0} \tag{15.8}$$

In order to calculate the entropy carried with the heat, the temperatures T respectively T_0 inside the system, where the heat crosses the system boundary are required. In this case, following the temperature slope, it is

$$T > T_0. \tag{15.9}$$

Obviously, cooling by $\dot{Q}_0$ is required in order to release entropy! Without releasing heat the balance of entropy can not be fulfilled. The cooling demand can be easily calculated and follows

$$\dot{Q}_0 = T_0 \left(\frac{\dot{Q}}{T} + \dot{S}_i \right). \tag{15.10}$$

The more entropy is produced within the machine, the more cooling is required. Furthermore, the larger the temperature of the cold reservoir, the more heat needs to be released. Now that the cooling requirement is known, the power output can be calculated with Eq. 15.6:

$$P = \dot{Q} - T_0 \left(\frac{\dot{Q}}{T} + \dot{S}_i \right). \tag{15.11}$$

In a different notation it is

$$P = \dot{Q} \left(1 - \frac{T_0}{T} \right) - \dot{S}_i T_0. \tag{15.12}$$

The efficiency of this machine is defined with the so-called thermal efficiency:

$$\eta_{th} = \frac{\text{Benefit}}{\text{Effort}}. \tag{15.13}$$

For this case it obeys

$$\eta_{th} = \frac{P}{\dot{Q}} \tag{15.14}$$

It follows, that

$$\boxed{\eta_{th} = \left(1 - \frac{T_0}{T} \right) - \dot{S}_i \frac{T_0}{\dot{Q}}} \tag{15.15}$$

Obviously, the best machine, i.e. the machine with the highest possible efficiency, does not have any internal entropy generation, $\dot{S}_i = 0$. Such a machine is denoted

---

[3]Mind, that generation of entropy is a *source term*, so it is going *into* the system!

Carnot machine with an efficiency of

$$\eta_{th, C} = 1 - \frac{T_0}{T}.$$ (15.16)

Obviously, the thermal engine (TE) works in-between a hot and cold reservoir. Due to the temperature gradient, according to the second law of thermodynamics, a heat flux occurs following the temperature slope of the two reservoirs, see Chap. 14. A thermal engine collects the heat flux and converts it partly into mechanical energy, while releasing a portion of energy to the cold reservoir.

15.1.2   Why Clockwise Cycle?

So far the thermal engine, that converts heat into work, has been presented in a black box notation. In this section it is clarified how a thermodynamic processes *within* the machine can be characterised. In the most simple case the process inside the machine consists of two consecutive changes of state, see Fig. 15.4. As the machine shall be operated under steady state conditions, the fluid needs to follow a thermodynamic loop. In case the fluid is in a closed system, e.g. a cylinder/piston system as shown in Fig. 15.2, the volume work can be illustrated in a p, V-diagram. To keep it simple, the process shown in this figure shall be reversible and outer energies are ignored, i.e. $W = W_V$. Instead of taking into account the power P it is continued with the work W. However, the conclusions are the same as

$$P = \frac{\delta W}{dt}.$$ (15.17)

Accordingly, the correlation between heat flux and heat is

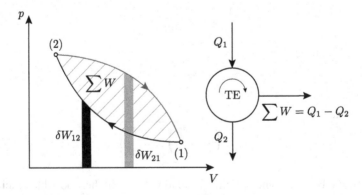

Fig. 15.4 Thermal engine in a p, V-diagram

$$\dot{Q} = \frac{\delta Q}{dt}. \tag{15.18}$$

Thus, the heat Q is shown in Fig. 15.4 instead of the heat flux $\dot{Q}$, that is illustrated in Fig. 15.3.

According to the p, V-diagram work is released by the fluid during expansion $(2) \rightarrow (1)$, i.e.

$$W_{21} = -\int_{2}^{1} p \, dV < 0. \tag{15.19}$$

For the compression $(1) \rightarrow (2)$ work is supplied, i.e.

$$W_{12} = -\int_{1}^{2} p \, dV > 0. \tag{15.20}$$

Work for compression respectively work, that is released, are both represented by the areas beneath the changes of state in a p, V-diagram. In order to achieve, that the overall process releases work, i.e. the benefit of a thermal engine, the change of state for expansion requires to be on a larger pressure level than the change of state for compression. According to Fig. 15.4 this only works out for a clockwise cycle, that releases more work than it consumes. The effective work is illustrated by the enclosed area of the changes of state in a p, V-diagram for a reversible operation with ignoring outer energies.

15.2 Formulation According to Clausius—Counterclockwise Cycle Processes

According to Clausius heat can not be shifted from a cold reservoir to a hot reservoir without any external efforts, see Fig. 15.5. If such a machine exists, the first law of thermodynamics, see Fig. 15.5a, obeys

$$Q_{in} = Q_{out}. \tag{15.21}$$

In steady state the energy being supplied to the system, needs to be equalised by the released energy. Thus, the first law of thermodynamics does not show any contradictions. As mentioned previously, the first law of thermodynamics does not restrict energy conversion. In order to evaluate the feasibility of such a process, a balance of entropy, see Fig. 15.5b needs to be applied. Again, in steady state operation, the entropy inside the machine needs to be constant over time, so that released entropy is balanced by supplied entropy:

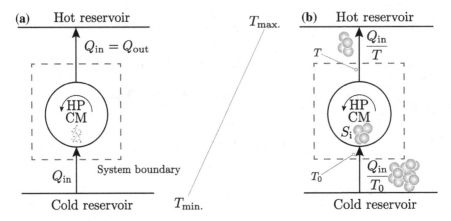

Fig. 15.5 Contradiction to the second law according to Clausius

$$\frac{Q_{in}}{T_0} + S_i = \frac{Q_{out}}{T} \tag{15.22}$$

This, in combination with the first law of thermodynamics, leads to

$$S_i = \frac{Q_{in}}{T} - \frac{Q_{in}}{T_0} < 0. \tag{15.23}$$

The entropy generation is negative, since, according to the temperature gradient between hot and cold reservoir, T is larger than T_0. Consequently, such a process is impossible! The amount of supplied entropy due to heat and internal generation of entropy is larger than the amount of entropy that is released by the heat. This results in the second law of thermodynamics according to Clausius:

Theorem 15.4 *Heat can never pass from a colder to a warmer reservoir by itself.*

15.2.1 The Cooling Machine/Heat Pump

Nevertheless, the question is, how a machine can be realised that takes thermal energy out of a cold reservoir and passes it to a warm reservoir. With respect to Fig. 15.5b the amount of entropy leaving the machine needs to be increased, so that the machine can be operated in steady state. According to the first law of thermodynamics in steady state, the heat respectively the entropy leaving the machine rise, if additional work is supplied, i.e.

$$Q_{in} + W = Q_{out} \tag{15.24}$$

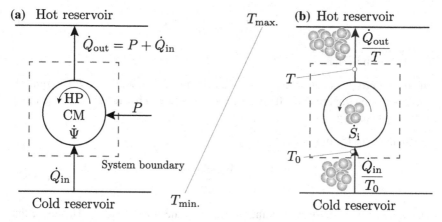

Fig. 15.6 Cooling machine/heat pump in black box notation

The derivative with respect to time leads to the fluxes, that are shown in Fig. 15.6a[4]:

$$\dot{Q}_{in} + P = \dot{Q}_{out} \tag{15.25}$$

However, applying the first law of thermodynamics leads to two unknowns (P and $\dot{Q}_{out}$) so that further information is required. This can be gained by an entropy balance in steady state:

$$\frac{dS}{dt} = 0 = \frac{\dot{Q}_{in}}{T_0} + \dot{S}_i - \frac{\dot{Q}_{out}}{T} \tag{15.26}$$

respectively, in an entropy-in-equals-entropy-out notation

$$\frac{\dot{Q}_{in}}{T_0} + \dot{S}_i = \frac{\dot{Q}_{out}}{T}. \tag{15.27}$$

Thus, the required heat to be released obeys

$$\dot{Q}_{out} = \frac{T}{T_0}\dot{Q}_{in} + \dot{S}_i T. \tag{15.28}$$

According to the first law of thermodynamics the required power for the operation then is

$$P = \dot{Q}_{out} - \dot{Q}_{in}. \tag{15.29}$$

Substituting the released heat results in

---

[4]Both notations are equivalent and lead to the same physical interpretations. Since the machine shall be operated in steady state, the indication of the fluxes is advantageous.

$$P = \dot{Q}_{\text{in}} \left(\frac{T}{T_0} - 1 \right) + \dot{S}_{\text{i}} T. \tag{15.30}$$

The conclusion is, that

- the more entropy is generated within the machine, the more power P is required.
- the larger the temperature difference between cold and hot reservoir is, the more power P is required.

The efficiency of the machine is

$$\varepsilon = \frac{\text{Benefit}}{\text{Effort}} \tag{15.31}$$

Usually, the efficiency of a machine is given by the greek letter η. However, the efficiency η by definition is in the range of $0 \ldots 1$. Since a heat pump, following Eq. 15.31, has an efficiency larger than 1, the coefficient of performance ε is used instead of an efficiency η. A distinction between cooling machine (refrigerator) and heat pump follows in the next section.

Cooling Machine

In case the machine is operated as a cooling machine, the purpose is to take heat from a cold reservoir.[5] Subjects in the fridge for instance release heat on a low temperature level. This heat is supplied to the cooling machine.[6] Consequently the coefficient of performance for a cooling machine (CM) is

$$\boxed{\varepsilon_{\text{CM}} = \frac{\dot{Q}_{\text{in}}}{P}} \tag{15.32}$$

Substitution of P leads to

$$\varepsilon_{\text{CM}} = \frac{\dot{Q}_{\text{in}}}{\dot{Q}_{\text{in}} \left(\frac{T}{T_0} - 1 \right) + \dot{S}_{\text{i}} T}. \tag{15.33}$$

The best machine works reversibly,[7] i.e. $\dot{S}_{\text{i}} = 0$, so that

$$\boxed{\varepsilon_{\text{CM,C}} = \frac{T_0}{T - T_0}} \tag{15.34}$$

---

[5]This machine is built for cooling!

[6]How this is realised will be clarified in part II.

[7]This is a Carnot-machine!

Heat Pump

In case the machine is operated as a heat pump, the purpose is to release heat to the hot reservoir.[8] However, the technical design of a heat pump is identical with a cooling machine. Consequently, the coefficient of performance for a heat pump (HP) obeys

$$\varepsilon_{\mathrm{HP}} = \frac{\dot{Q}_{\mathrm{out}}}{P} \qquad (15.35)$$

Dividing the first law of thermodynamics

$$\dot{Q}_{\mathrm{out}} = \dot{Q}_{\mathrm{in}} + P \qquad (15.36)$$

by the power P results in

$$\frac{\dot{Q}_{\mathrm{out}}}{P} = \frac{\dot{Q}_{\mathrm{in}}}{P} + 1. \qquad (15.37)$$

Introducing the coefficients of performance results in

$$\varepsilon_{\mathrm{HP}} = \varepsilon_{\mathrm{CM}} + 1 \qquad (15.38)$$

Hence, it is

$$\varepsilon_{\mathrm{HP}} = \frac{\dot{Q}_{\mathrm{in}}}{\dot{Q}_{\mathrm{in}}\left(\frac{T}{T_0} - 1\right) + \dot{S}_{\mathrm{i}}T} + 1. \qquad (15.39)$$

Again, the best machine works reversibly,[9] i.e. $\dot{S}_{\mathrm{i}} = 0$, so that:

$$\varepsilon_{\mathrm{HP,C}} = \frac{T}{T - T_0} \qquad (15.40)$$

15.2.2 Why Counterclockwise Cycle?

The explanation is the same as in Sect. 15.1.2. In this case the machine needs to be supplied with work, see Fig. 15.7. For visualisation the thermodynamic cycle is shown in a p, V-diagram and might be realised as a closed cylinder/piston system. To keep it simple, the process shown in this figure shall be reversible and outer energies are ignored, i.e. $W = W_{\mathrm{V}}$.

According to the p, V-diagram work is released by the fluid during expansion $(2){\rightarrow}(1)$, i.e.

---

[8]This machine is built for heating!
[9]This also is a Carnot-machine!

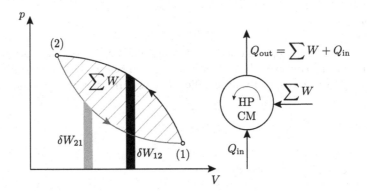

Fig. 15.7 Cooling machine/heat pump in a p, V-diagram

$$W_{21} = -\int_{2}^{1} p\,dV < 0. \qquad (15.41)$$

For the compression $(1) \rightarrow (2)$ work is supplied, i.e.

$$W_{12} = -\int_{1}^{2} p\,dV > 0. \qquad (15.42)$$

Work for compression respectively work, that is released, are both represented by the areas beneath the changes of state in a p, V-diagram. In order to achieve, that the overall process is supplied with work,[10] the change of state for expansion requires to be on a lower pressure level than the change of state for compression. According to Fig. 15.7 this only works out for a counterclockwise cycle, that always consumes more work than it releases work. The effective work is illustrated by the enclosed area of the changes of state in a p, V-diagram for a reversible operation with ignoring outer energies.

Problem 15.5 A heat pump uses ground water with a temperature of 8 °C as a heat source to heat a room with a temperature of 40 °C. The required thermal power for the heating purpose is 50 kW. The maximum electrical power consumption is 10 kW. Is the given task feasible? What is the heat flux that is extracted from the ground water?

---

[10]Only by doing so, heat can be *lifted* against the temperature gradient, as the entropy balance has shown!

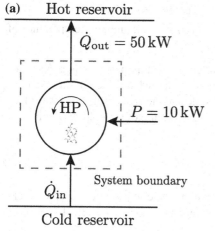

Fig. 15.8 Solution to Problem 15.5

Solution

According to Fig. 15.8a the first law of thermodynamics reads as

$$\dot{Q}_{in} + P = \dot{Q}_{out}. \tag{15.43}$$

The required heat from the ground water, i.e. the cold reservoir is

$$\dot{Q}_{in} = \dot{Q}_{out} - P = 40\,\text{kW}. \tag{15.44}$$

However, this does not answer the question, whether such a process is possible. Thus, a balance of entropy, see Fig. 15.8b is applied, i.e.

$$\frac{\dot{Q}_{in}}{T_0} + \dot{S}_i = \frac{\dot{Q}_{out}}{T}. \tag{15.45}$$

The entropy generation results in

$$\dot{S}_i = \frac{\dot{Q}_{out}}{T} - \frac{\dot{Q}_{in}}{T_0}. \tag{15.46}$$

This leads to

$$\dot{S}_i = 0.017395\frac{\text{kW}}{\text{K}} > 0. \tag{15.47}$$

Since $\dot{S}_i > 0$ the process is irreversible but possible!

Problem 15.6 The specific work $w = 100 \frac{kJ}{kg}$ is fed into a thermodynamic cycle. The machine releases a specific heat of $q_0 = 160 \frac{kJ}{kg}$ at an ambient temperature of $T_{env} = 290\,K$.

(a) Is the thermodynamic cycle a thermal engine, a heat pump or a cooling machine?
(b) What characteristic defines this cyclic process and what value does it reach?
(c) The thermodynamic cycle shall be reversible and the heat absorption takes place with a constant temperature. What is the temperature of the cold reservoir?

Solution

(a) Since work is effectively consumed by the machine, it must be a counterclockwise cycle. Heat is released at ambient temperature, so it is most likely a cooling machine.

(b) A counterclockwise cycle is characterised by the coefficient of performance. In this case it reads as

$$\varepsilon_{CM} = \frac{\text{Benefit}}{\text{Effort}} = \frac{q}{w}. \tag{15.48}$$

Applying the first law of thermodynamics results in

$$q + w = q_0. \tag{15.49}$$

Hence, it is

$$q = q_0 - w = 60 \frac{kJ}{kg}. \tag{15.50}$$

The coefficient of performance follows accordingly

$$\varepsilon_{CM} = \frac{q}{w} = 0.6. \tag{15.51}$$

(c) The balance of entropy for that case obeys

$$\frac{q}{T} + s_i = \frac{q_0}{T_{env}}. \tag{15.52}$$

Since the machine works reversibly, there is no generation of entropy, i.e. $s_i = 0$. Thus, it is

$$T = \frac{q}{q_0} T_{env} = 108.75\,K. \tag{15.53}$$

15.3 The Carnot-Machine

The recently discussed thermodynamic cycles have been introduced as *black boxes* so far. Clockwise cycles, i.e. thermal engines, convert heat into work. Counterclockwise cycles are heat pumps or cooling machines and lift heat from a lower temperature level upon a higher temperature level. The highest efficiency of machines like that can be achieved when being operated reversibly. In this section such a reversible machine, which is named after Nicolas Leonard Sadi Carnot (1796–1832), is investigated. The Carnot process shows, how heat with a certain temperature can be converted most efficiently into work by a thermal engine. It further clarifies how heat can pass most efficiently from a cold to a hot reservoir in terms of a cooling machine respectively heat pump.

The idealised cycle process according to Carnot consists of two reversible isothermal and two reversible adiabatic, i.e. isentropic, changes of state. Each change of state is reversible—including the heat transfer out of and into the reservoirs. In order to achieve this reversible heat transfer, the heat transfer from the machine respectively out of the machine must occur at the same temperature as the reservoirs, i.e. with a temperature potential $\Delta T \to 0$, see Chap. 14. Any temperature difference generates entropy during heat transfer and thus is irreversible. It has been shown, that the larger the temperature difference is, the more entropy is generated. However, heat transfer follows:

$$\dot{Q} = k A \underbrace{(T - T_{\text{env}})}_{\Delta T} \tag{15.54}$$

Hence, the heat flux is proportional to the temperature difference and to the surface where the heat transfer takes place. Consequently, for a reversible heat transfer ($\dot{S}_i \to 0$ at $\Delta T \to 0$) it follows that

$$k A \to \infty. \tag{15.55}$$

Consequently, the theoretical Carnot-machine needs to be infinite large!

15.3.1 The Carnot-Machine—Clockwise Cycle

First the focus is on a clockwise Carnot machine, i.e. a thermal engine, see Fig. 15.9a. In this figure the machine is given as a black box. It is important to emphasise, that the heat transfers between the machine and the two reservoirs run at $\Delta T = 0$ in order to keep the entire process reversible. The process internally does not generate entropy, i.e. $s_i = 0$, since the Carnot-machine follows a perfect, i.e. reversible, process. Hence, there is neither dissipation nor any other imperfections.

Now, it is investigated how the process runs *internally*, see also Fig. 15.9b and c, that show additionally a p, v-as well as a T, s-diagram. Obviously, the changes of state in a Carnot machine are as follows:

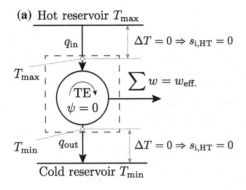

(a) Hot reservoir T_{max}

(b) $s = \text{const.}$

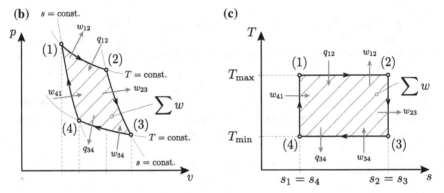

(c)

Fig. 15.9 Clockwise Carnot process—Thermal engine

- $(1) \rightarrow (2)$: Reversible, isothermal expansion

  In order to keep the temperature constant while expanding, heating is required. This is visible in the T, s-diagram, since the specific entropy rises. Since there is no dissipation the supplied specific heat is causing the increasing specific entropy. Following the p, v-diagram, the specific volume rises while the pressure decrease. Specific work is released by the machine.

- $(2) \rightarrow (3)$: Reversible, adiabatic expansion

  During this isentropic change of state, the specific entropy stays constant. The temperature drop is due to the released work during the expansion.

- $(3) \rightarrow (4)$: Reversible, isothermal compression

  In order to keep the temperature constant while compressing the fluid, cooling is required. This is visible in the T, s-diagram, since the specific entropy decreases. Since there is no dissipation the released specific heat is causing the decreasing

Fig. 15.10 Clockwise
Carnot process—T, s-
diagram

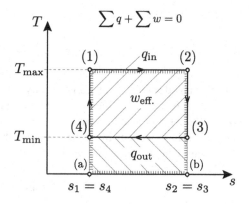

specific entropy. Following the p, v-diagram, the specific volume decreases while
the pressure rises. Specific work is supplied to the machine.

- $(4) \rightarrow (1)$: Reversible, adiabatic compression

During this isentropic change of state, the specific entropy stays constant. The
temperature rise is due to the supplied work during the compression.

Obviously, from $(1) \rightarrow (2) \rightarrow (3)$ specific work is released and can be illustrated by
the area beneath[11] the change of state in a p, v-diagram. The supplied specific work
is given by the area beneath $(3) \rightarrow (4) \rightarrow (1)$. Thus, the effective specific work is the
enclosed area in the p, v-diagram. The specific work gained follows from the first
law of thermodynamics for thermodynamic cycles, see Sect. 11.2:

$$\sum q + \sum w = 0. \tag{15.56}$$

Applied to the machine shown in Fig. 15.9 it reads as

$$q_{in} = w_{eff} + q_{out} \tag{15.57}$$

The specific heats q_{in} and q_{out} can be visualised in a T, s-diagram, see Fig. 15.10. As
the entire process, i.e. also the heat transfer, is reversible ($\psi = 0$), the areas beneath
the changes of state $(1) \rightarrow (2)$ and $(3) \rightarrow (4)$ show the transferred specific heats.
According to

$$w_{eff} = q_{in} - q_{in} \tag{15.58}$$

the enclosed hatched area in Fig. 15.10 represents the effective specific work. The
thermal efficiency in this case follows

$$\eta_{th} = \frac{w_{eff}}{q_{in}}. \tag{15.59}$$

---

[11]In case the Carnot machine is realised by a closed system!

Substitution of the effective specific work by Eq. 15.58 results in

$$\eta_{th} = \frac{q_{in} - q_{out}}{q_{in}}. \tag{15.60}$$

The exchanged specific heats can be easily calculated according to the T, s-diagram. Since all steps are reversible, the specific heats can be calculated by the rectangular areas. The specific heats in equation need to be positive since the sign has already been taken into account by the first law of thermodynamics, see Eq. 15.57. Thus, according to the T, s-diagram the specific heats are

$$q_{in} = T_{max}(s_2 - s_1) \tag{15.61}$$

respectively

$$q_{out} = T_{min}(s_3 - s_4). \tag{15.62}$$

Since $s_2 = s_3$ and $s_1 = s_4$ it follows

$$\eta_{th} = \frac{T_{max}(s_2 - s_1) - T_{min}(s_2 - s_1)}{T_{max}(s_2 - s_1)} \tag{15.63}$$

Rearranging results in

$$\boxed{\eta_{Carnot} = 1 - \frac{T_{min}}{T_{max}}} \tag{15.64}$$

This is exactly the same result that has already been derived with the black box approach, see Eq. 15.16.

15.3.2 The Carnot-Machine—Counterclockwise Cycle

The Carnot-machine can also be operated as a counterclockwise cycle, see Fig. 15.11. A reservoir releases heat Q_{in} at T_{min}, that is supplied to a Carnot-process. The machine releases a larger amount of heat at a higher level of temperature T_{max}. Thus, heat is lifted against the temperature gradient from a cold to a hot reservoir. In order to fulfil the steady state entropy balance, the released heat needs to be larger than the supplied heat. In specific notation it results in

$$q_{out} > q_{in}. \tag{15.65}$$

According to the first law of thermodynamics this can be achieved by supplying specific work

$$q_{out} = q_{in} + w_{eff}. \tag{15.66}$$

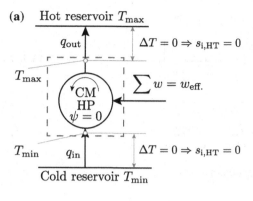

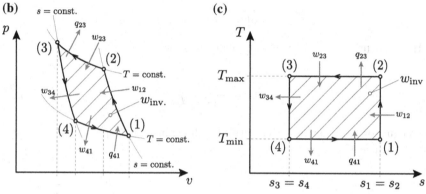

Fig. 15.11 Counterclockwise Carnot process—Cooling machine/heat pump

Following the principles of the previously discussed clockwise Carnot cycle, the specific heats can be calculated easily, i.e.

$$q_{in} = T_{min} (s_1 - s_4) \tag{15.67}$$

respectively

$$q_{out} = T_{max} (s_2 - s_3) . \tag{15.68}$$

Hence, the coefficients of performance can be calculated as follows:

• Cooling machine

$$\varepsilon_{CM} = \frac{q_{in}}{w_{eff}} = \frac{q_{in}}{q_{out} - q_{in}} \tag{15.69}$$

Substitution of the specific heats results in

$$\varepsilon_{CM} = \frac{T_{min} (s_1 - s_4)}{T_{max} (s_2 - s_3) - T_{min} (s_1 - s_4)} . \tag{15.70}$$

Hence, with $s_3 = s_4$ and $s_1 = s_2$ it finally reads as

$$\varepsilon_{\mathrm{CM}} = \frac{T_{\min}}{T_{\max} - T_{\min}} \tag{15.71}$$

- Heat pump

$$\varepsilon_{\mathrm{HP}} = \frac{q_{\mathrm{out}}}{w_{\mathrm{eff}}} = \frac{q_{\mathrm{out}}}{q_{\mathrm{out}} - q_{\mathrm{in}}} \tag{15.72}$$

Substitution of the specific heats results in

$$\varepsilon_{\mathrm{HP}} = \frac{T_{\max}(s_2 - s_3)}{T_{\max}(s_2 - s_3) - T_{\min}(s_1 - s_4)}. \tag{15.73}$$

Hence, with $s_3 = s_4$ and $s_1 = s_2$ it finally reads as

$$\varepsilon_{\mathrm{HP}} = \frac{T_{\max}}{T_{\max} - T_{\min}} \tag{15.74}$$

Both equations have already been derived with the black box approach as well, see Eqs. 15.32 and 15.40.

Chapter 16
Exergy

So far the law of energy conservation, i.e. the first law of thermodynamics, has been discussed, that quantified the changeability of energy. However, this changeability is limited, so that thermal energy can not be converted completely into mechanical energy in a steady state process. The second law of thermodynamics can be applied to evaluate the constraints of energy conversion. In Chap. 15 a clockwise Carnot machine was introduced: A machine that operates between two thermal reservoirs in order to convert thermal energy into maximum mechanical work. Its efficiency in best case is given by the minimum and maximum temperature the machine is working in-between. This principle is essential to understand the thermodynamic idea of exergy as maximum working capability of any form of energy. The significance of the exergy is presented in this chapter.

To get an idea of what the exergy is about, Fig. 16.1 provides a good introduction to the topic: Imagine two identical cups of tea, each filled with the same amount of tea at the same temperature, e.g. 50 °C, and the same pressure, so that their thermodynamic states are identical. Consequently, both of them have the same content of internal energy

$$U_{(a)} = U_{(b)}. \tag{16.1}$$

Anyhow, the cups are located in two different places: one at the North Pole, i.e. $-20\,°C$, and the other in the tropics, i.e. $50\,°C$. Now, the systems shall be evaluated in terms of their energy conversion potential, i.e. their working capability. For such an evaluation a thermal engine can be applied. In case 1, see Fig. 16.1a, the tea itself represents a hot reservoir, whereas the environment is the cold reservoir. Consequently, a heat flux arises due to the temperature difference between the two reservoirs. Hence, a thermal engine can be mounted to link the two reservoirs. Hence, the incoming heat flux can be collected to be converted into mechanical work, while releasing, due to the second law of thermodynamics, a part of this energy to the cold reservoir. Following Eq. 15.15 the conversion rate improves, the larger the temperature gap between the two reservoirs is.

© Springer Nature Switzerland AG 2019
A. Schmidt, *Technical Thermodynamics for Engineers*,
https://doi.org/10.1007/978-3-030-20397-9_16

Fig. 16.1 Evaluation of thermal energy sources

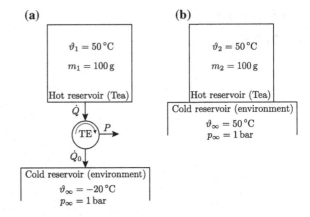

However, in case 2, see Fig. 16.1b, there is no temperature difference available, i.e. hot (tea) and cold (environment) reservoir are already in a thermodynamic equilibrium. Thus, it is not possible to operate a thermal engine in between. From a thermodynamic point of view the tea does not possess any working capability.

The maximum achievable working capability is called *exergy*. As the introducing example has shown, to evaluate the exergy of systems, the environment, i.e. the cold reservoir, needs be taken into account.

Definition 16.1 Exergy is the work, that can be gained maximally from the energy of a system when bringing it to a thermodynamic equilibrium with the environment.

Systems in *non-ambient* states posses exergy. Systems at ambient pressure and ambient temperature do not posses exergy and are worthless for any technical application. Strictly speaking, exergy is not a state value, since it depends on the environment's state as well. According to the definition made, energy E can be split into exergy E_x and an energetically worthless part, that can not be converted into work any further, i.e. the so-called anergy B_x. Thus, it reads as

$$E = E_x + B_x. \tag{16.2}$$

Unlimited convertible forms of energy solely consist of exergy, e.g. electrical energy or mechanical energy. Non convertible forms of energy solely consist of anergy, e.g. thermal energy at ambient temperature. Partially convertible energy consists of exergy and anergy. In contrast to energy,[1] exergy can be consumed and thus can decrease.[2] Consequently, exergy is no conservation quantity but it is an extensive property that can be balanced.[3] In the following sections exergy of several forms of energy are derived. This is required in order to balance exergy of open as well as closed systems. Furthermore, the cause for a loss of exergy is explained thermodynamically.

---

[1] The energy conservation principle claims, that energy must be constant.

[2] To keep the energy constant, anergy rises when exergy decreases!

[3] This balancing of exergy is usually performed with a *constant* environment.

Fig. 16.2 Gaining work
from heat

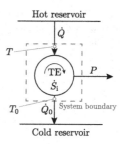

Hot reservoir

Cold reservoir

16.1 Exergy of Heat

In this first section the exergy of heat $E_{x,Q}$ is determined, i.e. how much work can
be gained maximally from thermal energy Q that is available at a temperature T.
In case a heat flux $\dot{Q}$ is of interest, there is a corresponding flux of exergy $\dot{E}_{x,Q}$.
The basic idea is to bring the heat into a thermal engine,[4] see Fig. 16.2. A thermal
engine splits thermal energy into mechanical and thermal energy. Consequently, a
thermal engine can be utilised as an *Energy-in-Exergy-and-Anergy-Decomposition-
Machine*. The maximum work can only be achieved in a reversible process, since
any irreversibility causes an increase of entropy and thus leads to a loss of working
capability. The efficiency of such a process has been already derived in Chap. 15:

$$\eta_{th} = \frac{P}{\dot{Q}} = 1 - \frac{T_0}{T} - \frac{T_0 \dot{S}_i}{\dot{Q}} \qquad (16.3)$$

With this finding the requirements for the thermal engine can be deduced:

- The introduced energy-in-exergy-and-anergy-decomposition-machine must therefore be free of any generation of entropy, i.e. $\dot{S}_i = 0$, in order to achieve the maximum work. Consequently, a Carnot machine is the best option to convert heat Q
  respectively a heat flux $\dot{Q}$ into work respectively power.
- To maximise the work, the released heat needs to be transferred at ambient temperature. Obviously, ambient temperature is the lower limit of the process, otherwise
  it would not be possible to transfer heat out of the machine into the environment.[5]
  In this theoretical limiting case no irreversibility occurs during the heat transfer,
  since the temperature difference reaches zero, see Sect. 15.3. As discussed previously the machine requires to be huge in size in order to transfer heat with a
  temperature difference of $\Delta T \rightarrow 0$.
- If the Carnot process releases heat above ambient temperature, working ability is
  wasted, since with the released heat above ambient temperature another Carnot

---

[4]A thermal engine has been introduced as a clockwise thermodynamic cycle, that converts heat into
work, see Chap. 15.

[5]In case the temperature at the cold end of the machine was *below* ambient temperature, there would
be a heat flux from environment *into* the machine!

Fig. 16.3 Energy-in-exergy-
and-anergy-decomposition-
machine

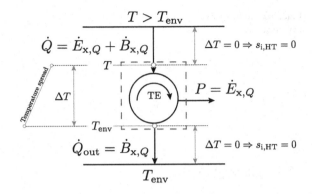

machine could be operated, since there still is a temperature potential against the environment.

Under these boundary conditions, see also Fig. 16.3, the power output is maximised and the exergy of the heat follows

$$\dot{E}_{x,Q} \overset{!}{=} P. \qquad (16.4)$$

The entering heat flux $\dot{Q}$ at a temperature $T > T_{env}$ consists of exergy $\dot{E}_{x,Q}$ and anergy $\dot{B}_{x,Q}$. Due to the second law of thermodynamics a heat flux $\dot{Q}_{out}$ must be released. Since the released heat flux is already at ambient temperature T_{env} it does not contain any exergy, i.e. no additional thermal engine can be operated due to the missing temperature gradient. Hence, $\dot{Q}_{out}$ is pure anergy. Under the discussed boundary conditions, the power output P is maximal and pure exergy as it is mechanical work. The principle, that heat has to be released at ambient temperature in order to achieve the maximum power output can be further explained with a T, s-diagram, see Fig. 16.4. In this figure a clockwise cycle[6] is shown. Specific heat at a constant temperature T, i.e. $q @ T$ has to be evaluated, is supplied to the thermal engine from (1) $\rightarrow$ (2). The amount of specific heat is given by the integral of this change of state. The released specific heat can be illustrated similarly: In this case the area beneath (3) $\rightarrow$ (4) represents the specific released heat. However, the enclosed area (1) (2) (3) (4) indicates the output of specific work. In order to determine the specific exergy this enclosed area has to be maximised. In Fig. 16.4a specific exergy is wasted, since the specific work is not yet maximised. Instead, Fig. 16.4b shows the maximum specific work. The upper limit of the enclosed area can not be touched, as the specific heat at that specific temperature T has to be investigated. Consequently, the specific heat release needs to be optimised. The maximum specific work is achieved, when the specific heat is released at ambient temperature. It is not possible to fall below ambient temperature since in that case it would be unfeasible to emit heat due to the

---

[6]The Energy-in-Exergy-and-Anergy-Decomposition-Machine must be a clockwise cycle!

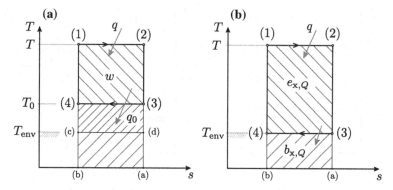

Fig. 16.4 Heat at constant temperature → evaluation of the exergy of the heat $q @ T$ with a Carnot-machine

second law of thermodynamics.[7] Under these conditions the specific exergy $e_{x,Q}$ can be illustrated as the hatched area (1) (2) (3) (4). Accordingly, the specific anergy $b_{x,Q}$ follows (3) (4) (a) (b), see Fig. 16.4b. The entire specific heat is the enclosed area (1) (2) (b) (a), i.e.

$$q = e_{x,Q} + b_{x,Q}. \tag{16.5}$$

Following the thermal efficiency of a Carnot-machine, see Eq. 15.64, the exergy can be calculated as follows:

$$\eta_{\text{Carnot}} = \frac{P_{\max}}{\dot{Q}} = 1 - \frac{T_{\text{env}}}{T}. \tag{16.6}$$

Solving for $P_{\max}$ results in

$$\dot{E}_{x,Q} \overset{!}{=} P_{\max} = \dot{Q} \underbrace{\left(1 - \frac{T_{\text{env}}}{T}\right)}_{=\eta_{\text{Carnot}}} \tag{16.7}$$

If not a flux is of interest but an amount of heat, it is accordingly

$$E_{x,Q} \overset{!}{=} W_{\max} = Q \left(1 - \frac{T_{\text{env}}}{T}\right) \tag{16.8}$$

In order to evaluate the exergy of heat, it is therefore necessary to know the temperature at which the heat is available. The larger the temperature compared to ambient temperature is, the more exergy the heat possess.

---

[7]The second law of thermodynamics states, that heat always follows a temperature gradient form hot to cold.

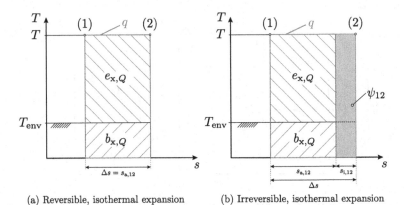

(a) Reversible, isothermal expansion   (b) Irreversible, isothermal expansion

Fig. 16.5 Exergy of heat at isothermal expansion according to [2]

Heat at Constant Temperature

For the special case of an isothermal expansion the exergy can be easily illustrated in a T, s-diagram, see Fig. 16.5. In case the change of state runs reversibly, the explanation is exactly as it was given above: The specific exergy is the hatched area between T and T_{env}, see Fig 16.5a. The integral from (1) to (2) represents the specific heat, since there is no dissipation. However in case b), i.e. an irreversible, isothermal expansion, the situation is slightly different, because specific dissipation occurs as well. Since the temperature is constant, the integral form (1) to (2) is re-ordered into a portion belonging to the specific heat q and another one belonging to the specific dissipation ψ_{12}: The entire change of specific entropy is due to one part caused by heating, i.e. $s_{a,12}$, and a second part due to dissipation, i.e. $s_{i,12}$. This approach can be applied, since specific heat q and specific dissipation ψ occur at the same temperature, that from (1) to (2) shall be constant.[8]

Heat at Variable Temperature

In case, heat is transferred at a variable temperature, e.g. the heat exchanger in Fig. 13.21, the exergy illustration in a T, s-diagram is more complex, see Fig. 16.6. If the change of state (1) → (2) runs reversibly, the supplied specific heat is the area beneath the change of state. In combination with the released specific heat at ambient temperature the specific exergy can be illustrated as hatched area as shown in Fig. 16.6a. This information can not be gained in case the change of state is irreversible, see Fig. 16.6b. Specific heat and specific dissipation occur *simultaneously*, a re-ordering as it has been done in Fig. 16.5b, can not be performed, as there is only one thermodynamic mean temperature for both, specific heat as well as specific dissipation. Even under these conditions the specific exergy of specific heat can be

---

[8]Thus, the thermodynamic mean temperature is constant as well. However, this is nontrivial if T is not constant, see Fig. 16.6.

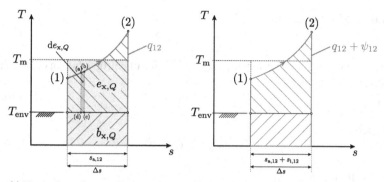

(a) Non-isothermal, reversible heat supply   (b) Non-isothermal, irreversible heat supply

Fig. 16.6 Non-isothermal heat supply (1) → (2)

calculated with a generic approach.[9] Since the temperature is variable a differential approach is followed. Equation 16.7 can be applied at any incremental, quasi-static change of state. Within this incremental step only a small portion of heat $\delta\dot{Q}$ is transferred at the current temperature T, see Fig. 16.6a:

$$d\dot{E}_{x,Q} = \left(1 - \frac{T_{env}}{T}\right)\delta\dot{Q}.$$
(16.9)

In order to calculate the entire exergy, an integration is done, i.e.

$$\dot{E}_{x,Q} = \int_1^2 d\dot{E}_{x,Q} = \int_1^2 \left(1 - \frac{T_{env}}{T}\right)\delta\dot{Q}.$$
(16.10)

Simplifying leads to

$$\dot{E}_{x,Q} = \dot{Q} - T_{env}\int_1^2 \frac{\delta\dot{Q}}{T}.$$
(16.11)

With the thermodynamic mean temperature, see Sect. 13.7.3, it results in

$$\dot{E}_{x,Q} = \dot{Q} - T_{env}\frac{\dot{Q}}{T_m}.$$
(16.12)

Hence, the generic approach for calculating the exergy of heat is

$$\boxed{\dot{E}_{x,Q} = \dot{Q}\left(1 - \frac{T_{env}}{T_m}\right)}$$
(16.13)

---

[9]No matter, if the change of state runs reversibly or irreversibly!

With Eq. 13.151 it reads as

$$\dot{E}_{x,Q} = \dot{Q} - T_{env}\dot{S}_a.$$ (16.14)

The flux of entropy $\dot{S}_a$ can be substituted by a balance of entropy. For an open system for instance it reads as

$$\dot{S}_a + \dot{S}_i = \dot{m}(s_2 - s_1)$$ (16.15)

respectively

$$\dot{S}_a = \dot{m}(s_2 - s_1) - \dot{S}_i.$$ (16.16)

Hence, finally it obeys

$$\boxed{\dot{E}_{x,Q} = \dot{Q} - \dot{m}T_{env}(s_2 - s_1) + T_{env}\dot{S}_i}$$ (16.17)

Obviously, this generic approach can be applied for reversible and irreversible changes of state!

In case an amount of heat is regarded,[10] it follows accordingly

$$\boxed{E_{x,Q} = Q\left(1 - \frac{T_{env}}{T_m}\right)}$$ (16.18)

With Eq. 13.164 it reads as

$$E_{x,Q} = Q - T_{env}S_a.$$ (16.19)

The entropy S_a might be substituted by a balance of entropy. For closed systems it obeys

$$S_a + S_i = S_2 - S_1.$$ (16.20)

Thus, it finally is

$$\boxed{E_{x,Q} = Q - T_{env}(S_2 - S_1) + T_{env}S_i}$$ (16.21)

Equation 16.11 respectively 16.18 show, that

- at large ambient temperature the exergy of heat is small and the anergy large. If ambient temperature is small, this is reverse!
- the larger the temperature T is at which heat occurs, the larger the exergy is!

This correlation is shown in Fig. 16.7. It is

$$Q = E_{x,Q} + B_{x,Q}.$$ (16.22)

---

[10] A closed system for example!

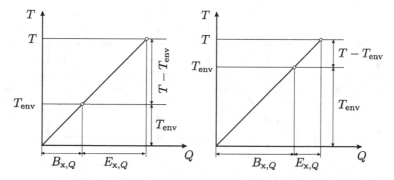

Fig. 16.7 Exergy of heat—influence of the environmental temperature according to [1]

For the anergy it follows

$$B_{x,Q} = Q - E_{x,Q}.$$ (16.23)

In combination with the exergy of heat these equations simplify to

$$E_{x,Q} = Q - Q\frac{T_{env}}{T}$$ (16.24)

and

$$B_{x,Q} = Q\frac{T_{env}}{T}.$$ (16.25)

Dividing these two equations results in so-called the lever rule, that is shown in Fig. 16.7.

$$\boxed{\frac{E_{x,Q}}{B_{x,Q}} = \frac{T - T_{env}}{T_{env}}}$$ (16.26)

Sign of the Exergy of Heat

Figure 16.8 shows the consequences of

$$\boxed{\dot{E}_{x,Q} = \dot{Q}\underbrace{\left(1 - \frac{T_{env}}{T}\right)}_{=\eta_{Carnot}}}$$ (16.27)

regarding the sign of exergy of heat, since the Carnot factor η_{Carnot} gets negative for temperatures $T < T_{env}$. In this figure the focus is on a body with a temperature T that is identified by a dashed line as system boundary. The following cases can occur:

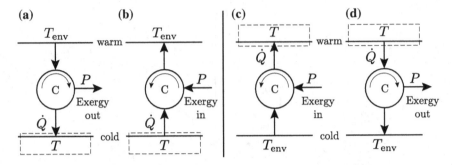

Fig. 16.8 Sign of exergy of heat

- Case (a): Heat supply, i.e. $\dot{Q} > 0$, at $T < T_{env}$

  In this case it is

$$\dot{E}_{x,Q} = \underbrace{\dot{Q}}_{>0} \underbrace{\left(1 - \frac{T_{env}}{T}\right)}_{<0} < 0 \qquad (16.28)$$

According to Fig. 16.8a exergy leaves, so that $\dot{E}_{x,Q} < 0$. Heat and exergy have opposite signs.

- Case (b): Heat release, i.e. $\dot{Q} < 0$, at $T < T_{env}$

  In this case it is

$$\dot{E}_{x,Q} = \underbrace{\dot{Q}}_{<0} \underbrace{\left(1 - \frac{T_{env}}{T}\right)}_{<0} > 0 \qquad (16.29)$$

According to Fig. 16.8b exergy is required,[11] so that $\dot{E}_{x,Q} > 0$. Heat and exergy have opposite signs.

- Case (c): Heat supply, i.e. $\dot{Q} > 0$, at $T > T_{env}$

  In this case it is

$$\dot{E}_{x,Q} = \underbrace{\dot{Q}}_{>0} \underbrace{\left(1 - \frac{T_{env}}{T}\right)}_{>0} > 0 \qquad (16.30)$$

According to Fig. 16.8c exergy is required, so that $\dot{E}_{x,Q} > 0$. Heat and exergy have the same sign.

- Case (d): Heat release, i.e. $\dot{Q} < 0$, at $T > T_{env}$

  In this case it is

$$\dot{E}_{x,Q} = \underbrace{\dot{Q}}_{<0} \underbrace{\left(1 - \frac{T_{env}}{T}\right)}_{>0} < 0 \qquad (16.31)$$

---

[11] To lift $\dot{Q}$ against the temperature gradient!

According to Fig. 16.8d exergy leaves, so that $\dot{E}_{x,Q} < 0$. Heat and exergy have the same sign.

16.2 Exergy of Fluid Flows

In this section it is clarified how much working capability a fluid flow contains. The maximum working capability is named exergy of fluid flows and is required to perform exergy balances with open systems. A fluid flow possesses working capability due to its mechanical energy, i.e. kinetic and potential energy, and due to its thermal energy when being in a thermodynamic imbalance with the environment,[12] see Fig. 16.9a. In order to get the maximal power output, the fluid is brought in a thermodynamic, i.e. mechanical and thermal, equilibrium with its environment, see Fig. 16.9b. This is achieved in a machine, that is supplied with the fluid in state (1) and that operates in steady state. Goal is to bring the fluid into equilibrium with the environment: Hence, the flow leaves the machine at ambient conditions, i.e. p_{env}, T_{env}, $c_{env} = 0$, $z_{env} = 0$. To bring the fluid from (1) to (env), technical work and heat can be exchanged with the environment. However, the task is to maximise the technical work as this is the exergy of the fluid flow[13] (1), i.e.

$$\dot{E}_{x,S,1} \overset{!}{=} -P_{t,rev}. \tag{16.32}$$

Obviously, the machine needs to work reversibly, since any dissipation reduces the power output, see Sect. 11.3.4. Thus, the fluid needs to be shifted reversibly, i.e. $\psi_{1env} = 0$, from state (1) to ambient state (env). For this change of state the first law of thermodynamics can be applied[14]:

$$\dot{Q}_{rev} + P_{t,rev} = \dot{m} \left[h_{env} - h_1 + \frac{1}{2} \left(c_{env}^2 - c_1^2 \right) + g \left(z_{env} - z_1 \right) \right] \tag{16.33}$$

Not only mechanical dissipation should be avoided, but heat transfer must be reversible as well, see Chap. 14. Thus, heat must be transferred with a $\Delta T \to 0$. Consequently, the heat transfer area needs to be huge in size. Since the environment is in rest and does not contain potential energy, the first law of thermodynamics results in

$$\dot{E}_{x,S,1} = \dot{m} \left[h_1 - h_{env} + \frac{1}{2} c_1^2 + g z_1 \right] + \dot{Q}_{rev}. \tag{16.34}$$

---

[12]It has been shown in the previous section, that thermal energy can be utilised in a thermal engine to gain work.

[13]Due to the sign convention a power output is negative. Thus, in order to get a positive exergy, a minus sign is applied in the definition.

[14]Following the *energy-in-is-balanced-by-energy-out* principle under steady state conditions.

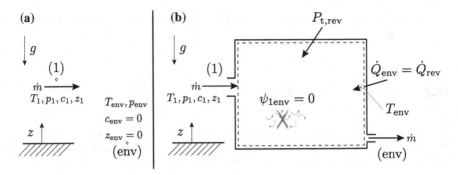

Fig. 16.9 Exergy of a fluid flow

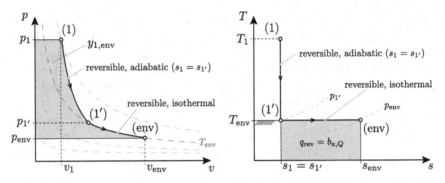

Fig. 16.10 Change of state (1) → (env) with $p_{1'} > p_{env}$, according to [2]

In the next step the reversible heat transfer $\dot{Q}_{rev}$ needs to be specified. The idea is illustrated in Fig. 16.10. The change of state is split into two sub-steps to get the maximal power output. At first, the change of state shall be reversible and adiabatic,[15] i.e. from (1) to (1'). If it is not adiabatic, heat is transferred at a temperature different than ambient temperature, so that exergy, i.e. working capability, is transferred as well. In order to avoid this, the first change of state to (1') is done until ambient temperature T_{env} is reached. In a second step, from (1') to (env) heat can then be transferred reversibly, since system and environment have the same temperature, i.e. $\Delta T \to 0$. Heat can be transferred reversibly in both directions, though an infinite heat transfer area would be required! According to the T, s-diagram in Fig. 16.10, the transferred heat then is pure anergy, so it does not affect our exergy considerations. With this approach the reversible heat can easily be calculated, i.e.

$$\mathrm{d}s = \delta s_a + \underbrace{\delta s_i}_{=0} = \delta s_a = \frac{\delta q_{rev}}{T} \tag{16.35}$$

---

[15]Reversible and adiabatic means the change of state is isentropic!

respectively

$$\delta q_{rev} = T \, ds. \tag{16.36}$$

Since $T = T_{env} = $ const. and $s_1 = s_{1'}$ it follows for the reversible transferred heat:

$$q_{rev} = \int\limits_{1'}^{env} T \, ds = T_{env} \left(s_{env} - s_1 \right). \tag{16.37}$$

The heat flux follows by multiplying with the mass flow rate $\dot{m}$ and obeys

$$\dot{Q}_{rev} = \dot{m} T_{env} \left(s_{env} - s_1 \right). \tag{16.38}$$

Thus, the exergy of the flux follows by substituting the heat in Eq. 16.34, i.e.

$$\dot{E}_{x,S,1} = \dot{m} \left[h_1 - h_{env} + \frac{1}{2}c_1^2 + gz_1 + T_{env} \left(s_{env} - s_1 \right) \right] \tag{16.39}$$

This equations shows, that kinetic as well as potential energies are pure exergy! For the anergy of a fluid flow it is

$$\dot{B}_{x,S,1} = \dot{m} \left[h_{env} - T_{env} \left(s_{env} - s_1 \right) \right] \tag{16.40}$$

The sum of exergy and anergy flux is the overall energy flux of state (1), i.e.

$$\dot{E}_{x,S,1} + \dot{B}_{x,S,1} = \dot{m} \left[h_1 + \frac{1}{2}c_1^2 + gz_1 \right] = \dot{E}_1. \tag{16.41}$$

However, let us have a closer look at Fig. 16.10, that shows how the machine works in a p, v-as well as in a T, s-diagram. The T, s-diagram has already been explained, as it shows the reversible heat at ambient temperature, i.e. pure anergy, that is required to reach ambient state (env). It is, in this example, obvious, that the fluid flow has to be heated, since the specific entropy from $(1')$ to (env) increases. Due to $\Delta T \to 0$ this can only be achieved with an infinite heat transfer area. In this borderline case, the environment supplies heat reversibly to the system. In order to interpret the p, v-diagram it makes sense to apply the partial energy equation for the open system from (1) to (env), i.e.

$$\dot{E}_{x,S,1} = -P_{t,rev} = -\dot{m} \left[\underbrace{\int\limits_{1}^{env} v \, dp + \underbrace{\psi_{1env}}_{=0}}_{=y_{1,env}} + \frac{1}{2} \left(c_{env}^2 - c_1^2 \right) + g \left(z_{env} - z_1 \right) \right]. \tag{16.42}$$

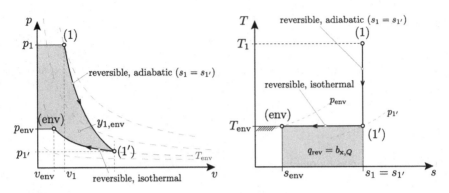

Fig. 16.11 Change of state (1) → (env) with $p_{1'} < p_{env}$, according to [2]

Thus, in the p, v-diagram, see Fig. 16.10, the specific pressure work $y_{1,env}$ is shown. Since the mechanical energy, i.e. pure exergy, is missing, the specific exergy is not illustrated in that diagram. Mind, that the specific pressure work for open systems is the projected area to the p-axis. In this example, the fluid releases specific pressure work from (1) to (1') and from (1') to (env). This is due to $p_{1'} > p_{env}$ after the isentropic change of state from (1) to (1').

Another example is shown in Fig. 16.11. In this case, after the isentropic change of state to (1'), it is $p_{1'} < p_{env}$. According to the T, s-diagram under these conditions a reversible cooling from (1') to (env) is required as the specific entropy needs to decrease. Once again, the released specific heat is pure anergy since it occurs at ambient temperature. The p, v-diagram implies, that pressure work is released from (1) to (1'), while pressure work needs to be supplied from (1') to (env). Hence, the resulting, effective specific pressure work is illustrated as grey area.

16.3 Exergy of Closed Systems

It has been shown how to evaluate the exergy of heat as well as of fluid flows. In this section the focus is on closed systems. It is investigated how much work can be gained maximally out of a closed system. This maximum effective work is named exergy of a closed system. According to Fig. 16.12 working capability is due to mechanical energy, i.e. kinetic and potential energy as the centre of gravity of a closed system can be in motion, and due to a thermodynamic imbalance with the environment. The gas inside a closed system can have a different state (1) than the environment (env), so that work can be released by the piston during the transient balancing process (1) → (env). However, the change of state from (1) to (env) needs to be reversible in

Fig. 16.12 Exergy of a closed system

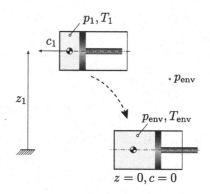

order to maximise the work. The exergy of the closed system in state (1) is defined as[16]:

$$E_{x,1} \overset{!}{=} -W_{eff,rev} = -W_{rev} - p_{env}m\left(v_{env} - v_1\right). \qquad (16.43)$$

The effective work[17] $W_{eff,rev}$ has been introduced in Sect. 9.2.3. It describes the work that can be utilised in a cylinder/piston system, when operated under ambient conditions. The environment supports the compression of a gas, while it reduces the released work at expansion. Thus, it is the effective work that comes along with the definition of exergy, as it takes the environment into account. The first law of thermodynamics for a reversible process reads as[18]

$$W_{rev} + Q_{rev} = U_{env} - U_1 - m\left(\frac{1}{2}c_1^2 + gz_1\right). \qquad (16.44)$$

Thus, in a first step the system is shifted to ambient temperature in an adiabatic, reversible change of state, see Fig. 16.13. In this first step, heat transfer must be avoided, since heat at a different temperature than environment contains exergy. In order to achieve the maximum output of work, transfer of exergy at that point is not wise. Furthermore, heat transfer is only reversible, in case $\Delta T \rightarrow 0$. Consequently, if heat transfer is required to achieve the ambient state (env), this is done in a second step, once ambient temperature is reached. Heat at ambient temperature is pure anergy. In case $p_{1'} > p_{env}$ the system needs to be heated, see T, s-diagram in Fig. 16.13. If $p_{1'} < p_{env}$, see T, s-diagram in Fig. 16.14, the system needs to be cooled in the second change of state from (1') to (env). Mind, that the heat transfer under these conditions is reversible[19] but it needs an infinite heat transfer area! According to the

---

[16]Due to the sign convention work release is negative. Hence, in order to get a positive exergy, a minus sign is applied in the definition.

[17]In Eq. 16.43 v_{env} denotes the specific volume the fluid has under ambient conditions, i.e. at T_{env} and p_{env}. It does not represent the specific volume of the environment!

[18]Mind, that the (env)-state is in rest!

[19]No entropy is generated, since $\Delta T = 0$ for the heat transfer!

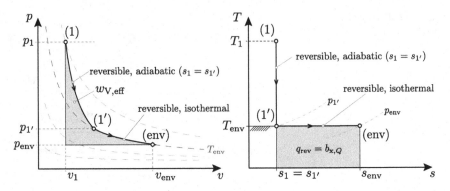

Fig. 16.13 Change of state (1) → (env) with $p_{1'} > p_{env}$

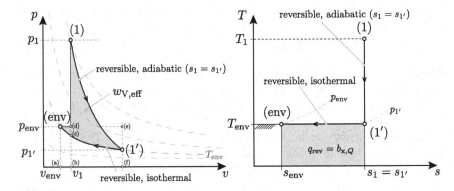

Fig. 16.14 Change of state (1) → (env) with $p_{1'} < p_{env}$

T, s-diagram the heat results in

$$Q_{rev} = m T_{env} (s_{env} - s_1). \tag{16.45}$$

Thus, the exergy of a closed system obeys

$$E_{x,1} = U_1 - U_{env} + m \left[\frac{1}{2} c_1^2 + g z_1 + T_{env} (s_{env} - s_1) - p_{env} (v_{env} - v_1) \right] \tag{16.46}$$

The anergy is

$$B_{x,1} = U_{env} + m \left[p_{env} (v_{env} - v_1) - T_{env} (s_{env} - s_1) \right] \tag{16.47}$$

since the sum of Eqs. 16.46 and 16.47 leads to

$$E_{x,1} + B_{x,1} = U_1 + m \left[\frac{1}{2} c_1^2 + g z_1 \right] = E_1. \tag{16.48}$$

However, let us have a closer look at Fig. 16.13, that shows the changes of state in p, v-as well as in a T, s-diagram. The T, s-diagram has already been explained, as it shows the reversible heat at ambient temperature, i.e. pure anergy, that is required to reach ambient state (env). Obviously, in case $p_{1'} > p_{env}$, the fluid flow has to be heated, since the specific entropy from (1') to (env) increases. Due to $\Delta T \rightarrow 0$ this can only be achieved with an infinite heat transfer area. In order to interpret the p, v-diagram it makes sense to apply the partial energy equation for a closed system from (1) to (env), i.e.

$$W_{rev} = W_V + \underbrace{\Psi}_{=0} + W_{mech} = -\int_1^{env} p\, dV + m\left[\frac{1}{2}\left(c_{env}^2 - c_1^2\right) + g\left(z_{env} - z_1\right)\right].$$

(16.49)

Thus, the effective work is[20]

$$W_{eff,rev} = W_{rev} + p_{env} m\left(v_{env} - v_1\right)$$

(16.50)

respectively

$$W_{eff,rev} = \underbrace{-\int_1^{env} p\, dV + p_{env} m\left(v_{env} - v_1\right)}_{w_{V,eff}} + m\left[\frac{1}{2}\left(c_{env}^2 - c_1^2\right) + g\left(z_{env} - z_1\right)\right].$$

(16.51)

Consequently, in the p, v-diagram, see Fig. 16.13, the specific effective volume work $w_{V,eff}$ is shown. Since the mechanical energy, i.e. pure exergy, is missing, the specific exergy is not illustrated in that diagram. Mind, that the specific effective volume work for closed systems is the projected area to the v-axis reduced by the work at the environment, see also Fig. 9.7. In this example, the fluid releases specific volume work from (1) to (1') and from (1') to (env), since the fluids expands. This is due to $p_{1'} > p_{env}$ after the isentropic change of state from (1) to (1').

Another example is shown in Fig. 16.14. In this case after the isentropic change of state to (1'), it is $p_{1'} < p_{env}$. According to the T, s-diagram under these conditions a reversible cooling from (1') to (env) is required as the specific entropy needs to decrease. Once again, the released specific heat is pure anergy since it occurs at ambient temperature. The specific effective volume work $w_{V,eff}$ in the p, v-diagram for that case is illustrated as grey area. Obviously, the specific effective volume work is composed like this:

---

[20] See Eq. 9.42.

- Let us first consider the change of state from (1) to (1′) to (c)
  From (1) to (1′) volume work is released (−), i.e. $\overline{1, 1', f, b, 1}$, which is reduced effectively by the environment (+), i.e. $\overline{b, d, e, f, b}$. From (1′) to (c) volume work is supplied (+), i.e. $\overline{1', f, b, c, 1'}$, which is supported effectively by the environment (−), i.e. $\overline{b, d, e, f, b}$. Thus, in total the environmental work $\overline{b, d, e, f, b}$ is neutralised. The specific effective volume work for this first step is $\overline{1, 1', c, 1}$.
- In the next step the final change of state from (c) to (env) is investigated
  From (c) to (env) volume work is supplied (+), i.e. $\overline{c, env, a, b, c}$, which is supported effectively by the environment (−), i.e. $\overline{a, b, d, env, a}$. According to the sketch in Fig. 16.14 the absolute value of the environmental work is larger than the supplied work for the compression. Consequently, the overall work that is released (−) from (c) to (env) results in $\overline{d, c, env, d}$!

Problem 16.2 Two equal, vertical standing pipes with a diameter of $D = 0.1$ m are connected with each other by a thin pipe and a valve, see Problem 13.14. One pipe is filled up to a height of $H = 10$ m with water, while the second one is empty. Water has a density of $\rho = 1000 \frac{\text{kg}}{\text{m}^3}$. Now the valve is opened. After a while a balance of the water volume develops. Calculate the exergy of the system in state (1) respect. in state (2)! How much exergy is lost during the change of state?

- Specific heat capacity (water): $c_p = 4, 18 \frac{\text{kJ}}{\text{kg K}}$
- Environmental state ($p_{env} = 1$ bar, $T_{env} = 293.15$ K)
- Initial temperature of the water $T_1 = 293.15$ K
- The system does not exchange any energy with the environment.

Solution

According to Eq. 16.46 the exergy in state (1) is

$$E_{x,1} = U_1 - U_{env} + m \left[\frac{1}{2} c_1^2 + g z_1 + T_{env}(s_{env} - s_1) - p_{env}(v_{env} - v_1) \right]. \quad (16.52)$$

It simplifies as follows, since state (1) is equal with state (env)

- Internal energy
$$U_1 - U_{env} = m c_p (T_1 - T_{env}) = 0 \quad (16.53)$$

- Outer energies
$$\frac{1}{2} c_1^2 + g z_1 = g \frac{H}{2} \quad (16.54)$$

- Entropy[21]

---

[21] Applying the caloric equation for incompressible liquids.

$$T_{\text{env}}(s_{\text{env}} - s_1) = T_{\text{env}}c_p \ln\frac{T_{\text{env}}}{T_1} = 0 \tag{16.55}$$

- Ambient work[22]

$$p_{\text{env}}(v_{\text{env}} - v_1) = 0 \tag{16.56}$$

Thus, the exergy in state (1) results in

$$E_{x,1} = mg\frac{H}{2} = 3852.4\,\text{J}. \tag{16.57}$$

The working capability only consists of the potential energy of the water, since the internal state is already in balance with the environment.

Now, let us investigate state (2). The exergy follows

$$E_{x,2} = U_2 - U_{\text{env}} + m\left[\frac{1}{2}c_c^2 + gz_2 + T_{\text{env}}(s_{\text{env}} - s_2) - p_{\text{env}}(v_{\text{env}} - v_2)\right]. \tag{16.58}$$

It can be simplified, since state (2) is also in rest and the fluid is still incompressible, i.e.

$$E_{x,2} = mc_p\left(T_2 - T_{\text{env}}\right) + mg\frac{H}{4} + mT_{\text{env}}c_p \ln\frac{T_{\text{env}}}{T_2}. \tag{16.59}$$

The temperature rise has already been calculated in Problem 12.8, i.e. $T_2 = 293.155867\,\text{K}$. Hence, the exergy in state (2) is

$$E_{x,2} = 1926.2\,\text{J}. \tag{16.60}$$

In state (2) the exergy is composed of mechanical energy as well as thermal energy, since the system's temperature is slightly larger than ambient temperature.

Obviously, since the system does not exchange any energy with the ambient,[23] there is a loss of exergy, i.e.

$$\Delta E_{x,V} = E_{x,1} - E_{x,2} = 1926.2\,\text{J}. \tag{16.61}$$

---

[22]Water is treated as an incompressible liquid, so that $v_{\text{env}} = v_1$.

[23]The system is adiabatic, so there is no exchange of exergy of heat! Furthermore, no electrical or mechanical work, that would be pure exergy, is transferred.

Fig. 16.15 Closed system: balance of exergy

16.4   Loss of Exergy

Problem 16.2 has shown, that a system can lose exergy. Obviously, there must be an internal reason for this loss of exergy, since the discussed system has not transferred any exergy across its border. Although exergy is not a conservation quantity such as mass, momentum and energy, it can be balanced. In this section the cause for a loss of exergy is investigated. To do so, a distinction between closed/open systems and thermodynamic cycles is made.

16.4.1   Closed System

The loss of exergy $\Delta E_{x,V}$ in a closed system can easily be determined by comparison of states (1) and (2). Starting from state (1) exergy increases by supply of exergy, i.e. due to heat and effective work. Work is pure exergy whereas heat contains partly exergy. It is postulated, that exergy is reduced by a loss of exergy as seen in Problem 16.2.

Thus, according to Fig. 16.15, for state (2) it is

$$E_{x,2} = E_{x,1} + W_{\text{eff}} + E_{x,Q} - \Delta E_{x,V}. \tag{16.62}$$

In other words, the exergy in state (2) is equal to the exergy initially available in state (1) plus the exergy that is supplied from (1) to (2) by heat and work, reduced by the loss of exergy.[24] Consequently, the loss of exergy follows

$$\boxed{\Delta E_{x,V} = E_{x,1} - E_{x,2} + W_{\text{eff.}} + E_{x,Q}} \tag{16.63}$$

The exergies for a closed system $E_{x,1}$ as well as $E_{x,2}$ can be substituted by the correlations, that have been deduced in Sect. 16.3. This results in

---

[24]Obviously, loss of exergy means a *sink*!

$$E_{x,1} - E_{x,2} = U_1 - U_2 +$$

$$+ m \left[\frac{1}{2} \left(c_1^2 - c_2^2 \right) + g \left(z_1 - z_2 \right) + T_{env} \left(s_2 - s_1 \right) - p_{env} \left(v_2 - v_1 \right) \right]. \tag{16.64}$$

The effective work reads as[25]:

$$W_{eff} = W_{12} + p_{env} \left(V_2 - V_1 \right). \tag{16.65}$$

Furthermore, the exergy of heat, see Sect. 16.1, reads as

$$E_{x,Q} = Q_{12} - T_{env} \left(S_2 - S_1 \right) + T_{env} S_{i,12}. \tag{16.66}$$

Applying the first law of thermodynamics, i.e.

$$W_{12} + Q_{12} = U_2 - U_1 + m \left[\frac{1}{2} \left(c_2^2 - c_1^2 \right) + g \left(z_2 - z_1 \right) \right] \tag{16.67}$$

results in the loss of exergy for a closed system:

$$\boxed{\Delta E_{x,V} = T_{env} S_{i,12}} \tag{16.68}$$

$S_{i,12}$ is the generated entropy caused by friction, mixing and balancing processes due to temperature-, pressure- or concentration gradients, see Chap. 14. Obviously, there is a correlation between dissipation, entropy generation and the loss of working capability, i.e.

$$\boxed{\Psi_{12} \rightarrow S_{i,12} \rightarrow \Delta E_{x,V}} \tag{16.69}$$

Dissipation should be avoided since it leads to entropy generation and finally to a loss of working capability!

Overview Closed System

Figure 16.16 gives an overview how to handle closed systems. It summarises what has been investigated so far:

- First law of thermodynamics, see Fig. 16.16a

$$Q_{12} + W_{12} = U_2 - U_1 + \underbrace{m \left[\frac{1}{2} \left(c_2^2 - c_1^2 \right) + g \left(z_2 - z_1 \right) \right]}_{=\Delta E_a} \tag{16.70}$$

---

[25]Mind, that W_{12} is the work that passes the system boundary of the fluid inside the cylinder, whereas W_{eff} is the work that can effectively be utilised at the piston!

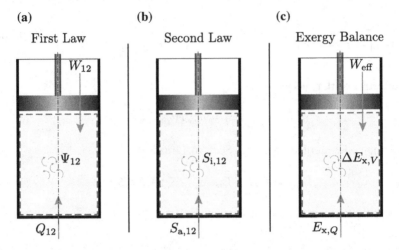

(a)   **(b)**   **(c)**

First Law   Second Law   Exergy Balance

Fig. 16.16 Overview closed system: balance of energy (**a**), entropy (**b**) and exergy (**c**)

– Partial energy equation

$$W_{12} = W_{12,\mathrm{V}} + \underbrace{W_{12,\mathrm{mech}}}_{=\Delta E_{\mathrm{a}}} + \Psi_{12} \qquad (16.71)$$

– Effective work

$$W_{\mathrm{eff}} = W_{12} + p_{\mathrm{env}}\,(V_2 - V_1) \qquad (16.72)$$

• Second law of thermodynamics, see Fig. 16.16b

$$S_2 - S_1 = m \cdot (s_2 - s_1) = S_{\mathrm{a},12} + S_{\mathrm{i},12} \qquad (16.73)$$

• Balance of exergy, see Fig. 16.16c

$$E_{\mathrm{x},2} = E_{\mathrm{x},1} + W_{\mathrm{eff}} + E_{\mathrm{x},Q} - \Delta E_{\mathrm{x},\mathrm{V}} \qquad (16.74)$$

with

$$\Delta E_{\mathrm{x},\mathrm{V}} = T_{\mathrm{env}}\,S_{\mathrm{i},12} \qquad (16.75)$$

16.4.2 Open System in Steady State Operation

The flux of loss of exergy $\Delta \dot{E}_{\mathrm{x},\mathrm{V}}$ in an open system can easily be determined by comparison of inlet state (1) and outlet state (2). Starting from state (1) exergy increases by supply of exergy, i.e. due to heat and technical power P_{t}. Mind, that

Fig. 16.17 Open system: balance of exergy

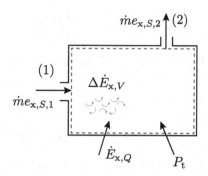

technical power is pure exergy. It is postulated, that exergy is reduced by a loss of exergy as seen in Problem 16.2.

Thus, according to Fig. 16.17, state (2) follows

$$\dot{E}_{x,S,2} = \dot{E}_{x,S,1} + P_t + \dot{E}_{x,Q} - \Delta\dot{E}_{x,V}. \tag{16.76}$$

In other words, the exergy in state (2) is equal to the exergy initially available in state (1) plus the exergy that is supplied from (1) to (2) by heat and work, reduced by the loss of exergy.[26] Rearranging this equation leads to

$$\boxed{\dot{E}_{x,S,1} + P_t + \dot{E}_{x,Q} = \dot{E}_{x,S,2} + \Delta\dot{E}_{x,V}} \tag{16.77}$$

Theorem 16.3 *Thus, the incoming flux of exergy is equal to the outgoing flux of exergy for steady state operation!*

The loss of exergy obeys

$$\boxed{\Delta\dot{E}_{x,V} = \dot{E}_{x,S,1} - \dot{E}_{x,S,2} + P_t + \dot{E}_{x,Q}} \tag{16.78}$$

The exergies for fluid flows $\dot{E}_{x,S,1}$ as well as $\dot{E}_{x,S,1}$ can be substituted by the correlations, that have been deduced in Sect. 16.2. This leads to

$$\dot{E}_{x,S,1} - \dot{E}_{x,S,2} = \dot{m}\left[h_1 - h_2 + \frac{1}{2}\left(c_1^2 - c_2^2\right) + g\left(z_1 - z_2\right) + T_{env}(s_2 - s_1)\right] \tag{16.79}$$

The flux of exergy due to heat flux obeys Eq. 16.17, i.e.

$$\dot{E}_{x,Q} = \dot{Q}_{12} - \dot{m}T_{env}\left(s_2 - s_1\right) + T_{env}\dot{S}_i \tag{16.80}$$

Applying the first law of thermodynamics, i.e.

---

[26]Exergy loss reduces the exergy, so it is an outgoing flux! It is treated as a *sink*.

$$P_t + \dot{Q}_{12} = \dot{m} \left[h_2 - h_1 + \frac{1}{2} \left(c_2^2 - c_1^2 \right) + g \left(z_2 - z_1 \right) \right] \qquad (16.81)$$

results in the loss of exergy for an open system in steady state:

$$\boxed{\Delta \dot{E}_{x,V} = T_{env} \dot{S}_i} \qquad (16.82)$$

$\dot{S}_i$ is the generated entropy caused by friction, mixing and balancing processes due to temperature-, pressure- or concentration gradients, see Chap. 14. Obviously, there is a correlation between dissipation, entropy generation and the loss of working capability, i.e.

$$\boxed{\dot{\Psi} \rightarrow \dot{S}_i \rightarrow \Delta \dot{E}_{x,V}} \qquad (16.83)$$

Dissipation should be avoided since it leads to entropy generation and finally to a loss of working capability!

Overview Open System

Figure 16.18 gives an overview how to handle open systems at steady state operation. It summarises what has been investigated so far:

- Law of mass conservation

$$\dot{m}_1 = \dot{m}_2 = \dot{m} \qquad (16.84)$$

$$\boxed{\text{Mass flux in is equal to mass flux out.}}$$

- First law of thermodynamics, see Fig. 16.18a

$$\dot{Q} + P_t = \dot{m} \cdot \left(h_2 - h_1 + \Delta e_{a,12} \right) \qquad (16.85)$$

$$\boxed{\text{Energy flux in is equal to energy flux out.}}$$

$$\dot{Q} + P_t + \dot{m} \cdot \left(h_1 + \Delta e_{a,1} \right) = \dot{m} \cdot \left(h_2 + \Delta e_{a,2} \right) \qquad (16.86)$$

- Partial energy equation

$$P_t = \dot{m} \cdot \left[\int_1^2 v \, dp + \psi_{12} + \Delta e_{a,12} \right] \qquad (16.87)$$

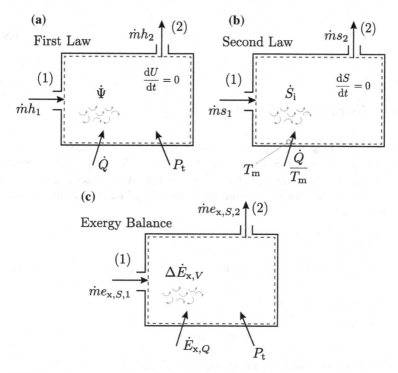

Fig. 16.18 Overview open system: balance of energy (**a**), entropy (**b**) and exergy (**c**)

- Second law of thermodynamics, see Fig. 16.18b

$$\frac{\mathrm{d}S}{\mathrm{d}t} = 0 = \dot{m} \cdot (s_1 - s_2) + \frac{\dot{Q}}{T_\mathrm{m}} + \dot{S}_\mathrm{i} \qquad (16.88)$$

> Entropy flux in is equal to entropy flux out.

$$\dot{m}s_1 + \underbrace{\frac{\dot{Q}}{T_\mathrm{m}} + \dot{S}_\mathrm{i}}_{=\dot{S}_\mathrm{a}} = \dot{m}s_2 \qquad (16.89)$$

- Balance of exergy, see Fig. 16.18c

$$\dot{m}e_{\mathrm{x},S,1} + \dot{E}_{\mathrm{x},Q} + P_\mathrm{t} = \Delta\dot{E}_{\mathrm{x},V} + \dot{m}e_{\mathrm{x},S,2} \qquad (16.90)$$

> Exergy flux in is equal to exergy flux out.

with

$$\Delta \dot{E}_{x,V} = T_{env} \dot{S}_i \tag{16.91}$$

16.4.3 Thermodynamic Cycles

In this section the loss of exergy in thermodynamic cycles is deduced. After this, an overview of the underlying concepts is given. Anyhow, a distinction is made between clockwise and counterclockwise cycles. Figure 16.19 shows the exergy flows for both cycles. As the cycle is operated in steady state, the exergy fluxes in need to be balanced by the fluxes out, i.e.

- Clockwise cycle, see Fig. 16.19a

$$\dot{E}_{x,Q} = P + \dot{E}_{x,Q_0} + \Delta \dot{E}_{x,V} \tag{16.92}$$

Hence, the loss of exergy is

$$\Delta \dot{E}_{x,V} = \dot{E}_{x,Q} - \dot{E}_{x,Q_0} - P. \tag{16.93}$$

Substitution of the exergies of heat, see Eq. 16.7, results in

$$\Delta \dot{E}_{x,V} = \dot{Q}\left(1 - \frac{T_{env}}{T}\right) - \dot{Q}_0\left(1 - \frac{T_{env}}{T_0}\right) - P. \tag{16.94}$$

Rearranging leads to

$$\Delta \dot{E}_{x,V} = \dot{Q} - \dot{Q}_0 - P - \dot{Q}\frac{T_{env}}{T} + \dot{Q}_0\frac{T_{env}}{T_0}. \tag{16.95}$$

Fig. 16.19 Balance of exergy: **a** Clockwise cycle, **b** Counterclockwise cycle

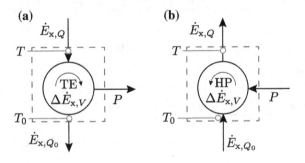

The first law of thermodynamics[27] obeys

$$\dot{Q} = \dot{Q}_0 + P.$$ (16.96)

Thus, Eq. 16.95 simplifies to

$$\Delta \dot{E}_{x,V} = T_{env} \left(\frac{\dot{Q}_0}{T_0} - \frac{\dot{Q}}{T} \right).$$ (16.97)

The second law of thermodynamics[28] reads as

$$\frac{\dot{Q}}{T} + \dot{S}_i = \frac{\dot{Q}_0}{T_0}$$ (16.98)

Hence, it finally is

$$\boxed{\Delta \dot{E}_{x,V} = T_{env} \dot{S}_i}$$ (16.99)

- Counterclockwise cycle, see Fig. 16.19b

$$\dot{E}_{x,Q_0} + P = \dot{E}_{x,Q} + \Delta \dot{E}_{x,V}$$ (16.100)

Hence, the loss of exergy is

$$\Delta \dot{E}_{x,V} = \dot{E}_{x,Q_0} - \dot{E}_{x,Q} + P.$$ (16.101)

Substitution of the exergies of heat, see Eq. 16.7, results in

$$\Delta \dot{E}_{x,V} = \dot{Q}_0 \left(1 - \frac{T_{env}}{T_0} \right) - \dot{Q} \left(1 - \frac{T_{env}}{T} \right) + P.$$ (16.102)

Rearranging leads to

$$\Delta \dot{E}_{x,V} = \dot{Q}_0 - \dot{Q} + P - \dot{Q}_0 \frac{T_{env}}{T_0} + \dot{Q} \frac{T_{env}}{T}.$$ (16.103)

The first law of thermodynamics obeys

$$\dot{Q} = \dot{Q}_0 + P.$$ (16.104)

Thus, Eq. 16.103 simplifies to

---

[27] Energy flux in is balanced by energy flux out.
[28] Entropy flux in is balanced by entropy flux out.

$$\Delta \dot{E}_{x,V} = T_{env} \left(\frac{\dot{Q}}{T} - \frac{\dot{Q}_0}{T_0} \right). \tag{16.105}$$

The second law of thermodynamics reads as

$$\frac{\dot{Q}_0}{T_0} + \dot{S}_i = \frac{\dot{Q}}{T}. \tag{16.106}$$

Hence, it finally is

$$\boxed{\Delta \dot{E}_{x,V} = T_{env} \dot{S}_i} \tag{16.107}$$

Overview Thermodynamic Cycles

Figure 16.20 gives an overview how to handle *clockwise cycles* in steady state operation. It summarises what has been investigated so far:

- First law of thermodynamics, see Fig. 16.20a

$$\frac{dU}{dt} = 0 = \dot{Q} - \dot{Q}_0 - P \tag{16.108}$$

Energy flux in is equal to energy flux out.

$$\dot{Q} = \dot{Q}_0 + P \tag{16.109}$$

- Second law of thermodynamics, see Fig. 16.20b

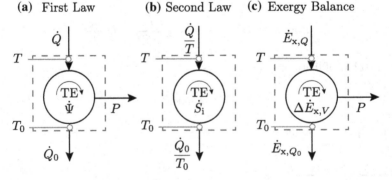

Fig. 16.20 Overview clockwise cycle: balance of energy (**a**), entropy (**b**) and exergy (**c**)

$$\frac{dS}{dt} = 0 = \frac{\dot{Q}}{T} - \frac{\dot{Q}_0}{T_0} + \dot{S}_i \tag{16.110}$$

> Entropy flux in is equal to entropy flux out.

$$\frac{\dot{Q}}{T} + \dot{S}_i = \frac{\dot{Q}_0}{T_0} \tag{16.111}$$

- Balance of exergy, see Fig. 16.20c

> Exergy flux in is equal to exergy flux out.

$$\dot{E}_{x,Q} = \dot{E}_{x,Q_0} + P + \Delta\dot{E}_{x,V} \tag{16.112}$$

respectively

$$P = \dot{E}_{x,Q} - \dot{E}_{x,Q_0} - \Delta\dot{E}_{x,V} \tag{16.113}$$

with

$$\Delta\dot{E}_{x,V} = T_{env}\dot{S}_i \tag{16.114}$$

- The exergetic efficiency is defined by the ratio of released power to the supplied flux of exergy. Thus, according to Fig. 16.20c it follows

$$\eta_{ex} = \frac{\text{released work}}{\text{supplied exergy}} = \frac{P}{\dot{E}_{x,Q}}. \tag{16.115}$$

Applying the balance of exergy, see Eq. 16.113, results in

$$\eta_{ex} = \frac{P}{\dot{E}_{x,Q}} = \frac{\dot{E}_{x,Q} - \dot{E}_{x,Q_0} - \Delta\dot{E}_{x,V}}{\dot{E}_{x,Q}} \tag{16.116}$$

Rearranging leads to

$$\eta_{ex} = 1 - \frac{\dot{E}_{x,Q_0}}{\dot{E}_{x,Q}} - \frac{\Delta\dot{E}_{x,V}}{\dot{E}_{x,Q}}. \tag{16.117}$$

- The term $\frac{\dot{E}_{x,Q_0}}{\dot{E}_{x,Q}}$ disappears in case the released heat flux $\dot{Q}_0$ does not contain any exergy, i.e. $T_0 = T_{env}$
- The term $\frac{\Delta\dot{E}_{x,V}}{\dot{E}_{x,Q}}$ disappears in case the machine works reversibly, i.e. $\Delta\dot{E}_{x,V} = 0$

In a theoretical borderline case, an exergetic efficiency of 100% is possible. Anyhow, the thermal efficiency is defined as

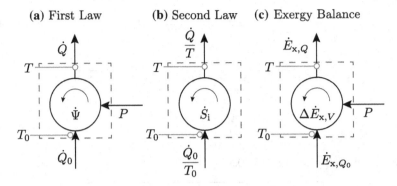

Fig. 16.21 Overview counterclockwise cycle: balance of energy (**a**), entropy (**b**) and exergy (**c**)

$$\eta_{th} = \frac{P}{\dot{Q}}. \tag{16.118}$$

Expanding this equation leads to

$$\eta_{th} = \underbrace{\frac{P}{\dot{E}_{x,Q}}}_{=\eta_{ex}} \frac{\dot{E}_{x,Q}}{\dot{Q}}. \tag{16.119}$$

Substitution of the exergy of heat $\dot{E}_{x,Q}$ results in

$$\boxed{\eta_{th} = \eta_{ex}\left(1 - \frac{T_{env}}{T}\right) = \eta_{ex}\eta_C} \tag{16.120}$$

The exergetic efficiency η_{ex} evaluates the *quality of the machine*.[29] In best, theoretical case it can reach 100% efficiency. To do so, it requires a reversible machine without any generation of entropy and a release of heat at ambient temperature. The Carnot efficiency η_C, however, characterises the *quality of the heat* that drives the machine. The larger the temperature at which the heat is available, the more exergy the heat contains and the more efficient the energy conversion runs.

Figure 16.20 gives an overview how to handle *counterclockwise cycles* at steady state operation. It summarises what has been investigated so far:

- First law of thermodynamics, see Fig. 16.21a

$$\frac{dU}{dt} = 0 = \dot{Q}_0 + P - \dot{Q} \tag{16.121}$$

---

[29]Thus, it is the efficiency that is relevant to evaluate the engineer's efforts!

> Energy flux in is equal to energy flux out.

$$\dot{Q}_0 + P = \dot{Q} \tag{16.122}$$

- Second law of thermodynamics, see Fig. 16.21b

$$\frac{dS}{dt} = 0 = \frac{\dot{Q}_0}{T_0} + \dot{S}_i - \frac{\dot{Q}}{T} \tag{16.123}$$

> Entropy flux in is equal to entropy flux out.

$$\frac{\dot{Q}_0}{T_0} + \dot{S}_i = \frac{\dot{Q}}{T} \tag{16.124}$$

- Balance of exergy, see Fig. 16.21c

> Exergy flux in is equal to exergy flux out.

$$\dot{E}_{x,Q_0} + P = \dot{E}_{x,Q} + \Delta \dot{E}_{x,V} \tag{16.125}$$

respectively

$$P = \dot{E}_{x,Q} - \dot{E}_{x,Q_0} + \Delta \dot{E}_{x,V} \tag{16.126}$$

with

$$\Delta \dot{E}_{x,V} = T_{env} \dot{S}_i \tag{16.127}$$

- The exergetic efficiency is:

$$\eta_{ex} = \frac{\text{Exergetic benefit}}{\text{Exergetic effort}} \tag{16.128}$$

  – In case of a *heat pump*

$$\eta_{ex,HP} = \frac{\dot{E}_{x,Q}}{P} \tag{16.129}$$

Applying the balance of exergy, see Eq. 16.126, results in

$$\eta_{ex,HP} = \frac{\dot{E}_{x,Q}}{\dot{E}_{x,Q} - \dot{E}_{x,Q_0} + \Delta \dot{E}_{x,V}} \tag{16.130}$$

  – In case of a *cooling machine*

$$\eta_{ex,CM} = \frac{\dot{E}_{x,Q_0}}{P} \tag{16.131}$$

Applying the balance of exergy, see Eq. 16.126, results in

$$\eta_{ex,CM} = \frac{\dot{E}_{x,Q_0}}{\dot{E}_{x,Q} - \dot{E}_{x,Q_0} + \Delta \dot{E}_{x,V}} \tag{16.132}$$

- The coefficient of performance is:

$$\varepsilon = \frac{\text{Benefit}}{\text{Effort}}. \tag{16.133}$$

  – In case of a *heat pump* it follows

$$\varepsilon_{HP} = \frac{\dot{Q}}{P}. \tag{16.134}$$

The exergy of heat is

$$\dot{E}_{x,Q} = \dot{Q}\left(1 - \frac{T_{env}}{T}\right). \tag{16.135}$$

Hence, it follows

$$\varepsilon_{HP} = \frac{T}{T - T_{env}} \cdot \frac{\dot{E}_{x,Q}}{P} = \frac{T}{T - T_{env}} \cdot \eta_{ex,HP}. \tag{16.136}$$

  – In case of a *cooling machine* it follows

$$\varepsilon_{CM} = \frac{\dot{Q}_0}{P}. \tag{16.137}$$

The exergy of heat is

$$\dot{E}_{x,Q_0} = \dot{Q}_0\left(1 - \frac{T_{env}}{T_0}\right). \tag{16.138}$$

Hence, it follows

$$\varepsilon_{CM} = \frac{T_0}{T_0 - T_{env}} \cdot \frac{\dot{E}_{x,Q_0}}{P} = \frac{T_0}{T_0 - T_{env}} \cdot \eta_{ex,CM}. \tag{16.139}$$

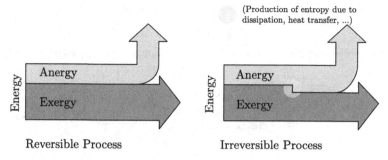

Fig. 16.22 Illustration of energy fluxes

16.5 Sankey-Diagram

Changes of state in thermodynamic cycles or systems can be visualised with so-called Sankey-diagrams,[30] that show the conversion of energy in a flow diagram. Furthermore, energy can be presented as exergy and anergy as well. The major idea is illustrated in Fig. 16.22. Exergy is that part of the energy, that can be converted in any other form of energy in particular into mechanical or electrical work. In contrast to the exergy, anergy is that part, which can not be converted any further and thus is technically useless. This is due to already being in equilibrium with the environment, so that there is no driving potential for a conversion available. Both, exergy and anergy form the energy that is a conservation quantity. It has been shown, that any irreversibility in a process leads to a loss of exergy and by the same amount to a rise of anergy. In reversible system, however, no loss of exergy occurs. This is visualised in Fig. 16.22. Energy enters a system and with it, exergy as well as anergy. Due to the first law of thermodynamics, energy being released in steady state operation is the same as energy being supplied. However, in case irreversibilities occur inside the process, exergy is lost and is converted to anergy.

16.5.1 Open System

An example of an open system in steady state operation is shown in Fig. 16.23. At the inlet (1) a fluid flow enters the system. The entire flux of energy consists of the exergy of the flow $\dot{m}e_{x,S,1}$, see Sect. 16.2, and anergy $\dot{m}b_{x,S,1}$.

Depending on the temperature at which the heat flux $\dot{Q}$ is supplied to the system, see Sect. 16.1, exergy $\dot{E}_{x,Q}$ as well as anergy $\dot{B}_{x,Q}$ enrich the system, see also Fig. 16.7. At that stage, the entire energy is the sum of the fluid flow's energy and the heat flux. In case technical power P_t is additionally supplied, the entire energy rises. So does

---

[30]Named after Matthew Henry Phineas Riall Sankey, *9 November 1853 in Nenagh, Ireland, †3 October 1925.

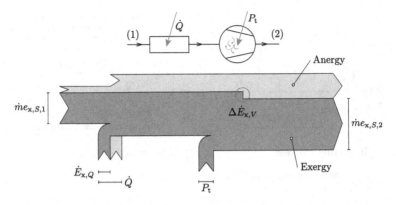

Fig. 16.23 Sankey-diagram of an open system

the exergy, as technical power is pure exergy. However, if there are imperfections internally, such as dissipation or other inhomogeneities, a loss of exergy appears. Thus, the exergy is reduced, while the anergy rises by that amount. Finally, the flux of energy leaving at the outlet (2) is exactly the same as the flux of energy, that has been supplied across the system boundary. However, this flux of energy is composed of exergy of the flow $\dot{E}_{x,S,2}$ and anergy $\dot{B}_{x,S,2}$.

16.5.2   Heat Transfer

Figure 16.24 shows an example of a steady state heat transfer through a conductive wall. This problem has been discussed several times as it is an excellent example of how entropy can be generated, though there is no dissipation, i.e. internal friction. Cause for a heat flux is a temperature difference, for instance $T_1 > T_2$ as shown in the figure.

From a thermodynamic point of view the wall is an imperfection as it inhibits to reach a thermodynamic equilibrium between left and right side, i.e. a homogeneous temperature distribution. In order to reach this equilibrium a heat flux occurs, that follows the temperature gradient according to the second law of thermodynamics. The heat flux at any horizontal position[31] must be the same. If this is not the case, the internal energy would locally vary by time. Thus, the problem would not be a steady state problem. Since the heat flux at the right border is equal to the heat flux leaving the system at the left border, the incoming entropy by heat is smaller than the outgoing entropy by heat. Consequently, inside the wall there must be a generation of entropy, which is the cause for the entire process being irreversible: Once the heat

---

[31] In case of one-dimensional heat transfer!

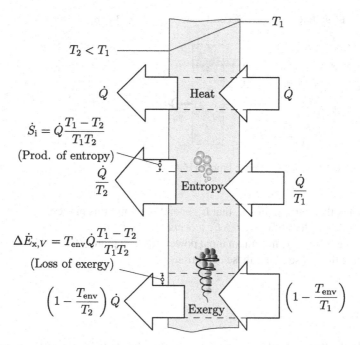

Fig. 16.24 Heat transfer (wall)

has left the system, there is no way back by itself.[32] The rate of entropy generation has been derived already and follows

$$\dot{S}_i = \dot{Q} \frac{T_1 - T_2}{T_1 T_2}. \tag{16.140}$$

This correlation shows, that the heat transfer mechanism is only reversible, if heat is theoretically transferred at $\Delta T \rightarrow 0$. Exergy of heat passes into the system on the right hand side. Due to the large temperature of the heat, the exergy on the right hand side is larger than the exergy of heat, that leaves on the left hand side at a small temperature. Obviously, the working capability of the heat has decreased while passing through the wall. That means, there is a loss of exergy internally, which obeys

$$\Delta \dot{E}_{x,V} = T_{env} \dot{Q} \frac{T_1 - T_2}{T_1 T_2} = T_{env} \dot{S}_i. \tag{16.141}$$

Problem 16.4 A heat flux of $\dot{Q} = 20 \, \text{kW}$ passes from the inside of a house ($\vartheta_1 = 22\,°C$) through the wall into the environment ($\vartheta_{env} = -5\,°C$). The process shall be investigated in steady state.

---

[32]Bringing back the heat *against* the gradient requires a heat pump, i.e. a counterclockwise cycle, that consumes technical power!

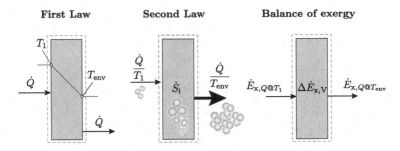

Fig. 16.25 Sketch for the solution to Problem 16.4

(a) What is the entropy flux $\dot{S}_i$, that is generated within this process?
(b) What flux of loss of exergy $\Delta \dot{E}_{x,V}$ occurs?
(c) What would be the minimum input power of a heat pump, that would be required to heat the house under these conditions?

Solution

Figure 16.25 shows the sketches that are required for parts (a) and (b).

(a) This part has already been solved in Problem 14.8. A balance of entropy in steady state leads to

$$\frac{\dot{Q}}{T_1} + \dot{S}_i = \frac{\dot{Q}}{T_{env}}. \tag{16.142}$$

For the generation of entropy we finally get:

$$\dot{S}_i = \dot{Q}\left(\frac{1}{T_{env}} - \frac{1}{T_1}\right) = 6.82 \ \frac{W}{K} > 0. \tag{16.143}$$

Hence, the process is irreversible, i.e. once the heat flux has left the building it does not return due to the temperature gradient.

(b) There are two possibilities to calculate the loss of exergy $\Delta \dot{E}_{x,V}$:

- Based on the generation of entropy:

$$\Delta \dot{E}_{x,V} = T_{env}\dot{S}_i = 1.8296 \ \text{kW} \tag{16.144}$$

- Based on a balance of exergy:

According to Fig. 16.25 the exergy balance in steady state[33] obeys

---

[33]Exergy *in* is balanced by exergy *out*. Mind, that the exergy loss reduces the exergy of a system. Thus, it counts as outgoing exergy!

$$\dot{E}_{x,Q@T_1} = \Delta\dot{E}_{x,V} + \dot{E}_{x,Q@T_{env}}. \tag{16.145}$$

The loss of exergy follows accordingly

$$\Delta\dot{E}_{x,V} = \dot{E}_{x,Q@T_1} - \dot{E}_{x,Q@T_{env}} \tag{16.146}$$

with the exergies of heat

$$\dot{E}_{x,Q@T_1} = \dot{Q}\left(1 - \frac{T_{env}}{T_1}\right) = 1.8296\,\text{kW} \tag{16.147}$$

and

$$\dot{E}_{x,Q@T_{env}} = \dot{Q}\left(1 - \frac{T_{env}}{T_{env}}\right) = 0. \tag{16.148}$$

Thus, the entire exergy of state (1) is lost, i.e.

$$\Delta\dot{E}_{x,V} = \dot{E}_{x,Q@T_1} = 1.8296\,\text{kW}. \tag{16.149}$$

(c) Again, there are several alternatives according to Fig. 16.26.

- Based on first and second law of thermodynamics:

$$\dot{Q}_{env} + P = \dot{Q}. \tag{16.150}$$

Hence, the minimum power (in case the heat pump is operated reversibly) follows

$$P = \dot{Q} - \dot{Q}_{env}. \tag{16.151}$$

In order to find $\dot{Q}_{env}$ the second law of thermodynamics is applied

$$\frac{\dot{Q}_{env}}{T_{env}} + \underbrace{\dot{S}_i}_{=0} = \frac{\dot{Q}}{T_1}. \tag{16.152}$$

Thus, the heat flux taken from the environment results in

$$\dot{Q}_{env} = T_{env}\frac{\dot{Q}}{T_1} = 18.17\,\text{kW}. \tag{16.153}$$

Consequently, the required power is

$$P = \dot{Q} - \dot{Q}_{env} = 1.8296\,\text{kW}. \tag{16.154}$$

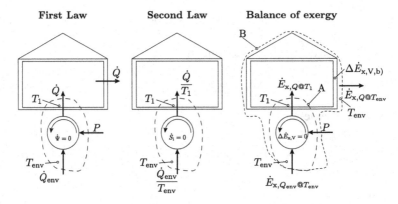

Fig. 16.26 Sketch for the solution to Problem 16.4

- Based on a balance of exergy according to system boundary A[34]:

$$\dot{E}_{x,Q_{env}@T_{env}} + P = \dot{E}_{x,Q@T_1} + \underbrace{\Delta\dot{E}_{x,V}}_{=0} \tag{16.155}$$

Since $\dot{Q}_{env}$ occurs at T_{env} it is

$$\dot{E}_{x,Q_{env}@T_{env}} = 0. \tag{16.156}$$

Hence, the required power P reads as

$$P = \dot{E}_{x,Q@T_1} = \dot{Q}\left(1 - \frac{T_{env}}{T_1}\right) = 1.8296\,\text{kW}. \tag{16.157}$$

- Based on a balance of exergy according to system boundary B:

$$\dot{E}_{x,Q_{env}@T_{env}} + P = \dot{E}_{x,Q@T_{env}} + \underbrace{\Delta\dot{E}_{x,V}}_{=0} + \Delta\dot{E}_{x,V,b)} \tag{16.158}$$

In this equation $\Delta\dot{E}_{x,V,b)}$ is the loss of exergy within the wall according to part (b), see Figs. 16.25 and 16.26. Since

$$\dot{E}_{x,Q@T_{env}} = 0 \tag{16.159}$$

respectively

$$\dot{E}_{x,Q_{env}@T_{env}} = 0 \tag{16.160}$$

---

[34]The loss of exergy in part (b) has nothing to do with the loss of exergy in part (b), since the system boundary is different!

the power finally results in

$$P = \Delta \dot{E}_{x,V,b}.$$ (16.161)

Problem 16.5 Air is compressed in an adiabatic compressor in a steady state process from state (1) ($p_1 = 1.05$ bar, $\vartheta_1 = \vartheta_{env} = 15\,°C$) to a pressure of $p_2 = 6.25$ bar in state (2). Air can be treated as an ideal gas with an individual gas constant of $R = 0.287\ \frac{kJ}{kg\,K}$ and a constant specific heat capacity of $c_p = 1.004\ \frac{kJ}{kg\,K}$. A specific technical work of $w_t = 230\ \frac{kJ}{kg}$ is required.

(a) What is the specific loss of exergy $\Delta e_{x,V,12}$?
(b) Sketch the change of state in a T, s-diagram and mark the specific dissipated energy by a vertical hatching and the specific exergy loss $\Delta e_{x,V,12}$ by a horizontal hatching.
(c) Calculate the specific dissipated energy.

Solution

Figure 16.27 shows the required balances.

(a) The specific loss of exergy $\Delta e_{x,V,12}$ follows

$$\Delta e_{x,V} = T_{env} s_{i,12}.$$ (16.162)

In order to calculate the specific generation of entropy, i.e. $s_{i,12}$, a balance of entropy is applied:

$$s_1 + s_{i,12} + \underbrace{s_{a,12}}_{=0} = s_2.$$ (16.163)

The specific generation of entropy obeys

$$s_{i,12} = s_2 - s_1 = c_p \ln\frac{T_2}{T_1} - R \ln\frac{p_2}{p_1}.$$ (16.164)

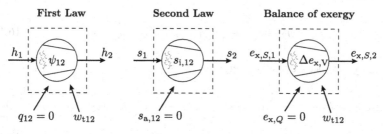

Fig. 16.27 Sketch for the solution to Problem 16.5

T_2 is unknown, so that the first law of thermodynamics is applied, i.e.

$$\underbrace{q_{12}}_{=0} + w_{t12} = h_2 - h_1 = c_p \left(T_2 - T_1\right).$$

(16.165)

This leads to the outlet temperature T_2, i.e.

$$T_2 = T_1 + \frac{w_{t12}}{c_p} = 517.23 \text{ K}.$$

(16.166)

For the specific generation of entropy it follows accordingly

$$s_{i,12} = c_p \ln\frac{T_2}{T_1} - R \ln\frac{p_2}{p_1} = 0.0754 \frac{\text{kJ}}{\text{kg K}}.$$

(16.167)

Thus, the loss of exergy finally is

$$\Delta ex_{,V} = T_{env} s_{i,12} = 21.726 \frac{\text{kJ}}{\text{kg}}.$$

(16.168)

Alternatively, a balance of exergy can be performed, i.e.

$$e_{x,S,1} + \underbrace{e_{x,Q}}_{=0} + w_{t12} = e_{x,S,2} + \Delta e_{x,V}.$$

(16.169)

Hence, that the specific loss of exergy follows

$$\Delta e_{x,V} = e_{x,S,1} - e_{x,S,2} + w_{t12}.$$

(16.170)

The specific exergies of the flow can be substituted by

$$e_{x,S,1} = h_1 - h_{env} + \frac{1}{2}c_1^2 + gz_1 + T_{env}\left(s_{env} - s_1\right)$$

(16.171)

respectively

$$e_{x,S,2} = h_2 - h_{env} + \frac{1}{2}c_2^2 + gz_2 + T_{env}\left(s_{env} - s_2\right).$$

(16.172)

Finally the specific loss of exergy is

$$\Delta e_{x,V} = h_1 - h_2 + \frac{1}{2}\left(c_1^2 - c_2^2\right) + g\left(z_1 - z_2\right) + T_{env}\left(s_2 - s_1\right) + \underbrace{\left(h_2 - h_1\right)}_{w_{t12}}.$$

(16.173)

Fig. 16.28 Sketch for the
solution to Problem 16.5

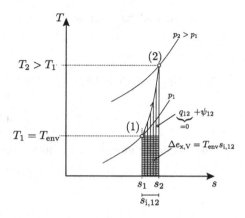

Consequently, the loss of exergy finally results in

$$\Delta e_{x,V} = T_{env} \left(s_2 - s_1 \right) = T_{env} \left[c_p \ln\frac{T_2}{T_1} - R \ln\frac{p_2}{p_1} \right] = 21.726 \frac{kJ}{kg}. \quad (16.174)$$

(b) Figure 16.28 shows the solution.
(c) The specific dissipated energy can be calculated with the partial energy equation, i.e.

$$w_{t12} = 230 \frac{kJ}{kg} = \int_1^2 v\,dp + \psi_{12} + \underbrace{\Delta e_{a12}}_{=0}. \quad (16.175)$$

Hence, the specific dissipation results in

$$\psi_{12} = w_{t12} - \underbrace{\int_1^2 v\,dp}_{y_{12}}. \quad (16.176)$$

The change of state is polytropic.[35] Thus, it can be described by a polytropic exponent n. The pressure work y_{12} for a polytropic change of state is

$$y_{12} = n\frac{RT_1}{n-1}\left[\frac{T_2}{T_1} - 1\right]. \quad (16.177)$$

The specific dissipation results in

$$\psi_{12} = w_{t12} - n\frac{RT_1}{n-1}\left[\frac{T_2}{T_1} - 1\right]. \quad (16.178)$$

---

[35]Mind, that any change of state is polytropic! In this case it is adiabatic and frictional.

In order to find the exponent n the polytropic change of state is investigated, i.e.

$$p_1 v_1^n = p_2 v_2^n.$$ (16.179)

Thus, the exponent n obeys

$$n = \frac{\ln \frac{p_1}{p_2}}{\ln \frac{v_2}{v_1}}.$$ (16.180)

The specific volumes v_1 and v_2 follow the thermal equation of state, i.e.

$$v_1 = \frac{RT_1}{p_1} = 0.7876 \frac{m^3}{kg}$$ (16.181)

respectively

$$v_2 = \frac{RT_2}{p_2} = 0.2375 \frac{m^3}{kg}.$$ (16.182)

Hence, the exponent n is

$$n = \frac{\ln \frac{p_1}{p_2}}{\ln \frac{v_2}{v_1}} = 1.488.$$ (16.183)

Finally, the specific dissipation results in

$$\psi_{12} = w_{t12} - n \frac{RT_1}{n-1} \left[\frac{T_2}{T_1} - 1 \right] = 29.5278 \frac{kJ}{kg}.$$ (16.184)

Chapter 17
Components and Thermodynamic Cycles

Although many technical components, that are required to run energy conversion processes, have already been discussed previously,[1] the focus in this chapter is on thermal turbo-machines as well as on heat exchangers. However, in this chapter the technical components are treated in steady state operation. In order to quantify the efficiency of turbine and compressor the so-called isentropic efficiency is defined. In addition, relevant thermodynamic cycles are introduced and discussed. Cyclic processes have been discussed in the previous chapters as well, but mostly in a black-box notation.[2] A distinction has been made between clockwise cycles, i.e. thermal engines, and counterclockwise cycles, i.e. cooling machines respectively heat pumps. This chapter concludes the first part of this book.

17.1 Components

17.1.1 Turbine

A turbine is a thermal turbo-machine, that converts enthalpy of a fluid into rotational energy, i.e. mechanical work is released, see Fig. 17.1. In doing so, the fluid's pressure decreases from p_1 to $p_2 < p_1$.

- The first law of thermodynamics for a turbine in steady state reads as[3]:

---

[1] Such as adiabatic throttle, compressor and turbine.

[2] Except for the Carnot-cycle, i.e. the technical benchmark, whose underlying changes of state, i.e. isentropic, isothermal, isentropic, isothermal, have been introduced as well.

[3] Following, that the energy flux *in* is balanced by the energy flux *out* in steady state!

© Springer Nature Switzerland AG 2019
A. Schmidt, *Technical Thermodynamics for Engineers*,
https://doi.org/10.1007/978-3-030-20397-9_17

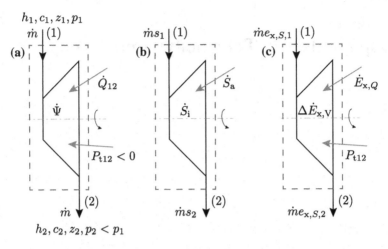

Fig. 17.1 Turbine: **a** First law, **b** Second Law, **c** Exergy

$$\dot{Q}_{12} + P_{t12} + \dot{m}\left[h_1 + \frac{1}{2}c_1^2 + gz_1\right] = \dot{m}\left[h_2 + \frac{1}{2}c_2^2 + gz_2\right]. \qquad (17.1)$$

In specific notation and with ignoring the potential energy, that is supposed to be comparable small in technical applications, it follows, that

$$q_{12} + w_{t12} = h_2 - h_1 + \frac{1}{2}\left(c_2^2 - c_1^2\right). \qquad (17.2)$$

With the so-called total enthalpy

$$h^+ = h + \frac{c^2}{2} \qquad (17.3)$$

the first law of thermodynamics finally obeys

$$\boxed{q_{12} + w_{t12} = h_2^+ - h_1^+} \qquad (17.4)$$

- The <u>partial energy equation</u> reads as

$$w_{t,12} = \int_1^2 v\,dp + \psi_{12} + \frac{1}{2}\left(c_2^2 - c_1^2\right) + g\left(z_2 - z_1\right). \qquad (17.5)$$

Ignoring the potential energy simplifies the partial energy equation, i.e.

$$w_{t,12} = \int\limits_1^2 v\,dp + \psi_{12} + \frac{1}{2}\left(c_2^2 - c_1^2\right) \qquad (17.6)$$

- The <u>second law of thermodynamics</u> in steady state[4] is, see Fig. 17.1b,

$$\dot{m}s_1 + \dot{S}_i + \dot{S}_a = \dot{m}s_2. \qquad (17.7)$$

In specific notation, by dividing with the mass flow rate, it is

$$s_2 - s_1 = s_{i12} + s_{a12} \qquad (17.8)$$

In differential notation it obeys

$$ds = \delta s_i + \delta s_a = \underbrace{\frac{\delta\psi}{T}}_{\geq 0} + \frac{\delta q}{T}. \qquad (17.9)$$

- The <u>balance of exergy</u> in steady state[5] reads as, see Fig. 17.1c,

$$\dot{m}e_{x,S,1} + \dot{E}_{x,Q} + P_{t12} = \dot{m}e_{x,S,2} + \Delta\dot{E}_{x,V} \qquad (17.10)$$

with

$$\Delta\dot{E}_{x,V} = T_{env}\dot{S}_i. \qquad (17.11)$$

Adiabatic Turbine

For an adiabatic turbine with neglected potential and kinetic energy the first law of thermodynamics simplifies to

$$w_{t12} = h_2 - h_1 < 0. \qquad (17.12)$$

Consequently, in order to make the turbine to release as much specific technical work as possible, the difference of specific enthalpies from (1) to (2) needs to be as large as possible. However, the second law of thermodynamics under these premises follows

$$s_2 - s_1 = s_{i12} \geq 0. \qquad (17.13)$$

---

[4]Entropy flux *in* is balanced by entropy flux *out*.
[5]Exergy flux *in* is balanced by exergy flux *out*.

Fig. 17.2 Adiabatic turbine:
Illustration in a T, s-diagram
(ideal and real gases)

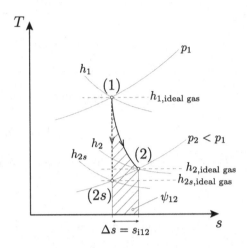

Thus, from state (1) to state (2) the specific entropy can either stay constant or it has to rise. Figure 17.2 shows the change of state in a T, s-diagram. As the pressure decreases from (1) to (2), state (2) lies on a lower isobar than state (1). The second law of thermodynamics has further shown, that since no heat is exchanged, the specific entropy needs to rise in case dissipation, i.e. $\psi_{12} \rightarrow s_{i12} \rightarrow \Delta e_{x,V}$, occurs. Consequently, state (2) is below and to the right of state (1). In best case, i.e. the change of state runs free of dissipation, the change of state is reversible and adiabatic, i.e. isentropic, so that the change of state runs vertical in a T, s-diagram. Under isentropic conditions, i.e. best case scenario, state (2s) is reached. The shown T, s-diagram additionally includes the isenthalps, i.e. curves of constant specific enthalpy. For an ideal gas isenthalps are horizontal, since the specific enthalpy is purely a function of temperature, i.e.

$$dh = c_p \, dT. \tag{17.14}$$

However, Fig. 17.2 as well contains the isenthalps for real fluids, that will be handled in part II. According to Eq. 12.94 the correlation between temperature T, specific entropy s and pressure p is more complex. Anyhow, the first law of thermodynamics has shown, that the specific technical work is a function of $\Delta h = h_2 - h_1$. Hence, Fig. 17.2 proves, that the thermodynamic process from (1) to (2) is not yet optimised, since Δh can be further increased. Obviously, the maximum technical work in an adiabatic turbine within a pressure range $p_1 \ldots p_2$ can be realised by an isentropic change of state $(1) \rightarrow (2s)$:

- $(1) \rightarrow (2)$: Adiabatic, technically realistic turbine, i.e. irreversible. The technical power is not maximised.
- $(1) \rightarrow (2s)$: Adiabatic, hypothetically best turbine, i.e. reversible.[6] Maximum technical power can be gained by this isentropic change of state!

---

[6]Thus, no dissipation or any other imperfections that cause generation of entropy.

It seems reasonable to compare these two adiabatic processes in terms of a so-called isentropic efficiency η_{sT}, i.e.

$$\eta_{sT} = \frac{\Delta h}{\Delta h_s} = \frac{h_2 - h_1}{h_{2s} - h_1} \qquad (17.15)$$

In case the gas is ideal, the definition simplifies to

$$\eta_{sT} = \frac{\Delta h}{\Delta h_s} = \frac{T_2 - T_1}{T_{2s} - T_1}. \qquad (17.16)$$

Thus, the turbine's power can be calculated as follows in case the turbine is adiabatic and kinetic/potential energies are ignored:

$$P_{t12} = \dot{m} w_{t12} = \dot{m}\, \Delta h = \dot{m}\eta_{sT}\, \Delta h_s < 0. \qquad (17.17)$$

Typical values for isentropic efficiencies are, see [4]:

- Steam turbine: $0.88\ldots0.94$
- Gas turbine: $0.90\ldots0.95$

17.1.2 Compressor

A compressor is a thermal turbo-machine, that compresses a fluid from a pressure p_1 to a pressure $p_2 > p_1$ by supply of work, see Fig. 17.3. The following balances can be conducted accordingly to the turbine:

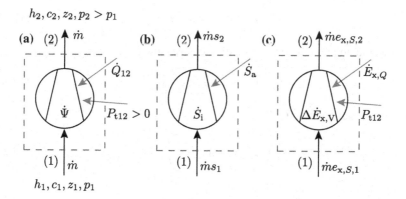

Fig. 17.3 Compressor: **a** First law, **b** Second Law, **c** Exergy

- The first law of thermodynamics for a compressor in steady state reads as[7]:

$$\dot{Q}_{12} + P_{t12} + \dot{m}\left[h_1 + \frac{1}{2}c_1^2 + gz_1\right] = \dot{m}\left[h_2 + \frac{1}{2}c_2^2 + gz_2\right]. \qquad (17.18)$$

In specific notation and with ignoring the potential energy, that is supposed to be comparable small in technical applications, it follows, that

$$q_{12} + w_{t12} = h_2 - h_1 + \frac{1}{2}\left(c_2^2 - c_1^2\right). \qquad (17.19)$$

With the total enthalpy the first law of thermodynamics finally obeys

$$\boxed{q_{12} + w_{t12} = h_2^+ - h_1^+} \qquad (17.20)$$

- The partial energy equation reads as:

$$w_{t,12} = \int_1^2 v\,dp + \psi_{12} + \frac{1}{2}\left(c_2^2 - c_1^2\right) + g\left(z_2 - z_1\right). \qquad (17.21)$$

Ignoring the potential energy simplifies the partial energy equation, i.e.

$$\boxed{w_{t,12} = \int_1^2 v\,dp + \psi_{12} + \frac{1}{2}\left(c_2^2 - c_1^2\right)} \qquad (17.22)$$

- The second law of thermodynamics in steady state[8] is, see Fig. 17.3b,

$$\dot{m}s_1 + \dot{S}_i + \dot{S}_a = \dot{m}s_2. \qquad (17.23)$$

In specific notation, by dividing with the mass flow rate, it is

$$\boxed{s_2 - s_1 = s_{i12} + s_{a12}} \qquad (17.24)$$

In differential notation it obeys

$$ds = \delta s_i + \delta s_a = \underbrace{\frac{\delta\psi}{T}}_{\geq 0} + \frac{\delta q}{T}. \qquad (17.25)$$

---

[7]Following, that the energy flux *in* is balanced by the energy flux *out* in steady state!
[8]Entropy flux *in* is balanced by entropy flux *out*.

- The balance of exergy in steady state[9] reads as, see Fig. 17.3c,

$$\boxed{\dot{m}e_{x,S,1} + \dot{E}_{x,Q} + P_{t12} = \dot{m}e_{x,S,2} + \Delta\dot{E}_{x,V}} \tag{17.26}$$

with

$$\Delta\dot{E}_{x,V} = T_{env}\dot{S}_i. \tag{17.27}$$

Adiabatic Compressor

For an adiabatic compressor with neglected potential and kinetic energy the first law of thermodynamics simplifies to

$$w_{t12} = h_2 - h_1 > 0. \tag{17.28}$$

Consequently, in order to minimise the required specific technical work, the difference of specific enthalpies from (1) to (2) needs to be as small as possible. However, the second law of thermodynamics under these premises follows

$$s_2 - s_1 = s_{i12} \geq 0. \tag{17.29}$$

Thus, from state (1) to state (2) the specific entropy can either stay constant or it has to rise. Figure 17.4 shows the change of state in a T, s-diagram. As the pressure rises from (1) to (2), state (2) lies on an upper isobar than state (1). The second law of thermodynamics has shown, that since not heat is exchanged, the specific entropy needs to increase in case dissipation, i.e. $\psi_{12} \to s_{i12} \to \Delta e_{x,V}$, occurs. Consequently, state (2) is above and to the right of state (1). In best case, i.e. the change of state runs free of dissipation, the change of state is reversible and adiabatic, i.e. isentropic, so that the change of state runs vertical in a T, s-diagram. Under isentropic conditions, i.e. best case scenario, state (2s) is reached. The shown T, s-diagram additionally includes the isenthalps. Figure 17.4 does not only show the isenthalps for an ideal gas but it contains the isenthalps for real fluids as well. However, the first law of thermodynamics has shown, that the specific technical work is a function of $\Delta h = h_2 - h_1$. Hence, Fig. 17.4 proves, that the thermodynamic process from (1) to (2) is not yet optimised, since Δh can be further decreased. Obviously, the minimum technical work in an adiabatic compressor within a pressure range $p_1 \ldots p_2$ can be realised by an isentropic change of state (1) $\to$ (2s):

- (1) $\to$ (2): Adiabatic, technically realistic compressor, i.e. irreversible. The technical power is not minimised.
- (1) $\to$ (2s): Adiabatic, hypothetically best compressor, i.e. reversible.[10] Minimum technical power can be achieved by this isentropic change of state!

---

[9]Exergy flux *in* is balanced by exergy flux *out*.

[10]Thus, no dissipation or any other imperfections that cause generation of entropy.

Fig. 17.4 Adiabatic
compressor: Illustration in a
T, s-diagram (ideal and real
gases)

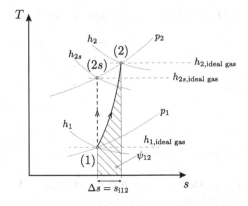

It seems reasonable to compare these two adiabatic processes in terms of a so-called
isentropic efficiency η_{sC}:

$$\boxed{\eta_{sC} = \frac{\Delta h_s}{\Delta h} = \frac{h_{2s} - h_1}{h_2 - h_1}} \tag{17.30}$$

In case the gas is ideal, the definition simplifies to

$$\eta_{sC} = \frac{\Delta h_s}{\Delta h} = \frac{T_{2s} - T_1}{T_2 - T_1}. \tag{17.31}$$

Thus, the compressor's power can be calculated as follows in case the compressor is
adiabatic and kinetic/potential energies are ignored:

$$P_{t12} = \dot{m} w_{t12} = \dot{m}\, \Delta h = \frac{\dot{m}\, \Delta h_s}{\eta_{sC}} > 0. \tag{17.32}$$

Typical values for isentropic efficiencies are in the range of $0.85 \dots 0.90$, see [4].

17.1.3   Thermal Turbo-Machines in a h, s-Diagram

Instead of using a T, s-diagram, the previous considerations have shown, that it
is advantageous to illustrate changes of state in a h, s-diagram. Hence, Fig. 17.5
summarises the changes of state in an adiabatic turbine respectively in an adiabatic
compressor. In case the change of state runs adiabatic the first law of thermodynamics
in specific notation reads as

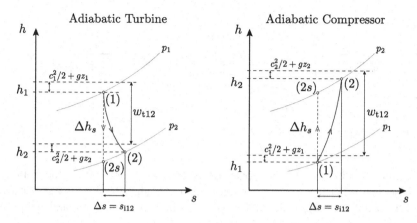

Fig. 17.5 Thermal turbo-machines in a h, s-diagram

$$w_{t12} = (h_2 - h_1) + \frac{1}{2}\left(c_2^2 - c_1^2\right) + g\left(z_2 - z_1\right) = \left(h_2^+ - h_1^+\right) + g\left(z_2 - z_1\right).$$
(17.33)

By applying the total enthalpy, see Eq. 17.3, it is

$$w_{t12} = \left(h_2^+ - h_1^+\right) + g\left(z_2 - z_1\right) \begin{cases} < 0 & \text{adiabatic turbine} \\ > 0 & \text{adiabatic compressor.} \end{cases}$$
(17.34)

Under these conditions, the specific technical work w_{t12} can be easily visualised in a h, s-diagram as shown in Fig. 17.5. If the kinetic respectively potential are taken into account as well, the specific enthalpy h has to be extended by $\left(0.5c^2 + gz\right)$, as done in Fig. 17.5. Generally speaking, in comparison with the caloric energy, the outer energies are often of a subordinate importance, thus they can be ignored for a steam power plant for instance. As mentioned before, for an ideal gas a h, s-diagram can be adapted from a T, s-diagram, since specific enthalpy h is proportional to temperature T. Consequently, isobars and isochors lines have the same mathematical shape as they have in a T, s-diagram. However, a h, s-diagram for non-ideal, i.e. real, gases/fluids will be introduced in part II.

17.1.4 Adiabatic Throttle

An adiabatic throttle is utilised in order to reduce the fluid's pressure. In contrast to a turbine, that also is applied to decrease the pressure, no technical work is released. Thus, a throttle is a passive, i.e. work-insulated, component. The working principle is illustrated in Fig. 17.6. A pressure drop is simply caused by a reduction of the cross section: In doing so turbulence, i.e. dissipation, is initiated. It has already been shown, that dissipation generates entropy. From a technical point of view, a throttle

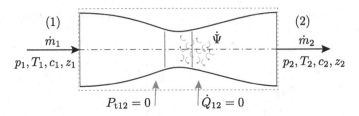

Fig. 17.6 Adiabatic throttle

is a component, that is built to dissipate energy, whereas other technical applications increase their efficiency by avoiding dissipation. An adiabatic throttle has already been calculated in Problem 12.10. However, the following assumptions and premises are made:

(1) No technical work is exchanged,[11] i.e. $w_{t12} = 0$
(2) System is adiabatic, i.e. $q_{12} = 0$
(3) System is in steady state, i.e. $\dot{m}_1 = \dot{m}_2 = \dot{m}$
(4) Change of potential energy is neglected, i.e. $z_1 = z_2$
(5) Change of kinetic energy is neglected, i.e. $c_1^2 = c_2^2$

- First law of thermodynamics

$$q_{12} + w_{t12} = h_2 - h_1 + \frac{1}{2}\left(c_2^2 - c_1^2\right) + g \cdot (z_2 - z_1) \qquad (17.35)$$

With the mentioned assumptions, the first law of thermodynamics simplifies to

$$\boxed{h_2 - h_1 = 0} \qquad (17.36)$$

In other words, an adiabatic throttle is isenthalpic, i.e. $dh = 0$.

- Partial energy equation[12]

$$w_{t12} = \int_1^2 v\,dp + \psi_{12} = 0 \qquad (17.37)$$

Rearranging leads to

$$\boxed{\int_1^2 v\,dp = -\psi_{12} < 0} \qquad (17.38)$$

---

[11] There are no mechanical/electrical parts within the component, that exchange work with the environment.
[12] Outer energies are ignored!

Thus, it is the dissipation that causes the pressure loss in an adiabatic throttle.

- Second law of thermodynamics

$$\dot{m}s_1 + \dot{S}_a + \dot{S}_i = \dot{m}s_2 \tag{17.39}$$

According to the second law of thermodynamics in steady state, the flux of entropy into the system needs to be balanced by the flux of entropy leaving the system. Since, the throttle is supposed to be adiabatic, the second law of thermodynamics obeys

$$\dot{m}s_1 + \dot{S}_i = \dot{m}s_2 \tag{17.40}$$

respectively

$$\boxed{\dot{S}_i = \dot{m}\,(s_2 - s_1)} \tag{17.41}$$

- Balance of exergy

$$\dot{m}e_{x,S,1} + \dot{E}_{x,Q} + \underbrace{P_{t12}}_{=0} = \dot{m}e_{x,S,2} + \Delta\dot{E}_{x,V} \tag{17.42}$$

In steady state, the flux of exergy into the system needs to be balanced by the flux being released. Mind, that the loss of exergy, that is due to the generation of entropy, from a thermodynamic point of view is a sink, whereas generation of entropy is a source. However, since the throttle is adiabatic, there is no exergy of heat, so that the balance simplifies to

$$\boxed{\dot{m}e_{x,S,1} = \dot{m}e_{x,S,2} + \Delta\dot{E}_{x,V}} \tag{17.43}$$

As mentioned above, the dissipation is the cause for the pressure loss in an adiabatic throttle. Thus, it is now derived how much energy is dissipated. Anyhow, there are two alternative approaches, that are both presented:

- Approach following the partial energy equation

$$w_{t12} = \int_1^2 v\,dp + \psi_{12} = 0 \tag{17.44}$$

The partial energy equation can be solved for the specific dissipation, i.e.

$$\psi_{12} = -\int_1^2 v\,dp. \tag{17.45}$$

In order the solve the integral, a correlation between pressure p and specific volume v is required. Since, the first law of thermodynamics has shown, that the change of state is isenthalpic, it follows for an ideal gas, that

$$h_2 = h_1 \Rightarrow T_2 = T_1. \tag{17.46}$$

Thus, applying the thermal equation of state reads as

$$pv = RT = \text{const.} \tag{17.47}$$

Substitution in the partial energy equation leads to

$$\psi_{12} = -\int_1^2 v\,dp = -p_1 v_1 \int_1^2 \frac{1}{p}\,dp = -RT \int_1^2 \frac{1}{p}\,dp. \tag{17.48}$$

Hence, it finally is

$$\boxed{\psi_{12} = -RT \ln \frac{p_2}{p_1}} \tag{17.49}$$

- Approach following the second law of thermodynamics

$$\dot{m}\,(s_1 - s_2) + \dot{S}_{a12} + \dot{S}_{i12} = 0 \tag{17.50}$$

Since the throttle is adiabatic, it follows in specific notation

$$s_2 - s_1 = s_{i12}. \tag{17.51}$$

Applying the caloric equation of state

$$s_2 - s_1 = c_p \cdot \ln \frac{T_2}{T_1} - R \cdot \ln \frac{p_2}{p_1}. \tag{17.52}$$

Since, for an ideal gas an isenthalpic change of state is isothermal as well, the second law of thermodynamics can be simplified, i.e.

$$s_2 - s_1 = -R \cdot \ln \frac{p_2}{p_1} = s_{i12}. \tag{17.53}$$

The specific generation of entropy obeys

$$s_{i12} = \int_1^2 \frac{\delta \psi}{T}. \tag{17.54}$$

Hence, it follows

$$-R \cdot \ln \frac{p_2}{p_1} = \int_1^2 \frac{\delta \psi}{T}. \tag{17.55}$$

Since temperature is constant, the integral can be simplified, i.e.

$$-R \cdot \ln \frac{p_2}{p_1} = \frac{1}{T} \int_1^2 \delta \psi = \frac{\psi_{12}}{T}. \tag{17.56}$$

Rearranging finally leads to the specific dissipation, i.e.

$$\boxed{\psi_{12} = -RT \ln \frac{p_2}{p_1}} \tag{17.57}$$

17.1.5 Heat Exchanger

A heat exchanger is a caloric apparatus, that is passed by two or even more fluids with different temperatures. In order to achieve a thermal equilibrium, heat is exchanged between the fluids. However, heat exchangers have already been treated in several problems before. There are many technical applications, e.g. heater, boiler, oil- and water coolers, that require heat exchangers. A differentiation can be made by the heat exchanger's design: plate heat exchanger, shell and tube heat exchanger, adiabatic wheel heat exchanger and several others more. The focus within this book is on heat exchangers that do not allow a mixing of the fluids. Anyhow, heat exchangers can be operated in parallel or countercurrent flow. Possible technical symbols for heat exchangers are shown in Fig. 17.7.

Energy Balancing Heat Exhangers

Let us have a closer look at the heat exchanger presented in Fig. 17.8, that consists of two flows in countercurrent operation. In this case the first law of thermodynamics in steady state, i.e. energy flux *in* is balanced by energy flux *out*, reads as

$$\dot{Q}_{\text{iso}} + P_{\text{t}} + \dot{m}_1 \left[h_{1,\text{E}} + c_{1,\text{E}}^2/2 + g \cdot z_{1,\text{E}} \right] + \dot{m}_2 \left[h_{2,\text{E}} + c_{2,\text{E}}^2/2 + g \cdot z_{2,\text{E}} \right]$$
$$= \dot{m}_1 \left[h_{1,\text{A}} + c_{1,\text{A}}^2/2 + g \cdot z_{1,\text{A}} \right] + \dot{m}_2 \left[h_{2,\text{A}} + c_{2,\text{A}}^2/2 + g \cdot z_{2,\text{A}} \right]. \tag{17.58}$$

Since the heat exchanger is a passive component, i.e. no work is transferred, the first law of thermodynamics can be simplified. Let us further assume, that the heat exchanger does not transfer heat with the environment and that potential energy can

Fig. 17.7 Symbols for heat exchangers

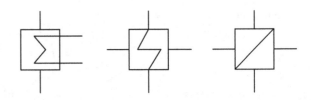

Fig. 17.8 Countercurrent heat exchanger

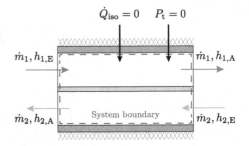

Fig. 17.9 Countercurrent heat exchanger

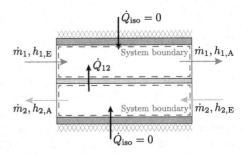

be ignored. Hence, the first law of thermodynamics obeys

$$0 = \dot{m}_1 \left[h_{1,A}^+ - h_{1,E}^+ \right] + \dot{m}_2 \left[h_{2,A}^+ - h_{2,E}^+ \right]. \tag{17.59}$$

Mind, that h^+ is the specific total enthalpy, according to Eq. 17.3, that includes the specific caloric enthalpy as well as the specific kinetic energy. Rearranging leads to

$$\dot{m}_1 \left[h_{1,A}^+ - h_{1,E}^+ \right] = \dot{m}_2 \left[h_{2,E}^+ - h_{2,A}^+ \right] \tag{17.60}$$

respectivley, if kinetic energy is ignored,

$$\boxed{\dot{m}_1 \left[h_{1,A} - h_{1,E} \right] = \dot{m}_2 \left[h_{2,E} - h_{2,A} \right]} \tag{17.61}$$

Obviously, the balance according to the system boundary given in Fig. 17.8 does not contain any information regarding the heat that is transferred between the fluids. Alternatively, the first law of thermodynamics is now applied for the system boundaries illustrated in Fig. 17.9. In doing so, the transferred heat $\dot{Q}_{12}$ between the two fluids is part of the balance. The first law of thermodynamics in steady state for fluid 1, see Fig. 17.9, with the same assumptions as before, is

$$\dot{Q}_{12} + \dot{m}_1 h_{1,E} = \dot{m}_1 h_{1,A}. \tag{17.62}$$

Applying the caloric equation of state for ideal gases results in

$$\dot{Q}_{12} = \dot{m}_1 c_{p,1} \left[T_{1,A} - T_{1,E} \right]. \tag{17.63}$$

The first law of thermodynamics in steady state for fluid 2, see Fig. 17.9 is

$$\dot{Q}_{12} + \dot{m}_2 h_{2,A} = \dot{m}_2 h_{2,E}. \tag{17.64}$$

Applying the caloric equation of state for ideal gases results in

$$- \dot{Q}_{12} = \dot{m}_2 c_{p,2} \left[T_{2,A} - T_{2,E} \right]. \tag{17.65}$$

However, combining Eqs. 17.62 and 17.64 leads to Eq. 17.61.

Countercurrent Versus Parallel Flow

So far, heat exchangers have been shown as countercurrent flows. For applying the first law of thermodynamics the flow direction does not have an impact as long as the rule, that the energy flux in is equalised by the energy flux out, is kept. Nevertheless, Fig. 17.10 shows a comparison between parallel and countercurrent flow.

In general, a countercurrent flow, operated under the same fluids' inlet temperatures as a parallel flow, is more efficient than a parallel flow. This is due to its larger averaged mean temperature difference compared to a parallel flow.

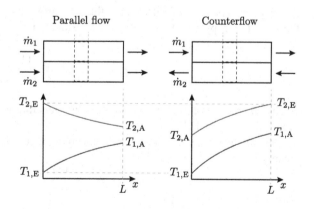

Fig. 17.10 Wall—Heat conduction (heat exchanger is supposed to be adiabatic to the environment)

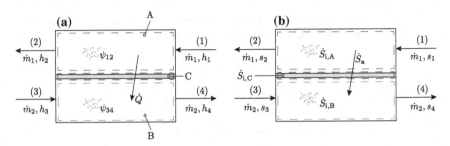

Fig. 17.11 Entropy balance (heat exchanger is supposed to be adiabatic to the environment)

Entropy Balancing Heat Exchangers

The entropy balance for heat transfer has already been investigated in Chap. 14 for a conductive wall. It has been shown, that any heat transfer at a temperature difference greater than zero comes along with generation of entropy. In this chapter, the entropy balance for an entire heat exchanger, see Fig. 17.11, is applied. In case, only system A is investigated, the generation of entropy within A is solely due to dissipation: If kinetic and potential energies are ignored, the partial energy equation simplifies to

$$ w_t = 0 = \int_1^2 v \, dp + \psi_{12}. \tag{17.66} $$

Thus, the pressure drop is an indicator for generated entropy, i.e.

$$ \psi_{12} = - \int_1^2 v \, dp. \tag{17.67} $$

A steady state balance of entropy, i.e. flux of entropy into the system is balanced by the flux of entropy out of the system, for system A is

$$ \dot{m}_1 s_1 + \dot{S}_{i,A} = \dot{m}_1 s_2 + \frac{\dot{Q}}{T_{m,A}}. \tag{17.68} $$

In case there is no pressure loss, there is no generation of entropy $\dot{S}_{i,A}$ under the premises being made.[13]

Nevertheless, the thermodynamic mean temperature for the system is, see Chap. 13,

$$ T_{m,A} = \frac{q_{12} + \psi_{12}}{s_2 - s_1}. \tag{17.69} $$

---

[13] Kinetic and potential energies ignored.

Applying the first law of thermodynamics

$$q_{12} = h_1 - h_2 \tag{17.70}$$

leads to

$$T_{m,A} = \frac{h_1 - h_2 + \psi_{12}}{s_2 - s_1}. \tag{17.71}$$

Thus, for steady state it is

$$\frac{dS_A}{dt} = 0 = \dot{S}_{i,A} - \frac{\dot{Q}}{T_{m,A}} - \dot{m}_1 (s_2 - s_1) \tag{17.72}$$

Hence, the entropy within system A is temporally constant, i.e. the flux in is balanced by the flux out. Now, let us focus on system B according to Fig. 17.11. The balance of entropy for system B obeys

$$\dot{m}_2 s_3' + \dot{S}_{i,B} + \frac{\dot{Q}}{T_{m,B}} = \dot{m}_2 s_4. \tag{17.73}$$

The thermodynamic mean temperature for system B follows

$$T_{m,B} = \frac{q_{34} + \psi_{34}}{s_4 - s_3}. \tag{17.74}$$

Applying the first law of thermodynamics

$$q_{34} = h_4 - h_3 \tag{17.75}$$

leads to

$$T_{m,B} = \frac{h_4 - h_3 + \psi_{34}}{s_4 - s_3}. \tag{17.76}$$

Thus, for steady state it is

$$\frac{dS_B}{dt} = 0 = \dot{S}_{i,B} + \frac{\dot{Q}}{T_{m,B}} - \dot{m}_2 (s_4 - s_3) \tag{17.77}$$

However, the overall system additionally consists of a third system C, i.e. a wall, that separates the two fluids, see Fig. 17.12. The balance of entropy for system C results in

$$\dot{S}_{i,C} + \frac{\dot{Q}}{T_{m,A}} = \frac{\dot{Q}}{T_{m,B}}. \tag{17.78}$$

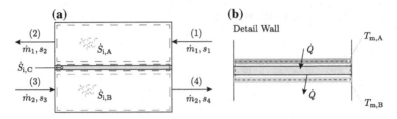

Fig. 17.12 Entropy balance (System C)

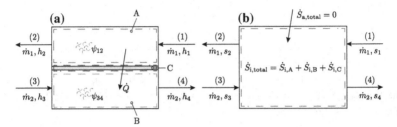

Fig. 17.13 Overall entropy balance

In steady state operation the heat flux into system C must be equal to the heat flux out of system C. However, the thermodynamic mean temperatures of systems A and B can be measured directly at the system boundary of the wall,[14] see Fig. 17.12b. Consequently, the entropy carried by heat can be easily calculated, see Eq. 17.78. The generated entropy in the wall, i.e. in system C, obeys[15]:

$$\dot{S}_{\mathrm{i,C}} = \frac{\dot{Q}}{T_{\mathrm{m,B}}} - \frac{\dot{Q}}{T_{\mathrm{m,A}}} \tag{17.79}$$

Thus, in an alternative notation it follows

$$\boxed{\frac{\mathrm{d}S_{\mathrm{C}}}{\mathrm{d}t} = 0 = \dot{S}_{\mathrm{i,C}} + \frac{\dot{Q}}{T_{\mathrm{m,A}}} - \frac{\dot{Q}}{T_{\mathrm{m,B}}}} \tag{17.80}$$

The overall system is the summation of the three sub-systems A, B and C, see Fig. 17.13. Once again in steady state the entropy within the overall system needs to be constant by time, so that the summation of systems A, B and C leads to

$$\frac{\mathrm{d}S_{\mathrm{total}}}{\mathrm{d}t} = 0 = \frac{\mathrm{d}S_{\mathrm{A}}}{\mathrm{d}t} + \frac{\mathrm{d}S_{\mathrm{B}}}{\mathrm{d}t} + \frac{\mathrm{d}S_{\mathrm{C}}}{\mathrm{d}t}. \tag{17.81}$$

---

[14]Systems A and B are supposed to be homogeneous.
[15]Generation of entropy in system C is caused by heat transfer, not by dissipation!

Substitution of the equations that have been derived before results in

$$0 = \underbrace{\dot{S}_{i,A} + \dot{S}_{i,B} + \dot{S}_{i,C}}_{=\dot{S}_{i,total}} - \dot{m}_1 (s_2 - s_1) - \dot{m}_1 (s_4 - s_3). \tag{17.82}$$

Thus, it finally is

$$\boxed{\dot{S}_{i,total} = \dot{m}_1 (s_2 - s_1) + \dot{m}_1 (s_4 - s_3)} \tag{17.83}$$

The same result can be gained much easier by an overall balance of the entire component, see Fig. 17.13b. Due to the internal heat exchange with a temperature difference between the fluids larger than zero, even then entropy is generated, when both flows are free of dissipation, i.e. no pressure drop occurs.

17.2 Thermodynamic Cycles

The focus in this section is on thermodynamic cycles. Thermodynamic cycles have already been discussed to motivate the second law of thermodynamics in Chap. 15. However, so far these cycles have been presented in black-box notation solely. Now, a technical realisation of cyclic processes is shown in detail, i.e. the technical integration of required components to form a machine. If a system after several changes of state finally reaches its initial state it is called a thermodynamic cycle. Thus, all state values Z reach their initial value, i.e.

$$\boxed{\oint dZ = 0} \tag{17.84}$$

Regarding thermodynamic cycles a distinction between clockwise and counterclockwise cycles can be made. Typical technical applications are:

- Thermal engines ("clockwise cycles")
  Closed gas turbine cycles
  Steam power plants
  Internal combustion engines (Otto- and Diesel)
  Open gas turbine plants
- Heat pump, fridge ("counterclockwise cycles")
  Absorption heat pump/fridge
  Compression heat pump/fridge

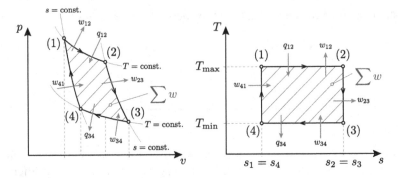

Fig. 17.14 Carnot process

17.2.1   Carnot Process

A Carnot-process is an idealised cyclic process, that consists of two reversible isothermal and two reversible adiabatic changes of state[16]:

- $(1) \rightarrow (2)$:
  Isothermal expansion
- $(2) \rightarrow (3)$:
  Reversible, adiabatic expansion
- $(3) \rightarrow (4)$:
  Isothermal compression
- $(4) \rightarrow (1)$:
  Reversible, adiabatic compression

Carnot's cycle has already been introduced in Chap. 15 in order to predict the maximum work that can be gained from heat at a certain temperature $T > T_{env}$. Mind, that a Carnot process reaches the maximum possible efficiency for a given temperature potential. Anyhow, the process is illustrated in a p, v- and a T, s-diagram, see Fig. 17.14. In this case the process is operated as a clockwise cycle, i.e. as a thermal engine.

The enclosed area in the p, v-diagram indicates the work, that can be gained effectively.[17] In contrast, the enclosed area in the T, s-diagram represents the effective heat, since the process comes along without any dissipation.[18] Furthermore, in a thermodynamic cycle, the first law of thermodynamics obeys, see Eq. 11.13,

$$\sum Q + \sum W = 0. \tag{17.85}$$

Thus, the enclosed area in the T, s-diagram is identical with the effective work.

---

[16] Since every single change of state is reversible, the entire process is reversible as well!

[17] This is due to no dissipation occurs and outer energies are ignored.

[18] Mind, that the area beneath a change of state in a T, s-diagram is the summation of specific heat and specific dissipation.

Fig. 17.15 Joule process—Layout clockwise cycle

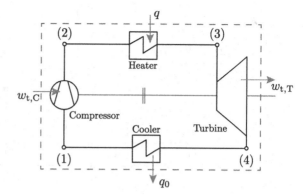

17.2.2 Joule Process

The Joule process typically is an ideal reference cycle for gas turbines respectively jet engines, i.e. for clockwise cycles. Its fluid is an ideal gas. Hence, the fluid does not underlie any phase change. The required components and the process layout are illustrated in Fig. 17.15. However, the process can also be operated as counterclockwise cycle.

Figure 17.16 shows the clockwise cycle in a p, v- and a T, s-diagram. It consists of the following changes of state:

- $(1) \rightarrow (2)$: Adiabatic, reversible compression
  The fluid is compressed isentropically in a compressor. Work is supplied. The fluid's pressure and temperature rise, whereas the specific volume decreases.
- $(2) \rightarrow (3)$: Isobaric heat transfer
  Since the heat exchanger does not have any pressure losses, the fluid's change of state needs to be reversible.[19] The supplied heat causes an increase of entropy, as can be seen in the T, s-diagram.
- $(3) \rightarrow (4)$: Adiabatic, reversible expansion
  The fluid expands isentropically in a turbine. Work is released. Pressure as well as temperature decrease, the specific volume rises.
- $(4) \rightarrow (1)$: Isobaric heat transfer
  Heat is released to the environment. Since the change of state is at constant pressure, there is no dissipation.[20] The entropy needs to decrease due to the release of heat.

---

[19] As long, as potential and kinetic energies are ignored! Anyhow, the *entire* heat exchanger generates entropy, since a temperature gradient is required to transfer heat. However, the T, s-diagram just shows the fluid, operated in the cycle. This fluid is supplied with heat, no matter regarding the source of the heat. The only cause for dissipation in the fluid is due to a (mechanical) pressure loss. Mind, that generation of entropy at heat transfer is at the interface of the two systems, i.e. imperfection in the wall caused by $\Delta T > 0$.

[20] As long, as potential and kinetic energies are ignored!

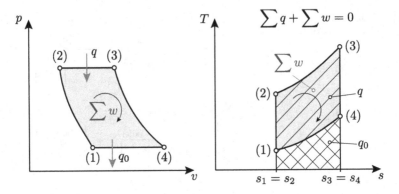

Fig. 17.16 Joule process

The enclosed areas in p, v- and T, s-diagram represent the effective work, that is released by the process. Mind, that the entire Joule process is free of dissipation, so that the enclosed area in the T, s-diagram does purely represent the effective heat. According to the first law of thermodynamics, see Eq. 17.85, it is identical with the effective work.

17.2.3 Clausius Rankine Process

The Clausius Rankine process is the ideal reference cycle for steam power plants. In contrast to the Joule process, the operating fluid is water, that undergoes phase changes. However, the basic principle for a Clausius Rankine and a Joule cycle is the same. The layout of the Clausius Rankine cycle is shown in Fig. 17.17. As the fluid changes its aggregate state, it can not be treated as in ideal gas. Thus, the process is investigated in detail in part II. However, the main aspects are introduced briefly. Figure 17.18 shows a corresponding p, v- and T, s-diagram. In contrast to ideal gases, both diagrams are characterised by limiting, bell-shaped curves, that separate liquid/wet-steam/gas region.

The following changes of state run in a Clausius Rankine process:

- $(1) \rightarrow (2)$: Isentropic compression
  The liquid fluid is compressed in a pump. Since the specific volume of liquids is rather small, the required specific pressure work

$$y_{12} = \int_1^2 v \, dp \qquad (17.86)$$

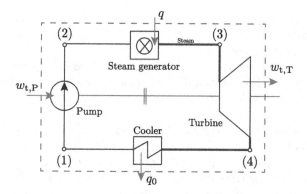

Fig. 17.17 Clausius Rankine process

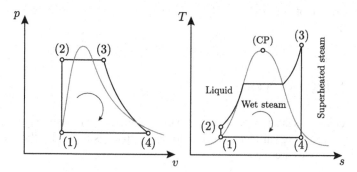

Fig. 17.18 Clausius Rankine process

is small as well.

- $(2) \rightarrow (3)$: Isobaric heat transfer + phase change
  The heat supply is supposed to be without pressure losses and thus reversible, since the change of potential and kinetic energies are ignored

$$w_t = 0 = \underbrace{\int_2^3 v \, dp}_{=0} + \psi_{23} + \underbrace{\Delta e_{a,23}}_{=0}. \tag{17.87}$$

Hence, it is

$$\psi_{23} = 0. \tag{17.88}$$

Due to the heating, first the temperature of the fluid rises.[21] Once the evaporation starts, the temperature is constant and the entire thermal energy is utilised for

---

[21] This is called sensible heat.

the phase change.[22] As soon as the phase change is completed, the temperature continues to rise.

- $(3) \rightarrow (4)$: Isentropic expansion
  The pressurised hot steam is now driving a steam turbine and expands. In doing so, technical work is released and the fluid's pressure decreases. The idealised Clausius Rankine process assumes, that the expansion runs adiabatically/reversibly, i.e. isentropically.
- $(4) \rightarrow (1)$: Isobaric heat transfer + phase change
  Finally, in order to reach initial state (1) the fluid needs to be cooled down. This is achieved by a cooler/condenser. Since the change of state is free of pressure losses, it needs to be reversible under the same premises as in $(2) \rightarrow (3)$. The entire condenser, i.e. fluid/housing/environment, is irreversible, since a temperature gradient is required for the heat transfer. However, the focus is on the fluid, so that generation of entropy can only be due to pressure losses, refer to Sect. 17.1.5 for details.

17.2.4 Seiliger Process

The Seiliger process is an ideal reference cycle for internal combustion engines. The fluid is supposed to be an ideal gas, i.e. no phase changes occur. Furthermore, the following premisses are made:

- The cylinder chamber is always completely filled with ideal gas
- The compression ratio ϵ is the same as in the real engine, i.e.

$$\epsilon = \frac{\text{Total volume before compression}}{\text{Rest volume after compression}} = \frac{V_{\text{swing}} + V_{\text{comp}}}{V_{\text{comp}}} \tag{17.89}$$

with

$$V_{\text{swing}} = \frac{\pi}{4} d^2 s. \tag{17.90}$$

The compression ratio in Otto engines is typically $10 : 1 \ldots 14 : 1$, in Diesel engines it is in the range of $19 : 1 \ldots 23 : 1$.

- The mass in the cylinder stays constant, i.e. the system is closed.
- The combustion process is substituted by heating. The amount of heat is equal to the heat, that is released during combustion.
- All walls are supposed to be adiabatic.

In fact, a distinction between two different combustion concepts needs to be done: Otto engines imply that the combustion, i.e. the heat supply, is at constant volume.

---

[22]This is called latent heat.

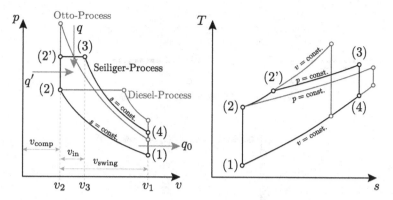

Fig. 17.19 Seiliger process

In contrast, Diesel engines are characterised by the pressure staying constant during combustion. The Seiliger process, however, is a mixed reference cycle, that takes into account both principles. Figure 17.19 shows p, v- and T, s-diagram for all three processes, i.e. Otto, Diesel and Seiliger.

The changes of state are as follows:

- $(1) \rightarrow (2)$: Isentropic compression
  Work is supplied, the specific volume decreases while the pressure rises.
- $(2) \rightarrow (2')$: Isochoric heat transfer
  Part one of the combustion, i.e. at constant volume according to the Otto engine, is performed. The pressure further rises. Combustion is substituted by supply of heat.
- $(2') \rightarrow (3)$: Isobaric heat transfer
  Part two of the combustion, i.e. at constant pressure according to the Diesel engine, is performed. The specific volume rise, so that work is released. Once again, the combustion is substituted by supply of heat.
- $(3) \rightarrow (4)$: Isentropic expansion
  The gas expands, i.e. expansion work is released. The specific volume rises, the pressure decreases.
- $(4) \rightarrow (1)$: Isochoric heat transfer
  According to the second law of thermodynamics, heat is released in order to reach the initial state (1). The heat transfer replaces the release of gas.

17.2.5 Stirling Process

The working principle of a Stirling process can be explained best with Fig. 17.20. In contrast to common cylinder/piston engines, a Stirling machine consists of two

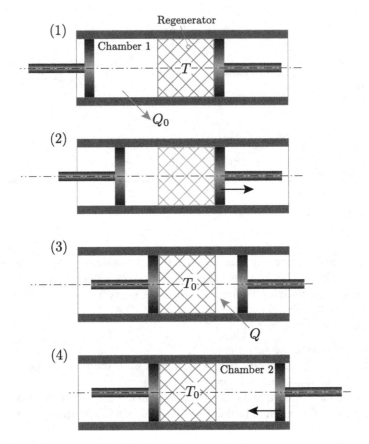

Fig. 17.20 Stirling process—Technical principle

pistons, that can move in parallel, i.e. the volume in-between the two pistons is constant, respectively that can move against each other, so that the volume can decrease respectively increase. The volume, that is limited by the two pistons, is filled with a gas. Additionally, a regenerator, i.e. a thermal storage, is mounted within the gas volume. Such a regenerator can easily be realised, e.g. by a metal mesh, that is characterised by a large thermal conductivity. The entire process is shown in a p, v- and T, s-diagram, see Fig. 17.21:

- $(1) \rightarrow (2)$: Isothermal compression
  The left piston starts to move to the right whereas the right piston is fixed. By doing so, the working fluid is compressed. In order to keep the temperature constant a heat release to the *outside* takes place. The regenerator is supposed to be hot.[23] Cooling leads to a decrease of entropy. Specific volume decreases, pressure rises.

---

[23] As will be seen at the end of the cycle!

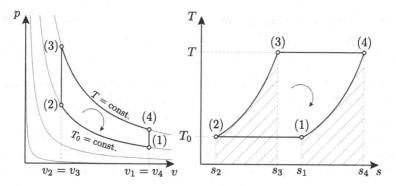

Fig. 17.21 Stirling process—p, v- and T, s-diagram

- $(2) \rightarrow (3)$: Isochoric heating
  In order to keep the volume constant, the two pistons move in parallel to the right. Consequently, the cold gas needs to pass the hot regenerator, so that heat is transferred from regenerator to the gas. This is an *internal* heat transfer, that cools down the regenerator. Heating is the cause for the rise of the entropy. The volume is fixed, so that the temperature increases.
- $(3) \rightarrow (4)$: Isothermal expansion
  The gas expands, while the right piston moves further to the right, while the left piston is fixed. In order to keep the temperature level constant, heat is supplied from the *outside*. The heat supply leads to an increase of entropy. Specific volume rises, pressure decreases.
- $(4) \rightarrow (1)$: Isochoric cooling
  The last change of state is required in order to reach the initial state (1) and thus for closing the cycle. The hot gas passes the cold regenerator by moving the two pistons in parallel to the left. This *internal* heat transfer cools down the gas and heats up the regenerator, i.e. the thermal storage is charged! Heat release causes a decrease of entropy. The volume is fixed, so that pressure and temperature decrease.

It is obvious, that heat is transferred in each change of state. However, a distinction between internal heat transfer and heat transfer to the outside needs to be made. Stirling's basic idea is, that the amount of internal heat supply is balanced by internal heat release: For the isochoric change of state $(2) \rightarrow (3)$ the first law of thermodynamics obeys

$$q_{23} + \underbrace{w_{23}}_{=0} = u_3 - u_2 = c_v(T_3 - T_2). \tag{17.91}$$

For the change of state $(4) \rightarrow (1)$ it follows accordingly, i.e.

$$q_{41} + \underbrace{w_{41}}_{=0} = u_1 - u_4 = c_v(T_1 - T_4). \tag{17.92}$$

Since both changes of state are isochoric, there is no work transferred. Consequently, it follows that

$$q_{23} = -q_{41}. \tag{17.93}$$

Thus, the thermal energy stored within the regenerator from $(4) \rightarrow (1)$ is identical with the thermal energy taken out of the regenerator from $(2) \rightarrow (3)$!

Now, let us focus on the *heat transfer with the environment* in order to derive the efficiency of the Stirling cycle. The supplied specific heat q follows from the first law of thermodynamics applied for the isothermal change of state $(3) \rightarrow (4)$, i.e.

$$w_{34} + q_{34} = u_4 - u_3 = 0. \tag{17.94}$$

Hence, the specific heat supply for an ideal gas is, see Fig. 13.13,

$$q = q_{34} = -w_{34} = RT \ln\left(\frac{v_4}{v_3}\right) > 0. \tag{17.95}$$

The specific heat release q_0 to the environment can be derived from the first law of thermodynamics from $(1) \rightarrow (2)$, i.e.

$$w_{12} + q_{12} = u_2 - u_1 = 0. \tag{17.96}$$

Thus, the specific heat release for an ideal gas is, see Fig. 13.13,

$$q_0 = q_{12} = -w_{12} = RT_0 \ln\left(\frac{v_2}{v_1}\right) < 0. \tag{17.97}$$

According to thermal engines, the gained work w can be calculated from released and supplied heat. The first law of thermodynamics for cycles claims, that the energy into the machine is balanced by the energy out of the machine. Thus, it is[24]

$$q = |q_0| + w. \tag{17.98}$$

The thermal efficiency is the ratio of benefit and effort, i.e.

$$\eta_{th} = \frac{w}{q} = \frac{q - |q_0|}{q} = 1 - \frac{|q_0|}{q}. \tag{17.99}$$

Substitution of the specific heats results in

---

[24]The notation of the applied first law of thermodynamics is, that the sign is taken into account by balancing in and out. This requires, that each energy is counted as absolute value. Thus, the heat release q_0, which is negative, needs to be taken as absolute value but on the side of the leaving energy.

Fig. 17.22 Compression heat pump—Layout

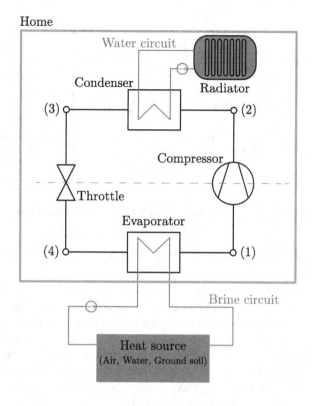

$$\eta_{\text{th}} = 1 - \frac{|q_0|}{q} = 1 - \frac{\left| RT_0 \ln\left(\frac{v_2}{v_1}\right) \right|}{RT \ln\left(\frac{v_4}{v_3}\right)} = 1 - \frac{RT_0 \ln\left(\frac{v_1}{v_2}\right)}{RT \ln\left(\frac{v_4}{v_3}\right)}. \tag{17.100}$$

Since $v_1 = v_4$ and $v_2 = v_3$, see Fig. 17.21, it follows that

$$\boxed{\eta_{\text{th}} = 1 - \frac{T_0}{T} = \eta_{\text{C}}} \tag{17.101}$$

The perfect Stirling process reaches the Carnot efficiency, i.e. the largest possible efficiency within a given temperature spread. Any irreversibility, e.g. internal friction, decreases the efficiency. Furthermore, the best efficiency can only be achieved in case the internal storage (regenerator) can be charged and discharged perfectly, i.e. the internal heat transfer works with $\Delta T \rightarrow 0$. However, in reality this can not be achieved, since a temperature potential $\Delta T > 0$ is required in order to transfer heat.

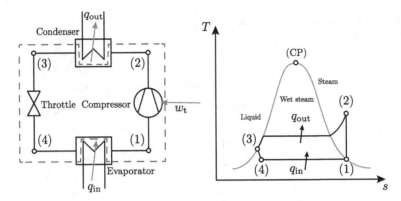

Fig. 17.23 Compression heat pump: Sketch and illustration in a T; *s*-diagram

17.2.6   Compression Heat Pump

So far clockwise cycles have been discussed. Now, the focus is on a heat pump, i.e. a counterclockwise cycle. However, these cycles are more complex as they are usually operated with real fluids, that underlie a phase change. This is why these cycles are discussed in detail in part II. In order to complete part I, the basic principle of a compression heat pump is explained briefly, see Fig. 17.22. The purpose of a heat pump is to utilise heat from the environment for heating, see Sect. 15.2.1, in which a heat pump was presented as black box. In doing so, the required power for heating, i.e. the effort, is reduced, since part of the heating demand is covered by the utilised heat from the environment. Ambient heat is supplied in the evaporator: The working fluid is operated at a low pressure, so that vaporisation takes place at low temperatures. Thus, heat can be transferred from the cold environment to the working fluid. The vaporised fluid is then compressed, so that pressure and temperature rise. In the condenser heat is released, e.g. to a water circuit for heating, so that the vaporised fluid cools down isobarically. This heat release causes a phase change, so that the fluid leaves the condenser as a liquid. In order to close the thermodynamic cycle, the fluid is finally throttled to the lower pressure, so that the cycle can start again. In the following, each change of state is summarised and displayed in a T, s-diagram, see Fig. 17.23.

For real fluids, i.e. fluids that can change its aggregate state, the T, s-diagram is split into three regions by a limiting, bell-shaped curve:

1. Liquid, left from the limiting curve
2. Wet steam,[25] within the enclosed area
3. Steam, right from the limiting curve

The T, s-diagram for real fluids is further discussed in the second part. However, the changes of state in a compression heat pump are:

---

[25]Liquid and vapour occur at the same time!

- $(4) \rightarrow (1)$

By outer heat transfer from the environment, e.g. air, water, ground soil, a working fluid is evaporated at a low temperature respectively pressure. In best case this is done without a pressure loss and thus reversibly.[26] The supplied specific heat is q_{in}.

- $(1) \rightarrow (2)$

The fluid's pressure is increased by a compressor. The required specific technical work is w_t. In the T, s-diagram this change of state is supposed to be adiabatic and reversible, i.e. isentropic.

- $(2) \rightarrow (3)$

Heat is released in a condenser e.g. to a room, that has to be heated. This takes place at a high temperature respectively pressure. By releasing heat isobarically, the fluid finally condenses. The released specific heat is supposed to be q_{out}.

- $(3) \rightarrow (4)$

By adiabatic throttling the fluid expands to the lower operating pressure level. While the specific enthalpy stays constant, if kinetic and potential energy are ignored, the entropy needs to rise. This is, since the internal friction within the adiabatic throttle is the cause for the required pressure loss.

17.2.7 Process Overview

Table 17.1 gives an overview of the previously discussed thermodynamic cycles. It shows, the scheme of the changes of state and illustrations in p, v- and T, s-digrams. Furthermore, it shows, how the process can be operated: *Left* means counterclockwise cycles, i.e. heat pump/fridges. *Right* stands for clockwise cycles, i.e. thermal engines.

Problem 17.1 A compressor plant consists of an adiabatic low-pressure compressor ($\eta_{sC,1} = 0.91$), an interstage cooler and an adiabatic high-pressure compressor ($\eta_{sC,2} = 0.91$) in serial connection. Air, that can be treated as an ideal gas with $R = 0.287 \frac{kJ}{kg\,K}$ and $c_p = 1.004 \frac{kJ}{kg\,K}$, is compressed in the low-pressure compressor from $p_1 = 1,0$ bar and $\vartheta_1 = 15\,°C$ to $p_2 = 6.1$ bar. After that, the air is cooled down in the interstage cooler to $\vartheta_3 = 25\,°C$ while the pressure sinks to $p_3 = 5.9$ bar. Finally, the air is compressed in the high-pressure compressor to $p_4 = 35$ bar. The volume flow of the air intake of the low-pressure compressor is $\dot{V}_1 = 10000 \frac{m^3}{h}$.

(a) Sketch the plant's layout and illustrate the changes of state in a p, v- diagram and in a T, s- diagram schematically.
(b) What is the power consumption P_{t12} of the low-pressure compressor?
(c) What is the heat flux $\dot{Q}_{23}$, that is released in the interstage cooler?
(d) What is the power consumption P_{t34} of the high-pressure compressor?

---

[26] In case kinetic and potential energies are ignored.

Table 17.1 Overview thermodynamic cycles

	Carnot	Seiliger	Stirling	Joule	CR
Scheme	T-s-T-s	s-v-p-s-v	T-v-T-v	s-p-s-p	s-p-s-p
p, v					
T, s					
Direction	left/right	right	left/right	left/right	left/right

Fig. 17.24 Sketch of the layout to Problem 17.1

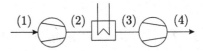

Solution

(a) The plant's layout is illustrated in Fig. 17.24. All three components are in serial connection. The process is documented in a p, v- as well as in a T, s-diagram, see Fig. 17.25.

(b) In order to calculate the power consumption P_{t12} of the low-pressure compressor, the first law of thermodynamics is applied,[27] i.e.

$$\underbrace{\dot{Q}_{12}}_{=0} + P_{t12} = \dot{m}\,(h_2 - h_1) = \dot{m}c_p\,(T_2 - T_1). \tag{17.102}$$

The mass flow rate follows the thermal equation of state

$$p_1 \dot{V}_1 = \dot{m} R T_1. \tag{17.103}$$

Mind, that in steady state the mass flux $\dot{m}$ is constant, whereas the volume flux is not! Solving for the mass flux leads to

---

[27] Since there is no information regarding potential and kinetic energy, both are neglected. Relevant information for the kinetic energy could be a volume flow rate in combination with a tube's cross section for instance.

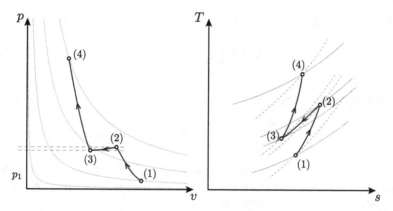

Fig. 17.25 p, v- and T, s-diagram to Problem 17.1

$$\dot{m} = \frac{p_1 \dot{V}_1}{R T_1} = 3.3589 \, \frac{\text{kg}}{\text{s}}. \tag{17.104}$$

The other unknown is the temperature T_2 after the low-pressure compressor. This can not be calculated directly, but the isentropic efficiency needs to be applied. In a first step a perfect adiabatic, i.e. isentropic, compressor is investigated. Thus, the fictitious change of state $(1) \rightarrow (2s)$ is isentropic, i.e.

$$T_{2s} = T_1 \left(\frac{p_2}{p_1} \right)^{\frac{\kappa - 1}{\kappa}}. \tag{17.105}$$

Mind, that this fictitious compressor is also working within the pressure range $p_1 \ldots p_2$. The isentropic exponent is

$$\kappa = \frac{c_p}{c_p - R} = 1.4003. \tag{17.106}$$

Temperature T_2 follows from the definition of the isentropic efficiency

$$\eta_{sC,1} = \frac{\Delta h_s}{\Delta h} = \frac{T_{2s} - T_1}{T_2 - T_1}. \tag{17.107}$$

Rearranging results in

$$T_2 = T_1 + \frac{T_{2s} - T_1}{\eta_{sC,1}} = 502.47 \, \text{K}. \tag{17.108}$$

Hence, the power consumption follows Eq. 17.102, i.e.

$$\boxed{P_{t12} = \dot{m} c_p \left(T_2 - T_1 \right) = 722.76 \, \text{kW}} \tag{17.109}$$

(c) The heat flux $\dot{Q}_{23}$ obeys the first law of thermodynamics for the heat exchanger, i.e.

$$\dot{Q}_{23} + \underbrace{P_{t23}}_{=0} = \dot{m}\,(h_3 - h_2)\,. \tag{17.110}$$

Applying the caloric equation of state results in

$$\boxed{\dot{Q}_{23} = \dot{m}c_p\,(T_3 - T_2) = -689.03\,\text{kW}} \tag{17.111}$$

(d) In order to calculate the power consumption P_{t34} of the high-pressure compressor, accordingly to b, the first law of thermodynamics in combination with the isentropic efficiency is applied, i.e.

$$\underbrace{\dot{Q}_{34}}_{=0} + P_{t34} = \dot{m}\,(h_4 - h_3) = \dot{m}c_p\,(T_4 - T_3)\,. \tag{17.112}$$

First, the fictitious isentropic change of state $(3) \rightarrow (4s)$ needs to be calculated, i.e.

$$T_{4s} = T_3 \left(\frac{p_4}{p_3}\right)^{\frac{\kappa-1}{\kappa}} = 495.97\,\text{K}. \tag{17.113}$$

Thus, temperature T_4 follows:

$$T_4 = T_3 + \frac{T_{4s} - T_1}{\eta_{sC,2}} = 515.53\,\text{K}. \tag{17.114}$$

Second, the power consumption results in

$$\boxed{P_{t34} = \dot{m}c_p\,(T_4 - T_3) = 733.09\,\text{kW}} \tag{17.115}$$

Alternatively, an overall balance including the entire plant $(1) \rightarrow (4)$ can be conducted. The first law of thermodynamics for this case is

$$\dot{Q}_{34} + P_{t12} + P_{t34} = \dot{m}\,(h_4 - h_1) = \dot{m}c_p\,(T_4 - T_1)\,. \tag{17.116}$$

Hence, the power consumption of the high-pressure compressor is

$$P_{t34} = \dot{m}c_p\,(T_4 - T_1) - \dot{Q}_{34} - P_{t12} = 733.09\,\text{kW}. \tag{17.117}$$

Problem 17.2 An open heat pump working with air, see Fig. 17.26, is built up of an adiabatic compressor ($\eta_{sC} = 0.89$), a heat exchanger, in which the compressed air releases heat to a water circuit, and an adiabatic turbine ($\eta_{sT} = 0.89$). Compressor and turbine are mechanically connected. The air is sucked out of the environment

Fig. 17.26 Sketch to
Problem 17.2

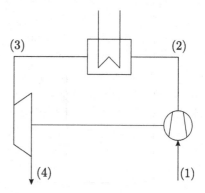

with $p_1 = p_{env} = 1\,bar$, $\vartheta_1 = \vartheta_{env} = 0\,°C$ and compressed to $p_2 = 4.2\,bar$. In the
heat exchanger the air is cooled down to $\vartheta_3 = 85\,°C$ while the pressure decreases
to $p_3 = 4.05\,bar$. In the turbine the air expands to $p_4 = p_{env} = 1\,bar$. Air can be
treated as an ideal gas with $c_p = 1.004\,\frac{kJ}{kg\,K} = $ const. and $R = 0.287\,\frac{kJ}{kg\,K}$. Within
the heat exchanger the water is heated up from $\vartheta_{We} = 70\,°C$ to $\vartheta_{Wa} = 90\,°C$. The
water's specific heat capacity is $c_W = 4.19\,\frac{kJ}{kg\,K} = $ const.. Water can be treated as an
incompressible liquid.

(a) Calculate temperature T_2.
(b) The water is supplied with a heat flux of $\dot{Q}_H = 1\,MW$ within the heat exchanger.
What are the mass flow $\dot{m}$ of the air and the mass flow $\dot{m}_W$ of the water?
(c) Sketch the temperature profile of the two fluid flows in the heat exchanger over
the specific enthalpy of the air. How is the heat exchanger operated?
(d) Calculate the power consumption P_{12} of the compressor and the released power
P_{34} of the turbine. What power P is effectively required in order to run the heat
pump? What is the coefficient of performance $\varepsilon = \frac{\dot{Q}_H}{P}$ of the heat pump?
(e) Calculate the flux of exergy $\dot{E}_{x,H}$, that the water absorbs, as well as the exergetic
efficiency $\varepsilon_{ex} = \frac{\dot{E}_{x,H}}{P}$ of the open heat pump.

Solution

(a) In order to determine temperature T_2 the fictitious isentropic change of state
$(1) \rightarrow (2s)$ is calculated first, see Fig. 17.27a, i.e.

$$T_{2s} = T_1 \left(\frac{p_2}{p_1} \right)^{\frac{\kappa-1}{\kappa}} = 411.68\,K. \qquad (17.118)$$

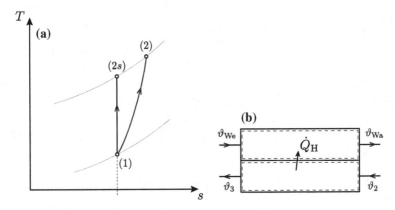

Fig. 17.27 Sketch to Problem 17.2: **a** Compressor in a p, v-diagram, **b** Heat exchanger

The definition of the isentropic efficiency is

$$\eta_{sC} = \frac{\Delta h_s}{\Delta h} = \frac{T_{2s} - T_1}{T_2 - T_1}. \tag{17.119}$$

Thus, temperature T_2 is

$$\boxed{T_2 = T_1 + \frac{T_{2s} - T_1}{\eta_{sC}} = 428.8 \text{ K}} \tag{17.120}$$

(b) Now the heat exchanger is investigated, see Fig. 17.27b. The energy balance for the water obeys[28]

$$\dot{m}_W h_{We} + \dot{Q}_H = \dot{m}_W h_{Wa}. \tag{17.121}$$

Thus, the mass flow rate of the water is

$$\dot{m}_W = \frac{\dot{Q}_H}{h_{Wa} - h_{We}}. \tag{17.122}$$

The caloric equation of state for water, i.e. incompressible liquid, is

$$h_{Wa} - h_{We} = c_W (T_{Wa} - T_{We}) + \underbrace{v_W (p_{Wa} - p_{We})}_{=0}. \tag{17.123}$$

Furthermore, there is no information, that a pressure loss within the heat exchanger on the *water's* side occurs! This brings the mass flow rate $\dot{m}_W$

---

[28] Energy *in* is balanced by energy *out* in steady state!

$$\boxed{\dot{m}_W = \frac{\dot{Q}_H}{c_W \left(T_{Wa} - T_{We}\right)} = 11.93 \, \frac{kg}{s}} \tag{17.124}$$

The air's mass flow rate follows accordingly by a balance for the air, see Fig. 17.27b, i.e.

$$\dot{m}h_2 = \dot{Q}_H + \dot{m}h_3. \tag{17.125}$$

Hence, the mass flow rate is

$$\boxed{\dot{m} = \frac{\dot{Q}_H}{h_2 - h_3} = \frac{\dot{Q}_H}{c_p \left(T_2 - T_3\right)} = 14.0973 \, \frac{kg}{s}} \tag{17.126}$$

(c) This question is intended to clarify how to operate the heat exchanger, i.e. in parallel or countercurrent flow. First, the enthalpy of the air is plotted over its temperature. At the air's inlet there is the maximum temperature and consequently the largest enthalpy, since enthalpy and temperature are proportional as long as the specific heat capacity is constant. Thus, air inlet and outlet are marked in the diagram, see Fig. 17.28. The position on the x-axis is not relevant in this sketch, but due to the enthalpy/temperature dependency, inlet and outlet temperature can be connected by a straight line. Once the air's profile is fixed, we can focus on the water's profile. In case of a countercurrent configuration, see Fig. 17.28a, the inlet for the water is on the opposite side of the air, i.e. where the air's enthalpy is the lowest, which is on the left. At this position, regarding h, the water's inlet temperature is marked. However, in this countercurrent configuration the water leaves on the right side, where the air's enthalpy is maximum. So the water's outlet temperature is marked at this position. Since c_W is supposed

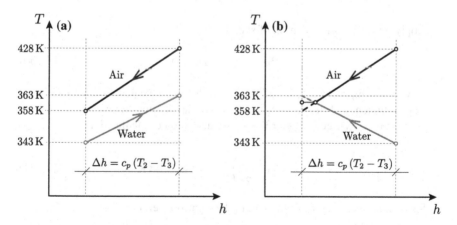

Fig. 17.28 Sketch to Problem 17.2c

to be constant and since there is no pressure loss on the water's side,[29] inlet and outlet temperature can be connected by a straight line as well. Figure 17.28a indicates that this configuration works, since the air's temperature is always larger than the temperature of the water. Hence, heat is transferred all way long from air to water.

Now, let us investigate, what happens if the heat exchanger is operated in parallel flow, see Fig. 17.28b. The air's profile is kept fix, whereas the water profile is just reversed. It is obvious, that in this configuration the two profiles intersect. Thus, starting from the right side, the heat exchanger work quite a while, i.e. the air temperature is larger than the temperature of the water. Consequently, air releases heat to the water. Once the curves intersect, there is no driving temperature potential left, so heat transfer stops, i.e. a thermal equilibrium has been reached! Thus, this heat exchanger is not capable to transfer the required thermal power of $\dot{Q}_H$.

(d) In order to calculate the power consumption of the adiabatic compressor P_{12} the first law of thermodynamics is applied, i.e.

$$P_{12} + \underbrace{\dot{Q}_{12}}_{=0} = \dot{m}(h_2 - h_1). \tag{17.127}$$

Applying the caloric equation of state leads to

$$\boxed{P_{12} = \dot{m}c_p(T_2 - T_1) = 2.2031 \times 10^3 \text{ kW}} \tag{17.128}$$

Accordingly, the power release of the turbine P_{34} can be calculated, i.e.

$$P_{34} + \underbrace{\dot{Q}_{34}}_{=0} = \dot{m}(h_4 - h_3). \tag{17.129}$$

Applying the caloric equation of state leads to

$$P_{34} = \dot{m}c_p(T_4 - T_3). \tag{17.130}$$

However, temperature T_4 is still unknown, but can be calculated with the isentropic efficiency of the turbine. In a first step the fictitious temperature T_{4s} from an isentropic change of state $(3) \rightarrow (4s)$ needs to be calculated, i.e.

$$T_{4s} = T_3 \left(\frac{p_4}{p_3}\right)^{\frac{\kappa-1}{\kappa}} = 240.1154 \text{ K}. \tag{17.131}$$

Second, temperature T_4 follows from the isentropic efficiency, i.e.

---

[29] Hence, the enthalpy of water is purely a function of temperature.

$$\eta_{sT} = \frac{\Delta h}{\Delta h_s} = \frac{T_4 - T_3}{T_{4s} - T_3}. \tag{17.132}$$

Hence, the required temperature T_4 results in

$$T_4 = T_3 + \eta_{sT} \left(T_{4s} - T_3 \right) = 253.0992 \, \text{K}. \tag{17.133}$$

The released power of the turbine is

$$\boxed{P_{34} = \dot{m} c_p \left(T_4 - T_3 \right) = -1.4869 \times 10^3 \, \text{kW}} \tag{17.134}$$

If the turbine runs the compressor, the technical effort is reduced. Hence, the required power to operate the turbine/compressor unit is

$$\boxed{P = P_{12} + P_{34} = 716.2074 \, \text{kW}} \tag{17.135}$$

(e) In order to calculated the exergy that is supplied to the water within the heat exchanger, there are two alternatives:

- Alternative 1
  The supplied exergy is the difference of the exergies of the fluid flow at outlet and inlet, see Sect. 16.2 and Fig. 17.29b, i.e.

$$\dot{E}_{x,H} = \dot{m}_W \left(e_{x,S,Wa} - e_{x,S,We} \right). \tag{17.136}$$

Applying Eq. 16.39 for inlet and outlet with ignoring kinetic and potential energies leads to

$$\dot{E}_{x,H} = \dot{m}_W \left[(h_{Wa} - h_{env}) - T_{env} (s_{Wa} - s_{env}) \right] + \\ -\dot{m}_W \left[(h_{We} - h_{env}) - T_{env} (s_{We} - s_{env}) \right]. \tag{17.137}$$

Simplifying results in

$$\dot{E}_{x,H} = \dot{m}_W \left[(h_{Wa} - h_{We}) - T_{env} (s_{Wa} - s_{We}) \right]. \tag{17.138}$$

With the caloric equations of state for incompressible liquids,[30] it finally is

$$\boxed{\dot{E}_{x,H} = \dot{m}_W \left[c_W (T_{Wa} - T_{We}) - T_{env} c_W \ln \frac{T_{Wa}}{T_{We}} \right] = 226.3258 \, \text{kW}} \tag{17.139}$$

---

[30] Without pressure loss!

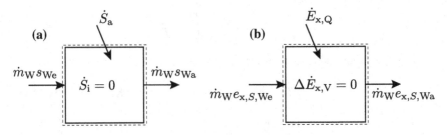

Fig. 17.29 Sketch to Problem 17.3e

Hence, the energetic efficiency of the plant is

$$\boxed{\varepsilon_{ex} = \frac{\text{Exergetic benefit}}{\text{Exergetic effort}} = \frac{\dot{E}_{x,H}}{P} = 0.316}$$ (17.140)

- Alternative 2
  The exergy balance of the system, see Fig. 17.29b reads as

$$\dot{m}_W e_{x,S,We} + \dot{E}_{x,Q} = \dot{m}_W e_{x,S,Wa} + \Delta\dot{E}_{x,V}.$$ (17.141)

Since there is no pressure loss on the water's side, the partial energy equation is

$$w_t = 0 = \underbrace{\int_1^2 v\,dp + \psi_{12} + \underbrace{\Delta e_{a12}}_{=0}}_{=0}.$$ (17.142)

Thus, there is no dissipation ($\psi_{12} = 0$). If there is no dissipation, there is no generation of entropy and no loss of exergy! Consequently, the exergy balance simplifies to

$$\underbrace{\dot{m}_W \left(e_{x,S,Wa} - e_{x,S,We} \right)}_{=\dot{E}_{x,H}} = \dot{E}_{x,Q}.$$ (17.143)

In other words, the exergy of the supplied heat is responsible for the increase of the water's exergy. The exergy of heat[31] obeys

$$\dot{E}_{x,Q} = \dot{Q}_H \left(1 - \frac{T_{env}}{T_m} \right).$$ (17.144)

---

[31] Mind, that T_m is the thermodynamic mean temperature of the water!

Rearranging results in

$$\dot{E}_{x,H} = \dot{E}_{x,Q} = \dot{Q}_H - T_{env} \underbrace{\frac{\dot{Q}_H}{T_m}}_{=\dot{S}_a}. \tag{17.145}$$

However, this requires a balance of entropy,[32] see Fig. 17.29a, i.e.

$$\dot{m}_W s_{We} + \dot{S}_a + \underbrace{\dot{S}_i}_{=0} = \dot{m}_W s_{Wa}. \tag{17.146}$$

Hence, the transferred entropy by heat is

$$\dot{S}_a = \dot{m}_W (s_{Wa} - s_{We}). \tag{17.147}$$

Applying the caloric equation of state for incompressible liquids results in

$$\dot{S}_a = \dot{m}_W c_W \ln\frac{T_{Wa}}{T_{We}}. \tag{17.148}$$

Substitution in Eq. 17.145 leads to

$$\dot{E}_{x,H} = \dot{E}_{x,Q} = \dot{Q}_H - T_{env} \dot{m}_W c_W \ln\frac{T_{Wa}}{T_{We}}. \tag{17.149}$$

Finally, the heat $\dot{Q}_H$ can be replaced according to part c), i.e.

$$\boxed{\dot{E}_{x,H} = \dot{m}_W \left[c_W (T_{Wa} - T_{We}) - T_{env} c_W \ln\frac{T_{Wa}}{T_{We}} \right] = 226.3258 \, \text{kW}} \tag{17.150}$$

Furthermore, the energetic efficiency of the plant is

$$\boxed{\varepsilon_{ex} = \frac{\text{Exergetic benefit}}{\text{Exergetic effort}} = \frac{\dot{E}_{x,H}}{P} = 0.316} \tag{17.151}$$

Problem 17.3 A tank with a volume of $V = 1 \, \text{m}^3$ contains an ideal gas ($c_v = 717 \frac{J}{kg \, K}$, $R = 287 \frac{J}{kg \, K}$.) In state (1) the pressure is $p_1 = 2$ bar, the temperature is $T_1 = 290$ K. The gas within the tank needs to be cooled down to a temperature of $T_2 = 250$ K. The released energy shall be transferred to a huge environment with a temperature of $T_{env} = 300$ K. In order to realise the heat transfer out of the tank

---

[32]Entropy *in* is balanced by entropy *out*. There is no generation of entropy, since no pressure loss occurs for the water!

respectively to the environment a temperature difference of $\Delta T = 5\,\text{K}$ is required at any point of time.

(a) Sketch the required machine as a black box!
(b) What is the mass of the gas within the tank and what pressure occurs in state (2)?
(c) How much heat needs to be released by the tank?
(d) What is the minimum work that needs to be supplied to the machine?
(e) What is the total loss of exergy $\Delta E_{x,V}$ for this process? What causes the loss of exergy?

Solution

(a) The problem treats a cooling machine. Its schematic is illustrated in Fig. 17.30.
(b) The mass in the tank is constant, since it is a closed system. Applying the thermal equation of states leads to

$$m = \frac{p_1 V}{RT_1} = 2.403\,\text{kg}. \tag{17.152}$$

The pressure p_2 in state (2) follows accordingly

$$p_2 = \frac{mRT_2}{V} = \frac{p_1 V R T_2}{RT_1 V} = \frac{p_1 T_2}{T_1} = 1.724\,\text{bar}. \tag{17.153}$$

(c) The heat Q_0 that needs to be released follows the first law of thermodynamics for the tank

$$Q_{12} + \underbrace{W_{12}}_{=0} = U_2 - U_1 = mc_v\,(T_2 - T_1)\,, \tag{17.154}$$

so that

$$Q_0 = Q_{12} = mc_v\,(T_2 - T_1) = \frac{p_1 V}{RT_1}c_v\,(T_2 - T_1) = -6.8914 \times 10^4\,\text{J}. \tag{17.155}$$

(d) The minimum work W can be calculated with the first law of thermodynamics for the cooling machine

$$|Q| = |Q_0| + W \rightarrow W = |Q| - |Q_0| \tag{17.156}$$

Obviously, there are two unknowns, i.e. W and $|Q|$, so that the second law of thermodynamics is applied. Since the tank's temperature is not constant, it needs to be done in differential notation, see Fig. 17.30, i.e.

Fig. 17.30 Sketch to Problem 17.3

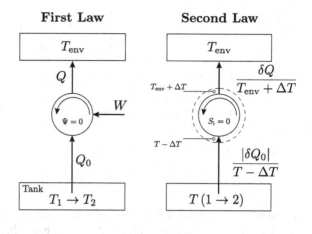

$$\frac{|\delta Q_0|}{T - \Delta T} = \frac{\delta Q}{T_{\text{env}} + \Delta T}. \tag{17.157}$$

This equation follows the balance of entropy, i.e. entropy in is balanced by entropy out. Since it is asked for the minimum work, the machine needs to be reversible, i.e. there is no entropy generation. Furthermore, the balance is given in steady state notation, i.e. no accumulation of entropy occurs. This is because any thermodynamic cycle follows Eq. 17.84. In this case it reads as

$$\oint dS = 0. \tag{17.158}$$

Anyhow, with

$$|\delta Q_0| = -mc_v \, dT \tag{17.159}$$

the balance of entropy 17.157 simplifies to

$$\frac{-mc_v \, dT}{T - \Delta T} = \frac{\delta Q}{T_{\text{env}} + \Delta T}. \tag{17.160}$$

The integration from state (1) to state (2) results in

$$-mc_v \int_1^2 \frac{dT}{T - \Delta T} = \int_1^2 \frac{\delta Q}{T_{\text{env}} + \Delta T} = \frac{Q}{T_{\text{env}} + \Delta T}. \tag{17.161}$$

Solving the integral leads to

$$-mc_v \ln (T - \Delta T) \Big|_{T_1}^{T_2} = \frac{Q}{T_{\text{env}} + \Delta T} \tag{17.162}$$

respectively

$$-mc_v \ln\frac{T_2 - \Delta T}{T_1 - \Delta T} = \frac{Q}{T_{\text{env}} + \Delta T}. \tag{17.163}$$

Hence, Q follows

$$Q = -mc_v (T_{\text{env}} + \Delta T) \ln\frac{T_2 - \Delta T}{T_1 - \Delta T} = 7.947 \times 10^4 \, \text{J}. \tag{17.164}$$

This finally results in

$$W = |Q| - |Q_0| = 1.0554 \times 10^4 \, \text{J}. \tag{17.165}$$

(e) In order to calculate the total loss of exergy $\Delta E_{\text{x,V}}$ two alternatives are possible. Both are sketched in Fig. 17.31. Mind, that both alternatives treat an extended system boundary, that includes the ΔT for the heat transfer!

- Alternative 1:
  A balance of exergy reads as

$$E_{\text{x},Q_0 @ T_m} + W = E_{\text{x},Q @ T_{\text{env}}} + \Delta E_{\text{x,V}}. \tag{17.166}$$

Since Q is transferred at ambient temperature, it follows

$$E_{\text{x},Q @ T_{\text{env}}} = 0. \tag{17.167}$$

The loss of exergy then is

$$\Delta E_{\text{x,V}} = E_{\text{x},Q_0 @ T_m} + W. \tag{17.168}$$

In order to solve this equation, the thermodynamic mean temperature of the tank is required, i.e.

$$T_m = \frac{Q_0}{S_{a12}} = \frac{Q_0}{mc_v \ln\frac{T_2}{T_1}} = \frac{mc_v (T_2 - T_1)}{mc_v \ln\frac{T_2}{T_1}} = \frac{(T_2 - T_1)}{\ln\frac{T_2}{T_1}} = 269.5 \, \text{K} \tag{17.169}$$

with[33]

$$\Delta S = \underbrace{S_{i12}}_{=0} + S_{a12} = m \left(c_v \ln\frac{T_2}{T_1} + \underbrace{R \ln\frac{V_2}{V_1}}_{=0} \right). \tag{17.170}$$

---

[33]The partial energy equation for the tank reads as $W_{12} = 0 = W_{12,\text{v}} + W_{12,\text{mech}} + \Psi_{12}$. Since there is no volume work respectively mechanical work, there is no dissipation and no generation of entropy within the tank!

Fig. 17.31 Sketch to Problem 17.3

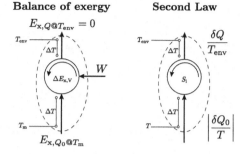

This leads to the loss of exergy, see Eq. 17.168, i.e.

$$\Delta E_{x,v} = E_{x,Q_0 @ T_m} + W = |Q_0| \left(1 - \frac{T_{env}}{T_m}\right) = 2.7557 \times 10^3 \, \text{J}. \quad (17.171)$$

- Alternative 2:

A balance of entropy for the *entire* system, see sketch 17.31 follows[34]

$$\frac{|\delta Q_0|}{T} + \delta S_i = \frac{\delta Q}{T_{env}}. \quad (17.172)$$

Substituting the heat results in

$$-\frac{mc_v \, dT}{T} + \delta S_i = \frac{\delta Q}{T_{env}}. \quad (17.173)$$

The integration leads to

$$-mc_v \int_{T_1}^{T_2} \frac{dT}{T} + S_{i,12} = \frac{Q}{T_{env}}. \quad (17.174)$$

Hence, the *entire* generation of entropy is

$$S_{i,12} = \frac{Q}{T_{env}} + mc_v \ln\frac{T_2}{T_1} = 9.1858 \, \frac{\text{J}}{\text{K}}. \quad (17.175)$$

The loss of exergy then is

$$\Delta E_{x,v} = S_{i,12} T_{env} = 2.7557 \times 10^3 \, \text{J}. \quad (17.176)$$

The loss of exergy is due to the irreversible heat transfer with $\Delta T > 0$!

---

[34] Again in differential notation, since the temperature in the tank is not constant! In this extended system there is dissipation due to the heat transfer, that is now part of the system boundary!

Part II
Real Fluids & Mixtures

Chapter 18
Single-Component Fluids

This chapter deals with single-component fluids, i.e. pure fluids not being mixed with other fluids. In Part I *idealised* single-component fluids have already been introduced: Ideal gases, that follow the well-known thermal and caloric equations of state, and incompressible liquids, whose specific volume remains constant. For incompressible liquids the caloric equations of state can be adapted easily. The behaviour of real fluids, however, deviates from the idealised models. In particular, real fluids can change their state of aggregation.

18.1 Ideal Gas Versus Real Fluids

So far, see Part I, in order to describe the thermodynamic state of a system an approach for ideal gases has been followed. Anyhow, ideal gases are subject to several model assumptions:

- Large distances between the particles[1]
  Interactions between the randomly moving particles are based on perfect elastic collisions. The larger the specific volume is, the more negligible these interactions are. Thus, at high temperatures and low pressures a gas behaves like an ideal gas. Intermolecular forces become less significant compared with its kinetic energy.
- The particles of an ideal gas have an insignificant expansion
  Particles are treated as mass points respectively punctiform. The empty space between molecules is dominant compared to the size of the particles.
- Ideal gases follow state equations
  The thermal equation of state, that has been derived in Chap. 6 and that can also be derived by means of statistical thermodynamics, reads as

$$pv = RT. \tag{18.1}$$

---

[1]Particle is a synonym for molecules respectively atoms!

© Springer Nature Switzerland AG 2019
A. Schmidt, *Technical Thermodynamics for Engineers*,
https://doi.org/10.1007/978-3-030-20397-9_18

The caloric equations of state, that have been deduced in Chap. 12, are:

$$du = c_v \, dT \tag{18.2}$$

and

$$dh = c_p \, dT \tag{18.3}$$

and

$$ds = \frac{c_v}{T} \cdot dT + \frac{R}{v} \cdot dv = \frac{c_p}{T} \cdot dT - \frac{R}{p} \cdot dp. \tag{18.4}$$

Many real gases behave as ideal gases under atmospheric conditions, i.e. standard temperature and standard pressure. Nitrogen, oxygen, hydrogen, noble gases and carbon dioxide can be treated as ideal gases for example with acceptable accuracy. However, at high pressures and low temperatures[2] real gases behave as follows:

- Spatial expansion of the particles and interacting forces can not be neglected.
- The thermal equation of state can not be applied any more. Thus, the deviation from the thermal equation of state for ideal gases is considered by the following modification

$$\boxed{pv = ZRT} \tag{18.5}$$

Z is a dimensionless compressibility factor, that equals one for ideal gases. Figure 18.1 shows the compressibility factor for air. Obviously, for real gases such a compressibility factor does not have to be constant. Generally, Z depends on pressure and temperature.

- Furthermore, at high pressures and low temperatures a liquefaction of real fluids is possible, i.e. a change in state of aggregation occurs.

Hence, real fluids can occur in different states of aggregation. In solid state the binding forces among the particles form solid structures. By supply of energy a liquid state can be reached, i.e. the binding forces are loosened by thermal induced movement of the particles. Further supply of energy can bring the fluid in gaseous state, so that the kinetic energy of the particles is enhanced. Possible changes of aggregation state are shown in Fig. 18.2. The change from solid to liquid is called *freezing* and its reverse is called *melting*. If a liquid becomes gas, this is named *vaporisation*. The way back is a *condensation*. *Sublimation* is a change of state, in which the solid phase passes to the gaseous phase directly without being liquid at any time. This occurs for instance when dry ice,[3] e.g. carbon dioxide, below the triple point[4] is supplied with heat: The dry ice sublimates, i.e. it leaves the solid aggregate state and becomes gaseous directly. The same can be observed when water in solid state[5] is in a cold and

---

[2]Hence, at small specific volumes!

[3]That does not contain any water.

[4]Which will be explained later in this chapter. However, for carbon dioxide the temperature at the triple point is at $-56.5\,°C$.

[5]Thus, ice!

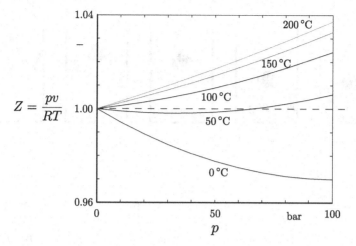

$$Z = \frac{pv}{RT}$$

Fig. 18.1 Compressibility factor Z for air according to [3]

Fig. 18.2 Changes of aggregation state

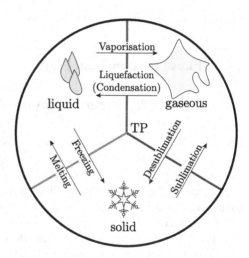

dry atmosphere, e.g. in winter time. The partial pressure, that will be explained in Chap. 19, of water in the atmosphere is rather small, so that the phase change, if heat is supplied, takes place below the water's triple point. That is the reason, why water sublimates under these conditions. Hence, hanging washed clothes in the outside in winter time is more efficient for drying than mounting it inside the house where the air might be already saturated with vapour. However, these two examples show what will be treated in Part II: First, it needs to be clarified how to quantify a phase change of real fluids. Second, by a change of state (solid ↔ liquid, liquid ↔ gaseous, solid ↔ gaseous) physical properties of a real fluid change jumpily, which needs to be taken into account by adapted equations of state. Third, thermodynamic approaches are required to handle the mixing of fluids, e.g. dry air and vapour.

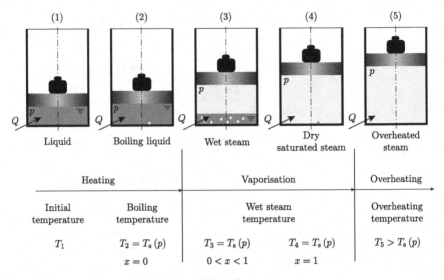

Fig. 18.3 Isobaric evaporation—schematic illustration

18.2   Phase Change Real Fluids

18.2.1   Example: Isobaric Vaporisation

Let us assume a closed system in state (1) is filled with a liquid, e.g. water, see Fig. 18.3. A freely movable piston closes the system. Hence, the system's pressure can easily be adjusted, since piston's mass and ambient pressure fix the pressure inside the cylinder in thermodynamic equilibrium, i.e. all forces acting on the system are balanced. It is further assumed, that there is no other substance within the system, so that it is free of any air for instance. After thermal energy is supplied a new equilibrium state is reached. Pressure is the same as before, since ambient pressure and mass of the piston are supposed to be constant. However, temperature rises. If the heat supply continues a state (2) is reached, at which boiling starts, i.e. the fluid begins to vaporise. At this point, the first vapour bubble appears, which starts to rise within the liquid. The temperature at which vaporisation starts depends on the fluid and its pressure, e.g. water at 1 bar starts boiling at almost 100 °C. During evaporation, the temperature remains constant despite further heat supply. Obviously, the supplied heat Q does not initiate a rise of temperature, but causes a change of the aggregate state.[6] State (3) shows the system at vaporisation: Though its temperature and pressure stay constant, its volume, and thus its specific volume, rises. The supplied energy is required for

---

[6]This is named *latent heat*, as the supplied heat does not cause a measurable caloric effect, i.e. rise of temperature, but a phase change that can not be detected by a thermocouple for instance. In contrast, supplied heat that causes a temperature rise, is called *sensible heat*.

the phase change and not for a further increase of temperature. State (3) can also exist in a thermodynamic equilibrium. However, a vapour ratio x is defined[7]:

$$x = \frac{\text{mass vapour}}{\text{total mass}} = \frac{m_V}{m_V + m_L} \qquad (18.6)$$

By definition, the vapour ratio in-between states (2) and (4) is in the range[8] of $0 \ldots 1$. Once, the vaporisation is finished, i.e. the last liquid has fully become vapour, see state (4) in Fig. 18.3, the fluid's temperature further increases if additional heat is supplied, see state (5).

The zone between states (2), i.e. boiling starts, and (4), i.e. phase change is completed, is called the *wet steam region*, because dry saturated steam and boiling liquid exist simultaneously.

In state (5) the steam is *overheated*, since its temperature is larger than the wet steam's temperature at the corresponding pressure.

Theorem 18.1 *While vaporisation the fluid's temperature depends on its pressure, i.e. an isobaric vaporisation is isothermal as well.*

Consequently, a state within the wet steam region can not be described unambiguously just by pressure p and temperature T as it was the case for ideal gases, see also Sect. 5.1.2.

Vaporisation of a Real Fluid

It is well-known from Part I, that a p, v-diagram is essential in thermodynamics as it, for example, illustrates the process value *work*. Hence, the focus now is to deduce a p, v-diagram for a vaporising liquid as shown in Fig. 18.3. In a first step the states (1) to (5) are marked in a p, v diagram, see Fig. 18.4a. The pressure for such a change of state is fix and supposed to be p_A, so that all states are located on a horizontal straight. States (1) and (2) are close to each other, since the increase of specific volume while heating up the liquid is rather small. State (3) is in the wet steam region, so that due to the ascending ratio of steam the specific volume increases rapidly. In the next steps, the pressure is increased to p_B respectively to p_C for instance by enlarging the mass of the piston. The measured states (1) to (5) are marked in the p, v-diagram again, see Fig. 18.4b, c. As before, state (2) indicates the boiling point and (4) represents the end of phase change. While conducting the experiment, the temperature was carefully determined as well, so that the isothermals can be added in the p, v-diagram, see Fig. 18.4d. The upper isothermals in the p, v-diagram indicate a larger temperature than the isothermals on a lower position. Thus, for any pressure state (1) has a lower temperature than state (2). Within the wet steam region the temperature is constant as long as the vaporisation takes place at a

---

[7]Unfortunately, x in Part II has multiple meaning!

[8]For states (1) and (5) it does not make sense to define a vapour ratio x!

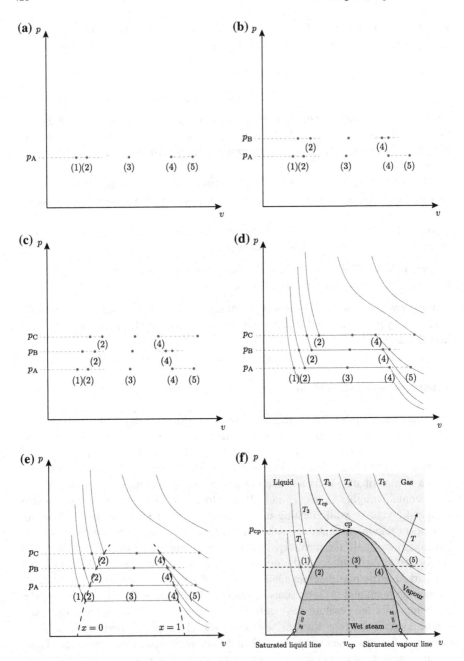

Fig. 18.4 Isobaric evaporation—schematic illustration in a p, v-diagram

constant pressure. However, compared to an ideal gas, refer to Fig. 7.2 for instance, the isothermals remind on a hyperbola that is interrupted for the wet steam region and provide a horizontal plateau in this section. Obviously, this region is an imperfection compared to an ideal gas. As mentioned before, pressure and temperature do not identify the thermodynamic state unambiguously.

In a next step, all initial boiling points, i.e. states (2), and all states that have just finished the phase change, i.e. states (4), are connected, see Fig. 18.4e. The left curve represents $x = 0$ and is called saturated liquid line. The right curve is characterised by $x = 1$ and is named saturated vapour line. The finalised p, v-diagram is shown in Fig. 18.4f. Both, saturated liquid line ($x = 0$) and saturated vapour line ($x = 1$), meet at the so-called critical point (cp), that is fluid specific and can be identified by p_{cp} and T_{cp}. In case the fluid's temperature is larger than T_{cp} the fluid can not be liquified any more, no matter how large its pressure is. In contrast, any vapour at a temperature below the critical temperature T_{cp} can be liquified by increasing the fluid's pressure. The region in-between the bell-shaped saturated liquid/vapour line is the wet steam, i.e. two phase, region. Within this region the temperature is solely a function of pressure. Left from the $x = 0$ line the fluid is in liquid state, right from the $x = 1$ line the fluid is in gaseous/vaporous state. By definition vapour respectively steam are gaseous fluids, whose thermodynamic state is close to the liquefaction point. In contrast, extremely overheated steams are called gases.

18.2.2 The p, v, T-State Space

In the previous section the phase change liquid to vapour has been explained. However, an isobaric phase change from liquid state to solid state and reverse shows a similar behaviour, i.e. the temperature stays constant. If water for instance is cooled down at 1 bar it reaches $0\,°C$. At this point the phase change liquid to solid starts and the first ice crystals occur. While further cooling down, the temperature still is constant at $0\,°C$, but the ratio of ice to liquid rises. Once the liquid has fully been converted into ice, thus the phase change is finished, temperature starts to drop, in case heat is further released.

In a so-called p, v, T state space the thermal state values can be visualised easily, see Fig. 18.5. For an ideal gas, as treated in Part I, the state space shows a smooth curve, since

$$p = f(v, T) = \frac{RT}{v}. \tag{18.7}$$

This graph is shown in Fig. 18.5b. From a thermodynamic point of view it does not show any imperfections or discontinuities. For real fluids, that underlie phase changes, the p, v, T-space, see Fig. 18.5a, shows a similar behaviour, but several discontinuities can be observed, each time a phase change takes place.

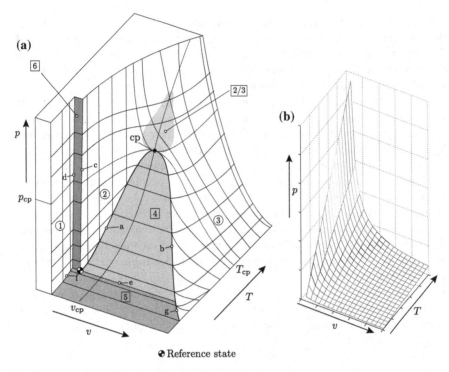

⊕ Reference state

Fig. 18.5 The p, v, T-diagram: **a** Real fluid, **b** Ideal gas

A p, v, T-diagram for real fluids illustrates the following characteristic:

- For small pressures p and small densities ρ, i.e. large specific volumes v, the ideal gas equation $pv = RT$ can be applied approximately for real gases, as in this region Fig. 18.5a, b show a similar behaviour.
- Approaching the saturated vapour line, indicated as curve b in Fig. 18.5a, the deviations from ideal gas law increase as a comparison between the two diagrams indicates.
- Region 1 represents the solid phase. This region is characterised by a rather small specific volume that remains almost constant, when temperature T and pressure p are varied.
- Liquid state is shown in region 2. However, liquid state and solid state are linked by region 6, i.e. the melting region. The transition from solid to liquid and vice versa at constant pressure runs isothermal while the specific volume varies slightly.
- Region 3 shows the gas phase.
- Region 4 in the diagram represents the wet steam region, that as been discussed previously. Within this region liquid and vapour occur coincidently, see Sect. 18.2.1.
- Region 5 is the sublimation region, i.e. solid and vapour exist at the same time.
- Region 2/3 represents the so-called overcritical gas/liquid region.

The limiting curves in the p, v, T state space have partly been discussed before and are named as follows:

- Curve a is the saturated liquid line, i.e. $x = 0$. It splits liquid from the wet steam.
- Curve b is the saturated liquid line, i.e. $x = 1$. It is the border line between wet steam region and gas phase.
- Curve c is the solidification line, that separates the melting region 6 and the liquid region 2.
- Curve d is the melting line, that separates the solid from the melting region.
- Curve e represents the triple point line, that links sublimation and wet steam region. Along this line the fluid can exist in all three states.
- Curve f is the sublimation line, that links sublimation region and solid region.
- Curve g is the sublimation line, that links sublimation region and gas region.
- The indicated reference point is required when phases of change are investigated energetically.

Examples

Isobaric Heating from (1a) → (1f)

In state (1a) a fluid is in supercooled solid state. Heating takes place at nearly constant specific volume until state (1b) is reached. At this time, melting starts, so that temperature is constant while the fluid's state changes from solid to liquid. However, the specific volume increases. The melting process is finished as soon as state (1c) is reached. Further supply of thermal energy leads to a temperature rise. The specific volume slightly rises as well. The liquid is supercooled as long as state (1d) is reached. At this point, vaporisation starts at a constant temperature. The vaporisation comes along with a large increase of specific volume. Vaporisation is finished in state (1e). Further supply of thermal energy results in an overheating of the vapour, i.e. temperature and specific volume rise.

Isobaric Heating from (2a) → (2b)

A so-called overcritical change of state occurs when $p > p_{cp}$. Initially, the fluid is in liquid state. Due to heating, i.e. supply of thermal energy, the fluid becomes gaseous without being liquid intermediately. In the states in-between (2a) and (2b) the fluid appears milky without any definite local phase boundaries, i.e. bubbles or droplets.

Isobaric Heating from (3a) → (3d)

Starting from state (3a), in which the fluid is in solid state, heating takes place at almost constant specific volume. When state (3b) is reached, the fluid starts to sublimate, i.e. it converts from liquid to gaseous state at constant temperature, without being liquid in-between. This change of state appears with a large increase of specific volume. In state (3c) sublimation is completed. Further supply of heat leads to an overheated gas (3d) while the specific volume rises continuously.

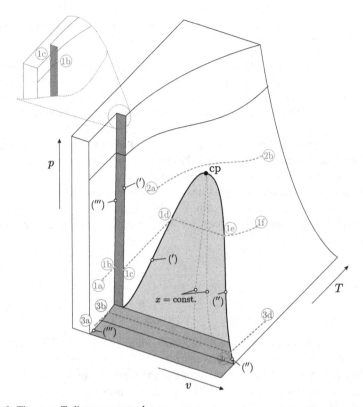

Fig. 18.6 The p, v, T-diagram: examples

Anomaly of Water

Figure 18.6 shows another interesting aspect. In contrast to most other fluids, the phase change of water from liquid state to solid state, e.g. freezing from state (1c) to state (1b), comes along with an increase of specific volume, i.e. a decrease of density. This is illustrated in the detailed view of Fig. 18.6. This is the reason why ice is floating on the surface of liquid water, since

$$\rho_{\text{ice}} < \rho_{\text{liquid}}. \tag{18.8}$$

Furthermore, a bottle of beer bursts in the deep-freezer compartment, as the specific volume rises during the freezing process. In contrast, most other fluids show the characteristic as given in Fig. 18.6, i.e. the change of state from (1c) to (1b) lead to a decrease of specific volume. This anomaly can be explained with the hydrogen bond. Due to this hydrogen bond the structure in solid state requires a larger space than in liquid state. Furthermore, water has its largest density at approx. 4 °C.

Projections of the p, v, T-Diagram

Obviously, the p, v, T-diagram provides a large quantity of information, so that it makes sense to concentrate this information. Once the fluid's behaviour as shown in Fig. 18.5 is available, it can be projected in order to derive standard x, y-plots. Possible projections are shown in Fig. 18.7. The derived p, v-diagram has previously been discussed, see Sect. 18.2.1. It is, as known from Part I, utilised to identify the *process value* work for thermodynamic changes of state. Furthermore, a p, T-diagram can be developed by a second projection. This diagram can be applied to identify the aggregate state of a fluid for a given pressure p as well as temperature T. However, the wet steam region in this diagram is hidden, since the saturated liquid and vapour line are congruent in this projection. It further includes the congruent solidification/melting lines as well as the sublimation curve. A correlation between pressure and temperature at vaporisation[9] can be gained easily. Finally, a T, v-diagram can be deduced as illustrated in Fig. 18.7. These three diagrams are discussed in the following sections.

18.2.3 p, T-Diagram

In a p, T-diagram the three different single-phase regions are separated by three phase limiting curves. These curves meet in the so called triple-point (TP). At that point, whose pressure as well as temperature are fluid-specific, the fluid can exist in all phases at the same time. The phase limiting curves are named as follows:

- Vapour pressure curve, separating liquid and steam
- Melting pressure curve, separating liquid and solid
- Sublimation pressure curve, separating solid and steam

Points (1) ... (5) indicated in Fig. 18.8 refer to the example given in Sect. 18.2.1. Obviously, states (2), (3) and (4), that are limiting the wet steam region respectively lie within this region, can not be distinguished in this p, T-diagram, as the wet steam information gets lost due to the projection. Figure 18.9 shows p, T-diagrams qualitatively for water (a) and non-water fluids (b). Obviously, the melting pressure curve differs. For water the phase change liquid to solid is almost independent from its pressure, i.e. it occurs at approximately $0\,°C$. Triple point as well as critical point have already been explained. For water these two states are as listed in Table 18.1.

18.2.4 T, v-Diagram

A T, v-diagram is shown in Fig. 18.10. Once again, the limiting curves $x = 0$, i.e. the saturated liquid line, and $x = 1$, i.e. vapour line, meet in the critical point (cp). As

---

[9]I.e. the vapour pressure curve.

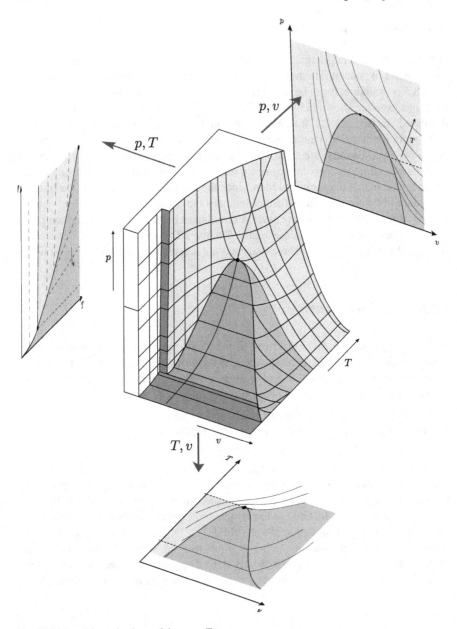

Fig. 18.7 Possible projections of the p, v, T-state space

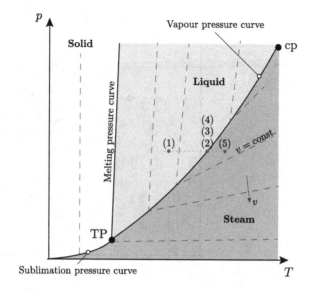

Fig. 18.8 The p, T-diagram

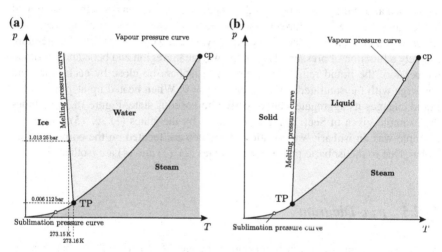

Fig. 18.9 The p, T-diagram: **a** Water, **b** Non-water fluid

Table 18.1 Triple/critical point of water

State value	Triple point	Critical point
p in [bar]	0.006117	220.64
ϑ in [°C]	0.01	373.95

Fig. 18.10 The
T, v-diagram

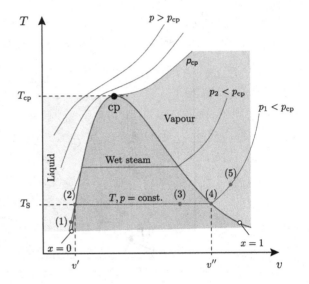

mentioned above the critical point is fluid-specific. The region beneath the saturated liquid/vapour curve is the already introduced wet steam, i.e. two-phase, region in which liquid and vapour occur coincidentally. In the wet steam region temperature is solely a function of pressure. Left of the wet steam region and beneath the critical temperature the liquid region is located. The isobars lie close by each other and converge with the saturated liquid line, i.e. $x = 0$. When heated up at $p > p_{cp}$ the liquid changes its aggregate state directly into gaseous state. Figure 18.10 includes the example given in Sect. 18.2.1 as indicated by the states (1) ... (5). Since this example was an isobaric vaporisation, the states are located on the corresponding isobar. Due to the isobaric phase change, states (2), (3) and (4) are isothermal.

18.2.5   p, v-Diagram

A p, v-diagram has already been introduced in Sect. 18.2.1 when the isobaric vaporisation has been discussed. However, Fig. 18.11 shows a p, v-diagram for water. Obviously, in the liquid region, i.e. left from the $x = 0$ curve, the specific volume changes only very slightly even at large pressure variations. This has been the reason, why in Part I liquids have been treated as incompressible. Since the specific volume increases strongly during vaporisation, it makes sense to use a logarithmic scale for the specific volume v. Saturated liquid line, i.e. $x = 0$, and saturated vapour line, i.e. $x = 1$, enclose the wet steam region, in which the fluid consists of boiling liquid and saturated steam at the same time. The ratio of vapour to total mass has been defined previously according to Eq. 18.6. Curves of constant vapour ratio x are represented in Fig. 18.11 as well. To the right of the $x = 1$ curve, the gaseous phase

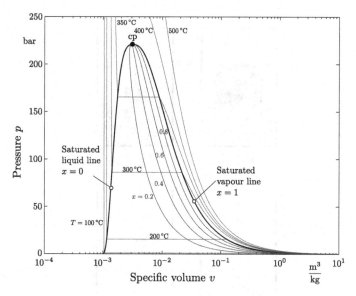

Fig. 18.11 The p, v-diagram for water

is given. Isothermals start from the top left and lead to the lower right part, similar to the isothermals, i.e. hyperbolas, of ideal gases. However, for real fluids, due to the phase change, the curves have a constant plateau within the wet steam region, where temperature is a function of pressure.

Figures 18.12 and 18.13 show an isothermal expansion of a closed system as an example. In state (1) the water is supposed to be an supercooled liquid at a pressure of 200 bar and a temperature of 100 °C. This state can be fixed in a p, v-diagram, see Fig. 18.13. However, the change of state follows the 100 °C-isothermal. Now, during the change of state the pressure is quasi-statically reduced, e.g .by decreasing the weight of the piston. Furthermore, temperature remains constant. This leads to state (2) and finally to state (3), i.e. the saturated liquid state, where vaporisation starts. For water at 100 °C this is approximately at 1 bar. Whereas the rise of specific volume from (1) to (2) to (3) was only very small, during vaporisation the increase of the specific volume is large. The process shall stop in state (4), when vaporisation is fully completed, i.e. $x = 1$ is achieved. One benefit of the p, v-diagram is, that the specific work can be visualised. Let us assume, that the process runs reversibly with the change of potential energy to be negligible, so that the entire work from (1) to (4) is purely volume work, i.e.

$$w_{14} = w_{V,14} + w_{mech,14} + \psi_{14} = w_{V,14} = -\int_1^4 p \, dv < 0. \qquad (18.9)$$

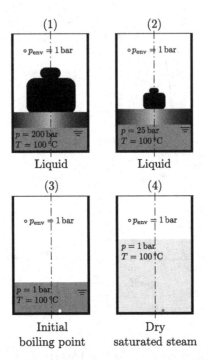

Fig. 18.12 Closed system—Isothermal expansion

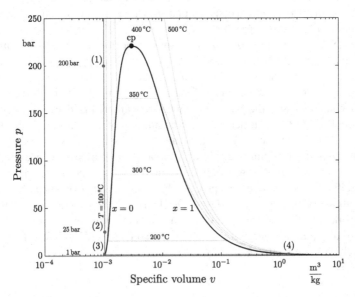

Fig. 18.13 Isothermal expansion (1) to (4) in a p, v-diagram

Obviously, the process is an expansion as the volume increases. Work is released and thus negative. The first law of thermodynamics from state (1) to state (4) obeys[10]

$$W_{14} + Q_{14} = U_4 - U_1. \tag{18.10}$$

In specific notation it is

$$w_{14} + q_{14} = u_4 - u_1. \tag{18.11}$$

Applying the steam table of water with the given states according to Fig. 18.12 results in

$$w_{14} + q_{14} = u_4 - u_1 = 496.3 \, \frac{kJ}{kg} \tag{18.12}$$

Hence, the specific heat is

$$q_{14} = u_4 - u_1 - w_{14} = 496.3 \, \frac{kJ}{kg} - \underbrace{w_{14}}_{<0} > 0. \tag{18.13}$$

Consequently, for an isothermal expansion, heat needs to be supplied to the system. In contrast to ideal gases, the internal energy of the fluid does not remain constant though the process is supposed to be isothermal. This, however, is due to the phase change. Obviously, according to Fig. 18.13, within the liquid region the specific volume of water changes only very slightly with large pressure variations, see phase change from state (1) to state (2) respectively (3). These three states are located on an almost vertical straight. Thus, liquids are often treated as incompressible fluids as it has been the case in Part I.

18.2.6 State Description Within the Wet-Steam Region

Section 18.2.1 has shown, that pressure and temperature within the wet steam region are not independent from each other. Thus, the state can not be unambiguously characterised by pressure p and temperature T, see also Sect. 5.1.2 as well as Fig. 5.1. To identify the state within the wet-steam region, the vapour ratio, see Eq. 18.6 has been introduced, that represents the ratio of vapour's mass to the total mass. Any extensive state value Z within the wet-steam region is composed of a vapour proportion and a liquid proportion, i.e.

$$Z = Z_V + Z_L \tag{18.14}$$

Introducing the specific state values according to Eq. 3.13 results in

---

[10]As mentioned before, outer energies shall be ignored!

$$m_{\text{total}}z = m_V z_V + m_L z_L. \tag{18.15}$$

The mass of the liquid m_L can be substituted by

$$m_L = m_{\text{total}} - m_V, \tag{18.16}$$

so that it is

$$m_{\text{total}}z = m_V z_V + (m_{\text{total}} - m_V) z_L. \tag{18.17}$$

A division by the total mass leads to

$$\boxed{z = xz_V + (1 - x) z_L} \tag{18.18}$$

z_V is the specific state value of the saturated vapour state, i.e.

$$z_V = z (x = 1) = z''. \tag{18.19}$$

z_L is the specific state value of the saturated liquid state,[11] i.e.

$$z_L = z (x = 0) = z'. \tag{18.20}$$

Hence, the specific state values within the wet-steam region, according to Eq. 18.18, obey

$$\boxed{v = \frac{V}{m} = xv'' + (1 - x)v'} \tag{18.21}$$

$$\boxed{u = \frac{U}{m} = xu'' + (1 - x)u'} \tag{18.22}$$

$$\boxed{h = \frac{H}{m} = xh'' + (1 - x)h'} \tag{18.23}$$

$$\boxed{s = \frac{S}{m} = xs'' + (1 - x)s'} \tag{18.24}$$

Mind, that $\rho \neq x\rho'' + (1 - x)\rho'$ since $\rho = \frac{1}{v}$!

The p, v-diagram shown in Fig. 18.14 indicates state (3) within the wet steam region, according to the example in Sect. 18.2.1. As explained before, the entire volume is composed of the volume of the dry saturated steam and the volume of the saturated liquid. The vapour ratio is

---

[11] The liquid is boiling!

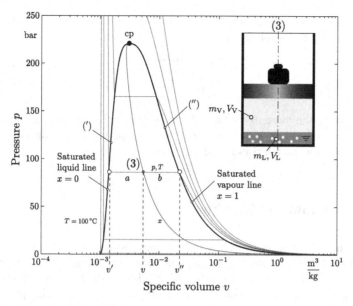

Fig. 18.14 Lever rule of the quantities

$$x = \frac{m_V}{m_V + m_L},$$ (18.25)

thus, the liquid ratio is

$$(1 - x) = \frac{m_L}{m_V + m_L}.$$ (18.26)

Combing these two equations results in

$$\frac{x}{1 - x} = \frac{m_V}{m_L} = \frac{m''}{m'}.$$ (18.27)

Solving Eq. 18.21 for the vapour ratio leads to

$$x = \frac{v - v'}{v'' - v'}.$$ (18.28)

Hence, it is

$$1 - x = \frac{v'' - v}{v'' - v'}.$$ (18.29)

Equations 18.28 and 18.29 lead to

$$\frac{x}{1 - x} = \frac{v - v'}{v'' - v} = \frac{a}{b}.$$ (18.30)

The distances a and b are indicated in Fig. 18.14. Combining Eqs. 18.27 and 18.30 result in the so-called lever rule, i.e.

$$\boxed{\frac{a}{b} = \frac{m''}{m'}} \tag{18.31}$$

In other words, the more dry saturated vapour the state contains, the larger a is. Consequently, state (3) moves to the right. In case, the vapour is reduced, thus the liquid ratio is increased, b rises and state (3) moves to the left.

18.3 State Values of Real Fluids

The previously discussed state diagrams are required to design technical components and machines. Thus, complex measurements need to be conducted to determine the fluid's properties. Such properties are the fundamentals for any thermodynamic calculation. Section 18.2.1 has shown, that the thermal state values, i.e. pressure p, temperature T and specific volume v, can be measured easily. However, in order to reduce the number of experiments, only a few grid points are investigated. Values in-between these grid points are derived by state equations:

- Ideal gases follow the thermal equation of state $pv = RT$.
- Real fluids show deviations from $pv = RT$, see Sect. 18.1, so that the thermal equation of state requires adaptions. Some of possible adaptions are listed in Table 18.2

Unfortunately, a determination of the caloric state values, such as specific entropy s, specific enthalpy h and specific internal energy u, can not be measured thus needs to be calculated.

Table 18.2 Further equations of state for real fluids

Name	Correlation	
Redlich–Kwong	$p = \frac{R_M T}{v_M - b} - \frac{a}{T^{0.5} v_M (v_M + b)}$	(18.23)
	$a = \frac{0.42748 R_M^2 T_{cp}^{2.5}}{p_{cp}}$	(18.33)
Peng Robinson	$p = \frac{R_M T}{v_M - b} - \frac{a\alpha}{v_M^2 + 2b v_M - b^2}$	(18.34)
Berthelot	$p = \frac{R_M T}{v_M - b} - \frac{a}{T v_M^2}$	(18.35)
Dieterici	$p = \frac{RT}{v_M - b} e^{-\frac{a}{v_M RT}}$	(18.36)
Virial	$\frac{p}{R_M T} = \frac{1}{v_M} + \frac{B(T)}{v_M^2} + \frac{C(T)}{v_M^3} + \frac{D(T)}{v_M^4} + \cdots$	(18.37)

18.3.1 Van der Waals Equation of State

Van der Waals[12] suggested the following equation of state in order to describe real fluids:

$$\left(p + \frac{a}{v_M^2}\right)(v_M - b) = R_M T. \tag{18.38}$$

With $a = 0$ and $b = 0$ this equation passes over to the thermal equation of state for ideal gases, see Eq. 6.26.

- b is the so-called co-volume. It describes the volume reduction due to the finite dimensions of the molecules. This reduces the space of free molecule movement.
- $\frac{a}{v_M^2}$ is the so-called cohesion pressure. It characterises the decrease of the pressure towards limiting walls due to intermolecular attraction.

Rearranging Eq. 18.38 results in

$$p = p(v_M) = \frac{R_M T}{v_M - b} - \frac{a}{v_M^2}. \tag{18.39}$$

The question is, how to derive the coefficients a as well as b for a real fluid: At the critical point the isothermal in a p, v-diagram shows a point of inflection with a horizontal tangent, see diagram 18.5 for instance. Above the critical temperate the isothermals approximate regular hyperbolas. Below the critical temperature the curves following Eq. 18.39 show a minimum and and a maximum. Hence, there needs to be a point of inflection, see Fig. 18.15, that shows a p, v-diagram for CO_2. Now, let us focus on the isothermal T_{cp}. However, for a point of inflection with a horizontal tangent the following mathematical conditions need to be fulfilled:

$$\left.\left(\frac{\partial p}{\partial v_M}\right)_T\right|_{cp} = 0 \tag{18.40}$$

and

$$\left.\left(\frac{\partial^2 p}{\partial v_M^2}\right)_T\right|_{cp} = 0 \tag{18.41}$$

with

$$p_{cp} = \frac{R_M T_{cp}}{v_{M,cp} - b} - \frac{a}{v_{M,cp}^2}. \tag{18.42}$$

The first derivative of Eq. 18.39 obeys

---

[12]Johannes Diderik van der Waals (∗23 November 1837 in Leiden, †8 March 1923 in Amsterdam).

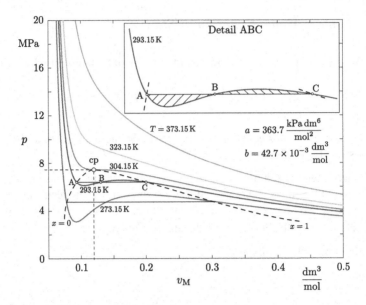

Fig. 18.15 Van der Waals approach for CO_2

$$\left(\frac{\partial p}{\partial v_M}\right)_T\bigg|_{cp} = -\frac{R_M T_{cp}}{\left(v_{M,cp} - b\right)^2} + \frac{2a}{v_{M,cp}^3} = 0 \tag{18.43}$$

and the second derivative follows

$$\left(\frac{\partial^2 p}{\partial v_M^2}\right)_T\bigg|_{cp} = \frac{2R_M T_{cp}}{\left(v_{M,cp} - b\right)^3} - \frac{6a}{v_{M,cp}^4} = 0. \tag{18.44}$$

Rearranging Eqs. 18.43 and 18.44 leads to

$$\frac{2a}{v_{M,cp}^3} = \frac{R_M T_{cp}}{\left(v_{M,cp} - b\right)^2} \tag{18.45}$$

respectively

$$\frac{6a}{v_{M,cp}^4} = \frac{2R_M T_{cp}}{\left(v_{M,cp} - b\right)^3}. \tag{18.46}$$

Dividing Eqs. 18.45 and 18.46 results in

$$\frac{1}{3}v_{M,cp} = \frac{1}{2}\left(v_{M,cp} - b\right). \tag{18.47}$$

Hence, the co-volume b can be calculated, in case the critical point is known, by

$$b = \frac{1}{3}v_{M,cp}.$$ (18.48)

Coefficient a follows from Eq. 18.45 with b being replaced by Eq. 18.48, i.e.

$$a = \frac{9}{8}R_M T_{cp} v_{M,cp}.$$ (18.49)

Rearranging results in

$$R_M T_{cp} = \frac{8}{9v_{M,cp}}.$$ (18.50)

The critical pressure follows from Eq. 18.42, i.e.

$$p_{cp} = \frac{3R_M T_{cp}}{2v_{M,cp}} - \frac{a}{v_{M,cp}^2}.$$ (18.51)

Substitution of $R_M T_{cp}$ by Eq. 18.50 brings

$$p_{cp} = \frac{a}{3v_{M,cp}^2}.$$ (18.52)

In other words, the cohesion pressure a obeys

$$a = 3p_{cp}v_{M,cp}^2$$ (18.53)

Figure 18.15 shows the van der Waals approach for CO_2. According to Eq. 18.39 the p, v-graphs can be calculated for several isothermals. Obviously, the curves fit the expected characteristic out of the wet-steam region as a comparison with Fig. 18.3 for instance shows. However, the wet-steam region can not be covered by van der Waals equation, since the isothermals do not show a horizontal line in a p, v-diagram. A fitting is done according to Maxwell,[13] so that

$$p\left(v_M'' - v_M'\right) = \int_A^C p \, dv_M.$$ (18.54)

This rule can be justified due to the fact that the area beneath the curve of a change of state in a p, v-diagram corresponds with the specific volume work. Thus, the work from A to B should be equal to the work from B to C, i.e. the two hatched areas in the detailed illustration (Fig. 18.15) need to have the same absolute value. Following this idea, the bell-shaped saturated liquid, i.e. $x = 0$, as well as vapour, i.e. $x = 1$, curves can be constructed. With this premise the calculated specific volume work

---

[13] James Clerk Maxwell (★13 June 1831 in Edinburgh, †5 November 1879 in Cambridge).

from A to C fits the actual specific volume work for an isobaric phase change from A to C.

18.3.2   Redlich–Kwong

Redlich–Kwong suggested the following correlation for real fluids, i.e.

$$p = \frac{R_M T}{v_M - b} - \frac{a}{T^{0.5} v_M (v_M + b)} \tag{18.55}$$

with

$$a = \frac{0.42748 R_M^2 T_{cp}^{2.5}}{p_{cp}} \tag{18.56}$$

and

$$b = \frac{0.08664 R_M T_{cp}}{p_{cp}}. \tag{18.57}$$

However, this correlation, which is similar to the van der Waals approach, is more accurate above critical temperature.

18.3.3   Peng–Robinson

An improvement of the van der Waals equation was developed by Peng–Robinson in 1976. Their approach follows

$$p = \frac{R_M T}{v_M - b} - \frac{a\alpha}{v_M^2 + 2b v_M - b^2} \tag{18.58}$$

with

$$a = \frac{0.457235 R_M T_{cp}^2}{p_{cp}} \tag{18.59}$$

and

$$b = \frac{0.077796 R_M T_{cp}}{p_{cp}}. \tag{18.60}$$

α can be calculated depending on the acentric factor[14] ω. For $\omega < 0.49$ it is

$$\alpha = \left[1 + \left(0.37464 + 1.54226\omega - 0.26992\omega^2 \right) \left(1 - T_r^{0.5} \right) \right]^2 \tag{18.61}$$

---

[14]The acentric factor is a measure of the non-sphericity (centricity) of molecules.

and for $\omega \geq 0.49$ it is

$$\alpha = \left[1 + (0.379642 + (1.48503 - (1.164423 - 1.016666\omega)\,\omega)\,\omega)\left(1 - T_r^{0.5}\right)\right]^2 .$$
(18.62)

T_r is the reduced temperature, that follows

$$T_r = \frac{T}{T_{\mathrm{cp}}}.$$
(18.63)

18.3.4 Berthelot

Berthelot suggested the following equation

$$p = \frac{R_{\mathrm{M}} T}{v_{\mathrm{M}} - b} - \frac{a}{T v_{\mathrm{M}}^2}$$
(18.64)

with

$$a = 3 T_{\mathrm{cp}} p_{\mathrm{cp}} v_{\mathrm{M,cp}}^2$$
(18.65)

and

$$b = \frac{v_{\mathrm{M,cp}}}{3}.$$
(18.66)

However, this correlation is rarely used.

18.3.5 Dieterici

According to Dieterici the following equation of state can be applied for real fluids, i.e.

$$p = \frac{RT}{v_{\mathrm{M}} - b} e^{-\frac{a}{v_{\mathrm{M}} R T}}.$$
(18.67)

The parameters a and b follow

$$a = 4 R_{\mathrm{M}} T_{\mathrm{cp}} \frac{v_{\mathrm{M,cp}}}{2}$$
(18.68)

and

$$b = \frac{v_{\mathrm{M,cp}}}{2}.$$
(18.69)

18.3.6 Virial Equation

Virial equations are extensions of the thermal equation of state for ideal gases and can be derived directly from statistical mechanics, i.e.

$$\frac{p}{R_M T} = \frac{1}{v_M} + \frac{B(T)}{v_M^2} + \frac{C(T)}{v_M^3} + \frac{D(T)}{v_M^4} + \dots \qquad (18.70)$$

They include series expansions of $\frac{1}{v_M}$. In case the series expansion is aborted after the first term, the virial equation turns into the thermal equation of state for ideal gases.

18.3.7 Steam Tables

Instead of calculating state values by state equations, the common procedure in technical thermodynamics is to apply so-called steam tables,[15] see Tables A.1, A.2, A.3, A.4, A.5, A.6, A.7, A.8 and A.9. These tables show state values for water: Table A.1 cover the limiting bell-shaped curve of the wet-steam region by denoting the saturated liquid state ($'$), i.e. $x = 0$, and the saturated steam state ($''$), i.e. $x = 1$. According to Sect. 18.2.1 pressure and temperature depend on each other, so that additionally the vapour ratio x identifies each specific state value

$$z = \frac{Z}{m} \qquad (18.71)$$

within the wet-steam region unambiguously, see Sect. 18.2.6, i.e.

$$z = x z'' + (1 - x) z'. \qquad (18.72)$$

Mind, that this equation can not be applied for the density ρ, but for the specific volume $v = \frac{1}{\rho} = \frac{V}{m}$. In addition to the wet-steam region, Tables A.2, A.3, A.4, A.5, A.6, A.7, A.8 and A.9 list the states value density ρ, specific enthalpy h, specific entropy s and specific heat capacity c_p of water for supercooled liquid as well as for overheated vapour. State values for water in this book are taken from steam tables and do not follow the previously described mathematical state functions.

Caloric state values require a reference point, that specific internal energy u, specific enthalpy h as well as specific entropy s can be compared with. The choice of a reference state is arbitrary but needs to be kept fix during thermodynamic calculations:

---

[15]The listed tables follow the International Association for the Properties of Water and Steam (IAPWS). Several digital tables are available that utilise the IAPWS guideline, e.g. citeXSteam. When applying state values from different sources, it needs to be checked if the reference points for caloric state values are identical. If not, they need to be corrected!

Definition 18.2 For the listed steam tables of water[16] the reference point is the saturated liquid state (') at the triple-point, i.e. at

$$p_0 = p_{TP} = 0.006117 \, \text{bar} \qquad (18.73)$$

and

$$\vartheta_0 = \vartheta_{TP} = 0.01 \, ^{\circ}\text{C}. \qquad (18.74)$$

The specific internal energy u is set to zero at reference level, i.e.

$$u_0 = 0 \, \frac{\text{kJ}}{\text{kg}}. \qquad (18.75)$$

Furthermore, the specific entropy s is set to zero as well, i.e.

$$s_0 = 0 \, \frac{\text{kJ}}{\text{kgK}}. \qquad (18.76)$$

Thus, the specific enthalpy h_0 at reference level follows

$$h_0 = u_0 + p_{TP} v_0. \qquad (18.77)$$

By taking the specific volume according to the steam table it follows

$$h_0 = 0 + 611.66 \, \frac{\text{N}}{\text{m}^2} \cdot 0.0010002 \, \frac{\text{m}^3}{\text{kg}} = 0.00061178 \, \frac{\text{kJ}}{\text{kg}}. \qquad (18.78)$$

By definition h_0 is the specific enthalpy at reference level, i.e. $h_0 = h'(p_{TP}, \vartheta_{TP})$, see Table A.1.

Example 18.3 (State functions versus steam table) In this example the specific enthalpy h' of saturated boiling water at $p = 1$ bar is to be investigated. It is already well known for real fluids, see Eq. 12.94 in Sect. 12.3.2, that

$$dh = c_p(p, T) \cdot dT + \left[-T \left(\frac{\partial v}{\partial T} \right)_p + v \right] \cdot dp. \qquad (18.79)$$

Since the specific enthalpy h' at 1 bar is requested, Eq. 18.79 has to be integrated. Reference level for the integration is the previously introduced saturated liquid state (') of the triple point. Thus, reference level and the demanded state are both liquid. Applying the assumption from Part I, that the liquid is almost incompressible, results in

---

[16]For other fluids the reference point has to be carefully checked!

$$\left(\frac{\partial v}{\partial T}\right)_p \approx 0. \tag{18.80}$$

Hence, the equation simplifies to

$$dh = c_p\,(p, T)\cdot dT + v\cdot dp. \tag{18.81}$$

Since v is regarded to be constant, an integration leads to

$$h'\,(1\,\text{bar}) - h_0 = \int_{T_0}^{T'} c_p\,dT + \int_{p_0}^{p} v\,dp = \int_{T_0}^{T'} c_p\,dT + v\,(p - p_0) \tag{18.82}$$

with

$$h_0 = h'\,(0.01\,^\circ\text{C}) = 0.00061178\,\frac{\text{kJ}}{\text{kg}}. \tag{18.83}$$

In other words

$$h'\,(1\,\text{bar}) = h_0 + \int_{T_0}^{T'} c_p\,dT + v\,(p - p_0)\,. \tag{18.84}$$

The temperature of the saturated liquid state (') at 1 bar is listed in Table A.1, i.e.

$$T' = 372.7559\,\text{K}. \tag{18.85}$$

Table A.1 further shows, that the fluids's specific heat capacity from the reference level, i.e. saturated liquid at the triple point, to (') at 1 bar is not constant. However, the averaged value is

$$c_p = \overline{c_p'}\Big|_{0.01\,^\circ\text{C}}^{99.6059\,^\circ\text{C}} = 4.192\,\frac{\text{kJ}}{\text{kgK}}. \tag{18.86}$$

The average specific volume according to Table A.1 is approximately constant, i.e.

$$v = \overline{v'}\Big|_{0.01\,^\circ\text{C}}^{99.6059\,^\circ\text{C}} = \overline{\frac{1}{\rho'}}\Big|_{0.01\,^\circ\text{C}}^{99.6059\,^\circ\text{C}} = 0.01\,\frac{\text{m}^3}{\text{kg}}. \tag{18.87}$$

This leads to

$$h'\,(1\,\text{bar}) = h_0 + c_p\,(T' - T_0) + v\,(p - p_0) = 418.5005\,\frac{\text{kJ}}{\text{kg}}. \tag{18.88}$$

The calculation has shown, that determining state values according to state functions is time consuming. Furthermore, a simplification of Eq. 18.79 regarding the compressibility has been applied. Thus, estimating state values with steam tables is much easier: In this example the requested specific enthalpy is $h' = 417.4365 \frac{kJ}{kg}$ which follows by interpolation of Table A.1. In this table the compressibility according to Eq. 18.79 is taken into account. However, in this case both approaches differ by less than 0.3%.

Problem 18.4 Determine the specific internal energy of saturated steam and saturated liquid at a pressure of 10 bar. Working fluid is water.

Solution

In order to find u' respectively u'' the steam Table A.1 can be applied. Since only the specific enthalpies h' and h'' are listed, these values need to be converted, i.e.

$$h' = u' + pv' \rightarrow u' = h' - pv' \qquad (18.89)$$

respectively

$$h'' = u'' + pv'' \rightarrow u'' = h'' - pv'. \qquad (18.90)$$

Let us start with the saturated liquid (') first. According to the steam table it is

$$h' = 762.65 \frac{kJ}{kg} \qquad (18.91)$$

and

$$v' = \frac{1}{\rho'} = 0.0011 \frac{m^3}{kg}. \qquad (18.92)$$

Hence, it is

$$u' = h' - pv' = 761.5556 \frac{kJ}{kg}. \qquad (18.93)$$

For the saturated vapour ('') it follows accordingly

$$h'' = 2777.1 \frac{kJ}{kg} \qquad (18.94)$$

and

$$v'' = \frac{1}{\rho''} = 0.1943 \frac{m^3}{kg}. \qquad (18.95)$$

Hence, it finally is

$$u'' = h'' - pv'' = 2582.8 \frac{kJ}{kg}. \qquad (18.96)$$

Problem 18.5 $8 \, \text{m}^3$ wet steam at a pressure of 9 bar has a vapour ratio of 35%. What is the mass of the wet steam and what is its enthalpy? Working fluid is water.

Solution

The wet-steam is composed of saturated liquid (') and saturated vapour ("), so that

$$V = V_L + V_V \tag{18.97}$$

and

$$m = m_L + m_V. \tag{18.98}$$

The specific volume of the wet-steam can be calculated of the (')- and (")-states, since its vapour ratio x is known:

$$v = v'(1 - x) + v''x. \tag{18.99}$$

Applying steam Table A.1 leads to

$$v' = v'(9 \, \text{bar}) = \frac{1}{\rho'(9 \, \text{bar})} = 0.0011211 \, \frac{\text{m}^3}{\text{kg}} \tag{18.100}$$

and

$$v'' = v''(9 \, \text{bar}) = \frac{1}{\rho''(9 \, \text{bar})} = 0.2149 \, \frac{\text{m}^3}{\text{kg}}. \tag{18.101}$$

Thus, the specific volume is

$$v = v'(1 - x) + v''x = 0.0759 \, \frac{\text{m}^3}{\text{kg}}. \tag{18.102}$$

It further is

$$v = \frac{V}{m}, \tag{18.103}$$

so that the entire mass m results in

$$m = m_L + m_V = \frac{V}{v} = 105.3539 \, \text{kg}. \tag{18.104}$$

According to

$$x = \frac{m_V}{m} \tag{18.105}$$

the mass of the saturated vapour is

$$m_V = xm = 36.8739 \, \text{kg} \tag{18.106}$$

and the mass of the saturated liquid is

$$m_L = m - m_V = 68.48 \, \text{kg}. \tag{18.107}$$

The entire enthalpy H of the wet-steam results in

$$H = m_L h' + m_V h''. \tag{18.108}$$

The specific enthalpies h' as well as h'' are listed in the steam table, i.e.

$$h' = h' \, (9 \, \text{bar}) = 742.7246 \, \frac{\text{kJ}}{\text{kg}} \tag{18.109}$$

respectively

$$h'' = h'' \, (9 \, \text{bar}) = 2773.0 \, \frac{\text{kJ}}{\text{kg}}. \tag{18.110}$$

Thus, it finally follows

$$H = m_L h' + m_V h'' = 153114.4 \, \text{kJ}. \tag{18.111}$$

Alternatively, the same result obeys

$$H = m \left[h' \, (1 - x) + h'' x \right] = 153114.4 \, \text{kJ}. \tag{18.112}$$

18.4 Energetic Consideration

18.4.1 Reversibility of Vaporisation

In a first step the reversibility of an isobaric vaporisation $(') \rightarrow ('')$, see Fig. 18.16, is investigated. The first law of thermodynamics for case (a) obeys

$$W_{12} + Q_{12} = U'' - U' \tag{18.113}$$

The partial energy equation follows

$$W_{12} = -p_{\text{env}} \, \Delta V = W_{V,12} + \Psi + \underbrace{W_{\text{mech},12}}_{=0}. \tag{18.114}$$

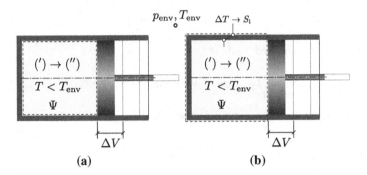

Fig. 18.16 Isobaric vaporisation: **a** reversible, **b** irreversible

The entire work, that is released by the system, is utilised to compress the environment by the piston. The mechanical work is zero, since the piston moves horizontally and states (') and (") are supposed to be in rest. For an isobaric change[17] of state the volume work can easily be calculated,[18] i.e.

$$W_{V,12} = - \int_{(')}^{('')} p \, \Delta V = -p_{env} \, \Delta V. \tag{18.115}$$

Thus, Eq. 18.114 results in

$$\boxed{\Psi = 0} \tag{18.116}$$

In other words, if the vaporisation is isobaric, there is no dissipation inside the system. Hence, the change of state as shown in Fig. 18.16a is reversible. This is the case, if the change of state runs quasi-statically and the system is homogeneous: Turbulence due to rising vapour bubbles, as sketched in the previous figures, needs to be neglected, since they cause dissipation. In such a case Eq. 18.114 reads as:

$$W_{12} = -p_{env} \, \Delta V = - \int_{(')}^{('')} p \, \Delta V + \Psi. \tag{18.117}$$

Hence, the dissipation is

$$\Psi = -p_{env} \, \Delta V + \int_{(')}^{('')} p \, \Delta V > 0. \tag{18.118}$$

---

[17]Thus, the change of state is supposed to run very slowly, see Example 7.24.
[18]For a quasi-static change of state the balance of forces results in $p = p_{env}$.

Consequently, for an expansion ($\Delta V > 0$) the pressure inside the system needs to be $p > p_{env}$ in case dissipation occurs.

Let us now have a closer look at Fig. 18.16b. Though, the process shall still be isobaric, i.e. there is no dissipation[19] inside the modified system, the choice of this new system boundary[20] now causes the process to be irreversible. Obviously, there is an imperfection regarding the temperature distribution, since for supplying heat, a temperature difference ΔT is required, see Fig. 18.16b. At the system boundary the temperature is T_{env}, inside the system it is $T = T_s(p)$. This imperfection generates entropy $S_{i,12}$, see Sect. 14.2 and Fig. 14.3. Thus, according to Fig. 18.16b the process including the environment now is irreversible, as the heat, that once is supplied to the system, is not able to follow the temperature gradient reversely for releasing heat. An entropy balance obeys

$$S'' - S' = S_{i,12} + S_{a,12}. \tag{18.119}$$

The supplied entropy due to the heat is[21]

$$S_{a,12} = \frac{Q_{12,(b)}}{T_{env}}. \tag{18.120}$$

Thus, it is

$$\boxed{S_{i,12} = \left(S'' - S'\right) - \frac{Q_{12,(b)}}{T_{env}} > 0} \tag{18.121}$$

18.4.2 Heat of Vaporisation

In a next step the required energy for vaporisation is investigated. An isobaric heating of a real fluid has been introduced in Sect. 18.2.1. According to Fig. 18.17, the change of state (2) $\rightarrow$ (4), i.e. from $(')\rightarrow ('')$, is analysed. The first law of thermodynamics obeys[22]

$$\delta q + \delta w = du. \tag{18.122}$$

With the partial energy equation[23]

$$\delta w = -p\,dv + \underbrace{\delta\psi}_{=0} + \underbrace{\delta w_{mech}}_{=0} \tag{18.123}$$

---

[19]Representing internal friction!

[20]The system now includes the cylinder wall!

[21]Mind, that the temperature inside the system, where the heat passes, is relevant. According to the boundary given in Fig. 18.16b it is T_{env}!

[22]The change of potential energy of the fluid is neglected!

[23]As described before the change of state is isobaric, so that no dissipation occurs.

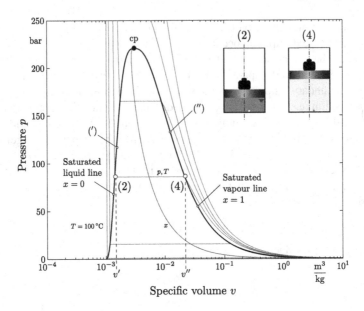

Fig. 18.17 Isobaric vaporisation of water

the first law of thermodynamics simplifies to

$$\delta q = \mathrm{d}u + p\,\mathrm{d}v. \tag{18.124}$$

Introducing the specific enthalpy

$$\mathrm{d}h = \mathrm{d}u + p\,\mathrm{d}v + v\,\mathrm{d}p \tag{18.125}$$

leads to the following notation of the first law of thermodynamics

$$\mathrm{d}h = \delta q + v\,\mathrm{d}p. \tag{18.126}$$

According to the second law of thermodynamics

$$\mathrm{d}s = \frac{\delta q}{T} + \underbrace{\frac{\delta \psi}{T}}_{=0} \tag{18.127}$$

the specific heat can be substituted by

$$\delta q = T\,\mathrm{d}s. \tag{18.128}$$

Thus, the fundamental equation of thermodynamics results in

$$dh = T \, ds + v \, dp. \tag{18.129}$$

For an isobaric vaporisation, i.e. $dp = 0$, it is

$$dh = T \, ds \tag{18.130}$$

as well as

$$\delta q = dh. \tag{18.131}$$

The integration from $(')$ → $('')$ results in

$$q = h'' - h' = \int_2^4 T \, ds. \tag{18.132}$$

An isobaric vaporisation from $(')$ → $('')$ is isothermal, i.e. $dT = 0$, as well. Thus, the specific enthalpy of vaporisation Δh_v can be defined, i.e.

$$\boxed{h'' - h' = \Delta h_v = T \left(s'' - s' \right)} \tag{18.133}$$

The specific enthalpy of vaporisation Δh_v is solely a function of temperature, i.e.

$$\Delta h_v = h'' - h' = f\left(T_s\right). \tag{18.134}$$

Figure 18.18 shows this correlation. Obviously, the specific enthalpy of vaporisation Δh_v decreases with rising temperature. For $\vartheta \to \vartheta_{cp}$ the specific enthalpy of vaporisation Δh_v is approaching zero. However, due to $h = u + pv$ it follows

$$\Delta h_v = h'' - h' = u'' - u' + p_s\left(v'' - v'\right). \tag{18.135}$$

Isobaric supply of specific enthalpy of vaporisation Δh_v increases the internal energy $\Delta u = u'' - u'$ though the temperature remains constant but the aggregate state changes. The second part $p_s\left(v'' - v'\right)$ represents the volume work, that is required during the isobaric expansion[24] of the fluid. According to Fig. 18.18 the change of the internal change dominates the volume work.

18.4.3 Caloric State Diagrams

Part I has shown, that the advantage of a p, v-diagram is, that work as a process value can be visualised. Furthermore, the T, s-diagram can be utilised to illustrate specific

---

[24]The specific volume increases largely during vaporisation!

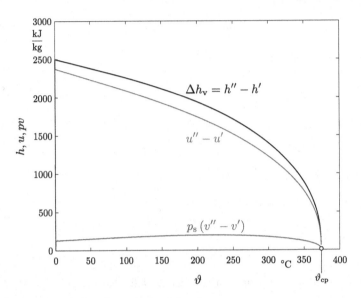

Fig. 18.18 Specific enthalpy of vaporisation Δh_{v} of water

heat as well as specific dissipation. Hence, in this section the focus is on relevant caloric state diagrams. The presented diagrams can be easily generated based on steam tables, e.g. steam tables of water according to Tables A.1, A.2, A.3, A.4, A.5, A.6, A.7, A.8 and A.9.

T, s-Diagram

The major benefit of the T, s-diagram is, that specific heat as well as specific dissipation can be visualised as area beneath the change of state. This has already been discussed in Part I. However, due to the phase change the T, s-diagram for real fluids differs from the diagrams for ideal gases. One example of a T, s-diagram for water is given with Fig. 18.19. A more detailed and magnified T, s-diagram for water can be found in Fig. C.2. The saturation lines, i.e. $x = 0$ and $x = 1$, form a bell-shaped curve with the critical point on its peak. Under this bell-shape curve the wet-steam region is located. Left to the wet steam region the liquid is supercooled, to the right the vapour state is covered. Isobars, i.e. $p = $ const., in the liquid region are approximately identical with the left saturation line ($x = 0$). As discussed at the beginning of this chapter within the wet steam region, i.e. $0 < x < 1$, isobars and isotherms, i.e. $T = $ const., are congruent. The specific entropy s is, according to our reference level, see Sect. 18.3.7, set to zero at the liquid state of the triple-point.

In case of an isobaric, reversible vaporisation, see Sect. 18.4.2, the specific heat of vaporisation

$$\Delta h_{\mathrm{v}} = h'' - h' = f(T_{\mathrm{s}}) \tag{18.136}$$

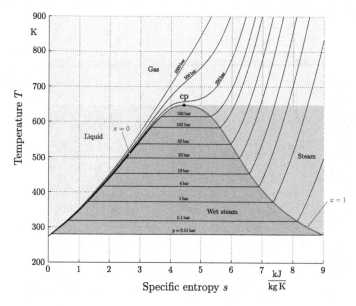

Fig. 18.19 T, s-diagram of water

can be illustrated as hatched, rectangular area, see Fig. 18.20, since the temperature is constant. Obviously, the rectangular area, i.e. the specific enthalpy of vaporisation, decreases with rising pressure respectively temperature. At the critical point the specific enthalpy of vaporisation is approaching zero.

h, s-Diagram

In addition to the T, s-diagram a h, s-diagram is beneficial, e.g. for visualising changes of state within thermal turbo-machines like compressor or turbine. An example for such a diagram is presented in Fig. 18.21 in order to characterise liquid/wet-steam/vapour regions. A more detailed diagram for water is given with Fig. C.3. Isobars, i.e. $p = $ const., within the wet steam region run as rising straights and cross the saturated vapour line, i.e. $x = 1$, steadily. Within the wet steam region isobars and isotherms, $T = $ const., are congruent. However, isobars run tangentially into the saturated liquid line ($x = 0$). Within the liquid region they are approximately identical with the saturated liquid line. Isotherms are almost horizontal straights within the overheated region and bend downwards when approaching the saturated vapour line ($x = 1$). By crossing the $x = 1$-curve a kink can be observed.

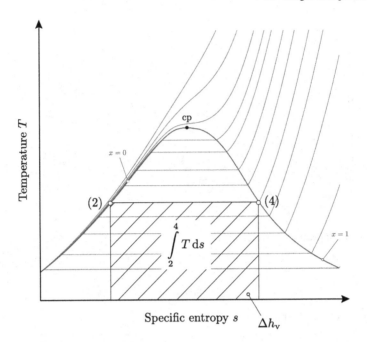

Fig. 18.20 T, s-diagram of water

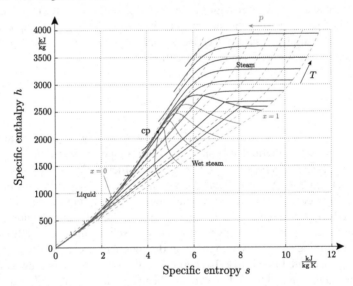

Fig. 18.21 h, s-diagram of water

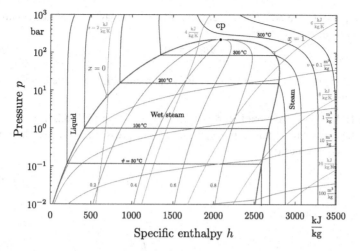

Fig. 18.22 log p, h-diagram of water

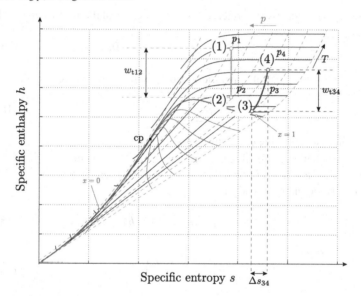

Fig. 18.23 Technical work in the h, s-diagram

log p, h-Diagram

This diagram is typically utilised for designing air-conditioning systems. Hence, it will be further discussed in Chap. 21. A log p, h-diagram for water is briefly shown in Fig. 18.22, a detailed diagram, however, is included in the appendix, see Fig. C.1.

Example 18.6 (*Thermal turbo-machines*) As mentioned above a h, s-diagram is beneficial to show changes of state in thermal turbo-machines, see Fig. 18.23. In case these components are adiabatic, according to the first law of thermodynamics for open systems in steady state[25]

$$\delta w_{t12} = dh \tag{18.137}$$

the specific technical work δw_t can be easily calculated. Let us have a closer look at two examples:

1. Case 1: Adiabatic, reversible turbine, see $(1) \rightarrow (2)$ in Fig. 18.23.
   Superheated steam expands from pressure p_1 to pressure $p_2 < p_1$. The first law of thermodynamics obeys

$$w_{t12} + q_{12} = h_2 - h_1. \tag{18.138}$$

With $q_{12} = 0$ it simplifies to

$$\boxed{w_{t12} = h_2 - h_1} \tag{18.139}$$

Obviously, technical work is released. The second law of thermodynamics follows

$$\dot{m}(s_1 - s_2) + \dot{S}_i + \dot{S}_a = 0. \tag{18.140}$$

With $\dot{S}_i = 0$, i.e. reversible operation, and $\dot{S}_a = 0$, i.e. adiabatic change of state, it is

$$\boxed{s_2 - s_1 = 0} \tag{18.141}$$

The change of state runs vertically in a h, s-diagram.

2. Case 2: Adiabatic, irreversible compressor, see $(3) \rightarrow (4)$ in Fig. 18.23.
   Superheated steam is compressed from pressure p_3 to pressure $p_4 > p_3$. The first law of thermodynamics reads as

$$w_{t34} + q_{34} = h_4 - h_3. \tag{18.142}$$

With $q_{34} = 0$ it is

$$\boxed{w_{t34} = h_4 - h_3} \tag{18.143}$$

Obviously, technical work is supplied. The second law of thermodynamics follows

$$\dot{m}(s_3 - s_4) + \dot{S}_i + \dot{S}_a = 0. \tag{18.144}$$

With $\dot{S}_i > 0$, since the change of state is supposed to be irreversible, and $\dot{S}_a = 0$ it results in

$$\boxed{s_4 - s_3 > 0} \tag{18.145}$$

---

[25]Note, that the change of kinetic as well as of potential energy is ignored! This is applied in the following two cases.

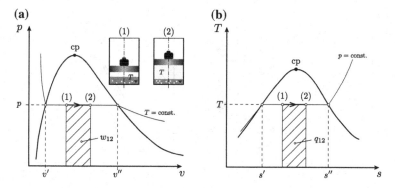

Fig. 18.24 Isobaric, isothermal, reversible change of state

The specific entropy rises from state (3) to state (4).

Example 18.7 (Isobaric, isothermal, reversible change of state) In this example a closed system, see Fig. 18.24, is investigated. The change of state (1) → (2) takes place within the wet-steam region.[26] The first law of thermodynamics reads as

$$w_{12} + q_{12} = u_2 - u_1. \tag{18.146}$$

Following the partial energy equation, the specific work is

$$w_{12} = \underbrace{w_{V,12} + w_{mech,12}}_{=0} + \underbrace{\psi_{12}}_{=0} = -p(v_2 - v_1). \tag{18.147}$$

Hence, the specific heat follows according to

$$q_{12} = u_2 - u_1 + p(v_2 - v_1) = h_2 - h_1. \tag{18.148}$$

The specific enthalpies h_1 and h_2 can be calculated with the vapour ratio x, see Sect. 18.2.6, i.e.

$$h_2 = (1 - x_2)h' + x_2 h'' \tag{18.149}$$

respectively

$$h_1 = (1 - x_1)h' + x_1 h''. \tag{18.150}$$

Thus, Eq. 18.148 simplifies to

$$\boxed{q_{12} = (x_2 - x_1)\,\Delta h_v} \tag{18.151}$$

---

[26]Changes of kinetic and potential energy shall be ignored.

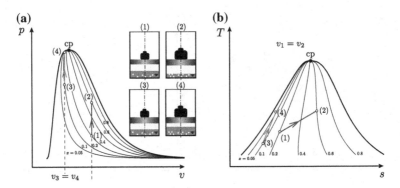

Fig. 18.25 Isochoric change of state

Alternatively, with the definition of the specific enthalpy of vaporisation, see Eq. 18.133, it is

$$\boxed{q_{12} = T\left(s'' - s'\right)(x_2 - x_1)}$$ (18.152)

Example 18.8 (Isochoric change of state) This example focuses on a closed system as well. A cylinder is filled with wet-steam and underlies a change of state while its volume remains constant. Hence, the change of state can be visualised easily in a p, v-diagram, i.e. Fig. 18.25a. Furthermore, the change of state can be shown in a T, s-diagram, see Fig. 18.25b. Obviously, two cases are conceivable:

1. Heating of wet-steam from $(1) \rightarrow (2)$: The vapour ratio x increases

$$x_2 > x_1.$$ (18.153)

2. Heating starting from state (3) with a low vapour ratio x_3 to state (4) the vapour ratio x can decrease, i.e.

$$x_4 < x_3.$$ (18.154)

Applying the first law of thermodynamics[27] results in

$$w_{12} + q_{12} = u_2 - u_1.$$ (18.155)

Following the partial energy equation, the specific work is

$$w_{12} = \underbrace{w_{V,12}}_{=0} + \underbrace{w_{\text{mech},12} + \psi_{12}}_{=0} = -p\left(v_2 - v_1\right) = 0.$$ (18.156)

Hence, the specific heat follows

---

[27]Note, that the change of kinetic as well as of potential energy is ignored!

$$q_{12} = u_2 - u_1 = u_2' - u_1' + x_2(u_2'' - u_2') - x_1(u_1'' - u_1'). \qquad (18.157)$$

Since the change of state is isochoric, the specific volume remains constant, i.e.

$$v_1 = v_2. \qquad (18.158)$$

Applying the vapour ratio leads to

$$(1 - x_1) v_1' + x_1 v_1'' = (1 - x_2) v_2' + x_2 v_2''. \qquad (18.159)$$

Solving for the vapour ratio in state (2) results in

$$\boxed{x_2 = x_1 \frac{v_1'' - v_1'}{v_2'' - v_2'} + \frac{v_1' - v_2'}{v_2'' - v_2'}} \qquad (18.160)$$

Example 18.9 (*Adiabatic change of state*) An adiabatic compression from p_1 to p_2 of a closed system is investigated, see Fig. 18.26. The following cases are possible:

- If a reversible adiabatic, i.e. isentropic, compression is conducted at

$$s > s_{cp}, \qquad (18.161)$$

the vapour ratio x increases, i.e. the vapour from (1) to (2) gets drier. Hence, it is

$$x_2 > x_1. \qquad (18.162)$$

- If an irreversible, i.e. $s_{i12'} > 0$, adiabatic compression is conducted from (1) $\rightarrow$ (2'), the specific entropy rises, i.e.

$$s_{2'} - s_1 = \underbrace{s_{a12'}}_{=0} + s_{i12'} > 0. \qquad (18.163)$$

- If a reversible adiabatic, i.e. isentropic compression is conducted at

$$s < s_{cp}, \qquad (18.164)$$

the vapour ratio decreases, i.e. the vapour from (3) to (4) gets wetter. Hence, it is

$$x_4 < x_3. \qquad (18.165)$$

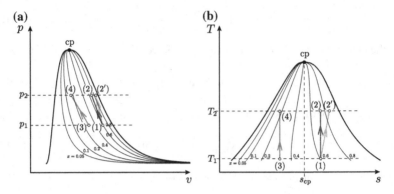

Fig. 18.26 Adiabatic change of state

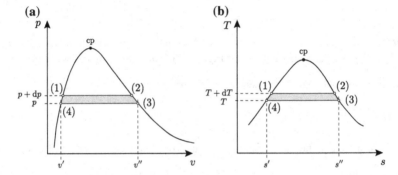

Fig. 18.27 Clausius–Clapeyron relation

18.4.4 Clausius–Clapeyron Relation

The Clausius–Clapeyron relation correlates pressure and temperature at vaporisation. It can be easily derived by the following thermodynamic cycle, that is briefly shown in Fig. 18.27. A real fluid in a closed system underlies the following changes of state:

- $(1) \to (2)$: Isobaric/isothermal vaporisation
- $(2) \to (3)$: Pressure rise by $\mathrm{d}p$
- $(3) \to (4)$: Isobaric/isothermal condensation
- $(4) \to (1)$: Pressure drop by $\mathrm{d}p$

The p, v-diagram in Fig. 18.27a proves that for small changes of pressure $\mathrm{d}p$ the specific volume at the changes of state $(2) \to (3)$ respectively $(4) \to (1)$ approximately is constant. Thus, these changes of state are approximately isochoric. The exchanged work can be calculated according to the partial energy equation, while the mechanical work is neglected:

- Isobaric vaporisation[28] $(1) \rightarrow (2)$

$$w_{12} = -(p + dp)(v'' - v')$$ (18.166)

- Isochoric change of state[29] $(2) \rightarrow (3)$

$$w_{23} = 0$$ (18.167)

- Isobaric condensation $(3) \rightarrow (4)$

$$w_{34} = -p(v' - v'')$$ (18.168)

- Isochoric change of state $(4) \rightarrow (1)$

$$w_{41} = 0$$ (18.169)

Thus, the overall work, i.e. the summation of Eqs. 18.166–18.169 results in

$$\sum w = -dp(v'' - v').$$ (18.170)

The T, s-diagram in Fig. 18.27b proves that for small changes of pressure dp, i.e. also small changes of dT, the specific entropy at the changes of state $(2) \rightarrow (3)$ respectively $(4) \rightarrow (1)$ approximately is constant. Thus, these changes of state are approximately isentropic. The exchanged specific heat follows:

- Isobaric vaporisation $(1) \rightarrow (2)$

$$q_{12} = (T + dT)(s'' - s')$$ (18.171)

- Isentropic change of state[30] $(2) \rightarrow (3)$

$$q_{23} = 0$$ (18.172)

- Isobaric condensation $(3) \rightarrow (4)$

$$q_{34} = T(s' - s'')$$ (18.173)

- Isentropic change of state $(4) \rightarrow (1)$

$$q_{41} = 0$$ (18.174)

---

[28]No dissipation due to isobaric, see Sect. 18.4.1!

[29]No work can be exchanged under these conditions, since the piston does not move.

[30]There is no fluid motion or dissipated electrical energy, so there is no dissipation, see Problem 11.4.

Thus, the overall heat, i.e. the summation of Eqs. 18.171–18.174 results in

$$\sum q = dT \left(s'' - s' \right).$$
(18.175)

For a thermodynamic cycle it is, see Eq. 11.13

$$\sum w = -\sum q.$$
(18.176)

Substituting Eqs. 18.170 and 18.175 leads to

$$dp \left(v'' - v' \right) = dT \left(s'' - s' \right).$$
(18.177)

With $\Delta h_v = T \left(s'' - s' \right)$ it is

$$\boxed{\frac{dp}{dT} = \frac{s'' - s'}{v'' - v'} = \frac{\Delta h_v}{T \left(v'' - v' \right)}}$$
(18.178)

The slope $\frac{dp}{dT}$ as well as its integration, i.e. $p_s = f(T)$, are shown in Fig. 18.28. Possible applications of the Clausius–Clapeyron relation are

- In a closed system the required specific enthalpy of vaporisation Δh_v can be measured for a specific pressure p respectively temperature T. Furthermore, the change of the specific volume at vaporisation, i.e. $\left(v'' - v' \right)$, can be estimated. In doing so, the slope $\frac{dp}{dT}$ according to Eq. 18.178 can be calculated. The procedure can be repeated at different temperatures, so that it finally is

$$\frac{dp}{dT} = f(T).$$
(18.179)

Fig. 18.28 Vapour pressure curve of water

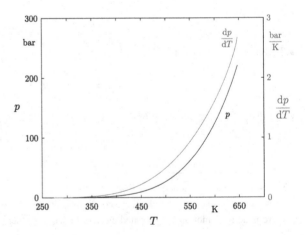

- In a closed system the slope of the vapour pressure curve and its according temperature T are measured. If additionally the change of specific volume at vaporisation $(v'' - v')$ is determined, the specific enthalpy of vaporisation Δh_v can be calculated according to

$$\Delta h_v = f(T) = T\left(v'' - v'\right) \frac{\mathrm{d}p}{\mathrm{d}T}. \tag{18.180}$$

18.5 Adiabatic Throttling—Joule–Thomson Effect

This section explains the different behaviour for an ideal and a real gas at adiabatic throttling. Adiabatic throttling has already been introduced in Part I: The fluid's pressure in a throttle is decreased by internal friction, see Fig. 18.29, so that

$$p_2 < p_1. \tag{18.181}$$

This can be realised easily by reducing the throttle's cross section. Anyhow, this causes to a rise of velocity and thus an increase of turbulence. However, the change of state is investigated under the following premises:

- Since no technical work is exchanged across the system boundary, the system is work-insulated, i.e. passive:

$$w_{t12} = 0. \tag{18.182}$$

- The system is supposed to be adiabatic. Thus, the heat exchange to the environment is neglected, i.e.

$$q_{12} = 0. \tag{18.183}$$

- The system is in steady state, so that the equation of continuity can be applied, i.e.

$$\dot{m}_1 = \dot{m}_2 \tag{18.184}$$

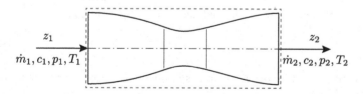

Fig. 18.29 Adiabatic throttling

- The change of mechanical energy is to be ignored,[31] i.e.

$$z_1 = z_2 \tag{18.185}$$

respectively

$$c_1^2 = c_2^2. \tag{18.186}$$

Thus, the first law of thermodynamics

$$q_{12} + w_{t12} = h_2 - h_1 + \frac{1}{2}\left(c_2^2 - c_1^2\right) + g \cdot (z_2 - z_1) \tag{18.187}$$

simplifies to

$$\boxed{h_2 - h_1 = 0} \tag{18.188}$$

Hence, the change of state is isenthalpic. In order to calculate the temperature, the caloric equation of state needs to be applied. According to Chap. 12 the specific enthalpy is a function of pressure and temperature, i.e.

$$h = h\left(p, T\right). \tag{18.189}$$

Its total differential follows

$$dh = \underbrace{\left(\frac{\partial h}{\partial T}\right)_p}_{c_p} dT + \left(\frac{\partial h}{\partial p}\right)_T dp. \tag{18.190}$$

18.5.1   Ideal Gas

Section 12.3.2 has shown, that for ideal gases it is

$$\left(\frac{\partial h}{\partial p}\right)_T = 0. \tag{18.191}$$

Thus, the specific enthalpy is solely a function of temperature, i.e.

$$dh = c_p \, dT. \tag{18.192}$$

Consequently, since the change of state is isenthalpic, it needs to be isothermal for ideal gases as well, i.e.

$$T_1 = T_2. \tag{18.193}$$

---

[31] The influence of the mechanical energies has been investigated in Problem 12.10.

18.5.2 Real Gas

However, for real gases the enthalpy does not only depend on the temperature but on the pressure as well. The change of enthalpy with respect to the pressure at constant temperature in Eq. 18.190 is described with the so-called isothermal throttle coefficient δ_T

$$\left(\frac{\partial h}{\partial p}\right)_T = \delta_T. \tag{18.194}$$

Hence, the specific enthalpy's total differential simplifies to

$$dh = c_p \, dT + \delta_T \, dp. \tag{18.195}$$

According to the first law of thermodynamics for an adiabatic throttling, see Eq. 18.188, it follows

$$dh = 0. \tag{18.196}$$

Equation 18.195 then reads as[32]

$$\boxed{\left(\frac{\partial T}{\partial p}\right)_h = -\frac{\delta_T}{c_p} = \delta_h} \tag{18.197}$$

Anyhow, this equation quantifies, how the temperature of a real gas changes when the pressure varies while the specific enthalpy remains constant. In contrast to ideal gases, a temperature change can occur for real gases at adiabatic throttling, i.e. when reducing the pressure. The sign of this temperature change, due to the specific heat capacity c_p being positive, solely depends on the Joule–Thomson coefficient respectively on the isenthalpic throttle coefficient δ_h, that is fluid specific. The adiabatic throttling is visualised in Fig. 18.30. The graph shows the isenthalps in a T, p-diagram. While for ideal gases the curves would run horizontally, since the specific enthalpy purely depends on the temperature, the graphs show a pressure dependency for real gases as well. However, throttling means reducing the pressure, so that the change of state in an adiabatic throttle moves from right to left along an isenthalp.[33] Obviously, three cases are possible:

- Case 1 (Temperature rises):
  According to Eq. 18.197 the temperature rises, i.e. $dT > 0$, when the pressure decreases, i.e. $dp < 0$, as long as

$$\left(\frac{\partial T}{\partial p}\right)_h = \delta_h < 0 \tag{18.198}$$

---

[32]The subscript h indicates, that the specific enthalpy is constant!

[33]It needs to run along an isenthalp, since the first law of thermodynamics has shown, that $dh =$ const.! However, Fig. 18.30 shows several isenthalps—depending on the inlet state of the fluid.

Fig. 18.30 Adiabatic throttling (Joule Thomson effect)

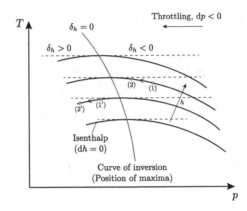

This case is shown in the figure as change of state $(1) \rightarrow (2)$.

- Case 2 (Temperature drops):
  According to Eq. 18.197 the temperature drops, i.e. $dT < 0$, when the pressure decreases, i.e. $dp < 0$, as long as

$$\left(\frac{\partial T}{\partial p}\right)_h = \delta_h > 0 \tag{18.199}$$

This case is shown in the figure as change of state $(1') \rightarrow (2')$.

- Case 3 (Temperature stays constant):
  According to Eq. 18.197 the temperature remains constant, i.e. $dT = 0$, when the pressure decreases, i.e. $dp < 0$, as long as

$$\left(\frac{\partial T}{\partial p}\right)_h = \delta_h = 0 \tag{18.200}$$

Example 18.10 In this example the adiabatic throttling of a real gas shall be investigated by applying a Van der Waals equation. In order to find the temperature characteristic while throttling the following equation needs to be solved:

$$\left(\frac{\partial T}{\partial p}\right)_h = \delta_h = -\frac{\delta_T}{c_p} = -\frac{1}{c_p}\left(\frac{\partial h}{\partial p}\right)_T. \tag{18.201}$$

The left hand side of this equation describes the change of temperature with the change of pressure while enthalpy remains constant. According to the first law of thermodynamics this is the case for an adiabatic throttling.[34] Following Eq. 12.93 the partial differential $\left(\frac{\partial h}{\partial p}\right)_T$ can be substituted, i.e.

---

[34]Changes of kinetic and potential energy are ignored!

$$\left(\frac{\partial T}{\partial p}\right)_h = \frac{T\left(\frac{\partial v}{\partial T}\right)_p - v}{c_p}. \tag{18.202}$$

Goal is to apply a van der Waals equation to handle real fluids. However, a van der Waals equation is denoted with molar state values, see Eq. 18.39, so that molar state values are introduced for Eq. 18.202 as well. With

$$v = \frac{v_M}{M} \tag{18.203}$$

and

$$c_p = \frac{c_{p,M}}{M} = \frac{C_p}{M} \tag{18.204}$$

it is

$$\left(\frac{\partial T}{\partial p}\right)_h = \frac{T\left(\frac{\partial v_M}{\partial T}\right)_p - v_M}{C_p}. \tag{18.205}$$

In order to solve $\left(\frac{\partial v_M}{\partial T}\right)_p$, a van der Waals equation is applied, see Sect. 18.3.1, i.e.

$$p = \frac{R_M T}{v_M - b} - \frac{a}{v_M^2}. \tag{18.206}$$

As long as $v_M \gg b$ this equation can be mathematically simplified to[35]

$$p = \frac{R_M T}{v_M}\left(1 + \frac{b}{v_M} - \frac{a}{v_M R_M T}\right). \tag{18.207}$$

Rearranging results in

$$p = \frac{R_M T}{v_M} + b\frac{R_M T}{v_M^2} - \frac{a}{v_M^2}. \tag{18.208}$$

Multiplying with $\frac{v_M}{p}$ leads to

$$v_M = \frac{R_M T}{p} + b\frac{R_M T}{v_M p} - \frac{a}{v_M p}. \tag{18.209}$$

For the correction terms including a and b the thermal equation of state $p v_M = R_M T$ is applied,[36] so that

$$v_M = \frac{R_M T}{p} + b - \frac{a}{R_M T}. \tag{18.210}$$

---

[35]However, this is the first approximation: Comparing Eqs. 18.206 and 18.207 leads to $v_M^2 = v_M^2 + b^2$, which is true for $v_M \gg b$!

[36]In fact, applying the correlation for ideal gases is the second approximation!

Hence, the partial differential $\left(\frac{\partial v_M}{\partial T}\right)_p$ can be solved, i.e.

$$\left(\frac{\partial v_M}{\partial T}\right)_p = \frac{R_M}{p} + \frac{a}{R_M T^2}. \tag{18.211}$$

Finally, this term can be substituted in Eq. 18.205, i.e.

$$\left(\frac{\partial T}{\partial p}\right)_h = \frac{\frac{R_M T}{p} + \frac{a}{R_M T} - v_M}{C_p}. \tag{18.212}$$

In this equation v_M can be replaced by Eq. 18.210, so that

$$\boxed{\left(\frac{\partial T}{\partial p}\right)_h = \frac{1}{C_p}\left[\frac{2a}{R_M T} - b\right]} \tag{18.213}$$

In order to solve this equation, the variables need to be separated:

$$\frac{\partial T}{\frac{2a}{R_M T} - b} = \frac{\partial p}{C_p}. \tag{18.214}$$

Performing an integration from state (1) to state (2) leads to[37]

$$\int_{T_1}^{T_2} \frac{\partial T}{\frac{2a}{R_M T} - b} = \int_{p_1}^{p_2} \frac{\partial p}{C_p} = \frac{p_2 - p_1}{C_p} \tag{18.215}$$

Solving results in

$$\left[-\frac{2a \ln(2a - b R_M T)}{b^2 R_M} - \frac{T}{b}\right]_{T_1}^{T_2} = \frac{p_2 - p_1}{C_p}. \tag{18.216}$$

Thus, it finally is

$$-\frac{1}{b}(T_2 - T_1) - \frac{2a}{b^2 R_M} \ln\frac{2a - b R_M T_2}{2a - b R_M T_1} = \frac{p_2 - p_1}{C_p}. \tag{18.217}$$

For a throttling from p_1 to $p_2 < p_1$ the outlet temperature T_2 can be calculated implicitly. However, it is much easier to solve Eq. 18.213 numerically. This can be done for air with the following van der Waals parameters for instance:

$$a = 135.8 \times 10^{-3} \frac{\text{Pa m}^6}{\text{mol}^2} \tag{18.218}$$

---

[37]It is assumed, that $C_p = $ const.

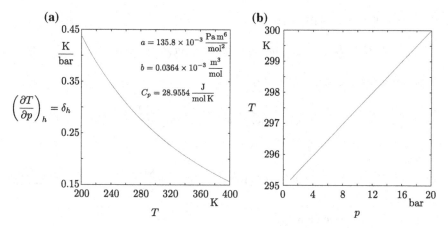

Fig. 18.31 Adiabatic throttling of air

and

$$b = 0.0364 \times 10^{-3} \frac{\text{m}^3}{\text{mol}} \qquad (18.219)$$

as well as

$$C_p = 28.9554 \frac{\text{J}}{\text{mol K}}. \qquad (18.220)$$

Figure 18.31a shows the temperature dependent isenthalpic throttle coefficient δ_h according to Eq. 18.213. Figure 18.31b, as a result of the numerical integration, visualises the throttling of air from an initial state (1), i.e. pressure $p_1 = 20$ bar and temperature $T_1 = 300$ K, to a pressure $p_2 = 1$ bar. The graph shows, that the outlet temperature is approximately $T_2 = 295.19$ K.

Problem 18.11 A steam vessel contains 10 t liquid water and 15 kg saturated vapour at a pressure of 10 bar. The vessel shall be treated as a closed system.

(a) Calculated the entire enthalpy of the vessel's content.
(b) What is the vessel's volume?
(c) What is the mass of the liquid water as well as of the vapour, if the pressure is decreased to 2 bar?
(d) What is the temperature within the vessel after the pressure has decreased?
(e) Sketch the change of state in a p, v- and T, s-diagram!

Solution

The vapour ratio can be calculated according to

$$x = \frac{m_V}{m_V + m_L} = 0.00149. \qquad (18.221)$$

(a) The extensive enthalpy H follows

$$H = H_V + H_L = m_V h_V + m_L h_L = m_V h'' + m_L h'. \qquad (18.222)$$

According to the steam table it is

$$h' = h' (10 \, \text{bar}) = 762.6828 \, \frac{\text{kJ}}{\text{kg}} \qquad (18.223)$$

respectively

$$h'' = h'' (10 \, \text{bar}) = 2777.1 \, \frac{\text{kJ}}{\text{kg}}. \qquad (18.224)$$

Thus, the extensive enthalpy is

$$H = m_V h'' + m_L h' = 7.6685 \times 10^6 \, \text{kJ}. \qquad (18.225)$$

Alternatively, one can follow the approach

$$H = m_{\text{total}} \left[(1 - x) \, h' + x h'' \right] = 7.6685 \times 10^6 \, \text{kJ}. \qquad (18.226)$$

(b) The volume is shared by vapour and liquid, i.e.

$$V = V_V + V_L = m_V v_V + m_L v_L = m_V v'' + m_L v'. \qquad (18.227)$$

According to the steam table it is

$$v' = v' (10 \, \text{bar}) = 0.0011 \, \frac{\text{m}^3}{\text{kg}} \qquad (18.228)$$

respectively

$$v'' = v'' (10 \, \text{bar}) = 0.1943 \, \frac{\text{m}^3}{\text{kg}}. \qquad (18.229)$$

Thus, the extensive volume is

$$V = m_V v'' + m_L v' = 14.1876 \, \text{m}^3. \qquad (18.230)$$

Alternatively, one can follow the approach

$$V = m_{\text{total}} \left[(1 - x) \, v' + x v'' \right] = 14.1876 \, \text{m}^3. \qquad (18.231)$$

(c) The vessel is supposed to be a closed system. Furthermore, we assume the vessel to be rigid, i.e. its volume is constant. At the new pressure $p_2 = 2$ bar it is

$$V = m_V v'' + m_L v' = 14.1876 \, \text{m}^3 \tag{18.232}$$

as well as

$$m_{\text{total}} = m_V + m_L = m'' + m' = 10015 \, \text{kg}. \tag{18.233}$$

According to the steam table it is

$$v' = v' \, (2 \, \text{bar}) = 0.0011 \, \frac{\text{m}^3}{\text{kg}} \tag{18.234}$$

respectively

$$v'' = v'' \, (2 \, \text{bar}) = 0.8857 \, \frac{\text{m}^3}{\text{kg}}. \tag{18.235}$$

Combining Eqs. 18.232 and 18.233 results in

$$m' = \frac{V - v'' m_{\text{total}}}{v' - v''} = 10011 \, \text{kg} \tag{18.236}$$

as well as

$$m'' = m_{\text{total}} - m' = 4 \, \text{kg}. \tag{18.237}$$

Thus, the vapour ratio x has decreased, i.e.

$$x_2 = \frac{m''}{m_{\text{total}}} = 4.0254 \times 10^{-4}. \tag{18.238}$$

Alternatively, it is

$$v = \frac{V}{m_{\text{total}}} = 0.0014 = \text{const.} \tag{18.239}$$

Since

$$v = (1 - x) \, v' + x v'' \tag{18.240}$$

it follows

$$x_2 = \frac{v - v' \, (2 \, \text{bar})}{v' \, (2 \, \text{bar}) - v'' \, (2 \, \text{bar})} = 4.0254 \times 10^{-4}. \tag{18.241}$$

Thus, the vapour's mass follows

$$m'' = x_2 m_{\text{total}} = 4 \, \text{kg}. \tag{18.242}$$

(d) A linear interpolation in the steam table results in a temperature $\vartheta_2 = 120.21 \, °\text{C}$.

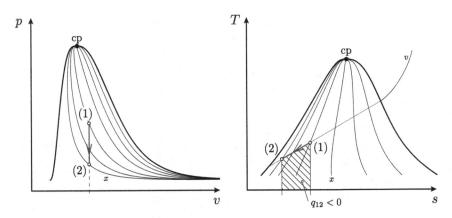

Fig. 18.32 Solution to Problem 18.11e

(e) The change of state is sketched in Fig. 18.32. According to the p, v-diagram no volume work is transferred as the volume remains constant. There shall also be no mechanical work, since the centre of gravity is fix. Thus, due to no work is exchanged, the dissipation follows from the partial energy equation, i.e.

$$W_{12} = 0 = \underbrace{W_{12,\mathrm{V}}}_{=0} + \underbrace{W_{12,\mathrm{mech}}}_{=0} + \Psi_{12}. \qquad (18.243)$$

Hence, the change of state is free of dissipation, i.e. $\Psi_{12} = 0$. Consequently, the hatched area in the T, s-diagram represents the specific heat. Obviously, the system needs to be cooled in order to achieve $p_2 = 2\,\mathrm{bar}$.

Problem 18.12   10 kg vapour at a pressure of 3 MPa and at a temperature of 290 °C (state 1) is trapped in a cylinder/piston system. By heat release the vapour is cooled down isochorically to 200 °C (state 2). Further heat is released isothermally until a pressure of 2.5 MPa (state 3) is reached.

(a) Localise the states in a p, v-diagram qualitatively.
(b) Determine the volume at each state as well as the vapour ratio.

The working fluid is water.

Solution

(a) Figure 18.33 shows the change of state in a p, v-diagram qualitatively.[38]

---

[38] In order to solve this problem, the knowledge of the isothermals in a p, v-diagram is required!

Fig. 18.33 Solution to
Problem 18.12

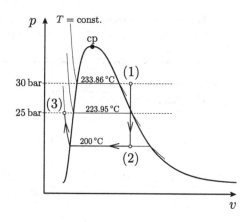

- State (1): It is known from state (1), that

$$p_1 = 3\,\text{MPa} = 30\,\text{bar} \tag{18.244}$$

respectively

$$\vartheta_1 = 290\,°\text{C}. \tag{18.245}$$

The vaporisation temperature at 30 bar is 233.86°C, see steam table. Thus, since $\vartheta_1 > \vartheta_s$ (30 bar) = 233.86 °C, state (1) is overheated.

- State (2):
The system is closed, so that the mass remains constant, i.e. $m_1 = m_2$. Furthermore, the change of state is isochoric, i.e. $V_1 = V_2$. Consequently, the specific volume follows

$$v_1 = v_2. \tag{18.246}$$

The change of state runs vertically in a p, v-diagram. The second information, that is required to fix the state thermodynamically unambiguously, is the temperature

$$\vartheta_2 = 200\,°\text{C}. \tag{18.247}$$

Unfortunately, it is not yet clear, if state (2) lies in the wet steam region, as shown in the p, v-diagram. That will be proven later on in part (b).

- State (3):
It is known from state (3), that

$$p_3 = 2.5\,\text{MPa} = 25\,\text{bar} \tag{18.248}$$

respectively, since the change of state runs isothermally,

$$\vartheta_3 = \vartheta_2 = 200\,°\text{C}. \tag{18.249}$$

The vaporisation temperature according to 25 bar is 223.96 °C, see steam table. Thus, since $\vartheta_3 < \vartheta_s$ (25 bar) = 223.96 °C, state (3) is supercooled.

(b) For the volume respectively the vapour ratio it follows:

- State (1) → overheated steam
  The steam table shows, that

$$v_1 = v\,(30\,\text{bar},\,290\,°\text{C}) = 0.0792\,\frac{\text{m}^3}{\text{kg}}. \qquad (18.250)$$

For volume V_1 it is

$$V_1 = mv_1 = 0.792\,\text{m}^3. \qquad (18.251)$$

- State (2)
  Since the change of state (1) → (2) is isochoric, it is $V_2 = V_1 = 0.792\,\text{m}^3$, i.e. $v_2 = v_1 = 0.0792\,\frac{\text{m}^3}{\text{kg}}$. Now, let us examine, whether state (2) lies in the wet steam region. According to the steam table it follows

$$v'\,(200\,°\text{C}) = 0.0012\,\frac{\text{m}^3}{\text{kg}} < v_2 < v''\,(200\,°\text{C}) = 0.1272\,\frac{\text{m}^3}{\text{kg}}. \qquad (18.252)$$

Thus, (2) is located in the wet steam region. Its vapour ratio is

$$x_2 = \frac{v_2 - v'\,(200\,°\text{C})}{v''\,(200\,°\text{C}) - v'\,(200\,°\text{C})} = 0.619. \qquad (18.253)$$

- State (3) → undercooled liquid:
  The steam table shows, that

$$v_3 = v\,(25\,\text{bar},\,200\,°\text{C}) = 0.00116\,\frac{\text{m}^3}{\text{kg}}. \qquad (18.254)$$

For volume V_3 it is

$$V_3 = mv_3 = 0.0116\,\text{m}^3. \qquad (18.255)$$

Chapter 19
Mixture of Gases

So far single component fluids, i.e. fluids of one chemical composition, have been treated. Part I has focused on ideal gases and incompressible liquids. In Chap. 18 fluids underlying a phase change have been investigated, i.e. a fluid can occur in solid, liquid and gaseous state—however, its chemical composition remains constant even when a phase change takes place. In this chapter it is clarified, how a mixture of several gases of different chemical compositions can be treated thermodynamically. Thus, mixtures of gases consist of two or more pure gases respectively components. Furthermore, these components must not react with each other chemically.[1] One example for a gaseous mixture is dry, atmospheric air. Its volume concentrations are as follows, see [21]:

- Nitrogen ($N_2 = 78.08$ Vol.-%)
- Oxygen ($O_2 = 20.95$ Vol.-%)
- Argon (Ar $= 0.93$ Vol.-%)
- Carbon dioxide ($CO_2 = 0.04$ Vol.-%)
- Neon Ne, helium He, methan CH_4, krypton Kr, hydrogen H_2, dinitrogen monoxide N_2O, xenon Xe in the ppm range[2]

Humid air, i.e. a mixture of dry air and water, i.e. vapour, liquid, solid, will be handled in Chap. 20.

19.1 Concentration Specifications

Section 5.1.3 has shown, that e.g. pressure p and temperature T are not sufficient in order to identify a thermodynamic system unambiguously. Further information according to the composition of the mixture is required. However, there are several possibilities to characterise the concentration of a component i within the mixture:

---

[1]Chemical reactions will be covered in Chaps. 23 and 24 though.
[2]1 ppm $\equiv 1 \times 10^{-4}$ Vol.-%.

© Springer Nature Switzerland AG 2019
A. Schmidt, *Technical Thermodynamics for Engineers*,
https://doi.org/10.1007/978-3-030-20397-9_19

- Mass concentration ξ_i

Defined as the ratio of the partial mass m_i of a component i to the total mass m_{total}, i.e.

$$\xi_i = \frac{m_i}{m_{\text{total}}}. \tag{19.1}$$

Obviously, it is

$$m_{\text{total}} = m_1 + m_2 + \cdots + m_n = \sum m_i, \tag{19.2}$$

so that the division by the total mass m_{total} results in

$$\frac{m_1}{m_{\text{total}}} + \frac{m_2}{m_{\text{total}}} + \cdots + \frac{m_n}{m_{\text{total}}} = \frac{\sum m_i}{m_{\text{total}}} = 1. \tag{19.3}$$

Thus, it follows, that

$$\boxed{\xi_1 + \xi_2 + \cdots + \xi_n = \sum \xi_i = 1} \tag{19.4}$$

- Molar concentration x_i

Defined as the ratio of the partial molar amount n_i of a component i to the total molar amount n_{total}, i.e.

$$x_i = \frac{n_i}{n_{\text{total}}}. \tag{19.5}$$

Obviously, it is

$$n_{\text{total}} = n_1 + n_2 + \cdots + n_n = \sum n_i, \tag{19.6}$$

so that the division by the total molar amount n_{total} results in

$$\frac{n_1}{n_{\text{total}}} + \frac{n_2}{n_{\text{total}}} + \cdots + \frac{n_n}{n_{\text{total}}} = \frac{\sum n_i}{n_{\text{total}}} = 1. \tag{19.7}$$

Thus, it follows, that

$$\boxed{x_1 + x_2 + \cdots + x_n = \sum x_i = 1} \tag{19.8}$$

- Volume concentration σ_i

Defined as the ratio of the partial volume V_i of a component i to the total volume V_{total}, i.e.

$$\sigma_i = \frac{V_i}{V_{\text{total}}}. \tag{19.9}$$

Obviously, it is

$$V_{\text{total}} = V_1 + V_2 + \cdots + V_n = \sum V_i,$$ (19.10)

so that the division by the total volume V_{total} results in

$$\frac{V_1}{V_{\text{total}}} + \frac{V_2}{V_{\text{total}}} + \cdots + \frac{V_n}{V_{\text{total}}} = \frac{\sum V_i}{V_{\text{total}}} = 1.$$ (19.11)

Thus, it follows, that

$$\boxed{\sigma_1 + \sigma_2 + \cdots + \sigma_n = \sum \sigma_i = 1}$$ (19.12)

- Partial pressure ratio π_i

Defined as the ratio of the partial pressure[3] p_i of a component i to the total pressure p_{total}, i.e.

$$\pi_i = \frac{p_i}{p_{\text{total}}}.$$ (19.13)

- Molarity c_i

Defined as the ratio of the partial molar amount n_i of a component i to the total volume V_{total}, i.e.

$$c_i = \frac{n_i}{V_{\text{total}}}.$$ (19.14)

- Volume ratio ψ_i

Defined as the ratio of the partial volume V_i of a component i to the partial volume V_k of a component k, i.e.

$$\psi_i = \frac{V_i}{V_k}.$$ (19.15)

- Partial density ρ_i

Defined as the ratio of the partial mass m_i of a component i to the total volume V_{total}, i.e.

$$\rho_i = \frac{m_i}{V_{\text{total}}} = \frac{n_i M_i}{V_{\text{total}}}.$$ (19.16)

---

[3] See Sect. 19.2.

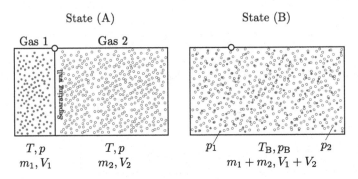

Fig. 19.1 Isobaric, isothermal mixing of two gases according to Dalton

19.2 Dalton's Law

In a mixture of ideal gases, each component acts as if it was alone, i.e. it covers the entire volume of the mixture V_{total}. Figure 19.1 shows the consequence for an adiabatic, isothermal, isobaric mixing of two ideal gases: Initially, two gases 1 and 2 are separated by a dividing wall, see left figure. In that state, both gases have the same pressure p and the same temperature T. Gas 1 possesses a mass m_1 and covers a volume of V_1. For gas 2 it is m_2 respectively V_2. Now, the wall is removed, so that both gases can share[4] the entire volume $V_{total} = V_1 + V_2$. According to Dalton, there is an equilibrium as soon as both gases have distributed across the entire volume.[5]

Temperature

Let us assume, that during mixing no energy is transferred with the environment.[6] Consequently, following the first law of thermodynamics

$$W + Q = 0 = m_1 c_{v,1} (T_B - T) + m_2 c_{v,2} (T_B - T) \qquad (19.17)$$

the temperature of the mixture is

$$\boxed{T_B = T} \qquad (19.18)$$

Hence, the mixture has the same temperature as the gases of initial state (A).

---

[4]This process is called diffusion.

[5]However, reaching this equilibrium takes a while.

[6]There is no heat crossing the system boundary, since the system is supposed to be adiabatic. Furthermore, no work passes the system boundary, see Fig. 19.1.

Pressure

Now, the focus is on the pressure of state (B). Since, left and right Fig. 19.1 represent equilibrium states (A) and (B) and both gases are supposed to be ideal, the thermal equation of state can be applied, i.e.

- Gas 1:

Before the mixing, i.e. state (A), it is

$$pV_1 = m_1 R_1 T. \tag{19.19}$$

After the mixing, i.e. state (B), it is

$$p_1 (V_1 + V_2) = m_1 R_1 T. \tag{19.20}$$

Obviously, in state (B) gas 1 occupies a larger volume than in state (A), while temperature and mass remain constant. Thus, the gas 1 in the mixture, i.e. state (B), has a smaller pressure[7] p_1 than in state (A). Combining these two equations brings

$$p_1 = \frac{V_1}{V_1 + V_2} p \tag{19.21}$$

- Gas 2:

Before the mixing it is

$$pV_2 = m_2 R_2 T. \tag{19.22}$$

After the mixing it is

$$p_2 (V_1 + V_2) = m_2 R_2 T. \tag{19.23}$$

Combining these two equations results in

$$p_2 = \frac{V_2}{V_1 + V_2} p. \tag{19.24}$$

Obviously, according to Eqs. 19.21 and 19.24, the pressures of gases 1 and 2 in the mixture are smaller than their initial pressures p in state (A).

Theorem 19.1 *Each component within a mixture has an individual pressure that is smaller than the total pressure of the mixture. This is called partial pressure p_i.*

With the partial pressures[8] according to Eqs. 19.21 and 19.24 the total pressure p_B of the mixture in state (B) follows

---

[7]This is called partial pressure!

[8]To get a better understanding of the partial pressure: A pressure sensor measures the kinetic collisions of the molecules on its surface. The more collisions, the larger the pressure within the

$$p_B = p_1 + p_2 = \frac{V_1}{V_1 + V_2} p + \frac{V_2}{V_1 + V_2} p. \tag{19.25}$$

Thus, it is

$$p_B = p_1 + p_2 = p. \tag{19.26}$$

In other words, the pressure of the mixture is the same as the initial pressures of the two components, i.e. under the given premises the change of state runs isobarically:

$$\boxed{p_B = p} \tag{19.27}$$

Based on Eq. 19.26 an extension for a mixture of n components can easily be made. Obviously, the total pressure is equal to the sum of all partial pressures:

$$\boxed{p_{total} = p = p_1 + p_2 + \cdots + p_n = \sum_i p_i} \tag{19.28}$$

Division by p_{total} brings

$$1 = \frac{p_1}{p_{total}} + \frac{p_2}{p_{total}} + \cdots + \frac{p_n}{p_{total}} \tag{19.29}$$

With the newly introduced partial pressure ratio, see Sect. 19.1, it finally is

$$\boxed{\sum \pi_i = 1} \tag{19.30}$$

Partial Energy Equation

It has already been shown that the temperature remains constant during the mixing process. This is due to no energy has passed the system boundary. However, both gases expand according to Fig. 19.1, so that temperature should decrease as discussed in part I: For a reversible expansion work used to be released, so that a heating was required to keep temperature constant. The partial energy equation,[9] see Eq. 11.21, can be applied to show how temperature remains constant for the mixing, i.e.

$$W = 0 = W_{V,1} + W_{V,2} + \Psi = - \int_A^B p_{Gas_1} \, dV_{Gas_1} - \int_A^B p_{Gas_2} \, dV_{Gas_2} + \Psi \tag{19.31}$$

---

mixture. However, the sensor does not distinguish between the molecules of the different gases—it always shows the entire pressure. Nevertheless, the different molecules have a specific ratio on the total pressure. The partial pressures p_i represent the part of the pressure that is induced by component i.

[9]The entire work $W = 0$, since no energy is exchanged with the environment!

Since the gases are ideal, the thermal equation of state can be applied, i.e.

$$0 = - \int_A^B \frac{m_1 R_1 T}{V_{\text{Gas}_1}} \, dV_{\text{Gas}_1} - \int_A^B \frac{m_2 R_2 T}{V_{\text{Gas}_2}} \, dV_{\text{Gas}_2} + \Psi. \tag{19.32}$$

The change of state is isothermal, i.e.

$$\Psi = m_1 R_1 T \ln \frac{V_1 + V_2}{V_1} + m_2 R_2 T \ln \frac{V_1 + V_2}{V_2}. \tag{19.33}$$

In other words, by applying the thermal equation of state, see Eqs. 19.19 and 19.22:

$$\boxed{\Psi = p V_1 \ln \frac{V_1 + V_2}{V_1} + p V_2 \ln \frac{V_1 + V_2}{V_2} > 0} \tag{19.34}$$

Obviously, it is the dissipation Ψ, that keeps the temperature constant. Dissipation is driven by the volume work of the expanding gases 1 and 2.

Entropy Generation

Applying the second law of thermodynamics leads to

$$S_B - S_A = \underbrace{S_a}_{=0} + S_i. \tag{19.35}$$

S_a disappears, since the system is adiabatic. A further simplification can be made according to

$$S_B - S_A = S_i = \int_A^B \frac{\delta \Psi}{T}. \tag{19.36}$$

Since $T = \text{const.}$ it follows

$$S_B - S_A = S_i = \int_A^B \frac{\delta \Psi}{T} = \frac{\Psi}{T}. \tag{19.37}$$

Finally, it is[10]

$$\boxed{S_i = m_1 R_1 \ln \frac{V_1 + V_2}{V_1} + m_2 R_2 \ln \frac{V_1 + V_2}{V_2}} \tag{19.38}$$

---

[10]The Gibb's paradox claims, that if two identical gases are mixed, no generation of entropy occurs, i.e. $S_i = 0$, see Sect. 19.3.4.

Respectively, with Eqs. 19.21 and 19.24, it results in

$$S_i = m_1 R_1 \ln\frac{p}{p_1} + m_2 R_2 \ln\frac{p}{p_2}$$

(19.39)

An extension to n components follows[11]

$$S_i = \sum_i^n m_i R_i \ln\frac{p}{p_i}$$

(19.40)

19.3   Laws of Mixing

19.3.1   Concentration, Thermal State Values

Concentration

Applying the thermal equation of state for a component i in state (A), see Fig. 19.1, leads to

$$p_{\text{total}} V_i = m_i R_i T.$$

(19.41)

The thermal equation of state for a component i in state (B) brings

$$p_i V_{\text{total}} = m_i R_i T.$$

(19.42)

Thus, its combination results in

$$\pi_i = \frac{p_i}{p_{\text{total}}} = \frac{V_i}{V_{\text{total}}} = \sigma_i.$$

(19.43)

Hence, the partial pressure ratio π_i is identical with the volume concentration σ_i.

Now let us generalise Eq. 19.42 for all components, i.e.

$$\sum_i p_i V_{\text{total}} = \sum_i m_i R_i T.$$

(19.44)

The component's mass m_i can be substituted by $n_i M_i$, i.e.

---

[11] The proof is simple, as instead of 2 components n components have to be taken into account starting with Eq. 19.31.

Fig. 19.2 Gas constant of a mixture of ideal gases

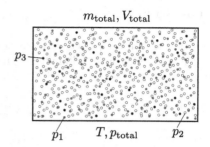

$$\sum_i p_i V_{total} = \sum_i n_i M_i R_i T = \sum_i n_i R_M T. \tag{19.45}$$

Thus, it is

$$V_{total} \sum_i p_i = \sum_i n_i M_i R_i T = R_M T \sum_i n_i. \tag{19.46}$$

Since

$$\sum_i p_i = p_{total} \tag{19.47}$$

respectively

$$\sum_i n_i = n_{total} \tag{19.48}$$

rearranging leads to

$$\frac{\sum_i p_i}{\sum_i n_i} = \frac{p_{total}}{n_{total}} = \frac{R_M T}{V_{total}}. \tag{19.49}$$

With

$$R_M T = \frac{p_i V_{total}}{n_i} \tag{19.50}$$

it finally results in

$$\pi_i = \frac{p_i}{p_{total}} = \frac{n_i}{n_{total}} = x_i. \tag{19.51}$$

In combination with Eq. 19.43 it is

$$\boxed{\frac{p_i}{p_{total}} = \frac{V_i}{V_{total}} = \frac{n_i}{n_{total}} = \pi_i = \sigma_i = x_i} \tag{19.52}$$

Hence, the partial pressure ratio π_i is identical with the volume concentration σ_i as well as with the molar concentration x_i.

Gas Constant

In this section a representative gas constant R_{total} for a mixture of ideal gases is deduced. For each gas within the mixture, see Fig. 19.2, it follows, see Eq. 19.42,

$$
\begin{aligned}
p_1 V_{total} &= m_1 R_1 T \\
p_2 V_{total} &= m_2 R_2 T \\
&\vdots \\
p_n V_{total} &= m_n R_n T
\end{aligned}
\tag{19.53}
$$

The summation of all components results in

$$
(p_1 + p_2 + \cdots + p_n)\, V_{total} = (m_1 R_1 + m_2 R_2 + \cdots + m_n R_n)\, T.
\tag{19.54}
$$

Now, a representative gas constant is defined as follows

$$
(m_1 R_1 + m_2 R_2 + \cdots + m_n R_n)\, T \stackrel{!}{=} m_{total} R_{total} T.
\tag{19.55}
$$

Thus, the thermal equation of state for the mixture reads as

$$
\boxed{p_{total} V_{total} = m_{total} R_{total} T}
\tag{19.56}
$$

Eq. 19.55 can be applied to derive R_{total}, i.e.

$$
\boxed{R_{total} = \frac{m_1 R_1 + m_2 R_2 + \cdots + m_n R_n}{m_{total}} = \sum_i \xi_i R_i}
\tag{19.57}
$$

Consequently, the representative gas constant of a mixture can be calculated with the mass-averaged individual gas constants of its single components.

Molar Mass

The molar mass of a mixture M_{total} shown in Fig. 19.2 is

$$
M_{total} = \frac{m_{total}}{n_{total}} = \sum_i \frac{m_i}{n_{total}} = \sum_i \frac{n_i}{n_{total}} M_i.
\tag{19.58}
$$

Hence, by introducing the molar concentration x_i, see Eq. 19.5, the molar mass of a mixture finally is

$$
\boxed{M_{total} = \sum_i (x_i M_i)}
\tag{19.59}
$$

Thus, the representative molar mass of a mixture can be calculated with the molar-averaged individual molar masses of its single components. R_{total} as well as M_{total} make it possible to apply the thermal equation of state easily for a mixture. The correlation between individual gas constant R_{total} and general gas constant R_M is

$$m_{total} R_{total} = n_{total} R_M.$$ (19.60)

Consequently, it finally is

$$\boxed{R_M = R_{total} M_{total}}$$ (19.61)

Conversion Mass/Molar Concentration

In this section a correlation between mass concentration ξ_i and molar concentration x_i is deduced. Following the definition of the mass concentration

$$\xi_i = \frac{m_i}{m_{total}}$$ (19.62)

the mass can be replaced by molar quantity and molar mass, i.e.

$$\xi_i = \frac{n_i M_i}{n_{total} M_{total}} = x_i \frac{M_i}{M_{total}}.$$ (19.63)

Hence, it results in

$$\boxed{\xi_i = x_i \frac{M_i}{M_{total}}}$$ (19.64)

19.3.2 Internal Energy, Enthalpy

Gibbs' theorem is an extension of Dalton's law, see [1]. It claims that the total internal energy of a mixture of ideal gases U_{total} is equal to the sum of all partial internal energies U_i, i.e.

$$U_{total} = U_1 + U_2 + \cdots + U_n = \sum_i U_i = \sum_i (m_i u_i).$$ (19.65)

Thus, the specific internal energy of a mixture u_{total} can be derived by division of the mixture's mass, i.e.

$$\boxed{u_{total} = \frac{U_{total}}{m_{total}} = \sum_i (\xi_i u_i)}$$ (19.66)

The enthalpy of a component i in a mixture of ideal gases follows according to the definition of the enthalpy, see part I, i.e.

$$H_i = U_i + p_i V_{\text{total}}. \tag{19.67}$$

In a mixture the component i possesses a partial pressure p_i and is distributed within the entire volume V_{total}. The total enthalpy of a mixture is

$$H_{\text{total}} = U_{\text{total}} + p_{\text{total}} V_{\text{total}}. \tag{19.68}$$

According to Gibbs/Dalton it is composed of its single components, i.e.

$$H_{\text{total}} = H_1 + H_2 + \cdots + H_n = \sum_i H_i = \sum_i U_i + V_{\text{total}} \sum_i p_i = \sum_i (m_i h_i). \tag{19.69}$$

Thus, the specific enthalpy of a mixture h_{total} can be derived by division of the mixture's mass, i.e.

$$\boxed{h_{\text{total}} = \frac{H_{\text{total}}}{m_{\text{total}}} = \sum_i (\xi_i h_i)} \tag{19.70}$$

Specific Heat Capacities of a Mixture of Ideal Gases

For ideal gases the internal energy is solely a function of temperature. Hence, it follows for a component i

$$dU_i = m_i c_v \, dT. \tag{19.71}$$

Thus, for a mixture it is

$$dU_{\text{total}} = \sum_i \left(m_i c_{v,i} \right) dT \stackrel{!}{=} m_{\text{total}} c_{v,\text{total}} \, dT. \tag{19.72}$$

Finally the isochoric specific heat capacity of a mixture $c_{v,\text{total}}$ reads as

$$\boxed{c_{v,\text{total}} = \sum_i \left(\frac{m_i}{m_{\text{total}}} c_{v,i} \right) = \sum_i \left(\xi_i c_{v,i} \right)} \tag{19.73}$$

The enthalpy of a component i within a mixture of ideal gases obeys

$$dH_i = m_i c_p \, dT. \tag{19.74}$$

Thus, for the mixture it is

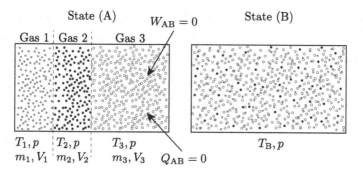

Fig. 19.3 Adiabatic mixing closed system—Exemplary for $n = 3$ components

$$dH_{\text{total}} = \sum_i \left(m_i c_{p,i} \right) dT \stackrel{!}{=} m_{\text{total}} c_{p,\text{total}} \, dT. \tag{19.75}$$

Finally the isobaric specific heat capacity of a mixture $c_{p,\text{total}}$ reads as

$$c_{p,\text{total}} = \sum_i \left(\frac{m_i}{m_{\text{total}}} c_{p,i} \right) = \sum_i \left(\xi_i c_{p,i} \right) \tag{19.76}$$

19.3.3 Adiabatic Mixing Temperature

Closed System

For a closed system, see Fig. 19.3, the first law of thermodynamics obeys[12]

$$Q_{\text{AB}} + W_{\text{AB}} = \Delta U. \tag{19.77}$$

Since the system is adiabatic and no work is exchanged, the first law of thermodynamics simplifies to

$$\Delta U = 0. \tag{19.78}$$

As shown in the last section, the change of internal energy of a mixture is composed of the changes of the internal energy of its individual components as follows[13]

---

[12]Kinetic as well as potential energies are ignored!

[13]Subject to the condition, that the specific heat capacity is temperature-independent!

Fig. 19.4 Adiabatic mixing
open system—Exemplary
for $n = 3$ components

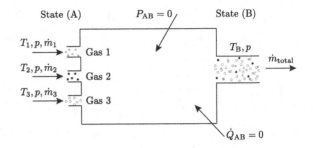

$$\Delta U_1 = m_1 c_{v,1}(T_B - T_1)$$
$$\Delta U_2 = m_2 c_{v,2}(T_B - T_2)$$
$$\vdots$$
$$\Delta U_n = m_n c_{v,n}(T_B - T_n).$$

$$(19.79)$$

Thus, the total change of internal energy U is the summation of its single components i, i.e.

$$\Delta U = \sum_i \Delta U_i = \sum_i m_i c_{v,i} (T_B - T_i) = 0. \qquad (19.80)$$

Rearranging results in

$$T_B \sum_i m_i c_{v,i} - \sum_i m_i c_{v,i} T_i = 0. \qquad (19.81)$$

This equation can be solved for the temperature T_B after mixing in state (B), i.e.

$$\boxed{T_B = \frac{\sum_i m_i c_{v,i} T_i}{\sum_i m_i c_{v,i}}} \qquad (19.82)$$

Open System

For an open system in steady state, see Fig. 19.4, the first law of thermodynamics obeys[14]

$$\dot{Q}_{AB} + P_{AB} = \Delta \dot{H}. \qquad (19.83)$$

Since the system is adiabatic and no power is exchanged, the first law of thermodynamics simplifies to

$$\Delta \dot{H} = 0. \qquad (19.84)$$

---

[14] Kinetic as well as potential energies are ignored!

As shown in the last section, the change of enthalpy of a mixture is composed of the changes of the enthalpies of its individual components as follows[15]

$$\Delta \dot{H}_1 = \dot{m}_1 c_{p,1}(T_B - T_1)$$
$$\Delta \dot{H}_2 = \dot{m}_2 c_{p,2}(T_B - T_2)$$
$$\vdots$$
$$\Delta \dot{H}_n = \dot{m}_n c_{p,n}(T_B - T_n).$$

(19.85)

Thus, the total change of enthalpy $\dot{H}$ is the summation of its single components i, i.e.

$$\Delta \dot{H} = \sum_i \Delta \dot{H}_i = \sum_i \dot{m}_i c_{p,i} (T_B - T_i) = 0.$$

(19.86)

Rearranging results in

$$T_B \sum_i \dot{m}_i c_{p,i} - \sum_i \dot{m}_i c_{p,i} T_i = 0.$$

(19.87)

This equation can be solved for the temperature T_B after mixing in state (B), i.e.

$$\boxed{T_B = \frac{\sum_i \dot{m}_i c_{p,i} T_i}{\sum_i \dot{m}_i c_{p,i}}}$$

(19.88)

19.3.4 Irreversibility of Mixing

Entropy

Though the irreversibility of mixing has already been derived based on the second law of thermodynamics in combination with the partial energy equation, see Sect. 19.2, an alternative approach is now presented.

In order to analyse the irreversibility of an adiabatic, isothermal mixing of ideal gases,[16] see Fig. 19.5, the generation of entropy from state (A) to state (B) $S_{i,12}$ is required. An analogous model is now postulated, that allows a hypothetical reversible mixing.[17] However, even in this hypothetical mixing the same state (B) shall be achieved as for the actual irreversible mixing, see state (B) in Fig. 19.5. A technique

---

[15]Subject to the condition, that the specific heat capacity is temperature-independent!

[16]In this case, to keep it simple, two ideal gases are mixed. However, this approach can easily be extended to n components.

[17]This is just hypothetic, since mixing of different gases is always irreversible!

State (A) $\xrightarrow{\text{Irreversible, adiabatic}}$ State (B)

$Q = 0$

$S_i > 0, S_a = 0$

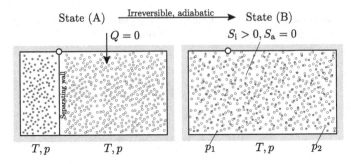

T, p $\qquad$ T, p $\qquad\qquad$ p_1 $\qquad$ T, p $\qquad$ p_2

Fig. 19.5 Irreversibility of mixing

State (A) $\xrightarrow{\text{Reversible, non-adiabatic}}$ State (B)

$Q > 0$

$S_i = 0, S_a > 0$

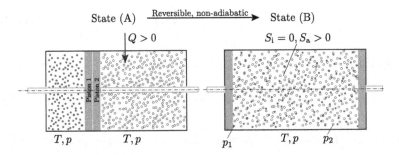

T, p $\qquad$ T, p $\qquad\qquad$ p_1 $\qquad$ T, p $\qquad$ p_2

Fig. 19.6 Irreversibility of mixing—Analogous model

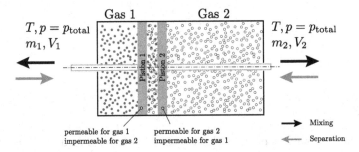

$T, p = p_{\text{total}}$
m_1, V_1

Gas 1 $\qquad$ **Gas 2**

$T, p = p_{\text{total}}$
m_2, V_2

permeable for gas 1 $\qquad$ permeable for gas 2
impermeable for gas 2 $\qquad$ impermeable for gas 1

$\longrightarrow$ Mixing
$\longleftarrow$ Separation

Fig. 19.7 Thought experiment for a reversible, isothermal mixing, pistons replace separating wall in Fig. 19.1

for this thought experiment is introduced with Fig. 19.6 as well as with Fig. 19.7. Instead of separating the ideal gases with a wall, the gases are now in two compartments that are limited by two pistons. Each of the pistons can move freely, i.e. without any friction. While the pistons slowly move to the cylinder walls, due to the diffusion of the gases, a mixing occurs as shown in Fig. 19.7: Gas 1 and gas 2 expand into the gap and further displace the pistons 1 and 2. Therefore, piston 1 is permeable for gas 1 and impermeable for gas 2. Piston 2 behaves vice versa, i.e. gas 2 can pass the piston and gas 1 can not.

Obviously, during this change of state the two gases expand,[18] thus the gases release work to the moving pistons: The left side of piston 1 lets gas 1 pass, so that gas 1 does not cause a force on piston 1. However, in the filling chamber, it is gas 2 that is responsible for causing a force on piston 1, since gas 2 can not pass piston 1. Consequently, work must be released due to the effective force and the distance piston 1 is moving. The force on piston 2 follows accordingly and is caused by gas 1. In sum, the volume work of gas 1 is transferred by piston 2, the volume work of gas 2 is transferred by piston 1. In case no irreversibilities[19] occur during the mixing in the cylinder, the amount of work during this expansion exactly covers the amount of work, that is required conversely for pushing the pistons in, i.e. compressing the two gases.[20] During such a compression the gases 1 and 2 are separated again due to the characteristics of the pistons. Hence, the entire process is reversible since state (A) has been re-reached and there are no changes in the environment. Now, let us focus on the change of state from (A) to (B): The entire work, which is released, is the isothermal volume work of the two gases and follows[21]

$$W_{\text{total}} = W_{V,1} + W_{V,2} = -p_{\text{total}} V_1 \ln\left(\frac{V_{\text{total}}}{V_1}\right) - p_{\text{total}} V_2 \ln\left(\frac{V_{\text{total}}}{V_2}\right) < 0, \quad (19.89)$$

since the change of state shall be isothermal, see also Sect. 19.2. To keep the change of state isothermal, according to the first law of thermodynamics, heat needs to be supplied, i.e.

$$W_{\text{total}} + Q = U_2 - U_1 = 0. \quad (19.90)$$

Obviously, in order to achieve state (B), the heat supply needs to follow

$$Q = -W_{\text{total}} > 0. \quad (19.91)$$

With the supplied heat the system additionally receives entropy. Thus, the reversible mixing follows

$$S_a = \frac{Q}{T} = -\frac{W_{\text{total}}}{T} = \frac{p_{\text{total}} V_{\text{total}}}{T}\left[\frac{V_1}{V_{\text{total}}} \ln\left(\frac{V_{\text{total}}}{V_1}\right) + \frac{V_2}{V_{\text{total}}} \ln\left(\frac{V_{\text{total}}}{V_2}\right)\right].$$
$$(19.92)$$

Hence, the entropy in state (B) obeys

$$S_B = S_A + S_a. \quad (19.93)$$

Premise is, that the irreversible mixing and the analogous model of hypothetic reversible mixing shall reach the same state (B). For the irreversible mixing there

---

[18]Expansion means release of volume work!

[19]This requires a quasi-static change of state!

[20]According to the partial energy equation.

[21]See Fig. 13.13.

was no heat exchange and thus no transport of entropy S_a. Consequently, in order to reach the same amount of entropy, there needs to be generation of entropy, i.e.

$$S_B = S_A + S_i. \tag{19.94}$$

The comparison of Eqs. 19.93 and 19.94 leads to $S_i = S_a$, so that

$$S_i = \frac{p_{total} V_{total}}{T} \left[\frac{V_1}{V_{total}} \ln\left(\frac{V_{total}}{V_1}\right) + \frac{V_2}{V_{total}} \ln\left(\frac{V_{total}}{V_2}\right) \right]. \tag{19.95}$$

Applying the thermal equation of state and Dalton's law, see Eqs. 19.21 and 19.24, the generation of entropy for an isothermal mixing of two ideal gases results in

$$\boxed{S_i = m_1 R_1 \ln\left(\frac{p_{total}}{p_1}\right) + m_2 R_2 \ln\left(\frac{p_{total}}{p_2}\right)} \tag{19.96}$$

Thus, the same result is achieved as with Eq. 19.39. Mind the so-called Gibbs paradox, which states, that if two identical ideal gases are mixed, no entropy of mixing occurs, i.e. $S_{i,Mix} = 0$. In case the two compartments contain the same ideal gas at pressure p and temperature T, there is a mixing when the sliding wall between the two compartments opens. When closing this wall again, each compartment then possesses the same state as before. Hence, from this perspective the change of state needs to be reversible. However, according to conventional thermodynamics there needs to be generation of entropy while the sliding wall opens and the two identical ideal gases mix with each other. Since the initial state is obviously achieved, when closing the wall, the amount of entropy needs to be the same as before. Thus, entropy must have been destroyed when closing the wall again,[22] i.e. the second law of thermodynamics is violated. In order to solve this problem a correction term needs to be defined that takes into account the permutation of identical particles by means of quantum mechanics. For further explanation see [22].

Based on the caloric equation of state with respect to entropy,[23] that has already been introduced in part I,

$$s_1 - s_0 = c_p \ln\left(\frac{T_1}{T_0}\right) - R \ln\left(\frac{p_1}{p_0}\right) \tag{19.97}$$

the entropy of state (A) follows accordingly:

---

[22]Otherwise, it is not possible to get the same amount of initial entropy.

[23]Mind, that s_0 is the specific entropy at an arbitrary reference level T_0, p_0.

- Entropy of component 1:

$$S_{A,1} = m_1 s_0 + m_1 c_{p,1} \ln\left(\frac{T}{T_0}\right) - m_1 R_1 \ln\left(\frac{p_{total}}{p_0}\right) \tag{19.98}$$

- Entropy of component 2:

$$S_{A,2} = m_2 s_0 + m_2 c_{p,2} \ln\left(\frac{T}{T_0}\right) - m_2 R_2 \ln\left(\frac{p_{total}}{p_0}\right) \tag{19.99}$$

Hence, the total entropy in state (A) is

$$S_A = S_{A,1} + S_{A,2}. \tag{19.100}$$

However, the entropy after adiabatic mixing in state (B) results in

$$S_B = S_A + S_i. \tag{19.101}$$

Combining Eq. 19.101 with Eqs. 19.100, 19.99, 19.98 and 19.96 leads to

$$\begin{aligned} S_B = & m_1 s_0 + m_1 c_{p,1} \ln\left(\frac{T}{T_0}\right) - m_1 R_1 \ln\left(\frac{p_{total}}{p_0}\right) \\ & + m_2 s_0 + m_2 c_{p,2} \ln\left(\frac{T}{T_0}\right) - m_2 R_2 \ln\left(\frac{p_{total}}{p_0}\right) \\ & + m_1 R_1 \ln\left(\frac{p_{total}}{p_1}\right) + m_2 R_2 \ln\left(\frac{p_{total}}{p_2}\right). \end{aligned} \tag{19.102}$$

Thus, rearranged

$$S_B = \underbrace{m_1 s_0 + m_1 c_{p,1} \ln\left(\frac{T}{T_0}\right) - m_1 R_1 \ln\left(\frac{p_1}{p_0}\right)}_{S_1} \tag{19.103}$$

$$+ \underbrace{m_2 s_0 + m_2 c_{p,2} \ln\left(\frac{T}{T_0}\right) - m_2 R_2 \ln\left(\frac{p_2}{p_0}\right)}_{S_2}.$$

Generally speaking, the entropy of a mixture of n ideal gases, i.e. in state (B), is the sum of its partial entropies, that are calculated with the particular partial pressures p_i:

$$\boxed{S_B = \sum_{i}^{n} S_i(p_i, T)} \tag{19.104}$$

In case a change of entropy is calculated, all reference values, i.e. s_0, T_0 and p_0, disappear!

Exergy

In this section it is investigated how much exergy the isothermal, isobaric mixing contains. Since the pure mixing is the focus, the system shall have already reached ambient temperature T_{env}, i.e. is at thermal equilibrium. The heat transfer for bringing the system from temperature T to ambient temperature T_{env} is not of interest. Due to keeping the temperature constant and due to the fixed masses of the gases within the volumes V_1 and V_2, the pressure of the system is also fixed. Hence, once the pressure is adjusted, e.g. by the amount of gases 1 and 2, it is not possible to bring the system to ambient pressure under the given premises. A hypothetic reversible mixing would release volume work, since the two gases expand from state (A) to state (B). Equation 19.89 states the maximum amount of work, that can be released by the system hypothetically, if the change of state would be reversible, i.e. this work is identical with the exergy[24] of the mixing:

$$E_{x,mix} = -W_{total} = +p_{total}V_1 \ln\left(\frac{V_{total}}{V_1}\right) + p_{total}V_2 \ln\left(\frac{V_{total}}{V_2}\right) > 0 \quad (19.105)$$

Rearranging and applying the thermal equation of state results in

$$E_{x,mix} = m_1 R_1 T_{env} \ln\left(\frac{V_{total}}{V_1}\right) + m_2 R_2 T_{env} \ln\left(\frac{V_{total}}{V_2}\right). \quad (19.106)$$

Applying Dalton's law, see Eqs. 19.21 and 19.24 finally leads to

$$\boxed{E_{x,mix} = m_1 R_1 T_{env} \ln\left(\frac{p_{total}}{p_1}\right) + m_2 R_2 T_{env} \ln\left(\frac{p_{total}}{p_2}\right)} \quad (19.107)$$

The realistic scenario, i.e. irreversible mixing, has shown, that dissipation occurs, see Sect. 19.2. The dissipation can be calculated according to Eq. 19.33, so that the generation of entropy was calculated to be

$$S_i = m_1 R_1 \ln\frac{p_{total}}{p_1} + m_2 R_2 \ln\frac{p_{total}}{p_2}. \quad (19.108)$$

Hence, the loss of exergy is

---

[24]Mind, that the effective work, see Sect. 9.2.3, does not play a role, since the pistons are encapsulated from the environment—except for the negligible cross sections of the two rods.

$$\Delta E_{x,V} = T_{env} S_i = m_1 R_1 T_{env} \ln\frac{p_{total}}{p_1} + m_2 R_2 T_{env} \ln\frac{p_{total}}{p_2} \qquad (19.109)$$

A comparison of Eqs. 19.109 and 19.107 shows, that the entire exergy gets lost during the mixing process, i.e.

$$\Delta E_{x,V} = E_{x,mix}. \qquad (19.110)$$

In case of a mixture of n ideal gases it follows accordingly

$$\Delta E_{x,V} = E_{x,mix} = \sum_i^n m_i R_i T_{env} \ln\left(\frac{p_{total}}{p_i}\right) \qquad (19.111)$$

Problem 19.2 A tank of a volume $V = 1\,m^3$ initially, i.e. state (A), contains air of $p_A = 1\,bar$ and $20\,°C$. Now, hydrogen (H_2) is filled in under pressure, so that the tank finally reaches state state (B) with $p_B = 4\,bar$ and $20\,°C$ (Fig. 19.8).

(a) What are the volume fractions of air and hydrogen in the tank?
(b) Determine the gas constant and molar mass of the mixture!
(c) How many moles of the mixture are in the tank?

Further information:

- Molar masses: $M_{H_2} = 2\frac{kg}{kmol}$, $M_{O_2} = 32\frac{kg}{kmol}$, $M_{N_2} = 28\frac{kg}{kmol}$
- General gas constant: $R_M = 8.3143\frac{kJ}{kmol\,K}$
- Composition of air: nitrogen 79 Vol.-%, oxygen 21 Vol.-%

Solution

(a) Volume fractions

- Air (1)

In state (A)[25] it is

$$p_A V = m_1 R_1 T. \qquad (19.112)$$

In state (B) it is

$$p_{B,1} V = m_1 R_1 T. \qquad (19.113)$$

Thus, it follows

$$p_{B,1} = p_A = 1\,bar. \qquad (19.114)$$

The partial pressure follows accordingly and is equal to the volume ratio

---

[25] In state (A) the total pressure is purely caused by air!

$$\boxed{\sigma_{B,1} = \pi_{B,1} = \frac{p_{B,1}}{p_B} = 0.25}$$

(19.115)

- Hydrogen (2)

  The total pressure is the summation of the partial pressures in state (B)

  $$p_{B,1} + p_{B,2} = p_B.$$

  (19.116)

  Hence, the partial pressure of the hydrogen is

  $$p_{B,2} = 3 \, \text{bar}.$$

  (19.117)

  The partial pressure follows accordingly and is equal to the volume ratio

  $$\boxed{\sigma_{B,2} = \pi_{B,2} = \frac{p_{B,2}}{p_B} = 0.75}$$

  (19.118)

(b) Gas constant and molar mass of the mixture

  The molar mass for hydrogen (2) is

  $$M_{H_2} = M_2 = 2 \frac{\text{kg}}{\text{kmol}}.$$

  (19.119)

  Its gas constant is

  $$R_{H_2} = R_2 = \frac{R_M}{M_2} = \frac{8.3143 \frac{\text{kJ}}{\text{kmol K}}}{2 \frac{\text{kg}}{\text{kJ}}} = 4.1571 \frac{\text{kJ}}{\text{kg K}}.$$

  (19.120)

  For air (1) with a composition of $\sigma_{1,O_2} = 0.21$, $\sigma_{1,N_2} = 0.79$ the molar mass follows

  $$M_1 = \sum_i x_i M_i = \sum_i \sigma M_i = (0.21 \cdot 32 + 0.79 \cdot 28) \frac{\text{kg}}{\text{kmol}} = 28.84 \frac{\text{kg}}{\text{kmol}}.$$

  (19.121)

  The gas constant for air (1) is

  $$R_1 = \frac{R_M}{M_1} = 0.28829 \frac{\text{kJ}}{\text{kg K}}.$$

  (19.122)

  Now, that we know the components R_i respectively M_i the mixture's properties can be determined. For the molar mass of the mixture M_B it follows

$$M_B = \sum_i x_i M_i = \sum_i \sigma_i M_i = \sum_i \pi_i M_i. \tag{19.123}$$

Thus, for state (B)

$$\boxed{M_B = \pi_{B,1} M_1 + \pi_{B,2} M_2 = 8.71 \frac{\text{kg}}{\text{kmol}}} \tag{19.124}$$

The gas constant for the mixture R_B reads as

$$\boxed{R_B = \frac{R_M}{M_B} = 954.56 \frac{\text{J}}{\text{kg K}}} \tag{19.125}$$

(c) Molar amount of the mixture

The thermal equation of state can be applied for state (B), i.e.

$$p_B V = m_B R_B T \tag{19.126}$$

respectively

$$p_B V = n_B M_B R_B T = n_B R_M T. \tag{19.127}$$

Thus, the molar quantity is

$$\boxed{n_B = \frac{p_B V}{R_M T} = 0.1641 \, \text{kmol}} \tag{19.128}$$

Problem 19.3 A mixture of 32 Mass-% ammonia ($R_{NH_3} = 0.4882 \frac{\text{kJ}}{\text{kg K}}$, $\kappa_{NH_3} = 1.31$) and 68 Mass-% nitrogen ($R_{N_2} = 0.2968 \frac{\text{kJ}}{\text{kg K}}$, $\kappa_{N_2} = 1.40$) with a temperature of 10 °C and a total mass of 6.3 kg is compressed reversibly and adiabatically from 0.98 bar to 2.94 bar.

(a) What is the molar mass of the mixture?
(b) Calculate the specific isochoric heat capacity c_v of the mixture!
(c) What is the change of enthalpy that is related with the compression?

General gas constant: $R_M = 8.3143 \frac{\text{kJ}}{\text{kmol K}}$

Solution

(a) The molar mass of the mixture follows

$$M = \frac{R_M}{R}. \tag{19.129}$$

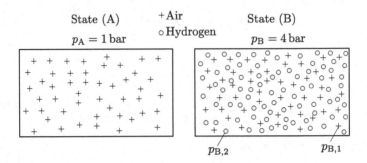

Fig. 19.8 Sketch to Problem 19.2

The gas constant of the mixture R can easily be calculated since the mass concentrations are given

$$R = \sum_i \xi_i R_i = (0.32 \cdot 0.4882 + 0.68 \cdot 0.2969)\, \frac{\text{kJ}}{\text{kg K}} = 0.358048\, \frac{\text{kJ}}{\text{kg K}}. \tag{19.130}$$

Thus, the molar mass is

$$\boxed{M = \frac{R_M}{R} = 23.22\, \frac{\text{kg}}{\text{kmol}}} \tag{19.131}$$

(b) The specific isochoric heat capacity c_v of the mixture can be calculated according to

$$c_v = \sum_i \xi_i c_{v,i}. \tag{19.132}$$

For the single components we know from part I, that

$$c_{v,i} = \frac{R_i}{\kappa_i - 1}. \tag{19.133}$$

Thus, for ammonia and nitrogen it is

$$c_{v,\text{NH}_3} = \frac{R_{\text{NH}_3}}{\kappa_{\text{NH}_3} - 1} = 1.5748\, \frac{\text{kJ}}{\text{kg K}} \tag{19.134}$$

respectively

$$c_{v,\text{N}_2} = \frac{R_{\text{N}_2}}{\kappa_{\text{N}_2} - 1} = 0.742\, \frac{\text{kJ}}{\text{kg K}}. \tag{19.135}$$

Hence, the specific isochoric heat capacity c_v of the mixture follows

$$c_v = \sum_i \xi_i c_{v,i} = \xi_{NH_3} c_{v,NH_3} + \xi_{N_2} c_{v,N_2} = 1.0085 \frac{kJ}{kg\,K} \qquad (19.136)$$

(c) In order to calculate the change of enthalpy, that is related with the compression, the temperature change needs to be determined first. Since the change of state is adiabatic and reversible, it is isentropic. For an isentropic change of state it is, see part I,

$$T_2 = T_1 \left(\frac{p_2}{p_1}\right)^{\frac{\kappa-1}{\kappa}}. \qquad (19.137)$$

Thus, the isentropic coefficient of the mixture κ needs to be calculated. For the mixture it is

$$c_p = R + c_v = 1.366548 \frac{kJ}{kg\,K}. \qquad (19.138)$$

Hence, the isentropic coefficient of the mixture follows

$$\kappa = \frac{c_p}{c_v} = 1.355\,03. \qquad (19.139)$$

The temperature after the compression is

$$T_2 = T_1 \left(\frac{p_2}{p_1}\right)^{\frac{\kappa-1}{\kappa}} = 377.59\,K. \qquad (19.140)$$

The temperature rise is

$$\Delta T = T_2 - T_1 = 94.445\,K. \qquad (19.141)$$

For the change of specific enthalpy the caloric equation of state[26] can be applied, i.e.

$$dh = c_p\,dT. \qquad (19.142)$$

Its integration brings[27]

$$\Delta h = c_p\,\Delta T = 129.06 \frac{kJ}{kg}. \qquad (19.143)$$

Multiplying with the total mass leads to the entire change of enthalpy

$$\boxed{\Delta H = m\,\Delta h = 813.1\,kJ} \qquad (19.144)$$

---

[26]Since the mixture is composed of ideal gases, the overall specific enthalpy purely is a function of temperature as well!

[27]Since no temperature dependencies are given!

Chapter 20
Humid Air

In the previous chapter it has been clarified how to handle mixtures of ideal gases. Definitions for characterising the mixture's composition have been introduced as well as the irreversibility of the mixing of different ideal gases. However, in this chapter the focus is on a technical relevant mixture of fluids: humid air. Its components at atmospheric conditions are as follows:

- Dry air, i.e. a mixture of N_2 and O_2 and with a small amount of CO_2. This type of mixture has been treated in Chap. 19. Mostly, these components are treated as ideal gases, so that the principles of Chap. 19 can easily been applied.
- Water within humid air can occur as liquid, i.e. fog or rain, as a solid, i.e. ice or snow, and as vapour. Chapter 18 has shown, that water as a real fluid can be subject to a phase change. Our everyday's experience shows, that this phase change happens at atmospheric conditions as it might rain or snow for instance. Hence, it needs to be clarified how the mixture can be described in case water condenses and forms a liquid.

Both, dry air and vapour, form a gaseous phase[1] and are commonly treated as ideal gases, see Fig. 20.1.[2] According to Dalton the total pressure of humid air is the summation of the partial pressures of dry air p_a and vapour p_v, i.e.

$$p = p_a + p_v. \tag{20.1}$$

Its deviation from non-ideal gas, i.e. real gas, behaviour is discussed in this chapter. Anyhow, this assumption leads to an error that is smaller than 3% for temperatures less than 100 °C. Usually, this error is acceptable for HVAC[3] systems.

---

[1] Usually, the partial pressure of vapour is significantly smaller than the total pressure, i.e. $p_v \ll p$.

[2] This figure shows air in a so-called unsaturated state.

[3] Heating, ventilation and air conditioning.

© Springer Nature Switzerland AG 2019
A. Schmidt, *Technical Thermodynamics for Engineers*,
https://doi.org/10.1007/978-3-030-20397-9_20

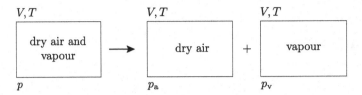

Fig. 20.1 Humid air—composition of the gaseous mixture of dry air and vapour

Furthermore, the water in non-vapour state[4] needs to be described thermodynamically as well as it occurs in addition to the gaseous mixture of dry air and vapour. Consequently, in this chapter it is investigated under what boundary conditions a phase change happens and how it influences the composition of the gaseous mixture of dry air and vapour.

Finally, it is discussed how and by what technical auxiliaries air can be conditioned in terms of temperature and humidity and what energy supply comes along with the conditioning.

20.1 Thermodynamic State

20.1.1 Concentration

The water content of air[5] x is defined as ratio of the mass of the water within the air to the mass of the *dry* air, i.e.

$$x = \frac{m_w}{m_a}. \tag{20.2}$$

Mind, not to mix it up with the molar concentration x, that has been introduced in Chap. 19, or the vapour ratio x, that has been defined in Chap. 18!

As the mass of the water m_w within the air can exist as vapour (v), liquid (liq) or as ice (i) it is

$$m_w = m_v + m_{liq} + m_{ice}. \tag{20.3}$$

Thus, the water content follows

$$\boxed{x = \frac{m_w}{m_a} = \frac{m_v}{m_a} + \frac{m_{liq}}{m_a} + \frac{m_{ice}}{m_a} = x_v + x_{liq} + x_{ice}} \tag{20.4}$$

---

[4]Rain or snow.

[5]Sometimes also named moisture grade.

Fig. 20.2 p, T-diagram of water

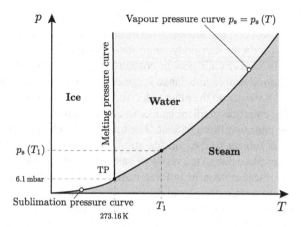

Thermodynamic states, in which water occurs as vapour, liquid and solid at the same time, are possible and will be discussed later.

The following examples show, that the water content x can take any positive number:

- Dry air

$$x = \frac{m_w}{m_a} = \frac{0}{m_a} = 0 \tag{20.5}$$

- Pure liquid water

$$x = \frac{m_w}{m_a} = \frac{m_{liq}}{0} \to \infty \tag{20.6}$$

20.1.2 Aggregate State of the Water

In order to decide on the state of aggregation of water in the air, both its pressure and temperature are required. Chapter 18 has shown, that these two state values fix the aggregate state. Hence, the water's state can then be visualised in a p, T-diagram[6] for instance, see Fig. 20.2. This diagram shows, that depending on pressure and temperature, water can occur as vapour respectively steam, as liquid or as solid. The vapour pressure curve $p_s = p_s(T)$ marks the border line between vapour/steam and liquid and is important to characterise the humid air. Obviously, for small pressures, the vaporisation already starts at low temperatures. However, the wet steam region, see Sect. 18.2, can not be seen in this p, T-projection as it is covered by the vapour

---

[6]Mind, that the p, T-diagram only takes into account the water, the air is not part of this diagram.

pressure line, see Sect. 18.2.2. Within this two-phase region, pressure is a function of temperature, i.e. another state information is required, e.g. the vapour ratio, in order to specify the thermodynamic state unambiguously.

In case dry air gets in contact with a reservoir of water, e.g. in liquid state, a thermodynamic imbalance is present: a large concentration of water on the one side and no water on the other side. According to the second law of thermodynamics that concentration gradient causes to an evaporisation[7] of the water in order to balance the concentration gradient. The transport mechanism is based on diffusion and has been subject to the considerations in the previous chapter. Vaporisation of water takes place due to the very low partial pressure of the water within the air right at the interface between air and water: At this interface water molecules can easy escape the liquid surface, the partial pressure of water in the surrounding air is small. According to Fig. 20.2, water thus vaporise at low temperatures. However, the required energy for the evaporisation reduces the temperature.[8] The energy balance, however, will be discussed later in this chapter. The more water vaporises, the larger the water's partial pressure and the lower the air's partial pressure are, i.e. the vapour concentration is rising. For the pressure Dalton's law can be applied, i.e.

$$p = p_a + p_v. \tag{20.7}$$

Let us assume, that air and water both have the same temperature T_1. Following the vapour pressure curve $p_s = p_s(T_s)$ in Fig. 20.2 the maximum vapour pressure is limited, i.e. the evaporisation at temperature T_1 stops as soon as the vapour's partial pressure reaches $p_s(T_1)$. Bringing any further water into the humid air means, that this water can not be carried as vapour, i.e. condensation leads to liquid water or even to ice.

The following definition is made for air with an exemplary temperature T_1:

- Unsaturated humid air:
  In case

$$p_v < p_s(T_1) \tag{20.8}$$

the air is still capable to carry a larger load of vapour. Thus, it is unsaturated.
- Saturated humid air:
  In case

$$p_v = p_s(T_1) \tag{20.9}$$

---

[7]It needs to be distinguished between vaporisation and evaporisation. Vaporisation means the phase change of a *single* fluid, e.g. water, that was handled in Chap. 18. Evaporisation, on the other hand, describes the vaporisation of one fluid, e.g water, *and* its diffusive transport into a second gas, e.g. air!

[8]Also known as evaporative cooling!

the maximal load of vapour is reached. Any further supply of water can not be carried as vapour but as liquid or as solid. However, even when further water is supplied, the gas phase still contains the maximal load of vapour belonging to $p_s(T_1)$.

20.1.3 Unsaturated Versus Saturated Air

Vapour Content and Partial Pressure

In order to derive a correlation between partial pressure and vapour content, the assumption is made, that both gases, dry air and vapour, can be treated as ideal. Applying the thermal equation of state for ideal gases for dry air

$$p_a V = m_a R_a T \rightarrow m_a = \frac{p_a V}{R_a} \qquad (20.10)$$

and vapour accordingly

$$p_v V = m_v R_v T \rightarrow m_v = \frac{p_v V}{R_v}. \qquad (20.11)$$

The vapour content of the air is by definition, see Sect. 20.1.1,

$$x_v = \frac{m_v}{m_a}. \qquad (20.12)$$

Combining Eqs. 20.10 and 20.11 with Eq. 20.12 results in

$$x_v = \frac{p_v R_a}{p_a R_v}. \qquad (20.13)$$

The partial pressure of air p_a follows from Dalton's law, see Eq. 20.7, and can be written as

$$p_a = p - p_v. \qquad (20.14)$$

Inserting Eq. 20.13 in Eq. 20.14 leads to

$$x_v = \frac{p_v R_a}{(p - p_v) R_v}. \qquad (20.15)$$

With the individual gas constants for dry air.[9]

---

[9] The molar masses for the air's components are $M_{O_2} = 31.998 \frac{kg}{kmol}$, $M_{N_2} = 28.0134 \frac{kg}{kmol}$, $M_{Ar} = 39.948 \frac{kg}{kmol}$ and $M_{CO_2} = 44.0087 \frac{kg}{kmol}$. The molar fractions follow according to Chap. 19.

$$R_a = \frac{R_M}{M_a} = 287.0409 \, \frac{J}{kg\,K} \tag{20.16}$$

with

$$M_a = x_{O_2} M_{O_2} + x_{N_2} M_{N_2} + x_{Ar} M_{Ar} + x_{CO_2} M_{CO_2} = 28.9656 \, \frac{kg}{kmol} \tag{20.17}$$

and vapour[10]

$$R_v = \frac{R_M}{M_W} = 461.5158 \, \frac{J}{kg\,K} \tag{20.18}$$

results in

$$\boxed{x_v = 0.622 \frac{p_v}{p - p_v}} \tag{20.19}$$

Respectively, re-arranged for the partial pressure of the vapour as function of the water content

$$\boxed{p_v = \frac{xp}{0.622 + x_v}} \tag{20.20}$$

Maximum Vapour Load of Humid Air

In case the humid air is saturated the vapour's partial pressure is equal to the saturation vapour pressure, see Fig. 20.2. All variables existing in saturated state are marked with a stroke (') from now on. Thus, it is

$$p_v' = p_s(T). \tag{20.21}$$

Hence, following Eq. 20.19 the maximum vapour content of humid air is

$$\boxed{x_v' = 0.622 \frac{p_v'}{p - p_v'}} \tag{20.22}$$

According to this equation, the maximum vapour content is a function of the total pressure p and the temperature T of the humid air, i.e. the larger the temperature is, the larger the maximum vapour load is.[11]

Vice versa, one gets the partial pressure in the saturated state from the maximum vapour content x_v', i.e.

---

[10]With the molar mass for water $M_W = 18.0152 \, \frac{kg}{kmol}$.

[11]This is due to the rising saturated vapour pressure with rising temperature, see Fig. 20.2.

$$p_v' = \frac{x_v' \cdot p}{0.622 + x_v'} \qquad (20.23)$$

Relative Saturation/Humidity

In case humid air is unsaturated it contains less vapour than potentially possible at the current temperature T. The ratio of the current vapour content to its maximum content at its temperature T is called degree of saturation or relative saturation ψ: i.e.

$$\psi = \frac{x_v}{x_v'} \qquad (20.24)$$

Another important measure for humid air is the relative humidity φ, that is defined as

$$\varphi = \frac{p_v}{p_v'}. \qquad (20.25)$$

Applying Eq. 20.20 finally results in

$$\varphi = \frac{p_v}{p_v'} = \frac{x_v}{x_v + 0.622} \cdot \frac{p}{p_v'} \qquad (20.26)$$

The correlation between degree of saturation ψ and the relative humidity φ follows by applying Eqs. 20.19 and 20.22, i.e.

$$\psi = \frac{x_v}{x_v'} = \frac{p_v}{p - p_v} \cdot \frac{p - p_v'}{p_v'} = \frac{p_v}{p_v'} \cdot \frac{p - p_v'}{p - p_v}. \qquad (20.27)$$

Thus, it finally is

$$\frac{\psi}{\varphi} = \frac{p - p_v'}{p - p_v} \qquad (20.28)$$

Absolute Humidity

The absolute humidity ρ_v is defined as the ratio of vapour's mass m_v to the total volume of the humid air V, that is shared by vapour and dry air, i.e.

$$\rho_v = \frac{m_v}{V}. \qquad (20.29)$$

By applying the thermal equation of state it results in

$$\boxed{\rho_v = \frac{p_v V}{R_v T} \cdot \frac{1}{V} = \frac{p_v}{R_v T}} \tag{20.30}$$

One gets the maximum absolute humidity ρ'_v, if the air is saturated, so that $p'_v = p_s(T)$

$$\rho'_v = \frac{p'_v}{R_v T}. \tag{20.31}$$

Hence, the maximum mass of the vapour within the saturated air follows

$$m'_v = \frac{p'_v}{R_v T} \cdot V. \tag{20.32}$$

Example 20.1 Starting from initial state (1), characterised by $p_{a,1} = 983$ mbar and $\vartheta_1 = 20\,°C$, humid air is cooled down at constant total pressure, i.e. $p = 1$ bar, and constant water content. What will happen?

According to Dalton, the vapour's partial pressure follows

$$p_{v,1} = p - p_{a,1} = 17\,\text{mbar}. \tag{20.33}$$

The maximum partial pressure of the vapour at $\vartheta_1 = 20\,°C$ can be taken from the steam table and is

$$p'_{v,1} = p_s(\vartheta_1) = 23.392\,\text{mbar}. \tag{20.34}$$

Hence, the humid air in state (1) is unsaturated, as it lies below the vapour pressure curve in the p, T-diagram and

$$p_{v,1} < p'_{v,1}. \tag{20.35}$$

However, when cooling down at constant total pressure, the air's capability of carrying vapour is reduced, see shape of the $p_s = p_s(T)$-function. First, the partial pressures of air as well as of vapour are constant, since air and vapour content do not vary. Consequently, the change of state follows a horizontal line in Fig. 20.3, because temperature reduces and partial pressure of the vapour[12] remains constant

$$p_v = p_{v,1}. \tag{20.36}$$

---

[12]Only the vapour's pressure is part of the p, T-diagram in Fig. 20.3, though the y-axis is named p. p does not represent the total pressure in this figure!

Fig. 20.3 Humid
air—Example 1/2

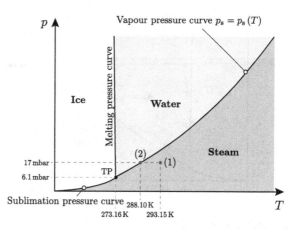

This works out until state (2) is reached, that still has the same vapour's partial
pressure as before, but which is in saturated state, i.e.

$$p_{v,2} = p_{v,1} = 17\,\text{mbar} = p_s\,(T_2)\,. \tag{20.37}$$

According to temperature T_2, the air now carries the maximum load of vapour. Following the steam table, Eq. 20.37 can be solved and one gets $\vartheta_2 = 14.95\,°\text{C}$.

Starting from state (2) isobaric[13] cooling continues. What will happen?

The change of state is illustrated in Fig. 18.11. State (1) used to be unsaturated,
whereas state (2) is saturated. While the temperature sinks during cooling down,
the capability of carrying vapour reduces as well. Consequently, the further the
temperature sinks, the more water starts to condense. In doing so, the gas phase contains a smaller amount of vapour as before, though the mass of the dry air remains
unchanged. Hence, the partial pressure of the vapour reduces, while the partial pressure of dry air rises.[14] Since, there is a rising reservoir of liquid water respectively
ice available when cooling down, the humid air always is in saturated state,[15] i.e.

$$p_{v,3} = p'_{v,3} = p_s\,(\vartheta_3) \tag{20.38}$$

respectively, when water starts to freeze

$$p_{v,4} = p'_{v,4} = p_s\,(\vartheta_4)\,. \tag{20.39}$$

---

[13] Thus, the total pressure remains constant!

[14] Otherwise, it would not be possible to keep the total pressure constant.

[15] The gas phase is in saturated state, with an extra amount of liquid water or ice, see Fig. 20.4.

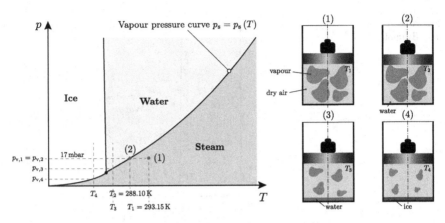

Fig. 20.4 Humid air—Example 2/2

20.2   Specific State Values

It is well know, that extensive state values Z can be referred to the mass of the system m, in order to gain mass-specific state values z, i.e.

$$z = \frac{Z}{m}.$$
(20.40)

In part I simplified fluids, such as ideal gas or incompressible liquids have been treated, so that m is the associated mass. However, if specific state values are calculated for humid air, another approach is followed. By definition, it is not the total mass of the gaseous mixture, but the *mass of the dry air* m_a, that is used for calculating mass-specific state values. This is denoted by the index[16] $1 + x$, i.e.

$$\boxed{z_{1+x} = \frac{Z}{m_a}}$$
(20.41)

This leads to the

- specific volume v_{1+x}
- specific density ρ_{1+x}
- specific internal energy u_{1+x}
- specific enthalpy h_{1+x}
- specific entropy s_{1+x}

for humid air.

---

[16]Mind, that the Z is the *total* extensive state value, i.e. of dry air and vapour!

20.2.1 Thermal State Values

Specific Volume of Unsaturated Humid Air

The specific volume of humid air is

$$v_{1+x} = \frac{V}{m_a}. \tag{20.42}$$

The partial volume of air V_a and of vapour V_v form the total volume V of the gaseous phase of unsaturated humid air. Thus, it follows

$$v_{1+x} = \frac{V_a}{m_a} + \frac{V_v}{m_a}. \tag{20.43}$$

By applying of the ideal gas law, i.e. $pV = mRT$, the following expressions can be substituted in Eq. 20.43

$$\frac{V_a}{m_a} = \frac{R_a T}{p} \tag{20.44}$$

and

$$V_v = m_v \frac{R_v T}{p}. \tag{20.45}$$

Hence, the specific volume of unsaturated humid air results in

$$v_{1+x} = \frac{R_a T}{p} + x_v \cdot \frac{R_v T}{p} = \frac{R_a T}{p} \left[1 + \frac{R_v}{R_a} \cdot x_v \right]. \tag{20.46}$$

The gas constants for dry air R_a and vapour R_v are constant and have been introduced in Sect. 20.1.3. Thus, Eq. 20.46 simplifies to

$$\boxed{v_{1+x} = \frac{R_a T}{p} [1 + 1.608 \cdot x_v]} \tag{20.47}$$

Following the definition of part I, i.e. with the total mass of the system, the specific volume is

$$v = \frac{V}{m}. \tag{20.48}$$

Re-arranging leads to

$$v = \frac{V}{m_a + m_v} = \frac{V}{m_a + m_v} \cdot \frac{m_a}{m_a} = \frac{V}{m_a} \cdot \frac{m_a}{m_a + m_v} = v_{1+x} \cdot \frac{1}{1 + x_v}. \tag{20.49}$$

Thus, the correlation between v and v_{1+x} is

$$v_{1+x} = (1 + x_v) \cdot v \tag{20.50}$$

Density of Unsaturated Humid Air

The density of humid air reads as

$$\rho_{1+x} = \frac{1}{v_{1+x}}. \tag{20.51}$$

Thus, it is

$$\rho_{1+x} = \frac{p}{R_a T} \cdot \left[\frac{1}{1 + 1.608 \cdot x_v} \right] \tag{20.52}$$

From

$$v_{1+x} = (1 + x_v) \cdot v \tag{20.53}$$

the correlation between ρ_{1+x} and ρ thus follows[17]:

$$\frac{1}{\rho_{1+x}} = \frac{(1 + x_v)}{\rho} \tag{20.54}$$

with

$$\rho = \frac{m_a + m_v}{V}. \tag{20.55}$$

Example 20.2 Why does humid air rise in an atmosphere of dry air?
The density of humid air is, see Eq. 20.54,

$$\rho = (1 + x_v) \cdot \rho_{1+x} = \frac{p}{R_a T} \cdot \left[\frac{1 + x_v}{1 + 1.608 \cdot x_v} \right]. \tag{20.56}$$

Hence, two cases can be investigated, i.e.

- density of dry air ($x_v = 0$):

$$\rho\,(x_v = 0) = \frac{p}{R_a T} \tag{20.57}$$

- density of humid air ($x_v > 0$):

$$\rho\,(x_v > 0) = \underbrace{\frac{p}{R_a T}}_{\rho(x_v=0)} \cdot \underbrace{\left[\frac{1 + x_v}{1 + 1.608 \cdot x_v} \right]}_{<1} \tag{20.58}$$

---

[17]Mind, that ρ considers the total mass!

Obviously, the density of humid air is smaller than the density of dry air

$$\rho(x_v > 0) < \rho(x_v = 0).$$ (20.59)

Thus humid air rises in a dry atmosphere!

Problem 20.3 In a closed room a temperature of $\vartheta = 18\,^\circ\text{C}$ and a relative humidity of $\varphi = 80\%$ are measured. The total pressure is approximately standard pressure, i.e. $p = 1.01325$ bar. The gas constant of dry air is $R_a = 0.28704\ \frac{\text{kJ}}{\text{kg K}}$.

(a) Calculate the mass m of $V = 1\,\text{m}^3$ of the air!
(b) What is the maximum mass of vapour m'_v the air with a volume of $V = 1\,\text{m}^3$ and a temperature of $\vartheta = 18\,^\circ\text{C}$ can carry without forming mist?

Solution

(a) Total mass of the humid air m

- Approach 1:
  Since the relative humidity is given, the vapour's partial pressure p_v can be calculated. According to the steam table it is

$$p_s(18\,^\circ\text{C}) = 20.6466\,\text{mbar}.$$ (20.60)

Thus, the vapour's partial pressure is

$$p_v = \varphi \cdot p_s(18\,^\circ\text{C}) = 16.5173\,\text{mbar}.$$ (20.61)

The vapour content x_v follows

$$x_v = 0.622 \frac{p_v}{p - p_v} = 0.010307 = 10.307\,\text{kg}_{\text{vapour}}/\text{kg}_{\text{dry air}}.$$ (20.62)

This leads to

$$\rho_{1+x} = \frac{p}{R_a T} \cdot \left[\frac{1}{1 + 1.608 \cdot x_v}\right] = 1.1927\ \frac{\text{kg}}{\text{m}^3}$$ (20.63)

and hence to

$$\rho = \rho_{1+x}(1 + x_v) = 1.2050\ \frac{\text{kg}}{\text{m}^3}.$$ (20.64)

Applying the thermal equation of state brings

$$\rho = \frac{m}{V},$$ (20.65)

so that the mass m finally follows

$$\boxed{m = \rho V = 1.205\,\text{kg}} \tag{20.66}$$

- Approach 2:
  The air's partial pressure follows from Dalton's law

$$p_a = p - p_v = 996.7372\,\text{mbar}. \tag{20.67}$$

Applying the thermal equation of state just for the dry air leads to

$$m_a = \frac{p_a V}{R_a T} = 1.1927\,\text{kg}. \tag{20.68}$$

From approach 1 we know, that

$$x_v = 0.010307, \tag{20.69}$$

so it follows

$$m_v = x_v m_a = 0.0123\,\text{kg}. \tag{20.70}$$

Finally, the entire mass is

$$\boxed{m = m_a + m_v = 1.205\,\text{kg}} \tag{20.71}$$

(b) Saturated state
Vapour shall be supplied to the closed system. Since the mass of the dry air is constant, the total pressure rises. For the partial pressure of the air it follows

$$p_a = \frac{m_a R_a T}{V} = \text{const.} = 996.7372\,\text{mbar}. \tag{20.72}$$

The vapour's partial pressure equals the saturated vapour pressure at $18\,°C$, i.e.

$$p_v' = p_s\,(18\,°C) = 20.6466\,\text{mbar}. \tag{20.73}$$

The maximum vapour load in saturated state is

$$x_v' = 0.622 \frac{p_v'}{p - p_v'} = 0.622 \frac{p_v'}{p_a} = 0.0128843. \tag{20.74}$$

Finally, the vapour's mass in saturated state follows

$$\boxed{m_v' = x_v' m_a = 0.0154\,\text{kg}} \tag{20.75}$$

The total pressure under these conditions is

$$p = p_a + p_s = 1017.4 \, \text{mbar}. \tag{20.76}$$

20.2.2 Caloric State Values

As will be seen in this chapter, the specific enthalpy plays an important role in technical applications. Hence, the relevance of this specific state value is motivated:

- In case, the system to be investigated is <u>open</u>, the first law of thermodynamics in differential notation reads as

$$\boxed{\delta q + \delta w_t = dh + de_a} \tag{20.77}$$

Obviously, the enthalpy is required to calculated open systems.
- Now, let us focus on <u>closed</u> systems. Its first law of thermodynamics in differential notation obeys

$$\delta q + \delta w = du + de_a. \tag{20.78}$$

With the partial energy equation

$$\delta w = \delta w_t + \delta w_{\text{mech}} + \delta \psi = -p \, dv + de_a + \delta \psi \tag{20.79}$$

the first law of thermodynamics simplifies to

$$\delta q - p \, dv + \delta \psi = du. \tag{20.80}$$

Introducing the entropy

$$dh = du + p \, dv + v \, dp \tag{20.81}$$

results in

$$\boxed{\delta q + \delta \psi = dh - v \, dp} \tag{20.82}$$

In the field of air conditioning most of the changes of state run at *constant pressure*, i.e. $dp = 0$. Thus, it finally is

$$\delta q + \delta \psi = dh. \tag{20.83}$$

Thus, even for closed systems, the specific enthalpy plays an important role. This is the reason, why in this chapter the focus is on the enthalpy, e.g. with the Mollier-diagram, see Sect. 20.3. However, if the specific energy u is required, it can easily be calculated by

$$u = h - pv. \tag{20.84}$$

20.2.3   Specific Enthalpy h_{1+x}

Since humid air is a multi-component mixture, i.e. it consists of water (vapour, liquid and ice) and dry air, the total extensive enthalpy is composed as follows, for reference see Sect. 19.3.2,

$$H = H_a + H_v + H_{liq} + H_{ice} = m_a \cdot h_a + m_v \cdot h_v + m_{liq} \cdot h_{liq} + m_{ice} \cdot h_{ice}. \tag{20.85}$$

As discussed before, the specific enthalpy h_{1+x} is referred to the mass of dry air m_a, so that it is

$$h_{1+x} = \frac{H}{m_a} = h_a + \frac{m_v}{m_a} \cdot h_v + \frac{m_{liq}}{m_a} \cdot h_{liq} + \frac{m_{ice}}{m_a} \cdot h_{ice}. \tag{20.86}$$

Substituting the water content x leads to

$$\boxed{h_{1+x} = h_a + x_v \cdot h_v + x_{liq} \cdot h_{liq} + x_{ice} \cdot h_{ice}} \tag{20.87}$$

h_a is the specific enthalpy of dry air and h_w the specific enthalpy of water.[18] Both are introduced in the following sections.

Specific Enthalpy of Dry Air h_a

Dry air can be treated as an ideal gas. Thus, the specific enthalpy of dry air h_a within the multi-component system follows

$$dh_a = c_{p,a} \cdot dT. \tag{20.88}$$

The integration of the caloric equation of state brings[19]

$$h_a = c_{p,a} \cdot (T - T_{0,a}) + h_{0,a}. \tag{20.89}$$

Thus, a reference level $h_0\left(T_{0,a}\right)$ is required and chosen[20] as

$$\boxed{h_{0,a}\left(T_{0,a} = 273.15\ \text{K}\right) = 0} \tag{20.90}$$

This reference level is beneficial, since it simplifies to calculate with the °C-scale, thus its handling is pretty simple, i.e.

---

[18] So far, no distinction has been made in what aggregate state the water exists, i.e. it can be solid, liquid or vapour.

[19] The specific heat capacity of air $c_{p,a}$ is regarded as constant.

[20] This choice is arbitrary but practicable!

$$h_a = c_{p,a} \cdot (T - 273.15\,\text{K}) = c_{p,a} \cdot \vartheta \tag{20.91}$$

The specific isobaric heat capacity for dry air is

$$c_{p,a} = 1.004\,\frac{\text{kJ}}{\text{kg K}}. \tag{20.92}$$

Specific Enthalpy of Water h_w

As known from Sect. 12.3 the caloric state value specific enthalpy of a real fluid is a function of two independent state values, e.g. pressure p and temperature T, so it follows[21]

$$dh_w = dh_w(p, T). \tag{20.93}$$

Hence, analogy with the specific enthalpy of air, a reference level for water is required. This reference level does not need to be the same as it was for the component air. Since for both components, their specific mass balance must be fulfilled and the first law of thermodynamics counts for the *change of state*, different reference levels for water as well as for air do not lead to a dilemma.[22] Thus, besides a reference temperature $\vartheta_{0,w}$ a reference pressure $p_{0,w}$ is required. Though the choice is arbitrary, the definition for a reference level for water within the multi-component system, follows the reference for the steam tables, see Sect. 18.3.7, i.e. it is the saturated liquid state (′) at the triple point, i.e.

$$h_{0,w}(T_{0,w} = 273.16\,\text{K},\ p_{0,w} = 0.006112\,\text{bar}) = 0.00061178\,\frac{\text{kJ}}{\text{kg}} \approx 0 \tag{20.94}$$

According to the steam table, the specific enthalpy at reference level is $0.00061178\,\frac{\text{kJ}}{\text{kg}}$, as it was the *specific internal energy* at reference level that was set to zero, i.e. $u_{0,w}(\vartheta_{0,w} = 0.01\,°\text{C},\ p_{0,w} = 0.006112\,\text{bar}) = 0$, see Sect. 18.3.7. However, the specific reference enthalpy is pragmatically set to zero,[23] see Eq. 20.94.

As already discussed, the water in the mixture humid air can occur in different aggregate states. In order to find out the water's thermodynamic state values, the steam table of water could be applied, but a more handy approach is usually preferred. This approach is discussed by making a distinction between the following possible cases:

---

[21] Under the premise, the fluid is an ideal gas, e.g. the vapour, this equation simplifies to $dh = dh(T)$, refer to Sect. 12.1.

[22] Mind, that the reference levels for water and for air must not be varied during a calculation!

[23] Actually, this simplification is not a mistake, but a comparison later on with the steam table, see Tables A.1 and A.9, it leads to minor deviations. The choice of a reference level is always arbitrary.

- Vapour

Most commonly, the vapour that occurs in humid air is considered to be an ideal gas. In order to determine its specific enthalpy the approach is divided into two steps

1. In the reference state, i.e. saturated liquid at the triple point (TP), the specific enthalpy of water is set to be zero, see Eq. 20.94. First, energy is supplied isobarically to vaporise the water.[24] This is done until saturated vapour at the triple point is reached. According to the steam table the heat of vaporisation

$$\Delta h_v(0.01\,°C) \approx 2500\,\frac{kJ}{kg} \tag{20.95}$$

is required. Thus, after step 1 vapour appears.
2. From now on, the steam table approach is no longer applied, but the vapour is treated as an *ideal gas*. Mind, that for ideal gases the specific enthalpy is purely a function of temperature, thus there is no further pressure dependency. In this second step, the heating respectively cooling of the vapour to adjust the temperature of the humid air is done[25] from $\vartheta_{0,w}$ to ϑ In this step the *ideal gas vapour* underlies the change of enthalpy of

$$\Delta h = c_{p,v}\,\Delta T = c_{p,v} \cdot (\vartheta - 0.01\,°C) \approx c_{p,v} \cdot \vartheta. \tag{20.96}$$

The specific isobaric heat capacity of vapour is

$$c_{p,v} = 1.86\,\frac{kJ}{kg\,K}. \tag{20.97}$$

Thus, the total specific enthalpy of vapour is the combination of steps 1 and 2:

$$h_v = \underbrace{\Delta h_v(0.01\,°C)}_{\text{Step 1}} + \underbrace{c_{p,v} \cdot \vartheta}_{\text{Step 2}} = 2500\,\frac{kJ}{kg} + 1.86\,\frac{kJ}{kg\,K} \cdot \vartheta \tag{20.98}$$

Example 20.4 The total pressure of humid air is $p = 1$ bar at a temperature of $\vartheta = 20\,°C$. The partial pressure of dry air is $p_a = 983$ mbar. Thus, the partial pressure of vapour is $p_v = 17$ mbar. What is the specific enthalpy of the vapour?

- Approach 1 (Steam table—real fluid):

$$h_v = h(20\,°C,\ 17\,mbar) = 2537.8\,\frac{kJ}{kg} \tag{20.99}$$

---

[24]The change of state (′) to (″) shall be isobaric and thus isothermal as well!
[25]In most cases the humid air does not have a temperature of 0.01 °C!

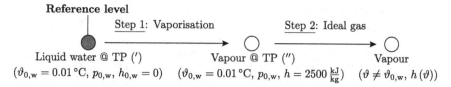

Reference level

Step 1: Vaporisation

Step 2: Ideal gas

Liquid water @ TP (′)
$(\vartheta_{0,w} = 0.01\,°C,\ p_{0,w},\ h_{0,w} = 0)$

Vapour @ TP (″)
$(\vartheta_{0,w} = 0.01\,°C,\ p_{0,w},\ h = 2500\,\frac{kJ}{kg})$

Vapour
$(\vartheta \neq \vartheta_{0,w},\ h\,(\vartheta))$

Fig. 20.5 How to treat the water in humid air—vapour

Fig. 20.6 How to treat the water in humid air—liquid

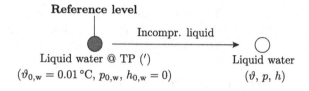

Reference level

Incompr. liquid

Liquid water @ TP (′)
$(\vartheta_{0,w} = 0.01\,°C,\ p_{0,w},\ h_{0,w} = 0)$

Liquid water
$(\vartheta,\ p,\ h)$

– Approach 2 (Ideal gas, according to Fig. 20.5):

$$h_v = 2500\,\frac{kJ}{kg} + 1.86 \cdot \vartheta = 2537.2\,\frac{kJ}{kg} \qquad (20.100)$$

That is, the error when using the ideal gas approach vs. real fluid approach is significant smaller than 0.1%!

- Liquid water
  Starting from reference point, see Fig. 20.6, that already is in liquid state, liquid water is now treated as an incompressible liquid. According to part I the change of enthalpy follows

$$h_{liq} = h_{0,w} + c_{liq}\underbrace{(\vartheta - 0.01\,°C)}_{\approx \vartheta} + v\,(p - p_{0,w}) \approx c_{liq}\vartheta + v\,(p - p_{0,w}).$$

$$(20.101)$$

Mind, that the pressure within the liquid p is equal to the pressure of the gaseous phase in thermodynamic equilibrium!.However, the term $v\,(p - p_{0,w})$ is usually ignored, so that the enthalpy of the liquid water follows

$$\boxed{h_{liq} = c_{liq}\vartheta} \qquad (20.102)$$

The specific heat capacity[26] of liquid water is

$$c_{liq} = 4.19\,\frac{kJ}{kg\ K}. \qquad (20.103)$$

---

[26]Mind, that for incompressible liquids there is no difference between c_p and c_v!

Reference level

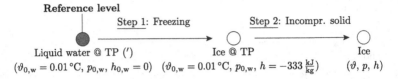

Fig. 20.7 How to treat the water in humid air—ice

Example 20.5 The total pressure of humid air is $p = 1$ bar at a temperature of $\vartheta = 20\,°\mathrm{C}$. The humid air is saturated and liquid water occurs. What is the specific enthalpy of the liquid water[27]?

– Approach 1 (according to Eq. 20.101):

$$
\begin{aligned}
h_{\mathrm{liq}} &= c_{\mathrm{liq}}\vartheta + v\left(p - p_{0,\mathrm{w}}\right) \\
&= 4.19\,\frac{\mathrm{kJ}}{\mathrm{kg\,K}} \cdot 20\,\mathrm{K} + 0.0010\,\frac{\mathrm{m}^3}{\mathrm{kg}}\,(1\,\mathrm{bar} - 0.006112\,\mathrm{bar}) \\
&= 83.8994\,\frac{\mathrm{kJ}}{\mathrm{kg}}
\end{aligned}
\tag{20.104}
$$

– Approach 2 (according to Eq. 20.102):

$$
h_{\mathrm{liq}} = c_{\mathrm{liq}}\vartheta = 83.8\,\frac{\mathrm{kJ}}{\mathrm{kg}}
\tag{20.105}
$$

Thus, the influence of the term $v\left(p - p_{0,\mathrm{w}}\right)$ is rather small, the deviation between the two approaches in this case is less than 0.12%.

• Solid water
  Starting from reference point, see Fig. 20.7, that is in liquid state, two successive steps are required:

1. In the reference state, i.e. saturated liquid at the triple point (TP), the specific enthalpy of water is set to be zero, see Eq. 20.94. First, energy is released isobarically to freeze the water.[28] This is done until the liquid is fully turned into ice at the triple point. According to [1] the specific heat of melting is

$$
\Delta h_{\mathrm{m}}(0.01\,°\mathrm{C}) \approx -333\,\frac{\mathrm{kJ}}{\mathrm{kg}}.
\tag{20.106}
$$

Thus, after step 1 ice at reference temperature is available.

---

[27]Mind, that the liquid water has the same pressure as the gaseous phase in thermodynamic equilibrium!
[28]During this phase change the pressure and the temperature remain constant!

2. From now on, the ice is treated as an incompressible solid. Thus, the change of enthalpy for the ice in step 2, until temperature ϑ is reached, follows

$$\Delta h = c_{ice} \underbrace{(\vartheta - 0.01\,°C)}_{\approx \vartheta} + v\,(p - p_{0,w}) \approx c_{ice}\vartheta + v\,(p - p_{0,w}). \quad (20.107)$$

The specific heat capacity[29] of ice is

$$c_{ice} = 2.05\,\frac{kJ}{kg\,K}. \quad (20.108)$$

Mind, that the pressure within the solid p is equal to the pressure of the gaseous phase in thermodynamic equilibrium. However, the term $v\,(p - p_{0,w})$ is usually ignored, so that the enthalpy of the solid follows

$$\Delta h = c_{ice}\vartheta. \quad (20.109)$$

Thus, the total specific enthalpy of a solid, i.e. ice, is the combination of steps 1 and 2, i.e.

$$h_{ice} = \underbrace{\Delta h_m(0.01\,°C)}_{Step\ 1} + \underbrace{c_{ice} \cdot \vartheta}_{Step\ 2} = -333\,\frac{kJ}{kg} + 2.05\,\frac{kJ}{kg\,K} \cdot \vartheta \quad (20.110)$$

20.2.4 Specific Entropy s_{1+x}

Since humid air is a multi-component mixture, i.e. it consists of water (vapour, liquid and ice) and dry air, the total extensive entropy is composed as follows, for reference see Sect. 19.3.2:

$$S = S_a + S_v + S_{liq} + S_{ice} = m_a \cdot s_a + m_v \cdot s_v + m_{liq} \cdot s_{liq} + m_{ice} \cdot s_{ice} \quad (20.111)$$

As discussed before, the specific entropy s_{1+x} is referred to the mass of dry air m_a, so that it is

$$s_{1+x} = \frac{S}{m_a} = s_a + \frac{m_v}{m_a} \cdot s_v + \frac{m_{liq}}{m_a} \cdot s_{liq} + \frac{m_{ice}}{m_a} \cdot s_{ice}. \quad (20.112)$$

Substituting the water content x leads to

$$s_{1+x} = s_a + x_v \cdot s_v + x_{liq} \cdot s_{liq} + x_{ice} \cdot s_{ice} \quad (20.113)$$

---

[29]Mind, that for incompressible solids there is no difference between c_p and c_v!

s_a is the specific enthalpy of dry air and s_w the specific enthalpy of water. Both are introduced in the following sections.

Specific Entropy of Dry Air s_a

Dry air can be treated as an ideal gas. However, similar as for the specific enthalpy, see Sect. 20.2.3, a reference level needs to be defined. Consequently, it is the same $T_{0,a}$ and $p_{0,a}$ as before, i.e.

$$\boxed{s_{0,a}\left(T_{0,a} = 273.15\,\text{K},\ p_{0,a} = 1\,\text{bar}\right) = 0} \qquad (20.114)$$

With this reference level the specific entropy for air is, see part I,

$$s_a - s_{0,a} = c_{p,a}\ln\frac{T}{T_{0,a}} - R_a\ln.\frac{p_a}{p_{0,a}} \qquad (20.115)$$

Finally, it is

$$\boxed{s_a = c_{p,a}\ln\frac{T}{T_{0,a}} - R_a\ln\frac{p_a}{p_{0,a}}} \qquad (20.116)$$

Specific Entropy of Water s_w

Analogous to the specific entropy of air, a reference level for water is required. As shown before, this reference level does not need to be the same as it was for the component air. Though the choice is arbitrary, the definition for a reference level for water within the multi-component system, follows the reference for the steam tables, see Sect. 18.3.7, i.e. it is the saturated liquid state (') at the triple point, i.e.

$$\boxed{s_{0,w}(T_{0,w} = 273.16\,\text{K},\ p_{0,w} = 0.006112\,\text{bar}) = 0} \qquad (20.117)$$

As already discussed, the water in the mixture humid air can occur in different aggregate states. In order to find out the water's thermodynamic state values, the steam table of water could be applied, but a more handy approach is usually preferred. This approach is discussed by making a distinction between the following possible cases:

- Vapour
  Most commonly, the vapour that occurs in humid air is considered to be an ideal gas. In order to determine its specific entropy the approach is divided into two steps

  1. In the reference state, i.e. saturated liquid at the triple point (TP), the specific entropy of water is set to be zero, see Eq. 20.117. First, energy is supplied

isobarically to vaporise the water.[30] This is done until saturated vapour at the triple point is reached. According to the steam table the heat of vaporisation

$$\Delta h_v(0.01\,°C) \approx 2500 \, \frac{kJ}{kg} \tag{20.118}$$

is required. The supply of thermal energy comes along with the supply of entropy, that can easily be calculated, since the change of state runs isothermally, i.e.

$$\Delta s = s_a + \underbrace{s_i}_{=0} = \frac{\Delta h_v(0.01\,°C)}{T_{0,w}}. \tag{20.119}$$

2. In this second step, the *ideal gas vapour* is adjusted from $\vartheta_{0,w}$ to ϑ respectively from $p_{0,w}$ to p_v.[31] Thus, the caloric equation of state as known from part I can be applied, i.e.

$$\Delta s = c_{p,v} \ln\frac{T}{T_{0,w}} - R_v \ln\frac{p_v}{p_{0,w}}. \tag{20.120}$$

Thus, the total specific entropy of vapour is the combination of steps 1 and 2, i.e.

$$\boxed{s_v = \underbrace{\frac{\Delta h_v(0.01\,°C)}{T_{0,w}}}_{\text{Step 1}} + \underbrace{c_{p,v} \ln\frac{T}{T_{0,w}} - R_v \ln\frac{p_v}{p_{0,w}}}_{\text{Step 2}}} \tag{20.121}$$

Example 20.6 The total pressure of humid air is $p = 1$ bar at a temperature of $\vartheta = 20\,°C$. The partial pressure of dry air is $p_a = 983$ mbar. Thus, the partial pressure of vapour is $p_v = 17$ mbar. What is the specific entropy of the vapour?

– Approach 1 (Steam table—real fluid):

$$s_v = s\,(20\,°C, \; 17\,\text{mbar}) = 8.8145 \, \frac{kJ}{kg\,K} \tag{20.122}$$

– Approach 2 (Ideal gas, according to Fig. 20.8):

$$s_v = \frac{\Delta h_v(0.01\,°C)}{T_{0,w}} + c_{p,v} \ln\frac{T}{T_{0,w}} - R_v \ln\frac{p_v}{p_{0,w}} = 8.8114 \, \frac{kJ}{kg\,K} \tag{20.123}$$

That is, the error when using the ideal gas approach vs. real fluid approach is significant smaller than 0.05%!

---

[30] The change of state (') to (") shall be isobaric and thus isothermal as well! It is considered to run reversible, since no pressure losses occur, see Sect. 18.4.1, i.e. $s_i = 0$.

[31] The vapour in the humid air possesses a temperature ϑ and its partial pressure is p_v!

Reference level

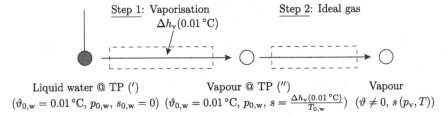

Fig. 20.8 How to treat the water in humid air—vapour

Fig. 20.9 How to treat the
water in humid air—liquid

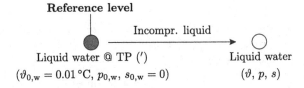

- Liquid water

  Starting from reference point, see Fig. 20.9, that already is in liquid state, liquid water is now treated as an incompressible liquid. According to part I the change of entropy for an incompressible liquid follows

  $$s_{\text{liq}} = s_{0,\text{w}} + c_{\text{liq}} \ln \frac{T}{T_{0,\text{w}}} = c_{\text{liq}} \ln \frac{T}{T_{0,\text{w}}} \tag{20.124}$$

- Solid water

  Starting from reference point, see Fig. 20.10, that is in liquid state, two successive steps are required:

  1. In the reference state, i.e. saturated liquid at the triple point (TP), the specific enthalpy of water is set to be zero, see Eq. 20.94. First, energy is released isobarically to freeze the water.[32] This is done until the liquid is fully turned into ice at the triple point. The specific heat of melting is

     $$\Delta h_{\text{m}}(0.01\,^{\circ}\text{C}) \approx -333\,\frac{\text{kJ}}{\text{kg}}. \tag{20.125}$$

     The release of thermal energy comes along with the release of entropy,[33] that can easily be calculated, since it runs isothermally, i.e.

---

[32]During this phase change the pressure and the temperature remain constant!

[33]The change of state (′) to ice shall be isobaric and thus isothermal as well! It is considered to run reversible, since no pressure losses occur, see Sect. 18.4.1, i.e. $s_{\text{i}} = 0$.

Reference level

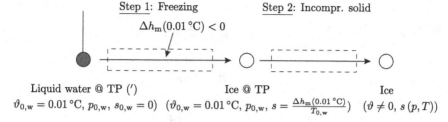

Fig. 20.10 How to treat the water in humid air—ice

$$\Delta s = s_a + \underbrace{s_i}_{=0} = \frac{\Delta h_m(0.01\,°C)}{T_{0,w}} < 0. \tag{20.126}$$

2. From now on, the ice is treated as an incompressible solid. Thus, the change of enthalpy for the ice in step 2, until temperature ϑ is reached, follows

$$s_{ice} = c_{ice}\ln\frac{T}{T_{0,w}}. \tag{20.127}$$

Consequently, the total specific entropy of the solid, i.e. ice, is the combination of steps 1 and 2, i.e.

$$s_{ice} = \underbrace{\frac{\Delta h_m(0.01\,°C)}{T_{0,w}}}_{\text{Step 1}} + \underbrace{c_{ice}\ln\frac{T}{T_{0,w}}}_{\text{Step 2}} \tag{20.128}$$

20.2.5 Overview Possible Cases

In this section an overview of possible states of humid air is given. Specific enthalpy as well as specific entropy are listed based on the investigations of Sects. 20.2.3 and 20.2.4. Mind, that in case the offer of liquid or solid water is sufficiently large, evaporisation takes place until the gaseous phase is saturated with vapour, i.e. the system has reached a thermodynamic equilibrium, see also Sect. 20.1.2! The required properties are listed in Table 20.1, see [1].

- Case 1: Unsaturated air
  The water exists solely as vapour, see sketch 20.11, so that

$$x = \frac{m_w}{m_a} = \frac{m_v}{m_a} = x_v. \tag{20.129}$$

Table 20.1 Properties—Humid air

Reference temperature air	$T_{0,a} = 273.15\,\text{K}$
Reference temperature water	$T_{0,w} = 273.16\,\text{K}$
Reference pressure air	$p_{0,a} = 1\,\text{bar}$
Reference pressure water	$p_{0,w} = 0.006112\,\text{bar}$
Gas constant dry air	$R_a = 287.0409\,\frac{\text{J}}{\text{kg}\,\text{K}}$
Gas constant vapour	$R_v = 461.5158\,\frac{\text{J}}{\text{kg}\,\text{K}}$
Specific isobaric heat capacity air	$c_{p,a} = 1.004\,\frac{\text{kJ}}{\text{kg}\,\text{K}}$
Specific isobaric heat capacity vapour	$c_{p,v} = 1.86\,\frac{\text{kJ}}{\text{kg}\,\text{K}}$
Specific heat capacity liquid water	$c_{\text{liq}} = 4.19\,\frac{\text{kJ}}{\text{kg}\,\text{K}}$
Specific heat capacity solid water	$c_{\text{ice}} = 2.05\,\frac{\text{kJ}}{\text{kg}\,\text{K}}$
Enthalpy of vaporisation at $T_{0,w}$	$\Delta h_v = \Delta h_v(0.01\,^\circ\text{C}) = 2500\,\frac{\text{kJ}}{\text{kg}}$
Enthalpy of melting at $T_{0,w}$	$\Delta h_m = \Delta h_m(0.01\,^\circ\text{C}) = -333\,\frac{\text{kJ}}{\text{kg}}$

Fig. 20.11 Unsaturated
humid air

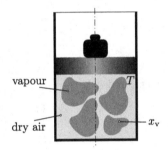

Its specific enthalpy for this case is

$$h_{1+x} = c_{p,a} \cdot \vartheta + x \cdot \left(\Delta h_v + c_{p,v} \cdot \vartheta\right)$$
(20.130)

The specific entropy obeys

$$s_{1+x} = c_{p,a} \ln\frac{T}{T_{0,a}} - R_a \ln\frac{p_a}{p_{0,a}} + x\left[c_{p,v} \ln\frac{T}{T_{0,w}} - R_v \ln\frac{p_v}{p_{0,w}} + \frac{\Delta h_v}{T_{0,w}}\right]$$
(20.131)

- Case 2: Saturated air + liquid water, i.e. $\vartheta > 0$
  In this case, the water is sufficiently available and exists as vapour and as liquid
  water,[34] see Fig. 20.12, so that

---

[34]Thus, according to diagram 18.9a, the temperature is $\vartheta > 0$! Actually, the temperature dependency
for freezing of water is rather small. Liquid water has the same pressure as the gaseous phase, so
that under atmospheric conditions it is almost $0\,^\circ\text{C}$ when freezing starts.

Fig. 20.12 Saturated humid
air and liquid water

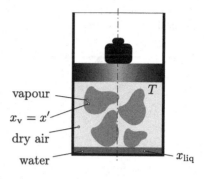

Fig. 20.13 Saturated humid
air and solid water

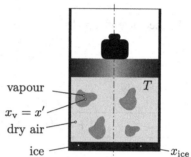

$$x = \frac{m_w}{m_a} = \frac{m_v + m_{liq}}{m_a} = x_v + x_{liq} = x' + x_{liq}. \tag{20.132}$$

Its specific enthalpy for this case is

$$h_{1+x} = h'_{1+x} + x_{liq}h_{liq} = c_{p,a} \cdot \vartheta + x' \cdot \left(\Delta h_v + c_{p,v} \cdot \vartheta\right) + \left(x - x'\right)c_{liq}\vartheta \tag{20.133}$$

The specific entropy[35] is

$$s_{1+x} = s'_{1+x} + x_{liq}s_{liq} = s'_{1+x} + \left(x - x'\right)c_{liq}\ln\frac{T}{T_{0,w}} \tag{20.134}$$

The volume of the liquid water is usually ignored in comparison with the volume
of the gaseous phase.

- Case 3: Saturated air + solid water, i.e. $\vartheta < 0$
  In this case, the water is sufficiently available and exists as vapour and as solid
  water, see Fig. 20.13, so that

$$x = \frac{m_w}{m_a} = \frac{m_v + m_{ice}}{m_a} = x_v + x_{ice} = x' + x_{ice}. \tag{20.135}$$

---

[35]Mind, that the pressure within the liquid is equal to the pressure of the gaseous phase!

Fig. 20.14 Saturated humid
air, liquid and solid water

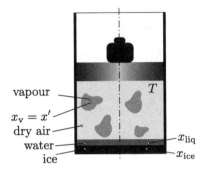

vapour
$x_v = x'$
dry air
water
ice

x_{liq}
x_{ice}

Its specific enthalpy for this case is

$$h_{1+x} = h'_{1+x} + x_{ice}h_{ice},$$ (20.136)

respectively

$$\boxed{h_{1+x} = c_{p,a} \cdot \vartheta + x' \cdot (\Delta h_v + c_{p,v} \cdot \vartheta) + (x - x')(\Delta h_m + c_{ice} \cdot \vartheta)}$$ (20.137)

The specific entropy[36] is

$$\boxed{s_{1+x} = s'_{1+x} + x_{ice}s_{ice} = s'_{1+x} + (x - x')\left(\frac{\Delta h_m}{T_{0,w}} + c_{ice}\ln\frac{T}{T_{0,w}}\right)}$$ (20.138)

The volume of the solid water is usually ignored in comparison with the volume of the gaseous phase.

- Case 4: Saturated air + liquid water + solid water, i.e. $\vartheta = 0$
  In this case, the water is sufficiently available and exists as vapour, as liquid water and as solid, see Fig. 20.14. This is under atmospheric conditions possible, when the temperature is $\vartheta = 0\,°C$, i.e. a three-phase mixture exists,[37] so that

$$x = \frac{m_w}{m_a} = \frac{m_v + m_{liq} + m_{ice}}{m_a} = x_v + x_{liq} + x_{ice} = x' + x_{liq} + x_{ice}.$$ (20.139)

Its specific enthalpy for this case is

$$\boxed{h_{1+x} = h'_{1+x} + x_{liq}h_{liq} + x_{ice}h_{ice}}$$ (20.140)

---

[36]Mind, that the pressure within the solid is equal to the pressure of the gaseous phase!

[37]See also Fig. 20.17! This state is not defined unambiguously and additional information is required, e.g. about the ratio of liquid and ice.

The specific entropy[38] is

$$\boxed{s_{1+x} = s'_{1+x} + x_{\text{liq}} s_{\text{liq}} + x_{\text{ice}} s_{\text{ice}}} \tag{20.141}$$

The volume of the liquid as well as solid water is usually ignored in comparison with the volume of the gaseous phase.

Problem 20.7 A cylinder contains $V_1 = 5\,\text{m}^3$ air as well as $0.119\,\text{kg}$ liquid water in thermodynamic equilibrium (1), see Fig. 20.16. The volume of the liquid water shall be ignored. The total pressure is $p_1 = 2.0\,\text{bar}$ at a temperature of $\vartheta_1 = 25\,°\text{C}$. The weights on top of the piston, that closes the cylinder tightly, are removed very slowly and step by step, so that the humid air expands isothermally and without any dissipation until state (2) is reached. In this state the liquid water is fully vaporised. Air and vapour can be treated as ideal gases ($R_a = 0.287\,\frac{\text{kJ}}{\text{kg K}}$, $R_v = 0.4615\,\frac{\text{kJ}}{\text{kg K}}$). The potential energy of the cylinder content can be disregarded. Ambient state is $\vartheta_{\text{env}} = \vartheta_1$ and $p_{\text{env}} = 0.8\,\text{bar}$. System and environment are perfectly thermally coupled.

(a) What is the mass of the dry air and the mass of the entire water within the cylinder?
(b) Calculate volume V_2 after the expansion? What is the pressure p_2?
(c) Calculate the released work W_{12} during the expansion.
(d) What amount of heat Q_{12} is be transferred with the environment?
(e) What is the change of entropy of the cylinder content during the process?
(f) What is the change of exergy $\Delta E_x = E_{x,2} - E_{x,1}$ of the closed system?

Solution

The following sketch 20.15 shows the change of state and the notation of the relevant variables, in order to solve this task. Since the system in state (1) is in thermodynamic equilibrium and the amount of liquid water is sufficiently large, the air is saturated with vapour (Fig. 20.16).

(a) Since the air is saturated the vapour's partial pressure can be taken out of the steam table, i.e.

$$p_{v_1} = p_s(\vartheta_1) = 0.0317\,\text{bar}. \tag{20.142}$$

According to Dalton the partial pressure of the air then follows

$$p_{a_1} = p_1 - p_{a_1} = 1.9683\,\text{bar}. \tag{20.143}$$

Applying the thermal equation for the air brings

$$m_{a_1} = \frac{p_{a_1} V_1}{R_a T_1} = 11.5013\,\text{kg} = \text{const.} = m_{a_2} = m_a. \tag{20.144}$$

---

[38] Mind, that the pressure within the solid and liquid is equal to the pressure of the gaseous phase!

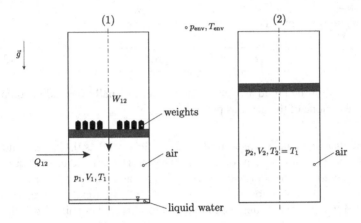

Fig. 20.15 Solution to Problem 20.7

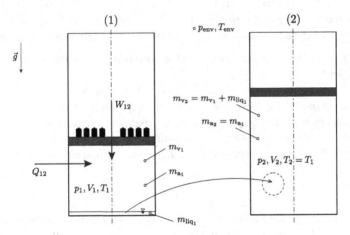

Fig. 20.16 Sketch to Problem 20.7

The mass of the vapour follows accordingly

$$m_{v_1} = \frac{p_{v_1} V_1}{R_v T_1} = 0.1152 \, \text{kg}. \tag{20.145}$$

Hence, the entire mass of the water is

$$\boxed{m_{w_1} = m_{v_1} + m_{liq_1} = 0.2342 \, \text{kg} = \text{const.} = m_{w_2} = m_w} \tag{20.146}$$

(b) In state (2) the liquid water has just turned into vapour, so that state (2) is saturated state. Thus, the vapour's partial pressure[39] is

---

[39] See steam table!

$$p_{v_2} = p_s \left(\vartheta_2 = \vartheta_1\right) = 0.0317 \, \text{bar} = p_{v_1} = p_v. \tag{20.147}$$

The entire water now is turned into vapour, i.e.

$$m_{v_2} = m_{w_1} = 0.2342 \, \text{kg}. \tag{20.148}$$

Applying the thermal equation of state for the vapour in state (2)

$$\boxed{V_2 = \frac{m_{v_2} R_v T_2}{p_{v_2}} = 10.1657 \, \text{m}^3} \tag{20.149}$$

The mass of the dry air remains constant, so that

$$p_{a_2} = \frac{m_{a_1} R_a T_2}{V_2} = 0.9681 \, \text{bar}. \tag{20.150}$$

The total pressure then follows

$$\boxed{p_2 = p_{a_2} + p_{v_2} = 0.9998 \, \text{bar}} \tag{20.151}$$

(c) The partial energy equation for the expansion reads as

$$W_{12} = W_{12,v} + \underbrace{W_{12,\text{mech}}}_{=0} + \underbrace{\Psi_{12}}_{=0} \tag{20.152}$$

As we know, the volume work $W_{12,v}$ counts for the *volume change*. In this case the entire volume of the cylinder content is composed of the volume of the gaseous phase *and* the volume of the liquid phase at *any point of time*. During the change of state the volume of the liquid phase shrinks to zero, whereas the volume of the gaseous phase rises. Accordingly, the volume work can be split into two parts[40]:

- gaseous phase $V_{1,\text{gas}} \rightarrow V_2$ with $p_1 \rightarrow p_2$
  - air $V_{1,\text{gas}} \rightarrow V_2$ with $p_{a_1} \rightarrow p_{a_2}$
  - vapour $V_{1,\text{gas}} \rightarrow V_2$ with $p_{v_1} = p_{v_2}$
- liquid phase $V_{\text{liq}_1} \rightarrow 0$ with $p_{\text{liq}_1} = p_1 \rightarrow p_{v_2}$. Mind, that $V_{\text{liq}_1} \approx 0$!

Hence, the work W_{12} follows

$$W_{12} = W_{12,v} = W_{12,v,\text{gas}} + W_{12,v,\text{liq}}. \tag{20.153}$$

---

[40]It is the changing *volume*, that causes work, since $\delta W_v = -p A \, dx = -p \, dV$. Thus, the mass is not involved *directly*. However, the mass is implied due to the thermal equation of state. Consequently, it is the volume change of the gaseous phase, i.e. air and vapour, as well as the liquid's volume change, that needs to be taken into account. The mass of the vaporising liquid is the cause for the vapour's partial pressure in the gaseous phase to be constant.

Introducing the volume work and applying Dalton's law results in

$$
W_{12} = -\underbrace{\int_{V_{1,\text{gas}}}^{V_2} p\,dV_{\text{gas}}}_{\text{gaseous phase}} - \underbrace{\int_{V_{\text{liq}_1}}^{0} p_{\text{liq}}\,dV_{\text{liq}}}_{=0,\text{ since } V_{\text{liq}_1}\approx 0} = -\int_{V_{1,\text{gas}}}^{V_2} (p_a + p_v)\,dV_{\text{gas}} - 0
$$

$$
= -\int_{V_{1,\text{gas}}}^{V_2} p_a\,dV_{\text{gas}} - \int_{V_{1,\text{gas}}}^{V_2} p_v\,dV_{\text{gas}}.
$$

(20.154)

As mentioned above, the volume of the liquid can be ignored, so that $V_{1,\text{gas}} = V_1$. Since the vapour's pressure is constant, one gets

$$
W_{12} = -\int_{V_1}^{V_2} p_a\,dV_{\text{gas}} - p_v\,(V_2 - V_1)\,.
$$

(20.155)

Applying the isothermal volume work for air, treated as ideal gas, results in

$$
\boxed{W_{12} = -m_a R_a T_1 \ln\frac{p_{a_1}}{p_{a_2}} - p_v\,(V_2 - V_1) = -7.1471 \times 10^5\,\text{J}}
$$

(20.156)

(d) The first law of thermodynamics leads to the exchanged heat Q_{12}. The change of the internal energy is due to the change of the internal energy of the air m_a, the change of the internal energy of the initial vapour m_{v_1} and the change of the internal energy of the liquid m_{liq_1}, see also Fig. 20.15, i.e.

$$
W_{12} + Q_{12} = \Delta U_{\text{total}} = \Delta U_a + \Delta U_v + \Delta U_{\text{liq}}.
$$

(20.157)

Applying the caloric equation of state, with air and vapour to be treated as an ideal gas, results in

$$
\begin{aligned}
W_{12} + Q_{12} &= \Delta U_{\text{total}} \\
&= m_a c_{v,a}\,(T_2 - T_1) + m_{v_1} c_{v,v}\,(T_2 - T_1) + m_{\text{liq}_1}\,\Delta u.
\end{aligned}
$$

(20.158)

Since, the change of state is isothermal one gets

$$
Q_{12} = m_{\text{liq}_1}\,\Delta u - W_{12} = m_{\text{liq}_1}\left(u_{v_2} - u_{\text{liq}_1}\right) - W_{12}.
$$

(20.159)

With

$$
u_{v_2} = h_{v_2} - p_{v_2} v_{v_2} = 2500\,\frac{\text{kJ}}{\text{kg}} + c_{p,v}\vartheta_2 - p_{v_2}\frac{V_2}{m_w} = 2408.9\,\frac{\text{kJ}}{\text{kg}}
$$

(20.160)

respectively

$$u_{\text{liq}_1} = h_{\text{liq}_1} - p_1 v_{\text{liq}_1} = c_{\text{liq}} \vartheta_1 - p_1 \underbrace{v(p_1, T_1)}_{\text{Steam table}} = 104.5494 \, \frac{\text{kJ}}{\text{kg}} \quad (20.161)$$

one finally gets

$$\boxed{Q_{12} = m_{\text{liq}_1} \Delta u - W_{12} = m_{\text{liq}_1} \left(u_{\text{v}_2} - u_{\text{liq}_1} \right) - W_{12} = 9.8893 \times 10^5 \, \text{J}}$$
$$(20.162)$$

(e) There are two alternatives in order to calculate the change of entropy:

- Alternative 1: Process values
  The change of entropy for this reversible process follows

$$\Delta S = S_2 - S_1 = \underbrace{S_{i,12}}_{=0} + S_{a,12}. \quad (20.163)$$

Since the change of state is isothermal, it simplifies to

$$\boxed{\Delta S = S_2 - S_1 = S_{a,12} = \frac{Q_{12}}{T_1} = 3.317 \times 10^3 \, \frac{\text{J}}{\text{K}}} \quad (20.164)$$

- Alternative 2: State values
  The entire change of entropy is due to the change of entropy for air, vapour and liquid, accordingly to the internal energy in part d), i.e.

$$\Delta S = \Delta S_a + \Delta S_v + \Delta S_{\text{liq}} \quad (20.165)$$

with
- Air

$$\Delta S_a = S_{a_2} - S_{a_1}$$
$$= m_a \left[\left(c_{p,a} \ln \frac{T_2}{T_{0,a}} - R_a \ln \frac{p_{a_2}}{p_{0,a}} \right) - \left(c_{p,a} \ln \frac{T_1}{T_{0,a}} - R_a \ln \frac{p_{a_1}}{p_{0,a}} \right) \right]$$
$$= 2.3422 \times 10^3 \, \frac{\text{J}}{\text{K}}$$
$$(20.166)$$

- Vapour

$$\Delta S_v = S_{v_2} - S_{v_1}$$

$$= m_{v_1} \left[\frac{\Delta h_v}{T_{0,w}} + c_{p,v} \ln \frac{T_2}{T_{0,w}} - R_v \ln \frac{p_{v_2}}{p_{0,w}} \right] +$$

$$- m_{v_1} \left[\frac{\Delta h_v}{T_{0,w}} + c_{p,v} \ln \frac{T_1}{T_{0,w}} - R_v \ln \frac{p_{v_1}}{p_{0,w}} \right]$$

$$= 0$$

(20.167)

This is, since the temperature and the partial pressure of the vapour remain constant!

– Liquid

$$\Delta S_{liq} = S_{liq_2} - S_{liq_1}$$

$$= m_{liq_1} \left[\left(\frac{\Delta h_v}{T_{0,w}} + c_{p,v} \ln \frac{T_2}{T_{0,w}} - R_v \ln \frac{p_{v_2}}{p_{0,w}} \right) - c_{liq} \ln \frac{T_1}{T_{0,w}} \right]$$

$$= 974.4384 \, \frac{J}{K}$$

(20.168)

Thus, the entire change of entropy results in

$$\boxed{\Delta S = \Delta S_a + \Delta S_v + \Delta S_{liq} = 3.317 \times 10^3 \, \frac{J}{K}}$$
(20.169)

The deviation of approach 1 and 2 is less than 0.01% and most likely a result of inaccuracies in the fluid's properties with respect to the steam table.

(f) The balance of exergy leads to

$$E_{x,2} = E_{x,1} + W_{\mathrm{eff}} + E_{x,Q} - \Delta E_{x,V}.$$
(20.170)

The heat transfer occurs at ambient temperature, so that

$$E_{x,Q} = 0.$$
(20.171)

Furthermore, the change of state is reversible, i.e.

$$\Delta E_{x,V} = 0.$$
(20.172)

So, one gets

$$\boxed{\Delta E_x = E_{x,2} - E_{x,1} = W_{\mathrm{eff}} = W_{12} + p_{\mathrm{env}} (V_2 - V_1) = -3.0145 \times 10^5 \, \mathrm{J}}$$
(20.173)

20.3 The h_{1+x}, x-Diagram According to Mollier

Obviously, humid air is a two- respectively three-phase mixture and the relevant correlations to describe its thermodynamic states have been introduced in the previous sections. It has been clarified under what conditions liquid or even solid water occur in humid air. In order to handle changes of state, it has been beneficial to visualise equilibrium states, e.g. in p, v- or T, s-diagrams. This has been done in part I, so that process values for instance can be easily illustrated. Especially in the field of HVAC technology the application of the so-called h_{1+x}, x-diagram according to Mollier,[41] that is introduced in this section, is advantageous. A schematic h_{1+x}, x-diagram is shown in Fig. 20.17. A detailed diagram can be found in the appendix, see Fig. D.1. Actually, all correlations discussed so far are gathered and visualised in this diagram. The water content x is plotted on the x-axis, the specific enthalpy h_{1+x} is shown on the y-axis. Figure 20.17 visualises the previously discussed regions, i.e. it shows the state of the water. The so-called saturation curve, i.e. the relative humidity is $\varphi = 1$, is characteristic for the h_{1+x}, x-diagram. Above that curve, the humid air is unsaturated, so that the relative humidity φ is less than 100%. Below the $\varphi = 1$-curve the humid air is saturated and liquid or solid water occurs additionally. The following examples summarise possible states:

- State (1)
  Unsaturated humid air, single-phase region, i.e. vapour and dry air are in gaseous state
- State (2)
  Saturated humid air, the humid air has its maximum vapour load with respect to its temperature
- State (3)
  Saturated humid air (gaseous phase composed of dry air and vapour) and ice (solid phase), i.e. two-phase region
- State (4)
  Saturated humid air (gaseous phase composed of dry air and vapour) and ice (solid phase) and water (liquid phase), i.e. three-phase region
- State (5)
  Saturated humid air (gaseous phase composed of dry air and vapour) and water (liquid phase), i.e. two-phase region

However, there are a few characteristics related with the h_{1+x}, x-diagram that need further explanation:

- Lines of constant enthalpy h_{1+x} are inclined in a way, that the $\vartheta = 0\,°C$-isothermal for unsaturated air, i.e. in the single-phase region, runs horizontally, see Fig. 20.17. However, the other isothermals in the unsaturated region are not perfectly horizontal.

---

[41] Richard Mollier (∗30 November 1863 in Triest, †13 March 1935 in Dresden). At the 1923 Thermodynamics Conference held in Los Angeles it was decided to name all state diagrams, that have the specific enthalpy h as one of its axis, after Mollier.

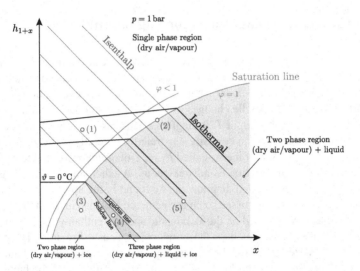

Fig. 20.17 Mollier h_{1+x}, x-diagram—schematic

- At the saturation line ($\varphi = 1$) isothermals kink, so that their gradient becomes significant smaller than in the unsaturated region:

  – Single-phase region

$$\left.\frac{\partial h_{1+x,1P}}{\partial x}\right|_{\vartheta=\text{const.}} = \left.\frac{\partial \left[c_{p,a} \cdot \vartheta + x \cdot \left(\Delta h_v + c_{p,v} \cdot \vartheta\right)\right]}{\partial x}\right|_\vartheta \qquad (20.174)$$
$$= \Delta h_v + c_{p,v} \cdot \vartheta$$

  – Two-phase region (liquid)

$$\left.\frac{\partial h_{1+x,2P}}{\partial x}\right|_{\vartheta=\text{const.}} = \left.\frac{\partial \left[c_{p,a}\vartheta + x'\left(\Delta h_v + c_{p,v}\vartheta\right) + \left(x - x'\right)c_{\text{liq}}\vartheta\right]}{\partial x}\right|_\vartheta$$
$$= c_{\text{liq}} \cdot \vartheta$$

$$(20.175)$$

Obviously, it follows, that

$$\left.\frac{\partial h_{1+x,2P}}{\partial x}\right|_{\vartheta=\text{const.}} < \left.\frac{\partial h_{1+x,1P}}{\partial x}\right|_{\vartheta=\text{const.}} \qquad (20.176)$$

Thus, the isothermals kink at the saturation curve. Though, lines of constant enthalpy h_{1+x} run without kink through the different regions.

- As mentioned before, three phases can occur at $\vartheta = 0\,°\text{C}$, i.e. gaseous, liquid and solid phase. Hence, this isothermal splits at the saturation curve $\varphi = 1$ into liquidus and solidus line, see Fig. 20.17. According to Eq. 20.175, the gradient of the liquidus line is

$$\left.\frac{\partial h_{1+x}}{\partial x}\right|_{\vartheta=0\,°C} = c_{\mathrm{liq}} \cdot \vartheta = 0 \tag{20.177}$$

Thus, the liquidus line runs congruent to its isenthalp!
For the solidus line it is

$$\left.\frac{\partial h_{1+x}}{\partial x}\right|_{\vartheta=0\,°C} = \left.\frac{\partial\left[c_{p,\mathrm{a}}\vartheta + x'\left(\Delta h_{\mathrm{v}} + c_{p,\mathrm{v}}\vartheta\right) + \left(x - x'\right)\left(\Delta h_{\mathrm{m}} + c_{\mathrm{ice}}\vartheta\right)\right]}{\partial x}\right|_{\vartheta=0\,°C}$$

$$= \Delta h_{\mathrm{m}} + c_{\mathrm{ice}} \cdot \vartheta = \Delta h_{\mathrm{m}}$$

$$< 0 \tag{20.178}$$

- Usually, the h_{1+x}, x-diagram is given for a total pressure $p = 1$ bar. A conversion to different pressures is given later on.

20.4 Changes of State for Humid Air

The handling of the h_{1+x}, x-diagram is demonstrated in this section by showing relevant changes of states of humid air for HVAC technology. However, the focus is on the thermodynamic balances on the one hand side. These balances include mass conservation for water and air as well as the first law of thermodynamics. Each change of state is further illustrated in the h_{1+x}, x-diagram on the other hand, so that its beneficial usage becomes obvious.

20.4.1 Heating and Cooling at Constant Water Content

This change of state is shown exemplary in Fig. 20.18. As the entire water content remains constant,[42] i.e. $x = $ const., the change of state follows a vertical line. In order to perform the relevant balances, a distinction is made between open and closed systems:

- Open system
  The first law of thermodynamics[43] in steady state reads as

$$\dot{Q}_{12} = \dot{m} \cdot (h_2 - h_1). \tag{20.179}$$

Introducing the specific enthalpy of humid air $h_{1+x,2}$, that is referred to the mass of the dry air, brings

---

[42] This is due to no water is added or remove. Only thermal energy is supplied respectively released.
[43] Kinetic as well as potential energy are ignored.

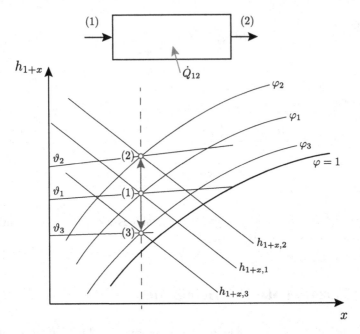

Fig. 20.18 Heating (1) → (2) and cooling (1) → (3) at constant water content

$$\dot{Q}_{12} = \dot{m}_a \cdot \left(h_{1+x,2} - h_{1+x,1} \right)$$
(20.180)

Thus, a heating, i.e. $\dot{Q} > 0$ leads to an increase of specific enthalpy, a cooling, i.e. $\dot{Q} < 0$ to a decrease of the specific enthalpy.
The mass conservation for the water is

$$\dot{m}_{1,w} = \dot{m}_{2,w} = \dot{m}_w$$
(20.181)

respectively for the air

$$\dot{m}_{1,a} = \dot{m}_{2,a} = \dot{m}_a.$$
(20.182)

With the entire mass

$$\dot{m} = \dot{m}_a + \dot{m}_w = \text{const.},$$
(20.183)

so that

$$\dot{m}_a = \frac{\dot{m}}{1 + x}$$
(20.184)

- Closed system

  The first law of thermodynamics in differential notation[44] is

  $$\delta Q + \delta W = dU. \tag{20.185}$$

  Applying the partial energy equation[45]

  $$\delta W = -p \, dV \tag{20.186}$$

  and the definition for the enthalpy brings

  $$\delta Q - p \, dV = dH - p \, dV - V \, dp. \tag{20.187}$$

  As mentioned before, in HVAC applications, mostly isobaric changes of state, i.e. $dp = 0$, occur, so that

  $$Q_{12} = m \cdot (h_2 - h_1) \tag{20.188}$$

  Introducing the specific enthalpy for humid air $h_{1+x,2}$

  $$\boxed{Q_{12} = m_a \cdot \left(h_{1+x,2} - h_{1+x,1}\right)} \tag{20.189}$$

  The mass conservation for the water is

  $$m_{1,w} = m_{2,w} = m_w \tag{20.190}$$

  respectively for the air

  $$m_{1,a} = m_{2,a} = m_a. \tag{20.191}$$

  With the entire mass

  $$m = m_a + m_w = \text{const.}, \tag{20.192}$$

  so that

  $$\boxed{m_a = \frac{m}{1 + x}} \tag{20.193}$$

  Obviously, according to Fig. 20.18 two options are possible:

- Heating $(1) \rightarrow (2)$

  - Water content x stays constant
  - Enthalpy h_{1+x} increases, see first law of thermodynamics
  - Temperature ϑ increases

---

[44]Let us assume, the change of state is reversible and kinetic as well as potential energy can be ignored.

[45]No mechanical work, since kinetic and potential energy are ignored!

– Relative humidity φ decreases

- Cooling (1)$\rightarrow$(3)

  – Water contant x stays constant
  – Enthalpy h_{1+x} decreases, see first law of thermodynamics
  – Temperature ϑ decreases
  – Relative humidity φ increases
  – State (3) might even be located in the two-phase region in case cooling is suffi-
    cient

20.4.2 Dehumidification

This change of state is essential in air conditioning systems, e.g. for operation during
summer period. As already known, air of large temperature potentially carries a
high load of vapour. Hence, it can be required to dehumidify the air, in order to
prevent condensing of water, when the air is cooled down. This example is shown
in Fig. 20.19 for an open system in steady state. Obviously, three components are
required to design an air conditioning system: a cooler, a water separator as well as
a heater:

- Within the cooler the humid air (1) is cooled down, while the water content x
  remains constant, since no water is released or supplied. Hence, the change of
  state runs vertically downwards in the h_{1+x}, x-diagram. At state (1′) the humid air
  reaches saturated state. A further release of thermal energy leads to liquid water,
  while the gaseous phase still is saturated, i.e. the two-phase region is reached and
  cooling stops at state (2).
  The first law of thermodynamics[46] reads as

$$\dot{Q}_{12} = \dot{m} \cdot (h_2 - h_1) = \dot{m}_a \cdot \left(h_{1+x,2} - h_{1+x,1} \right) \tag{20.194}$$

with the specific enthalpies

$$h_{1+x,1} = c_{p,a}\vartheta_1 + x_1 \cdot \left(\Delta h_v + c_{p,v}\vartheta_1 \right) \tag{20.195}$$

respectively, since the air with a temperature of ϑ_2 can just carry $x_{2'}$ as vapour load

$$h_{1+x,2} = c_{p,a}\vartheta_2 + x_{2'} \cdot \left(\Delta h_v + c_{p,v}\vartheta_2 \right) + (x_2 - x_{2'}) \cdot c_{\text{liq}} \cdot \vartheta_2. \tag{20.196}$$

The dry air balance reads as

$$m_a = \text{const.} = m_{1,a} = m_{2,a}. \tag{20.197}$$

---

[46]Let us assume, kinetic as well as potential energy can be ignored.

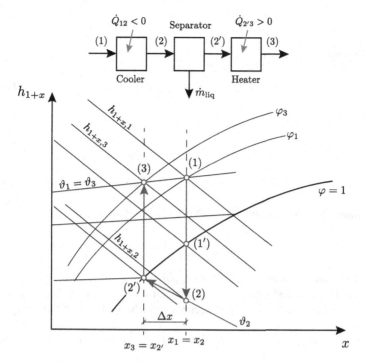

Fig. 20.19 Dehumidification of air

The water balance reads as

$$m_w = \text{const.} = m_{1,w} = m_{2,w}. \tag{20.198}$$

In state (1) the water is in vapour state, i.e.

$$m_{1,w} = m_{1,v} = x_1 m_a \tag{20.199}$$

and in state (2) it occurs as vapour[47] and liquid

$$m_{2,w} = m_{2,v} + m_{2,\text{liq}} = x_2' m_a + m_{2,\text{liq}}. \tag{20.200}$$

Thus, the mass of the liquid water results in

$$m_{2,\text{liq}} = x_1 m_a - x_2' m_a = m_a \left(x_1 - x_2' \right) = m_a \, \Delta x. \tag{20.201}$$

- In the second step, the liquid water is released by a separator. Let us assume, the separator works adiabatically, so the first law of thermodynamics reads as

---

[47] In saturated state!

$$\dot{m}_a h_{1+x,2} = \dot{m}_a h_{1+x,2'} + m_{2,\text{liq}} h_{2,\text{liq}} \qquad (20.202)$$

Dividing by $\dot{m}_a$ results in

$$h_{1+x,2} = h_{1+x,2'} + \Delta x\, h_{2,\text{liq}} \qquad (20.203)$$

with

$$h_{2,\text{liq}} = c_{\text{liq}} \cdot \vartheta_2 \qquad (20.204)$$

and

$$h_{1+x,2'} = c_{p,a}\vartheta_{2'} + x_{2'} \cdot \left(\Delta h_v + c_{p,v}\vartheta_{2'}\right). \qquad (20.205)$$

Combining Eqs. 20.203, 20.204, 20.205 and 20.196 brings

$$\vartheta_{2'} = \vartheta_2. \qquad (20.206)$$

Thus, the water separator works isothermally.
- Now the dehumidified air is heat up again until $\vartheta_3 = \vartheta_1$ is reached. The required thermal energy can be calculated according to the first law of thermodynamics, i.e.

$$\dot{Q}_{2'3} = \dot{m} \cdot (h_3 - h_{2'}) = \dot{m}_a \cdot \left(h_{1+x,3} - h_{1+x,2'}\right) \qquad (20.207)$$

with the specific enthalpy of state (3)

$$h_{1+x,3} = c_{p,a}\vartheta_1 + x_{2'} \cdot \left(\Delta h_v + c_{p,v}\vartheta_1\right). \qquad (20.208)$$

According to Fig. 20.19 state (3) finally has the same temperature as the initial state (1), but the air has been dehumidified. The relative humidity φ as well as the water content x have been reduced. If a further dehumidification is desired, the air needs a further temperature sink within the cooler.

20.4.3   Adiabatic Mixing of Humid Air

In this section the adiabatic mixing of humid air is investigated. This change of state is essential, when an air conditioning system is operated in bypass-mode, i.e. fresh air is mixed with recirculating air. Its principle is shown in Fig. 20.20: An air flow (1) is mixed adiabatically with a second flow (2) and the mixed air (3) is leaving the system boundary. As already mentioned, a mass balance is required for the dry air as well as for the water, following the idea of mass conservation. However, furthermore an energy balance is required. First, the mathematical correlation is derived. In the second step, the change of state is shown in a h_{1+x}, x-diagram.

Fig. 20.20 Adiabatic mixing of humid air

Let us start with the mass balance for dry air, i.e.

$$\dot{m}_{a,3} = \dot{m}_{a,1} + \dot{m}_{a,2}. \tag{20.209}$$

The mass conservation for water[48] reads as

$$\dot{m}_{w,3} = \dot{m}_{w,1} + \dot{m}_{w,2} \tag{20.210}$$

With introducing the water content x one gets:

$$\dot{m}_{a,3}x_3 = \dot{m}_{a,1}x_1 + \dot{m}_{a,2}x_2. \tag{20.211}$$

These equations can be solved for the water content of the mixed humid air x_3, i.e.

$$\boxed{x_3 = \frac{\dot{m}_{a,1}x_1 + \dot{m}_{a,2}x_2}{\dot{m}_{a,3}} = \frac{\dot{m}_{a,1}x_1 + \dot{m}_{a,2}x_2}{\dot{m}_{a,1} + \dot{m}_{a,2}}} \tag{20.212}$$

Once, the water content x_3 is know, the following equations can be derived easily, i.e.

$$\begin{aligned} x_1 - x_3 &= \frac{x_1\left(\dot{m}_{a,1} + \dot{m}_{a,2}\right) - \dot{m}_{a,1}x_1 - \dot{m}_{a,2}x_2}{\dot{m}_{a,1} + \dot{m}_{a,2}} \\ &= \frac{\dot{m}_{a,2}\left(x_1 - x_2\right)}{\dot{m}_{a,1} + \dot{m}_{a,2}} \end{aligned} \tag{20.213}$$

respectively

$$\begin{aligned} x_3 - x_2 &= \frac{\dot{m}_{a,1}x_1 + \dot{m}_{a,2}x_2 - x_2\left(\dot{m}_{a,1} + \dot{m}_{a,2}\right)}{\dot{m}_{a,1} + \dot{m}_{a,2}} \\ &= \frac{\dot{m}_{a,1}\left(x_1 - x_2\right)}{\dot{m}_{a,1} + \dot{m}_{a,2}}. \end{aligned} \tag{20.214}$$

---

[48] At this stage no distinction is made regarding the state of the water!

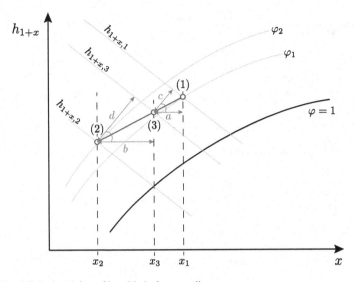

Fig. 20.21 Adiabatic mixing of humid air, h_{1+x}, x-diagram

Thus, it finally is

$$\boxed{\frac{x_1 - x_3}{x_3 - x_2} = \frac{\dot{m}_{a,2}}{\dot{m}_{a,1}} = \frac{a}{b}}$$ (20.215)

The distances a and b are indicated in the h_{1+x}, x-diagram, see Fig. 20.21.

Now, the first law of thermodynamics for the adiabatic mixing in steady state[49] is applied, i.e.

$$\dot{H}_1 + \dot{H}_2 = \dot{H}_3.$$ (20.216)

Introducing the specific enthalpy leads to

$$\dot{m}_{a,1} h_{1+x,1} + \dot{m}_{a,2} h_{1+x,2} = \left(\dot{m}_{a,1} + \dot{m}_{a,2}\right) h_{1+x,3}.$$ (20.217)

Hence, the specific enthalpy of the mixed air (3) follows

$$\boxed{h_{1+x,3} = \frac{\dot{m}_{a,1} h_{1+x,1} + \dot{m}_{a,2} h_{1+x,2}}{\dot{m}_{a,1} + \dot{m}_{a,2}}}$$ (20.218)

Once, the water content $h_{1+x,3}$ is known, the following equations can be derived easily

---

[49]I.e. energy flux in is balanced by the energy flux out. Kinetic as well as potential energies are neglected.

$$h_{1+x,1} - h_{1+x,3} = \frac{h_{1+x,1}\left(\dot{m}_{a,1} + \dot{m}_{a,2}\right) - \dot{m}_{a,1}h_{1+x,1} - \dot{m}_{a,2}h_{1+x,2}}{\dot{m}_{a,1} + \dot{m}_{a,2}}$$

$$= \frac{\dot{m}_{a,2}\left(h_{1+x,1} - h_{1+x,2}\right)}{\dot{m}_{a,1} + \dot{m}_{a,2}} \tag{20.219}$$

respectively

$$h_{1+x,3} - h_{1+x,2} = \frac{\dot{m}_{a,1}h_{1+x,1} + \dot{m}_{a,2}h_{1+x,2} - h_{1+x,2}\left(\dot{m}_{a,1} + \dot{m}_{a,2}\right)}{\dot{m}_{a,1} + \dot{m}_{a,2}}$$

$$= \frac{\dot{m}_{a,1}\left(h_{1+x,1} - h_{1+x,2}\right)}{\dot{m}_{a,1} + \dot{m}_{a,2}}. \tag{20.220}$$

Hence, it finally is

$$\boxed{\frac{h_{1+x,1} - h_{1+x,3}}{h_{1+x,3} - h_{1+x,2}} = \frac{\dot{m}_{a,2}}{\dot{m}_{a,1}} = \frac{c}{d}} \tag{20.221}$$

The distances c and d are indicated in the h_{1+x}, x-diagram,[50] see Fig. 20.21.

A comparison of Eqs. 20.215 and 20.221 leads to

$$\frac{h_{1+x,1} - h_{1+x,3}}{h_{1+x,3} - h_{1+x,2}} = \frac{x_1 - x_3}{x_3 - x_2} \tag{20.222}$$

Rearranging this equation brings

$$\frac{h_{1+x,3} - h_{1+x,1}}{x_3 - x_1} = \frac{h_{1+x,2} - h_{1+x,3}}{x_2 - x_3}. \tag{20.223}$$

These terms represent the gradient of the curves $\overline{31}$ and $\overline{23}$:

$$\boxed{\left.\frac{\Delta h}{\Delta x}\right|_{\overline{31}} = \frac{c}{a} = \frac{d}{b} = \left.\frac{\Delta h}{\Delta x}\right|_{\overline{23}}} \tag{20.224}$$

Thus, the gradients in the *inclined* h_{1+x}, x-diagram are identical. Consequently, the points (1), (2) and (3) need to lie on a straight line. Figure 20.22 summarises the change of state for an adiabatic mixing of two flows. Equation 20.215 can be extended: Since all points are locate on one straight line it also follows,[51] see Fig. 20.22:

$$\boxed{\frac{\dot{m}_{a,2}}{\dot{m}_{a,1}} = \frac{x_1 - x_3}{x_3 - x_2} = \frac{\overline{13}}{\overline{23}}} \tag{20.225}$$

---

[50]Mind, that it is not yet proven, that (1), (2) and (3) lie on a straight line! The prove follows now.

[51]This is due to $(x_3 - x_2) = \overline{23} \cdot \cos\alpha$ and $(x_1 - x_3) = \overline{13} \cdot \cos\alpha$.

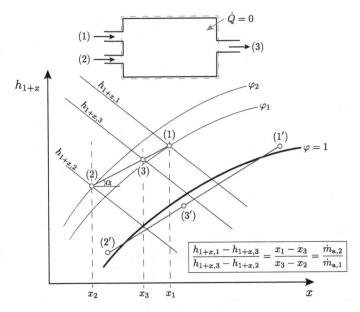

Fig. 20.22 Adiabatic mixing of humid air, h_{1+x}, x-diagram

Theorem 20.8 *A mass flux (1) of humid air is mixed adiabatically with a mass flux (2) of humid air. The new state (3) lies on a straight line between (1) and (2). This line is called mixing straight. The larger the amount of dry air, the smaller the distance to the mixing point:*

- $\dot{m}_{a,1} = \dot{m}_{a,2}$: The mixing state (3) is located right in the middle of the line $\overline{12}$.
- $\dot{m}_{a,1} > \dot{m}_{a,2}$: The mixing state (3) moves closer to (1).
- $\dot{m}_{a,1} < \dot{m}_{a,2}$: The mixing state (3) moves closer to (2).

Mind, that no restrictions have been made regarding the region of the mixing states. Thus, this principle even works, if the mixing states are located in the two/three-phase region. It is even possible, that mixing of two unsaturated flows of humid air (1') and (2') leads to a state, where liquid water (3') occurs, as shown in Fig. 20.22. This can be observed in winter time, when the ambient air is cold and does not contain much vapour, see state (2'). When a person breathes out, this state is rather warm and carries vapour, see state (1'). The mixing then often leads to condensing of the breathing air, see state (3').

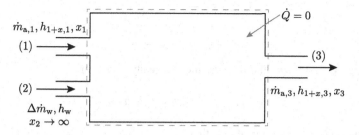

Fig. 20.23 Humidification of air with pure water

20.4.4 Humidification of Air

Section 20.4.2 has shown, that a dehumidification of air might be required, e.g in summer time, when air potentially carries a large load of vapour due to its temperature. However, in winter time, characterised by cold and dry air, a major task of HVAC systems is to humidify air. This can be achieved with liquid, solid or vaporous water, that is injected to the rather dry air. The consequences are discussed within this section thermodynamically. However, the principle is shown in Fig. 20.23. The air, that needs to be humidified, shall be of state (1). The flow of pure water[52] is state (2) and the humidified air is state (3). Unfortunately, due to

$$x_2 = \frac{\Delta \dot{m}_\mathrm{w}}{\dot{m}_{\mathrm{a},2}} \to \infty \tag{20.226}$$

the mixing-straight-approach, see Sect. 20.4.3, does not work any more.[53] However, the mass balance for air is

$$\dot{m}_{\mathrm{a},3} = \dot{m}_{\mathrm{a},1} = \dot{m}_\mathrm{a} = \text{const.} \tag{20.227}$$

The mass balance for water reads as

$$\dot{m}_{\mathrm{w},3} = \dot{m}_{\mathrm{w},1} + \Delta \dot{m}_\mathrm{w}. \tag{20.228}$$

Introducing the water content x brings

$$\dot{m}_\mathrm{a} x_3 = \dot{m}_\mathrm{a} x_1 + \Delta \dot{m}_\mathrm{w}. \tag{20.229}$$

Hence, rearranging leads to

---

[52]It is assumed, that the water is added purely, i.e. without any carrying flow of dry air, i.e. $\dot{m}_{\mathrm{a},2} = 0$.
[53]Due to $x_2 \to \infty$ state (2) can not be fixed in the h_{1+x}, x-diagram.

$$x_3 - x_1 = \Delta x = \frac{\Delta \dot{m}_w}{\dot{m}_a}$$

(20.230)

Applying the first law of thermodynamics in steady state[54] brings

$$\dot{m}_a h_{1+x,1} + \Delta \dot{m}_w h_w = \dot{m}_a h_{1+x,3}.$$

(20.231)

This equation can be rearranged, so that

$$h_{1+x,3} - h_{1+x,1} = \Delta h_{1+x} = \frac{\Delta \dot{m}_w}{\dot{m}_a} h_w$$

(20.232)

Dividing Eqs. 20.230 and 20.232 leads to

$$\frac{\Delta h_{1+x}}{\Delta x} = \frac{h_{1+x,3} - h_{1+x,1}}{x_3 - x_1} = h_w$$

(20.233)

Obviously, this gradient $\frac{\Delta h_{1+x}}{\Delta x}$ shows the *direction a change of state* takes in the h_{1+x}, x-diagram. According to Eq. 20.233 this gradient is identical with the specific enthalpy of the water that is added. Starting from the initial (1) the increase of the water content x can be calculated according to Eq. 20.230. Thus, the mass of the added water needs to be known. However, the knowledge of Δx is not sufficient in order to fix state (3), since it can be at any vertical position corresponding with the new water content

$$x_3 = x_1 + \Delta x.$$

(20.234)

The second information to fix state (3) comes from the gradient information. Figure 20.24 shows how this information can be applied in a h_{1+x}, x-diagram. Starting from a so-called *pole* the $\frac{\Delta h_{1+x}}{\Delta x}$ can be grabbed with the rulers along the edges of the h_{1+x}, x-diagram. Once the gradient has been fixed by scale and pole, it can be shifted parallel through state (1). The point of intersection with the vertical line containing the Δx information gives the new state (3).

Example 20.9 The following examples show how to handle the h_{1+x}, x-diagram for air humidification. Let us assume the initial state (1) is well known and can be illustrated in the diagram. Each of the following examples can be solved mathematically by applying the derived correlations. However, the focus is on a graphical solution. In most cases this procedure is beneficial, since the change of state can be investigated fast compared to a mathematical solution. The following steps are required:

- Step 1: Determine Δx
  This can be done easily by applying

---

[54]Kinetic as well as potential energy is ignored! The incoming energy is balanced by the outgoing energy.

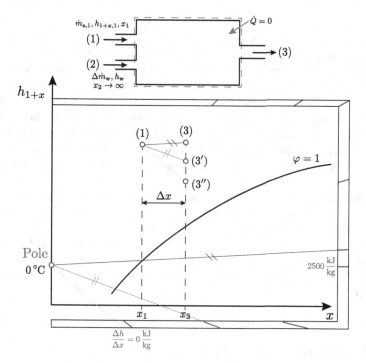

Fig. 20.24 Humidification of air with pure water

$$\Delta x = \frac{\Delta \dot{m}_w}{\dot{m}_a} \rightarrow x_3 = x_1 + \Delta x. \tag{20.235}$$

Thus, the vertical x_3 line can be drawn in the h_{1+x}, x-diagram.

- **Step 2:** Determine the gradient $\frac{\Delta h_{1+x}}{\Delta x}$
  Depending on the state of water the gradient follows:

  – Case (a): Vapour (1) → (3)

$$\frac{\Delta h_{1+x}}{\Delta x} = h_w = h_v = \Delta h_v + c_{p,v} \cdot \vartheta \tag{20.236}$$

  – Case (b): Liquid water (1) → (3')

$$\frac{\Delta h_{1+x}}{\Delta x} = h_w = h_{liq} = c_{liq} \cdot \vartheta \tag{20.237}$$

  – Case (c): Soild water, i.e. ice (1) → (3'')

$$\frac{\Delta h_{1+x}}{\Delta x} = h_w = h_{ice} = \Delta h_m + c_{ice} \cdot \vartheta < 0 \tag{20.238}$$

- Step 3: Graphical construction in the h_{1+x}, x-diagram

  The three cases are exemplary shown in Fig. 20.24. Starting from the Δx line, the states can be fixed in the diagram by parallel shifting of the according gradients, see step 2, and marking the point of intersection. Obviously, the mixing of water affects the temperature of the final state. Case (3″) brings the lowest temperature: The initial state (1) is unsaturated, so that it still can carry a larger loaf of vapour as it currently has. Water is available, but in solid state. Consequently, the aggregate state needs to be changed from solid to liquid to vapour. As already known, these phase changes require energy, that is taken from the surrounding air (1). Thus, the initial temperature of state (1) sinks, since the energy is utilised for the phase change.

  The same effect occurs with liquid water, that is added in the example (1) → (3′). However, the required energy is less compared with (1) → (3″), since only one phase change from liquid to vapour needs to be done.

  Finally, the example (1) → (3), i.e. vapour is added, does not require phase change energy,[55] but sensible heat to bring vapour and air in a thermal equilibrium. Mind, that the temperature of the vapour can be adjusted by its pressure, see steam Tables A.1–A.12.

20.4.5   Adiabatic Saturation Temperature

Let us assume, unsaturated humid air (1) flows over a surface, that is fully covered with liquid water (2), see Fig. 20.25. When having the surface passed (3), the air is in thermodynamic equilibrium[56] with the reservoir of water, so that

$$\vartheta_3 = \vartheta_2 \tag{20.239}$$

and

$$\varphi_3 = 1. \tag{20.240}$$

Since the air, shown exemplary in Fig. 20.25, is assumed to be unsaturated, it can take a further load of vapour. In order to do so, liquid water at its surface starts to vaporise due to its low partial pressure. The required energy for the phase change is taken from the surrounding air, that cools down to a lower temperature, i.e.

$$\vartheta_3 < \vartheta_1 \tag{20.241}$$

Actually this problem comes along with two possible questions:

- What temperature on the humid surfaces ϑ_2 can be achieved in steady state?

  State (3) is the result of the mixing of state (1) and state (2), that is pure liquid

---

[55] Known as latent heat.

[56] Assuming, that the contact time of air and water is sufficiently long!

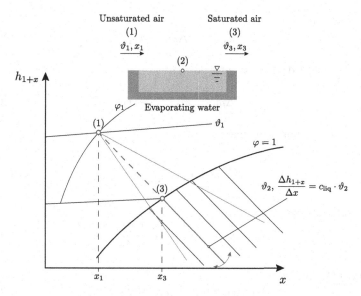

Fig. 20.25 Adiabatic saturation temperature

water, see Sect. 20.4.4. Hence, state (2), i.e. $x_2 \to \infty$, can not be fixed in a h_{1+x}, x-diagram. In case, that state (1) is well known, e.g. by temperature and relative humidity, it can be localised in a h_{1+x}, x-diagram. Now, the following requirements need to be satisfied:

1. State (3) needs to have the same temperature as (2) due to thermodynamic equilibrium. State (2) is somewhere in the two-phase region.
2. Furthermore, (3) needs to be on the saturation line, i.e. $\varphi_3 = 1$.
3. It is further known, see Sect. 20.4.4, that the direction from (1) to (3) needs to follow

$$\frac{\Delta h_{1+x}}{\Delta x} = h_{\mathrm{w}} = h_{\mathrm{liq}} = c_{\mathrm{liq}} \cdot \vartheta_2. \tag{20.242}$$

However, in the two-phase region this gradient is identical with the isothermal ϑ_2, see Eq. 20.175. Thus, one can solve this problem graphical by fixing one end of a ruler in state (1) and start rotating the ruler as long as one finds a straight line, that covers (1), (2) and (3) and that is congruent with an isothermal in the two-phase region. Such a line is shown in Fig. 20.25 and fulfils all three requirements.

In the following, the mathematical strategy is introduced, in case no h_{1+x}, x-diagram is present. Starting point is state (1) given by ϑ_1 and φ_1:

1. Determination of the water content x_1, i.e.

$$x_1 = 0.622 \frac{\varphi_1 p_{\mathrm{v}}'(\vartheta_1)}{p - \varphi_1 p_{\mathrm{v}}'(\vartheta_1)}. \tag{20.243}$$

2. Water balance

$$\dot{m}_{a,1} x_1 + \Delta \dot{m}_w = \dot{m}_{a,1} x_3 \tag{20.244}$$

Thus, the supplied water is

$$\Delta \dot{m}_w = \dot{m}_{a,1} (x_3 - x_1) . \tag{20.245}$$

With state (3) being saturated, i.e.

$$x_3 = x_3' = 0.622 \frac{p_v'(\vartheta_2)}{p - p_v'(\vartheta_2)} . \tag{20.246}$$

Thus, one gets

$$\Delta \dot{m}_w = \dot{m}_{a,1} \left(0.622 \frac{p_v'(\vartheta_2)}{p - p_v'(\vartheta_2)} - x_1 \right) . \tag{20.247}$$

3. Energy balance

$$\dot{m}_{a,1} h_{1+x,1} + \Delta \dot{m}_w h_{liq} = \dot{m}_{a,1} h_{1+x,3} \tag{20.248}$$

Substitution of $\Delta \dot{m}_w$ and reducing of $\dot{m}_{a,1}$ results in

$$h_{1+x,1} + \left(0.622 \frac{p_v'(\vartheta_2)}{p - p_v'(\vartheta_2)} - x_1 \right) h_{liq} = h_{1+x,3} . \tag{20.249}$$

The according specific enthalpies are

$$h_{liq} = c_{liq} \vartheta_2 \tag{20.250}$$

and

$$h_{1+x,1} = c_{p,a} \cdot \vartheta_1 + x_1 \cdot \left(\Delta h_v + c_{p,v} \cdot \vartheta_1 \right) \tag{20.251}$$

and

$$h_{1+x,3} = c_{p,a} \cdot \vartheta_2 + x_3' \cdot \left(\Delta h_v + c_{p,v} \cdot \vartheta_2 \right) . \tag{20.252}$$

Equations 20.243 to 20.252 can be solved numerically. The solution is presented in Fig. 20.26.

On the x-axis the humid air temperature ϑ_1 of the flow is shown, the y-axis shows the adiabatic saturation temperature ϑ_2, that can be achieved in thermodynamic equilibrium. Obviously, with $\varphi_1 = 100\%$ it is not possible to cool the surface down, since the air flow (1) is already saturated with vapour, so that no extra water can evaporate at the liquid surface, i.e. the system has already reached its equilibrium. Thus, if no phase change takes place, no energy is consumed, that causes a temperature drop.[57]

---

[57] As long as no water is transferred into the air flow, the energy balance is not affected, see first law of thermodynamics, Eq. 20.248. However, in case the liquid surface has a lower temperature than

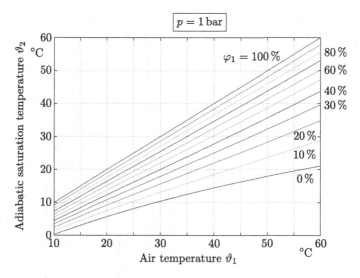

Fig. 20.26 Adiabatic saturation temperature

The further the relative humidity sinks, the larger the cooling potential is, since the amount of water, that evaporates and then is carried by the air flow, rises. However, that principle can be utilised for cooling purpose.

- What is the water content of the unsaturated air (1)?
  Alternatively, if the temperature of the water ϑ_2 is measured and if the temperature of the unsaturated air ϑ_1 is well-known, the principle, described with question one, can be utilised to estimate the water content x_1 respectively the relative humidity φ_1 of the unsaturated air. The previously given requirements are still valid. State (3), however, can easily be fixed in the h_{1+x}, x-diagram, since $\vartheta_3 = \vartheta_2$ and $\varphi_3 = 1$. Now, the ϑ_3-isothermal needs to be extended from the two-phase region to the one-phase region, see Fig. 20.25. Its intersection with the ϑ_1-isothermal in the one-phase region marks state (1). Finally, x_1 can easily be determined with a h_{1+x}, x-diagram. A possible experimental set-up, also known as psychrometer, is shown in Fig. 20.27. The wet-bulb can be realised by placing the thermometer in some wet cloth for instance.

20.4.6 The h_{1+x}, x-Diagram for Varying Total Pressure

Usually, the h_{1+x}, x-diagram is given for a fixed total pressure of 1 bar. So the question rises, how to handle the diagram for varying pressures 1 bar. Let us assume a

---

the surrounding air, there is convective heat transfer. This phenomena is not part of this investigation and thus neglected!

Fig. 20.27 Psychrometer
(wet-and-dry-bulb
thermometer)

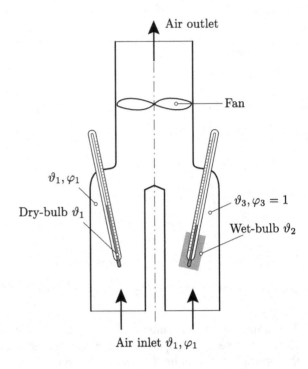

Air outlet

Fan

ϑ_1, φ_1

Dry-bulb ϑ_1

$\vartheta_3, \varphi_3 = 1$

Wet-bulb ϑ_2

Air inlet ϑ_1, φ_1

state (1) with ϑ_1, φ_1 and $p_1 \neq p_{ref} = 1$ bar shall be visualised in a standard h_{1+x}, x-diagram, that is valid for $p = p_{ref} = 1$ bar.

Obviously, the relative humidity of a state

$$\varphi = \frac{x}{x + 0.622} \cdot \frac{p}{p'_v} \tag{20.253}$$

depends on the total pressure. In order to convert from $p \to p_{ref}$ only the pressure is varied whereas temperature ϑ and water content x stay constant. The varying terms are separated to the left side of the equation, so that

$$\frac{\varphi}{p} = \frac{x}{x + 0.622} \cdot \frac{1}{p'_v} = \text{const.} \tag{20.254}$$

Obviously, the ratio $\frac{\varphi}{p}$ is constant, i.e. applied for state (1) it is

$$\boxed{\varphi_{1,ref} = \frac{p_{ref}}{p_1} \cdot \varphi_1} \tag{20.255}$$

Fig. 20.28 The h_{1+x}, x-diagram - $p \neq 1$ bar

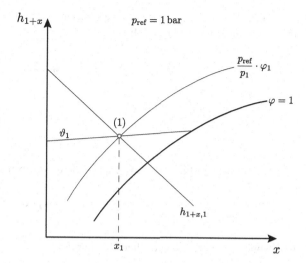

Finally, this equation can be applied to mark the state with a total pressure of $p \neq p_{\text{ref}} = 1$ bar in a standard h_{1+x}, x-diagram with $p = 1$ bar. The only variable that needs to be converted is the relative humidity which is a function of the total pressure. This conversion is shown in Fig. 20.28.

Problem 20.10 In the heat exchanger of an air conditioning system an air flow of $500 \, \frac{\text{kg}}{\text{h}}$ humid air with a relative humidity of 40% and a temperature of 32.9 °C is cooled down. The process runs isobarically at a pressure of 1 bar. The cooling power is $13.773 \, \frac{\text{MJ}}{\text{h}}$.

(a) Sketch the change of state schematically in a h_{1+x}, x-diagram.
(b) Calculate the air's water content at the inlet of the heat exchanger.
(c) Calculate the mass flow rate of the dry air.
(d) Determine the specific enthalpy, the temperature as well as the water content of the air when leaving the heat exchanger.
(e) How much liquid water needs to be removed adiabatically and isobarically in order to get saturated humid air?

Solution

(a) Sketch of the change of state in a h_{1+x}, x-diagram:
(b) The water content x_1 of state (1) can easily be determined with a h_{1+x}, x-diagram, i.e.

$$x_1 = 0.0127. \qquad (20.256)$$

In order to *calculate* x_1 the following equation needs to be applied

$$x_1 = 0.622 \frac{\varphi_1 p_s (\vartheta_1)}{p - \varphi_1 p_s (\vartheta_1)} = 0.0127. \qquad (20.257)$$

According to the steam table, the saturated pressure is $p_s(\vartheta_1) = 0.0501$ bar.

(c) The total mass flow rate follows

$$\dot{m} = \dot{m}_1 = \dot{m}_{a,1} + \dot{m}_{v,1} = 500\,\frac{kg}{h}. \tag{20.258}$$

Substitution of the vapour's mass

$$\dot{m} = \dot{m}_1 = \dot{m}_{a,1} + x_1\dot{m}_{a,1} \tag{20.259}$$

and solving for the mass flow rate of the dry air brings

$$\dot{m}_{a,1} = \frac{\dot{m}}{1+x_1} = 493.83\,\frac{kg}{h}. \tag{20.260}$$

(d) Applying the first law of thermodynamics[58]

$$\dot{Q}_{12} + \underbrace{P_{t,12}}_{=0} = \dot{m}_a\left(h_{1+x,2} - h_{1+x,1}\right) \tag{20.261}$$

leads to the specific enthalpy of state (2), i.e.

$$h_{1+x,2} = h_{1+x,1} + \frac{\dot{Q}_{12}}{\dot{m}_a} \tag{20.262}$$

$h_{1+x,1}$ is unknown and can be determined either with the h_{1+x}, x-diagram or calculated, i.e.

$$h_{1+x,1} = c_{p,a}\vartheta_1 + x_1\left(\Delta h_v + c_{p,v}\vartheta_1\right) = 65.5\,\frac{kJ}{kg}. \tag{20.263}$$

This leads to the specific enthalpy of state (2) according to Eq. 20.262

$$h_{1+x,2} = h_{1+x,1} + \frac{\dot{Q}_{12}}{\dot{m}_a} = 37.6\,\frac{kJ}{kg}. \tag{20.264}$$

The water content x_1, however, is constant, since no water is removed or added, so that

$$x_2 = x_1 = 0.0127. \tag{20.265}$$

x_2 and $h_{1+x,2}$ fix state (2) unambiguously in the h_{1+x}, x-diagram (Fig. 20.29), so that its temperature can be easily determined, i.e.

---

[58]There is no technical power $P_{t,12}$ to be transferred in the heat exchanger and the change of kinetic as well as potential energy shall be ignored. Mind, that a cooling is done, so that $\dot{Q}_{12} = -13.773\,\frac{MJ}{h}$.

$$x_2, h_{1+x,2} \rightarrow \vartheta_2 = 13.4\,°C. \tag{20.266}$$

Alternatively, temperature ϑ_2 can be calculated. The specific enthalpy in state (2) is

$$h_{1+x,2} = 37.6\,\frac{kJ}{kg} = c_{p,a}\vartheta_2 + x_2'\left(\Delta h_v + c_{p,v}\vartheta_2\right) + \left(x_2 - x_2'\right)c_{liq}\vartheta_2. \tag{20.267}$$

Solving for the temperature ϑ_2

$$\vartheta_2 = \frac{h_{1+x,2} - x_2'\Delta h_v}{c_{p,a} + x_2'c_{p,v} + \left(x_2 - x_2'\right)c_{liq}} \tag{20.268}$$

with

$$x_2' = 0.622\frac{p_s\,(\vartheta_2)}{p - p_s\,(\vartheta_2)}. \tag{20.269}$$

Obviously, Eq. 20.268 needs to be solved numerically, since $x_2' = f\,(\vartheta_2)$. The iteration leads to

$$\vartheta_2 = 13.4\,°C. \tag{20.270}$$

(e) The liquid water that needs to be removed follows

$$\dot{m}_{liq} = \dot{m}_a x_2 - \dot{m}_a x_2' = \dot{m}_a\left(x_2 - x_2'\right). \tag{20.271}$$

The water content x_2' is

$$x_2' = 0.622\frac{p_s\,(\vartheta_2)}{p - p_s\,(\vartheta_2)} = 0.0097. \tag{20.272}$$

According to the steam table, the saturated pressure is $p_s\,(\vartheta_2) = 0.015$ bar. Thus, the mass flow rate of the liquid water is

$$\dot{m}_{liq} = \dot{m}_a\left(x_2 - x_2'\right) = 1.48\,\frac{kg}{h}. \tag{20.273}$$

Problem 20.11 A room needs to be supplied with $\dot{V} = 20000\,\frac{m^3}{h}$ humid air ($\vartheta_4 = 20\,°C$, $\varphi_4 = 0.4$). This state of air is achieved by adiabatic mixing of heated fresh air (3) and used air (1). State (1) is characterised by $\vartheta_1 = 25\,°C$ and $\varphi_1 = 0.6$. Fresh air of state (2) has a temperature of $\vartheta_2 = 0\,°C$ and a relative humidity of $\varphi_2 = 0.8$. Figure 20.30 shows the technical layout. The total pressure shall be $p = 1$ bar.

(a) Sketch the changes of state schematically in a h_{1+x}, x-diagram.
(b) Determine state (3) to which the fresh air (2) needs to be conditioned.

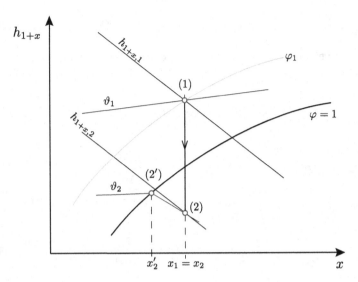

Fig. 20.29 Solution to Problem 20.10

Fig. 20.30 Technical layout
to Problem 20.11

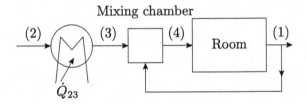

(c) To what ratio needs the fresh air and used air to be mixed?
(d) What heat flux needs to be supplied to the fresh air?

Solution

(a) Sketch of the change of state in a h_{1+x}, x-diagram; see Fig. 20.31.
(b) There are two alternatives to fix state (3):

- Alternative 1: Graphical solution

1. Since the change of state from (2) to (3) is pure heating, the water content remains constant, i.e.

$$x_2 = x_3. \tag{20.274}$$

Thus, state (3) needs to be located vertically below state (2).
2. States (1), (3) and (4) need to be located on a straight, i.e. mixing straight, due to the adiabatic mixing, see Sect. 20.4.3. However, states (1) and (4) are

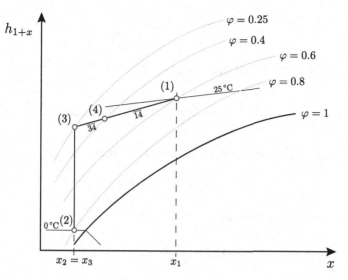

Fig. 20.31 Solution to Problem 20.11

given in the task description, so that the straight has to be extended to the left.

3. The graphical construction of 1. and 2. leads to state (3), which is the point of intersection. It can be fixed in the h_{1+x}, x-diagram, i.e.

$$\vartheta_3 = 17.7\,°C \tag{20.275}$$

and

$$\varphi_3 = 25\%. \tag{20.276}$$

- Alternative 2: Numerical solution
  The mixing chamber's mass balance for air reads as

$$\dot{m}_{a,3} + \dot{m}_{a,1} = \dot{m}_{a,4}. \tag{20.277}$$

For water accordingly

$$\dot{m}_{a,3}x_3 + \dot{m}_{a,1}x_1 = \dot{m}_{a,4}x_4. \tag{20.278}$$

The first law of thermodynamics in steady state for the mixing chamber with the common premises is

$$\dot{H}_1 + \dot{H}_3 = \dot{H}_4. \tag{20.279}$$

Introducing the specific enthalpy brings

$$\dot{m}_{a,3} h_{1+x,3} + \dot{m}_{a,1} h_{1+x,1} = \underbrace{\left(\dot{m}_{a,3} + \dot{m}_{a,1} \right)}_{=\dot{m}_{a,4}} h_{1+x,4}. \qquad (20.280)$$

Hence, the specific enthalpy of state (3) is

$$h_{1+x,3} = \frac{\left(\dot{m}_{a,3} + \dot{m}_{a,1} \right) h_{1+x,4} - \dot{m}_{a,1} h_{1+x,1}}{\dot{m}_{a,3}}. \qquad (20.281)$$

For state (1) it is known from the steam table, that

$$p_s \left(\vartheta_1 \right) = 0.0317 \, \text{bar}. \qquad (20.282)$$

The water content follows

$$x_1 = 0.622 \frac{\varphi_1 p_s \left(\vartheta_1 \right)}{p - \varphi_1 p_s \left(\vartheta_1 \right)} = 0.0121. \qquad (20.283)$$

Its specific enthalpy is

$$h_{1+x,1} = c_{p,a} \vartheta_1 + x_1 \left(\Delta h_v + c_{p,v} \vartheta_1 \right) = 55.81 \, \frac{\text{kJ}}{\text{kg}}. \qquad (20.284)$$

For state (2) it is known from the steam table, that

$$p_s \left(\vartheta_2 \right) = 0.0061 \, \text{bar}. \qquad (20.285)$$

The water content follows

$$x_2 = 0.622 \frac{\varphi_2 p_s \left(\vartheta_2 \right)}{p - \varphi_2 p_s \left(\vartheta_2 \right)} = 0.0031. \qquad (20.286)$$

We already know, that for state (3) it is

$$x_3 = x_2 = 0.0031. \qquad (20.287)$$

For state (4) it is known from the steam table, that

$$p_s \left(\vartheta_4 \right) = 0.0234 \, \text{bar}. \qquad (20.288)$$

The water content follows

$$x_4 = 0.622 \frac{\varphi_4 p_s \left(\vartheta_4 \right)}{p - \varphi_4 p_s \left(\vartheta_4 \right)} = 0.0059. \qquad (20.289)$$

Its specific enthalpy is

$$h_{1+x,4} = c_{p,a}\vartheta_4 + x_4 \left(\Delta h_v + c_{p,v}\vartheta_4\right) = 34.98 \, \frac{\text{kJ}}{\text{kg}}. \tag{20.290}$$

The mass flow rate $\dot{m}_{a,4}$ of the dry air can be calculated with

$$v_{1+x,4} = \frac{\dot{V}}{\dot{m}_{a,4}} \tag{20.291}$$

so that

$$\dot{m}_{a,4} = \frac{\dot{V}}{v_{1+x,4}}. \tag{20.292}$$

The specific volume of state (4) results in

$$v_{1+x,4} = \frac{R_a T_4}{p} \left[1 + 1.608 x_4\right]. \tag{20.293}$$

Hence, the mass flow rate is

$$\dot{m}_{a,4} = \frac{p\dot{V}}{R_a T_4 \left[1 + 1.608 x_4\right]} = 6.539 \, \frac{\text{kg}}{\text{s}}. \tag{20.294}$$

Rearranging the air's mass balance, see Eq. 20.277, brings

$$\dot{m}_{a,3} = \dot{m}_{a,4} - \dot{m}_{a,1}. \tag{20.295}$$

Substitution in the water's mass balance, see Eq. 20.278

$$\left(\dot{m}_{a,4} - \dot{m}_{a,1}\right) x_3 + \dot{m}_{a,1} x_1 = \dot{m}_{a,4} x_4 \tag{20.296}$$

and solving for

$$\dot{m}_{a,1} = \dot{m}_{a,4} \frac{x_4 - x_3}{x_1 - x_3} = 2.0479 \, \frac{\text{kg}}{\text{s}}. \tag{20.297}$$

The mass flow rate of state (3) follows accordingly

$$\dot{m}_{a,3} = \dot{m}_{a,4} - \dot{m}_{a,1} = 4.4913 \, \frac{\text{kg}}{\text{s}}. \tag{20.298}$$

Hence, finally the specific enthalpy can be calculated, see Eq. 20.281,

$$h_{1+x,3} = \frac{\left(\dot{m}_{a,3} + \dot{m}_{a,1}\right) h_{1+x,4} - \dot{m}_{a,1} h_{1+x,1}}{\dot{m}_{a,3}} = 25.49 \, \frac{\text{kJ}}{\text{kg}}. \tag{20.299}$$

It is further known, that

$$h_{1+x,3} = c_{p,a}\vartheta_3 + x_3\left(\Delta h_v + c_{p,v}\vartheta_3\right) = 25.49\,\frac{kJ}{kg}. \qquad (20.300)$$

This equation can be solved for the temperature in state (3), i.e.

$$\vartheta_3 = \frac{h_{1+x,3} - x_3\Delta h_v}{c_{p,a} + x_3 c_{p,v}} = 17.68\,°C. \qquad (20.301)$$

The relative humidity is

$$\varphi_3 = \frac{x_3}{x_3 + 0.622}\frac{p}{p_s\left(\vartheta_3\right)} = 24.15\% \qquad (20.302)$$

with

$$p_s\left(\vartheta_3\right) = 0.0202\,bar \qquad (20.303)$$

taken from the steam table.
A comparison of alternative 1 and 2 clearly shows how beneficial the graphical solution is.

(c) Once again, there are two alternatives:

- Alternative 1: Graphical solution
  According to Sect. 20.4.3, the ratio can be fixed with the length of the lever arms, i.e.

$$\frac{\dot{m}_{a,3}}{\dot{m}_{a,1}} = \frac{\overline{14}}{\overline{34}} = 2.14 \qquad (20.304)$$

  This result has been achieved by measuring the length of the lever arms in a h_{1+x}, x-diagram.
- Alternative 2: Numerical solution
  The mass flow rates have already been calculated in part (b), i.e.

$$\frac{\dot{m}_{a,3}}{\dot{m}_{a,1}} = \frac{4.4913\,\frac{kg}{s}}{2.0479\,\frac{kg}{s}} = 2.193. \qquad (20.305)$$

(d) The heat flux follows by applying the first law of thermodynamics with the common premises for the heater

$$\dot{Q}_{23} + \underbrace{P_{t,23}}_{=0} = \dot{m}_{a,2}\left(h_{1+x,3} - h_{1+x,2}\right) \qquad (20.306)$$

with

$$h_{1+x,2} = c_{p,a}\vartheta_2 + x_2 \left(\Delta h_v + c_{p,v}\vartheta_2\right) = 7.64 \, \frac{\text{kJ}}{\text{kg}}. \tag{20.307}$$

Thus, finally the heat flux is

$$\dot{Q}_{23} = \dot{m}_{a,2} \left(h_{1+x,3} - h_{1+x,2}\right) = 80.19 \, \text{W}. \tag{20.308}$$

Chapter 21
Steady State Flow Processes

Steady state flow processes have already been discussed when introducing the first law of thermodynamics for open systems, see Sect. 11.3. Figure 21.1 shows an example for a simple open system with a single inlet and a single outlet. Characteristic for open systems in steady state is, that the mass inside the system remains constant with respect to time, so that the mass flux into the system needs to be balanced by the mass flux out of the system, i.e.

$$\boxed{\dot{m}_1 = \dot{m}_2} \tag{21.1}$$

Furthermore, the *state* inside the system must not vary with respect to time, so that the first law of thermodynamics needs to follow

$$\frac{dE}{dt} = 0 = \dot{E}_{in} - \dot{E}_{out}. \tag{21.2}$$

In other words, the energy flux into the system must be equal to the energy flux out of the system, i.e.

$$\dot{m}_1\left(h_1 + \frac{1}{2}c_1^2 + gz_1\right) + \dot{Q} + P_t - \dot{m}_2\left(h_2 + \frac{1}{2}c_2^2 + gz_2\right) \tag{21.3}$$

respectively in specific notation

$$\boxed{q + w_t = h_2 - h_1 + \frac{1}{2}\left(c_2^2 - c_1^2\right) + g\left(z_2 - z_1\right)} \tag{21.4}$$

and in differential notation

$$\boxed{\delta q + \delta w_t = dh + c\,dc + g\,dz} \tag{21.5}$$

© Springer Nature Switzerland AG 2019
A. Schmidt, *Technical Thermodynamics for Engineers*,
https://doi.org/10.1007/978-3-030-20397-9_21

First Law   **Entropy - Second Law**

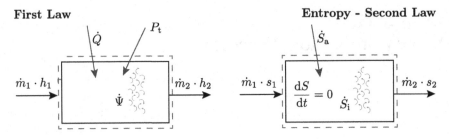

Fig. 21.1 Steady state flow process

The specific technical work w_t can be calculated with the partial energy equation, see Sect. 11.3.4, and follows

$$w_t = \int_1^2 v \, dp + \frac{1}{2} \left(c_2^2 - c_1^2 \right) + g \left(z_2 - z_1 \right) + \psi_{12} \qquad (21.6)$$

Since no state value is allowed to vary by time in a steady state open system, the entropy needs to follow accordingly

$$\frac{dS}{dt} = 0 = \dot{S}_{in} - \dot{S}_{out}. \qquad (21.7)$$

In other words, the entropy flux into the system must be equal to the entropy flux out of the system, i.e.

$$\dot{S}_i + \dot{S}_a + \dot{m}_1 s_1 = \dot{m}_2 s_2 \qquad (21.8)$$

respectively in specific notation

$$s_i + s_a = s_2 - s_1 \qquad (21.9)$$

and in differential notation

$$\delta s_i + \delta s_a = ds \qquad (21.10)$$

The process values δs_i and δs_a can be substituted, so that

$$\frac{\delta \psi}{T} + \frac{\delta q}{T} = ds \qquad (21.11)$$

In Part I these equations have been discussed and practised in detail. So far no restrictions have been made regarding the fluid, so the given correlation are generally valid. They can be applied for ideal gases as well as for real fluids. In this chapter the focus is further on steady state flow processes including supersonic flows.

21.1 Incompressible Flows

Figure 21.2a shows an open system in steady state. Let us assume, that the fluid is incompressible, e.g. an incompressible liquid as treated in Part I. Incompressible fluids obey

$$\rho = \frac{1}{v} = \text{const.} \rightarrow v = \text{const.} \tag{21.12}$$

Under this premise the partial energy Eq. 21.6 can be simplified as follows

$$w_{t12} = v\,(p_2 - p_1) + \psi_{12} + \frac{c_2^2 - c_1^2}{2} + g \cdot (z_2 - z_1). \tag{21.13}$$

For a work-insulated flow process, i.e. no technical work is transferred across the system boundary, it follows:

$$w_{t12} = 0 = v\,(p_2 - p_1) + \psi_{12} + \frac{c_2^2 - c_1^2}{2} + g \cdot (z_2 - z_1). \tag{21.14}$$

Multiplication with the density leads to

$$0 = (p_2 - p_1) + \rho\psi_{12} + \rho\frac{c_2^2 - c_1^2}{2} + g\rho \cdot (z_2 - z_1). \tag{21.15}$$

Rearranging finally brings

$$\boxed{\left(p + \frac{\rho}{2}c^2 + g\rho z\right)_2 - \left(p + \frac{\rho}{2}c^2 + g\rho z\right)_1 = -\rho\psi_{12} \le 0} \tag{21.16}$$

Now, let us further assume, the flow is frictionless, i.e.

$$\psi_{12} = 0. \tag{21.17}$$

Under this premise the partial energy equation can be simplified as follows

$$\left(p + \frac{\rho}{2}c^2 + g\rho z\right)_2 - \left(p + \frac{\rho}{2}c^2 + g\rho z\right)_1 = 0. \tag{21.18}$$

In other words

$$\boxed{\left(p + \frac{\rho}{2}c^2 + g\rho z\right) = \text{const.}} \tag{21.19}$$

This equation is known in fluid dynamics as Bernoulli equation. It claims, that in a frictionless flow with no technical work being transferred, see Fig. 21.2b, the *total*

(a)
$c_1, z_1, p_1, A_1, \dots$

(b)
$c_1, z_1, p_1, A_1, \dots$

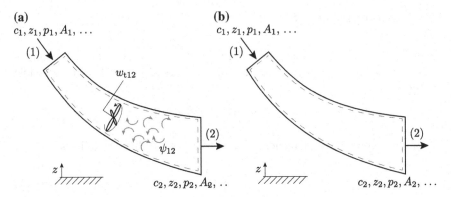

Fig. 21.2 Incompressible flow

pressure[1] $\left(p + \frac{\varrho}{2}c^2 + g\rho z\right)$ remains constant. If friction occurs, the total pressure along the flow is reduced by $\rho\psi_{12}$, see Eq. 21.16.

21.2 Adiabatic Flows

In this section adiabatic flows, as shown in Fig. 21.3, are investigated in steady state. The basic principles have already been derived in Part I and are presented briefly. Since the flow is adiabatic, there is no exchange of thermal energy with with environment, i.e.

$$q_{12} = 0. \tag{21.20}$$

Furthermore, a work-insulated flow process without any exchange of technical work across the system boundary is assumed, so that

$$w_{t12} = 0. \tag{21.21}$$

Under these conditions the first law of thermodynamics for steady state conditions obeys

$$q_{12} + w_{t12} = 0 = h_2 - h_1 + \frac{c_2^2 - c_1^2}{2} + g \cdot (z_2 - z_1). \tag{21.22}$$

With the assumption, that potential energies[2] can be ignored, i.e. $z_2 = z_1$, it simplifies to

---

[1]The total pressure is composed of the static pressure p, that a person would be faced who is following the flow with its velocity c, the dynamic pressure $\frac{\varrho}{2}c^2$ and the hydraulic pressure $g\rho z$.
[2]Generally speaking, potential energies can be neglected as long as the vertical distance between inlet and outlet is not huge in size.

$$h_2 + \frac{c_2^2}{2} = h_1 + \frac{c_1^2}{2}. \tag{21.23}$$

Introducing the so-called total enthalpy $h^+ = h + \frac{c^2}{2}$ results in

$$\boxed{h_2^+ = h_1^+} \tag{21.24}$$

The second law of thermodynamics obeys

$$\dot{m}s_1 + \dot{S}_i + \dot{S}_a = \dot{m}s_2. \tag{21.25}$$

Since the flow is adiabatic, the exchanged entropy by heat across the system boundary is zero, i.e. $\dot{S}_a = 0$. This leads to

$$\dot{m}(s_1 - s_2) + \dot{S}_i = 0. \tag{21.26}$$

Dividing by the mass flow rate means

$$s_2 - s_1 = s_{i12} = \int_1^2 \frac{\delta\psi}{T} \geq 0 \tag{21.27}$$

respectively in differential notation

$$\boxed{ds = \delta s_i = \frac{\delta\psi}{T} \geq 0} \tag{21.28}$$

In other words, a decrease of entropy is not possible for an adiabatic flow. The partial energy equation for the technical work obeys

$$w_{t,12} = 0 = \int_1^2 v\,dp + \psi_{12} + \frac{c_2^2 - c_1^2}{2} + g \cdot (z_2 - z_1). \tag{21.29}$$

With ignoring the change of potential energy, see above, it simplifies to

$$\boxed{\int_1^2 v\,dp + \psi_{12} + \frac{c_2^2 - c_1^2}{2} = 0} \tag{21.30}$$

Figure 21.4 shows an adiabatic flow process in a h, s-diagram. Let state (1) be the initial start, due to Eq. 21.27, the entropy can not sink, so that the outlet state must move to the right or vertically. The first law of thermodynamics has shown, that

Fig. 21.3 Adiabatic flow
with $A \neq$ const.

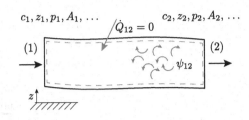

Fig. 21.4 Adiabatic flow
with $A \neq$ const.

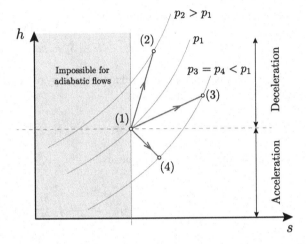

$$h^+ = h + \frac{c^2}{2} = \text{const.} \tag{21.31}$$

However, two cases are possible:

- Case 1: Accelerated flow, i.e. $dc > 0$
  If the specific enthalpy decreases, the velocity needs to increase, in order to keep
  the total enthalpy h^+ constant. According to the partial energy equation, i.e.

$$\int_1^2 v \, dp = \underbrace{-\psi_{12}}_{<0} + \underbrace{\frac{c_1^2 - c_2^2}{2}}_{<0} < 0, \tag{21.32}$$

  this case always comes along with a pressure loss.

- Case 2: Decelerated flow, i.e. $dc < 0$
  If the specific enthalpy increases, the velocity needs to decrease, in order to keep
  the total enthalpy h^+ constant. For this case the partial energy equation, i.e.

Fig. 21.5 Adiabatic diffusor

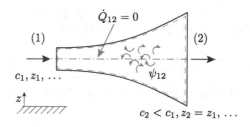

$$\int\limits_1^2 v\,\mathrm{d}p = \underbrace{-\psi_{12}}_{<0} + \underbrace{\frac{c_1^2 - c_2^2}{2}}_{>0}, \tag{21.33}$$

shows, that a pressure rise is possible, when

$$\psi_{12} < \frac{c_1^2 - c_2^2}{2} \tag{21.34}$$

and a pressure loss, when

$$\psi_{12} > \frac{c_1^2 - c_2^2}{2}. \tag{21.35}$$

In case that

$$\psi_{12} = \frac{c_1^2 - c_2^2}{2} \tag{21.36}$$

the pressure remains constant.

21.2.1 Adiabatic Diffusor

Figure 21.5 shows an adiabatic diffusor, that commonly is utilised to reduce the flow velocity by a cross-section expansion. Thus, the outlet velocity is smaller than the inlet velocity, i.e.

$$c_2 < c_1. \tag{21.37}$$

Section 21.2 has shown, that, in steady state and for ignoring potential energies, the first law of thermodynamics obeys

$$h^+ = h + \frac{c^2}{2} = \text{const.} \tag{21.38}$$

so that

$$h_1 + \frac{c_1^2}{2} = h_2 + \frac{c_2^2}{2}. \tag{21.39}$$

Since the flow is decelerated the specific enthalpy needs to rise in order to keep the total specific enthalpy constant, i.e.

$$h_2 > h_1. \tag{21.40}$$

The outlet velocity c_2 can easily be calculated and results in

$$\boxed{c_2 = \sqrt{c_1^2 + 2\,(h_1 - h_2)}} \tag{21.41}$$

According to the second law for thermodynamics for adiabatic flows, see Sect. 21.2, reads as

$$\boxed{\mathrm{d}s = \delta s_{\mathrm{i}} = \frac{\delta \psi}{T} \geq 0} \tag{21.42}$$

so that the entropy needs to increase. Since no technical work is transferred across the system boundary, the partial energy equation for the technical work is

$$\boxed{w_{\mathrm{t},12} = 0 = \int_1^2 v\,\mathrm{d}p + \psi_{12} + \frac{c_2^2 - c_1^2}{2}} \tag{21.43}$$

This leads to a correlation for the change of pressure, i.e.

$$\int_1^2 v\,\mathrm{d}p = -\psi_{12} + \frac{c_1^2 - c_2^2}{2}. \tag{21.44}$$

However, two cases are possible:

- Case 1: Pressure increase, $(1) \rightarrow (2)$

$$\psi_{12} < \frac{c_1^2 - c_2^2}{2} \tag{21.45}$$

- Case 2: Pressure decrease, $(1) \rightarrow (3)$

$$\psi_{13} > \frac{c_1^2 - c_3^2}{2} \tag{21.46}$$

These considerations are summarised in a h, s-diagram, see Fig. 21.6. An isentropic diffusor efficiency is defined as

Fig. 21.6 Adiabatic diffusor in a h, s-diagram

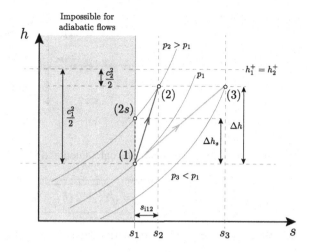

$$\eta_{sD} = \frac{\Delta h_s}{\Delta h} = \frac{h_{2s} - h_1}{h_2 - h_1} = \frac{0.5 \left(c_1^2 - c_{2s}^2 \right)}{0.5 \left(c_1^2 - c_2^2 \right)} \tag{21.47}$$

i.e. the reference is an isentropic change of state $(1) \rightarrow (2s)$.

21.2.2 Adiabatic Nozzle

Figure 21.7 shows an adiabatic nozzle, that is utilised to increase the flow velocity by a cross-section reduction. Thus, the outlet velocity is larger than the inlet velocity, i.e.

$$c_2 > c_1. \tag{21.48}$$

Section 21.2 has shown, that, in steady state and for ignoring potential energies, the first law of thermodynamics obeys (Fig. 21.8)

$$h^+ = h + \frac{c^2}{2} = \text{const.} \tag{21.49}$$

so that

$$h_1 + \frac{c_1^2}{2} = h_2 + \frac{c_2^2}{2}. \tag{21.50}$$

Since the flow is accelerated the specific enthalpy needs to sink in order to keep the total specific enthalpy constant, i.e.

$$h_2 < h_1. \tag{21.51}$$

Fig. 21.7 Adiabatic nozzle

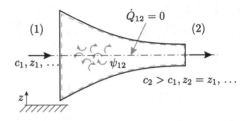

The outlet velocity c_2 can easily be calculated and is

$$c_2 = \sqrt{c_1^2 + 2\left(h_1 - h_2\right)} \tag{21.52}$$

According to the second law for thermodynamics for adiabatic flows, see Sect. 21.2, reads as

$$ds = \delta s_i = \frac{\delta \psi}{T} \geq 0 \tag{21.53}$$

so that the entropy needs to increase. Since no technical work is transferred across the system boundary, the partial energy equation for the technical work is

$$w_{t,12} = 0 = \int_1^2 v\,dp + \psi_{12} + \frac{c_2^2 - c_1^2}{2} \tag{21.54}$$

This leads to a correlation for the change of pressure, i.e.

$$\int_1^2 v\,dp = \underbrace{-\psi_{12}}_{<0} + \underbrace{\frac{c_1^2 - c_2^2}{2}}_{<0} < 0. \tag{21.55}$$

Thus, for an adiabatic nozzle the pressure needs to decrease. These considerations are summarised in a h, s-diagram, see Fig. 21.6. An isentropic diffusor efficiency is defined as

$$\eta_{sS} = \frac{\Delta h}{\Delta h_s} = \frac{h_2 - h_1}{h_{2s} - h_1} = \frac{0.5\left(c_1^2 - c_2^2\right)}{0.5\left(c_1^2 - c_{2s}^2\right)} \tag{21.56}$$

i.e. the reference is an isentropic change of state $(1) \rightarrow (2s)$.

Problem 21.1 A mass flux of $\frac{kg}{s}$ vapour with an initial state of $p_1 = 22$ bar and $\vartheta_1 = 500\,°C$ shall expand polytropically with a constant exponent of $n = 1.2$ in adiabatic nozzle to a pressure of $p_2 = 15$ bar. The vapour's inlet velocity is $c_1 = 10\,\frac{m}{s}$. Changes of the potential energy can be ignored. Please calculate:

Fig. 21.8 Adiabatic nozzle in a h, s-diagram

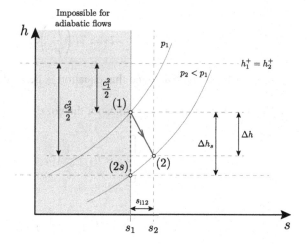

(a) the specific volume v_2 and the temperature ϑ_2 of the vapour in state (2),
(b) the outlet velocity c_2 of the vapour,
(c) the specific pressure work $y_{12} = \int_1^2 v \, dp$ and the specific dissipation ψ_{12},
(d) the thermodynamic mean temperature,
(e) the cross-sectional area of inlet and outlet.

Solution

(a) According to the steam table state (1) is overheated steam with

$$v_1 = 0.1595 \, \frac{m^3}{kg}. \tag{21.57}$$

For a polytropic change of state it is[3]

$$pv^n = \text{const.} \tag{21.58}$$

This leads to

$$v_2 = v_1 \left(\frac{p_1}{p_2} \right)^{\frac{1}{n}} = 0.2194 \, \frac{m^3}{kg}. \tag{21.59}$$

According to the steam table one gets

$$\vartheta_2 = f(v_2, p_2) = 450.72 \,^\circ\text{C}. \tag{21.60}$$

---

[3] No matter what fluid is used!

Mind, that

$$T_2 = T_1 \left(\frac{p_2}{p_1} \right)^{\frac{n-1}{n}} \tag{21.61}$$

can not been applied, since this equation is derived by applying the ideal gas law.

(b) The first law of thermodynamics for the nozzle reads as

$$\dot{Q}_{12} + P_{t,12} = 0 = \dot{m} \left[h_2 - h_1 + \frac{1}{2} \left(c_2^2 - c_1^2 \right) + \underbrace{g \left(z_2 - z_1 \right)}_{=0} \right]. \tag{21.62}$$

Thus, it follows

$$c_2 = \sqrt{2 \left(h_1 - h_2 \right) + c_1^2}. \tag{21.63}$$

The specific enthalpies follow from the steam table, i.e.

$$h_1 = h \left(22 \, \text{bar}, 500\,°C \right) = 3465.9 \, \frac{\text{kJ}}{\text{kg}} \tag{21.64}$$

and

$$h_2 = h \left(15 \, \text{bar}, 450.72\,°C \right) = 3366.2 \, \frac{\text{kJ}}{\text{kg}}. \tag{21.65}$$

Finally the outlet velocity is[4]

$$c_2 = \sqrt{2 \left(h_1 - h_2 \right) + c_1^2} = 446.57 \, \frac{\text{m}}{\text{s}}. \tag{21.66}$$

(c) Part (a) has shown, that for a polytropic change of state it always follows

$$v = v_1 \left(\frac{p_1}{p} \right)^{\frac{1}{n}} = 0.2194 \, \frac{\text{m}^3}{\text{kg}}. \tag{21.67}$$

Thus, the pressure works can be calculated accordingly

$$y_{12} = \int_1^2 v \, dp = v_1 p_1 \frac{n}{n-1} \left[\left(\frac{p_2}{p_1} \right)^{\frac{n-1}{n}} - 1 \right] = -130.16 \, \frac{\text{kJ}}{\text{kg}}. \tag{21.68}$$

The technical work then is

---

[4]Mind the units! The specific enthalpy should be converted in $\frac{\text{J}}{\text{kg}}$.

$$w_{t,12} = \int_1^2 v\,dp + \psi_{12} + \frac{1}{2}\left(c_2^2 - c_1^2\right) = 0. \tag{21.69}$$

Mind, that no technical work is transferred[5] in the nozzle across the system boundary. Thus, the dissipation is

$$\psi_{12} = \frac{1}{2}\left(c_1^2 - c_2^2\right) - \int_1^2 v\,dp = h_2 - h_1 - \int_1^2 v\,dp = 30.49 \frac{kJ}{kg}. \tag{21.70}$$

(d) The second law of thermodynamics obeys

$$\dot{m}s_1 + \dot{S}_i + \dot{S}_a = \dot{m}s_2. \tag{21.71}$$

Since the nozzle is adiabatic it simplifies to

$$\dot{S}_i = \dot{m}\left(s_2 - s_1\right). \tag{21.72}$$

In specific notation it results in

$$s_i = s_2 - s_1 = \int_1^2 \frac{\delta\psi}{T} = \frac{1}{T_m}\int_1^2 \delta\psi = \frac{\psi_{12}}{T_m}. \tag{21.73}$$

Hence, the thermodynamic mean temperature follows accordingly

$$T_m = \frac{\psi_{12}}{s_2 - s_1}. \tag{21.74}$$

The specific entropies follow from the steam table, i.e.

$$s_1 = s(22\,bar, 500\,°C) = 7.3873 \frac{kJ}{kg\,K} \tag{21.75}$$

and
$$s_2 = s(15\,bar, 450.72\,°C) = 7.4281 \frac{kJ}{kg\,K}. \tag{21.76}$$

Finally the thermodynamic mean temperature is

$$T_m = \frac{\psi_{12}}{s_2 - s_1} = 748.36\,K = 475.21\,°C. \tag{21.77}$$

---

[5]A nozzle is a work-insulated, i.e. passive, system.

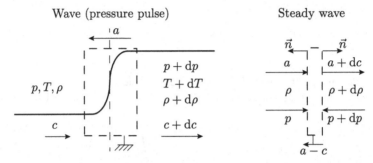

Wave (pressure pulse)   Steady wave

(a) Fixed coordinate system   (b) Moving coordinate system

Fig. 21.9 Velocity of sound a

(e) The mass flux needs to be constant in steady state, so that

$$\dot{m} = \rho \dot{V} = \rho A c = \frac{Ac}{v} = \text{const.} \tag{21.78}$$

This leads to

$$A_1 = \frac{\dot{m} v_1}{c_1} = 0.1595 \, \text{m}^2 \tag{21.79}$$

and

$$A_2 = \frac{\dot{m} v_2}{c_2} = 49.13 \, \text{cm}^2. \tag{21.80}$$

21.3   Velocity of Sound

For the following sections it is required to know the velocity of sound a. The velocity
of sound describes the speed at which a pressure pulse moves through a fluid. In order
to derive this velocity, a fluid is assumed, that is moving unhindered with a velocity c,
see Fig. 21.9. At a certain point of time a pressure pulse is initiated, that causes a wave
moving with a speed a. This situation is shown in Fig. 21.9a with a fixed coordinate
system. Left from the pulse, the fluid is undisturbed and enters the coordinate system
with a pressure p, a temperature T, a density ρ and a velocity c. However, each of
these state values will have varied when leaving the control volume, i.e. the pressure
is $p + dp$, the temperature is $T + dT$, the density $\rho + d\rho$ and the velocity is $c + dc$.

In a next step the coordinate system starts moving, see Fig. 21.9. The velocity of
the moving coordinate system can be adjusted,[6] so that the fluid enters the coordinate

---

[6]Imagine a pressure pulse is traveling with a velocity $a = 100 \, \frac{\text{km}}{\text{h}}$ from right to left, facing a
headwind of $c = 20 \, \frac{\text{km}}{\text{h}}$. Thus, the fluid hits the pulse front with a velocity of $120 \, \frac{\text{km}}{\text{h}}$. However, the
fluid's velocity behind the pulse shall be slightly higher, e.g. $c + dc = 22 \, \frac{\text{km}}{\text{h}}$, so that the velocity

system on the left with the velocity a, i.e. according to the sketch the moving velocity of the coordinate system needs to be $a - c$. This results to the inlet and outlet velocities of Fig. 21.9b:

- Left of the control volume:

$$\underbrace{c}_{a)} + \underbrace{(a - c)}_{\text{Velocity of co-system}} = \underbrace{a}_{b)} \qquad (21.81)$$

- Right of the control volume:

$$\underbrace{c + dc}_{a)} + \underbrace{(a - c)}_{\text{Velocity of co-system}} = \underbrace{a + dc}_{b)} \qquad (21.82)$$

The equation of continuity, i.e. principle of mass conservation, for the moving coordinate system, Fig. 21.9b, reads as

$$A\rho a = A\,(\rho + d\rho)\,(a + dc). \qquad (21.83)$$

Ignoring the non-linear term of second order $d\rho\,dc$ results in

$$\rho\,dc = -a\,d\rho. \qquad (21.84)$$

The conservation of momentum for the moving coordinate system is, see [23],

$$\frac{dI}{dt} = \int_A \rho \vec{c}\,(\vec{c} \cdot \vec{n})\,dA = \sum \vec{F} = \int_A -\vec{n}\,p\,dA. \qquad (21.85)$$

In this case the problem is one-dimensional, so that

$$\underbrace{A\,(\rho + d\rho)\,(a + dc)}_{A\rho a}\,(a + dc) - A\rho aa = pA - (p + dp)\,A. \qquad (21.86)$$

In combination with the mass conservation it follows

$$A\rho a\,(a + dc) - A\rho aa = pA - (p + dp)\,A. \qquad (21.87)$$

Rearranging brings

$$\rho\,dc = -\frac{dp}{a}. \qquad (21.88)$$

---

rises by $dc = 2\,\frac{km}{h}$. If we want the air to face the pulse front with exactly the velocity $a = 100\,\frac{km}{h}$, the coordinate system needs to go with a velocity of $a - c = 80\,\frac{km}{h}$ to the left. In this case the velocity of the air leaving the coordinate system on the right hand side is $a + dc = 102\,\frac{km}{h}$. Mind, that for balancing any coordinate system can be applied. The solution must never depend on the choice of the coordinate system!

Combining equation of continuity and conservation of momentum results in

$$a^2 = \frac{\mathrm{d}p}{\mathrm{d}\rho}. \tag{21.89}$$

The momentum equation has been applied without any friction, i.e. the flow is reversible. If it is further assumed, that the pulse moves so fast, that there is no time for heat exchange, the flow is adiabatic as well. Hence, the process is isentropic. Hence, the velocity of sound a is

$$a = \sqrt{\left(\frac{\mathrm{d}p}{\mathrm{d}\rho}\right)_s}. \tag{21.90}$$

For an isentropic change of state of an ideal gas it is known that

$$pv^\kappa = \frac{p}{\rho^\kappa} = \text{const.} \rightarrow p = C\rho^\kappa. \tag{21.91}$$

The derivative for an isentropic change follows

$$\frac{\mathrm{d}p}{\mathrm{d}\rho} = C\kappa\rho^{\kappa-1}. \tag{21.92}$$

The constant C can be substituted by, see Eq. 21.91,

$$C = \frac{p}{\rho^\kappa}. \tag{21.93}$$

This finally leads to

$$\frac{\mathrm{d}p}{\mathrm{d}\rho} = \frac{p}{\rho^\kappa}\kappa\rho^{\kappa-1} = \kappa\frac{p}{\rho} = \kappa pv. \tag{21.94}$$

Hence, the velocity of sound follows

$$\boxed{a = \sqrt{\kappa pv} = \sqrt{\kappa RT}} \tag{21.95}$$

The Mach-number Ma is defined by the ratio

$$\boxed{\mathrm{Ma} = \frac{c}{a}} \tag{21.96}$$

The same correlation as Eq. 21.94 can be derived with the second law of thermodynamics. In differential notation it obeys

$$\mathrm{d}s = \delta s_\mathrm{i} + \delta s_\mathrm{a} = 0 = c_v\frac{\mathrm{d}T}{T} + R\frac{\mathrm{d}v}{v}. \tag{21.97}$$

With the thermal equation of state

$$T = \frac{pv}{R} \rightarrow dT = \frac{p}{R}\,dv + \frac{v}{R}\,dp \tag{21.98}$$

one gets with $c_p = R + c_v$

$$0 = c_p\frac{dv}{v} + c_v\frac{dp}{p}. \tag{21.99}$$

It is further known, that

$$v = \rho^{-1} \rightarrow dv = -\frac{1}{\rho^2}d\rho. \tag{21.100}$$

Thus, for an isentropic pressure pulse the ratio of pressure change to density change needs to follow

$$\frac{dp}{d\rho} = \kappa\frac{p}{\rho} = \kappa pv. \tag{21.101}$$

The pressure pulse in Fig. 21.9b represents an imperfection, that usually is related with generation of entropy, see Part I. However, it is assumed, that this imperfection moves so fast, that the system is not able to respond to that imperfection. This is the cause, why no generation of entropy was assumed and why the conservation of momentum has been applied frictionless.

21.4   Fanno Correlation

Let us assume an adiabatic but frictional flow in a tube with a constant cross-section A, see Fig. 21.10. The equation of continuity states, that the mass flux is constant in steady state, i.e.

$$\dot{m} = \rho A c = \frac{Ac}{v} = \text{const.} \tag{21.102}$$

In other words the mass flux density is constant as well, since the cross-section does not vary

$$\boxed{\frac{c}{v} = \frac{\dot{m}}{A} = \dot{m}'' = \text{const.}} \tag{21.103}$$

The first law of thermodynamics for steady state conditions obeys[7]

$$q + w_t = 0 = h_2 - h_1 + \frac{1}{2}\left(c_2^2 - c_1^2\right) \tag{21.104}$$

with the so-called specific total enthalpy h^+ it reads as

---

[7]Ignoring potential energies!

$$h^+ = h + \frac{c^2}{2} = \text{const.} = h_1^+.$$ (21.105)

The specific total enthalpy of the inlet shall be given[8] and is h_1^+. However, the combination of first law of thermodynamics and mass conservation results in

$$\boxed{h + \frac{1}{2}v^2\dot{m}''^2 = \text{const.} = h_1^+}$$ (21.106)

According to the second law of thermodynamics for adiabatic flow it is

$$\boxed{ds = \delta s_i = \frac{\delta\psi}{T} \geq 0}$$ (21.107)

Thus, in this case the entropy needs to rise due to friction. A reduction of entropy is not possible, since the flow is supposed to be adiabatic. Furthermore, the partial energy equation under these conditions is

$$w_t = 0 = \int_1^2 v\,dp + \frac{1}{2}\left(c_2^2 - c_1^2\right) + \psi_{12}.$$ (21.108)

Unfortunately, this equation can not be applied any further, since the friction is unknown!

The idea now is to visualise the change of state for an ideal gas in a h, s-diagram:

1. Inlet state (1) is fixed by two independent state values, e.g. T_1 and p_1. Thus, the specific volume is

$$v_1 = \frac{RT_1}{p_1}.$$ (21.109)

The specific enthalpy[9] is

$$h_1 = h_{\text{ref}} + c_p\left(T_1 - T_{\text{ref}}\right).$$ (21.110)

The specific entropy obeys

$$s_1 = s_{\text{ref}} + c_p \ln\frac{T_1}{T_{\text{ref}}} - R \ln\frac{p_1}{p_{\text{ref}}}.$$ (21.111)

Hence, state (1) can be fixed unambiguously in a h, s-diagram.
2. The velocity c_1 is chosen next, so that the mass flux density is defined as

---

[8]For instance by its internal state p_1, T_1 and its velocity c_1.

[9]T_{ref}, p_{ref}, h_{ref} and s_{ref} represent a arbitrary reference state!

Fig. 21.10 Adiabatic, frictional tube flow (A = const.)

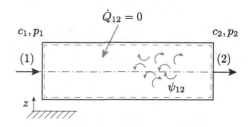

$$\frac{c_1}{v_1} = \frac{\dot{m}}{A}\bigg|_1 = \dot{m}''_{(a)} = \text{const.} \tag{21.112}$$

3. Now, the specific volume v is varied continuously while keeping the mass flux density constant. Thus, a new velocity can be calculated according to

$$c(v) = v\frac{c_1}{v_1} = v\dot{m}''_{(a)}. \tag{21.113}$$

The updated specific enthalpy then follows, see Eq. 21.105,

$$h(v) = h_1^+ - \frac{1}{2}c^2 = h_1 + \frac{c_1^2}{2} - \frac{c^2}{2} = h_1 + \frac{\dot{m}''^2_{(a)}}{2}\left(v_1^2 - v^2\right). \tag{21.114}$$

The temperature is

$$T(v) = T_{\text{ref}} + \frac{h(v) - h_{\text{ref}}}{c_p} \tag{21.115}$$

and the pressure follows

$$p(v) = \frac{RT}{v}. \tag{21.116}$$

Finally, the new specific entropy is

$$s(v) = s_{\text{ref}} + c_p \ln\frac{T(v)}{T_{\text{ref}}} - R \ln\frac{p(v)}{p_{\text{ref}}}. \tag{21.117}$$

4. The new state (h, s) can be fixed in the h, s-diagram for varying specific volumes.
5. The previous steps 2–4 can now be repeated for a different velocity c_1, i.e. for a different mass flux density $\dot{m}''_{(b)}$.

By following these steps the corresponding curves in a h, s-diagram are called Fanno-curves, see Figs. 21.11 and 21.13. However, a distinction regarding the inlet condition (1) is made:

- Subsonic inlet, i.e. $\text{Ma}_1 < 1$
  Under these conditions the given algorithm, see steps 1–5, leads to the curves as shown in Fig. 21.11. Starting at the inlet state (1), the specific entropy starts rising.

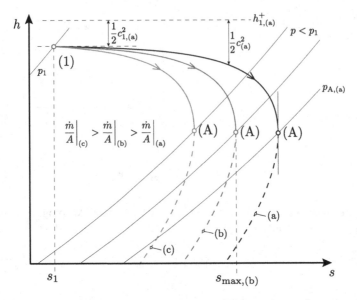

Fig. 21.11 Subsonic Fanno-curves, ideal gas with $\kappa = $ const.

This characteristic makes sense, as while passing the tube, the friction leads to an increase of specific entropy. Furthermore, the pressure decreases as proven by the isobars in the h, s-diagram, see Fig. 21.11. Figure 21.12 additionally proves, that the velocity as well as the specific volume start to rise.[10] At a point (A) the flow finally reaches the speed of sound, i.e. Ma $= 1$. However, any further acceleration would then lead to a decrease of entropy. This would violate the second law of thermodynamics, so the dashed line might be possible from a mathematical point of view, but it is not from a thermodynamic perspective.

Theorem 21.2 *Due to friction the velocity in an adiabatic flow with constant cross-section and subsonic inlet conditions rises. The maximum velocity that can be achieved is the speed of sound, i.e. Ma $= 1$.*

The outlet pressure can not be reduced under its critical pressure p_A in state (A), no matter what ambient pressure rules. Thus, to further reduce the pressure, the mass flux density needs to be lowered and the Fanno-curves shifts to the right until the desired outlet pressure occurs. Normal shocks, as described in Sect. 21.6, can not occur, since the flow is subsonic.

So far no statement has been made regarding the length of the tube. However, for a subsonic flow, there is a critical length, where Ma $= 1$ is achieved. In case the tube is now extended beyond the critical length, at which the specific entropy can no further rise, the new outlet velocity still is Ma $= 1$. However, this comes along with

---

[10]Compare with Problem 12.10!

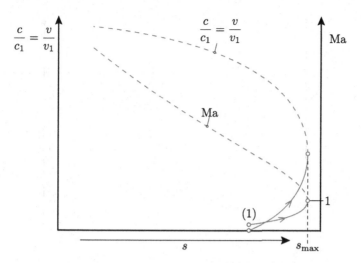

Fig. 21.12 Subsonic Fanno-curves for one mass flux density, ideal gas with $\kappa = $ const.

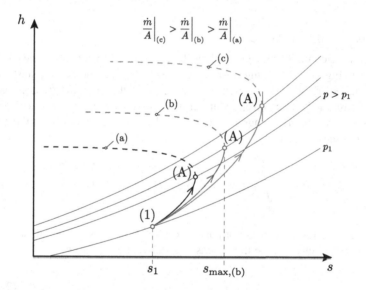

Fig. 21.13 Supersonic Fanno-curves, ideal gas with $\kappa = $ const.

a reduction of the mass flow rate, i.e. also with a reduction of the initial velocity. The flow is blocked by frictional effects ("choking"), the Fanno curves are shifted to the right. The frictional blocking leads to a decrease of the effective cross-section. Thus, a supersonic state can not be achieved, as this would require a cross-section expansion.

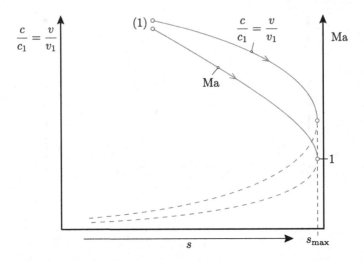

Fig. 21.14 Supersonic Fanno-curves for one mass flux density, ideal gas with κ = const.

- Supersonic inlet, i.e. $Ma_1 > 1$

  Under these conditions the given algorithm, see steps 1 to 5, leads to the curves as shown in Fig. 21.13. Starting at the inlet state (1), the specific entropy starts rising. This characteristic makes sense, as while passing the tube, the friction leads to an increase of specific entropy. Furthermore, the pressure increases as proven by the isobars in the h, s-diagram, see Fig. 21.13. According to the partial energy equation, see Eq. 21.108,

$$\int\limits_1^2 v\,\mathrm{d}p = -\frac{1}{2}\left(c_2^2 - c_1^2\right) - \psi_{12} > 0 \qquad (21.118)$$

  the change of pressure does not purely depend on the dissipation but also on the change of velocity, i.e. the flow needs to decelerate $c_2 < c_1$, i.e.

$$c_2 < \sqrt{c_1^2 - 2\psi_{12}}. \qquad (21.119)$$

  Figure 21.14 additionally proves, that the velocity as well as the specific volume decrease. At a point (A) the flow finally has decelerated to the speed of sound, i.e. $Ma = 1$. However, any further deceleration would then lead to a decrease of entropy. This would violate the second law of thermodynamics, so the dashed line might be possible from a mathematical point of view, but it is not from a thermodynamic perspective.

Theorem 21.3 *Due to friction the velocity in an adiabatic flow with constant cross-section and supersonic inlet conditions drops. The minimum velocity that can be achieved is the speed of sound, i.e. Ma = 1.*

From these considerations the outlet pressure can not exceed the critical pressure in state (A). Thus, to further increase the pressure, the mass flux density needs to be increased and the Fanno-curves shift to the right until the desired outlet pressure occurs. Depending on the ambient pressure a normal shock can occur, see Sect. 21.6, so that the supersonic flow compresses in a *unsteady change of state* to a higher pressure level.

In case the flow is supersonic, there is Ma = 1 after a critical length due to frictional effects. Let's now extend the tube beyond this critical length: The flow now adjusts with a normal shock, so that the flow becomes subsonic, see Sect. 21.6. Starting from that new subsonic state the flow now is accelerated, so that it again reaches Ma = 1 at its end, see also [24].

Both cases have shown, that the velocity in point (A) is the velocity of sound, which is now proven. However, Figs. 21.11 and 21.13 indicate that the Fanno-curves have a vertical tangent at point (A), i.e. $ds_A = 0$. At that point the entropy can not further rise. Hence, the fundamental equation of thermodynamics, see Sect. 12.3.2, in point (A) reads as

$$T \, ds = dh - v \, dp = 0. \tag{21.120}$$

In differential notation the first law of thermodynamics obeys

$$dh + c \, dc = 0. \tag{21.121}$$

The equation of continuity $c\rho = $ const. leads to

$$d(c\rho) = \rho \, dc + c \, d\rho = 0. \tag{21.122}$$

Combining these three equation leads to the velocity in point (A)

$$\boxed{c = \sqrt{\left(\frac{\partial p}{\partial \rho}\right)_s} = a} \tag{21.123}$$

Thus, the velocity in point (A) needs to be the velocity of sound. Since no restrictions have been made regarding the inlet conditions, it is valid for subsonic as well as for supersonic inlet states (1).

Following the first law of thermodynamics state (A) obeys

$$h_A + \frac{1}{2}c_A^2 = h_1 + \frac{1}{2}c_1^2 \rightarrow h_1 - h_A + \frac{1}{2}c_1^2 = \frac{1}{2}c_A^2. \tag{21.124}$$

With the caloric equation of state it results in

$$c_p (T_1 - T_A) + \frac{1}{2}c_1^2 = \frac{1}{2}c_A^2.$$

(21.125)

The velocity in (A) is the velocity of sound, i.e.

$$c_A = \sqrt{\kappa R T_A}$$

(21.126)

Combining both equations leads to

$$T_A = \frac{c_p T_1 + \frac{1}{2}c_1^2}{\frac{1}{2}\kappa R + c_p} = \frac{c_p T_1 + \frac{1}{2}\dot{m}''^2 v_1^2}{\frac{1}{2}\kappa R + c_p} \rightarrow \boxed{c_A = \sqrt{\kappa R T_A}}$$

(21.127)

The conservation of mass brings

$$v_A = c_A \frac{v_1}{c_1} = \frac{v_1}{c_1}\sqrt{\kappa R T_A} = \frac{\sqrt{\kappa R T_A}}{\dot{m}''}$$

(21.128)

The pressure follows the thermal equation of state

$$p_A = \frac{R T_A}{v_A}$$

(21.129)

The caloric state values can then be calculated

$$s_A = s_{ref} + c_p \ln\frac{T_A}{T_{ref}} - R \ln\frac{p_A}{p_{ref}}$$

(21.130)

and

$$h_A = h_{ref} + c_p (T_A - T_{ref})$$

(21.131)

Each Fanno-curve is a function of an initial state (1) and a mass flux density $\dot{m}''$, i.e. its velocity c_1.

21.5 Rayleigh Correlation

Let us now assume a non-adiabatic but frictionless tube flow with a constant cross-section A, see Fig. 21.15. The equation of continuity states, that the mass flux is constant in steady state, i.e.

$$\dot{m} = \rho A c = \frac{Ac}{v} = \text{const.} \tag{21.132}$$

In other words the mass flux density is constant as well, since the cross-section does not vary, i.e.

$$\boxed{\frac{c}{v} = \frac{\dot{m}}{A} = \dot{m}'' = \text{const.}} \tag{21.133}$$

The first law of thermodynamics for steady state conditions obeys

$$q + \underbrace{w_t}_{=0} = h_2 - h_1 + \frac{1}{2}\left(c_2^2 - c_1^2\right). \tag{21.134}$$

However, for a non-adiabatic flow this does not help, since the specific heat q is not specified. According to the second law of thermodynamics for diabatic flows it is

$$\boxed{ds = \delta s_a = \frac{\delta q}{T} \lessgtr 0} \tag{21.135}$$

Thus, in this case the entropy can rise due to heating or it can sink due to cooling.[11] The partial energy equation under these conditions is

$$w_t = 0 = \int_1^2 v\,dp + \frac{1}{2}\left(c_2^2 - c_1^2\right) + \underbrace{\psi_{12}}_{=0}. \tag{21.136}$$

This equation, in contrast to the derivation of the Fanno-correlation, can now be applied, since the flow is supposed to be frictionless. In differential notation it reads as

$$v\,dp + c\,dc = 0. \tag{21.137}$$

Dividing by the specific volume brings

$$dp + \frac{c}{v}\,dc = dp + \dot{m}''\,dc = 0. \tag{21.138}$$

The integration leads to

$$p_2 - p_1 + \dot{m}''\left(c_2 - c_1\right) = 0 \tag{21.139}$$

respectively the equation for constant total pressure

$$\boxed{p_2 + \dot{m}''^2 v_2 = p_1 + \dot{m}''^2 v_1 = \text{const.}} \tag{21.140}$$

---

[11] Sure, it can remain constant, but then there would be no driver for any changes of the flow.

Since the cross-section is constant this equation can easily be interpreted as conservation of momentum.

Same as before with the Fanno-curves, the idea now is to visualise the change of state for an ideal gas in a h, s-diagram:

1. Inlet state (1) is fixed by two independent state values, e.g. T_1 and p_1. Thus, the specific volume is

$$v_1 = \frac{RT_1}{p_1}. \tag{21.141}$$

The specific enthalpy[12] is

$$h_1 = h_{ref} + c_p \left(T_1 - T_{ref} \right). \tag{21.142}$$

The specific entropy obeys

$$s_1 = s_{ref} + c_p \ln \frac{T_1}{T_{ref}} - R \ln \frac{p_1}{p_{ref}}. \tag{21.143}$$

Hence, state (1) can be fixed unambiguously in a h, s-diagram.
2. The velocity c_1 is chosen next, so that the mass flux density is defined as

$$\frac{c_1}{v_1} = \left. \frac{\dot{m}}{A} \right|_1 = \dot{m}''_{(a)} = \text{const.} \tag{21.144}$$

3. Now, the velocity c is varied continuously while keeping the mass flux density $\dot{m}''_{(a)}$ constant. Thus, a new specific volume can be calculated according to

$$v(c) = c \frac{v_1}{c_1} = \frac{c}{\dot{m}''_{(a)}}. \tag{21.145}$$

The updated pressure then follows, see Eq. 21.140,

$$p(c) = p_1 - \dot{m}''^2_{(a)} \left(v(c) - v_1 \right). \tag{21.146}$$

The temperature is

$$T(c) = \frac{p v(c)}{R} \tag{21.147}$$

and the specific enthalpy follows

$$h(c) = h_{ref} + c_p \left(T(c) - T_{ref} \right). \tag{21.148}$$

---

[12] $T_{ref}, p_{ref}, h_{ref}$ and s_{ref} represent a arbitrary reference state!

Fig. 21.15 Non-adiabatic, frictionless tube flow ($A = $ const.)

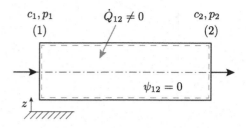

Finally, the new specific entropy is

$$s\left(c\right) = s_{\text{ref}} + c_p \ln\frac{T\left(c\right)}{T_{\text{ref}}} - R\ln\frac{p\left(c\right)}{p_{\text{ref}}}. \tag{21.149}$$

4. The new state (h, s) can be fixed in the h, s-diagram for varying velocities.
5. The previous steps 2–4 can now repeated for a different velocity c_1, i.e. for a different mass flux density $\dot{m}''_{(b)}$.

By following these steps the corresponding curves in a h, s-diagram are called Rayleigh-curves. However, a distinction regarding the inlet condition (1) is made:

- Subsonic inlet, i.e. $\text{Ma}_1 < 1$
  Under these conditions the given algorithm, see steps 1–5, leads to the curves as shown in Fig. 21.16. Starting at the inlet state (1), the specific entropy rises when heat is supplied. According to the isobars in Fig. 21.16 the pressure decreases under these conditions. Figure 21.17 additionally proves, that the velocity rises while heat is supplied. At a point (A) the flow finally has reached the speed of sound, i.e. $\text{Ma} = 1$. However, any further acceleration would then lead to a decrease of entropy, i.e. it is impossible for a heated tube.

Theorem 21.4 *The maximum heat supply is limited. In order to increase the heating, the mass flux density needs to be decreased. This results in a reduced pressure.*

Heating can lead to a thermal blocking of the subsonic flow, i.e. the flow causes an unsteady effect similar to a reduction of the cross section, so that its initial state respectively the mass flow rate is adjusted, see [24]. It is not possible to accelerate the fluid *thermally* beyond state (A), see [25]. Interestingly, close to the critical point (A)[13] the maximal specific enthalpy is reached before the maximal specific entropy is reached, see Fig. 21.16. In other words, once h_{max} is reached, a further heating leads to a temperature reduction due to the large expansion of the fluid. The expansion comes along with a fluid acceleration, see Eq. 21.133. Hence, the first law of thermodynamics in differential notation is

---

[13]Thus, at large subsonic velocities!

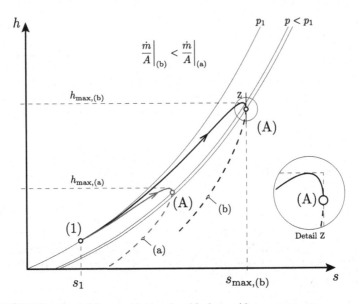

Fig. 21.16 Subsonic Rayleigh-curves for heating, ideal gas with $\kappa = $ const.

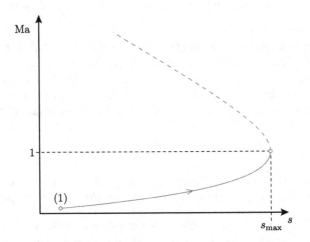

Fig. 21.17 Subsonic Rayleigh-curves for one mass flux density and heating, ideal gas with $\kappa = $ const.

$$\delta q = \mathrm{d}h + c\,\mathrm{d}c. \tag{21.150}$$

The closer one gets to state (A) the larger $c\,\mathrm{d}c$ gets. Combining the first law of thermodynamics with the partial energy equation, see Eq. 21.137, results in

$$\delta q = \mathrm{d}h - v\,\mathrm{d}p \tag{21.151}$$

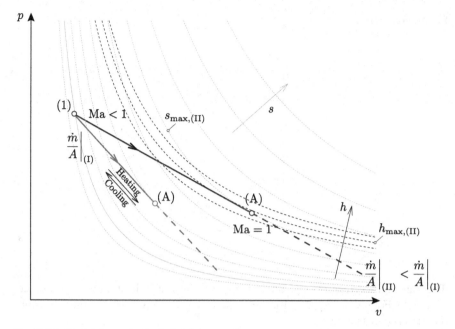

Fig. 21.18 Rayleigh correlation (subsonic), p, v-diagram

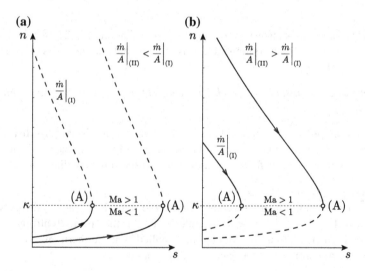

Fig. 21.19 Polytropic exponent—**a** subsonic, **b** supersonic

respectively

$$\delta q = \mathrm{d}u + v\,\mathrm{d}p + p\,\mathrm{d}v - v\,\mathrm{d}p = c_v\mathrm{d}T + p\,\mathrm{d}v. \tag{21.152}$$

Fig. 21.20 Supersonic
Rayleigh-curves for heating,
ideal gas with $\kappa = $ const.

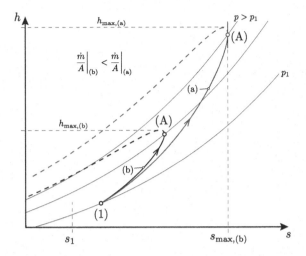

The Rayleigh-correlation can additionally be illustrated in a p, v-diagram, see
Fig. 21.18 Following Eq. 21.140, the slope of the Rayleigh-correlation in a p, v-
diagram is constant, i.e.

$$\frac{\mathrm{d}p}{\mathrm{d}v} = -\dot{m}''^2 = \text{const.} \tag{21.153}$$

Though the focus has been on heating so far, the p, v-diagram Fig. 21.18 also contains
information regarding the cooling case. It is known, that cooling leads to a decrease
of specific enthalpy, i.e.

Theorem 21.5 *A subsonic Rayleigh flow is accelerated by heating and decelerated
by cooling.*

Obviously, the polytropic exponent can not be constant in this case. Its distribution
is shown in Fig. 21.19a. Once, state (A) is achieved, it is $n = \kappa$ since no heat can be
supplied any more and the flow is frictionless, i.e. isentropic conditions occur.

- Supersonic inlet, i.e. $\mathrm{Ma}_1 > 1$
  Under these conditions the given algorithm, see steps 1–5, leads to the curves as
  shown in Fig. 21.20. Starting at the inlet state (1), the specific entropy rises as
  entropy is carried into the system by the supplied heat. Furthermore, the pressure
  increases. According to the partial energy equation

$$\int_1^2 v \, \mathrm{d}p = -\frac{1}{2} \left(c_2^2 - c_1^2 \right) > 0 \tag{21.154}$$

the rise of pressure is because the flow decelerates, i.e. $c_2 < c_1$. Figure 21.21 additionally proves, that the velocity decreases. At a point (A) the flow finally has

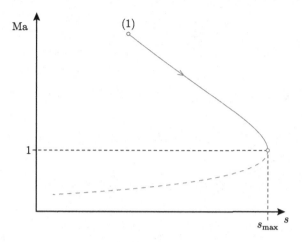

Fig. 21.21 Supersonic Rayleigh-curves for one mass flux density and heating, ideal gas with $\kappa = \text{const.}$

decelerated to the speed of sound, i.e. Ma = 1. However, any further deceleration would then require the entropy to decrease, i.e. it is impossible for a heated tube.

Theorem 21.6 *The maximum heat supply is limited. In order to increase the heating, the mass flux density needs to be increased. This leads to an enlarged pressure.*

In case Ma = 1 is reached and heat is further supplied, a normal shock takes place, see Sect. 21.6. By this normal shock the flow is then subsonic and starts to accelerate again to Ma = 1, see [24]. In case heating is further increased the normal shock moves upstream until the entire tube has subsonic velocity. The Rayleigh-correlation can additionally be illustrated in a p, v-diagram, see Fig. 21.22. Same as for the subsonic flow, following Eq. 21.140, the slope of the Rayleigh-correlation in a p, v-diagram is constant, i.e.

$$\frac{\mathrm{d}p}{\mathrm{d}v} = -\dot{m}''^2 = \text{const.} \tag{21.155}$$

Though the focus has been on heating so far, the p, v-diagram 21.22 also contains information regarding the cooling case. It is known, that cooling leads to a decrease of specific enthalpy, i.e.

Theorem 21.7 *A supersonic Rayleigh flow is accelerated by cooling and decelerated by heating.*

Obviously, the polytropic exponent can not be constant in this case. Its distribution is shown in Fig. 21.19b.

Both cases have shown, that the velocity in point (A) is the velocity of sound, which is now proven. However, Figs. 21.16 and 21.20 indicate that the Rayleigh-curves

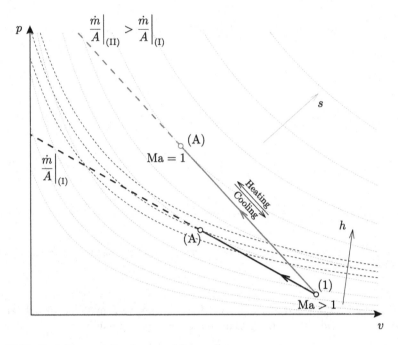

Fig. 21.22 Rayleigh correlation (supersonic), p, v-diagram

have a vertical tangent at point (A), i.e. $ds_A = 0$. Hence, the fundamental equation of thermodynamics, see Sect. 12.3.2, in point (A) reads as

$$T \, ds = dh - v \, dp = 0. \tag{21.156}$$

In differential notation the first law of thermodynamics obeys

$$dh + c \, dc = \delta q. \tag{21.157}$$

Combining with the second law of thermodynamics

$$ds = \frac{\delta q}{T} + \underbrace{\frac{\delta \psi}{T}}_{=0} \tag{21.158}$$

the first law of thermodynamics is

$$dh + c \, dc = T \, ds = 0. \tag{21.159}$$

The equation of continuity $c\rho = \text{const.}$ leads to

$$d\,(c\rho) = \rho\,dc + c\,d\rho = 0. \tag{21.160}$$

Combining these equation leads to the velocity in point (A), i.e.

$$\boxed{c = \sqrt{\left(\frac{\partial p}{\partial \rho}\right)_s} = a} \tag{21.161}$$

Thus, the velocity in point (A) needs to be the velocity of sound. Since no restrictions have been made regarding the inlet conditions, it is valid for subsonic as well as for supersonic inlet states (1).

Following the partial energy equation state (A) can be described with

$$p_A = p_1 - \dot{m}_1''^2\,(v_A - v_1). \tag{21.162}$$

With the conservation of mass it is

$$\frac{c_A}{v_A} = \frac{c_1}{v_1} = \dot{m}_1''. \tag{21.163}$$

The velocity in (A) is the velocity of sound, i.e.

$$c_A = \sqrt{\kappa R T_A} = \sqrt{\kappa p_A v_A}. \tag{21.164}$$

Combining these equations leads to

$$\boxed{p_A = \frac{p_1 + \dot{m}_1''^2 v_1}{1 + \kappa}} \tag{21.165}$$

respectively

$$\boxed{v_A = \left(\frac{v_1}{c_1}\right)^2 \kappa p_A = \frac{\kappa p_A}{\dot{m}_1''^2}} \rightarrow \boxed{c_A = \sqrt{\kappa R T_A}} \tag{21.166}$$

and

$$\boxed{T_A = \frac{p_A v_A}{R}} \tag{21.167}$$

The caloric state values can then be calculated

$$\boxed{s_A = s_{\mathrm{ref}} + c_p \ln\frac{T_A}{T_{\mathrm{ref}}} - R \ln\frac{p_A}{p_{\mathrm{ref}}}} \tag{21.168}$$

and

$$\boxed{h_A = h_{\mathrm{ref}} + c_p\,(T_A - T_{\mathrm{ref}})} \tag{21.169}$$

Each Rayleigh-curve is a function of an initial state (1) and a mass flux density $\dot{m}_1''$, i.e. its velocity c_1.

21.6 Normal Shock

Let us now investigate what can happen in a tube flow with constant cross-section under supersonic but adiabatic conditions. The considerations in Sect. 21.4 have shown, that the Fanno-correlation must be fulfilled. Based on the first law of thermodynamics the energy needs to be constant, i.e.

$$h_1 + \frac{1}{2}v_1^2\dot{m}''^2 = h_2 + \frac{1}{2}v_2^2\dot{m}''^2 \tag{21.170}$$

Since no heat is exchanged the first law of thermodynamics simplifies to a constant specific total enthalpy. It has been shown, that due to friction the flow decelerates until $Ma = 1$ is reached in state (A). Thus, as shown in Fig. 21.23, the lower, supersonic branch of the Fanno-curve has to be followed. Under this premise a pressure p_2, which is for instance fixed by a huge environment, can not be reached in case

$$p_2 > p_A. \tag{21.171}$$

However, a so-called normal shock can occur. According to [26] the fluid compresses in an unsteady change of state, i.e. it is non-quasi-static, from a lower, supersonic pressure of state (1) to the fixed pressure of subsonic state (2), that is ruled by the environment for instance. This normal shock ("jump") is supposed to be linear, i.e. its spatial expansion is neglected. In case the spatial expansion is low, there is no friction. Thus, for the change of state from (1) to (2) a frictionless tube flow can be assumed, that needs to follow the Rayleigh correlation. This correlation has shown, that in case no friction occurs, the total pressure remains constant, i.e.

$$p_2 - p_1 + \dot{m}''^2 (v_2 - v_1) = 0 \tag{21.172}$$

In other words, the states (1) and (2) under these specific conditions need to fulfil both, the Fanno and the Rayleigh-correlation, as shown in Fig. 21.23. Consequently, both states, i.e. (1) and (2) are on a par. Once, state (2) is reached, the further process takes place under subsonic conditions and is friction-controlled again.

The change of state from (1) to (2) can be visualised in a h, s-diagram as follows:

- A Fanno-curve is calculated by the given mass flux density and any supersonic state in front of the normal shock, e.g. state (I)
- State (2) is then unambiguously fixed by the Fanno-curve and the isobar p_2
- A Rayleigh-curve is calculated by the given mass flux density and state (2)
- A second point of intersection between Fanno- and Rayleigh-correlation occurs, i.e. state (1)

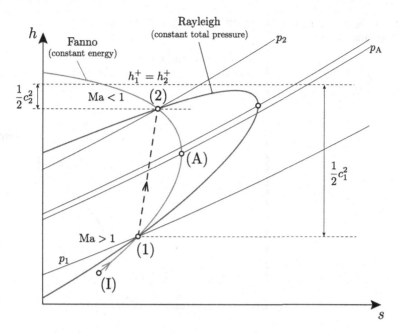

Fig. 21.23 Normal shock, adiabatic tube flow

- The normal shock takes place from (1) to (2)

Due to the unsteady jump and its non-quasi-static character, the normal shock comes along with a rise of entropy, see Fig. 21.23, so it does not violate the second law of thermodynamics. According to the second law of thermodynamics entropy can not be destroyed. This is the reason why a normal shock can not be achieved from (2) to (1), as this would require a destruction of entropy, as shown in Fig. 21.23.

The entropy increase has been treated as a discontinuity—otherwise it has not been possible to get from state (1) to state (2). However, a normal shock in reality is spread over several free path lengths of the particles, so that it can be treated as a steady effect. Thus, the rise of entropy then can be explained with frictional effects and heat fluxes within the normal shock.

21.7 Supersonic Flows

21.7.1 Flow of a Converging Nozzle

Before starting with supersonic flows in a so-called Laval-nozzle, the focus first is on a much simpler problem as sketch in Fig. 21.24. A huge tank contains an ideal gas *in*

Fig. 21.24 Steady flow from
vessels according to [1]

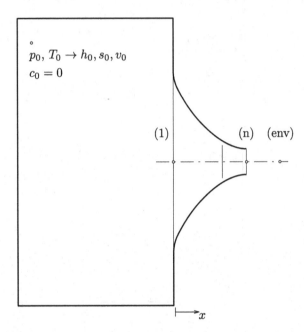

$$p_0, T_0 \rightarrow h_0, s_0, v_0$$
$$c_0 = 0$$

(1)   (n)   (env)

rest, i.e. its velocity is $c_0 = 0$, given by its state values pressure p_0 and temperature T_0.

A small nozzle connects the container with the environment (env). However, the nozzle is converging, i.e. in flow direction its cross-section decreases with a narrowest cross-section (n) at its end, see sketch 21.24. In case

$$p_0 > p_{env} \tag{21.173}$$

there is a flow from the vessel into the environment. Nozzle and vessel are supposed to be adiabatic. Since the tank is supposed to be huge in size, its internal state is time invariant, so that a steady state shall be investigated.

First Law of Thermodynamics

In order to derive the outflow velocity c_n the first law of thermodynamics with the made premises is applied from state (0) to state (n) and results in

$$q_{0n} + w_{t,0n} = 0 = h_n + h_0 + \frac{1}{2} \left(c_n^2 - c_0 \right). \tag{21.174}$$

Applying the caloric equation of state and taking into account, that $c_0 = 0$ leads to

$$c_n^2 = 2c_p \left(T_0 - T_n\right) = 2c_p T_0 \left(1 - \frac{T_n}{T_0}\right). \tag{21.175}$$

In case the change of state is isentropic, i.e. adiabatic and frictionless, it follows for an ideal gas, that

$$\frac{T_n}{T_0} = \left(\frac{p_n}{p_0}\right)^{\frac{\kappa-1}{\kappa}}. \tag{21.176}$$

Thus, the outflow velocity is

$$c_n^2 = 2c_p \left(T_0 - T_n\right) = 2c_p T_0 \left[1 - \left(\frac{p_n}{p_0}\right)^{\frac{\kappa-1}{\kappa}}\right]. \tag{21.177}$$

The tank temperature T_0 follows from the thermal equation of state, i.e.

$$T_0 = \frac{p_0 v_0}{R} \tag{21.178}$$

and it further is

$$c_p = \frac{\kappa}{\kappa - 1} R. \tag{21.179}$$

This finally leads to the outflow velocity

$$\boxed{c_n = \sqrt{2 \frac{\kappa}{\kappa - 1} p_0 v_0 \left[1 - \left(\frac{p_n}{p_0}\right)^{\frac{\kappa-1}{\kappa}}\right]}} \tag{21.180}$$

In case the flow would not be frictionless, i.e. non-isentropic, the outflow velocity can be derived with a polytropic change of state with $n < \kappa$ and lead to

$$c_{n,\text{fric}} = \sqrt{2 \frac{\kappa}{\kappa - 1} p_0 v_0 \left[1 - \left(\frac{p_n}{p_0}\right)^{\frac{n-1}{n}}\right]} < c_n. \tag{21.181}$$

However, the focus now is on the ideal case of an isentropic change of state as a benchmarking change of state.

Conservation of Mass

The continuity equation for any cross-section of the nozzle, i.e. also in the narrowest cross-section, in steady state reads as

$$\dot{m} = \frac{cA}{v} = \frac{c_n A_n}{v_n} = \dot{m}_n = \text{const.} \tag{21.182}$$

Substituting the velocity c_n of the narrowest cross-section brings

$$\dot{m}_n = \frac{A_n}{v_n} \sqrt{2 \frac{\kappa}{\kappa - 1} p_0 v_0 \left[1 - \left(\frac{p_n}{p_0} \right)^{\frac{\kappa - 1}{\kappa}} \right]}. \tag{21.183}$$

Due to the isentropic change of state it is

$$v_n = v_0 \left(\frac{p_0}{p_n} \right)^{\frac{1}{\kappa}}. \tag{21.184}$$

Hence, it follows accordingly

$$\dot{m}_n = A_n \left(\frac{p_n}{p_0} \right)^{\frac{1}{\kappa}} \sqrt{2 \frac{\kappa}{\kappa - 1} \frac{p_0}{v_0} \left[1 - \left(\frac{p_n}{p_0} \right)^{\frac{\kappa - 1}{\kappa}} \right]}. \tag{21.185}$$

Now, a so-called flow function Ω_n for the narrowest cross-section is defined, so that

$$\dot{m}_n = A_n \Omega_n \sqrt{2 \frac{p_0}{v_0}} = A_n \Omega_n \sqrt{2 p_0 T_0}. \tag{21.186}$$

Thus, the dimensionless flow function is

$$\Omega_n = \sqrt{\frac{\kappa}{\kappa - 1} \left[1 - \left(\frac{p_n}{p_0} \right)^{\frac{\kappa - 1}{\kappa}} \right] \left(\frac{p_n}{p_0} \right)^{\frac{1}{\kappa}}} \tag{21.187}$$

respectively with the pressure ratio

$$\pi_n = \frac{p_n}{p_0} \tag{21.188}$$

it is

$$\Omega_n = \sqrt{\frac{\kappa}{\kappa - 1} \left[1 - \pi_n^{\frac{\kappa - 1}{\kappa}} \right]} \pi_n^{\frac{1}{\kappa}}. \tag{21.189}$$

Since the mass flow rate needs to be constant in steady state, a flow function can be defined at any position within the nozzle, i.e. from state (0) to (n), so that the general notation is

$$\boxed{\dot{m}_n = \dot{m} = A \Omega \sqrt{2 \frac{p_0}{v_0}} = A \Omega \sqrt{2 p_0 T_0}} \tag{21.190}$$

with

$$\Omega = \Omega(\pi) = \sqrt{\frac{\kappa}{\kappa - 1} \left[1 - \pi^{\frac{\kappa - 1}{\kappa}} \right]} \pi^{\frac{1}{\kappa}} \tag{21.191}$$

and

$$\pi = \frac{p}{p_0} \tag{21.192}$$

Now, the maximum mass flow rate that can occur in a converging nozzle is investigated. Unfortunately, the mathematical function of the converging shape, i.e. the local cross-section A, is not given. Anyhow, the local flow function Ω, see Eq. 21.191, can be plotted over the local pressure ratio π, see Fig. 21.25. Since the mass flux needs to be constant in steady state, the corresponding cross-section can easily be added qualitatively in the plot as well, following

$$A = \frac{\dot{m}}{\Omega \sqrt{2\rho_0 T_0}} \propto \frac{1}{\Omega}. \tag{21.193}$$

Thus, the larger the local Ω is, the smaller the local cross-section A becomes. The flow starts within the tank in state (0), i.e. at $\pi = 1$ at the right position of Fig. 21.25. On its way to (n) the cross-section gets smaller, since the nozzle is assumed to be converging. Correspondingly, the flow function Ω rises and reaches a maximum. Due to the constant mass flow rate, the cross-section then needs to have a minimum, that has been introduced as A_n. A further movement to the left beyond that point is not possible, since then the nozzle would need to diverge, i.e. the dashed lines can not be reached with a converging geometry. The pressure ratio at (n) can be calculated by finding the maximum of $\Omega(\pi)$, i.e.

$$\frac{d\Omega}{d\pi} = \frac{2\pi^{\frac{1}{\kappa}} - (\kappa + 1)\pi}{2(\kappa - 1)\pi \sqrt{\frac{\kappa}{1 - \kappa} \left(\pi^{\frac{\kappa - 1}{\kappa}} - 1 \right)}} = 0. \tag{21.194}$$

Thus, the maximum, which is at the narrowest cross-section, of the curve is at

$$\pi_{\min} = \left(\frac{\kappa + 1}{2} \right)^{\frac{\kappa}{1 - \kappa}}. \tag{21.195}$$

The maximum flow function at that point follows according to

$$\Omega_{\max} = \sqrt{\frac{\kappa}{\kappa - 1} \left[1 - \left(\frac{2}{\kappa + 1} \right) \right]} \left(\frac{\kappa + 1}{2} \right)^{\frac{1}{1 - \kappa}} \tag{21.196}$$

respectively

$$\Omega_{\max} = \left(\frac{2}{\kappa+1}\right)^{\frac{1}{\kappa-1}} \sqrt{\frac{\kappa}{\kappa+1}}. \tag{21.197}$$

However, that point is called critical point of the nozzle[14] and its state values then follow

$$p_{\mathrm{crit}} = p_0 \left(\frac{\kappa+1}{2}\right)^{\frac{\kappa}{1-\kappa}} \rightarrow \pi_{\mathrm{crit}} = \left(\frac{\kappa+1}{2}\right)^{\frac{\kappa}{1-\kappa}} \tag{21.198}$$

and due to the isentropic change of state

$$v_{\mathrm{crit}} = v_0 \left(\frac{p_{\mathrm{n}}}{p_0}\right)^{-\frac{1}{\kappa}} = v_0 \left(\frac{\kappa+1}{2}\right)^{\frac{1}{\kappa-1}} \tag{21.199}$$

respectively

$$T_{\mathrm{crit}} = T_0 \left(\frac{p_{\mathrm{n}}}{p_0}\right)^{\frac{\kappa-1}{\kappa}} = T_0 \frac{2}{\kappa+1}. \tag{21.200}$$

Now that we know the critical state values, the velocity can now be calculated according to Eq. 21.180, i.e.

$$c_{\mathrm{crit}} = \sqrt{2\frac{\kappa}{\kappa-1} p_0 v_0 \left[1 - \left(\frac{p_{\mathrm{crit}}}{p_0}\right)^{\frac{\kappa-1}{\kappa}}\right]}. \tag{21.201}$$

With the other critical state values as derived before one gets

$$c_{\mathrm{crit}} = \sqrt{\kappa p_{\mathrm{crit}} c_{\mathrm{crit}}} = \sqrt{\kappa R T_{\mathrm{crit}}} = a_{\mathrm{crit}}. \tag{21.202}$$

Theorem 21.8 *Hence, the maximum velocity at the narrowest cross-section of a converging nozzle is velocity of sound when the pressure reaches its critical value. The pressure can not fall below that critical value.*

In Fig. 21.25 it has been assumed that π can freely vary from $0 \ldots 1$ and the consequences have been investigated mathematically. From a thermodynamic point of view the range of π was further limited from π_{crit}, in best case at the narrowest cross-section, to 1 at state (0). The range from $0 \ldots \pi_{\mathrm{crit}}$ is impossible for a converging shape. Now, several case studies are discussed with the focus on the question what happens, if π_{crit} is not achieved in the narrowest cross-section, e.g. due to the counter ambient pressure. Hence, Figs. 21.26 and 21.27 show cases (a), (b), (c) and (d), that need to be discussed further and that show, that the full range of π can not be covered. It is known that the cause for any flow is a pressure difference: The driver for a flow through the converging nozzle is the pressure gap between vessel p_0 and ambient

---

[14]For a given p_0 it is not possible to reach a pressure smaller than p_{crit} in a converging nozzle, see sketch 21.25.

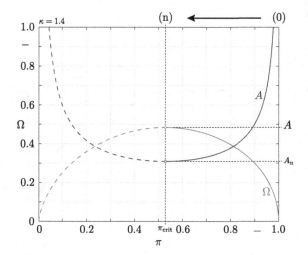

Fig. 21.25 Local flow function Ω for $\kappa = 1.4$, state (0) in rest

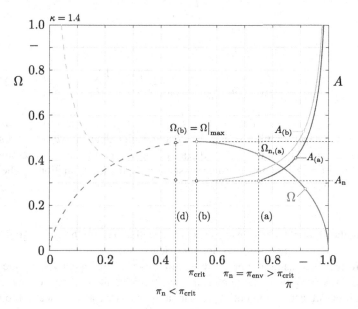

Fig. 21.26 Converging nozzle scenarios I/II

pressure p_{env}, that acts as a kind of counter pressure. So, let us increase the vessel's pressure p_0 step by step and fix the ambient pressure:

(a) The vessel's pressure p_0 is larger than ambient pressure. As shown in Fig. 21.27 the flow starts, but the critical pressure at the narrowest cross-section, i.e. at the outlet is not reached, i.e.

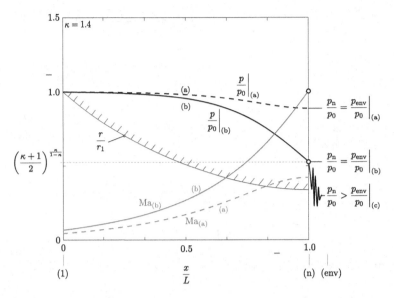

Fig. 21.27 Converging nozzle scenarios II/II

$$\frac{p_{\mathrm{n}}}{p_0} = \left.\frac{p_{\mathrm{env}}}{p_0}\right|_{(a)} > \frac{p_{\mathrm{crit}}}{p_0} = \left(\frac{\kappa+1}{2}\right)^{\frac{\kappa}{1-\kappa}}. \tag{21.203}$$

The pressure at the outlet (n) is ambient pressure. The according outlet velocity, see Mach-number in Fig. 21.27, is smaller than the velocity of sound. Figure 21.26 additionally shows the according shape of the cross-section as a function of π. At the narrowest cross-section A_{n} the flow function reaches

$$\Omega_{\mathrm{n,(a)}} < \Omega_{\mathrm{max}}. \tag{21.204}$$

Thus, the maximal mass flow rate according to the shape of the converging nozzle and the vessel's pressure p_0 is not yet reached.

(b) By further increasing the pressure within the tank, the velocity rises due to the larger pressure potential, which is the driver for the flow. In case (b) the chosen tank pressure causes the outlet pressure at the narrowest cross-section to be

$$\frac{p_{\mathrm{n}}}{p_0} = \left.\frac{p_{\mathrm{env}}}{p_0}\right|_{(a)} = \frac{p_{\mathrm{crit}}}{p_0} = \left(\frac{\kappa+1}{2}\right)^{\frac{\kappa}{1-\kappa}}. \tag{21.205}$$

Consequently, the velocity reaches its maximum value, i.e. $\mathrm{Ma} = 1$, see Fig. 21.27. According to Fig. 21.26, the flow function Ω is maximised at the narrowest cross-section, i.e.

$$\Omega_{\mathrm{n,(b)}} = \Omega_{\mathrm{max}}. \tag{21.206}$$

Hence, the mass flow rate is maximal and can not be exceeded for the given state (0) and the shape of the nozzle. The outlet pressure is equal with ambient pressure. This case is the design case of a perfect converging nozzle.

(c) In case the tank pressure p_0 is further increased, the critical pressure p_{crit} rises as well, but according to Fig. 21.25 the critical pressure ratio is still fixed to

$$\pi_{crit} = \frac{p_{crit}}{p_0} = \left(\frac{\kappa + 1}{2}\right)^{\frac{\kappa}{1-\kappa}}. \tag{21.207}$$

Thus, the dimensionless function $\pi = f\left(\frac{x}{L}\right)$ is the same as in case (b). Same as in case (b), the flow function reaches is maximal value, i.e.

$$\Omega_{n,(c)} = \Omega_{max}. \tag{21.208}$$

Consequently, the Mach-number at the outlet still is $Ma = 1$. However, at the outlet the ambient pressure is not reached, i.e.

$$p_n = p_{crit} > p_{env}. \tag{21.209}$$

According to [1] the nozzle is still left with

$$p_n = p_{crit}. \tag{21.210}$$

The flow starts to expand periodically when leaving the nozzle and finally reaches ambient pressure. However, let us have a look at the mass flow rate. According to Eq. 21.186 it follows

$$\dot{m}_n = A_n \Omega_n \sqrt{2\frac{p_0}{v_0}}. \tag{21.211}$$

The narrowest cross-section and the local flow function at this spot are the same as in (b). However, the pressure p_0 has been increased, so that v_0 decreases if the tank's temperature T_0 has not been modified compared to (b), i.e.

$$v_0 = \frac{RT_0}{p_0}. \tag{21.212}$$

Consequently, the mass flow rate is larger than in (b) but still maximised for the given geometry and the initial pressure p_0.

(d) On the other hand a smaller $\pi_n < \pi_{crit}$ is not possible, since this would require the cross-section to increase again—which violates the assumption of a converging nozzle. This case is shown in Fig. 21.26.

The mass flow rate through the nozzle in critical state is maximal, see case (b) respectively (c) in Figs. 21.26 and 21.27. The mass flow rate is proportional to the nozzle's cross-section at the narrowest cross-section A_n and its flow function at that point Ω_n, see Eq. 21.193. The cross-section at the narrowest point is fixed, but only in case (b) respectively (c), the flow function reaches its maximum!

The curves in Fig. 21.27 have been calculated by solving the following equations:

1. The first law of thermodynamics for adiabatic flows, i.e.

$$h^+ = \text{const.} = h + \frac{1}{2}c^2 = h_0 = h_1 + \frac{1}{2}c_1^2 = h_n + \frac{1}{2}c_n^2 \qquad (21.213)$$

2. Isentropic change of state, i.e.

$$pv^\kappa = p_0 v_0^\kappa = p_1 v_1^\kappa = p_n v_n^\kappa \qquad (21.214)$$

3. Conservation of mass, i.e.

$$\frac{cA}{v} = \frac{c_1 A_1}{v_1} = \frac{c_n A_n}{v_n} \qquad (21.215)$$

4. Thermal equation of state, i.e.

$$pv = RT \qquad (21.216)$$

5. Caloric equation of state, i.e.

$$h = c_p \left(T - T_{ref} \right) + h_{ref} \qquad (21.217)$$

Since the flow is isentropic it further is

$$ds = 0. \qquad (21.218)$$

21.7.2 Laval-Nozzle

The previous section has shown, that the maximum outlet velocity from a vessel through a converging nozzle is the velocity of sound in case the pressure in the narrowest cross-section reaches its critical value p_{crit}. This critical value is fluid specific and has been derived in the previous section—depending on the rest pressure in the tank and the counter pressure this critical value can be achieved. Within this section it is now investigated how to reach a supersonic state, i.e. Mach-numbers greater than one. In order to do so, a further theoretical approach follows, i.e. a correlation between cross-sectional area and mass flow density through a nozzle is investigated in a differential approach. Due to the mass conservation in a steady state nozzle with variable cross-section the mass flow rate needs to be constant, i.e.

$$\dot{m} = c\rho A = \text{const.} \qquad (21.219)$$

The derivation brings

$$d(c\rho A) = 0. \tag{21.220}$$

Rearranging leads to

$$A d(c\rho) + \rho c\, dA = 0. \tag{21.221}$$

Hence, the change of cross-section follows

$$\frac{dA}{A} = -\frac{d(c\rho)}{\rho c} = -\frac{c\, d\rho + \rho\, dc}{\rho c} = -\frac{d\rho}{\rho} - \frac{c\, dc}{c^2}. \tag{21.222}$$

Equation 21.222 can also be written with the mass flux density, i.e.

$$\dot{m}'' = \frac{\dot{m}}{A} = \rho c \tag{21.223}$$

namely

$$\frac{dA}{A} = -\frac{d\dot{m}''}{\dot{m}''}. \tag{21.224}$$

A physical interpretation is, that the cross-sectional area A has a minimum, i.e. the first derivative is $dA = 0$ and the second derivative is $d^2 A > 0$, at the point where the mass flow density $\dot{m}''$ is maximal, i.e. first derivative is $d\dot{m}'' = 0$, second derivative is $d^2\dot{m}'' < 0$. This narrowest cross-section is important further on. However, the first derivative at this location is

$$\frac{dA}{A} = -\frac{d\dot{m}''}{\dot{m}''} = 0 \tag{21.225}$$

and the second derivative at this location is

$$\underbrace{\frac{d^2 A}{A}}_{>0} = -\underbrace{\frac{d^2\dot{m}''}{\dot{m}''}}_{<0}. \tag{21.226}$$

This interpretation of a maximum mass flux density at the narrowest cross-section is congruent with the flow function Ω that has been introduced in Sect. 21.7.1, see Fig. 21.25 for instance.

Assuming an adiabatic and frictionless, i.e. isentropic, flow with negligible changes of potential energies, the partial energy equation for technical work[15] reads as

---

[15]This is equivalent with a constant total pressure. Mind, that no dissipation occurs.

$$w_t = 0 = \int_1^2 v\,dp + \frac{1}{2}\left(c_2^2 - c_1^2\right).$$ (21.227)

In differential notation

$$v\,dp + c\,dc = 0.$$ (21.228)

Combining with the equation of continuity, see Eq. 21.222, results in

$$\frac{dA}{A} = -\frac{d\rho}{\rho} + \frac{v\,dp}{c^2}.$$ (21.229)

When deducing the velocity of sound it has already been proven for an isentropic change of state, see Eq. 21.90, that

$$d\rho = \frac{dp}{a^2}.$$ (21.230)

Hence, the following correlation can be found for the cross-sectional area A of the nozzle

$$\boxed{\frac{dA}{A} = \left(\frac{1}{c^2} - \frac{1}{a^2}\right) v\,dp}$$ (21.231)

Introducing the Mach-number

$$\text{Ma} = \frac{c}{a}$$ (21.232)

and applying Eq. 21.228 one gets the so-called Rankine–Hugoniot equation, i.e.

$$\boxed{\frac{dc}{c} = -\frac{dA}{A\left(1 - \text{Ma}^2\right)}}$$ (21.233)

Respectively, with Eq. 21.228

$$\boxed{\frac{dp}{\rho c^2} = \frac{dA}{A\left(1 - \text{Ma}^2\right)}}$$ (21.234)

The consequences of these two equations are shown in Fig. 21.28:

- Subsonic flows, i.e. Ma < 1
  According to the Rankine–Hugoniot-equation a nozzle, i.e. accelerating a fluid so that $dc > 0$, requires a converging geometry, i.e. $dA < 0$. The pressure while passing the nozzle decreases. A diffusor, however, characterised by decelerating a fluid so that $dc < 0$, requires a diverging geometry with $dA > 0$. The pressure within the diffusor rises.

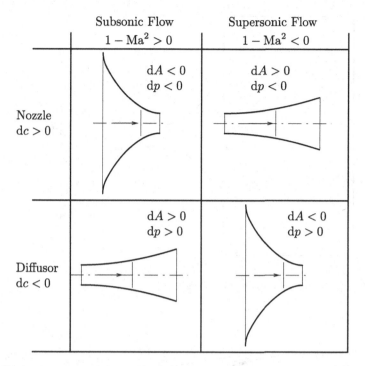

Fig. 21.28 Overview subsonic/supersonic flows

- Supersonic flows, i.e. Ma > 1

According to the Rankine–Hugoniot-equation a nozzle, i.e. $dc > 0$ requires a diverging geometry, i.e. $dA > 0$. The pressure while passing the nozzle decreases. A diffusor, however, characterised by $dc < 0$ requires a converging geometry with $dA < 0$. The pressure within the diffusor rises.

With this knowledge the generation of a supersonic flow with subsonic inlet conditions can be derived: A converging/diverging nozzle shape is required and shown in Fig. 21.29. This type of nozzle is a so-called Laval-nozzle. The setup shall be the same as before: A tank contains an ideal gas in rest state (0), i.e. the pressure shall be p_0, its temperature is T_0 and the velocity is $c_0 = 0$. The other state values follow accordingly, while the potential energy being ignored. State (1) is the inlet state in the converging part of the Laval-nozzle, state (n) is the state at the narrowest cross-section and state (2) indicates the outlet state. Outside shall be ambient pressure p_{env}. Obviously, the driver for a flow is the pressure difference between tank and ambient. In case the ambient pressure is equal to the pressure within the tank, no flow develops. Let us now start increasing the pressure p_0 within the tank while fixing the ambient pressure p_{env}.

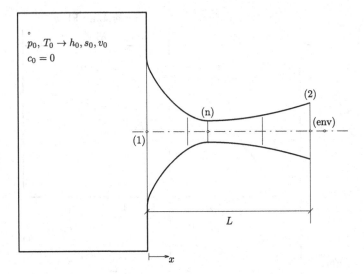

Fig. 21.29   Steady flow from vessels using a Laval-nozzle

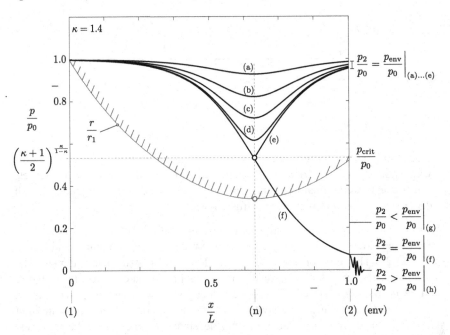

Fig. 21.30   Pressure characteristic Laval-nozzle

The scenarios (a)–(h) are shown in Figs. 21.30 and 21.31. Mind, that it is not sufficient just to regard the pressure difference between tank and environment, but it is required to take the *level of the ambient pressure* into account as well, since this controls the flow inside the Laval-nozzle:

(a) The pressure within the tank p_0 is larger than the ambient pressure p_{env}. A flow develops, but the pressure in state (n) is smaller than the critical pressure, i.e.

$$\pi_n < \pi_{crit}. \qquad (21.235)$$

The flow leaves with

$$\pi_2 = \pi_{env} \rightarrow p_2 = p_{env}. \qquad (21.236)$$

The counter pressure p_{env} is too large to reach a supersonic state within the nozzle. Velocity of sound is not reached and a pressure distribution as shown in Fig. 21.30 is measured. The flow leaves with subsonic velocity and the mass flow rate is not maximised according to the geometry of the Laval-nozzle. The temperature profile as well as the Mach-number profile are additionally shown in Fig. 21.31.

(b) The same as in (a) while p_n further approximates p_{crit}.
(c) The same as in (b) while p_n further approximates p_{crit}.
(d) The same as in (c) while p_n further approximates p_{crit}.
(e) In this case, the mass flux is sufficiently high, so that

$$\pi_n = \pi_{crit}. \qquad (21.237)$$

However, it is

$$\pi_n = \pi_{crit} < \pi_2 = \pi_{env} \rightarrow p_n = p_{crit} < p_2 = p_{env}. \qquad (21.238)$$

The counter pressure ratio π_{env} has been lowered compared to the previous cases but it is still too large for a supersonic flow in the diverging sector. According to [1] a normal shock, see Sect. 21.6, takes place at the narrowest cross-section.

(f) If the design fits to the geometry ("adapted Laval-nozzle") and the initial condition in the tank, i.e. state (0), the outlet pressure is ambient pressure, i.e.

$$\pi_2 = \pi_{env} \rightarrow p_2 = p_{env} \qquad (21.239)$$

and the pressure in the narrowest cross-section fulfils

$$\pi_n = \pi_{crit} > \pi_2 = \pi_{env} \rightarrow p_n = p_{crit} > p_2 = p_{env}. \qquad (21.240)$$

The mass flow rate is maximised for p_0 and the nozzle's geometry. Mind, that in this case π_2 is purely a function of the nozzle's geometry.

(g) A supersonic state is achieved within the diverging part of the nozzle, i.e.

$$\pi_n = \pi_{crit} \rightarrow p_n = p_{crit}. \tag{21.241}$$

However, the ambient pressure information can not get upstream, since the flow moves faster than velocity of sound. In this case the ambient pressure is in the range

$$\pi_{env,(g)} = \pi_{env,(e)} \cdots \pi_{env,(f)} \rightarrow p_{env,(g)} = p_{env,(e)} \cdots p_{env,(f)}. \tag{21.242}$$

Normal shocks within the nozzle occur, i.e. the fluid compresses unsteadily from the lower pressure to the larger pressure at the environment. Mind, that this characteristic within the nozzle is not sketched in Fig. 21.30.

(h) A supersonic state is achieved within the diverging part of the nozzle. Same as before, the ambient pressure information can not get upstream, since the flow moves faster than velocity of sound. However, the pressure at the outlet p_2 can be larger than the ambient pressure p_{env}, i.e.

$$\pi_2 > \pi_{env} \rightarrow p_2 > p_{env}, \tag{21.243}$$

while the critical pressure at (n) is still reach. It further is

$$\pi_{env} < \pi_{env,f}. \tag{21.244}$$

The fluid then starts to expand as soon as it leaves the nozzle. Periodic jet expansions and constrictions occur, see [1]. However, this scenario might happen if p_0 is further increased beyond the design case according to (f). Any further increase of p_0 leads to the same dimensionless characteristic

$$\pi = \frac{p}{p_0} = f\left(\frac{x}{L}\right) \tag{21.245}$$

as shown in Fig. 21.30f. This function purely depends on the geometry of the Laval-nozzle, see Fig. 21.25, i.e. the same π_{crit} and the same π_2 is reached. Increasing the pressure within the tank p_0 leads to a larger mass flow rate. However, though the function $\pi = \frac{p}{p_0} = f\left(\frac{x}{L}\right)$ is fixed for supersonic flows, the ratio $\frac{p_{env}}{p_0}$ decreases with rising p_0 while ambient pressure being constant.

A more detailed explanation regarding the scenarios can be found in [1, 24]. The shown graphs have been calculated with the same approach as in Sect. 21.7.1, i.e. solving Eqs. 21.213–21.217. Figure 21.31 compares cases (a), i.e. a subsonic flow, with the design case (f) of a Laval-nozzle. It shows the dimensionless shape of the Laval-nozzle, as well as the dimensionless temperature and pressure distributions. In case (f) the critical state at the narrowest cross-section is reached, so that

$$p_{crit} = p_0\left(\frac{\kappa+1}{2}\right)^{\frac{\kappa}{1-\kappa}} \tag{21.246}$$

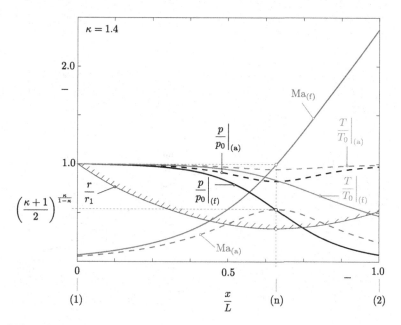

Fig. 21.31 (*a*) Venturi-nozzle, (*f*) Supersonic flow

respectively

$$T_{\text{crit}} = T_0 \left(\frac{p_\text{n}}{p_0}\right)^{\frac{\kappa-1}{\kappa}} = T_0 \frac{2}{\kappa+1}. \tag{21.247}$$

Consequently, the Mach-number is

$$\text{Ma} = 1. \tag{21.248}$$

For this shown example with the applied geometry the pressure ratio at the outlet fulfils

$$\pi_2 = \frac{p_2}{p_0} = 0.0712 \tag{21.249}$$

Case (a) is also called a Venturi-nozzle. The entire flow remains subsonic but is supposed to be adiabatic and frictionless, i.e. isentropic. Thus, it should not be confused with an adiabatic throttle, that has a similar geometrical shape but is based on friction in order to reduce the pressure.

Adiabatic/Polytropic Laval-Nozzle

Figure 21.32 shows a comparison between an isentropic Laval-nozzle, as it has been investigated so far, and an adiabatic/polytropic, i.e. frictional, operated Laval-

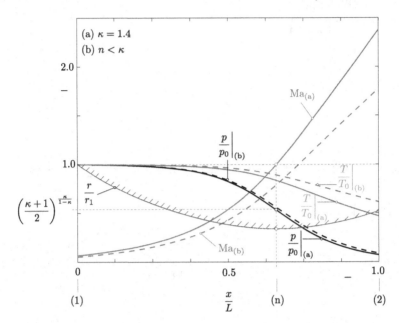

Fig. 21.32 Supersonic exit Laval-nozzle (a) isentropic, (b) polytropic/adiabatic, i.e. with friction

nozzle. In this case the algorithm for calculating this flow is the same as before, see
Eqs. 21.213–21.217, but the change of state is treated polytropically with $n < \kappa$, see
Eq. 21.214, that then reads as

$$pv^n = p_0 v_0^n = p_1 v_1^n = p_n v_n^n = p_2 v_2^n. \tag{21.250}$$

As indicated by the graph in Fig. 21.32 the supersonic flow does not reach its critical
pressure in the narrowest cross-section, but in the diverging part of the nozzle. The
Ma-number follows accordingly and thus is shifted to the right, i.e. Ma = 1 is in
the diverging section. Figure 21.33 shows the isentropic as well as the frictional
Laval-nozzle in a T, s-diagram. Obviously, the polytropic change of state with $n < \kappa$
causes the specific entropy to rise. This increase correlates with the friction, that
can be visualised by the area beneath the change of state in the T, s-diagram. The
temperature in the frictional case is larger than for isentropic conditions.

Laval-Diffusor

If the convergent/divergent shape of a Laval-nozzle is operated with a supersonic
inlet, see Fig. 21.34, the characteristic follows a diffusor. According to the Rankine–
Hugoniot-equation

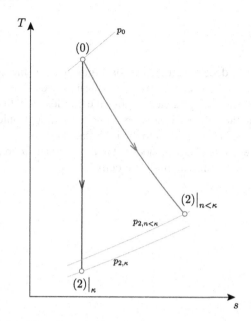

Fig. 21.33 Supersonic exit Laval-nozzle, T, s-diagram

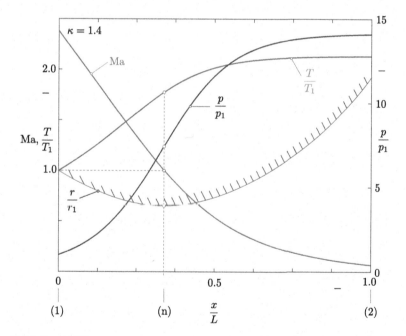

Fig. 21.34 Laval-diffusor

$$\frac{dc}{c} = -\frac{dA}{A\left(1 - \text{Ma}^2\right)} \tag{21.251}$$

the supersonic flow is decelerated in the converging part of the geometry. Under isentropic conditions the flow reaches Ma $= 1$ in the narrowest cross-section and then further decelerates in the diverging part. Finally, the flow leaves with a larger pressure as at the inlet, while the velocity is smaller. Mind, that dimensionless pressure and temperature in Fig. 21.34 are referred to the inlet state of the nozzle (1). A state (0), that used to be the rest state of a tank, does not make sense any more, since now state (1) is given and the system has an initial velocity.

Chapter 22
Thermodynamic Cycles with Phase Change

Though thermodynamic cycles have already been introduced in Sect. 17.2, this chapter covers cycles, in which the working fluid is subject to phase changes. If a system after several changes of state finally reaches the initial state, it is called a thermodynamic cycle. Thus, all state values reach their initial value, i.e.

$$\oint dZ = 0. \tag{22.1}$$

A distinction has been made regarding the application of such processes: *Clockwise cycles* convert heat into work, while with respect to the second law of thermodynamics a thermal efficiency of 100% is not possible, i.e. heat needs to be passed to the environment. A typical representative of such a cycle is the Joule process. However, the working fluids so far used to be ideal gases. In this chapter the focus is on clockwise cycles with working fluids, that typically change their aggregate state, e.g. a steam power process. *Counterclockwise cycle* shift heat from a lower temperature to a higher level. In order to do so, according to the second law of thermodynamics, work needs to be supplied. This working principle can be utilised for heat pumps as well as for cooling machines. However, these cycles have only been discussed briefly in part I, so they are focus in this chapter.

The Carnot-process is an idealised benchmarking cycle process, that consists of two reversible isothermal and two reversible adiabatic, i.e. isentropic, changes of state. Thus, each change of state is reversible and the process achieves its maximum efficiency. This process has already been discussed in part I and is shown exemplary for a clockwise cycle in Fig. 22.1. The process consists of the following changes of state:

- $(1) \rightarrow (2)$: Isothermal expansion
- $(2) \rightarrow (3)$: Reversible, adiabatic expansion
- $(3) \rightarrow (4)$: Isothermal compression
- $(4) \rightarrow (1)$: Reversible, adiabatic compression.

© Springer Nature Switzerland AG 2019
A. Schmidt, *Technical Thermodynamics for Engineers*,
https://doi.org/10.1007/978-3-030-20397-9_22

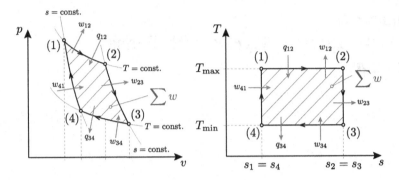

Fig. 22.1 Clockwise Carnot process: thermal engine

The thermal efficiency for such a process solely depends on the minimum and maximum operation temperature, i.e.

$$\eta_C = 1 - \frac{T_{\min}}{T_{\max}} \tag{22.2}$$

22.1   Steam Power Process

22.1.1   Clausius-Rankine Process

The Clausius-Rankine process is an ideal[1] reference cycle for steam power plants, i.e. it covers phase changes of the working fluid, that usually takes place in liquid and gaseous state. Its principal layout is shown in Fig. 22.2 and is similar to the already introduced Joule-process, see part I. The process consists of the following changes of state, that are illustrated in a p, v- as well as T, s-diagram, see Fig. 22.3:

- $(1) \rightarrow (2)$: Adiabatic, frictionless compression
  The required compression to reach a high pressure level is idealised and supposed to be adiabatic and frictionless. It is advantageous to increase the pressure of a liquid fluid instead of a gaseous fluid, since the technical work under these conditions, with ignoring the kinetic and potential energies, follows

$$w_{t,12} = \int_1^2 v \, dp > 0. \tag{22.3}$$

---

[1]Idealised in this case means, that there is no generation of entropy!

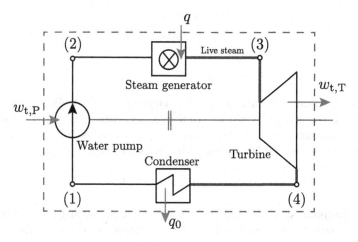

Fig. 22.2 Clausius-Rankine cycle: sketch

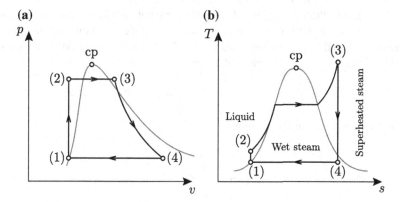

Fig. 22.3 Clausius-Rankine cycle: **a** p, v-diagram, **b** T, s-diagram

In contrast to gases/steams the specific volume of liquids is rather small. The first law of thermodynamics obeys

$$w_{t,12} = w_{t,P} = h_2 - h_1 > 0. \tag{22.4}$$

In the illustrated example it is assumed, that state (1) is a $(')$-state.

- $(2) \rightarrow (3)$: Isobaric heat transfer + phase change
  In case the heat supply is isobaric, the partial energy equation with the same premise as for the change of state from (1) to (2) reads as

$$w_{t,23} = \int_2^3 v \, dp + \underbrace{\psi_{23}}_{=0} = 0. \tag{22.5}$$

Thus, no dissipation occurs, so that this change of state is idealised as well. The rise of specific entropy is just due to the supplied heat. The initial liquid fluid becomes saturated liquid in $(')$-state, fully vaporises and is further overheated to reach state (3). The first law of thermodynamics results in

$$q_{23} = q = h_3 - h_2 > 0. \tag{22.6}$$

Due to the chosen system boundary,[2] the entropy generation that comes along with the heat transfer from outside to the fluid is not part of the system, see Sect. 17.1.5 for details.

- (3) → (4): Adiabatic, frictionless expansion
  In this step the fluid, being on a high pressure level and due to the heating on a high energetic level, now releases technical work in the adiabatic and frictionless turbine, so that the pressure level decreases. The first law of thermodynamics, with ignoring kinetic as well as potential energies, reads as

$$w_{t,34} = -w_{t,T} = h_4 - h_3 < 0. \tag{22.7}$$

The partial energy equation follows

$$w_{t,34} = \int_3^4 v \, dp < 0. \tag{22.8}$$

- (4) → (1): Isobaric heat transfer + phase change
  In case the heat release is isobaric, the partial energy equation with the same premise as for the change of state from (1) to (2) reads as

$$w_{t,41} = \int_4^1 v \, dp + \underbrace{\psi_{41}}_{=0} = 0. \tag{22.9}$$

Thus, no dissipation occurs, so that this change of state is idealised as well. The wet steam condenses and finally reaches state (1). The first law of thermodynamics follows

$$q_{41} = -q_0 = h_1 - h_4. \tag{22.10}$$

---

[2]Only the fluid flow through the heat exchanger is part of the system boundary. The environment being on a larger temperature level, and thus forming an imperfection with the fluid, is not covered.

It is assumed, that state (4) after the turbine has reached the wet-steam region.

The graphs in Fig. 22.3 prove, that the Clausius-Rankine cycle is a clockwise cycle. Thus, effectively work is released, i.e. the work released by the turbine is larger than the work supplied at the water pump

$$|w_{t,34}| > |w_{t,12}|. \tag{22.11}$$

The overall first law of thermodynamics obeys

$$\sum w_t + \sum q = 0 \rightarrow \underbrace{w_{t,12} + w_{t,34}}_{=w_{eff}} + \underbrace{q_{23} + q_{41}}_{=q_{eff}} = 0. \tag{22.12}$$

The effective work

$$w_{eff} = w_{t,12} + w_{t,34} \tag{22.13}$$

is represented as enclosed area in the p, v-diagram.[3] Since the entire process is free of generation of entropy, it is the specific heat that is represented as area beneath each change of state in the T, s-diagram. Consequently, the enclosed area in the T, s-diagram represents the effective heat

$$q_{eff} = q_{23} + q_{41}. \tag{22.14}$$

According to the first law of thermodynamics it is equivalent with

$$w_{eff} = -q_{eff}. \tag{22.15}$$

22.1.2 Steam Power Plant

In this section a non-ideal process is discussed, that in contrast to the Clausius-Rankine process generates entropy. Before doing so, the heat supply in a steam power process is formally split, see Fig. 22.4. In the first step, i.e. from state (2) the pre-heating takes place until the super-cooled liquid reaches the (3) = (′)-state. Secondly, the fluid vaporises fully from state (3) to state (4) = (″). Finally, the overheating takes place from state (4) to state (5). The according T, s-diagram is shown in Fig. 22.5. Since the specific entropy rises within pump and turbine, the process is not ideal, i.e. it differs from a perfect Clausius-Rankine process. It is beneficial to illustrate a steam power process in a h, s-diagram, as been done in Fig. 22.6, since the specific works as well as the specific heats can easily been identified under the premise, that potential as well as kinetic energies are ignored. Thus, it is further assumed, that pump and turbine are adiabatic. For the turbine the first law of thermodynamics

---

[3]Since potential and kinetic energies have been ignored. Furthermore, no dissipation occurs.

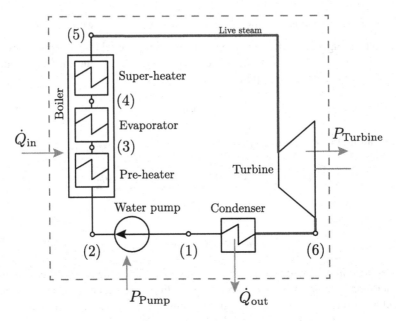

Fig. 22.4 Steam power process: boiler

exemplary reads as

$$w_{t,56} + \underbrace{q_{56}}_{=0} = h_6 - h_5 < 0. \tag{22.16}$$

For the overall heat supply it is

$$\underbrace{w_{t,25}}_{=0} + q_{25} = h_5 - h_2 > 0. \tag{22.17}$$

With the assumption that the turbine is adiabatic, the rise of specific entropy from (5) to (6) is due to frictional entropy generation, see Fig. 22.6, i.e.

$$ds = \delta s_i + \underbrace{\delta s_a}_{=0} \rightarrow \Delta s_{56} = s_6 - s_5 = \Delta s_{i,T}. \tag{22.18}$$

In order to calculated a steam power process, the required caloric state values can be taken from steam tables. However, the shown expansion from state (5) to state (6) is not chosen perfectly, since the expanded vapour still is on a high temperature respectively enthalpy level. Thus, it is very unlikely to expand the fluid into a super-heated state, but into the wet-steam region. According to the h, s-diagram, see Fig. 22.6, the lower pressure level p_1 would be further decreased. Typical steam parameters are $\vartheta_5 = 550\,°C$ and $p_5 = 250\,bar$. The work effectively gained in the plant is

Fig. 22.5 Steam power
process: T, s-diagram

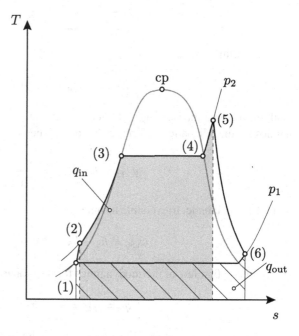

Fig. 22.6 Steam power
process: h, s-diagram

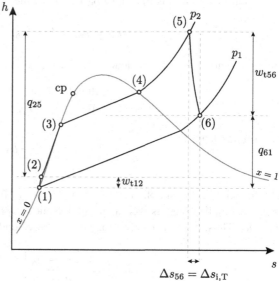

$$w_{t,\text{eff}} = \sum w_t = w_{t,12} + w_{t,56}. \qquad (22.19)$$

Technical work needs to be supplied to the pump, whereas the turbine releases technical work. The first law of thermodynamics for the overall power plant leads to

$$\sum w_t + \sum q = 0. \tag{22.20}$$

Thus, according to Fig. 22.4 it is

$$w_{t,\text{eff}} = -(q_{\text{in}} - |q_{\text{out}}|). \tag{22.21}$$

With introducing the isentropic efficiency, see Sect. 17.1, the first law of thermodynamics for the adiabatic and irreversible pump obeys

$$w_{t,12} = h_2 - h_1 = \frac{\Delta h_s}{\eta_{s,P}} \tag{22.22}$$

and for the adiabatic, irreversible turbine

$$w_{t,56} = h_6 - h_5 = \Delta h_s \eta_{s,T}. \tag{22.23}$$

Applying the first law of thermodynamics for the isobaric boiler brings

$$q_{\text{in}} = q_{25} = h_5 - h_2. \tag{22.24}$$

The condenser as well is treated isobaric, so that the first law of thermodynamics is

$$q_{\text{out}} = q_{61} = h_1 - h_6. \tag{22.25}$$

Hence, it is now possible to calculate the thermal efficiency of the steam power plant, i.e.

$$\eta_{\text{th}} = \frac{\text{Benefit}}{\text{Effort}} = \frac{|w_{t,\text{eff}}|}{q_{\text{in}}}. \tag{22.26}$$

Substitution leads to

$$\eta_{\text{th}} = \frac{|h_2 - h_1 + h_6 - h_5|}{h_5 - h_2}. \tag{22.27}$$

The technical work supplied to the pump $w_{t,12}$ is often ignored compared to the released/supplied heat as well as to the technical work within the turbine, so that $h_2 \approx h_1$. Thus, the thermal efficiency under these conditions obeys

$$\boxed{\eta_{\text{th}} = \frac{h_5 - h_6}{h_5 - h_1}} \tag{22.28}$$

Reheating

Figure 22.5 has shown, that it is beneficial not to let the turbine expand into the superheated region, but to bring the fluid in wet-steam state. By doing so, the released technical work can be maximised, see Fig. 22.7. However, in case the vapour ratio x is small the wet-steam carries liquid droplets, that can cause, when hitting the blades of the turbine, erosion. According to Fig. 22.7, this threat can be avoided by further shifting state (5) up to larger temperatures. However, as mentioned before, the maximum temperature is limited and increasing the lower operation pressure, i.e. bringing state (6) closer to the $x = 1$ curve, would decrease the power output of the turbine. A solution is the so-called reheating, as shown in Fig. 22.8: After partial expansion in a first high-pressure turbine, the fluid is reheated in a heat exchanger. In order to achieve the lower operation pressure the fluid then is expanded in the low pressure turbine. The entire cycle follows:

- (1) → (2): Polytropic compression
  As in a regular steam power process, the pressure of the liquid fluid is increased. In most cases, the pump is assumed to be adiabatic, so that the increase of entropy is caused by frictional effects. Thus, the change of state is polytropic.
- (2) → (3): Pre-heating
  By heat supply the super-cooled[4] liquid is heated-up to reach the (')-state.
- (3) → (4): Vaporisation
  Due to further heat supply the fluid changes from (') to (")-state. i.e. from the left $x = 0$ curve to the $x = 1$ curve of the wet-steam region.
- (4) → (5): Overheating
  Once, the vaporisation is finished, further supply of thermal energy leads to sensible heating, i.e. the vapour temperature rises.
- (5) → (6): High pressure expansion
  The fluid partially expands from the maximum operation pressure p_{HD} to a medium pressure p_{MD}. Thus, the minimum operation pressure of the process is not achieved in the HP-turbine. State (6) is super-heated fluid.
- (6) → (7): Isobaric reheating
  The super-heated fluid (6) is further heated up by external heat supply. Figure 22.9 shows, that this change of state is isobaric, i.e. in case kinetic and potential energies are ignored it is frictionless and thus idealised. However, in common steam power plants a pressure drop occurs. State (7) is now further away from the $x = 1$ curve.
- (7) → (8): Low pressure expansion
  In this step the final expansion is done in the LP-turbine, in order to reach the minimum operation pressure. Since the state (7) has been shifted further to the right, the expansion now can be done without running into danger to cause droplet erosion at the turbine's blades.
- (8) → (1): Condensation
  In the last change of state the fluid now is condensed to reach state (1) again. The thermodynamic cycle is closed. Once again it is assumed, that the condensation runs isobarically. The same has been done for the heat supply from (2) to (5).

---

[4]Super-cooled means, that the fluid's temperature is lower than the boiling temperature at this pressure.

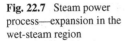**Fig. 22.7** Steam power process—expansion in the wet-steam region

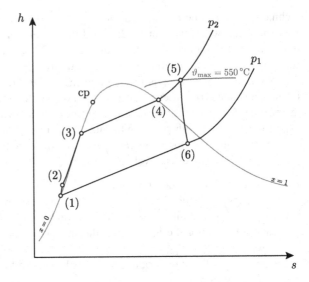

Figure 22.9 shows the advantage of the reheating: Damage at the turbine blades due to erosion can be prevented. Furthermore, the efficiency of such a steam process is increased, since the entire heat $q_{in,1} + q_{in,2}$ is supplied on a higher average temperature level. This phenomenon is called *carnotisation*. According to the ideal Carnot process, its efficiency is purely a function of minimum and maximum operation temperature. The larger the temperature is, at which the heat is supplied, the larger the efficiency is. However, this is utilised in a steam process with reheating, since the specific heat $q_{in,2}$ is supplied at a high temperature level. The thermal efficiency of the shown process is, while ignoring the rather small technical work of the pump,

$$\eta_{th} = \frac{\text{Benefit}}{\text{Effort}} = \frac{|w_{t,56} + w_{t,78}|}{q_{25} + q_{67}}. \tag{22.29}$$

Regenerative Feed Water Preheating

The idea of the *carnotisation* can be further developed. Figure 22.10a shows a steam power process as discussed before in a T, s-diagram. Obviously, heat is supplied from state (2) to state (5). Consequently, the averaged temperature $\overline{T}_{\dot{Q}_{in}}$, at which heat is supplied *externally*, needs to be somewhere in-between temperature T_2 and temperature T_5. According to Carnot this temperature should be as high as possible. Unfortunately, it is the pre-heating that reduces the averaged temperature of external heat supply, since in comparison with states (3), (4) and (5) temperature T_2 is rather small. In order to maximise the thermal efficiency only the externally supplied heat is relevant, so that the idea of a *carnotisation* is to realise the feed water pre-heating

Fig. 22.8 Steam power process—reheating

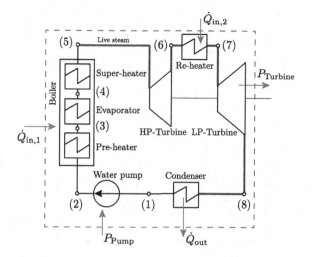

Fig. 22.9 Steam power process—reheating

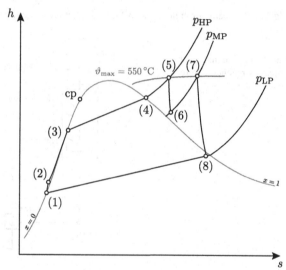

inside the process, i.e. without transferring thermal energy from outside to the process. Thus, the required energy is taken from the process itself, i.e. by regenerative means. The major advantage is shown in Fig. 22.10b: The averaged temperature $\overline{T}_{\dot{Q}_{in}}$, at which *external* heat is supplied, rises and increases the thermal efficiency.

Figure 22.11 shows, how regenerative[5] feed water preheating from state (2) to state (3) can be realised. Partially expanded steam from the turbine is extracted, state (6) respectively state (7), that still is on a high temperature level. Hence, this extracted mass flow can be utilised to heat up the feed water after the water pump, state (2).

---

[5]The required energy comes from the process itself!

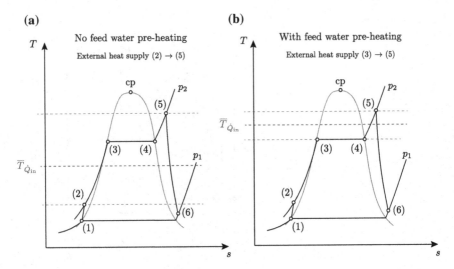

(a)

No feed water pre-heating

External heat supply (2) → (5)

(b)

With feed water pre-heating

External heat supply (3) → (5)

Fig. 22.10 Steam power process—**a** No feed water preheating, **b** Regenerative feed water preheating

Fig. 22.11 Steam power process—regenerative feed water preheating

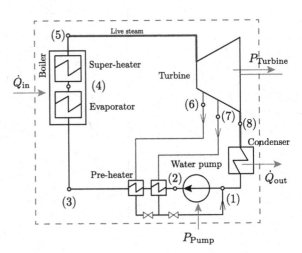

After leaving the pre-heaters the pressure needs to be adjusted before mixing the extracted flow back to the main flow (1). This pressure adjustment can easily be done by throttles, since the extracted flow still is on a high pressure level, since it has been taken out of the turbine before fully being expanded. The sketched system boundary in Fig. 22.11 clearly indicates, that the water pre-heating now is achieved with internal means and no external energy is transferred across the system boundary to heat up the water. However, the externally supplied heat flux $\dot{Q}_{in}$ is required for the vaporisation as well as for the super-heating of the fluid.

22.2 Heat Pump and Cooling Machine

In this section the focus is on counterclockwise cycles, i.e. thermodynamic cycles that shift heat from a cold reservoir to a hot reservoir, see Fig. 15.5. As already known, according to the second law of thermodynamics, work needs to be supplied. On the one hand side the benefit of such a process can be the heat, that is taken out of the cold tank. If this is the case, the machine is called a cooling machine. Alternatively, the purpose can be the heating of a room for instance, so that the heat, that is shifted to the hot tank is the benefit. In such a case, the machine is called a heat pump. However, cooling machine and heat pump have the same technical components and follow the same thermodynamic principle. It is just the definition of its benefit, that makes the difference between them.

22.2.1 Mechanical Compression

In order to realise a counterclockwise cycle, the working fluid needs to be compressed from a lower working pressure to a higher working pressure, see Sect. 15.2.1. This can be achieved with a compressor in case the fluid is in gaseous state. Figure 22.12 shows exemplary the layout of a heat pump. Heat is transferred from the environmental heat source, that is on a low temperature level, to the working fluid. The fluid is compressed and releases heat on a larger temperature level. This might be an external water circuit in order to operate a radiator for heating purpose. To close the cycle the fluid then needs to be expanded to the lower level. The changes of state are as follows and are illustrated in a T, s-diagram, see Fig. 22.13:

- $(1) \rightarrow (2)$: Pressure increase
  The fluid is supposed to be super-heated, see step (4) to (1). Thus, the required specific technical work follows

$$
w_{t,12} = w_t = \int_1^2 v \, dp + \psi_{12} > 0.
\tag{22.30}
$$

Kinetic as well as potential energies are as usual ignored. In case the compressor is adiabatic, the indicated increase of specific entropy in the T, s-diagram, see Fig. 22.13, is due to frictional entropy generation. After the compression the fluid is on a larger pressure and, according to the first law of thermodynamics, on a larger level of specific enthalpy, i.e.

$$
w_{t,12} + \underbrace{q_{12}}_{=0} = h_2 - h_1 > 0.
\tag{22.31}
$$

This enthalpy increase leads to an increase of temperature.

- $(2) \rightarrow (3)$: Isobaric heat release

  Now the fluid is on a large temperature level, heat can be released to a hot reservoir, e.g. a room, that needs to be heated. Due to the heat release the fluid condenses and leaves the heat exchanger as super-cooled liquid. However, the heat release was supposed to be isobaric, i.e. in case kinetic and potential energies are ignored there is no dissipation and thus no entropy generation. The first law of thermodynamics reads as

$$q_{23} = q_{\text{out}} = h_3 - h_2 < 0. \tag{22.32}$$

- $(3) \rightarrow (4)$: Adiabatic throttling

  In order to reduce the pressure level an adiabatic throttle is utilised. As already mentioned, no technical work is released in a throttle and the pressure drop is purely based on dissipation, i.e.

$$w_{t,34} = 0 = \int_3^4 v \, dp + \psi_{34} \rightarrow \int_3^4 v \, dp = -\psi_{34}. \tag{22.33}$$

Consequently, the specific entropy needs to rise, while the specific enthalpy remains constant, i.e.

$$\underbrace{w_{t,34}}_{=0} + \underbrace{q_{34}}_{=0} = h_4 - h_3 = 0. \tag{22.34}$$

- $(4) \rightarrow (1)$: Isobaric heat supply

  For closing the thermodynamic cycle, heat needs to be supplied. As can be seen in the T, s-diagram, see Fig. 22.13, this needs to be realised on a low temperature level and a low corresponding pressure level as well. The thermal energy shall be taken from the environment, e.g. air, water or ground soil, that already is on a low temperature level. For the heat transfer a ΔT is required, i.e. the fluid's temperature in the thermodynamic process needs to be by ΔT colder than ambient temperature. Real fluids, i.e. fluids that can underlie a phase change, start to vaporise at a low temperature in case the pressure is low. This happens in the change of state from (4) to (1): Though ambient temperature is low, heat is supplied to the even colder fluid in the circuit. The fluid starts to vaporise, while, if the change of state runs isobarically, the temperature remains constant until the $(")$-state is reached. If the ΔT is still sufficient, the fluid is further super-heated until state (1) is finally reached. The first law of thermodynamics obeys

$$q_{41} = q_{\text{in}} = h_1 - h_4 > 0. \tag{22.35}$$

Obviously, according to the T, s-diagram in Fig. 22.13 the process is counterclockwise. Since heat supply and heat release have been assumed to be isobaric, no entropy is generated. Consequently, both, the supplied specific heat q_{in} and the released heat q_{out} can be identified in the T, s-diagram. According to the heat pump's layout, see

Fig. 22.12 Compression heat pump

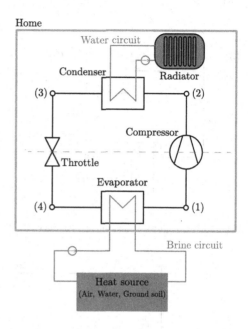

system boundary in Fig. 22.13, technical work w_t is supplied to the machine. Thus, the overall first law of thermodynamics obeys

$$|q_{out}| = q_{in} + w_t. \tag{22.36}$$

In HVAC applications the counterclockwise cycle is often illustrated in a log p, h-diagram, that as been introduced in Sect. 18.4.3. Such a diagram is exemplary shown in Fig. 22.14 for the discussed cycle. With the made premises[6] the specific heats and specific technical work can easily be visualised by horizontal distances, see Fig. 22.14. Similar to a clockwise cycle, the efficiency of counterclockwise cycles can be calculated by the coefficient of performance, i.e. the ration of benefit and effort. For a heat pump it is

$$\varepsilon_{HP} = \frac{|q_{out}|}{w_{t,12}} = \frac{h_2 - h_3}{h_2 - h_1} \tag{22.37}$$

For the cooling machine it is

$$\varepsilon_{CM} = \frac{q_{in}}{w_{t,12}} = \frac{h_1 - h_4}{h_2 - h_1} \tag{22.38}$$

---

[6]Ignoring the kinetic and potential energies!

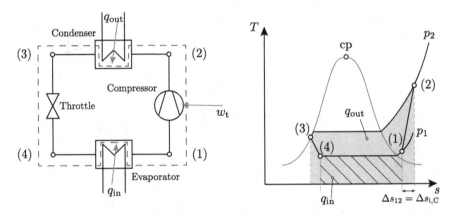

Fig. 22.13 Compression heat pump—layout (left), T, s-diagram (right)

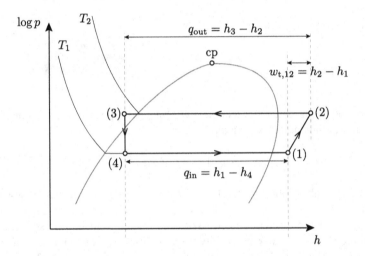

Fig. 22.14 Compression heat pump—log p, h-diagram

22.2.2 Thermal Compression

In Sect. 22.2.1 it has been shown, that the pressure increase is realised with a super-heated fluid. However, according to the partial energy equation

$$w_{t,12} = w_t = \int_1^2 v\,dp + \psi_{12} > 0 \qquad (22.39)$$

it would be beneficial to increase the pressure of a liquid fluid, since the specific volume of a liquid is smaller than the specific volume of a gaseous fluid. Consequently, replacing the mechanical compression is the goal of an absorption heat pump, see sketch 22.15. Accept of the compressor, the components are the same as before for the mechanical compression according to Fig. 22.13. Hence, the condenser is still applied to release heat on the high temperature level, the evaporator receives the heat on a low temperature level. The pressure decrease is realised by a throttle. The grey-boxed components replace the former mechanical compressor: Super-heated vapour reaches the so-called absorber, that is filled with water as solvent respectively absorbent for instance. The gaseous working fluid, also called refrigerant, is then absorbed, i.e. the working fluid forms a *liquid mixture* with the absorbent. While doing so, heat is released on a low temperature level, so that its exergy content is rather low. The major advantage of this absorption technique compared to compression heat pumps is, that now a pump can be applied to achieve the higher pressure level of this liquid mixture. Thus, the technical work is much lower than for a gaseous fluid, see Eq. 22.39. Once, the mixture is on the upper working pressure, it needs to be separated again, i.e. the refrigerant must be released. In general the boiling of the refrigerant starts earlier than the boiling of the solvent. This principle is utilised in the expeller, where heat is supplied on a large temperature level, so that the refrigerant is separated from the liquid mixture. Refrigerant in gaseous state leaves the expeller on the high pressure level and the circuit can be continued as before with the mechanical compression. The remaining solvent in the expeller can be throttled and led back to the absorber, see Fig. 22.15. Thus, the technical effort of an absorption heat pump is not mechanical work but thermal energy. This is why the grey-boxed components in Fig. 22.15 are called *thermal compressor*. Figure 22.16 shows an example of the so-called solar cooling. As mentioned before, cooling machine and heat pump follow the same technical principles. The absorption-based counterclockwise cycle requires thermal energy on a high temperature as a driver. In case solar energy is available, this technique can be utilised for a cooling machine.

Problem 22.1 The high pressure (HP) turbine of a steam power plant expands vapour from state 1 ($\vartheta_1 = 540\,°C$, $p_1 = 170\,bar$) to state 2 ($\vartheta_2 = 320\,°C$, $p_2 = 35\,bar$). The fluid is re-heated to $\vartheta_3 = 540\,°C$ and finally expands in the low pressure (LP) turbine to $p_4 = 0.1\,bar$. The vapour ratio is $x_4 = 0.96$. Due to friction the pressure drop within the re-heater is $\Delta p = 2.5\,bar$. All changes of state are polytropic with a constant polytropic exponent. Changes of kinetic and potential energies can be ignored. Both turbines are adiabatic (Fig. 22.17).

(a) Sketch the process in a h, s-diagram.
(b) Calculate the specific technical work in both turbines as well as the transferred specific heat in the re-heater.
(c) What is the polytropic exponent n in the re-heater?
(d) Calculate the specific dissipation in the re-heater.
(e) What is the specific loss of exergy of the entire plant, in case the heat supply is realised with a reservoir of $\vartheta_R = 650\,°C$? The ambient temperature is $\vartheta_{env} = 17\,°C$.

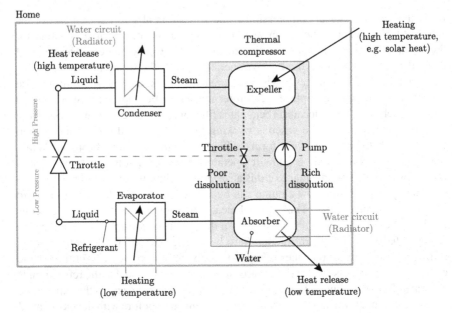

Fig. 22.15 Absorption heat pump

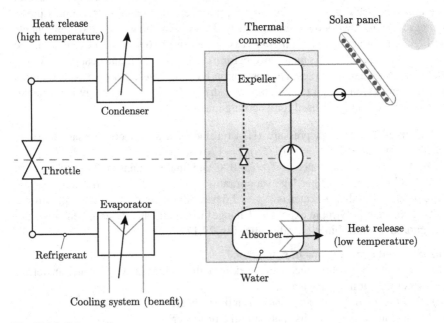

Fig. 22.16 Solar cooling

Fig. 22.17 Sketch to
Problem 22.1

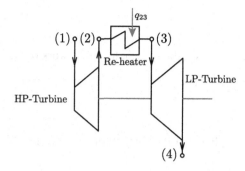

Fig. 22.18 h, s-diagram to
Problem 22.1

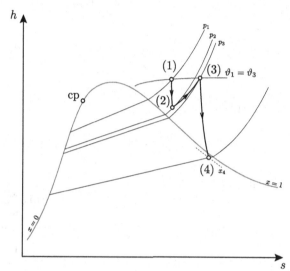

Solution

(a) Figure 22.18 shows the process in a h, s-diagram.

(b) In order to calculate the specific technical work of the turbines the first law of thermodynamics is applied. For the high pressure turbine it reads as

$$\underbrace{q_{12}}_{=0} + w_{t,12} = h_2 - h_1. \tag{22.40}$$

The specific enthalpy can be taken from the steam table, i.e.

$$h_1 = h\,(p_1,\,\vartheta_1) = 3400.9\,\frac{\text{kJ}}{\text{kg}} \tag{22.41}$$

and

$$h_2 = h\,(p_2,\,\vartheta_2) = 3030.5\,\frac{\text{kJ}}{\text{kg}}. \tag{22.42}$$

Hence, the specific technical work of the HP-turbine is

$$\boxed{w_{t,12} = h_2 - h_1 = -370.4\,\frac{\text{kJ}}{\text{kg}}} \tag{22.43}$$

For the LP-turbine it follows accordingly

$$\underbrace{q_{34}}_{=0} + w_{t,34} = h_4 - h_3 \tag{22.44}$$

with

$$h_3 = h\,(p_3,\,\vartheta_3) = 3544.6\,\frac{\text{kJ}}{\text{kg}}. \tag{22.45}$$

State (4) follows

$$h_4 = h_4' \cdot (1 - x_4) + h_4'' x_4 = 2488.2\,\frac{\text{kJ}}{\text{kg}} \tag{22.46}$$

with

$$h_4'\,(p_4) = 191.8123\,\frac{\text{kJ}}{\text{kg}} \tag{22.47}$$

and

$$h_4''\,(p_4) = 2583.9\,\frac{\text{kJ}}{\text{kg}}. \tag{22.48}$$

Hence, the specific technical work of the LP-turbine is

$$\boxed{w_{t,34} = h_4 - h_3 = -1056.4\,\frac{\text{kJ}}{\text{kg}}} \tag{22.49}$$

The specific heat in the re-heater obeys the first law of thermodynamics

$$\boxed{q_{23} + \underbrace{w_{t,23}}_{=0} = h_3 - h_2 = 514.14\,\frac{\text{kJ}}{\text{kg}}} \tag{22.50}$$

(c) A polytropic change of state is defined by

$$pv^n = \text{const.} \tag{22.51}$$

Thus, it is

$$p_2 v_2^n = p_3 v_3^n.$$ (22.52)

Solving for n brings

$$n = \frac{\log\left(\frac{p_2}{p_3}\right)}{\log\left(\frac{v_3}{v_2}\right)} = 0.1635$$ (22.53)

with

$$v_3 = 0.1131 \, \frac{m^3}{kg}$$ (22.54)

and

$$v_2 = 0.0719 \, \frac{m^3}{kg}.$$ (22.55)

Both specific volumes have been taken from the steam table. Mind, that

$$\frac{T_3}{T_2} = \left(\frac{p_3}{p_2}\right)^{\frac{n-1}{n}}$$ (22.56)

can not be applied, since the vapour does not obey the ideal gas law.

(d) In order to calculate the specific dissipation in the re-heater, the partial energy equation is applied, i.e.

$$w_{t,23} = 0 = \int_2^3 v \, dp + \psi_{23} + \underbrace{\Delta e_{a,23}}_{=0}.$$ (22.57)

Solving for the specific dissipation[7] leads to

$$\psi_{23} = -\int_2^3 v \, dp = -v_2 p_2 \frac{n}{n-1}\left[\left(\frac{p_3}{p_2}\right)^{\frac{n-1}{n}}\right] = 22.67 \, \frac{kJ}{kg}.$$ (22.58)

(e) The specific loss of exergy of the entire plant follows

$$\Delta e_{x,V} = T_{env} s_{i,14}.$$ (22.59)

Thus, a balance of entropy according to Fig. 22.19 needs to be done. The incoming entropy must be balanced by the outgoing entropy in steady state. Hence, the second law of thermodynamics reads as

---

[7]See Fig. 13.13!

$$\dot{m}s_1 + \dot{S}_{i,14} + \dot{S}_a = \dot{m}s_4, \tag{22.60}$$

so that the specific generation of entropy is

$$s_{i,14} = s_4 - s_1 - s_{a,23}. \tag{22.61}$$

For the specific entropy, that is carried with the heat it follows[8]:

$$s_{a,23} = \frac{q_{23}}{T_R} = 0.5569 \, \frac{\text{kJ}}{\text{kg K}}. \tag{22.62}$$

Mind, that the specific generation of entropy $s_{i,14}$ includes the specific production of entropy due to dissipation within both turbines, the re-heater and the imperfection of the heat transfer.[9] The specific entropy of state (1) can be taken out of the steam table, i.e.

$$s_1 = s\,(p_1, \vartheta_1) = 6.4106 \, \frac{\text{kJ}}{\text{kg K}}. \tag{22.63}$$

For state (4) it is

$$s_4 = s_4' \cdot (1 - x_4) + s_4'' x_4 = 7.8489 \, \frac{\text{kJ}}{\text{kg K}} \tag{22.64}$$

with

$$s_4'\,(p_4) = 0.6492 \, \frac{\text{kJ}}{\text{kg K}} \tag{22.65}$$

and

$$s_4''\,(p_4) = 8.1489 \, \frac{\text{kJ}}{\text{kg K}}. \tag{22.66}$$

Hence, the specific generation of entropy is

$$s_{i,14} = 0.8813 \, \frac{\text{kJ}}{\text{kg K}}. \tag{22.67}$$

The specific loss of exergy finally is

$$\boxed{\Delta e_{x,V} = T_{env} s_{i,14} = 255.7 \, \frac{\text{kJ}}{\text{kg}}} \tag{22.68}$$

---

[8]This is since the temperature of the heat reservoir is constant!

[9]This imperfection is due to the ΔT between fluid and reservoir.

Problem 22.2 A compression cooling machine operated with ammonia as working fluid shall keep a room at a temperature of $\vartheta_0 = -12\,°C$. To do so, the cooling machine receives a cooling power of $\dot{Q} = 135.0\,kW$. The vaporisation pressure is $p_0 = 2.264\,bar$. The vapour being in saturated state ($''$) is compressed in an adiabatic compressor ($\eta_{s,V} = 0.785$) to a pressure of $p = 11.666\,bar$. After being fully condensed isobarically, i.e. to the ($'$) state, it is adiabatically throttled to a pressure of p_0. The condenser is operated with cooling water, that enters with a temperature of $\vartheta_{env} = 15\,°C$. The cooling water shall be treated as an incompressible liquid. There shall be no pressure drop for the cooling water. Changes of kinetic as well as potential energies shall be disregarded.

(a) Sketch the layout of the cooling machine and illustrate the changes of state in a T, s-diagram. Additionally mark the temperatures T_0 and T_{env}. Clearly identify the states (1) to (4).
(b) Calculate the mass flux of the ammonia $\dot{m}_A$.
(c) What is the technical power of the compressor? Determine the coefficient of performance of the cooling machine ε! What is the benchmarking coefficient of performance?
(d) The maximum temperature of the cooling water shall not exceed $\vartheta_W = 27\,°C$. What is the required mass flux of the cooling water $\dot{m}_W$? The specific heat capacity of the water is $c_W = 4.187\,\frac{kJ}{kg\,K}$).
(e) What is the loss of exergy $\Delta\dot{E}_{x,V}$ in the condenser?

For ammonia in saturated state it is

$p[bar]$	$\vartheta_s[°C]$	$h[\frac{kJ}{kg}]$	$s[\frac{kJ}{kg\,K}]$
2.264	−16.0	$h' = 288.5,\ h'' = 1604.1$	$s' = 1.2877,\ s'' = 6.4038$
11.666	30.0	$h' = 503.6,\ h'' = 1648.1$	$s' = 2.0512,\ s'' = 5.8267$

For ammonia at $p = 11.666\,bar$ it is

$\vartheta[°C]$	$h[\frac{kJ}{kg}]$	$s[\frac{kJ}{kg\,K}]$
100	1840.1	6.3988
110	1865.2	6.4653
120	1890.1	6.5297
130	1914.9	6.5919

Solution

(a) The layout of this process is shown in Fig. 22.20a. Figure 22.20b illustrates the changes of state in a T, s-diagram. Mind, that ambient temperature T_{env} needs

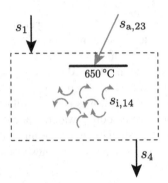

Fig. 22.19 Entropy balance to Problem 22.1

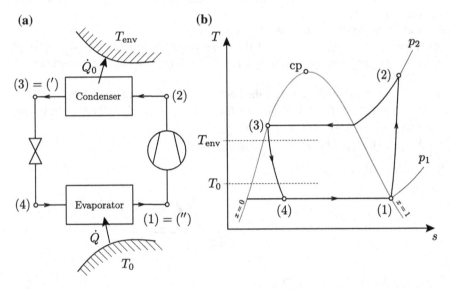

Fig. 22.20 **a** Sketch of the plant, **b** T, s-diagram to Problem 22.2

to be smaller than the temperatures T_2 respectively T_3, since the fluid needs to release heat to the environment. T_0 accordingly needs to be larger than $T_1 = T_4$ in order to supply heat to the working fluid.

(b) The mass flux $\dot{m}_A$ follows by applying the first law of thermodynamics for the evaporator, i.e.

$$\dot{Q} = 135 \, \text{kW} = \dot{m}_A \, (h_1 - h_4) \,. \tag{22.69}$$

The throttle is adiabatic and kinetic respectively potential energies are to be ignored, so that the change of state is isenthalpic, i.e.

$$h_4 = h_3 = h' = 503.6 \, \frac{\text{kJ}}{\text{kg}} \,. \tag{22.70}$$

For state (1) it is known, that

$$h_1 = h'' = 1604.1 \, \frac{\text{kJ}}{\text{kg}}. \tag{22.71}$$

Thus, the mass flux follows

$$\boxed{\dot{m}_A = \frac{\dot{Q}}{h_1 - h_3} = 0.1227 \, \frac{\text{kg}}{\text{s}}} \tag{22.72}$$

(c) The technical power of the compressor follows from the first law of thermodynamics, i.e.

$$P_{t,12} = \dot{m}_A \, (h_2 - h_1). \tag{22.73}$$

By introducing the isentropic efficiency $\eta_{s,V}$ it results in

$$P_{t,12} = \dot{m}_A \frac{(h_{2s} - h_1)}{\eta_{s,V}}. \tag{22.74}$$

For state (1) it is known, that

$$h_1 = h'' = 1604.1 \, \frac{\text{kJ}}{\text{kg}} \tag{22.75}$$

and

$$s_1 = s'' = 6.4038 \, \frac{\text{kJ}}{\text{kg K}}. \tag{22.76}$$

The hypothetic change of state from (1) to (2s) is isentropic, i.e.

$$s_{2s} = s_1 = 6.4038 \, \frac{\text{kJ}}{\text{kg K}}. \tag{22.77}$$

Together with the pressure $p_{2s} = 11.666 \, \text{bar}$, the specific enthalpy follows by linear interpolation from the steam data, i.e.

$$h_{2s} = 1842.0 \, \frac{\text{kJ}}{\text{kg}}. \tag{22.78}$$

Finally, the technical power of the compressor obeys

$$\boxed{P_{t,12} = \dot{m}_A \frac{(h_{2s} - h_1)}{\eta_{s,V}} = 37.17 \, \text{kW}} \tag{22.79}$$

The specific enthalpy of state (2) then is

$$h_2 = h_1 = \frac{P_{t,12}}{\dot{m}_A} = 1907.1 \, \frac{kJ}{kg}. \tag{22.80}$$

The coefficient of performance for the cooling machine is

$$\boxed{\varepsilon = \frac{\dot{Q}}{P_{t,12}} = 3.6313} \tag{22.81}$$

A perfect (=Carnot) process would lead to an efficiency of

$$\boxed{\varepsilon_C = \frac{T_{min}}{T_{max} - T_{min}} = 9.6722} \tag{22.82}$$

(d) The mass flux for the cooling water $\dot{m}_W$ follows by applying he first law of thermodynamics for the condenser, see Fig. 22.21a, i.e.

$$\dot{m}_A h_2 + \dot{m}_W h_{env} = \dot{m}_A h_3 + \dot{m}_W h_W. \tag{22.83}$$

Re-arranging results in

$$\dot{m}_A h_2 - \dot{m}_A h_3 = \dot{m}_W h_W - \dot{m}_W h_{env}. \tag{22.84}$$

Applying the caloric equation of state for the cooling water[10]

$$\dot{m}_A h_2 - \dot{m}_A h_3 = \dot{m}_W c_W \left(\vartheta_W - \vartheta_{env} \right). \tag{22.85}$$

Thus, the mass flow rate of the cooling water is

$$\boxed{\dot{m}_W = \frac{\dot{m}_A \left(h_2 - h_3 \right)}{\vartheta_W - \vartheta_{env}} = 3.4268 \, \frac{kg}{s}} \tag{22.86}$$

(e) The loss of exergy $\Delta \dot{E}_{x,V}$ in the condenser follows

$$\Delta \dot{E}_{x,V} = T_{env} \dot{S}_i. \tag{22.87}$$

Hence, a balance of entropy is required in order to calculate the generation of entropy, see Fig. 22.21b. This balance leads to

$$\dot{m}_A s_2 + \dot{m}_W s_{env} + \dot{S}_i = \dot{m}_A s_3 + \dot{m}_W s_W. \tag{22.88}$$

Hence, the entropy generation obeys

---

[10]Cooling water is regarded as an incompressible liquid. There shall be no pressure drop for the cooling water!

(a)

(b)

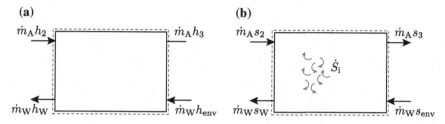

Fig. 22.21 **a** Energy balance, **b** Entropy balance for the condenser

$$\dot{S}_i = \dot{m}_A \left(s_3 - s_2 \right) + \dot{m}_W \left(s_W - s_{env} \right). \tag{22.89}$$

Applying the caloric equation of state for the cooling water, that is supposed to be incompressible, leads to

$$\dot{S}_i = \dot{m}_A \left(s_3 - s_2 \right) + \dot{m}_W c_W \log \frac{T_W}{T_{env}}. \tag{22.90}$$

The specific entropy of state (3) is

$$s_3 = s' = 2.0512 \, \frac{kJ}{kg \, K}. \tag{22.91}$$

For state (2) it follows

$$s_2 = s \left(h_2, \, p_2 \right) = 6.5724 \, \frac{kJ}{kg \, K}. \tag{22.92}$$

Thus, the generation of entropy[11] is

$$\dot{S}_i = \dot{m}_A \left(s_3 - s_2 \right) + \dot{m}_W c_W \log \frac{T_W}{T_{env}} = 0.0308 \, \frac{kW}{K}. \tag{22.93}$$

Finally, the loss of exergy is

$$\boxed{\Delta \dot{E}_{x,V} = T_{env} \dot{S}_i = 8.8701 \, kW} \tag{22.94}$$

---

[11] The generation of entropy is due to the heat transfer from ammonia to the cooling water. Since there is no pressure drop for the ammonia respectively the cooling water, there is no dissipation.

Part III
Reactive Systems

Chapter 23
Combustion Processes

In Part I ideal gases and incompressible liquids have been introduced and the basic thermodynamic correlations have been derived. Part II has shown, that real fluids can be subject to a change of aggregate state and that many thermodynamic cycles are based on phase change, e.g. a steam power plant. Furthermore, Part II covered mixture of fluids, e.g. humid air or mixtures of ideal gases. However, within these mixture each component was stable and not part of a chemical reaction, i.e. decomposition of the present atoms and molecules. In Part III the focus now is on chemical reacting systems: First, the stoichiometry of a chemical reaction is investigated, i.e. the principle of mass conservation is applied to reactants and products of a chemical reaction. In doing so it is possible to predict the composition of the products. This is important for instance, when the composition of an exhaust gas of a combustion process needs to fulfil technical thresholds. Second, a chemical reacting system is investigated in terms of an energy balance. In this chapter the heating-value approach is followed with focus on technical combustions, i.e. combustions based on fossil fuels. In Chap. 24 another energetic approach based on absolute enthalpy respectively entropy is introduced. Major advantage of this method is, that the irreversibility of chemical reactions can be quantified.

23.1 Fossil Fuels

A combustion is a chemical reaction of fuels with oxygen, i.e. oxidation. Oxygen is mostly taken from atmospheric air. In case the reaction is exothermal, heat is released and can be utilised, e.g. in a steam power process, see Fig. 23.2. Obviously, the fuel's chemical bonded energy is converted into thermal energy. As mentioned above, in this chapter the focus is on fossil fuels. Fossil fuels are in particular characterised by carbon, hydrogen and oxygen. Beyond this, sulphur can be converted by oxidation:

© Springer Nature Switzerland AG 2019
A. Schmidt, *Technical Thermodynamics for Engineers*,
https://doi.org/10.1007/978-3-030-20397-9_23

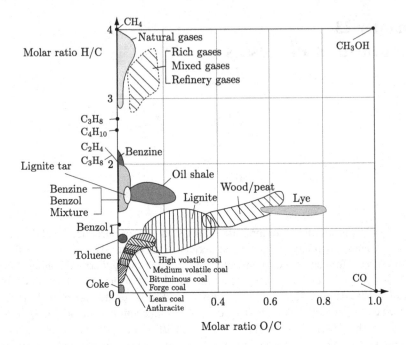

Fig. 23.1 Fossil fuels according to [5]

- Carbon C → CO$_2$
- Hydrogen[1] H$_2$ → H$_2$O
- Sulphur S → SO$_2$

In general, fuels consist of the listed combustible components (C, H$_2$, S) and non-flammable components, e.g. O$_2$, N$_2$, CO$_2$, H$_2$O, noble gases, ashes. For solid fossil fuels the ratio of carbon and hydrogen to oxygen content varies with the age of the fuel, see Fig. 23.1. For instance this degree of carbonisation is much larger for coke than for lignite. However, fossil fuels contain many other components, e.g. among others water, ashes and nitrogen N$_2$, but in this frame they are treated as to be inert, i.e. not chemically reactive. Nevertheless, they have a thermal impact on the chemical reaction. In case of the nitrogen, the reactive Part Is much more complex and subject of many other books, see [5, 27].

   In case the supply of oxygen is too low, an oxidation is incomplete, i.e. the exhaust or flue gas still contains reactants, that could be further converted. This can be the fossil fuel itself or not yet fully oxidised products like carbon monoxide CO for instance, that could be further oxidised in case oxygen would be available. In this chapter the focus is on stoichiometric reactions with sufficient oxygen for a full conversion of the fuel's components carbon, hydrogen and sulphur.

---

[1]Hydrogen in its stable form under atmospheric conditions is molecular.

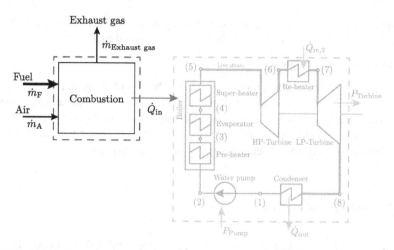

Fig. 23.2 Combustion process

Anyhow, fossil fuels can occur in different states of aggregation at standard conditions, i.e.

- Solid fuels: Hard coal, lignite, biomass
- Liquid fuels: oil, diesel, hydrocarbons etc.
- Gaseous fuels:
  Hydrogen, gaseous hydrocarbons (methane CH_4, natural gas, biogas)
  Flammable gases: carbon monoxide (CO), hydrogen (H_2), methane (CH_4), ethane (C_2H_6), ethylene (C_2H_4), acetylene (C_2H_2), further hydrocarbons (C_xH_y)

To run an oxidation the ignition temperature, i.e. the temperature at which the reaction continues by itself under release of heat, needs to be reached. Furthermore, a sufficient amount of oxygen must be available. For gaseous fuels, the mixture of fuel and oxygen is required to be within the ignition limit, since only then the released heat is sufficient to keep the temperature above ignition temperature. The lower respectively upper limit for these mixtures is the minimum respectively maximum ratio of gaseous fuel in the fuel/air-mixture, e.g. for a mixture of natural gas and air it is 4.2 … 16.5 Vol.-%, for a mixture of hydrogen and air it is 4 … 77 Vol.-%. In case a liquid or solid fuel shall be oxidised, the fuel needs to evaporate first.

Figure 23.2 shows a combustion process in black box notation. A mass flow rate of fuel as well as of air/oxygen are led into the combustion chamber. The exhaust gas, i.e. the product of the chemical reaction leaves the system boundary. Released heat can be utilised to run a steam power plant. In order to calculate such a process, two steps are followed:

- Step 1: Stoichiometry

Based on the chemical reaction scheme, material balances can be performed. It is clarified, what amount of oxygen is required for a total oxidation of the fuel.

Furthermore the composition of the exhaust gas is calculated. In order to do so, the composition of the reactants needs to be determined first.

- Step 2: Energetic balancing

Similar to the non-reacting open systems in Part I and II, an energy balance is performed. The energetic aspect of the material decomposition, i.e. oxidation of the fuel, needs to be taken into account. Such a balance allows to calculate the exhaust gas temperature respectively the amount of released heat. It is further possible to predict if water condenses out of the exhaust gas.

23.2 Fuel Composition

In order to perform mass and energy balances of the combustion process, the fuel composition is required first. However, a distinction is made based on in which aggregate state the fuel is.

23.2.1 Solid Fuels

The composition of solid and liquid fuels is usually given by mass fractions with regard to the fuel's mass m_F, i.e.

- Carbon

$$\xi_C = \frac{m_C}{m_F} = c \tag{23.1}$$

- Hydrogen

$$\xi_{H_2} = \frac{m_{H_2}}{m_F} = h \tag{23.2}$$

- Sulphur

$$\xi_S = \frac{m_S}{m_F} = s \tag{23.3}$$

- Oxygen

$$\xi_{O_2} = \frac{m_{O_2}}{m_F} = o \tag{23.4}$$

- Nitrogen

$$\xi_{N_2} = \frac{m_{N_2}}{m_F} = n \tag{23.5}$$

- Ashes

$$\xi_{Ash} = \frac{m_{Ash}}{m_F} = a \tag{23.6}$$

- Water

$$\xi_W = \frac{m_W}{m_F} = w \tag{23.7}$$

The fuel's total mass m_F follows

$$m_F = m_C + m_{H_2} + m_S + m_{O_2} + m_{N_2} + m_{Ash} + m_W. \tag{23.8}$$

The devision by m_F leads to

$$\boxed{c + h + s + o + n + a + w = 1} \tag{23.9}$$

23.2.2 Liquid Fuels

Liquid fuels primarily consist of hydrocarbons. They contain carbon C, hydrogen H_2, oxygen O_2 and sulphur S. Its composition as usually given in mass fractions as well, i.e.

- Carbon

$$\xi_C = \frac{m_C}{m_F} = c \tag{23.10}$$

- Hydrogen

$$\xi_{H_2} = \frac{m_{H_2}}{m_F} = h \tag{23.11}$$

- Sulphur

$$\xi_S = \frac{m_S}{m_F} = s \tag{23.12}$$

- Oxygen

$$\xi_{O_2} = \frac{m_{O_2}}{m_F} = o \tag{23.13}$$

Due to

$$m_F = m_C + m_{H_2} + m_S + m_{O_2} \tag{23.14}$$

it is

$$\boxed{c + h + s + o = 1} \tag{23.15}$$

23.2.3 Gaseous Fuels

Usually gaseous fuels are mixtures of gases, whose components need to be determined by a elemental analysis. The concentration of a component k of the mixture is given in volume fractions with regard to the entire volume of the fuel V_F, i.e.

$$\sigma_K = \frac{V_K}{V_F}. \tag{23.16}$$

In case the components can be treated as ideal gases, the volume fraction in a gaseous mixture is equal with its molar fraction, see Part II Sect. 19.3.1, i.e.

$$\sigma_K = x_K = \frac{n_K}{n_F}. \tag{23.17}$$

23.3 Stoichiometry

The following equations show the relevant chemical reactions of the *combustible* components,[2] i.e. C, H, S, of fossil fuels and oxygen under stoichiometric[3] conditions, i.e.

- Carbon:

$$C + O_2 \rightarrow CO_2 \tag{23.18}$$

- Hydrogen:

$$H_2 + \frac{1}{2}O_2 \rightarrow H_2O \tag{23.19}$$

- Sulphur:

$$S + O_2 \rightarrow SO_2 \tag{23.20}$$

The chemical equations are no quantity equations in the mathematical sense, but stoichiometric equations, i.e. they count for the mass conservation. The number of atoms of each element on the side of the reactants need to be equal to the number of atoms of the same element on the product side. Equation 23.18 for instance has one C-atom on the left hand side and one C-atom on the right hand side. The number of O-atoms is two on the left and two on the right hand side. Thus, the mass conservation of Eq. 23.18 is fulfilled. Actually, the stoichiometry shows in which numerical ratio the elements within the reaction exist.

---

[2]The other components like nitrogen, water and ashes are treated as inert, i.e. not chemically reactive.
[3]Thus, each combustible element is fully oxidised. The oxygen on the left hand side of the equation is fully consumed, so that there is no residual oxygen on the right hand side.

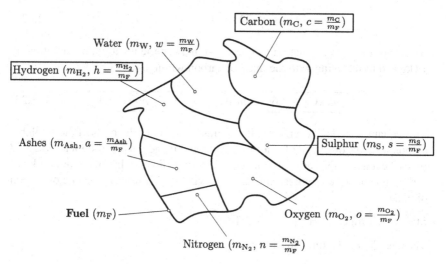

Carbon $(m_C, c = \frac{m_C}{m_F})$

Water $(m_W, w = \frac{m_W}{m_F})$

Hydrogen $(m_{H_2}, h = \frac{m_{H_2}}{m_F})$

Ashes $(m_{Ash}, a = \frac{m_{Ash}}{m_F})$

Sulphur $(m_S, s = \frac{m_S}{m_F})$

Fuel (m_F)

Oxygen $(m_{O_2}, o = \frac{m_{O_2}}{m_F})$

Nitrogen $(m_{N_2}, n = \frac{m_{N_2}}{m_F})$

Fig. 23.3 Example of a solid fossil fuel

23.3.1 Solid/Liquid Fuels

Air/Fuel Composition

Oxygen Need

The minimum need for oxygen is investigated in this section, i.e. according to the chemical Eqs. 23.18–23.20 it is clarified how much oxygen is required to fully oxidise all combustible components of a solid/liquid fossil fuel. The composition of this fuel is visualised in Fig. 23.3. Mind, that the combustible elements are marked by a box. In order to determine the minimum oxygen need, each combustible element is analysed step by step, starting with the carbon:

- Step 1 (Carbon): $C + O_2 \rightarrow CO_2$

  The chemical reaction obeys

$$1 \text{ atom } C + 1 \text{ molecule } O_2 \rightarrow 1 \text{ molecule } CO_2 \qquad (23.21)$$

  Multiplied with the Avogadro constant, i.e. 6.022×10^{26} particles represent 1 kmol:

$$1 \text{ kmol } C + 1 \text{ kmol } O_2 \rightarrow 1 \text{ kmol } CO_2 \qquad (23.22)$$

  Introducing the molar masses[4] it finally is

---

[4] 1 kmol of C has a mass of 12 kg. 1 kmol of O has a mass of 16 kg. 1 kmol of CO_2 has a mass of $1 \cdot 12 \text{ kg} + 2 \cdot 16 \text{ kg} = 44 \text{ kg}$. See also Table 3.2.

$$12 \, kg \, C + 32 \, kg \, O_2 \rightarrow 44 \, kg \, CO_2 \qquad (23.23)$$

By rule of three, this equation can be given in a generic notation, i.e. dividing by 12 kg and multiplying with the mass of carbon within the fuel m_C, i.e.

$$\boxed{m_C(C) + 2.667 \cdot m_C(O_2) \rightarrow 3.667 \cdot m_C(CO_2)} \qquad (23.24)$$

This equation reads like a recipe: It requires 2.667 times the mass of the available carbon as mass of oxygen for a full oxidation. Oxygen and carbon lead to 3.667 times the mass of the carbon as mass of the carbon dioxide. In other words: 1 kg of carbon for instance requires 2.667 kg of oxygen and reacts to 3.667 kg of carbon dioxide.

- Step 2 (Hydrogen): $2H_2 + O_2 \rightarrow 2H_2O$

The chemical reaction reads as

$$2 \text{ molecules } H_2 + 1 \text{ molecule } O_2 \rightarrow 2 \text{ molecules } H_2O \qquad (23.25)$$

Analogous to step 1, the multiplication with the Avogadro constant results in

$$2 \, kmol \, H_2 + 1 \, kmol \, O_2 \rightarrow 2 \, kmol \, H_2O \qquad (23.26)$$

And the multiplication with the molar masses leads to the mass balance

$$4 \, kg \, H_2 + 32 \, kg \, O_2 \rightarrow 36 \, kg \, H_2O \qquad (23.27)$$

A generalisation based on the rule of three finally brings the equation

$$\boxed{m_{H_2}(H_2) + 8 \cdot m_{H_2}(O_2) \rightarrow 9 \cdot m_{H_2}(H_2O)} \qquad (23.28)$$

It requires 8 times the mass of the available hydrogen as mass of oxygen for a full oxidation. Oxygen and hydrogen lead to 9 times the mass of the hydrogen as mass of the water.

- Step 3 (Sulphur): $S + O_2 \rightarrow SO_2$

The chemical reaction obeys

$$1 \text{ atom } S + 1 \text{ molecule } O_2 \rightarrow 1 \text{ molecule } SO_2 \qquad (23.29)$$

Multiplying with the Avogadro constant

$$1 \, kmol \, S + 1 \, kmol \, O_2 \rightarrow 1 \, kmol \, SO_2 \qquad (23.30)$$

and multiplying with the molar masses brings

$$32 \, \text{kg S} + 32 \, \text{kg O}_2 \rightarrow 64 \, \text{kg SO}_2 \qquad (23.31)$$

Once again, a generalisation based on the rule of three finally leads to equation

$$\boxed{m_S(S) + m_S(O_2) \rightarrow 2 \cdot m_S(SO_2)} \qquad (23.32)$$

It requires the mass of the available sulphur as mass of oxygen for a full oxidation. Oxygen and sulphur lead to twice the mass of the sulphur as mass of the sulphur dioxide.

Now, by steps 1–3 *all the combustible elements* are oxidised. Combining Eqs. 23.24, 23.28 and 23.32 gives the minimum need for oxygen. The feeding amount of oxygen is reduced by the mass of oxygen m_{O_2}, that might be bonded in the fuel itself, see Fig. 23.3. Thus, the minimum oxygen need $m_{O_2,min}$ obeys

$$m_{O_2,min} = 2.667 \cdot m_C + 8 \cdot m_{H_2} + m_S - m_{O_2}. \qquad (23.33)$$

Division by the mass of the fuel m_F as reference leads to

$$\frac{m_{O_2,min}}{m_F} = 2.667 \cdot \frac{m_C}{m_F} + 8 \cdot \frac{m_{H_2}}{m_F} + \frac{m_S}{m_F} - \frac{m_{O_2}}{m_F}. \qquad (23.34)$$

With the already introduced abbreviations, see Sect. 23.2, the mass-specific minimum oxygen need results in

$$\boxed{o_{min} \equiv \frac{m_{O_2,min}}{m_F} = 2.667 \cdot c + 8 \cdot h + s - o} \qquad (23.35)$$

Air Need

Usually the reaction is not operated with pure oxygen, but with atmospheric air. Dry, atmospheric air mainly consists of nitrogen N_2 and oxygen O_2, see Chap. 19. Its composition is now simplified to consists solely out of these two components, i.e. ignoring for instance carbon dioxide and argon. Hence, the following Table 23.1 summarises mass fractions as well as molar-fractions. A conversion from molar to mass fractions has been introduced in Chap. 19 and follows, see Eq. 19.64,

Table 23.1 Simplified composition of dry, atmospheric air

	N_2	O_2
Mass fraction ξ_i	0.77	0.23
Molar fraction x_i	0.79	0.21

$$\xi_i = x_i \frac{M_i}{M_{\text{total}}}.$$ (23.36)

Thus, the minimum need for air, referred to the mass of the fuel, can be calculate as follows

$$l_{\min} = \frac{m_{a,min}}{m_F} = \frac{m_{a,min}}{m_{O_2,min}} \cdot \frac{m_{O_2,min}}{m_F} = \frac{1}{\xi_{O_2,\text{dry air}}} \cdot o_{\min}$$ (23.37)

This equation indicates, that the mass of air is larger than the mass of oxygen, since the air is not purely composed of oxygen, but carries a large amount of nitrogen as well. If humid air is used instead of dry air, the required amount of air further rises due to the water the air contains, i.e.

$$m_{A,min} = \underbrace{m_{a,min}}_{\text{dry air}} + \underbrace{m_w}_{\text{water}} = m_{a,min} + x \cdot m_{a,min} = m_{a,min}(1 + x)$$ (23.38)

Mind, that x is the water content of air, see Chap. 20.

Air-Fuel Equivalence Ratio

So far, the minimum oxygen/air need has been derived. Often, in order to ensure, that the oxidisation is complete, the amount of air is further increased. Hence, the air-fuel equivalence ratio λ is defined. In fact, it is the ratio of fed air mass to the required minimum air mass for a stoichiometric combustion, i.e. full oxidisation:

$$\lambda = \frac{m_A}{m_{a,min}} = \frac{l}{l_{\min}}$$ (23.39)

However, the range of λ is defined as follows

$$\lambda \begin{cases} < 1, \text{ lack of air, rich mixture} \\ = 1, \text{ stoichiometric combustion} \\ > 1, \text{ excess air, lean mixture} \end{cases}$$ (23.40)

Diesel-engines e.g. are operated with $\lambda = 1.1 \ldots 1.15$, gasoline-engines with catalytic converter with $\lambda = 0.99 \ldots 1.00$. Gas turbines are operated with $\lambda > 5$, in order to reduce the flame temperature, see Sect. 23.4. In case humid air is used instead of dry air, the required amount of air has to be corrected by $(1 + x)$, since

$$m_A = m_a + m_w = m_a + x \cdot m_a = m_a(1 + x) = \lambda \cdot m_{a,min}(1 + x)$$ (23.41)

Rich mixtures are not covered by this book, since the mechanisms of incomplete oxidisation are rather complicated and part of several other textbooks, see [5, 28]. Hence, the focus here is on oxidisations with $\lambda \geq 1$.

Exhaust Gas Composition

Now, that we know how to mix air and fuel in order to achieve a full oxidation, the exhaust gas composition is investigated in a next step. It is assumed, that the combustion air purely consists of nitrogen and oxygen. According to the deduced Eqs. 23.24, 23.28 and 23.32 for the combustible elements the composition can be derived. Step 1 has shown, that

$$m_C(C) + 2.667 \cdot m_C(O_2) \rightarrow 3.667 \cdot m_C(CO_2). \tag{23.42}$$

Step 2 led to

$$m_{H_2}(H_2) + 8 \cdot m_{H_2}(O_2) \rightarrow 9 \cdot m_{H_2}(H_2O) \tag{23.43}$$

an step 3 to

$$m_S(S) + m_S(O_2) \rightarrow 2 \cdot m_S(SO_2). \tag{23.44}$$

According to these equations it follows for the exhaust gas:

- Carbon dioxide CO_2

$$m_{EG,CO_2} = 3.667 \cdot m_C \tag{23.45}$$

This mass can be referred to the mass of the fuel, so that one gets the mass-specific exhaust gas composition of carbon dioxide, i.e.

$$\boxed{\mu_{EG,CO_2} = \frac{m_{EG,CO_2}}{m_F} = 3.667 \cdot \frac{m_C}{m_F} = 3.667 \cdot c} \tag{23.46}$$

- Sulphur dioxide SO_2

$$m_{EG,SO_2} = 2 \cdot m_S \tag{23.47}$$

This mass can be referred to the mass of the fuel, so that one gets the mass-specific exhaust gas composition of sulphur dioxide, i.e.

$$\boxed{\mu_{EG,SO_2} = \frac{m_{EG,SO_2}}{m_F} = 2 \cdot \frac{m_S}{m_F} = 2 \cdot s} \tag{23.48}$$

- Water H_2O

There are several reasons for water being in the exhaust gas. Due to the hydrogen of fossil fuels, water is formed by oxidisation as described by Eq. 23.28. However, the fossil fuel might also carry water, see Fig. 23.3, that is inert but that leaves the combustion process as well. Last, the combustion might be operated with humid air, so that this water, carried by the air, additionally leaves with the exhaust gas without being chemically reactive, i.e.

$$m_{EG,H_2O} = 9 \cdot m_{H_2} + \underbrace{m_W}_{\text{fuel water}} + \underbrace{\lambda \cdot m_{a,min} \cdot x}_{\text{water from the comb. air}} \quad . \tag{23.49}$$

This mass can be referred to the mass of the fuel, so that one gets the mass-specific exhaust gas composition of water, i.e.

$$\boxed{\mu_{EG,H_2O} = \frac{m_{EG,H_2O}}{m_F} = 9 \cdot h + w + \lambda \cdot l_{min} \cdot x} \tag{23.50}$$

- Nitrogen (inert) N_2

Actually, there are two sources for the nitrogen: One part comes from the nitrogen, that is part of the fossil fuel, see Fig. 23.3. The other part is from the air, that is required for the oxidation, i.e.

$$m_{EG,N_2} = \underbrace{m_{N_2}}_{\text{fuel nitrogen}} + \underbrace{\lambda \cdot m_{a,min} \cdot \xi_{N_2,\text{dry air}}}_{\text{nitrogen from the comb. air}} \cdot \tag{23.51}$$

This mass can be referred to the mass of the fuel, so that one gets the mass-specific exhaust gas composition of nitrogen, i.e.

$$\boxed{\mu_{EG,N_2} = \frac{m_{EG,N_2}}{m_F} = n + \lambda \cdot l_{min} \cdot \xi_{N_2,\text{dry air}}} \tag{23.52}$$

- Oxygen O_2

The cause for oxygen being in the exhaust gas is, that the combustion is performed with a lean-mixture, i.e. $\lambda > 1$. Oxygen in the exhaust gas is the difference between the total supplied oxygen on the reactant side reduced by the minimum required oxygen need in order to run the combustion, i.e.

$$m_{EG,O_2} = \underbrace{\lambda \cdot m_{a,min} \cdot \xi_{O_2,\text{dry air}}}_{\text{oxygen from the comb. air}} - \underbrace{m_{a,min} \cdot \xi_{O_2,\text{dry air}}}_{\text{oxygen req. for comb.}} \tag{23.53}$$

$$= m_{a,min} \cdot \xi_{O_2,\text{dry air}}(\lambda - 1).$$

This mass can be referred to the mass of the fuel, so that one gets the mass-specific exhaust gas composition of oxygen, i.e.

$$\boxed{\mu_{EG,O_2} = \frac{m_{EG,O_2}}{m_F} = l_{min} \cdot \xi_{O_2,\text{dry air}} \cdot (\lambda - 1) = o_{min} \cdot (\lambda - 1)} \tag{23.54}$$

However, this equation shows, that there is no exhaust gas oxygen in case the combustion runs stoichiometrically, i.e. $\lambda = 1$.

In case the combustion air is not simplified to be a composition of just nitrogen and oxygen, the composition of the exhaust gas contains other components as well, i.e. the calculation needs to be adjusted.

Exhaust Gas Concentration

In the previous section the exhaust gas composition $\mu_{EG,i}$ has been derived. Since it refers to the fuel's mass m_F, it is possible to achieve $\mu_{EG,i} > 1$. Thus, the exhaust gas composition is not to be mixed up with an exhaust gas concentration. Such a concentration is required to define the caloric as well as thermal properties of the exhaust gas, that shall be treated as a mixture of ideal gases, see Chap. 19. The *mass-fraction* of a exhaust gas component i is referred to the mass of the exhaust gas and thus follows

$$\xi_{EG,i} = \frac{m_{EG,i}}{m_{EG}} = \frac{m_F \cdot \mu_{EG,i}}{m_F \cdot \mu_{EG}} = \frac{\mu_{EG,i}}{\mu_{EG}}. \tag{23.55}$$

Mind, that the entire mass of the exhaust gas is the sum over the individual masses of its components, i.e.

$$m_{EG} = \sum_i m_{EG,i}. \tag{23.56}$$

So dividing by the fuel's mass leads to

$$\mu_{EG} = \sum_i \mu_{EG,i}. \tag{23.57}$$

Thus, the components concentration simplifies to

$$\xi_{EG,i} = \frac{\mu_{EG,i}}{\sum_i \mu_{EG,i}}. \tag{23.58}$$

Thus, applied to the concentrations of the exemplary exhaust gas it is:

- Carbon dioxide CO_2

$$\xi_{EG,CO_2} = \frac{m_{EG,CO_2}}{m_{EG}} = \frac{m_F \cdot \mu_{EG,CO_2}}{m_F \cdot \sum_i \mu_{EG,i}} = \frac{\mu_{EG,CO_2}}{\sum_i \mu_{EG,i}} \tag{23.59}$$

- Sulphur dioxide SO_2

$$\xi_{EG,SO_2} = \frac{m_{EG,SO_2}}{m_{EG}} = \frac{m_F \cdot \mu_{EG,SO_2}}{m_F \cdot \sum\limits_i \mu_{EG,i}} = \frac{\mu_{EG,SO_2}}{\sum\limits_i \mu_{EG,i}} \tag{23.60}$$

- Water H_2O

$$\xi_{EG,H_2O} = \frac{m_{EG,H_2O}}{m_{EG}} = \frac{m_F \cdot \mu_{EG,H_2O}}{m_F \cdot \sum\limits_i \mu_{EG,i}} = \frac{\mu_{EG,H_2O}}{\sum\limits_i \mu_{EG,i}} \tag{23.61}$$

- Oxygen O_2

$$\xi_{EG,O_2} = \frac{m_{EG,O_2}}{m_{EG}} = \frac{m_F \cdot \mu_{EG,O_2}}{m_F \cdot \sum\limits_i \mu_{EG,i}} = \frac{\mu_{EG,O_2}}{\sum\limits_i \mu_{EG,i}} \tag{23.62}$$

- Nitrogen N_2

$$\xi_{EG,N_2} = \frac{m_{EG,N_2}}{m_{EG}} = \frac{m_F \cdot \mu_{EG,N_2}}{m_F \cdot \sum\limits_i \mu_{EG,i}} = \frac{\mu_{EG,N_2}}{\sum\limits_i \mu_{EG,i}} \tag{23.63}$$

Once the concentration of each component within the exhaust gas is known, the exhaust gas properties can be estimated, i.e. $c_{p,EG}$, R_{EG} and M_{EG}. Thus, all principles following Chap. 19 can be applied. Mind, that mass fractions can be converted into molar fractions according to Eq. 19.64.

23.3.2 Gaseous Fuels

Air/Fuel Composition

Oxygen Need

The minimum need for oxygen is investigated in this section, i.e. according to the chemical Eqs. 23.18–23.20 it is clarified how much oxygen is required to fully oxidise all combustible components of a gaseous fossil fuel. The procedure is analogue to the approach we followed for solid fuels. However, the *exemplary* composition of this gaseous fuel is visualised in Fig. 23.4. Mind, that the combustible elements in this example are marked by a box. In order to determine the minimum oxygen need, each combustible element is analysed step by step, starting with the hydrogen:

- Step 1 (Hydrogen): n_{H_2}

    The chemical reaction reads as

$$2H_2 + O_2 \rightarrow 2H_2O \tag{23.64}$$

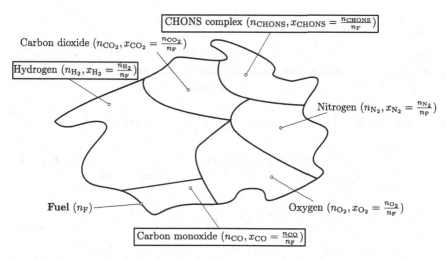

Fig. 23.4 Exemplary gaseous fuel

Multiplying with the Avogadro constant

$$2 \, \text{kmol} \, H_2 + 1 \, \text{kmol} \, O_2 \rightarrow 2 \, \text{kmol} \, H_2O \tag{23.65}$$

A generalisation based on the rule of three finally leads to the equation

$$n_{H_2}(H_2) + 0.5 \cdot n_{H_2}(O_2) \rightarrow 1 \cdot n_{H_2}(H_2O) \tag{23.66}$$

- Step 2 (Carbon dioxide): $n_{CO_2} \rightarrow$ inert, no further reaction
- $\overline{\text{Step 3}}$ (Oxygen): $n_{O_2} \rightarrow$ lowers the required feed of oxygen
- $\overline{\text{Step 4}}$ (Nitrogen): $n_{N_2} \rightarrow$ inert, no further reaction
- $\overline{\text{Step 5}}$ (Carbon monoxide): n_{CO}

The chemical reaction obeys

$$2CO + O_2 \rightarrow 2CO_2 \tag{23.67}$$

Multiplying with the Avogadro constant

$$2 \, \text{kmol} \, CO + 1 \, \text{kmol} \, O_2 \rightarrow 2 \, \text{kmol} \, CO_2 \tag{23.68}$$

A generalisation based on the rule of three finally leads to the equation

$$n_{CO}(CO) + 0.5 \cdot n_{CO}(O_2) \rightarrow 1 \cdot n_{CO}(CO_2) \tag{23.69}$$

- Step 6 (Chemical bonds in the form of $C_aH_bO_zN_pS_r$):

There might be several complexes of hydrocarbons[5] part of the gaseous fuel as well. Goal is to find a generic chemical reaction scheme, that allows to handle any of these complexes. Such a generic chemical reaction obeys

$$C_aH_bO_zN_pS_r + \left(a + \frac{b}{4} - \frac{z}{2} + r\right) \cdot O_2 \rightarrow a \cdot CO_2 + \frac{b}{2} \cdot H_2O + r \cdot SO_2 + \frac{p}{2} \cdot N_2$$

$$(23.70)$$

By adjusting the stoichiometric parameters a, b, z, p, r any component can be covered by this approach.[6] Multiplying with the Avogadro constant

$$1 \text{ kmol } C_aH_bO_zN_pS_r + \left(a + \frac{b}{4} - \frac{z}{2} + r\right) \cdot \text{kmol } O_2$$

$$(23.71)$$

$$\rightarrow a \cdot \text{kmol } CO_2 + \frac{b}{2} \cdot \text{kmol } H_2O + r \cdot \text{kmol } SO_2 + \frac{p}{2} \cdot \text{kmol } N_2$$

A generalisation based on the rule of three finally leads to the equation

$$n_{C_aH_bO_zN_pS_r}(C_aH_bO_zN_pS_r) + (a + \frac{b}{4} - \frac{z}{2} + r) \cdot n_{C_aH_bO_zN_pS_r}(O_2) \rightarrow$$

$$a \cdot n_{C_aH_bO_zN_pS_r}(CO_2) +$$

$$+ \frac{b}{2} \cdot n_{C_aH_bO_zN_pS_r}(H_2O) + \qquad (23.72)$$

$$+ r \cdot n_{C_aH_bO_zN_pS_r}(SO_2) +$$

$$+ \frac{p}{2} \cdot n_{C_aH_bO_zN_pS_r}(N_2)$$

Now, each of the exemplary components, see sketch Fig. 23.4, has been investigated. For the fuel mixture it is:

$$n_F = n_{H_2} + n_{CO_2} + n_{O_2} + n_{N_2} + n_{CO} + \sum_i n_{C_aH_bO_zN_pS_r,i}. \qquad (23.73)$$

In order to oxidise the exemplary fuel mixture completely, analogue to the solid fuel the oxygen need is the summation of the oxygen need of each individual component, see steps 1 to 6:

$$n_{O_2,min} = 0.5 \cdot n_{H_2} + 0.5 \cdot n_{CO} + \sum_i \left(a_i + \frac{b_i}{4} - \frac{z_i}{2} + r_i\right) \cdot n_{C_aH_bO_zN_pS_r,i} - n_{O_2}$$

$$(23.74)$$

---

[5]Such as methane, propane, butane, ethanol or others.
[6]Actually, steps 1–5 could be skipped, since the former reactions can be handled by this generic approach as well!

The right hand side of the equation represents the molar amounts of the fuel's individual components. n_{O_2} is the molar amount of oxygen in the fuel mixture, see Fig. 23.4, so the overall oxygen need is reduced by this amount. Dividing by the molar amount of the entire fuel mixture n_F leads to the molar-specific minimum oxygen need of the fuel, i.e.

$$O_{min} = \frac{n_{O_2,min}}{n_F}. \tag{23.75}$$

Thus, it is

$$O_{min} = 0.5 \cdot x_{H_2} + 0.5 \cdot x_{CO} + \sum_i \left(a_i + \frac{b_i}{4} - \frac{z_i}{2} + r_i\right) \cdot x_{C_aH_bO_zN_pS_{r,i}} - x_{O_2} \tag{23.76}$$

Mind, that not only one hydrocarbon complex might be part of the gaseous fuel, but several. Each of it has its own specific oxygen need to fully oxidise.

Air Need

Usually the reaction is not operated with pure oxygen, but with atmospheric air. Dry, atmospheric air mainly consists of nitrogen N_2 and oxygen O_2, see Chap. 19. Its composition is now simplified to consists solely out of these two components, i.e. ignoring for instance carbon dioxide and argon, see e.g. Table 23.1. The minimum need for air, referred to the molar amount of the fuel, can be calculate as follows

$$L_{min} = \frac{n_{a,min}}{n_F} = \frac{n_{a,min}}{n_{O_2,min}} \cdot \frac{n_{O_2,min}}{n_F} = \frac{1}{x_{O_2,dry\,air}} \cdot O_{min} \tag{23.77}$$

This equation indicates, that the molar amount of air is larger than the molar amount of oxygen, since the air is not purely composed of oxygen, but carries a large amount of nitrogen as well. If humid air is used instead of dry air, the required amount of air further rises due to the water the air contains, i.e.

$$m_{A,min} = m_{a,min} + m_v = m_{a,min} + x \cdot m_{a,min} = m_{a,min}(1 + x). \tag{23.78}$$

The correlation for the molar amount follows accordingly:

- Dry air

$$m_{a,min} = n_{a,min} M_a \tag{23.79}$$

so, that

$$n_{a,min} = \frac{m_{a,min}}{M_a} \tag{23.80}$$

- Water

$$m_w = n_w M_w = x \cdot m_{a,min} = x \cdot n_{a,min} M_a \tag{23.81}$$

so, that

$$n_w = x \cdot n_{a,min} \frac{M_a}{M_w} = 1.6022 \cdot x \cdot n_{a,min} \tag{23.82}$$

Mind, that the ratio $\frac{M_a}{M_w} = 1.6022$ has been derived in Chap. 20 and only counts for a standard atmosphere.[7]

Thus, if the combustion air is humid, the minimum molar amount of humid air is

$$\boxed{n_{A,min} = n_{a,min} + n_w = n_{a,min}(1 + 1.6022 \cdot x)} \tag{23.83}$$

Air-Fuel Equivalence Ratio

The air-fuel equivalence ratio λ has already been introduced previously with the solid fuels and is now adapted for gaseous fuels. It is the ratio of fed air mass to the required minimum air mass for a stoichiometric combustion, i.e. full oxidisation:

$$\boxed{\lambda = \frac{m_a}{m_{a,min}} = \frac{n_a M_a}{n_{a,min} M_a} = \frac{n_a}{n_{a,min}} = \frac{L}{L_{min}}} \tag{23.84}$$

However, the range of λ is defined as follows

$$\lambda \begin{cases} < 1, \text{ lack of air, rich mixture} \\ = 1, \text{ stoichiometric combustion} \\ > 1, \text{ excess air, lean mixture} \end{cases} \tag{23.85}$$

In case humid air is used instead of dry air, the required amount of air has to be

$$n_A = \lambda \cdot n_{A,min}. \tag{23.86}$$

In combination with Eq. 23.83 it is

$$n_A = \lambda \cdot n_{A,min} = \lambda \cdot n_{a,min}(1 + 1.6022 \cdot x). \tag{23.87}$$

Exhaust Gas Composition

Now, that we know how to mix air and fuel in order to achieve a full oxidation, the exhaust gas composition is investigated in a next step. It is assumed, that the com-

---

[7]The value 1.6022 is taken here, though the standard atmosphere has been reduced to nitrogen and oxygen! The related error shall be ignored.

bustion air purely consists of nitrogen and oxygen. According to the deduced equations 23.66–23.72 for the exemplary gaseous fuel the composition can be derived.

- Carbon dioxide CO_2

In this example the carbon dioxide is a result of the fuel's carbon monoxide, the fuel's carbon dioxide, that does not react any further, and a result of the oxidisation of the hydrocarbon complexes, see Fig. 23.4, i.e.

$$n_{EG,CO_2} = 1 \cdot n_{CO} + \sum_i a_i \cdot n_{C_aH_bO_zN_pS_{r,i}} + n_{CO_2}. \tag{23.88}$$

This molar amount can be referred to the molar quantity of the fuel, so that one gets the molar-specific exhaust gas composition of carbon dioxide, i.e.

$$\boxed{\nu_{EG,CO_2} = \frac{n_{EG,CO_2}}{n_F} = x_{CO} + \sum_i a_i \cdot x_{C_aH_bO_zN_pS_{r,i}} + x_{CO_2}} \tag{23.89}$$

- Sulphur dioxide SO_2

The sulphur dioxide in the exhaust gas is due to the oxidisation of the hydrocarbon complexes, i.e.

$$n_{EG,SO_2} = \sum_i r_i \cdot n_{C_aH_bO_zN_pS_{r,i}}. \tag{23.90}$$

This molar amount can be referred to the molar quantity of the fuel, so that one gets the molar-specific exhaust gas composition of sulphur dioxide, i.e.

$$\boxed{\nu_{EG,SO_2} = \frac{n_{EG,SO_2}}{n_F} = \sum_i r_i \cdot x_{C_aH_bO_zN_pS_{r,i}}} \tag{23.91}$$

- Water H_2O

The water is a result of the combustion of the fuel's hydrogen, the fuel's hydrocarbon complexes, see Fig. 23.4, and the water, that is part of the oxidisation air, i.e.

$$n_{EG,H_2O} = n_{H_2} + \sum_i \frac{b_i}{2} \cdot n_{C_aH_bO_zN_pS_{r,i}} + \underbrace{1.6022 \cdot \lambda \cdot n_{a,min} \cdot x}_{\text{water from the comb. air}}. \tag{23.92}$$

This molar amount can be referred to the molar quantity of the fuel, so that one gets the molar-specific exhaust gas composition of water, i.e.

$$\nu_{EG,H_2O} = \frac{n_{EG,H_2O}}{n_F} = x_{H_2} + \sum_i \frac{b_i}{2} \cdot x_{C_aH_bO_zN_pS_{r,i}} + 1.6022 \cdot \lambda \cdot L_{min} \cdot x$$

$$(23.93)$$

- Nitrogen N_2

The nitrogen is a result of the combustion of the fuel's nitrogen, the fuel's hydro-carbon complexes, see Fig. 23.4, and the nitrogen, that is part of the oxidisation air, i.e.

$$n_{EG,N_2} = \underbrace{n_{N_2}}_{\text{fuel nitrogen}} + \underbrace{\sum_i \frac{p_i}{2} \cdot n_{C_aH_bO_zN_pS_{r,i}}}_{\text{from the CHONS-complexes}} + \underbrace{\lambda \cdot n_{a,min} \cdot x_{N_2,\text{dry air}}}_{\text{nitrogen from the comb. air}} \cdot \qquad (23.94)$$

This molar amount can be referred to the molar quantity of the fuel, so that one gets the molar-specific exhaust gas composition of nitrogen, i.e.

$$\nu_{EG,N_2} = \frac{n_{EG,N_2}}{n_F} = x_{N_2} + \sum_i \frac{p_i}{2} \cdot x_{C_aH_bO_zN_pS_{r,i}} + \lambda \cdot L_{min} \cdot x_{N_2,\text{dry air}} \qquad (23.95)$$

- Oxygen O_2

The cause for oxygen being in the exhaust gas is, that the combustion is performed with a lean-mixture, i.e. $\lambda > 1$. Oxygen in the exhaust gas is the difference between the total supplied oxygen on the reactant side reduced by the minimum required oxygen need in order to run the combustion, i.e.

$$n_{EG,O_2} = \underbrace{\lambda \cdot n_{a,min} \cdot x_{O_2,\text{dry air}}}_{\text{oxygen from comb. air}} - \underbrace{n_{a,min} \cdot x_{O_2,\text{dry air}}}_{\text{oxygen req. for comb.}}$$

$$= n_{a,min} \cdot x_{O_2,\text{dry air}}(\lambda - 1). \qquad (23.96)$$

This molar amount can be referred to the molar quantity of the fuel, so that one gets the molar-specific exhaust gas composition of oxygen, i.e.

$$\nu_{EG,O_2} = \frac{n_{EG,O_2}}{n_F} = L_{min} \cdot x_{O_2,\text{dry air}} \cdot (\lambda - 1) = O_{min} \cdot (\lambda - 1) \qquad (23.97)$$

In case the combustion air is not simplified to be a composition of just nitrogen and oxygen, the composition of the exhaust gas contains other components as well, i.e. the calculation needs to be adjusted. The same is for gaseous fuels, that do have a different composition than shown in Fig. 23.4.

Exhaust Gas Concentration

In the previous section the exhaust gas composition $\nu_{EG,i}$ has been derived. Since it refers to the fuel's molar quantity n_F, it is possible to achieve $\nu_{EG,i} > 1$. Thus, the exhaust gas composition is not to be mixed up with an exhaust gas concentration. Such a concentration is required to define the caloric as well as thermal properties of the exhaust gas, that shall be treated as a mixture of ideal gases, see Chap. 19. The *molar-fraction* of a exhaust gas component i is referred to the molar quantity of the exhaust gas and thus follows

$$x_{EG,i} = \frac{n_{EG,i}}{n_{EG}} = \frac{n_F \cdot \nu_{EG,i}}{n_F \cdot \nu_{EG}} = \frac{\nu_{EG,i}}{\nu_{EG}}. \tag{23.98}$$

Mind, that the entire molar quantity of the exhaust gas is the sum over the individual molar quantities of its components, i.e.

$$n_{EG} = \sum_i n_{EG,i}. \tag{23.99}$$

Hence, dividing by the fuel's molar quantity leads to

$$\nu_{EG} = \sum_i \nu_{EG,i}. \tag{23.100}$$

Consequently, the components concentration simplifies to

$$x_{EG,i} = \frac{\nu_{EG,i}}{\sum_i \nu_{EG,i}}. \tag{23.101}$$

Thus, applied to the concentrations of the exemplary exhaust gas it is:

- Carbon dioxide CO_2

$$x_{EG,CO_2} = \frac{n_{EG,CO_2}}{n_{EG}} = \frac{n_F \cdot \nu_{EG,CO_2}}{n_F \cdot \sum_i \nu_{EG,i}} = \frac{\nu_{EG,CO_2}}{\sum_i \nu_{EG,i}} \tag{23.102}$$

- Sulphur dioxide SO_2

$$x_{EG,SO_2} = \frac{n_{EG,SO_2}}{n_{EG}} = \frac{n_F \cdot \nu_{EG,SO_2}}{n_F \cdot \sum_i \nu_{EG,i}} = \frac{\nu_{EG,SO_2}}{\sum_i \nu_{EG,i}} \tag{23.103}$$

- Water H_2O

$$x_{EG,H_2O} = \frac{n_{EG,H_2O}}{n_{EG}} = \frac{n_F \cdot \nu_{EG,H_2O}}{n_F \cdot \sum_i \nu_{EG,i}} = \frac{\nu_{EG,H_2O}}{\sum_i \nu_{EG,i}} \tag{23.104}$$

- Oxygen O_2

$$x_{EG,O_2} = \frac{n_{EG,O_2}}{n_{EG}} = \frac{n_F \cdot \nu_{EG,O_2}}{n_F \cdot \sum_i \nu_{EG,i}} = \frac{\nu_{EG,O_2}}{\sum_i \nu_{EG,i}} \qquad (23.105)$$

- Nitrogen N_2

$$x_{EG,N_2} = \frac{n_{EG,N_2}}{n_{EG}} = \frac{n_F \cdot \nu_{EG,N_2}}{n_F \cdot \sum_i \nu_{EG,i}} = \frac{\nu_{EG,N_2}}{\sum_i \nu_{EG,i}} \qquad (23.106)$$

Once the concentration of each component within the exhaust gas is known, the exhaust gas properties can be estimated, i.e. $c_{p,EG}$, R_{EG} and M_{EG}. Thus, all principles following Chap. 19 can be applied. Mind, that molar fractions can be converted into mass fractions according to Eq. 19.64.

23.3.3 Mass Conservation

Sure, the mass conservation needs to be fulfilled and obeys in steady state

$$\underbrace{m_F}_{\text{fuel}} + \underbrace{\lambda \cdot m_{a,min} \cdot (1 + x)}_{\text{comb. air}} = \underbrace{m_F \cdot \sum_i \mu_{EG,i}}_{\text{exhaust gas}} + \underbrace{m_A}_{\text{fuel ashes}} . \qquad (23.107)$$

The left side of this equation summarises the fluxes in. They need to be balanced by the fluxes out that are given on the right hand side of this equation. Mind, that usually the ashes are not part of the exhaust gas, but are treated separately. Dividing by the mass of the fuel m_F brings

$$1 + \lambda \cdot l_{min} \cdot (1 + x) = \sum_i \mu_{EG,i} + a. \qquad (23.108)$$

Re-arranging results in

$$\boxed{\sum_i \mu_{EG,i} = (1 - a) + \lambda \cdot l_{min} \cdot (1 + x)} \qquad (23.109)$$

Each stoichiometric calculation should be checked carefully regarding mass conservation.

23.3.4 Conversions

The major difference between the handling of solid/liquid and gaseous fuels obviously is, that solid/liquid fuels are most likely treated in mass-specific notation, whereas gaseous fuels are treated in molar-specific notation. However, with the knowledge of Chap. 19 a conversion between these two approaches is simple:

- Molar fraction → mass fraction

$$\boxed{\xi_{EG,i} = x_{EG,i} \cdot \frac{M_{EG,i}}{M_{EG}}} \tag{23.110}$$

- Air need (mass) → Air need (molar amount)

$$L = \frac{n_{air}}{n_F} = \frac{m_{air} M_F}{m_F M_{air}} = l \cdot \frac{M_F}{M_{air}} \tag{23.111}$$

- Exhaust gas composition (molar amount) → Exhaust gas composition (mass)

$$\mu_{EG,i} = \frac{m_{EG,i}}{m_F} \tag{23.112}$$

$$\mu_{EG,i} = \frac{n_{EG,i} M_{EG,i}}{n_F M_F} \tag{23.113}$$

$$\boxed{\mu_{EG,i} = \nu_{EG,i} \cdot \frac{M_{EG,i}}{M_F}} \tag{23.114}$$

Elemental Analysis of Fuels

Single-component Fuel

If the structural chemical formula of a one-component gaseous fuel is known, e.g. in the form of $C_a H_b O_z N_p S_r$, an elemental analysis, in order to determine the mass fractions h, c, o, n, s of the fuel, can be performed easily. The major advantage is, that the oxygen need as well as the composition of the exhaust gas can then be calculated according to the steps shown in Sect. 23.3.1. So let's investigate the gaseous fuel $C_a H_b O_z N_p S_r$. The molar amount of the fuel shall be n_F:

- Mass fraction carbon c

The fuel contains $a \cdot n_F$ carbon atoms, see structural chemical formula, i.e.

$$c = \xi_C = \frac{m_C}{m_F} = \frac{n_C M_C}{n_F M_F} = \frac{a n_F M_C}{n_F M_{C_a H_b O_z N_p S_r}} = \frac{a M_C}{M_{C_a H_b O_z N_p S_r}} \tag{23.115}$$

- Mass fraction hydrogen h

  The fuel contains $b \cdot n_F$ hydrogen atoms, see structural chemical formula, i.e.

  $$h = \xi_H = \frac{bM_H}{M_{C_aH_bO_zN_pS_r}} \tag{23.116}$$

- Mass fraction oxygen n

  The fuel contains $z \cdot n_F$ oxygen atoms, see structural chemical formula, i.e.

  $$o = \xi_O = \frac{zM_O}{M_{C_aH_bO_zN_pS_r}} \tag{23.117}$$

- Mass fraction nitrogen n

  The fuel contains $p \cdot n_F$ nitrogen atoms, see structural chemical formula, i.e.

  $$n = \xi_N = \frac{pM_N}{M_{C_aH_bO_zN_pS_r}} \tag{23.118}$$

- Mass fraction sulphur s

  The fuel contains $r \cdot n_F$ sulphur atoms, see structural chemical formula, i.e.

  $$s = \xi_S = \frac{rM_S}{M_{C_aH_bO_zN_pS_r}} \tag{23.119}$$

Multi-component Fuel (Mixture)

In case the fuel is a mixture of several components i each with well-known structural chemical formula, e.g. methane and propane, an elemental analysis can be performed as well. The mass of the fuel component i shall be m_i, so that

$$\sum_i m_i = m_F. \tag{23.120}$$

The mass fractions of carbon of the fuel component i shall be c_i, see Fig. 23.5. The other elements follow accordingly. Hence, for the mixture of fuels the elemental analysis is:

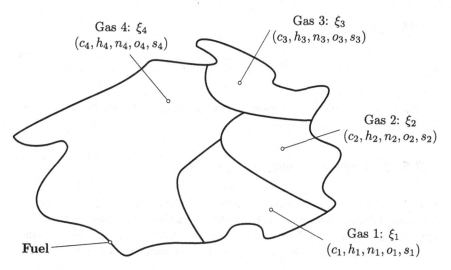

Fig. 23.5 Exemplary gaseous fuel

- Mass fraction carbon c

$$c = \frac{\sum_i c_i m_i}{\sum_i m_i} = \frac{\sum_i c_i m_i}{m_F} = \sum_i c_i \xi_i \qquad (23.121)$$

- Mass fraction hydrogen h

$$h = \frac{\sum_i h_i m_i}{\sum_i m_i} = \sum_i h_i \xi_i \qquad (23.122)$$

- Mass fraction nitrogen n

$$n = \frac{\sum_i n_i m_i}{\sum_i m_i} = \sum_i n_i \xi_i \qquad (23.123)$$

- Mass fraction oxygen o

$$o = \frac{\sum_i o_i m_i}{\sum_i m_i} = \sum_i o_i \xi_i \qquad (23.124)$$

- Mass fraction sulphur s

$$s = \frac{\sum_i s_i m_i}{\sum_i m_i} = \sum_i s_i \xi_i \qquad (23.125)$$

The mass fractions of the components can be calculated with Eqs. 23.115–23.119.

23.3.5 Setting up a Chemical Equation

Instead of applying the approach for CHONS-complexes, see Sect. 23.3.2, the chemical equation can be set up alternatively. Mind, that a chemical equation does not represent the *mass* ratio of the involved chemical bonds,[8] but it shows the *molar* ratio of them. This approach is shown exemplary for methanol with $\lambda > 1$ step by step[9]:

- Stoichiometric with oxygen

$$1\,CH_3OH + \frac{3}{2}\,O_2 \rightarrow 1\,CO_2 + 2\,H_2O \qquad (23.126)$$

Left and right hand side of this chemical equation are balanced by one carbon atom, four hydrogen atoms and four oxygen atoms.

- Stoichiometric with air

$$1\,CH_3OH + \frac{3}{2}\,O_2 + \frac{79}{21}\cdot\frac{3}{2}\,N_2 \rightarrow 1\,CO_2 + 2\,H_2O + \frac{79}{21}\cdot\frac{3}{2}\,N_2 \qquad (23.127)$$

The ratio $\frac{79}{21}$ is the molar ratio $\frac{x_{N_2,dry\,air}}{x_{O_2,dry\,air}}$. It shows how much more nitrogen than oxygen is in the combustion air. Since the mount of nitrogen on the left side is equal to the amount of nitrogen on the right side, nitrogen is inert, i.e. chemically not reactive in this combustion.

- With excess air (dry air)

$$\boxed{\begin{aligned} CH_3OH + \lambda\cdot\frac{3}{2}\,O_2 + \lambda\cdot\frac{79}{21}\cdot\frac{3}{2}\,N_2 \\ \rightarrow CO_2 + 2\,H_2O + \lambda\cdot\frac{79}{21}\cdot\frac{3}{2}\,N_2 + (\lambda-1)\cdot\frac{3}{2}\,O_2 \end{aligned}}$$
$$(23.128)$$

In case $\lambda = 1$ there is no oxygen in the exhaust gas, so that the combustion then is stoichiometric.

---

[8]Sure, the mass balance needs to be fulfilled!

[9]Applying a simplified standard atmosphere containing purely nitrogen and oxygen.

According to this chemical Eq. 23.128 oxygen need as well as exhaust gas composition can be determined instantaneously:

- Oxygen need

It requires $\lambda \cdot \frac{3}{2}$ times as much molar amount than molar amount of the fuel, i.e.

$$n_{O_2} = \lambda \cdot \frac{3}{2} \cdot n_F. \tag{23.129}$$

Thus, it is by division of n_F

$$\boxed{O = \lambda \cdot \frac{3}{2}} \tag{23.130}$$

- Exhaust gas composition

The molar amount of each exhaust gas component with reference to the molar amount of the fuel can be taken out of Eq. 23.128:

− Carbon dioxide

Comparing the stoichiometric factors of carbon dioxide and methanol in Eq. 23.128 leads to

$$n_{EG,CO_2} = 1 \cdot n_F. \tag{23.131}$$

Thus, it is by division of n_F

$$\boxed{\nu_{EG,CO_2} = 1} \tag{23.132}$$

This means, the molar amount of fuel and carbon dioxide is identical.

− Water

Comparing the stoichiometric factors of water and methanol in Eq. 23.128 leads to

$$n_{EG,H_2O} = 2 \cdot n_F. \tag{23.133}$$

Thus, it is by division of n_F

$$\boxed{\nu_{EG,H_2O} = 2} \tag{23.134}$$

That means, the molar amount of water is twice as much as the molar amount of fuel.

− Nitrogen

Comparing the stoichiometric factors of nitrogen and methanol in Eq. 23.128 leads to

$$n_{EG,N_2} = \lambda \cdot \frac{79}{21} \cdot \frac{3}{2} \cdot n_F. \tag{23.135}$$

Thus, it is by division of n_F

$$\boxed{\nu_{EG,N_2} = \lambda \cdot \frac{79}{21} \cdot \frac{3}{2}} \qquad (23.136)$$

– Oxygen

Comparing the stoichiometric factors of product-oxygen and methanol in Eq. 23.128 leads to

$$n_{EG,O_2} = (\lambda - 1) \cdot \frac{3}{2} \cdot n_F. \qquad (23.137)$$

Thus, it is by division of n_F

$$\boxed{\nu_{EG,O_2} = (\lambda - 1) \cdot \frac{3}{2}} \qquad (23.138)$$

This means, in case $\lambda = 1$ there is no oxygen in the exhaust gas.

23.3.6 Dew Point of the Exhaust Gas

The exhaust gas investigation has shown, that fossil fuels deliver water, depending on the amount of hydrogen respectively with the amount of water which is carried by the fuel. As already known water tends to change its aggregate state, i.e. vapour can condense and bring out liquid. In this section it is analysed under what conditions water condenses and how much water turns into liquid state. In order to do so, the gaseous exhaust gas is treated as a mixture of *ideal* gases. Once the concentration of the exhaust gas is known, the partial pressure of the water in the gaseous mixture can be calculated, see Chap. 19. Actually, based on the assumption the entire product water is in vapour state two alternatives can be followed:

• Case 1: Molar concentration $x_{EG,i}$ is well-known

This case is simple, since the given correlation can be applied for ideal mixtures, i.e.

$$x_i = \pi_i = \frac{p_i}{p}. \qquad (23.139)$$

This equation states, that molar-fraction and partial pressure fraction are equal, see Chap. 19. Hence, the partial pressure of the vapour is

$$p_v = p_{EG,H_2O} = x_{EG,H_2O} \cdot p. \qquad (23.140)$$

• Case 2: Mass concentration $\xi_{EG,i}$ is well-known

A conversion from mass-fractions to molar-fractions $x_{EG,i}$ is required, i.e.

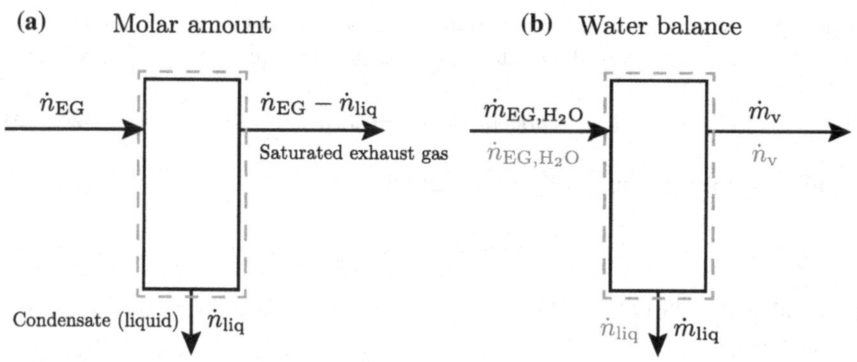

(a) Molar amount **(b)** Water balance

Fig. 23.6 Condensing product water

$$x_{EG,i} = \xi_{EG,i} \cdot \frac{M_{EG}}{M_{EG,i}}. \tag{23.141}$$

Mind, that M_{EG} is the molar mass of the exhaust gas, i.e. the gaseous mixture. It can be determined following the principles of Chap. 19. $M_{EG,i}$, however, is the molar mass of the specific component with the given concentration.[10] Hence, the vapour's partial pressure is

$$p_v = p_{EG,H_2O} = x_{EG,H_2O} \cdot p = \xi_{EG,H_2O} \cdot \frac{M_{EG}}{M_{EG,H_2O}} \cdot p. \tag{23.142}$$

Once the partial pressure of the vapour p_v in the exhaust gas is known, the related *saturated steam temperature* (=dew point temperature ϑ_τ) can be looked up, e.g. in the steam table, i.e.

$$\boxed{\vartheta_\tau = f(p_v)} \tag{23.143}$$

In other words, this is the minimum temperature which is required to fulfil the assumption, that the entire water occurs as vapour. In case this temperature is fallen short of, the vapour starts to condense. Let us now analyse, how much water condenses if the temperature decreases. This scenario is sketched in Fig. 23.6.

Once the condensed water has been fallen out and is separated from the gaseous mixture, the remaining exhaust gas is *saturated* with vapour according to its temperature. Hence, for the exhaust gas, see Fig. 23.6a, it is

$$\pi_s = \frac{p_s(T_{EG})}{p} = \frac{\dot{n}_v}{\dot{n}_{EG} - \dot{n}_{liq}} = x_s. \tag{23.144}$$

Obviously, the total molar amount leaving the separator is reduced by the molar amount of the liquid water that is removed. The water in the remaining exhaust gas

---

[10] In this section it is the component *water*.

is in vapour state. It is well-known from Chap. 19, that molar concentration equals partial pressure ratio. Now, let us state the mass balance according to Fig. 23.6b, i.e.

$$\dot{m}_{EG,H_2O} = \dot{m}_v + \dot{m}_{liq}. \tag{23.145}$$

Dividing by the molar mass of water results in

$$\dot{n}_{EG,H_2O} = \dot{n}_v + \dot{n}_{liq}. \tag{23.146}$$

Combining Eqs. 23.144 and 23.146 leads to the molar quantity of the liquid, i.e. condensed water,

$$\dot{n}_{liq} = \frac{\dot{n}_{EG,H_2O} - x_s \dot{n}_{EG}}{1 - x_s}. \tag{23.147}$$

Conversion into mass fluxes leads to

$$\dot{m}_{liq} = \frac{\dot{m}_{EG,H_2O} - x_s \dot{m}_{EG} \frac{M_{H_2O}}{M_{EG}}}{1 - x_s}. \tag{23.148}$$

With reference to the fuel's mass it follows

$$\boxed{\frac{\dot{m}_{liq}}{\dot{m}_F} = \mu^*_{EG,H_2O} = \frac{\mu_{EG,H_2O} - x_s \mu_{EG} \frac{M_{H_2O}}{M_{EG}}}{1 - x_s}} \tag{23.149}$$

with x_s being of function of total pressure p and the saturated pressure according to the exhaust gas temperature, see also Eq. 23.144, i.e.

$$\boxed{x_s = \frac{p_s(T_{EG})}{p}} \tag{23.150}$$

Problem 23.1 Coal is oxidised with an air fuel equivalence ratio of $\lambda = 1.35$. The air shall be dry. Please calculate the minimum oxygen need, the overall air need and the concentrations $\xi_{EG,i}$ of the exhaust gas. The composition of the coal is as follows: $c = 0.75$, $h = 0.05$, $s = 0.01$, $o = 0.06$, $n = 0.01$, $w = 0.06$, $a = 0.06$.

Solution

The minimum oxygen need obeys

$$o_{min} = 2.667c + 8h + s - o = 2.3502 \, kg_{O_2}/kg_F. \tag{23.151}$$

The mass fraction of oxygen in a simplified standard atmosphere is $\xi_{O_2,air} = 0.23$ and $\xi_{N_2,air} = 0.77$, so that the minimum air need is

$$l_{min} = \frac{o_{min}}{\xi_{O_2,air}} = 10.2185. \tag{23.152}$$

The combustion is performed with excess air, so that the overall air need follows

$$l = \lambda l_{min} = 13.7949. \tag{23.153}$$

The exhaust gas composition is

- Carbon dioxide:

$$\mu_{EG,CO_2} = 3.667c = 2.7502 \tag{23.154}$$

- Water:

$$\mu_{EG,H_2O} = 9h + w = 0.51 \tag{23.155}$$

- Sulphur dioxide:

$$\mu_{EG,SO_2} = 2s = 0.02 \tag{23.156}$$

- Nitrogen:

$$\mu_{EG,N_2} = n + \lambda l_{min}\xi_{N_2,air} = 10.6321 \tag{23.157}$$

- Oxygen:

$$\mu_{EG,O_2} = o_{min}(\lambda - 1) = 0.8226 \tag{23.158}$$

The mass conservation needs to be fulfilled[11]:

$$\underbrace{\sum_i \mu_{EG,i}}_{=\mu_{EG}=14.7349} = \underbrace{(1-a) + \lambda l_{min}(1+x)}_{=14.7349} \tag{23.159}$$

Hence, the concentration of the exhaust gas is

- Carbon dioxide:

$$\xi_{EG,CO_2} = \frac{\mu_{EG,CO_2}}{\mu_{EG}} = 0.1866 \tag{23.160}$$

- Water:

$$\xi_{EG,H_2O} = \frac{\mu_{EG,H_2O}}{\mu_{EG}} = 0.0346 \tag{23.161}$$

- Sulphur dioxide:

$$\xi_{EG,SO_2} = \frac{\mu_{EG,SO_2}}{\mu_{EG}} = 0.0014 \tag{23.162}$$

---

[11] The air is supposed to be dry, so that its water content $x = 0$!

- Nitrogen:

$$\xi_{EG,N_2} = \frac{\mu_{EG,N_2}}{\mu_{EG}} = 0.7216 \qquad (23.163)$$

- Oxygen:

$$\xi_{EG,O_2} = \frac{\mu_{EG,O_2}}{\mu_{EG}} = 0.0558 \qquad (23.164)$$

Problem 23.2 Pure hydrogen is oxidised with $\lambda = 2.9$. Air shall be treated simplified as a two component mixture of N_2/O_2. The mass-fraction of oxygen shall be $\xi_{O_2,air} = 0.23$.

(a) Calculate the minimum air need l_{min} and the overall air need l.
(b) What is the total specific exhaust gas composition μ_{EG} and the concentration of the exhaust gas $\xi_{EG,i}$?
(c) Determine the dew point ϑ_τ of the exhaust gas at a pressure of $p = 1060\,\text{mbar}$.

Solution

(a) Since the fuel composition is $h = 1$, the minimum oxygen need obeys

$$o_{min} = 2.667c + 8h + s - o = 8. \qquad (23.165)$$

The mass fraction of oxygen in a simplified standard atmosphere is $\xi_{O_2,air} = 0.23$ and $\xi_{N_2,air} = 0.77$, so that the minimum air need is

$$l_{min} = \frac{o_{min}}{\xi_{O_2,air}} = 34.7826. \qquad (23.166)$$

The combustion is performed with excess air, so that the overall air need results in

$$l = \lambda l_{min} = 100.8696. \qquad (23.167)$$

(b) The exhaust gas composition is

- Water:

$$\mu_{EG,H_2O} = 9h + w = 9 \qquad (23.168)$$

- Nitrogen:

$$\mu_{EG,N_2} = n + \lambda l_{min}\xi_{N_2,air} = 77.6696 \qquad (23.169)$$

- Oxygen:

$$\mu_{EG,O_2} = o_{min}(\lambda - 1) = 15.2000 \qquad (23.170)$$

The summation leads to

$$\mu_{EG} = \sum_i \mu_{EG,i} = 101.8696. \tag{23.171}$$

Hence, the concentration of the exhaust gas is

- Water:

$$\xi_{EG,H_2O} = \frac{\mu_{EG,H_2O}}{\mu_{EG}} = 0.0883 \tag{23.172}$$

- Nitrogen:

$$\xi_{EG,N_2} = \frac{\mu_{EG,N_2}}{\mu_{EG}} = 0.7624 \tag{23.173}$$

- Oxygen:

$$\xi_{EG,O_2} = \frac{\mu_{EG,O_2}}{\mu_{EG}} = 0.1492 \tag{23.174}$$

(c) In order to calculate the dew point ϑ_τ of the exhaust gas the partial pressure of the product water is required. The molar fraction x_{EG,H_2O} follows

$$x_{EG,H_2O} = \xi_{EG,H_2O} \frac{M_{EG}}{M_{H_2O}}. \tag{23.175}$$

The molar mass of the exhaust gas is

$$M_{EG} = \frac{R_M}{R_{EG}} \tag{23.176}$$

with

$$R_{EG} = \sum_i \xi_{EG,i} R_i. \tag{23.177}$$

The gas constants of the exhaust gas components are:

- Water

$$R_{H_2O} = \frac{R_M}{M_{H_2O}} = 461.9 \, \frac{J}{kg \, K} \tag{23.178}$$

- Nitrogen

$$R_{N_2} = \frac{R_M}{M_{N_2}} = 296.94 \, \frac{J}{kg \, K} \tag{23.179}$$

- Oxygen

$$R_{O_2} = \frac{R_M}{M_{O_2}} = 259.82 \, \frac{J}{kg \, K} \tag{23.180}$$

Hence, the gas constant of the exhaust gas is

$$R_{EG} = \sum_i \xi_{EG,i} R_i = 305.9755 \, \frac{J}{kg \, K}. \tag{23.181}$$

Its molar mass is

$$M_{EG} = \frac{R_M}{R_{EG}} = 27.1731 \, \frac{kg}{kmol}. \tag{23.182}$$

The molar fraction of the water in the exhaust gas is

$$x_{EG,H_2O} = \xi_{EG,H_2O} \frac{M_{EG}}{M_{H_2O}} = 0.1334. \tag{23.183}$$

As known from Chap. 19 it is

$$x_{EG,H_2O} = \frac{p_{EG,H_2O}}{p}. \tag{23.184}$$

Hence, the partial pressure of the vapour in the exhaust gas results in

$$p_{EG,H_2O} = x_{EG,H_2O} \cdot p = 0.1414 \, bar. \tag{23.185}$$

The corresponding saturated temperature to this pressure can be taken from the steam table, i.e.

$$\vartheta_\tau = f\left(p_{EG,H_2O}\right) = 52.75 \, °C. \tag{23.186}$$

Problem 23.3 Natural gas is oxidised with dry air and an air fuel equivalence ratio $\lambda = 1.3$. What are the minimum oxygen need O_{min}, the overall air need L and the exhaust gas composition $\nu_{RG,i}$? What are the exhaust gas concentrations in molar- and mass-fractions? Natural gas shall be treated as a mixture of the following components:

- Methane CH_4: $x_{CH_4} = 0.8$
- Ethane C_2H_6: $x_{C_2H_6} = 0.02$
- Propane C_3H_8: $x_{C_3H_8} = 0.01$
- Carbon dioxide CO_2: $x_{CO_2} = 0.03$
- Nitrogen N_2: $x_{N_2} = 0.14$

Solution

In order to calculate the minimum oxygen need Eq. 23.76 for the exemplary fuel, see sketch Fig. 23.4, is applied and modified, i.e.

$$O_{min} = 0.5 \cdot x_{H_2} + 0.5 \cdot x_{CO} + \sum_i (a_i + \frac{b_i}{4} - \frac{z_i}{2} + r_i) \cdot x_{C_a H_b O_z N_p S_{r,i}} - x_{O_2}.$$

(23.187)

Methane, ethane and propane are representatives for the CHONS-complexes. Since the fuel does not contain hydrogen H_2, oxygen O_2 and carbon monoxide CO the equation simplifies to

$$O_{min} = \sum_i (a_i + \frac{b_i}{4} - \frac{z_i}{2} + r_i) \cdot x_{C_a H_b O_z N_p S_{r,i}}.$$

(23.188)

Hence, the CHONS-complexes are handled as follows[12]:

$$O_{min} = \left(1 + \frac{4}{4}\right) \cdot x_{CH_4} + \left(2 + \frac{6}{4}\right) \cdot x_{C_2 H_6} + \left(3 + \frac{8}{4}\right) \cdot x_{C_3 H_8} = 1.72 \quad (23.189)$$

Hence, in order to oxidise 1 kmol of fuel 1.72 kmol oxygen are required. The minimum air need is larger, since air is composed to 21 Vol.-% of oxygen, i.e.

$$L_{min} = \frac{O_{min}}{x_{O_2,air}} = 8.1905.$$

(23.190)

The overall air need is

$$L = \lambda \cdot L_{min} = 10.6476.$$

(23.191)

The exhaust gas composition can be calculated according to the exemplary fuel, see Fig. 23.4:

- Carbon dioxide CO_2

Starting from the exemplary mixture, i.e.

$$\nu_{EG,CO_2} = x_{CO} + \sum_i a_i \cdot x_{C_a H_b O_z N_p S_{r,i}} + x_{CO_2}$$

(23.192)

it follows for the given gaseous mixture, i.e.

$$\boxed{\nu_{EG,CO_2} = 1 \cdot x_{CH_4} + 2 \cdot x_{C_2 H_6} + 3 \cdot x_{C_3 H_8} + x_{CO_2} = 0.9}.$$

(23.193)

- Water H_2O

The example of the gaseous fuel according to Fig. 23.4 has shown

---

[12] According to their chemical composition: $CH_4 \rightarrow a = 1, b = 4$, $C_2H_6 \rightarrow a = 2, b = 6$ and $C_3H_8 \rightarrow a = 3, b = 8$.

$$\nu_{EG,H_2O} = x_{H_2} + \sum_i \frac{b_i}{2} \cdot x_{C_aH_bO_zN_pS_{r,i}} + 1.6022 \cdot \lambda \cdot L_{min} \cdot x. \qquad (23.194)$$

Since the combustion air is dry and the water purely comes from the fuel's bonded hydrogen, it is

$$\boxed{\nu_{EG,H_2O} = 2 \cdot x_{CH_4} + 3 \cdot x_{C_2H_6} + 4 \cdot x_{C_3H_8} = 1.7} \qquad (23.195)$$

- Nitrogen N_2

For the nitrogen it is

$$\nu_{EG,N_2} = \frac{n_{EG,N_2}}{n_F} = x_{N_2} + \sum_i \frac{p_i}{2} \cdot x_{C_aH_bO_zN_pS_{r,i}} + \lambda \cdot L_{min} \cdot x_{N_2,dry\ air}. \qquad (23.196)$$

Since the CHONS-complexes do not contain any nitrogen, it simplifies to

$$\boxed{\nu_{EG,N_2} = x_{N_2} + \lambda \cdot L_{min} \cdot x_{N_2,dry\ air} = 8.5516} \qquad (23.197)$$

- Oxygen O_2

The example of the gaseous fuel according to Fig. 23.4 has shown

$$\nu_{EG,O_2} = \frac{n_{EG,O_2}}{n_F} = L_{min} \cdot x_{O_2,dry\ air} \cdot (\lambda - 1) = O_{min} \cdot (\lambda - 1). \qquad (23.198)$$

Applied to the composition of the given mixture it is

$$\boxed{\nu_{EG,O_2} = L_{min} \cdot x_{O_2,dry\ air} \cdot (\lambda - 1) = O_{min} \cdot (\lambda - 1) = 0.5160} \qquad (23.199)$$

In a next step the exhaust gas concentration in molar fractions is calculated. To do, so overall ν_{EG} is required, i.e.

$$\nu_{EG} = \sum_i \nu_{EG,i} = 11.6676. \qquad (23.200)$$

Hence the concentration is

- Carbon dioxide:

$$x_{EG,CO_2} = \frac{\nu_{EG,CO_2}}{\nu_{EG}} = 0.0771 \qquad (23.201)$$

- Water:

$$x_{EG,H_2O} = \frac{\nu_{EG,H_2O}}{\nu_{EG}} = 0.1457 \qquad (23.202)$$

- Nitrogen:

$$x_{EG,N_2} = \frac{\nu_{EG,N_2}}{\nu_{EG}} = 0.7329 \qquad (23.203)$$

- Oxygen:

$$x_{EG,O_2} = \frac{\nu_{EG,O_2}}{\nu_{EG}} = 0.0442 \qquad (23.204)$$

To convert these concentration into mass-fractions, the following equation is applied

$$\xi_{EG,i} = x_{EG,i} \cdot \frac{M_{EG,i}}{M_{EG}}. \qquad (23.205)$$

In order to do so, the molar mass of the exhaust gas is required, i.e.

$$\begin{aligned}
M_{EG} &= \sum_i x_{EG,i} \cdot M_i \\
&= x_{EG,CO_2} M_{CO_2} + x_{EG,H_2O} M_{H_2O} + x_{EG,N_2} M_{N_2} + x_{EG,O_2} M_{O_2} \\
&= 27.9541 \frac{kg}{kmol}.
\end{aligned} \qquad (23.206)$$

Thus, the mass fractions of the exhaust gas follow accordingly

- Carbon dioxide:

$$\xi_{EG,CO_2} = x_{EG,CO_2} \cdot \frac{M_{EG,CO_2}}{M_{EG}} = 0.1214 \qquad (23.207)$$

- Water:

$$\xi_{EG,H_2O} = x_{EG,H_2O} \cdot \frac{M_{EG,H_2O}}{M_{EG}} = 0.0938 \qquad (23.208)$$

- Nitrogen:

$$\xi_{EG,N_2} = x_{EG,N_2} \cdot \frac{M_{EG,N_2}}{M_{EG}} = 0.7341 \qquad (23.209)$$

- Oxygen:

$$\xi_{EG,O_2} = x_{EG,O_2} \cdot \frac{M_{EG,O_2}}{M_{EG}} = 0.0506 \qquad (23.210)$$

Problem 23.4 A mixture of gaseous fuels contains 75 Vol.-% ethane (C_2H_6) and 25 Vol.-% methane (CH_4). Please perform a elemental analysis and estimate c and h. What is the minimum oxygen need o_{min}?

Solution

First, the elemental analysis of each component, i.e. ethane and methane, is done

- Ethane:

  Carbon

$$c_{C_2H_6} = \xi_C = \frac{a M_C}{M_{C_a H_b O_z N_p S_r}} = \frac{2 \cdot M_C}{M_{C_2H_6}} = \frac{24}{30} = 0.8 \qquad (23.211)$$

and hydrogen

$$h_{C_2H_6} = \xi_H = \frac{b M_H}{M_{C_a H_b O_z N_p S_r}} = \frac{6 \cdot M_H}{M_{C_2H_6}} = \frac{6}{30} = 0.2 \qquad (23.212)$$

- Methane:

  Carbon

$$c_{CH_4} = \xi_C = \frac{a M_C}{M_{C_a H_b O_z N_p S_r}} = \frac{1 \cdot M_C}{M_{CH_4}} = \frac{12}{16} = 0.75 \qquad (23.213)$$

and hydrogen

$$h_{CH_4} = \xi_H = \frac{b M_H}{M_{C_a H_b O_z N_p S_r}} = \frac{4 \cdot M_H}{M_{CH_4}} = \frac{4}{16} = 0.25 \qquad (23.214)$$

The molar mass of the gaseous fuel is[13]

$$M_F = \sum_i x_{F,i} \cdot M_i = x_{C_2H_6} M_{C_2H_6} + x_{CH_4} M_{CH_4} = 26.5 \, \frac{kg}{kmol} \qquad (23.215)$$

Now, the composition of the gaseous mixture is converted into mass-fractions, i.e.

$$\xi_{C_2H_6} = x_{C_2H_6} \cdot \frac{M_{C_2H_6}}{M_F} = 0.8491 \qquad (23.216)$$

and

$$\xi_{CH_4} = x_{CH_4} \cdot \frac{M_{CH_4}}{M_F} = 0.1509. \qquad (23.217)$$

Hence, it finally is for the gaseous mixture

---

[13] Mind, that $\sigma_i = x_i$, i.e. molar fractions are identical with volume fractions!

$$c = \sum_i c_i \xi_i = c_{C_2H_6} \xi_{C_2H_6} + c_{CH_4} \xi_{CH_4} = 0.7925 \tag{23.218}$$

and

$$h = \sum_i h_i \xi_i = h_{C_2H_6} \xi_{C_2H_6} + h_{CH_4} \xi_{CH_4} = 0.2075. \tag{23.219}$$

Once the elemental analysis is done, the minimum oxygen need can be calculated easily

$$o_{min} = 2.667c + 8h = 3.7738. \tag{23.220}$$

23.4 Energetic Balancing

Now that it has been clarified how much air for an oxidisation of fossil fuels is required and how the composition of the exhaust gas looks like, the focus in this section is on the energy balance. Hence, a correlation of released heat and corresponding temperature of the exhaust gas is deduced. However, this correlation depends on the fuel/air mixture. Figure 23.7 shows a combustion chamber as an open system, that is assumed to be in steady state. Obviously, fuel and air enter the system, whereas ashes and exhaust gas leave the system, so that in steady state the mass balance obeys

$$\dot{m}_F + \dot{m}_a = \dot{m}_{EG} + \dot{m}_{Ash}. \tag{23.221}$$

Ashes might be disregarded in case the fuel is gaseous, so that

$$\dot{m}_F + \dot{m}_a = \dot{m}_{EG}. \tag{23.222}$$

The first law of thermodynamics for steady state simplifies, so that the energy flux into the system is balanced by the energy flux out of the system, i.e.

$$\dot{Q} + P + \sum_{in,k} \dot{m}_k \cdot \left[h_k + \frac{c_k^2}{2} + g \cdot z_k \right] = \sum_{out,i} \dot{m}_i \cdot \left[h_i + \frac{c_i^2}{2} + g \cdot z_i \right]. \tag{23.223}$$

Obviously, according to Fig. 23.7 no technical work[14] crosses the system boundary. Let us further ignore the ashes as well as kinetic and potential energies, so that the first law of thermodynamics follows

$$\dot{Q} + \dot{m}_F \cdot h_F(T_F) + \dot{m}_a \cdot h_a(T_a) = \dot{m}_{EG} \cdot h_{EG}(T_{EG}). \tag{23.224}$$

Rearranging and replacing the mass of the exhaust gas according to Eq. 23.222 results in

---

[14]Mechanical or electrical energy!

Fig. 23.7 Energy
balance—Combustion
chamber

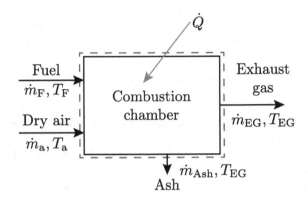

$$\dot{Q} = \dot{m}_F \left[h_{EG}(T_{EG}) - h_F(T_F) \right] + \dot{m}_a \left[h_{EG}(T_{EG}) - h_a(T_a) \right]. \qquad (23.225)$$

Specific enthalpy shall purely be a function of temperature, because fuel, air and
exhaust gas are treated simplified as ideal fluids. T_F is the fuel's temperature, T_a
the temperature of the combustion air and T_{EG} the exhaust gas temperature. Open
systems have been balanced in the previous parts several times and one expects to
substitute the specific enthalpies by applying a caloric equation of states, such as

$$h_2 - h_1 = c_p \left(T_2 - T_1 \right). \qquad (23.226)$$

Unfortunately, an application of the caloric equation of state fails in Eq. 23.225,
since in the first bracket the released exhaust gas and the supplied fuel are compared,
respectively in the second bracket exhaust gas and air.[15] So far, the enthalpy included
the caloric effect, i.e. the temperature dependency on the supplied heat,[16] chemical
bonded energies are not part of the specific energy though. Consequently, the recently
applied approaches are limited. However, this dilemma is solved in Chap. 24, in
which the absolute enthalpies are introduced that not just include the caloric part but
the chemical bonded energy as well. In this chapter another pragmatic approach is
followed, that splits the combustion process in three conceptual, consecutive steps.

23.4.1 Lower Heating Value

The major problem of the specific enthalpy as used so far is, that the material conversion is energetically not covered. Systems treated in Part I/II used to be ideal

---

[15]The enthalpy differences can not be substituted by temperature differences, as the exhaust gas
and fuel/air have different specific heat capacities.

[16]Supplied thermal energy increases temperature and thus enthalpy, heat release decreases temperature and enthalpy.

Fig. 23.8 Definition of the lower heating value H_U

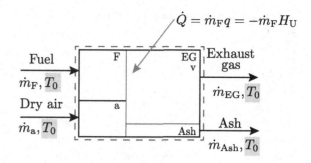

$$\dot{Q} = \dot{m}_F q = -\dot{m}_F H_U$$

fluids that are chemically not active. Consequently, the chemical reaction and thus the material conversion needs to be described alternatively, see Fig. 23.8.

Fuel as well as air enter the combustion chamber with the same *reference* temperature T_0. During the oxidation the temperature increases, so that under adiabatic conditions the exhaust gas leaves with a larger temperature than inlet temperatures. Now, let's drop this adiabatic condition and apply a cooling. The cooling shall be as effective, that the exhaust gas leaves with the same reference temperature T_0 as the reactants have been supplied. The corresponding cooling power is

$$\dot{Q} = \dot{m}_F q < 0. \tag{23.227}$$

Obviously, the released heat flux depends on the fluid, so that the so-called lower heating value $H_U(T_0) > 0$ is defined as follows

$$\dot{Q} = \dot{m}_F q \stackrel{!}{=} -\dot{m}_F H_U(T_0) < 0. \tag{23.228}$$

Definition 23.5 The lower heating value $H_U(T_0)$ of fuels is the specific amount of heat, that is needed to cool down the exhaust gas back to a reference temperature T_0 of the supplied fuel and air. By definition the entire product water in this case is in vapour state.

In order to estimate the lower heating value experimentally it must be guaranteed, that the oxidisation is complete, i.e. the highest oxidation state is reached. Hence, the experiments are performed with an air-fuel equivalence ratio of $\lambda > 1$. The excess air, that is not chemically reactive, does not have an energetic impact on the heating value, since air inlet and outlet temperature are identical T_0 and thus have no caloric effect. With the same premise humid, unsaturated air can be supplied, since the included humidity, i.e. vapour, has the same inlet and outlet temperature.[17] Tables 23.2 and 23.3 show lower heating values of relevant technical fossil fuels.

---

[17]Hence, Fig. 23.8 also works out with *unsaturated* humid air $\dot{m}_A$. If liquid water would be supplied with the combustion air, thermal energy would be required to vaporise the liquid water and turn it into vapour. Thus, the specific lower heating value would decrease.

Table 23.2 Lower heating value H_U (Solids and Liquids) at $\vartheta_0 = 25\,°C$, see [7]

Solid fuels		Liquid fuels	
Fuel	H_U in $\frac{MJ}{kg}$	Fuel	H_U in $\frac{MJ}{kg}$
Wood, dry	14.65 ... 16.75	Ethanol	26.9
Turf, dry	11.72 ... 15.07	Benzol	40.15
Raw lignite	8.37 ... 11.30	Toluol	40.82
Brown coal briquettes	19.68 ... 20.10	Naphtalin	38.94
Hard coal	27.31 ... 34.12	Pentane	45.43
Anthracite	32.66 ... 33.91	Octane	44.59
Coke	27.84 ... 30.35	Benzine	42.7

Table 23.3 Lower heating value H_U (Gases) at $\vartheta_0 = 25\,°C$, see [7]

Gaseous fuels	
Fuel	H_U in $\frac{MJ}{kg}$
Hydrogen	119.97
Carbon monoxide	10.10
Methane	50.01
Ethane	47.49
Propane	46.35
Ethylene	47.15
Acetylene	48.22

23.4.2 Conceptual 3-Steps Combustion

As discussed, the chemical reactive part is energetically handled with the lower heating value. However, inlet and outlet temperature usually differ from the reference temperature T_0 the lower heating value is related with, see Fig. 23.7. In order to calculate any combustion process with varying temperatures a pre- and post-conditioning is, in addition to the lower heating value, required. This concept is further discussed in this section.

Step 1—Pre-conditioning

Fuel and air enter the combustion chamber with any temperature $T_F \neq T_0$ respectively $T_a \neq T_0$. In the chamber both fluids are conditioned to the reference temperature T_0, so that in step 2 the lower heating value approach can be followed. Thus, in step 1, see Fig. 23.9, no chemical reactions run and the energy balance is purely based on caloric effects. Consequently, the first law of thermodynamics obeys

$$\dot{Q}_1 + \dot{m}_F h_F (T_F) + \dot{m}_a h_a (T_a) = \dot{m}_F h_F (T_0) + \dot{m}_a h_a (T_0). \qquad (23.229)$$

Fig. 23.9 Combustion—step 1

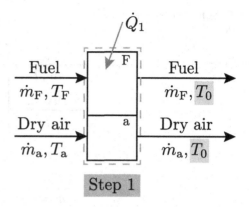

Solving for the required heat flux results in

$$\dot{Q}_1 = -\dot{m}_F\left[h_F\left(T_F\right) - h_F\left(T_0\right)\right] - \dot{m}_a\left[h_a\left(T_a\right) - h_a\left(T_0\right)\right].$$
(23.230)

In specific notation with reference to the mass flux of the fuel it is

$$\boxed{q_1 = -\left[h_F\left(T_F\right) - h_F\left(T_0\right)\right] - \lambda l_{\min}\left[h_a\left(T_a\right) - h_a\left(T_0\right)\right]}$$
(23.231)

with

$$\lambda l_{\min} = l = \frac{\dot{m}_a}{\dot{m}_F}.$$
(23.232)

This equation can actually be calculated with the means of Part I: the caloric equation of state can be applied.[18] The specific heat flux q_1 can be positive, negative or zero depending on the temperatures T_F and T_a in comparison with the reference temperature T_0.

Step 2—Chemical Reaction

Now, that the fluids have both been conditioned to reference temperature T_0, the chemical reaction can be energetically realised by the lower heating value approach, see Fig. 23.10. While doing so, the exhaust gas leaves with reference temperature as well. Thus, the cooling power follows, see Sect. 23.4.1,

$$\dot{Q}_2 = -\dot{m}_F H_U\left(T_0\right).$$
(23.233)

In specific notation with reference to the mass flux of the fuel it is

---

[18]The first bracket in Eq. 23.231 contains purely fuel, the second purely air.

Fig. 23.10 Combustion—step 2

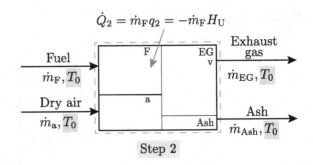

$$\dot{Q}_2 = \dot{m}_F q_2 = -\dot{m}_F H_U$$

Step 2

$$\boxed{q_2 = -H_U (T_0)} \qquad (23.234)$$

Mind, that the exhaust gas leaves step 2 with reference temperature T_0. Due to the definition of the lower heating value, the product water is in gaseous state when leaving step 2.

Step 3—Post-conditioning

The exhaust gas leaves the combustion chamber with a temperature T_{EG} which usually is unequal to the reference temperature T_0, so that a post-conditioning must be done, see Fig. 23.11. Since there is not chemical reaction this is purely a caloric step. The first law of thermodynamics in this case reads in steady state

$$\dot{Q}_3 + \dot{m}_{EG} h_{EG} (T_0) + \dot{m}_{Ash} h_{Ash} (T_0) = \dot{m}_{EG} h_{EG} (T_{EG}) + \dot{m}_{Ash} h_{Ash} (T_{EG}). \qquad (23.235)$$

Solving for the heat flux

$$\dot{Q}_3 = \dot{m}_{EG} [h_{EG} (T_{EG}) - h_{EG} (T_0)] + \dot{m}_{Ash} [h_{Ash} (T_{EG}) - h_{Ash} (T_0)]. \qquad (23.236)$$

In specific notation with reference to the mass flux of the fuel it results in

$$\boxed{q_3 = \sum_i \mu_{EG,i} \cdot [h_{EG} (T_{EG}) - h_{EG} (T_0)] + a \cdot [h_{Ash} (T_{EG}) - h_{Ash} (T_0)]} \qquad (23.237)$$

with

$$\sum_i \mu_{EG,i} = \mu_{EG} = \frac{\dot{m}_{EG}}{\dot{m}_F} \qquad (23.238)$$

and

$$a = \frac{\dot{m}_{Ash}}{\dot{m}_F}. \qquad (23.239)$$

Fig. 23.11 Combustion—
step 3

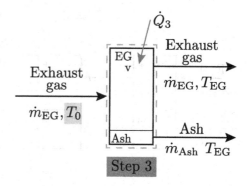

This energy balance can actually be calculated with the means of Part I: the caloric equation of state can be applied in case the specific heat capacity of the exhaust gas is known.

Conclusion

Steps 1–3 can be summarised, so that the entire combustion process is covered with the following energy balance, that is a combination of Eqs. 23.231, 23.234 and 23.237:

$$
\begin{aligned}
q = & q_1 + q_2 + q_3 = \\
& \mu_{EG} \cdot [h_{EG}(T_{EG}) - h_{EG}(T_0)] + a \cdot [h_{Ash}(T_{EG}) - h_{Ash}(T_0)] + \\
& - [h_F(T_F) - h_F(T_0)] - \lambda l_{min} \cdot [h_a(T_a) - h_a(T_0)] - H_U(T_0).
\end{aligned}
\tag{23.240}
$$

A superposition of the combustion process can be visualised according to Fig. 23.12. Thermal energy $Q = Q_1 + Q_2 + Q_3$ is transferred, that includes caloric as well as chemical effects. The overall system boundary leads to the following mass fluxes:

- <u>Inlet</u>: Fuel @ T_F and dry air @ T_a
- <u>Outlet</u>: Exhaust gas @ T_{EG}, Ashes[19] @ T_{EG}

Though the combustion air was assumed to be dry, water can occur within the exhaust gas, e.g. water that is bonded in the fuel or hydrogen that is oxidised, i.e. due to w and h. Hence, the exhaust gas shown in Fig. 23.12 might contain water, but this water is in vapour state, since this is the premise of step 2: Applying the lower heating value approach means, that the product water is in vapour state. This vapour in step 3 has just been treated in terms of sensible heat, no latent heat, related with a phase change, has been applied. However, the exhaust gas temperature can be below the dew point

---

[19]Ashes usually occur for solid fuels! The impact of the ashes on the energy balance disappear for $a = 0$.

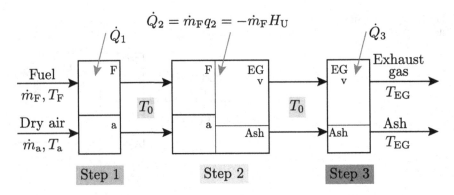

Fig. 23.12 Combustion—summary

Fig. 23.13 Condensation of water

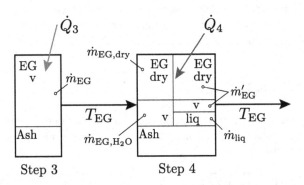

and liquid water occurs, see Sect. 23.3.6. The energy balance then could be covered by a step 4, see Fig. 23.13.

Ash as well as exhaust gas[20] leave step 3. Let us now split the exhaust gas leaving step 3 in two parts: one contains the entire water in vapour state $\dot{m}_{EG,H_2O}$, the other one all the other components except for the water

$$\dot{m}_{EG,dry} = \dot{m}_{EG} - \dot{m}_{EG,H_2O}. \tag{23.241}$$

The amount of liquid water $\dot{m}_{liq}$, when the exhaust gas temperature is below the dew point, can be determined according to Sect. 23.3.6. The rest of the water is still in vapour state, so that the remaining exhaust gas has a mass of

$$\dot{m}'_{EG} = \dot{m}_{EG} - \dot{m}_{liq}, \tag{23.242}$$

that contains the remaining vapour and dry exhaust gas. Let us now apply the first law of thermodynamics in steady state following the system boundary of Fig. 23.13, that includes the phase change due to the condensation. Water in vapour state as well

---

[20]With its water in vapour state!

as a heat flux $\dot{Q}_4$ enter the system, wheres liquid water and water in vapour state leave the system, i.e.

$$\dot{m}_{EG,H_2O}h_v(T_{EG}) + \dot{Q}_4 + \dot{m}_{EG,dry}h_{EG}(T_{EG}) + \dot{m}_{Ash}h_{Ash}(T_{EG}) =$$
$$\dot{m}_{liq}h_{liq}(T_{EG},\ p) + \left(\dot{m}_{EG,H_2O} - \dot{m}_{liq}\right)h_v(T_{EG}) + \qquad (23.243)$$
$$+ \dot{m}_{EG,dry}h_{EG}(T_{EG}) + \dot{m}_{Ash}h_{Ash}(T_{EG})\,.$$

The specific enthalpies of vapour and liquid can be taken from the steam table. Thus, the condensation heat $\dot{Q}_4$ follows[21]:

$$\dot{Q}_4 = \dot{m}_{liq}\left[h_{liq}(T_{EG},\ p) - h_v(T_{EG})\right] \approx -\dot{m}_{liq}\,\Delta h_v(T_{EG})\,. \qquad (23.244)$$

Hence, since step 4 is isothermal only the phase change of the water is energetically relevant. In specific notation the first law of thermodynamics for this step obeys

$$\boxed{q_4 = -\frac{\dot{m}_{liq}}{\dot{m}_F}\,\Delta h_v(T_{EG})} \qquad (23.245)$$

Thus, the entire heat when condensation occurs is

$$\boxed{\begin{aligned} q =& q_1 + q_2 + q_3 + q_4 = \\ & \mu_{EG} \cdot [h_{EG}(T_{EG}) - h_{EG}(T_0)] + a \cdot [h_{Ash}(T_{EG}) - h_{Ash}(T_0)] + \\ & - [h_F(T_F) - h_F(T_0)] - \lambda l_{min} \cdot [h_a(T_a) - h_a(T_0)] - H_U(T_0) + \\ & - \frac{\dot{m}_{liq}}{\dot{m}_F}\,\Delta h_v(T_{EG}) \end{aligned}} \qquad (23.246)$$

The mass of the liquid $\dot{m}_{liq}$ follows according to Sect. 23.3.6.

Caloric Equations of State

In order to solve Eq. 23.240 the enthalpies need to be calculated. On the assumption that all components are ideal, the specific enthalpies follow:

- Dry Air

$$h_a(T_a) - h_a(T_0) = \overline{c_{p,a}}\big|_{\vartheta_0}^{\vartheta_a} \cdot (\vartheta_a - \vartheta_0) = \overline{c_{p,a}}\big|_0^{\vartheta_a} \cdot \vartheta_a - \overline{c_{p,a}}\big|_0^{\vartheta_0} \cdot \vartheta_0 \qquad (23.247)$$

---

[21] The enthalpy difference between super-cooled liquid and saturated liquid is assumed to be negligible small. An evaluation regarding this aspect is done in Sect. 23.4.3.

Table 23.4 Averaged specific heat capacity $\overline{c_p}|_0^{\vartheta}$ in $\frac{kJ}{kg\,K}$, according to [4]

ϑ [°C]	Air	N_2	O_2	CO_2	H_2O	SO_2
−25	1.0034	1.0393	0.9135	0.8035	1.8567	0.6010
0	1.0037	1.0394	0.9147	0.8173	1.8589	0.6079
25	1.0042	1.0395	0.9163	0.8307	1.8615	0.6149
50	1.0048	1.0397	0.9182	0.8437	1.8646	0.6219
75	1.0055	1.0400	0.9204	0.8563	1.8682	0.6289
100	1.0064	1.0404	0.9229	0.8684	1.8724	0.6359
150	1.0087	1.0416	0.9288	0.8914	1.8820	0.6495
200	1.0116	1.0435	0.9354	0.9128	1.8931	0.6626
400	1.0285	1.0567	0.9649	0.9856	1.9466	0.7078
600	1.0498	1.0763	0.9925	1.0427	2.0083	0.7418
800	1.0712	1.0976	1.0158	1.0885	2.0744	0.7672
1000	1.0910	1.1179	1.0350	1.1257	2.1421	0.7867
2000	1.1614	1.1914	1.0991	1.2378	2.4425	0.8410

- Exhaust gas

$$h_{EG}(T_{EG}) - h_{EG}(T_0) = \overline{c_{p,EG}}|_0^{\vartheta_{EG}} \cdot \vartheta_{EG} - \overline{c_{p,EG}}|_0^{\vartheta_0} \cdot \vartheta_0 \qquad (23.248)$$

Exhaust gas shall be treated as a mixture of ideal gases, whose concentrations are well known. Chapter 19 has shown, that the specific heat capacity of such a mixture follows

$$c_{p,total} = \sum_i \left(\xi_i c_{p,i} \right). \qquad (23.249)$$

Hence, for the exhaust gas it is

$$\overline{c_{p,EG}}|_0^{\vartheta} = \sum_i \left(\xi_{EG,i} \cdot \overline{c_{p,i}}|_0^{\vartheta} \right). \qquad (23.250)$$

Thereby $\xi_{EG,i}$ is the mass fraction of component i of the exhaust gas and $\overline{c_{p,i}}|_0^{\vartheta}$ is the temperature-averaged specific heat capacity of the component.

- Fuel

$$h_F(T_F) - h_F(T_0) = \overline{c_{p,F}}|_0^{\vartheta_F} \cdot \vartheta_F - \overline{c_{p,F}}|_0^{\vartheta_0} \cdot \vartheta_0 \qquad (23.251)$$

- Ashes

$$h_{Ash}(T_{EG}) - h_{Ash}(T_0) = \overline{c_{p,Ash}}|_0^{\vartheta_{EG}} \cdot \vartheta_{EG} - \overline{c_{p,Ash}}|_0^{\vartheta_0} \cdot \vartheta_0 \qquad (23.252)$$

The specific heat capacities are usually temperature-dependent, see Sect. 12.4.4. Table 23.4 gives an overview of the most relevant components.

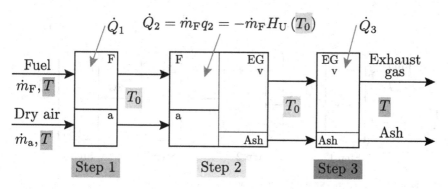

Fig. 23.14 $H_U(T_0) \rightarrow H_U(T)$

Conversion $H_U(T_0) \rightarrow H_U(T)$

The specific lower heating value is temperature dependent, so that a reference temperature T_0 is defined at which its value is given. In Table 23.2 it is a reference temperature of $\vartheta_0 = 25\,°C$, that usually is chosen. However, from time to time it is required to convert a given specific lower heating value from one reference temperature T_0 to another temperature T, see Fig. 23.14.

By definition of the lower heating value inlet and outlet temperature are equal, i.e. in this case T. Thus, the converted specific lower heating value follows according to Fig. 23.14

$$- \dot{m}_F H_U(T) \stackrel{!}{=} \dot{Q}_1 + \dot{Q}_2 + \dot{Q}_3. \tag{23.253}$$

Applying Eq. 23.240 leads to

$$\begin{aligned}
-\dot{m}_F H_U(T) = & - \dot{m}_F [h_F(T) - h_F(T_0)] - \dot{m}_a [h_a(T) - h_a(T_0)] + \\
& - \dot{m}_F H_U(T_0) + \dot{m}_F \mu_{EG} \cdot [h_{EG}(T) - h_{EG}(T_0)] + \\
& + \dot{m}_F a \cdot [h_{Ash}(T) - h_{Ash}(T_0)].
\end{aligned} \tag{23.254}$$

Solving for $H_U(T)$ brings the specific lower heating value at the new temperature T, i.e.

$$\boxed{\begin{aligned}
H_U(T) = & H_U(T_0) + [h_F(T) - h_F(T_0)] + \lambda l_{min} [h_a(T) - h_a(T_0)] + \\
& - \mu_{EG} [h_{EG}(T) - h_{EG}(T_0)] - a [h_{Ash}(T) - h_{Ash}(T_0)]
\end{aligned}} \tag{23.255}$$

Combustion with Humid Air

In this section it is distinguished between unsaturated humid combustion air and combustion air that includes water in non-vapour state.

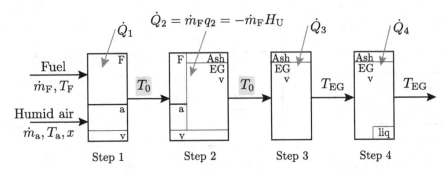

Fig. 23.15 Combustion with unsaturated humid air—summary

- Humid, unsaturated air:

This case is shown in Fig. 23.15 under the premise, that the humid air before and after begin pre-conditioned in step 1 is unsaturated, i.e. the entire water shall be in vapour state for the entire combustion process. The first law of thermodynamics, including steps 1 to 3, then reads based on Eq. 23.240

$$\dot{Q} = \dot{Q}_1 + \dot{Q}_2 + \dot{Q}_3 =$$
$$\dot{m}_{EG} \cdot [h_{EG}(T_{EG}) - h_{EG}(T_0)] + \dot{m}_{Ash} \cdot [h_{Ash}(T_{EG}) - h_{Ash}(T_0)] +$$
$$- \dot{m}_F [h_F(T_F) - h_F(T_0)] - \dot{m}_a \cdot [h_{1+x}(T_a, x) - h_{1+x}(T_0, x)] +$$
$$- \dot{m}_F H_U(T_0).$$

(23.256)

Mind, that the specific enthalpy of the humid air h_{1+x} follows Chap. 20, i.e. it is referred to the mass of the dry air $\dot{m}_a$ and not to the mass of the humid air $\dot{m}_A = (1 + x) \dot{m}_a$. In specific notation with reference to the mass flux of the fuel the first law of thermodynamics then obeys

$$q = q_1 + q_2 + q_3 =$$
$$\mu_{EG} \cdot [h_{EG}(T_{EG}) - h_{EG}(T_0)] + a \cdot [h_{Ash}(T_{EG}) - h_{Ash}(T_0)] +$$
$$- [h_F(T_F) - h_F(T_0)] - \lambda l_{min} \cdot [h_{1+x}(T_a, x) - h_{1+x}(T_0, x)] +$$
$$- H_U(T_0).$$

(23.257)

In case after step 3 the temperature is below the dew point temperature another subsequent step 4 according to Fig. 23.13 must be added to take the condensation into the calculation, i.e.

$$\dot{Q}_4 = -\dot{m}_{liq} \Delta h_v (T_{EG}) < 0.$$

(23.258)

Thus, the entire energy balance follows

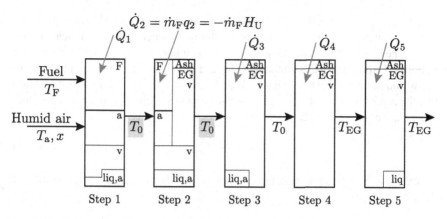

Fig. 23.16 Combustion with saturated humid air—summary

$$\dot{Q} = \dot{Q}_1 + \dot{Q}_2 + \dot{Q}_3 + \dot{Q}_4 =$$
$$\dot{m}_{EG} \cdot [h_{EG}(T_{EG}) - h_{EG}(T_0)] + \dot{m}_{Ash} \cdot [h_{Ash}(T_{EG}) - h_{Ash}(T_0)] +$$
$$- \dot{m}_F [h_F(T_F) - h_F(T_0)] - \dot{m}_a \cdot [h_{1+x}(T_a, x) - h_{1+x}(T_0, x)] +$$
$$- \dot{m}_F H_U(T_0) - \dot{m}_{liq} \, \Delta h_v (T_{EG}) .$$

(23.259)

In specific notation with reference to the mass flux of the fuel the first law of thermodynamics then obeys

$$q = q_1 + q_2 + q_3 + q_4 =$$
$$\mu_{EG} \cdot [h_{EG}(T_{EG}) - h_{EG}(T_0)] + a \cdot [h_{Ash}(T_{EG}) - h_{Ash}(T_0)] +$$
$$- [h_F(T_F) - h_F(T_0)] - \lambda l_{min} \cdot [h_{1+x}(T_a, x) - h_{1+x}(T_0, x)] +$$
$$- H_U(T_0) - \frac{\dot{m}_{liq}}{\dot{m}_F} \, \Delta h_v (T_{EG}) .$$

(23.260)

Mind, that the enthalpy difference $[h_{EG}(T_{EG}) - h_{EG}(T_0)]$ is related to step 3, so that the exhaust gas composition of step 3 is relevant.

- Humid, saturated air with liquid water:

This case is shown in Fig. 23.16. Fuel as well as humid air enter step 1. Humid air contains dry air (a), vapour (v) as well as liquid water (liq). During the pre-conditioning in step 1, the entire mass of the water remains constant, but mass of the vapour can decrease whereas the mass of the liquid water increases.[22] As usual the chemical reaction is performed in step 2: The lower heating value approach just includes the heat that is realised during oxidisation with the product water

---

[22]E.g. lowering the temperature from T_a to T_0 decreases the capability of air to carry vapour, see Chap. 20. In case $T_a < T_0$ the amount of liquid water sinks and the amount of vapour rises.

being in vapour state, i.e. it is based on the reaction of fuel (F) and the dry air (a). The available vapour (v) is energetically inert, since it enters and leaves with the same temperature. The liquid water (liq) is ignored by the lower heating value approach, so that it enters and leaves step 2. Excess air, i.e. $\lambda > 1$, and vapour are now counted to the exhaust gas (EG) and enter step 3. In step 3 the vaporisation of the liquid water, that has been brought into the system with the air, is realised: Hence, the mass of the liquid that is vaporised in step 3 is

$$\dot{m}_{\text{liq,a}} = \dot{m}_{\text{a}} x_{\text{liq}} \tag{23.261}$$

with x_{liq} being the liquid water content of the air after step 1. The vaporisation of that liquid comes along with a supply of heat, i.e. $\dot{Q}_3 > 0$,

$$\dot{Q}_3 = \dot{m}_{\text{liq,a}} \left[h_{\text{v}}(T_0) - h_{\text{liq}}(T_0, p) \right] \approx \dot{m}_{\text{liq,a}} \Delta h_{\text{v}}(T_0). \tag{23.262}$$

In step 4 the post-conditioning is done: Ashes (Ash) and exhaust gas are post-conditioned to the exhaust gas temperature as in the previous examples. Hence, the exemplary first law of thermodynamics then obeys

$$\dot{Q} = \dot{Q}_1 + \dot{Q}_2 + \dot{Q}_3 + \dot{Q}_4 =$$
$$\dot{m}_{\text{EG}} \cdot [h_{\text{EG}}(T_{\text{EG}}) - h_{\text{EG}}(T_0)] + \dot{m}_{\text{Ash}} \cdot [h_{\text{Ash}}(T_{\text{EG}}) - h_{\text{Ash}}(T_0)] +$$
$$- \dot{m}_{\text{F}} [h_{\text{F}}(T_{\text{F}}) - h_{\text{F}}(T_0)] - \dot{m}_{\text{a}} \left[h_{1+x}(T_{\text{a}}, x) - h_{1+x}(T_0, x) \right] +$$
$$- \dot{m}_{\text{F}} H_{\text{U}}(T_0) + \dot{m}_{\text{a}} x_{\text{liq}}(T_0) \Delta h_{\text{v}}(T_0). \tag{23.263}$$

In specific notation with reference to the mass flux of the fuel the first law of thermodynamics then is

$$\boxed{\begin{aligned} q &= q_1 + q_2 + q_3 + q_4 = \\ &\mu_{\text{EG}} \cdot [h_{\text{EG}}(T_{\text{EG}}) - h_{\text{EG}}(T_0)] + a \cdot [h_{\text{Ash}}(T_{\text{EG}}) - h_{\text{Ash}}(T_0)] + \\ &- [h_{\text{F}}(T_{\text{F}}) - h_{\text{F}}(T_0)] - \lambda l_{\min} \left[h_{1+x}(T_{\text{a}}, x) - h_{1+x}(T_0, x) \right] + \\ &- H_{\text{U}}(T_0) + \lambda l_{\min} x_{\text{liq}}(T_0) \Delta h_{\text{v}}(T_0). \end{aligned}} \tag{23.264}$$

In case after step 4 the temperature is below the dew point temperature another subsequent step 5 according to Fig. 23.13 must be added to take the condensation into the calculation, i.e.

$$\dot{Q}_5 = \dot{m}_{\text{liq}} \left[h_{\text{liq}}(T_{\text{EG}}, p) - h_{\text{v}}(T_{\text{EG}}) \right] \approx -\dot{m}_{\text{liq}} \Delta h_{\text{v}}(T_{\text{EG}}). \tag{23.265}$$

Hence, the entire energy balance in specific notation is

$$
\begin{aligned}
q = & q_1 + q_2 + q_3 + q_4 + q_5 = \\
& \mu_{EG} \cdot [h_{EG}(T_{EG}) - h_{EG}(T_0)] + a \cdot [h_{Ash}(T_{EG}) - h_{Ash}(T_0)] + \\
& - [h_F(T_F) - h_F(T_0)] - \lambda l_{min} [h_{1+x}(T_a, x) - h_{1+x}(T_0, x)] + \\
& - H_U(T_0) + \lambda l_{min} x_{liq}(T_0) \Delta h_v(T_0) - \frac{\dot{m}_{liq}}{\dot{m}_F} \Delta h_v(T_{EG})
\end{aligned}
\tag{23.266}
$$

Mind, that the enthalpy difference $[h_{EG}(T_{EG}) - h_{EG}(T_0)]$ is related to step 4, so that the exhaust gas composition of step 4 is relevant. The mass of the condensing water follows, see Sect. 23.3.6,

$$
\frac{\dot{m}_{liq}}{\dot{m}_F} = \mu^*_{EG,H_2O} = \frac{\mu_{EG,H_2O} - x_s \mu_{EG} \frac{M_{H_2O}}{M_{EG}}}{1 - x_s}
\tag{23.267}
$$

with x_s being of function of total pressure p and the saturated pressure according to the exhaust gas temperature, see also Eq. 23.144, i.e.

$$
x_s = \frac{p_s(T_{EG})}{p}.
\tag{23.268}
$$

23.4.3 Upper Heating Value

The lower heating value is the specific heat release of an oxidising fuel with constant educt and product temperature T_0 under the premise, that the possible product water[23] is in gaseous state, i.e. no water condenses, see Sect. 23.4.1. As the previous sections have shown, the released thermal energy of a combustion increases, when gaseous water condenses and turns into liquid state. Based on this phase change energy the specific upper heating value is defined, see Fig. 23.17. In contrast to the specific lower heating value, the specific upper heating value $H_0(T_0)$ represents the specific amount of heat, that is isothermally released at T_0, if the produced water is *entirely* in liquid state. Thus, after step 1, at which the chemical reaction is done at T_0 and the product water is gaseous, i.e. applying the lower heating value approach, a step 2 follows, that covers the condensation. The released heat in step 1 is already known and follows

$$
\dot{Q}_1 = -\dot{m}_F H_U(T_0) < 0.
\tag{23.269}
$$

In step 2 the exhaust gas is split into dry exhaust gas $\dot{m}_{EG,dry}$, that does not contain any water, and the remaining water. Water in vapour state $\dot{m}_{EG,H_2O}$ enters step 2 and liquid water $\dot{m}_{liq}$ occurs at the outlet. The mass balance for the water is

---

[23] According to Fig. 23.17 the water results from the fuel: it can contain water or water can be a result of the oxidisation of hydrogen.

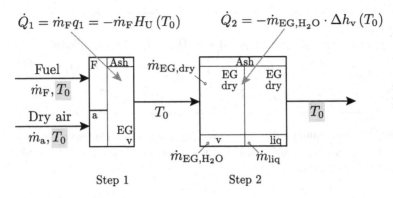

$$\dot{Q}_1 = \dot{m}_F q_1 = -\dot{m}_F H_U(T_0) \qquad\qquad \dot{Q}_2 = -\dot{m}_{EG,H_2O} \cdot \Delta h_v(T_0)$$

Step 1   Step 2

Fig. 23.17 Definition of the upper heating value H_0

$$\dot{m}_{EG,H_2O} = \dot{m}_{liq}. \tag{23.270}$$

The first law of thermodynamics in steady state for step 2 obeys

$$\dot{Q}_2 + \dot{m}_{EG,H_2O} h_v(T_0) + \dot{m}_{EG,dry} h_{EG,dry}(T_0) + \dot{m}_{Ash} h_{Ash}(T_0) =$$
$$\dot{m}_{EG,dry} h_{EG,dry}(T_0) + \dot{m}_{Ash} h_{Ash}(T_0) + \dot{m}_{liq} h_{liq}(T_0, p). \tag{23.271}$$

Solving for $\dot{Q}_2$ results in

$$\dot{Q}_2 = \dot{m}_{EG,H_2O} \left[h_{liq}(T_0, p) - h_v(T_0) \right]. \tag{23.272}$$

The specific enthalpies of vapour and liquid can be taken from the steam table. With a reference temperature for the heating value of $\vartheta_0 = 25\,°C$ for instance and a pressure[24] of $p = 1$ bar it is

- Vapour (treated as an ideal gas)

$$h_v(\vartheta_0) = h''(\vartheta_0) = 2546.5\,\frac{kJ}{kg} \tag{23.273}$$

Or alternatively following Chap. 20

$$h_v(\vartheta_0) = \Delta h_v(0\,°C) + c_{p,v} \cdot \vartheta_0 = 2546.5\,\frac{kJ}{kg} \tag{23.274}$$

- Super-cooled liquid

$$h_{liq}(\vartheta_0, p) = 104.9281\,\frac{kJ}{kg} \tag{23.275}$$

---

[24]Mind, that the pressure of the liquid is equal to the pressure of the gaseous atmosphere above the water!

Table 23.5 Upper heating values H_0 at $\vartheta_0 = 25\,°C$, according to [7]

Solid fuels		Liquid fuels	
Fuel	H_0 in $\frac{MJ}{kg}$	Fuel	H_0 in $\frac{MJ}{kg}$
Wood, dry	15.91...18.0	Ethanol	29.73
Turf, dry	13.83...16.33	Benzol	41.87
Raw lignite	10.47...12.98	Toluol	42.75
Brown coal briquettes	20.93...21.35	Naphtalin	40.36
Hard coal	29.31...35.17	Pentane	49.19
Anthracite	33.49...34.75	Octane	48.15
Coke	28.05...30.56	Benzine	46.05

Thus, it is

$$\left[h_v(T_0) - h_{liq}(T_0, p)\right] = 2441.6\,\frac{kJ}{kg} \approx \Delta h_v(\vartheta_0) = 2441.7\,\frac{kJ}{kg}. \tag{23.276}$$

With this simplification[25] the heat release in step 2 is

$$\dot{Q}_2 = \dot{m}_{EG,H_2O}\left[h_{liq}(T_0, p) - h_v(T_0)\right] \approx -\dot{m}_{EG,H_2O}\Delta h_v(\vartheta_0) < 0. \tag{23.277}$$

By definition the upper heating value is

$$\dot{Q}_1 + \dot{Q}_2 \overset{!}{=} -\dot{m}_F H_0(T_0). \tag{23.278}$$

Solving for $H_0(T_0)$ leads to

$$H_0(T_0) = H_U(T_0) + \frac{\dot{m}_{EG,H_2O}}{\dot{m}_F}\Delta h_v(T_0). \tag{23.279}$$

With $\mu_{EG,H_2O} = \frac{\dot{m}_{EG,H_2O}}{\dot{m}_F}$ one finally gets

$$\boxed{H_0(T_0) = H_U(T_0) + \mu_{EG,H_2O} \cdot \Delta h_v(T_0)} \tag{23.280}$$

Mind, that the product water is due to w and h, i.e. it purely comes from the fuel and not from humidity of the combustion air, see [8]. In case humid air would be utilised, the amount of water in the exhaust gas would rise and its condensation leads to an increase of the specific upper heating value. This is why Fig. 23.17 emphasises, that dry air $\dot{m}_a$ is used. Specific upper heating values for technically relevant fuels are shown in Tables 23.5 and 23.6.

---

[25]The enthalpy difference between saturated liquid and super-cooled liquid is assumed to be negligible small!

Gaseous fuels	
Fuel	H_0 in $\frac{MJ}{kg}$
Hydrogen	141.80
Carbon monoxide	10.10
Methane	55.50
Ethane	51.88
Propane	50.35
Ethylene	50.28
Acetylene	49.91

Table 23.6 Upper heating values H_0 (Gases) at $\vartheta_0 = 25\,°C$, according to [7]

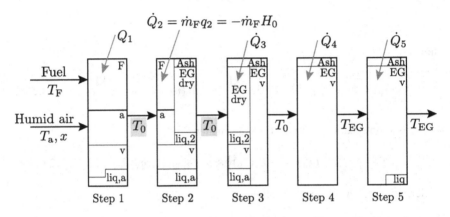

Fig. 23.18 Combustion with saturated humid air—applying the specific upper heating value

Example

The most generic case of a combustion with saturated humid air has been intro-duced based on the specific lower heating value, see Fig. 23.16 and the correspond-ing Eq. 23.266. However, it can also be handled with the specific *upper* heating value, see Fig. 23.18. When doing so, step 2 differs from the approach with the lower heating value: The combustion based on the specific upper heating value requires a combustion with dry air (a) and the product water is entirely in liquid state, i.e.

$$\dot{m}_{liq,2} = \dot{m}_{EG,H_2O}.\qquad (23.281)$$

Energetically, the released heat in step 2 under these conditions then is

$$\dot{Q}_2 = -\dot{m}_F H_0\,(T_0).\qquad (23.282)$$

In step 3 the entire liquid water, i.e. (liq, 2) and (liq, a), is vaporised by heat supply $\dot{Q}_3$, i.e.

$$\dot{Q}_3 = \left(\dot{m}_{\text{liq},2} + \dot{m}_{\text{liq,a}}\right) \Delta h_v \left(T_0\right) = \left(\dot{m}_{\text{EG,H}_2\text{O}} + \dot{m}_{\text{liq,a}}\right) \Delta h_v \left(T_0\right) \qquad (23.283)$$

with, see Eq. 23.261,

$$\dot{m}_{\text{liq,a}} = \dot{m}_a x_{\text{liq}} \left(T_0\right) . \qquad (23.284)$$

The following steps are the same as before shown in Fig. 23.16. In step 4 the exhaust gas is lifted to the exhaust gas temperature T_{EG} with the water being in gaseous state. In case the exhaust gas temperature is below the dew point step 5 is needed, in order to describe the condensation. Hence, the overall energy balance is a result of steps 1–5 and follows

$$
\begin{aligned}
q = & q_1 + q_2 + q_3 + q_4 + q_5 = \\
& \mu_{\text{EG}} \cdot [h_{\text{EG}}(T_{\text{EG}}) - h_{\text{EG}}(T_0)] + a \cdot [h_{\text{Ash}} (T_{\text{EG}}) - h_{\text{Ash}} (T_0)] + \\
& - [h_{\text{F}}(T_{\text{F}}) - h_{\text{F}}(T_0)] - \lambda l_{\min} \left[h_{1+x}(T_a, x) - h_{1+x}(T_0, x)\right] + \\
& \underbrace{-H_0(T_0) + \mu_{\text{EG,H}_2\text{O}} \Delta h_v \left(T_0\right)}_{-H_{\text{U}}(T_0)} + \lambda l_{\min} x_{\text{liq}}(T_0) \Delta h_v \left(T_0\right) + \qquad (23.285) \\
& - \frac{\dot{m}_{\text{liq}}}{\dot{m}_{\text{F}}} \Delta h_v \left(T_{\text{EG}}\right) .
\end{aligned}
$$

Hence, Eqs. 23.266 and 23.285 lead to the same result.

23.4.4 Molar and Volume Specific Lower/Upper Heating Value

So far the heating values $H_{\text{U}} \left(T_0\right)$ respectively $H_0 \left(T_0\right)$ have been given with reference to the fuel's mass, i.e. in $[\frac{\text{kJ}}{\text{kg}}]$. Additionally, it makes sense, to give the specific heating values with reference to the molar quantity, i.e. in $\frac{\text{kJ}}{\text{kmol}}$, respectively with reference to the volume, i.e. in $[\frac{\text{kJ}}{\text{m}^3}]$:

- Molar specific lower/upper heating value in $[\frac{\text{kJ}}{\text{kmol}}]$

The overall heating value is

$$m_{\text{F}} \cdot H_{\text{U}} \left(T_0\right) = n_{\text{F}} H_{\text{UM}} \left(T_0\right) , \qquad (23.286)$$

so that its conversion from the mass specific lower heating value $H_{\text{U}} \left(T_0\right)$ to the molar specific lower heating value $H_{\text{UM}} \left(T_0\right)$ follows

$$\boxed{H_{\text{UM}} \left(T_0\right) = M_{\text{F}} \cdot H_{\text{U}} \left(T_0\right)} \qquad (23.287)$$

The molar specific upper heating value $H_{0\text{M}} \left(T_0\right)$ follows accordingly

$$H_{0M}(T_0) = M_F \cdot H_0(T_0)$$ (23.288)

In ideal fuel mixtures the molar lower/upper heating values apply according to the rule of mixtures, i.e.

$$H_{UM}(T_0) = \sum_i \frac{n_i}{n_F} \cdot H_{UM,i}(T_0)$$ (23.289)

respectively

$$H_{0M}(T_0) = \sum_i \frac{n_i}{n_F} \cdot H_{0M,i}(T_0)$$ (23.290)

- Lower/upper heating value with reference to the volume in $[\frac{kJ}{m^3}]$

  The overall heating value is

$$n_F \cdot H_{UM}(T_0) = V_F H_{Uv}(T_0),$$ (23.291)

so that its conversion from the molar specific lower heating value $H_{UM}(T_0)$ to the volume specific lower heating value $H_{Uv}(T_0)$ follows for standard conditions

$$H_{Uv}(T_0) = \frac{H_{UM}(T_0)}{v_M}$$ (23.292)

respectively

$$H_{0v}(T_0) = \frac{H_{0M}(T_0)}{v_M}$$ (23.293)

Mind, that the standard volume[26] is $v_M = 22.41 \frac{m^3}{kmol}$.

23.4.5   Combustion Temperature

(I) Combustion Temperature Above Dew Point ϑ_τ

In Sect. 23.4.2 the following energy balance for a combustion with unsaturated humid air has been deduced, see Eq. 23.257. In case the combustion temperature T_{EG} is above the dew point,[27] i.e. no condensation occurs, it is

---

[26] At standard conditions!

[27] No step 4 is required!

$$q = \mu_{EG} \cdot [h_{EG}(T_{EG}) - h_{EG}(T_0)] + a \cdot [h_{Ash}(T_{EG}) - h_{Ash}(T_0)] +$$
$$- [h_F(T_F) - h_F(T_0)] - \lambda l_{min} \cdot \left[h_{1+x}(T_a, x) - h_{1+x}(T_0, x)\right] - H_U(T_0).$$
$$(23.294)$$

With the following abbreviations

$$h_{in}(\vartheta_a, \vartheta_F, \lambda) = [h_F(T_F) - h_F(T_0)] +$$
$$+ \lambda l_{min} \cdot \left[h_{1+x}(T_a, x) - h_{1+x}(T_0, x)\right] + H_U(T_0)$$
$$(23.295)$$

and

$$h_{out}(\vartheta_{EG}, \lambda) = \mu_{EG} \cdot [h_{EG}(T_{EG}) - h_{EG}(T_0)] + a \cdot [h_{Ash}(T_{EG}) - h_{Ash}(T_0)]$$
$$(23.296)$$

the specific released heat $-q$ is:

$$\boxed{-q = h_{in}(\vartheta_a, \vartheta_F, \lambda) - h_{out}(\vartheta_{EG}, \lambda)} \qquad (23.297)$$

(II) Combustion Temperature Below Dew Point ϑ_τ

In Sect. 23.4.2 the following energy balance for a combustion with unsaturated humid air has been deduced, see Eq. 23.260, under the premise, that the combustion temperature T_{EG} is below the dew point, i.e. vapour partially condenses[28]:

$$q = \mu_{EG} \cdot [h_{EG}(T_{EG}) - h_{EG}(T_0)] + a \cdot [h_{Ash}(T_{EG}) - h_{Ash}(T_0)] +$$
$$- [h_F(T_F) - h_F(T_0)] - \lambda l_{min} \cdot \left[h_{1+x}(T_a, x) - h_{1+x}(T_0, x)\right] - H_U(T_0) +$$
$$- \frac{\dot{m}_{liq}}{\dot{m}_F} \Delta h_v(T_{EG}).$$
$$(23.298)$$

With the following abbreviations

$$h_{in}(\vartheta_a, \vartheta_F, \lambda) = [h_F(T_F) - h_F(T_0)] + \lambda l_{min} \cdot \left[h_{1+x}(T_a, x) - h_{1+x}(T_0, x)\right] +$$
$$+ H_U(T_0)$$
$$(23.299)$$

and

$$h_{out}(\vartheta_{EG}, \lambda) = \mu_{EG} \cdot [h_{EG}(T_{EG}) - h_{EG}(T_0)] +$$
$$+ a \cdot [h_{Ash}(T_{EG}) - h_{Ash}(T_0)] - \frac{\dot{m}_{liq}}{\dot{m}_F} \Delta h_v(T_{EG})$$
$$(23.300)$$

---

[28] This case requires a step 4!

the specific released heat $-q$ is:

$$\boxed{-q = h_{\text{in}}\left(\vartheta_a, \vartheta_F, \lambda\right) - h_{\text{out}}\left(\vartheta_{EG}, \lambda\right)} \tag{23.301}$$

The mass of the condensing water $\dot{m}_{\text{liq}}$ has been investigated in Sect. 23.3.6. It is

$$\frac{\dot{m}_{\text{liq}}}{\dot{m}_F} = \mu^*_{\text{EG,H}_2\text{O}} = \frac{\mu_{\text{EG,H}_2\text{O}} - x_s \mu_{\text{EG}} \frac{M_{\text{H}_2\text{O}}}{M_{\text{EG}}}}{1 - x_s}. \tag{23.302}$$

with x_s being of function of total pressure p and the saturated pressure according to the exhaust gas temperature, see also Eq. 23.144,

$$x_s = \frac{p_s(T_{EG})}{p}. \tag{23.303}$$

The smaller the exhaust gas temperature is, the more water condenses.

The h, ϑ-Diagram

The equation, that has just been derived in the previous sections[29]

$$h_{\text{in}}(\vartheta_a, \vartheta_F, \lambda) - h_{\text{out}}(\vartheta_{EG}, \lambda) = -q > 0 \tag{23.304}$$

can be visualised in a so-called h, ϑ-diagram, that is exemplary shown in Fig. 23.19. Enthalpy differences of the fuel usually can be ignored compared to the lower heating value, so that h_{in} simplifies to

$$h_{\text{in}}(\vartheta_a, \vartheta_F, \lambda) \approx h_{\text{in}}(\vartheta_a, \lambda) = \lambda l_{\text{min}} \cdot \left[h_{1+x}(\vartheta_a) - h_{1+x}(\vartheta_0)\right] + H_U(\vartheta_0). \tag{23.305}$$

The functions of h_{in} and h_{out} can be calculated and plotted as a function of temperature with fuel-air equivalence ratio and humidity being parameters. A combustion, shown in a h, ϑ-diagram 23.19, exemplary starts at (Start), i.e. with air being supplied at

$$\vartheta_a = \vartheta_F = \vartheta_0. \tag{23.306}$$

Thus, the enthalpy into the system h_{in} is fixed. In case (1) the combustion releases a specific heat of $q < 0$ and the exhaust gas temperature is $\vartheta_{EG,1}$. The remaining enthalpy is released with the exhaust gas and the ashes. In case there are no ashes it reads as

$$h_{\text{out}}(\vartheta_{EG}, \lambda) = \mu_{EG} \cdot \left[h_{EG}(\vartheta_{EG}) - h_{EG}(\vartheta_0)\right] = \Delta h_{EG}. \tag{23.307}$$

---

[29]This equations counts for combustion temperatures below and above the dew point!

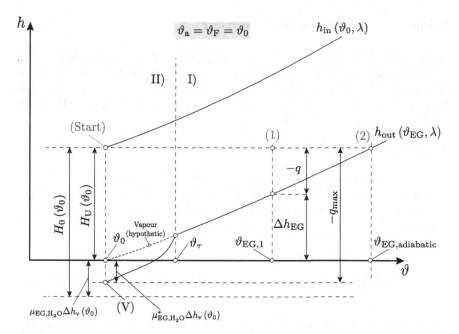

Fig. 23.19 h, ϑ-diagram

Obviously, the exhaust gas temperature is above the dew point, so that no water condenses. The larger the exhaust gas temperature gets, the less heat is released during the combustion. Finally, in case (2) the combustion does not release any heat and the exhaust gas temperature is maximised. This temperature is called adiabatic flame temperature $\vartheta_{EG, \text{adiabat}}$ and the entire energy is released as exhaust gas/ash enthalpy. At the dew point the h_{out}-curve bends. A comparison of Eqs. 23.296 and 23.300 proves, that this is due to the condensation of water, i.e. $\frac{\dot{m}_{liq}}{\dot{m}_F} \Delta h_v (\vartheta_{EG})$, that occurs when the temperature falls below the dew point.

According to Fig. 23.19, in case the combustion runs isothermally at reference temperature ϑ_0, with the *entire* water being hypothetic in vapour state, the released specific heat is $H_U(\vartheta_0)$, i.e. it fulfils the definition of the lower heating value, see Sect. 23.4.1. However, if the *entire* product water under these conditions condenses and turns into liquid, the released heat would be enlarged by the specific enthalpy of vaporisation $\mu_{EG,H_2O} \Delta h_V(\vartheta_0)$, see sketch Fig. 23.19. By definition, see Sect. 23.4.3, this released heat is the specific upper heating value. Realistically, according to the p, T-diagram, see Fig. 18.8, it is not possible to condense the entire water, but, depending on the exhaust gas temperature, water still is in vapour state. Thus, only the scaled specific enthalpy of vaporisation $\mu^*_{EG,H_2O} \Delta h_V(\vartheta_0)$ is released. This is represented by point (V) in Fig. 23.19. Following this consideration, the maximum specific heat, that can be released, is smaller than the specific upper heating value und marked in the h, ϑ-diagram as q_{max}.

Calculation of the Exhaust Gas Temperature

Now, that we know how to determine the dependency between released heat and exhaust gas temperature graphically, the exhaust gas temperature is now calculated. As the thermal properties of the exhaust gas are temperature dependent numerical methods are required to solve the correlation between released heat and exhaust gas temperature. This is shown for two cases:

- Adiabatic flame temperature (dry air)

  In case the combustion air is dry, see sketch Fig. 23.20, the first law of thermodynamics follows Eq. 23.240 under the premise that the adiabatic flame temperature is above dew point. Since the combustion is supposed to be adiabatic, the specific heat is $q = 0$. Hence, the first law of thermodynamics obeys

$$0 = - [h_F(T_F) - h_F(T_0)] - \lambda l_{min} \cdot [h_a(T_a) - h_a(T_0)] - H_U(T_0) +$$
$$+ \sum_i \mu_{EG,i} \cdot [h_{EG}(T_{EG}) - h_{EG}(T_0)] + a \cdot [h_{Ash}(T_{ad}) - h_{Ash}(T_0)]^{\cdot}$$

$$(23.308)$$

Applying caloric equations of state for the specific enthalpies requires averaged specific heat capacities, see Eqs. 23.247–23.252. Thus, the first law of thermodynamics can be solved for ϑ_{ad}, i.e.

$$\boxed{\vartheta_{ad} = \vartheta_0 + \frac{H_U(T_0) + \overline{c_{p,F}}|_{\vartheta_0}^{\vartheta_F} \cdot (\vartheta_F - \vartheta_0) + \lambda \cdot l_{min} \cdot \overline{c_{p,a}}|_{\vartheta_0}^{\vartheta_a} \cdot (\vartheta_a - \vartheta_0)}{\mu_{EG} \cdot \overline{c_{p,EG}}|_{\vartheta_0}^{\vartheta_{ad}} + a \cdot \overline{c_{p,Ash}}|_{\vartheta_0}^{\vartheta_{ad}}}}$$

$$(23.309)$$

Unfortunately, this is an implicit equation, since the specific heat capacities of the exhaust gas and the ashes also depend on the adiabatic flame temperature. Hence, an iterative, numerical solving strategy needs to be followed:

(1) Start with an initial temperature $\vartheta_{ad,0}$
(2) Calculate specific heat capacities for exhaust gas and ashes at this temperature
(3) Calculate $\vartheta_{ad,n+1}$ following Eq. 23.309
(4) Repeat steps 2 and 3 until the deviation $(\vartheta_{ad,n+1} - \vartheta_{ad,n})$ his sufficiently small

Alternatively, the adiabatic flame temperature can be estimated graphically in a h, ϑ-diagram, see Fig. 23.19.

- Exhaust gas temperature—generic approach

  A generic combustion process follows Fig. 23.16. Humid air is utilised that additionally carries liquid water into the combustion chamber. The first law of thermodynamic has been derived with Eq. 23.266 and thus obeys

Fig. 23.20 Adiabatic flame temperature (dry air)

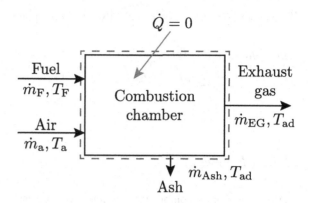

$$
\begin{aligned}
q =\ & \mu_{EG} \cdot [h_{EG}(T_{EG}) - h_{EG}(T_0)] + a \cdot [h_{Ash}(T_{EG}) - h_{Ash}(T_0)] + \\
& - [h_F(T_F) - h_F(T_0)] - \lambda l_{min} \left[h_{1+x}(T_a, x) - h_{1+x}(T_0, x) \right] + \\
& - H_U(T_0) + \lambda l_{min} x_{liq}(T_0) \Delta h_v(T_0) - \underbrace{\frac{\dot{m}_{liq}}{\dot{m}_F} \Delta h_v(T_{EG})}_{\text{for } T_{EG} < T_\tau} .
\end{aligned}
\tag{23.310}
$$

Mind, that the heat of vaporisation with the last term only occurs, if the combustion temperature falls below the dew point, that is corresponding with the exhaust gas composition, see Sect. 23.3.6. The function $q = f(T_{EG})$ can be calculated straight forward, taking the temperature dependencies of the thermal properties into account and distinguish if water condenses finally. However, the inverse function

$$
T_{EG} = f(q)
\tag{23.311}
$$

is rather complex, since equation can not be explicitly solved for T_{EG}, since

– the specific enthalpies of exhaust gas,
– the specific enthalpies of ashes and, in case water condenses in step 5,
– the specific heat of vaporisation

depend on the exhaust gas temperature. Hence, numerical methods have to be applied, to solve Eq. 23.311.

23.5 Combustion Chamber

A combustion chamber is shown exemplary in Fig. 23.21. Fuel as well as air are supplied, exhaust gas and ashes[30] are released. The utilised specific thermal power shall be $q_E = \frac{\dot{Q}_E}{\dot{m}_F}$. However, due to insulation losses the specific heat release $q_{iso} =$

---

[30]In case of solid fuels.

$\frac{\dot{Q}_{iso}}{\dot{m}_F}$ can not be further used technically. The corresponding energy balance[31] obeys

$$q_E + q_{iso} = -\left[h_F(T_F) - h_F(T_0)\right] - \lambda l_{min} \cdot \left[h_{1+x}(T_a) - h_{1+x}(T_0)\right] - H_U(T_0) +$$
$$+ \sum_i \mu_{EG,i} \cdot \left[h_{EG}(T_{EG}) - h_{EG}(T_0)\right] + a \cdot \left[h_{Ash}(T_{EG}) - h_{Ash}(T_0)\right].$$

$$(23.312)$$

23.5.1 Efficiency

The efficiency of the combustion is defined with the usable thermal energy according to

$$\eta_K = \frac{|q_E|}{H_U}. \tag{23.313}$$

However, the overall combustion efficiency follows

$$\eta_F = \frac{|q_E + q_{iso}|}{H_U}. \tag{23.314}$$

The specific enthalpy, that is released with the exhaust gas, results in

$$\Delta h_{EG} = \sum_i \mu_{EG,i} \cdot \left[h_{EG}(EG) - h_{EG}(T_0)\right] \tag{23.315}$$

with

$$\sum_i \mu_{EG,i} = \mu_{EG} = (1 - a) + \lambda \cdot l_{min} \cdot (1 + x). \tag{23.316}$$

In terms of maximising the thermal output of the combustion chamber, the exhaust gas temperatures should be small.

23.5.2 Operation

In this section the influence of air preheating as well as of the fuel-air equivalence ratio λ on the combustion temperature is discussed. Figure 23.22, based on the previously introduced h, ϑ-diagram, indicates, that in case the air is preheated, the supplied specific enthalpy h_{in} rises, i.e.

---

[31]In case the combustion air is humid but unsaturated before and after step 1 and in case the combustion temperature is above the dew point!

Fig. 23.21 Combustion chamber

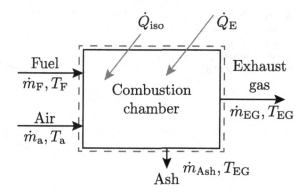

Fig. 23.22 Combustion chamber—variation of λ

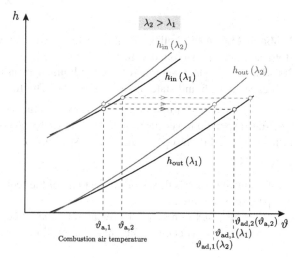

$$h_{in}(\vartheta_a, \vartheta_F, \lambda) = [h_F(T_F) - h_F(T_0)] + \\ + \lambda l_{min} \cdot [h_{1+x}(T_a, x) - h_{1+x}(T_0, x)] + H_U(T_0). \tag{23.317}$$

Consequently, the adiabatic flame temperature rises. Figure 23.22 additionally shows, when the fuel-air equivalence ratio λ is increased, the adiabatic flame temperature sinks. This is caused by the increasing mass of the reactants. The chemical bonded energy of the fuel is required to heat up a larger mass. This leads to a reduced temperature increase. The suppled specific enthalpy h_{in}, according to Eq. 23.317, as well as the released specific enthalpy h_{out}

$$h_{out}(\vartheta_{EG}, \lambda) = \mu_{EG} \cdot [h_{EG}(T_{EG}) - h_{EG}(T_0)] + a \cdot [h_{Ash}(T_{EG}) - h_{Ash}(T_0)] \tag{23.318}$$

run with a larger gradient with rising λ. Figure 23.23 shows the influence of both parameters exemplary for the combustion of oil: Preheating the combustion air increases the exhaust gas temperature and increasing the amount of air reduces it.

Fig. 23.23 Influence of fuel-air equivalence ratio and air preheating on the adiabatic flame temperature for oil according to [6]

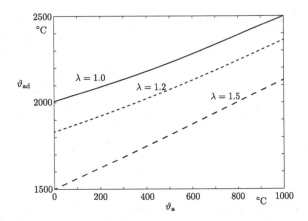

Problem 23.6 A fuel cell is operated stoichiometrically with hydrogen and oxygen. Both reactants are supplied with $\vartheta = 25\,°C$, the hydrogen's inlet pressure is $p_F = 200\,kPa$. The product water is released in liquid state at $\vartheta_W = 60\,°C$ with a specific heat capacity in liquid state of $c_W = 4.19\,\frac{kJ}{kg\,K}$. The fuel cell delivers a DC-voltage of $U_{el} = 52\,V$ and a current of $I_{el} = 118\,A$ with a fuel consumption of $V_B = 2.2\,\frac{m^3}{h}$. The fuel's specific lower heating voltage at a reference temperature of $\vartheta_0 = 25\,°C$ follows $H_U = 119.946\,\frac{MJ}{kg}$. Water has a specific heat of vaporisation at $25\,°C$ of $\Delta h_V = 2442\,\frac{kJ}{kg}$.

(a) Calculate the mass fluxes of the fuel $\dot{m}_F$, of the required oxygen $\dot{m}_{O_2}$ and of the product water $\dot{m}_W$.
(b) What heat flux $\dot{Q}$ needs to be released by the fuel cell?
(c) Calculate the fuel cell's efficiency, that shall follow $\eta = \frac{-P_{el}}{\dot{m}_F H_U}$.

Solution

(a) Hydrogen is treated as an ideal gas, so that the mass fluxes of the fuel $\dot{m}_F$ follows the thermal equation of state

$$p_F \dot{V}_F = \dot{m}_F R_F T_0. \tag{23.319}$$

The individual gas constant is

$$R_F = \frac{R_M}{M_F} = 4157.1\,\frac{J}{kg\,K}. \tag{23.320}$$

Hence, the mass flux of hydrogen is

$$\dot{m}_F = \frac{p_F \dot{V}_F}{R_F T_0} = 9.8610 \times 10^{-5}\,\frac{kg}{s}. \tag{23.321}$$

Since the fuel is pure hydrogen its elemental analysis results in $h = 1$. Hence, the minimum oxygen need at stoichiometric conditions obeys

$$o_{min} = 2.667c + 8h = 8. \tag{23.322}$$

The mass flux of the oxygen follows accordingly

$$\dot{m}_{O_2} = o_{min}\dot{m}_F = 7.8888 \times 10^{-4}\frac{kg}{s}. \tag{23.323}$$

Hydrogen and oxygen react and water occurs with

$$\dot{m}_W = \dot{m}_{O_2} + \dot{m}_F = 8.8749 \times 10^{-4}\frac{kg}{s}. \tag{23.324}$$

(b) The energy balance follows Fig. 23.24 and is split in three consecutive steps.

- Step 1—Reaction

  A preconditioning is not required in this case, since hydrogen and oxygen are supplied at reference temperature. Within step 1 the (electrochemical) reaction is done, so that the entire chemical bonded energy is released as thermal energy and electrical energy, i.e.

$$\dot{Q}_1 + P_{el} = -\dot{m}_F H_U. \tag{23.325}$$

The released electrical power follows

$$P_{el} = -U_{el}I_{el}. \tag{23.326}$$

Hence, the released heat is

$$\dot{Q}_1 = -\dot{m}_F H_U + U_{el}I_{el} = -5.6918\,kW. \tag{23.327}$$

Mind, that the product water leaving step 1 is in gaseous state, since the lower heating value has been applied.

- Step 2—Latent heat

The first law of thermodynamics for step 2 obeys

$$\dot{Q}_2 + \dot{m}_W h_v(T_0) = \dot{m}_W h_{liq}(T_0). \tag{23.328}$$

The released heat flux follows

$$\dot{Q}_2 = \dot{m}_W\left[h_{liq}(T_0) - h_v(T_0)\right] = -\dot{m}_W \Delta h_v(T_0) = -2.1672\,kW. \tag{23.329}$$

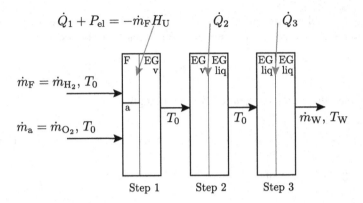

$$\dot{Q}_1 + P_{el} = -\dot{m}_F H_U \qquad \dot{Q}_2 \qquad \dot{Q}_3$$

Fig. 23.24 Solution to Problem 23.4b

- Step 3—Sensible heat

  In step 3 the liquid water is post-conditioned from 25 °C to 60 °C. Since no phase change occurs, this is pure sensible heat and the first law of thermodynamics is[32]

  $$\dot{Q}_3 + \dot{m}_W h_{liq}(T_0) = \dot{m}_W h_{liq}(T_W). \tag{23.330}$$

  Solving for the heat flux and applying the caloric equation of state for incompressible liquids[33] leads to

  $$\dot{Q}_3 = \dot{m}_W \left[h_{liq}(T_W) - h_{liq}(T_0) \right] = \dot{m}_W c_W (T_W - T_0) = 0.1302\,\text{kW}. \tag{23.331}$$

  The overall thermal energy flux is

  $$\dot{Q} = \dot{Q}_1 + \dot{Q}_2 + \dot{Q}_3 = -7.7289\,\text{kW}. \tag{23.332}$$

(c) The fuel cell's efficiency follows

$$\eta = \frac{-P_{el}}{\dot{m}_F H_U} = 0.5188. \tag{23.333}$$

Thus, 51.88 % of the chemical bonded energy are converted in electricity. The rest is released as heat.

Problem 23.7 A fossil fuel ($c = 0.85$, $h = 0.15$) is fired with air, that is composed of O_2, N_2 and CO_2, see Table 23.7. The mass flux of the air is $\dot{m}_a = 8\,\frac{\text{kg}}{\text{s}}$.

---

[32]Step 3 is supposed to be isobar.
[33]And further assuming, that step 3 is isobar.

Table 23.7 Composition of the combustion air

	O_2	N_2	CO_2
ξ_i	0.12	0.70	0.18

Table 23.8 Averaged specific heat capacities $\overline{c_p}|_0^\vartheta$ in $\frac{kJ}{kg\,K}$

ϑ [°C]	O_2	N_2	CO_2	H_2O
0	0.9147	1.0394	0.8173	1.8589
25	0.9163	1.0395	0.8307	1.8615
100	0.9229	1.0404	0.8684	1.8724
200	0.9354	1.0435	0.9128	1.8931
1100	1.0434	1.1254	1.1413	2.1744
1200	1.0511	1.1343	1.1560	2.2075
1300	1.0583	1.1426	1.1693	2.2399

(a) What is the maximum mass flux of the fossil fuel, that can be oxidised?

(b) Calculate the mass concentrations of the exhaust gas $\xi_{EG,i}$.

(c) Calculate the adiabatic flame temperature if the air is supplied with 100 °C and the fuel with 25 °C. The lower heating value of the fuel is $H_U(25\,°C) = 42.6\,\frac{MJ}{kg}$. Required averaged specific heat capacities can be taken from Table 23.8.

Solution

(a) The minimum oxygen need is

$$o_{min} = \frac{\dot{m}_{O_2,min}}{\dot{m}_F} = 2.667c + 8h + s - o. \tag{23.334}$$

The available mass flux of oxygen is given by the concentration of the combustion air, i.e.

$$\dot{m}_{O_2,min} = \xi_{O_2}\dot{m}_a. \tag{23.335}$$

Combining both equations leads to

$$\dot{m}_F = \frac{\xi_{O_2}\dot{m}_a}{2.667c + 8h + s - o} = 0.2769\,\frac{kg}{s}. \tag{23.336}$$

(b) The composition of the exhaust gas $\mu_{EG,i}$ is

- Carbon monoxide

$$\mu_{EG,CO_2} = \underbrace{3.667c}_{\text{comb.}} + \underbrace{\frac{\xi_{CO_2}\dot{m}_a}{\dot{m}_F}}_{\text{air}} = 8.3174 \tag{23.337}$$

- Water

$$\mu_{EG,H_2O} = 9h = 1.35 \tag{23.338}$$

- Oxygen

$$\mu_{EG,O_2} = 0 \tag{23.339}$$

There is no oxygen in the exhaust gas, since[34] $\lambda = 1$.
- Nitrogen

$$\mu_{EG,N_2} = \frac{\xi_{N_2}\dot{m}_a}{\dot{m}_F} = 20.2239 \tag{23.340}$$

Hence, for the exhaust gas it follows

$$\mu_{EG} = \sum_i \mu_{EG,i} = 29.8912. \tag{23.341}$$

According to that the concentrations are
- Carbon monoxide

$$\xi_{EG,CO_2} = \frac{\mu_{EG,CO_2}}{\mu_{EG}} = 0.2783 \tag{23.342}$$

- Water

$$\xi_{EG,H_2O} = \frac{\mu_{EG,H_2O}}{\mu_{EG}} = 0.0452 \tag{23.343}$$

- Oxygen

$$\xi_{EG,O_2} = \frac{\mu_{EG,O_2}}{\mu_{EG}} = 0 \tag{23.344}$$

- Nitrogen

$$\xi_{EG,N_2} = \frac{\mu_{EG,N_2}}{\mu_{EG}} = 0.6766 \tag{23.345}$$

(c) The equation for the adiabatic flame temperature has been derived in Sect. 23.4.5 and follows, see Eq. 23.309,

---

[34] It has been asked for the *maximum* mass flux of fuel in a.

Table 23.9 Iterations

Iteration	$\vartheta_{ad,i}$ in °C	$\overline{c_{p,a}}\big|_{\vartheta_0}^{\vartheta_a}$ in $\frac{kJ}{kg\,K}$	$\vartheta_{ad,i+1}$ in °C
0	1300	1.2031	1245.4
1	1245.4	1.1973	1251.5
2	1251.5	1.1979	1250.8
3	1250.8	1.1979	1250.9

$$\vartheta_{ad} = \vartheta_0 + \frac{H_U(T_0) + \overline{c_{p,F}}\big|_{\vartheta_0}^{\vartheta_F} \cdot (\vartheta_F - \vartheta_0) + \lambda \cdot l_{min} \cdot \overline{c_{p,a}}\big|_{\vartheta_0}^{\vartheta_a} \cdot (\vartheta_a - \vartheta_0)}{\mu_{EG} \cdot \overline{c_{p,EG}}\big|_{\vartheta_0}^{\vartheta_{ad}} + a \cdot \overline{c_{p,Ash}}\big|_{\vartheta_0}^{\vartheta_{ad}}}.$$

$$(23.346)$$

The fuel is supplied at reference temperature, there are no ashes and it is a stoichiometric combustion with $\lambda = 1$, so that the equation simplifies to

$$\vartheta_{ad} = \vartheta_0 + \frac{H_U(T_0) + l_{min} \cdot \overline{c_{p,a}}\big|_{\vartheta_0}^{\vartheta_a} \cdot (\vartheta_a - \vartheta_0)}{\mu_{EG} \cdot \overline{c_{p,EG}}\big|_{\vartheta_0}^{\vartheta_{ad}}} \qquad (23.347)$$

with

$$l = \frac{\dot{m}_a}{\dot{m}_F} = 28.8912 = \mu_{EG} - 1. \qquad (23.348)$$

According to Eq. 12.183 the averaged specific heat capacity is

$$\overline{c_{p,EG}}\big|_{\vartheta_0}^{\vartheta_{ad}} = \frac{1}{\vartheta_{ad} - \vartheta_0}\left[\overline{c_{p,EG}}\big|_0^{\vartheta_{ad}}\vartheta_{ad} - \overline{c_{p,EG}}\big|_0^{\vartheta_0}\vartheta_0\right] \qquad (23.349)$$

with

$$c_{p,EG} = \sum \xi_{EG,i} c_{p,EG,i}. \qquad (23.350)$$

For the air it follows accordingly

$$\overline{c_{p,a}}\big|_{\vartheta_0}^{\vartheta_a} = \frac{1}{\vartheta_a - \vartheta_0}\left[\overline{c_{p,a}}\big|_0^{\vartheta_a}\vartheta_a - \overline{c_{p,a}}\big|_0^{\vartheta_0}\vartheta_0\right] = 0.9981\,\frac{kJ}{kg\,K} \qquad (23.351)$$

with

$$c_{p,a} = \sum \xi_i c_{p,i}. \qquad (23.352)$$

The averaged specific heat capacities can be taken from Table 23.8. Obviously, Eq. 23.346 can be solved numerically, see iterations documented in Table 23.9.

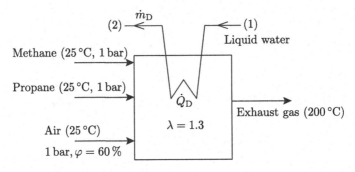

Fig. 23.25 Sketch to Problem 23.8

After 3 iterations the solution does not change significantly, so that the adiabatic flame temperature is $\vartheta_{ad} = 1250.8\,°C$.

Problem 23.8 In a combustion chamber methane (CH_4) and propane (C_3H_8) are fired isobarically with a fuel-air equivalence ratio of $\lambda = 1.3$, see Fig. 23.25. Both fuels are supplied with a temperature of 25 °C and a pressure of 1 bar. The humid combustion air has a temperature of 25 °C and a relative humidity of $\varphi = 60\,\%$. The mass fractions of the dry air are $\xi_{O_2,dry} = 0.23$ and $\xi_{N_2,dry} = 0.77$. A heat flux of $\dot{Q}_D$ is released, so that the exhaust gas temperature is 200 °C. The pressure shall be 1 bar. It further is

- Methane: $\dot{V}_{CH_4} = 0.5\,\frac{m^3}{h}$, lower heating value $H_{U,CH_4}(25\,°C) = 50.01\,\frac{MJ}{kg}$
- Propane: $\dot{V}_{C_3H_8} = 0.3\,\frac{m^3}{h}$, lower heating value $H_{U,C_3H_8}(25\,°C) = 46.35\,\frac{MJ}{kg}$
- Environment: $\vartheta_{env} = 25\,°C$, $p = 1\,bar$

Isolation losses as well as changes of the kinetic as well potential energies can be ignored.

(a) Calculate the mass fluxes of methane and propane. Both shall be treated as ideal gases.
(b) Estimate the overall mass fractions of the mixture of methane and propane c and h.
(c) What is the minimum oxygen o_{min}? Calculate the mass flux of the humid air.
(d) Calculate the concentrations of the exhaust gas $\xi_{EG,i}$.
(e) What is the dew point of the exhaust gas?
(f) What is the overall lower heating value $H_{U,F}(25\,°C)$ of the fuel mixture?
(g) Calculate the released heat flux $\dot{Q}_D$?

Solution

(a) The mass fluxes follow from the thermal equation of state, since both gases shall be treated as ideal gases, i.e.

$$\dot{m} = \frac{p\dot{V}}{RT}. \tag{23.353}$$

The individual gas constant can by estimated with the molar masses

$$M_{CH_4} = 16 \; \frac{kg}{kmol} \tag{23.354}$$

and

$$M_{C_3H_8} = 44 \; \frac{kg}{kmol}. \tag{23.355}$$

So, the mass fluxes follow

$$\dot{m}_{CH_4} = \frac{p\dot{V}_{CH_4} M_{CH_4}}{R_M T} = 8.9645 \times 10^{-5} \frac{kg}{s} \tag{23.356}$$

respectively

$$\dot{m}_{C_3H_8} = \frac{p\dot{V}_{C_3H_8} M_{C_3H_8}}{R_M T} = 1.4791 \times 10^{-4} \frac{kg}{s}. \tag{23.357}$$

(b) The individual mass fractions of the fuels are

- Methane

$$c_{CH_4} = \frac{1 M_C}{M_{CH_4}} = 0.75 \tag{23.358}$$

and

$$h_{CH_4} = \frac{4 M_H}{M_{CH_4}} = 0.25 \tag{23.359}$$

- Propane

$$c_{C_3H_8} = \frac{3 M_C}{M_{C_3H_8}} = 0.8182 \tag{23.360}$$

and

$$h_{C_3H_8} = \frac{8 M_H}{M_{C_3H_8}} = 0.1818 \tag{23.361}$$

So for the mixture of both fuels it is

$$c = \frac{\dot{m}_{CH_4} c_{CH_4} + \dot{m}_{C_3H_8} c_{C_3H_8}}{\dot{m}_{CH_4} + \dot{m}_{C_3H_8}} = 0.7925 \tag{23.362}$$

and

$$h = \frac{\dot{m}_{CH_4} h_{CH_4} + \dot{m}_{C_3H_8} h_{C_3H_8}}{\dot{m}_{CH_4} + \dot{m}_{C_3H_8}} = 0.2075 \tag{23.363}$$

(c) The minimum oxygen is

$$o_{min} = 2.667c + 8h = 3.7738. \tag{23.364}$$

The minimum need of dry air is

$$l_{min} = \frac{o_{min}}{\xi_{O_2,dry}} = 16.408. \tag{23.365}$$

The overall need of dry air obeys

$$l = \lambda l_{min} = 21.33. \tag{23.366}$$

The mass flux of dry air follows

$$\dot{m}_a = l \underbrace{\left(\dot{m}_{CH_4} + \dot{m}_{C_3H_8}\right)}_{=\dot{m}_F}. \tag{23.367}$$

In order to find the mass flux of the humid air, the vapour content x of the combustion air is needed,[35] i.e.

$$x = 0.622\frac{\varphi \cdot p_s(25\,°C)}{p - \varphi \cdot p_s(25\,°C)} = 0.0121. \tag{23.368}$$

Hence, the mass flux of the humid air is

$$\dot{m}_A = (1+x)\dot{m}_a = (1+x)l\left(\dot{m}_{CH_4} + \dot{m}_{C_3H_8}\right) = 18.4621\frac{kg}{h}. \tag{23.369}$$

(d) The composition of the exhaust gas $\mu_{EG,i}$ is

- Carbon monoxide

$$\mu_{EG,CO_2} = 3.667c = 2.9059 \tag{23.370}$$

- Water

$$\mu_{EG,H_2O} = 9h + \lambda l_{min}x = 2.1252 \tag{23.371}$$

- Oxygen

$$\mu_{EG,O_2} = o_{min}(\lambda - 1) = 1.1322 \tag{23.372}$$

---

[35] According to the steam table it is $p_s(25\,°C) = 0.0317\,bar$.

- Nitrogen

$$\mu_{EG,N_2} = \lambda l_{min} \xi_{N_2,dry} = 16.4244 \qquad (23.373)$$

Hence, for the exhaust gas it follows

$$\mu_{EG} = \sum_i \mu_{EG,i} = 22.5877. \qquad (23.374)$$

Hence, the concentrations are

- Carbon monoxide

$$\xi_{EG,CO_2} = \frac{\mu_{EG,CO_2}}{\mu_{EG}} = 0.1287 \qquad (23.375)$$

- Water

$$\xi_{EG,H_2O} = \frac{\mu_{EG,H_2O}}{\mu_{EG}} = 0.0941 \qquad (23.376)$$

- Oxygen

$$\xi_{EG,O_2} = \frac{\mu_{EG,O_2}}{\mu_{EG}} = 0.0501 \qquad (23.377)$$

- Nitrogen

$$\xi_{EG,N_2} = \frac{\mu_{EG,N_2}}{\mu_{EG}} = 0.7271 \qquad (23.378)$$

(e) In case condensation starts to occur, it is

$$x_{EG,H_2O} = \frac{p_s(\vartheta_\tau)}{p}. \qquad (23.379)$$

Consequently, the mass fraction of the water needs to be converted into molar fractions, i.e.

$$x_{EG,H_2O} = \xi_{EG,H_2O} \frac{M_{EG}}{M_{H_2O}} \qquad (23.380)$$

with

$$M_{EG} = \frac{R_M}{R_{EG}}. \qquad (23.381)$$

The gas constant of the exhaust gas is

$$R_{EG} = \sum_i \xi_{RG,i} R_i = \sum_i \xi_{RG,i} \frac{R_M}{M_i} = 296.7 \frac{J}{kg\,K}. \tag{23.382}$$

So, the molar mass is

$$M_{EG} = \frac{R_M}{R_{EG}} = 28.02 \frac{kg}{kmol}. \tag{23.383}$$

The mass fraction of the product water is

$$x_{EG,H_2O} = \xi_{EG,H_2O} \frac{M_{EG}}{M_{H_2O}} = 0.1465. \tag{23.384}$$

The critical partial pressure obeys

$$p_s(\vartheta_\tau) = x_{EG,H_2O}\,p = 0.1465\,bar. \tag{23.385}$$

According to the steam table the dew point temperature finally is

$$\vartheta_\tau = 53.48\,°C. \tag{23.386}$$

(f) The overall lower heating value $H_{U,F}(25\,°C)$ of the fuel mixture obeys

$$\dot{m}_F H_{U,F}(25\,°C) \overset{!}{=} \dot{m}_{CH_4} H_{U,CH_4}(25\,°C) + \dot{m}_{C_3H_8} H_{U,C_3H_8}(25\,°C). \tag{23.387}$$

Solving for $H_{U,F}(25\,°C)$ brings

$$H_{U,F}(25\,°C) = \frac{\dot{m}_{CH_4} H_{U,CH_4}(25\,°C) + \dot{m}_{C_3H_8} H_{U,C_3H_8}(25\,°C)}{\dot{m}_{CH_4} + \dot{m}_{C_3H_8}} = 47.731 \frac{MJ}{kg}. \tag{23.388}$$

(g) The heat flux for the combustion with unsaturated humid air is,[36] see Eq. 23.256,

$$\dot{Q}_D = \dot{m}_{EG} \cdot [h_{EG}(T_{EG}) - h_{EG}(T_0)] + \dot{m}_{Ash} \cdot [h_{Ash}(T_{EG}) - h_{Ash}(T_0)] +$$
$$- \dot{m}_F [h_F(T_F) - h_F(T_0)] - \dot{m}_a \cdot [h_{1+x}(T_a, x) - h_{1+x}(T_0, x)] +$$
$$- \dot{m}_F H_{U,F}(T_0). \tag{23.389}$$

Fuel and air are supplied at reference temperature and no ashes occur, so that the energy balance simplifies to

$$\dot{Q}_D = -\dot{m}_F H_{U,F}(T_0) + \dot{m}_{EG} \cdot [h_{EG}(T_{EG}) - h_{EG}(T_0)]. \tag{23.390}$$

Applying the caloric equation of state for the exhaust gas leads to

---

[36]The exhaust gas temperature is above the dew point.

$$\dot{Q}_D = -\dot{m}_F H_{U,F}(T_0) + \dot{m}_{EG} \cdot \left[\overline{c_{p,EG}} \big|_0^{200\,°C} 200\,°C - \overline{c_{p,EG}} \big|_0^{25\,°C} 25\,°C \right]$$

$$(23.391)$$

with

$$\overline{c_{p,EG}} \big|_0^{200\,°C} = \sum_i \xi_{EG,i} \overline{c_{p,i}} \big|_0^{200\,°C} = 1.1004 \, \frac{kJ}{kg\,K} \qquad (23.392)$$

and

$$\overline{c_{p,EG}} \big|_0^{25\,°C} = \sum_i \xi_{EG,i} \overline{c_{p,i}} \big|_0^{25\,°C} = 1.0837 \, \frac{kJ}{kg\,K}. \qquad (23.393)$$

Finally, the heat flux is

$$\dot{Q}_D = -\dot{m}_F H_{U,F}(T_0) + \dot{m}_{EG} \cdot \left[\overline{c_{p,EG}} \big|_0^{200\,°C} 200\,°C - \overline{c_{p,EG}} \big|_0^{25\,°C} 25\,°C \right]$$
$$= -10.3\,kW.$$

$$(23.394)$$

Chapter 24
Chemical Reactions

This chapter examines chemical reactions more generally than the previous chapters. In Chap. 23 the focus has been on the combustion of fossil fuels and the lower/upper heating value has been introduced to handle a chemical decomposition energetically. However, this method used to be rather impractical, since the oxidisation has to be split into several parts and a distinction has been required whether condensation occurs. Now, a more straightforward method is preferred by introducing the so-called absolute specific enthalpy/entropy. In doing so, the specific enthalpy does not only imply a caloric effect, as it has been done in part I and II, but it includes the specific chemical bonded energy as well.

24.1 Mass Balance

For a chemical reaction the mass balances are derived generically first, i.e. how much of each educt is needed to achieve a certain amount of products. Actually, this step has recently been done in Chap. 23 as well. However, for doing so, in a first step a chemical reaction scheme is required, i.e.

$$\gamma_A A + \gamma_B B \rightarrow \gamma_C C + \gamma_D D \tag{24.1}$$

Components A and B chemically react and finally lead to the products C and D. The stoichiometric factors γ_i actually show, how particles, i.e. atoms or molecules, are consumed and rearranged by the chemical reaction. The stoichiometric factors can be found under the premise, that the number of any atom on the educt side needs to be balanced by the number of the same atom on the product side. According to the rule of three it can easily be linked with the molar amount n_i. Thus, the reaction schemes is referred to 1 particle of component A, i.e.

© Springer Nature Switzerland AG 2019
A. Schmidt, *Technical Thermodynamics for Engineers*,
https://doi.org/10.1007/978-3-030-20397-9_24

$$1 + \frac{\gamma_B}{\gamma_A} \rightarrow \frac{\gamma_C}{\gamma_A} + \frac{\gamma_D}{\gamma_A} \qquad (24.2)$$

Multiplying with n_A particles then brings

$$n_A + \frac{\gamma_B}{\gamma_A} n_A \rightarrow \frac{\gamma_C}{\gamma_A} n_A + \frac{\gamma_D}{\gamma_A} n_A \qquad (24.3)$$

If a molar amount of n_A particles of A shall be converted by a chemical reaction with B, the $\frac{\gamma_B}{\gamma_A}$-fold amount of B is needed and the $\frac{\gamma_C}{\gamma_A}$-fold amount of C occurs. With the molar mass of each component

$$M_i = \frac{m_i}{n_i}, \qquad (24.4)$$

the converted masses can be easily calculated, i.e.

$$n_A M_A + \frac{\gamma_B}{\gamma_A} n_A M_B \rightarrow \frac{\gamma_C}{\gamma_A} n_A M_C + \frac{\gamma_D}{\gamma_A} n_A M_D \qquad (24.5)$$

Analogous to the previous chapter the molar specific conversions rate can be defined with respect to component A for instance, i.e.

$$\nu_i = \frac{n_i}{n_A} \qquad (24.6)$$

The mass specific conversion rate follows accordingly

$$\mu_i = \frac{m_i}{m_A} \qquad (24.7)$$

This can be applied for the educts

$$\nu_B = \frac{\gamma_B}{\gamma_A} \left[\frac{\text{mol}_B}{\text{mol}_A} \right] \qquad (24.8)$$

and

$$\mu_B = \frac{\gamma_B}{\gamma_A} \frac{M_B}{M_A} \left[\frac{\text{kg}_B}{\text{kg}_A} \right]. \qquad (24.9)$$

For the products it is

$$\nu_C = \frac{\gamma_C}{\gamma_A} \left[\frac{\text{mol}_C}{\text{mol}_A} \right] \qquad (24.10)$$

and

$$\mu_C = \frac{\gamma_C}{\gamma_A} \frac{M_C}{M_A} \left[\frac{\text{kg}_C}{\text{kg}_A} \right] \qquad (24.11)$$

respectively

$$\nu_D = \frac{\gamma_D}{\gamma_A} \left[\frac{mol_D}{mol_A} \right] \tag{24.12}$$

and

$$\mu_D = \frac{\gamma_D}{\gamma_A} \frac{M_D}{M_A} \left[\frac{kg_D}{kg_A} \right]. \tag{24.13}$$

The product composition ("Pr") can be specified by molar- or mass concentration, i.e.

$$\boxed{x_{Pr,i} = \frac{n_{Pr,i}}{\sum n_{Pr,i}}} \tag{24.14}$$

and

$$\boxed{\xi_{Pr,i} = \frac{m_{Pr,i}}{\sum m_{Pr,i}}} \tag{24.15}$$

with

$$\sum_i x_{Pr,i} = 1 \tag{24.16}$$

and

$$\sum_i \xi_{Pr,i} = 1 \tag{24.17}$$

According to our example it results in

$$x_{Pr,C} = \frac{\gamma_C}{\gamma_C + \gamma_D} \tag{24.18}$$

and

$$x_{Pr,D} = \frac{\gamma_D}{\gamma_C + \gamma_D} \tag{24.19}$$

respectively

$$\xi_{Pr,C} = \frac{\gamma_C M_C}{\gamma_C M_C + \gamma_D M_D} \tag{24.20}$$

and

$$\xi_{Pr,D} = \frac{\gamma_D M_D}{\gamma_C M_C + \gamma_D M_D}. \tag{24.21}$$

Example: Combustion of Methanol with $\lambda > 1$

In this example methanol CH_3OH shall be oxidised. The reaction scheme is now derived step by step:

- Stoichiometric with oxygen

  In this case the reaction scheme follows

$$1 \underbrace{CH_3OH}_{\text{``F''}} + \frac{3}{2} O_2 \rightarrow 1 CO_2 + 2 H_2O. \qquad (24.22)$$

Mind, that the stoichiometric coefficients fulfil the conservation of mass: The number of C-atoms on left and right side is 1, the H-atoms are 4 and the O-atoms are 4 as well on both sides.

- Stoichiometric with air[1]

  If air is supplied it not just delivers the required oxygen, but nitrogen as well, that needs to be included by the reaction scheme. As in the chapters before it shall be treated inert, i.e. it enters and leaves without being chemically reactive, i.e.

$$1 CH_3OH + \frac{3}{2} O_2 + \frac{79}{21} \cdot \frac{3}{2} N_2 \rightarrow 1 CO_2 + 2 H_2O + \frac{79}{21} \cdot \frac{3}{2} N_2 \qquad (24.23)$$

- Excess air

  A fuel-air equivalence ratio of $\lambda > 1$ leads to excess air, that is not consumed by the reaction. Consequently, this unused air leaves on the product side.

$$1 CH_3OH + \lambda \cdot \frac{3}{2} O_2 + \lambda \cdot \frac{79}{21} \cdot \frac{3}{2} N_2 \rightarrow 1 CO_2 + 2 H_2O + \lambda \cdot \frac{79}{21} \cdot \frac{3}{2} N_2 + (\lambda - 1) \cdot \frac{3}{2} O_2 \qquad (24.24)$$

The fuel-air equivalence ration is defined as

$$\lambda = \frac{n_{\text{Air}}}{n_{\text{Air, min}}} = \frac{m_{\text{Air}}}{m_{\text{Air, min}}}, \qquad (24.25)$$

see Sect. 23.3.1. Mind, that if $\lambda = 1$ the product side is free of oxygen.

Oxygen/Air Need

The molar oxygen need can now be taken from the reaction scheme, i.e.

---

[1] The ratio $\frac{79}{21}$ is the molar ratio of atmospheric air $\frac{x_{N_2,\text{Air}}}{x_{O_2,\text{Air}}}$. Thus, it shows how much more nitrogen compared to oxygen is in the combustion air.

$$n_{O_2} = \lambda \cdot \frac{3}{2} \cdot n_F. \tag{24.26}$$

In other words, it requires the $\lambda \cdot \frac{3}{2}$-fold amount of fuel as oxygen, see reaction scheme 24.24. The molar-specific oxygen need then is

$$\boxed{\nu_{O_2} = \frac{n_{O_2}}{n_F} \equiv O = \lambda \cdot O_{min} = \lambda \cdot \frac{3}{2}} \tag{24.27}$$

The mass consumption of oxygen follows accordingly with the molar masses

$$m_{O_2} = \lambda \cdot \frac{3}{2} \frac{M_{O_2}}{M_F} \cdot m_F \tag{24.28}$$

respectively in specific notation

$$\boxed{\mu_{O_2} = \frac{m_{O_2}}{m_F} \equiv o = \lambda \cdot o_{min} = \lambda \cdot \frac{3}{2} \frac{M_{O_2}}{M_F}} \tag{24.29}$$

In a next step the air need can be calculated

$$n_a = \lambda \cdot \frac{3}{2} \cdot n_F \cdot \frac{1}{x_{O_2,Air}} \tag{24.30}$$

respectively in specific notation

$$\boxed{\nu_a = \frac{n_a}{n_F} \equiv L = \lambda \cdot L_{min} = \lambda \cdot \frac{3}{2} \frac{1}{x_{O_2,Air}}} \tag{24.31}$$

The mass consumption of air follows accordingly with the molar masses

$$m_a = \lambda \cdot \frac{3}{2} \frac{M_{O_2}}{M_F} \cdot m_F \cdot \frac{1}{\xi_{O_2,Air}} \tag{24.32}$$

respectively in specific notation

$$\boxed{\mu_a = \frac{m_a}{m_F} \equiv l = \lambda \cdot l_{min} = \lambda \cdot \frac{3}{2} \frac{M_{O_2}}{M_F} \cdot \frac{1}{\xi_{O_2,Air}}} \tag{24.33}$$

Product Composition

The product composition is as follows and can be taken directly from the chemical Eq. 24.24

- Carbon dioxide

$$n_{Pr,CO_2} = 1 \cdot n_F \tag{24.34}$$

respectively in fuel-specific notation

$$\boxed{\nu_{Pr,CO_2} = \frac{n_{Pr,CO_2}}{n_F} = 1} \tag{24.35}$$

- Water

$$n_{Pr,H_2O} = 2 \cdot n_F \tag{24.36}$$

respectively in fuel-specific notation

$$\boxed{\nu_{Pr,H_2O} = \frac{n_{Pr,H_2O}}{n_F} = 2} \tag{24.37}$$

- Nitrogen

$$n_{Pr,N_2} = \lambda \cdot \frac{79}{21} \cdot \frac{3}{2} \cdot n_F \tag{24.38}$$

respectively in fuel-specific notation

$$\boxed{\nu_{Pr,N_2} = \frac{n_{Pr,N_2}}{n_F} = \lambda \cdot \frac{79}{21} \cdot \frac{3}{2}} \tag{24.39}$$

- Oxygen

$$n_{Pr,O_2} = (\lambda - 1) \cdot \frac{3}{2} \cdot n_F \tag{24.40}$$

respectively in fuel-specific notation

$$\boxed{\nu_{Pr,O_2} = \frac{n_{Pr,O_2}}{n_F} = (\lambda - 1) \cdot \frac{3}{2}} \tag{24.41}$$

24.2 Energy Balance

In Chap. 23 the energy balance of the oxidisation has been described with the specific lower heating value of the fuel H_U. Fuel and air are supplied at a reference temperature, the exhaust has leaves the system at the same reference temperature with the product water being *entirely* in vapour state. Under these conditions, the specific lower heating value represents the specific heat that needs to be released. Though it used to be a simple approach, several steps have been required to fully describe the energetics of the combustion and possible condensation. The specific enthalpy of any involved component covered the caloric part, i.e. it has been the specific sensible/latent heat that influenced the specific enthalpy. However, it has not been possible to count for entropic effects, so the irreversibility of such a combustion could not be investigated.

Within this section a new approach is followed, that extends the specific enthalpy to include chemical effects has well. This is the key to perform energy balances as it used to be done in part I and II: In steady state the energy flux into an open system is balanced by the energy flux out of this system. A classification into several sub-steps, see Sect. 23.4, then is obsolete. Furthermore, the specific entropy is extended to chemical effects as well, so that a entropy balance for chemical reactions is possible. However, these chemical effects require a unified reference level for any chemical substance. This new approach is based on the so-called specific absolute enthalpy/entropy.

24.2.1 Caloric Equations of State

Enthalpy of Formation, Absolute Enthalpy

The molar specific enthalpy of formation[2] $\Delta_B^0 H_m$ is defined at standard conditions for any chemical substance and covers the chemical bonded energy to form this component, see also [2]. Actually, the standard conditions are arbitrary but usually defined as

$$T_0 = 298.15 \, \text{K} \tag{24.42}$$

and

$$p_0 = 1 \, \text{bar}. \tag{24.43}$$

The molar specific enthalpy of formation of the *elements i* in their stable state at standard conditions, e.g.

---

[2]Though the absolute specific enthalpy is a specific state value, a capital letter H_m is used instead of h_M, to emphasise that the *absolute* specific enthalpy is meant. The same counts for the absolute specific entropy S_m.

- hydrogen: H_2,
- nitrogen: N_2,
- oxygen: O_2,
- carbon: C,
- sulphur: S_2,

is referenced to $\Delta_B^0 H_{m,i} (T_0) = 0$. For any other chemical bond the specific enthalpy of formation is determined with reference to the standard conditions. For varying temperatures and pressures from standard state the caloric effect needs to be taken into account as well, so that the absolute specific enthalpy $H_{m,i}$ is defined:

- **Ideal gas**

In case the component is gaseous it shall follow an ideal gas, i.e. the absolute specific enthalpy $H_{m,i} (T)$ is purely a function of temperature. With the molar specific heat capacity C_p it can be applied for ideal gases:

$$H_{m,i} (T) = \underbrace{\Delta_B^0 H_{m,i} (T_0)}_{\text{chemical}} + \underbrace{\overline{C_p}\Big|_{T_0}^{T} (T - T_0)}_{\text{caloric}}. \tag{24.44}$$

The first term on the right hand side of Eq. 24.44 represents the chemical energy of formation and the second the caloric effect, in case the temperature differs from standard temperature. For the absolute specific internal energy it then follows

$$H_{m,i} = U_{m,i} + p v_{M,i}. \tag{24.45}$$

With the thermal equation of state, i.e.

$$p v_{M,i} = R_M T, \tag{24.46}$$

the absolute specific internal energy follows

$$U_{m,i} (T) = H_{m,i} (T) - R_M T. \tag{24.47}$$

- **Incompressible liquid**

For an incompressible liquid the absolute enthalpy is, see Eq. 12.124 in part I,

$$H_{m,i} (T, p) = \underbrace{\Delta_B^0 H_{m,i} (T_0, p_0)}_{\text{chemical}} + \underbrace{\overline{C_p}\Big|_{T_0}^{T} (T - T_0)}_{\text{caloric, } T} + \underbrace{v_{M,i} (p - p_0)}_{\text{caloric, } p}. \tag{24.48}$$

The last term of Eq. 24.48 can usually be ignored and is excluded for liquids in Appendix B. However, the absolute specific internal energy follows

$$U_{m,i}(T, p) = H_{m,i}(T, p) - p v_{M,i}. \tag{24.49}$$

$H_{m,i}$ is the so-called absolute specific enthalpy of a component i. It is listed for several elements and bonds in Appendix B. All components are supposed to be ideal.

Absolute Entropy—Third Law of Thermodynamics

The third law of thermodynamics according to Nernst claims:

Theorem 24.1 *The entropy of each homogeneous, perfectly crystallised substance is zero at* 0 K.

The reference at 0 K is required, since substances at T_0, p_0 can occur in different aggregate states and thus have different levels of entropy. Hence, 0 K is an absolute reference level for any substance. However, the absolute specific entropy $S_m^0(T_0, p_0)$ can be determined for each substance at the same standard conditions as for the absolute specific enthalpy, i.e. at

$$T_0 = 298.15 \, \text{K} \tag{24.50}$$

and

$$p_0 = 1 \, \text{bar}. \tag{24.51}$$

At T_0, p_0 according to Nernst the absolute specific entropy for any substance is larger than zero. In Appendix B absolute molar specific entropies are listed for idealised components, i.e. for ideal gases as well as for incompressible liquids:

- Ideal gas

  For varying states T and p_i the caloric equation of state, see Eq. 12.110, can be applied. Mind, that if component i is part of a gaseous mixture, the partial pressure p_i of that component must be applied, see also Sect. 19.3.4, i.e.

  $$S_m(T, p_i) = S_m^0(T_0, p_0) + \overline{C_p}\bigg|_{T_0}^{T} \ln\frac{T}{T_0} - R_m \ln\frac{p_i}{p_0} \tag{24.52}$$

  The temperature influence is already covered by the listed data in Appendix B. Hence, the tables show $S_m(T, p_0)$, i.e.

  $$S_m(T, p_i) = \underbrace{S_m^0(T_0, p_0) + \overline{C_p}\bigg|_{T_0}^{T} \ln\frac{T}{T_0}}_{S_m(T, p_0)} - R_m \ln\frac{p_i}{p_0}. \tag{24.53}$$

  Consequently, for varying pressures the following correction is required

$$S_m(T, p_i) = \underbrace{S_m(T, p_0)}_{\text{listed}} - R_m \ln \frac{p_i}{p_0}. \tag{24.54}$$

- Incompressible liquid

For incompressible liquids, the specific entropy purely is a function of temperature, see Chap. 12. Hence, a pressure correction is not required and the absolute specific entropy obeys

$$S_m(T) = S_m^0(T_0, p_0) + \overline{C}\Big|_{T_0}^{T} \ln \frac{T}{T_0} = \underbrace{S_m(T, p_0)}_{\text{listed}}. \tag{24.55}$$

24.2.2 Open Systems

First Law of Thermodynamics

Since the absolute specific enthalpy $H_{m,i}(T, p)$ now includes the chemical energy as well, the first law of thermodynamics for a chemical reaction can easily be handled, see Fig. 24.1. In contrast to Sect. 23.4 no standardisation as in "Step 2" is required. For a steady state process the energy flux into the system needs to be equal to the energy flux out, i.e. the first law of thermodynamics obeys[3]

$$\dot{Q} + \sum_{j,\text{Educts}} \dot{n}_j H_{m,j}(T_j) = \sum_{i,\text{Products}} \dot{n}_i H_{m,i}(T_i) \tag{24.56}$$

or solved for the heat flux

$$\boxed{\dot{Q} = \sum_{i,\text{Products}} \dot{n}_i H_{m,i}(T_i) - \sum_{j,\text{Educts}} \dot{n}_j H_{m,j}(T_j)} \tag{24.57}$$

The absolute enthalpies $H_{m,i}(T, p)$ can be found in Appendix B. Mind, that the listed data is given as a function of T, thus it can be directly applied for ideal gases. In case an incompressible liquid is relevant a correction of the absolute specific enthalpy according to Eq. 24.48 is needed.

If a reaction occurs at standard conditions, i.e. products as well as educts have reference temperature T_0, see Fig. 24.2, the absolute specific enthalpies just include the chemical energy. Thus, the fist law of thermodynamics reads as

---

[3]Changes of kinetic/potential energies shall be ignored!

Fig. 24.1 Enthalpy of reaction

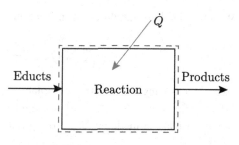

Fig. 24.2 Enthalpy of reaction at T_0

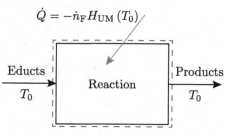

$$\dot{Q} = \dot{n}_F \Delta_R^0 H_m (T_0) = \sum_{i,\text{Products}} \dot{n}_i \Delta_B^0 H_{m,i} (T_0) - \sum_{j,\text{Educts}} \dot{n}_j \Delta_B^0 H_{m,j} (T_0) \quad (24.58)$$

$\Delta_R^0 H_m (T_0)$ is called specific enthalpy of reaction at standard conditions. If the product water is gaseous, the released heat correspond to the lower heating value, i.e.

$$\dot{Q} = -\dot{n}_F H_{UM} (T_0). \quad (24.59)$$

Thus, it is

$$\Delta_R^0 H_m (T_0) = \sum_{i,\text{Products}} \frac{\dot{n}_i}{\dot{n}_F} \Delta_B^0 H_{m,i} (T_0) - \sum_{j,\text{Educts}} \frac{\dot{n}_j}{\dot{n}_F} \Delta_B^0 H_{m,j} (T_0)$$
$$= -H_{UM} (T_0). \quad (24.60)$$

Example 24.2 (*Lower/upper heating value of methane*) As an example the specific enthalpy of reaction at standard conditions of methane is calculated. The chemical equation follows

$$CH_4 + 2O_2 \rightarrow CO_2 + 2H_2O. \quad (24.61)$$

According to Eq. 24.60 with the assumption that <u>water occurs as vapour</u> it is

$$\Delta_R^0 H_m (T_0) = 2 \cdot \Delta_B^0 H_{m,H_2O,g} (T_0) + 1 \cdot \Delta_B^0 H_{m,CO_2} (T_0) + \quad (24.62)$$
$$- 2 \cdot \Delta_B^0 H_{m,O_2} (T_0) - 1 \cdot \Delta_B^0 H_{m,CH_4} (T_0). \quad (24.63)$$

With the absolute enthalpies taken from the tables in Appendix B it follows, that

$$\Delta_R^0 H_m (T_0) = -802.562 \frac{kJ}{mol}. \tag{24.64}$$

Hence, following Eq. 24.60 the molar specific lower heating value is $H_{UM} = 802.562 \frac{kJ}{mol}$. It can be converted into a mass specific lower heating value by using the molar mass of methane $M = 16 \frac{kg}{kmol}$, i.e.

$$\boxed{H_U (25\,°C) = 50.1 \frac{MJ}{kg}} \tag{24.65}$$

In case the product water is in liquid state it is

$$\Delta_R^0 H_m (T_0) = 2 \cdot \Delta_B^0 H_{m,H_2O,liq} (T_0) + 1 \cdot \Delta_B^0 H_{m,CO_2} (T_0) + \tag{24.66}$$
$$- 2 \cdot \Delta_B^0 H_{m,O_2} (T_0) - 1 \cdot \Delta_B^0 H_{m,CH_4} (T_0). \tag{24.67}$$

With the absolute enthalpies taken from the tables in Appendix B it follows, that

$$\Delta_R^0 H_m (T_0) = -890.57 \frac{kJ}{mol}. \tag{24.68}$$

Hence, the molar specific upper heating value is: $H_{OM} = 890.57 \frac{kJ}{mol}$. It can be converted into the mass specific upper heating value by using the molar mass of methane $M = 16 \frac{kg}{kmol}$, i.e.

$$\boxed{H_O (25\,°C) = 55.6 \frac{MJ}{kg}} \tag{24.69}$$

Second Law of Thermodynamics

A major disadvantage regarding the lower heating value approach, is the impossibility to determine the degree of irreversibility, i.e. applying the second law of thermodynamics. With the method of absolute entropy it is now feasible to conduct a balance of entropy and hence to calculate the generation of entropy. A steady state entropy balance for a chemical reaction means, that incoming fluxes of entropy need to be balanced by the outgoing fluxes of entropy. According to Fig. 24.3 the balance then obeys

$$\dot{S}_a + \dot{S}_i + \dot{S}_I = \dot{S}_{II}. \tag{24.70}$$

State (I) represents the inlet, i.e. the educts, state (II) represents the outlet, i.e. the products. Since several educts respectively products might be present, the convective

Fig. 24.3 Entropy and
irreversibility of a reaction

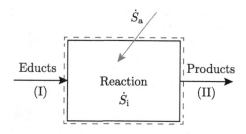

entropy fluxes follow

$$\dot{S}_{\mathrm{II}} - \dot{S}_{\mathrm{I}} = \sum_{i,\text{Products}} \dot{n}_i S_{\mathrm{m},i}\,(T_i, p_i) - \sum_{j,\text{Educts}} \dot{n}_j S_{\mathrm{m},j}\,\left(T_j, p_j\right). \tag{24.71}$$

Mind, that $\dot{S}_{\mathrm{i}}$ represents the generation of entropy due to irreversibilites and $\dot{S}_{\mathrm{a}}$ counts
for the entropy fluxes due to heat transfer. Thus, it is

$$\boxed{\dot{S}_{\mathrm{i}} + \dot{S}_{\mathrm{a}} = \sum_{i,\text{Products}} \dot{n}_i S_{\mathrm{m},i}\,(T_i, p_i) - \sum_{j,\text{Educts}} \dot{n}_j S_{\mathrm{m},j}\,\left(T_j, p_j\right)} \tag{24.72}$$

Hence, for an adiabatic reaction it is $\dot{S}_{\mathrm{a}} = 0$. The molar specific entropy of a reaction
$\Delta_{\mathrm{R}} S_{\mathrm{m}}$ is defined as

$$\dot{n}_{\mathrm{F}} \Delta_{\mathrm{R}} S_{\mathrm{m}} \stackrel{!}{=} \dot{S}_{\mathrm{II}} - \dot{S}_{\mathrm{I}} = \sum_{i,\text{Products}} \dot{n}_i S_{\mathrm{m},i}\,(T_i, p_i) - \sum_{j,\text{Educts}} \dot{n}_j S_{\mathrm{m},j}\,\left(T_j, p_j\right). \tag{24.73}$$

$$\Rightarrow \boxed{\Delta_{\mathrm{R}} S_{\mathrm{m}} = \sum_{i,\text{Products}} \frac{\dot{n}_i}{\dot{n}_{\mathrm{F}}} S_{\mathrm{m},i}\,(T_i, p_i) - \sum_{j,\text{Educts}} \frac{\dot{n}_j}{\dot{n}_{\mathrm{F}}} S_{\mathrm{m},j}\,\left(T_j, p_j\right)} \tag{24.74}$$

With this definition the balance of entropy for a steady state reaction obeys

$$\dot{n}_{\mathrm{F}} \Delta_{\mathrm{R}} S_{\mathrm{m}} = \dot{S}_{\mathrm{a}} + \dot{S}_{\mathrm{i}}. \tag{24.75}$$

24.2.3 Closed Systems

First Law of Thermodynamics

Let us now assume, the chemical reaction runs in a closed system, see Fig. 24.4: state
(I) includes the educts that lead to the products in state (II). State (I) and state (II)
shall be in thermodynamic equilibrium, i.e. the educts have a common temperature

Fig. 24.4 First law of thermodynamics reactive closed system

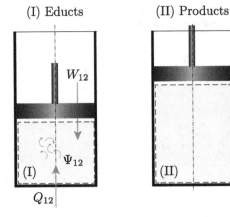

(I) Educts   (II) Products

T_I, the products a common temperature T_II. The first law of thermodynamics obeys

$$Q_{12} + W_{12} = U_\text{II} - U_\text{I} + \Delta E_{a,\text{I,II}}. \tag{24.76}$$

For the change of internal energy it is

$$U_\text{II} - U_\text{I} = \sum_{i,\text{Products}} n_i U_{m,i} (T_\text{II}) - \sum_{j,\text{Educts}} n_j U_{m,j} (T_\text{I}). \tag{24.77}$$

Since the specific absolute internal energies $U_{m,i}$ and $U_{m,j}$ are not listed in Appendix B, they have to be calculated according to Eqs. 24.47 and 24.49 depending on the fluid's aggregate state:

- Ideal gas

$$U_{m,i} (T_\text{II}) = H_{m,i} (T_\text{II}) - R_M T_\text{II} \tag{24.78}$$

$$U_{m,j} (T_\text{I}) = H_{m,j} (T_\text{I}) - R_M T_\text{I} \tag{24.79}$$

- Incompressible liquid

$$U_{m,i} (T_\text{II}, p_\text{II}) = H_{m,i} (T_\text{II}, p_\text{II}) - p_\text{II} v_{M,i} \tag{24.80}$$

$$U_{m,j} (T_\text{I}, p_\text{I}) = H_{m,j} (T_\text{I}, p_\text{I}) - p_\text{I} v_{M,j} \tag{24.81}$$

Second Law of Thermodynamics

Figure 24.5 shows the entropy balance for a closed, reactive system. The second law of thermodynamics obeys

$$S_{a,12} + S_{i,12} = S_{II} - S_{I}. \tag{24.82}$$

State (I) represents the educts, state (II) represents the products. Since several educts respectively products might be present, the entropy balance follows

$$S_{II} - S_{I} = \sum_{i,\text{Products}} n_i S_{m,i} \left(T_{II}, p_{II,i} \right) - \sum_{j,\text{Educts}} n_j S_{m,j} \left(T_{I}, p_{I,j} \right). \tag{24.83}$$

The entropy being transferred with the heat is

$$S_{a,12} = \int_{I}^{II} \frac{\delta Q}{T}. \tag{24.84}$$

Problem 24.3 In an electrolysis system $\dot{m}_W = 18 \, \frac{kg}{s}$ water is chemically decomposed into hydrogen and oxygen under isothermal and isobaric conditions (25 °C, $p = 20 \, \text{bar}$), see Fig. 24.6.

(a) What are the molar fluxes $\dot{n}_{H_2O}$, $\dot{n}_{H_2}$, $\dot{n}_{O_2}$ under steady state conditions?
(b) What are the enthalpy and entropy fluxes at inlet/outlet (water is supposed to be liquid)?
(c) What is the required power P_t at reversible operation, if heat is exchanged with the environment ($T_{env} = 298.15 \, \text{K}$)?

Solution

(a) The chemical reaction follows

$$2 H_2O \rightarrow 2 H_2 + 1 O_2 \tag{24.85}$$

The molar flux of water is

$$\dot{n}_{H_2O} = \frac{\dot{m}_{H_2O}}{M_{H_2O}} = 1 \, \frac{kmol}{s}. \tag{24.86}$$

Thus, according to the chemical equation it is

$$\dot{n}_{H_2} = \dot{n}_{H_2O} = 1 \, \frac{kmol}{s} \tag{24.87}$$

respectively

$$\dot{n}_{O_2} = \frac{1}{2}\dot{n}_{H_2O} = 0.5 \, \frac{kmol}{s}. \tag{24.88}$$

(b) The enthalpy and entropy fluxes are

- Inlet (I)@25 °C, liquid

  According to Appendix B it is

  $$H_{m,I,H_2O} = -285.83 \, \frac{kJ}{mol} \tag{24.89}$$

  and

  $$S_{m,I,H_2O} = 69.942 \, \frac{kJ}{kmol \, K}. \tag{24.90}$$

  The fluxes are

  $$\dot{H}_I = \dot{n}_{H_2O} H_{m,I,H_2O} = -285.83 \times 10^3 \, kW \tag{24.91}$$

  and

  $$\dot{S}_I = \dot{n}_{H_2O} S_{m,I,H_2O} = 69.942 \, \frac{kW}{K}. \tag{24.92}$$

- Outlet (II)

  At the outlet (II) oxygen and hydrogen occur. Following Appendix B results in

  $$H_{m,II,O_2} = 0 \, \frac{kJ}{mol} \tag{24.93}$$

  and

  $$H_{m,II,H_2} = 0 \, \frac{kJ}{mol}. \tag{24.94}$$

  The entire enthalpy flux is

  $$\dot{H}_{II} = \dot{n}_{H_2} H_{m,II,H_2} + \dot{n}_{O_2} H_{m,II,O_2} = 0 \, kW. \tag{24.95}$$

  For the entropy it is

  $$S_{m,II,O_2} = 205.149 \, \frac{kJ}{kmol \, K} - 8.3143 \, \frac{kJ}{kmol \, K} \ln\frac{20}{1} = 180.24 \, \frac{kJ}{kmol \, K} \tag{24.96}$$

  and

  $$S_{m,II,H_2} = 130.681 \, \frac{kJ}{kmol \, K} - 8.3143 \, \frac{kJ}{kmol \, K} \ln\frac{20}{1} = 105.77 \, \frac{kJ}{kmol \, K}. \tag{24.97}$$

  The entire entropy flux is

Fig. 24.5 Second law of thermodynamics reactive closed system

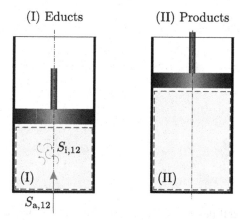

(I) Educts   (II) Products

$$\dot{S}_{II} = \dot{n}_{H_2} S_{m,II,H_2} + \dot{n}_{O_2} S_{m,II,O_2} = 195.89 \, \frac{\text{kW}}{\text{K}}. \tag{24.98}$$

(c) The first law of thermodynamics obeys

$$\dot{Q} + P_{t,rev} = \dot{H}_{II} - \dot{H}_{I} + \underbrace{\Delta \dot{E}_a}_{=0}. \tag{24.99}$$

Obviously, there are two unknowns ($\dot{Q}$ and $P_{t,rev}$), so further information can be gained by the second law of thermodynamics

$$\dot{S}_a + \dot{S}_i + \dot{S}_I = \dot{S}_{II}. \tag{24.100}$$

The reaction shall be reversible, so that $\dot{S}_i = 0$. Since heat is exchanged with a constant ambient temperature, it follows

$$\dot{S}_a = \frac{\dot{Q}}{T_{env}}. \tag{24.101}$$

Hence, the heat flux is

$$\dot{Q} = T_{env} \left(\dot{S}_{II} - \dot{S}_I \right). \tag{24.102}$$

Finally, the reversible technical power that has to be supplied is

$$P_{t,rev} = \dot{H}_{II} - \dot{H}_{I} - T_{env} \left(\dot{S}_{II} - \dot{S}_I \right) = 248.27 \, \text{MW}. \tag{24.103}$$

Problem 24.4 A fuel cell is operated stoichiometrically with hydrogen and oxygen. Both reactants are supplied with $\vartheta = 25 \, °\text{C}$, the hydrogen's inlet pressure is $p_F = 200 \, \text{kPa}$. The product water is released in liquid state at $\vartheta_w = 60 \, °\text{C}$. The fuel

Fig. 24.6 Sketch to
Problem 24.3

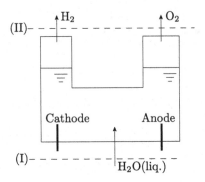

cell delivers a DC-voltage of $U_{el} = 52\,V$ and a current of $I_{el} = 118\,A$ with a fuel
consumption of $V_B = 2.2\,\frac{m^3}{h}$.

Mind, that this problem is identical with Problem 23.6. Anyhow, it is now solved
with the absolute enthalpy/entropy approach.

(a) Calculate the mass fluxes of the fuel $\dot{m}_F$, of the required oxygen $\dot{m}_{O_2}$ and of the
product water $\dot{m}_W$.
(b) What heat flux $\dot{Q}$ needs to be released by the fuel cell?

Solution

(a) Hydrogen is treated as an ideal gas, so that the mass fluxes of the fuel $\dot{m}_F$ follows
the thermal equation of state

$$p_F \dot{V}_F = \dot{m}_F R_F T_0. \tag{24.104}$$

The individual gas constant is

$$R_F = \frac{R_M}{M_F} = 4157.1\,\frac{J}{kg\,K}. \tag{24.105}$$

Hence, the mass flux of hydrogen is

$$\dot{m}_F = \frac{p_F \dot{V}_F}{R_F T_0} = 9.8610 \times 10^{-5}\,\frac{kg}{s} \tag{24.106}$$

Since the fuel is pure hydrogen its elemental analysis is $h = 1$. Hence, the minimum oxygen need at stoichiometric conditions obeys

$$o_{min} = 2.667c + 8h = 8. \tag{24.107}$$

The mass flux of the oxygen follows accordingly

$$\dot{m}_{O_2} = o_{\min}\dot{m}_F = 7.8888 \times 10^{-4}\,\frac{\text{kg}}{\text{s}}. \tag{24.108}$$

Hydrogen and oxygen react and water occurs with

$$\dot{m}_W = \dot{m}_{O_2} + \dot{m}_F = 8.8749 \times 10^{-4}\,\frac{\text{kg}}{\text{s}}. \tag{24.109}$$

The molar fluxes are

$$\dot{n}_W = \frac{\dot{m}_W}{M_W} = 4.9305 \times 10^{-5}\,\frac{\text{kmol}}{\text{s}} \tag{24.110}$$

and

$$\dot{n}_F = \frac{\dot{m}_F}{M_F} = 4.9305 \times 10^{-5}\,\frac{\text{kmol}}{\text{s}} \tag{24.111}$$

and

$$\dot{n}_{O_2} = \frac{\dot{m}_{O_2}}{M_{O_2}} = 2.4652 \times 10^{-05}\,\frac{\text{kmol}}{\text{s}}. \tag{24.112}$$

(b) The first law of thermodynamics obeys

$$P_{el} + \dot{Q} + \dot{H}_I = \dot{H}_{II} \tag{24.113}$$

with

$$P_{el} = -U_{el}I_{el}. \tag{24.114}$$

Thus, the heat follows

$$\dot{Q} = \dot{H}_{II} - \dot{H}_I + U_{el}I_{el}. \tag{24.115}$$

According to Appendix B the enthalpies are

$$\dot{H}_I = \dot{n}_{O_2} H_{m,O_2}(25\,°\text{C}) + \dot{n}_F H_{m,H_2}(25\,°\text{C}) = 0 \tag{24.116}$$

and

$$\dot{H}_{II} = \dot{n}_W H_{m,H_2O,\text{liq}}(60\,°\text{C}) = \dot{n}_W - 283.2105\,\frac{\text{kJ}}{\text{mol}} = -13.9637\,\text{kW}. \tag{24.117}$$

Hence, the heat flux finally is

$$\dot{Q} = \dot{H}_{II} - \dot{H}_I + U_{el}I_{el} = -7.8277\,\text{kW}. \tag{24.118}$$

24.3 Gibbs Enthalpy

24.3.1 Definition

A new state value, the specific Gibbs enthalpy, is defined as a combination of other state values

$$\boxed{g = g\,(T,\,p) = h - Ts = u + pv - Ts} \tag{24.119}$$

It has already been introduced in Chap. 12, in which the caloric state equations have been derived. Hence, its differential is given by

$$dg = dh - T\,ds - s\,dT = du + v\,dp + p\,dv - T\,ds - s\,dT \tag{24.120}$$

Applying the fundamental equation of thermodynamics, see Eq. 13.28, it is

$$dg = dh - T\,ds - s\,dT = -s\,dT + v\,dp. \tag{24.121}$$

However, this total differential also needs to follow

$$dg = \left(\frac{\partial g}{\partial T}\right)_p dT + \left(\frac{\partial g}{\partial p}\right)_T dp. \tag{24.122}$$

A comparison of Eqs. 24.121 and 24.122 leads to the change of Gibbs enthalpy with respect to temperature

$$\boxed{\left(\frac{\partial g}{\partial T}\right)_p = -s} \tag{24.123}$$

and to the change of Gibbs enthalpy with respect to pressure

$$\boxed{\left(\frac{\partial g}{\partial p}\right)_T = v} \tag{24.124}$$

24.3.2 Molar Gibbs Enthalpy

The extensive Gibbs enthalpy follows from its definition, see Eq. 24.119,

$$G = mg\,(T,\,p) = H - TS. \tag{24.125}$$

Replacing the mass with the molar quantity and the molar mass leads to

$$G = nMg\,(T,\,p) = H - TS. \tag{24.126}$$

Thus, the molar specific Gibbs enthalpy is

$$\boxed{G_\mathrm{m} = \frac{G}{n} = M g\,(T, p) = H_\mathrm{m} - T\,S_\mathrm{m}} \tag{24.127}$$

According to that the molar specific Gibbs enthalpy for any substance can be calculated following Tables B.1, B.2, B.3, B.4, B.5, B.6, B.7, B.8, B.9, B.10, B.11, B.12, B.13, B.14, B.15, B.16, B.17, B.18, B.19, B.20, B.21, B.22, B.23, B.24, B.25 and B.26 in Appendix B.

24.3.3 Motivation

Closed System

Imagine a closed system, in which an *isobar/isothermal* process takes place. According to Fig. 24.7, the question is, how much non-volume work $\delta w_\mathrm{non\text{-}v}$ can be released by such a system. However, this non-volume work covers any mechanical, electrical or chemical related work. Thus, also the dissipation. The first law of thermodynamics obeys[4]

$$\delta q + \delta w = \mathrm{d}u. \tag{24.128}$$

The partial energy equation reads as

$$\delta w = \delta w_\mathrm{non\text{-}v} + \delta w_\mathrm{V}. \tag{24.129}$$

Applying the second law of thermodynamics

$$\delta q + \delta \psi = T\,\mathrm{d}s \tag{24.130}$$

finally brings the released specific work $\delta w_\mathrm{non\text{-}v}$

$$\delta w_\mathrm{non\text{-}v} = \mathrm{d}u + p\,\mathrm{d}v - T\,\mathrm{d}s + \delta\psi. \tag{24.131}$$

Combining with the definition of the Gibbs enthalpy, see Eq. 24.120, results in

$$\mathrm{d}g = \mathrm{d}u + v\,\mathrm{d}p + p\,\mathrm{d}v - T\,\mathrm{d}s - s\,\mathrm{d}T = \delta w_\mathrm{non\text{-}v} - \delta\psi + v\,\mathrm{d}p - s\,\mathrm{d}T. \tag{24.132}$$

With $\mathrm{d}p = 0$, $\mathrm{d}T = 0$ it is

$$\boxed{\delta w_\mathrm{non\text{-}v} = \mathrm{d}g + \delta\psi} \tag{24.133}$$

---

[4]Kinetic/potential energies ignored!

Fig. 24.7 Gibbs
enthalpy—Closed system

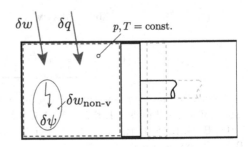

In case the change of state is reversible, the Gibbs enthalpy is the maximum non-volume work, i.e.

$$\boxed{\delta w_{\text{non-v,rev}} = dg}$$
(24.134)

Open System

Imagine an open system, in which an *isobar/isothermal* process takes place. According to Fig. 24.8, the question is, how much non-pressure work $\delta w_{\text{non-p}}$ can be released by such a system. However, this non-volume work covers any mechanical, electrical or chemical related work. Thus, also the dissipation. The first law of thermodynamics obeys[5]

$$\delta q + \delta w_{\text{t}} = dh.$$
(24.135)

The partial energy equation reads as

$$\delta w_{\text{t}} = \delta w_{\text{non-p}} + \delta y = \delta w_{\text{non-p}} + v \, dp,$$
(24.136)

i.e. the technical work is the sum of pressure work and non-pressure work. Applying the second law of thermodynamics

$$\delta q + \delta \psi = T \, ds$$
(24.137)

finally brings the released specific work $\delta w_{\text{non-p}}$

$$\delta w_{\text{non-p}} = dh - v \, dp - T \, ds + \delta \psi.$$
(24.138)

Combining with the definition of the Gibbs enthalpy, see Eq. 24.120, results in

$$dg = dh - T \, ds - s \, dT = \delta w_{\text{non-p}} - \delta \psi + v \, dp - s \, dT.$$
(24.139)

---

[5]Kinetic/potential energies ignored!

Fig. 24.8 Gibbs enthalpy—Open system

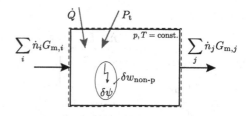

With $\mathrm{d}p = 0$, $\mathrm{d}T = 0$ it is

$$\boxed{\delta w_{\text{non-p}} = \mathrm{d}g + \delta \psi}. \tag{24.140}$$

In case the change of state is reversible, the Gibbs enthalpy is the maximum non-volume work, i.e.

$$\boxed{\delta w_{\text{non-p,rev}} = \mathrm{d}g} \tag{24.141}$$

Thus, in order to calculate the maximum power, that can be released according to Fig. 24.8, the isobaric/isothermal reaction needs to be reversible, i.e. $\delta \psi = 0$:

$$
\begin{aligned}
P_{\text{rev}} &= \sum_j \dot{n}_j G_{\text{m},j} - \sum_i \dot{n}_i G_{\text{m},i} \\
&= \sum_j \dot{n}_j \left(H_{\text{m},j} - T S_{\text{m},j} \right) - \sum_i \dot{n}_i \left(H_{\text{m},i} - T S_{\text{m},i} \right).
\end{aligned} \tag{24.142}
$$

24.4 Exergy of a Fossil Fuel

Let us have a closer look at a combustion process of a fossil fuel. For the following investigations any changes of kinetic and potential energies shall be ignored. The combustion shall be performed isothermally at ambient temperature $T_{\text{env}} = 298.15\ \text{K}$ and isobarically at ambient pressure $p_{\text{env}} = 1\ \text{bar}$, see Fig. 24.9. This represents a standard atmosphere. Since not only the thermomechanic properties of this standard atmosphere are relevant, its standard chemical composition is defined as well according to [29], see Table 24.1. This standard atmosphere is in thermodynamic equilibrium, i.e. it is saturated with water. It further occurs liquid water in this saturated atmosphere with a pressure of $p_{\text{env}} = 1\ \text{bar}$. The combustion shall be performed with oxygen, that is taken from this atmosphere. The exhaust gas (PR) is released to the huge atmosphere without changing its composition significantly. In order to find the exergy of the fossil fuel, the process must be reversible, see Sect. 24.3.3.

The following balances can be performed for steady state:

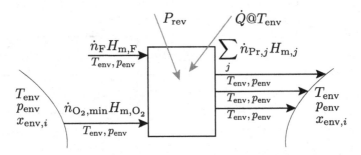

Fig. 24.9 Exergy of a fossil fuel—Energy balance

Table 24.1 Chemical composition of the standard atmosphere

x_{env,N_2}	x_{env,O_2}	x_{env,H_2O}	$x_{env,Ar}$	x_{env,CO_2}
0.756 08	0.2084	0.031 71	0.009 06	0.000 31

- First law of thermodynamics, see Fig. 24.9

$$\dot{Q} + P_{rev} + \dot{n}_{O_2,min} H_{m,O_2} + \dot{n}_F H_{m,F} = \sum_j \dot{n}_{Pr,j} H_{m,j} \qquad (24.143)$$

Rearranging brings

$$\dot{Q} + P_{rev} = \sum_j \dot{n}_{Pr,j} H_{m,j} - \dot{n}_{O_2,min} H_{m,O_2} - \dot{n}_F H_{m,F}. \qquad (24.144)$$

In case the product water is in liquid state, the heat flux and the release power are equal to the specific upper heating value, i.e.

$$\dot{Q} + P_{rev} = -\dot{n}_F H_{0M} \left(T_{env} \right). \qquad (24.145)$$

- Second law of thermodynamics, see Fig. 24.10

$$\dot{S}_i + \dot{S}_a + \dot{n}_{O_2,min} S_{m,O_2} + \dot{n}_F S_{m,F} = \sum_j \dot{n}_{Pr,j} S_{m,j} \qquad (24.146)$$

Since, the released technical power shall be maximised, the reaction needs to be reversible, so that

$$\dot{S}_i = 0. \qquad (24.147)$$

The exchanged entropy with the heat $\dot{S}_a$ can easily be substituted, since the combustion is isothermal, i.e.

$$\dot{S}_a = \frac{\dot{Q}}{T_{env}}. \qquad (24.148)$$

Hence, the balance of entropy is

$$\frac{\dot{Q}}{T_{\text{env}}} = \sum_j \dot{n}_{\text{Pr},j} S_{\text{m},j} - \dot{n}_{O_2,\text{min}} S_{\text{m},O_2} - \dot{n}_F S_{\text{m},F}. \tag{24.149}$$

The heat flux then is

$$\dot{Q} = T_{\text{env}} \left[\sum_j \dot{n}_{\text{Pr},j} S_{\text{m},j} - \dot{n}_{O_2,\text{min}} S_{\text{m},O_2} - \dot{n}_F S_{\text{m},F} \right]. \tag{24.150}$$

In combination with the first law of thermodynamics, see Eq. 24.144, it results in

$$P_{\text{rev}} = \sum_j \dot{n}_{\text{Pr},j} H_{\text{m},j} - \dot{n}_{O_2,\text{min}} H_{\text{m},O_2} - \dot{n}_F H_{\text{m},F} +$$

$$- T_{\text{env}} \left[\sum_j \dot{n}_{\text{Pr},j} S_{\text{m},j} - \dot{n}_{O_2,\text{min}} S_{\text{m},O_2} - \dot{n}_F S_{\text{m},F} \right]. \tag{24.151}$$

With the newly introduced Gibbs enthalpy it is

$$P_{\text{rev}} = \sum_j \dot{n}_{\text{Pr},j} G_{\text{m},j} - \dot{n}_{O_2,\text{min}} G_{\text{m},O_2} - \dot{n}_F G_{\text{m},F} \tag{24.152}$$

respectively

$$\frac{P_{\text{rev}}}{\dot{n}_F} = \sum_j \nu_{\text{Pr},j} G_{\text{m},j} - O_{\text{min}} G_{\text{m},O_2} - G_{\text{m},F}. \tag{24.153}$$

- Exergy balance, see Fig. 24.11

$$\dot{E}_{\text{x},Q} + P_{\text{rev}} + \dot{n}_F E_{\text{xm},F} + \dot{n}_{O_2,\text{min}} E_{\text{xm},O_2} = \Delta \dot{E}_{\text{x},V} + \sum_j \dot{n}_{\text{Pr},j} E_{\text{xm},j} \tag{24.154}$$

Since the heat $\dot{Q}$ is transferred at environmental temperature T_{env} it is

$$\dot{E}_{\text{x},Q} = 0. \tag{24.155}$$

The process shall further be reversible, so that $\Delta \dot{E}_{\text{x},V} = 0$. Solving for the exergy $E_{\text{xm},F}$ of the fuel leads to

$$\dot{n}_F E_{\text{xm},F} = -P_{\text{rev}} + \sum_j \dot{n}_{\text{Pr},j} E_{\text{xm},j} - \dot{n}_{O_2,\text{min}} E_{\text{xm},O_2} \tag{24.156}$$

respectively

$$E_{xm,F} = -\frac{P_{rev}}{\dot{n}_F} + \sum_j \nu_{Pr,j} E_{xm,j} - O_{min} E_{xm,O_2}. \qquad (24.157)$$

The exergy of the flows has been treated in Sect. 16.2. A flow in part I used to carry exergy due to a thermal and mechanical[6] imbalance with the environment. The flow of carries would have the following exergy

$$E_{xm,O_2} = H_{m,O_2} - H_{m,O_2,env} + T_{env} \left[S_{m,O_2,env} - S_{m,O_2} \right]. \qquad (24.158)$$

The first term disappears, since the oxygen enters with ambient temperature. The oxygen is supplied the combustion process with the reference pressure of p_{env}. Thus, it contains exergy, since the partial pressure of the oxygen is

$$p_{env,O_2} = p_{env} x_{env,O_2}. \qquad (24.159)$$

The exergy of the flow of oxygen obeys

$$E_{xm,O_2} = 0 + T_{env} \left[0 - R_M \log \frac{p_{env} x_{env,O_2}}{p_{env}} \right] = T_{env} R_M \log \frac{1}{x_{env,O_2}}. \qquad (24.160)$$

The products can be treated accordingly: However, during the combustion the exhaust gas components would mix with one another and leave the combustion process as a mixture of gases with a total pressure of p_{env}. Each component in this mixture then has a partial pressure that correlates with its concentration. Unfortunately, such a mixture is related with generation of entropy and thus with a loss of exergy, see Sect. 19.3.4. In order to find the exergy, i.e. maximum working capability of the fuel, any irreversibility has to be avoided. Consequently, in our approach each product component leaves the system boundary with ambient temperature and ambient pressure p_{env}, i.e. no mixing occurs, see also Chap. 19. Hence, the exergy, each component carries, is due to the different partial pressure of this component in the standard atmosphere, i.e. due to a thermomechanical imbalance. The individual partial pressure of a component j in the environment is

$$p_{env,j} = p_{env} x_{env,j}. \qquad (24.161)$$

The exergy then obeys

---

[6]In case kinetic/potential energies are counted as well!

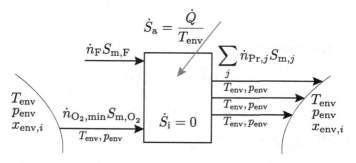

Fig. 24.10 Exergy of a fossil fuel—Entropy balance

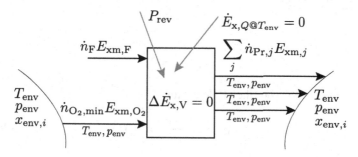

Fig. 24.11 Exergy of a fossil fuel—Exergy balance

$$E_{xm,j} = H_{m,j} - H_{m,j,env} + T_{env}\left[S_{m,j,env} - S_{m,j}\right]$$
$$= 0 + T_{env}\left[0 - R_M \log\frac{p_{env}x_{env,j}}{p_{env}}\right] \quad (24.162)$$
$$= T_{env}R_M \log\frac{1}{x_{env,j}}.$$

For the molar specific exergy of the fuel it then is

$$E_{xm,F} = -\frac{P_{rev}}{\dot{n}_F} + T_{env}R_M\left[\sum_j \nu_{Pr,j}\log\frac{1}{x_{env,j}} - O_{min}\log\frac{1}{x_{env,O_2}}\right] \quad (24.163)$$

In case the fuel contains sulphur, so that SO_2 is produced, the according reference atmosphere needs to be extended to x_{env,SO_2}.

Problem 24.5 Calculate the molar specific exergy of methane at standard conditions. i.e. $T_{env} = 298.15\,K$, $p_{env} = 1\,bar$ and a composition according to Table 24.1. The process is isobaric/isothermal.

Solution

The chemical equation reads as

$$CH_4 + 2O_2 \rightarrow CO_2 + 2H_2O \tag{24.164}$$

According to the method introduced in Sect. 23.3.5 it follows:

- Minimum oxygen need:

$$O_{min} = 2 \tag{24.165}$$

- Product composition:

$$\nu_{Pr,CO_2} = 1 \tag{24.166}$$

and

$$\nu_{Pr,H_2O} = 2 \tag{24.167}$$

The molar concentrations then follow with $\nu_{Pr} = 3$

$$x_{Pr,CO_2} = \frac{\nu_{Pr,CO_2}}{\nu_{Pr}} = 0.333 \tag{24.168}$$

and

$$x_{Pr,H_2O} = \frac{\nu_{Pr,H_2O}}{\nu_{Pr}} = 0.667. \tag{24.169}$$

Thus, almost the entire product water is in liquid state, since $x_{Pr,H_2O} > \frac{p_s(T_{env})}{p_{env}}$!

The molar specific Gibbs enthalpies are, refer to Tables B.1, B.2, B.3, B.4, B.5, B.6, B.7, B.8, B.9, B.10, B.11, B.12, B.13, B.14, B.15, B.16, B.17, B.18, B.19, B.20, B.21, B.22, B.23, B.24, B.25 and B.26 in Appendix B:

- Methane

$$G_{m,CH_4} = H_{m,CH_4} - T_{env}S_{m,CH_4} = -1.3017 \times 10^5 \frac{kJ}{kmol} \tag{24.170}$$

- Oxygen

$$G_{m,O_2} = H_{m,O_2} - T_{env}S_{m,O_2} = -6.1165 \times 10^4 \frac{kJ}{kmol} \tag{24.171}$$

- Carbon dioxide

$$G_{m,CO_2} = H_{m,CO_2} - T_{env}S_{m,CO_2} = -4.5725 \times 10^5 \frac{kJ}{kmol} \tag{24.172}$$

- Water (liquid)

$$G_{m,H_2O} = H_{m,H_2O} - T_{env} S_{m,H_2O} = -3.0668 \times 10^5 \frac{kJ}{kmol} \qquad (24.173)$$

Thus, it is

$$\frac{P_{rev}}{\dot{n}_F} = \sum_j \nu_{Pr,j} G_{m,j} - O_{min} G_{m,O_2} - G_{m,F} = -8.1812 \times 10^5 \frac{kJ}{kmol}. \qquad (24.174)$$

Finally, the molar specific exergy of the fuel is

$$E_{xm,F} = -\frac{P_{rev}}{\dot{n}_F} + T_{env} R_M \left[\sum_j \nu_{Pr,j} \log \frac{1}{x_{env,j}} - O_{min} \log \frac{1}{x_{env,O_2}} \right]. \qquad (24.175)$$

In this case the product water is assumed to be *entirely* liquid, it already is in equilibrium with the liquid water in the atmosphere, that also has a pressure of $p_{env} = 1$ bar, i.e.

$$E_{xm,F} = -\frac{P_{rev}}{\dot{n}_F} + T_{env} R_M \left[\nu_{Pr,CO_2} \log \frac{1}{x_{env,CO_2}} - O_{min} \log \frac{1}{x_{env,O_2}} \right]$$
$$= 830\,240 \frac{kJ}{kmol}. \qquad (24.176)$$

Mind, that the molar specific upper heating value is

$$H_{0M}(T_{env}) = -\sum_j \nu_{Pr,j} H_{m,j} + O_{min} H_{m,O_2} + H_{m,F} = 890\,570 \frac{kJ}{kmol} \qquad (24.177)$$

and the molar specific lower heating value with the water being in gaseous state

$$H_{UM}(T_{env}) = -\sum_j \nu_{Pr,j} H_{m,j} + O_{min} H_{m,O_2} + H_{m,F} = 802\,562 \frac{kJ}{kmol}. \qquad (24.178)$$

24.5 Chemical Potential

24.5.1 Multi-component Systems

Now, a system is considered in which there are k different substances. Let n_1 be the number of moles of substance 1, n_2 of substance 2 and so on. For a single component the Gibbs enthalpy G only depends on T and p, see Sect. 24.3. However, for variable compositions it then is

$$G = G\left(T, p, n_1, n_2, \ldots, n_k\right). \tag{24.179}$$

Thus, the total differential follows

$$dG = \left(\frac{\partial G}{\partial T}\right)_{p,n_i} dT + \left(\frac{\partial G}{\partial p}\right)_{T,n_i} dp + \sum_{i=1}^{k}\left(\frac{\partial G}{\partial n_i}\right)_{T,p,n_j} dn_i. \tag{24.180}$$

This expression can be interpreted as follows:

• Temperature potential

$$\left(\frac{\partial G}{\partial T}\right)_{p,n_i} dT \rightarrow \text{driver for heat transfer} \tag{24.181}$$

• Pressure potential

$$\left(\frac{\partial G}{\partial p}\right)_{T,n_i} dp \rightarrow \text{driver for exchange of work} \tag{24.182}$$

• Chemical potential

$$\sum_{i=1}^{k}\left(\frac{\partial G}{\partial n_i}\right)_{T,p,n_j} dn_i \rightarrow \text{driver for mass transfer} \tag{24.183}$$

A chemical potential μ_i is defined by

$$\boxed{\mu_i = \left(\frac{\partial G}{\partial n_i}\right)_{T,p,n_j}} \tag{24.184}$$

As we know from the single system it is

$$\left(\frac{\partial G}{\partial T}\right)_{p,n_i} = -S \tag{24.185}$$

and

$$\left(\frac{\partial G}{\partial p}\right)_{T,n_i} = V. \tag{24.186}$$

Hence, it follows

$$dG = V\,dp - S\,dT + \sum_{i=1}^{k}\mu_i dn_i. \tag{24.187}$$

The chemical potential μ_i is required for systems in which there are changes of composition. As already known, temperature is the driver for a thermal balancing process, pressure, however, is the driver for a mechanical balancing process. Consequently, the chemical potential is the driver for a chemical balancing process.

For any extensive state variable Z in a mixture under isothermal/isobaric conditions it is

$$Z(T, p, x_i) = n \sum_i x_i z_{m,i}(T, p) = \sum_i n_i z_{m,i}(T, p) =$$
$$= n_1 z_{m,1}(T, p) + n_2 z_{m,2}(T, p) + \cdots + n_k z_{m,k}(T, p) \tag{24.188}$$

with x_i being the mole fraction of each component and $z_{m,i}$ the molar state variable. This can be easily shown with the following example

$$V(T, p) = n_1 V_{m,1}(T, p) + n_2 V_{m,2}(T, p) + \cdots \tag{24.189}$$

Thus, it is

$$H(T, p) = n_1 H_{m,1}(T, p) + n_2 H_{m,2}(T, p) + \cdots \tag{24.190}$$

The entropy has to be calculated with care, in order to take into account the irreversibility of the mixing process. As shown in Chap. 19 one has to calculate with the partial pressure p_i of each component in the mixture, i.e.

$$S(T, p) = n_1 S_{m,1}(T, p_1) + n_2 S_{m,2}(T, p_2) + \cdots \tag{24.191}$$

For isothermal/isobaric conditions it results in

$$\left(\frac{\partial Z}{\partial n_i}\right)_{p,T} = Z_{m,i}. \tag{24.192}$$

Thus, substituting Z with G and the molar variable $z_{m,i}$ with $G_{m,i}$ leads to

$$\mu_i = \left(\frac{\partial G}{\partial n_i}\right)_{T,p,n_j} = G_{m,i} = H_{m,i} - T S_{m,i}. \tag{24.193}$$

The temperature dependency of μ_i is

$$\left(\frac{\partial \mu_i}{\partial T}\right)_{p,n_i} = \left(\frac{\partial^2 G}{\partial T \partial n_i}\right)_{p,n_i} = \left(\frac{\partial}{\partial n_i}\frac{\partial G}{\partial T}\right)_{p,n_i} = -\left(\frac{\partial}{\partial n_i}S\right)_{p,n_i} = -S_{m,i}. \tag{24.194}$$

The pressure dependency of μ_i is

$$\left(\frac{\partial \mu_i}{\partial p}\right)_{T,n_i} = \left(\frac{\partial^2 G}{\partial p \partial n_i}\right)_{T,n_i} = \left(\frac{\partial}{\partial n_i}\frac{\partial G}{\partial p}\right)_{T,n_i} = \left(\frac{\partial}{\partial n_i}V\right)_{T,n_i} = V_{m,i} \quad (24.195)$$

Hence, if $T = $ const. it follows

$$d\mu_i = V_{m,i}\,dp. \quad (24.196)$$

Based on the ideal gas equation $p_i V = n_i R_m T$ and a mixture of ideal gases, i.e. $\frac{p_i}{p} = \frac{V_i}{V}$, it is

$$V_i = \frac{p_i}{p}V = \frac{n_i R_m T}{p} \Rightarrow V_{m,i} = \frac{R_m T}{p}. \quad (24.197)$$

The integration from standard state (0) to partial pressure p_i yields

$$\boxed{\mu_i\,(T) = \mu_i^0\,(T) + R_m T \ln\frac{p_i}{p_0}} \quad (24.198)$$

Note, that the standard state is different at different temperatures.
    As seen before it is

$$\mu_i = \left(\frac{\partial G}{\partial n_i}\right)_{T,p,n_j} = G_{m,i} = H_{m,i} - T S_{m,i}. \quad (24.199)$$

Finally, the Gibbs-enthalpy for isothermal/isobaric conditions results in

$$\boxed{G = \sum_i n_i G_{m,i} = \sum_i n_i \mu_i\,(T, p, x_i) = \sum_i n_i\left(H_{m,i} - T S_{m,i}\right)} \quad (24.200)$$

The same result occurs if

$$dG = V\,dp - S\,dT + \sum_{i=1}^{k}\mu_i dn_i \quad (24.201)$$

is integrated while p and T remain constant.

Example 24.6 Consider the mixing of n_A moles A with n_B moles of B both at the same temperature T and pressure p, see Fig. 24.12. According to Dalton it is

$$p_A + p_B = p. \quad (24.202)$$

The first law of thermodynamic obeys

$$W_{12} + Q_{12} = 0 = U_2 - U_1$$
$$= n_A \left[U_{m,A}(T_2) - U_{m,A}(T_1) \right] + n_B \left[U_{m,B}(T_2) - U_{m,B}(T_1) \right].$$
(24.203)

Thus, it is $T_2 = T_1$. Consequently, it follows

$$\Delta H = n_A \left[H_{m,A}(T_2) - H_{m,A}(T_1) \right] + n_B \left[H_{m,B}(T_2) - H_{m,B}(T_1) \right] = 0 \quad (24.204)$$

respectively

$$\Delta H_m = \frac{\Delta H}{n_A} = 0. \tag{24.205}$$

According to the <u>second law of thermodynamics</u> it is

$$\Delta S = S_{i,12} + S_{a,12} = S_{i,12} > 0. \tag{24.206}$$

The rise of entropy is due to entropy generation at the transient balancing process. Since the system is supposed to be adiabatic, no entropy is exchanged with the environment. The entropy for the mixture then is

$$\Delta S = n_A \left[S_{m,A}(T_2, p_A) - S_{m,A}(T_1, p) \right] + n_B \left[S_{m,B}(T_2, p_B) - S_{m,B}(T_1, p) \right] > 0 \tag{24.207}$$

respectively

$$\Delta S_m = \frac{\Delta S}{n_A} > 0. \tag{24.208}$$

The <u>Gibbs enthalpy</u> for the mixing reads as

$$\Delta G = G_2 - G_1 \tag{24.209}$$

with

$$G_1 = n_A \left[H_{m,A,1} - T S_{m,A,1} \right] + n_B \left[H_{m,B,1} - T S_{m,B,1} \right] \tag{24.210}$$

and

$$G_2 = n_A \left[H_{m,A,2} - T S_{m,A,2} \right] + n_B \left[H_{m,B,2} - T S_{m,B,2} \right]. \tag{24.211}$$

Since $T = $ const. it follows $H_{m,A,1} = H_{m,A,2}$ and $H_{m,B,1} = H_{m,B,2}$. Thus, it results in

$$\Delta G = -n_A \left[S_{m,A,2} - S_{m,A,1} \right] T - n_B \left[S_{m,B,2} - S_{m,B,1} \right] T \tag{24.212}$$

with

$$S_{m,A,1} = S_{m,A}^0 - R_m \ln \frac{p_A}{p_0} \tag{24.213}$$

ans

$$S_{m,A,2} = S_{m,A}^0 - R_m \ln \frac{p}{p_0}. \tag{24.214}$$

B is analogue, i.e.

$$\Delta G = n_A R_m T \ln\frac{p_A}{p} + n_B R_m T \ln\frac{p_B}{p} \tag{24.215}$$

Alternatively, another approach can be applied:
The change of Gibbs enthalpy is

$$\Delta G_{mix} = G_2 - G_1. \tag{24.216}$$

With the following Gibbs-enthalpies:

$$G_2 = n_A \mu_{A,2} + n_B \mu_{B,2} \tag{24.217}$$

and

$$G_1 = n_A \mu_{A,1} + n_B \mu_{B,1}. \tag{24.218}$$

The chemical potentials are as follows. For gas A it is for the mixing

$$\mu_{A,1} = \mu_A^0 + R_m T \ln\frac{p}{p^0} \tag{24.219}$$

and after the mixing it is

$$\mu_{A,2} = \mu_A^0 + R_m T \ln\frac{p_A}{p^0}. \tag{24.220}$$

For gas B it follows accordingly

$$\mu_{B,1} = \mu_B^0 + R_m T \ln\frac{p}{p^0} \tag{24.221}$$

and

$$\mu_{B,2} = \mu_B^0 + R_m T \ln\frac{p_B}{p^0}. \tag{24.222}$$

Hence, the change of the Gibbs-enthalpy results in

$$\Delta G_{mix} = n_A R_m T \ln\frac{p_A}{p} + n_B R_m T \ln\frac{p_B}{p}. \tag{24.223}$$

Since

$$\frac{n_i}{n} = x_i = \pi_i = \frac{p_i}{p} \tag{24.224}$$

it finally is

$$\boxed{\Delta G_{mix} = n R_m T \left(x_A \ln x_A + x_B \ln x_B \right)} \tag{24.225}$$

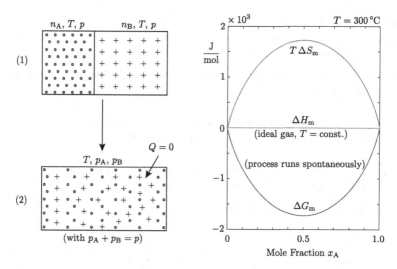

Fig. 24.12 Mixing of two ideal gases

As indicated by Fig. 24.12 the change of Gibbs enthalpy ΔG is negative, so that G decreases when gases are mixed. A negative ΔG indicates, that the process runs spontaneously without impact from outside. In case the molar quantity of A is the same as of B, so entropy generation is maximised and the Gibbs enthalpy minimised. However, according to Eq. 24.134 no work can be gained, since the entire Gibbs enthalpy is consumed by the dissipation. If the system would be reversible, see also Chap. 19, the Gibbs enthalpy, see Eq. 24.225, would be equal to Eq. 19.89. Anyhow, Eq. 24.206 has shown, that the mixing always is irreversible. Once the mixing is complete, i.e. a thermodynamic equilibrium is reached in state (2), ΔG is zero.

24.5.2 Chemical Reactions

Let there be a dynamic chemical reaction under isothermal/isobaric conditions according to

$$\gamma_A \, A + \gamma_B \, B \leftrightharpoons \gamma_C \, C + \gamma_D \, D \qquad (24.226)$$

The reaction can go back and forth. In case a *dynamic* equilibrium is reached the reactions back and forth run with the same velocity.[7] However, the Gibbs enthalpy G reaches a minimum, so that

$$dG = 0 \qquad (24.227)$$

---

[7]Hence, it is no *static* equilibrium. In a static equilibrium it would obey $dn_i = 0$.

Macroscopically, the molar quantities of all components remain constant, though chemical reactions run in both directions. In states (1) and (2) the molar quantities of each component are measured, see Fig. 24.13. The following cases can occur:

- If $\Delta G = G_2 - G_1 < 0$, the process runs spontaneously. The formation of products is dominating.

- If $dG = G_2 - G_1 = 0$, the process has already reached a chemical equilibrium, i.e. the Gibbs enthalpy reaches a minimum, see also Example 24.7.

- If $\Delta G = G_2 - G_1 > 0$, the process does not run spontaneously. Energy is required. The formation of educts is dominating.

Hence, in thermodynamic equilibrium it is

$$dG = \left(\frac{\partial G}{\partial T}\right)_{p,n_i} dT + \left(\frac{\partial G}{\partial p}\right)_{T,n_i} dp + \sum_{i=1}^{k} \underbrace{\left(\frac{\partial G}{\partial n_i}\right)_{T,p,n_j}}_{\mu_i} dn_i = 0. \quad (24.228)$$

Since the reaction shall be isothermal/isobaric the equation simplifies to

$$\sum_{i=1}^{k} \mu_i dn_i = 0. \quad (24.229)$$

Applied to the given example it is

$$\mu_A \, dn_A + \mu_B \, dn_B + \mu_C \, dn_C + \mu_D \, dn_D = 0. \quad (24.230)$$

According to the chemical equation it is

-
$$n_A = -\frac{\gamma_A}{\gamma_D} n_D \rightarrow dn_A = -\frac{\gamma_A}{\gamma_D} dn_D \quad (24.231)$$

-
$$n_B = -\frac{\gamma_B}{\gamma_D} n_D \rightarrow dn_B = -\frac{\gamma_B}{\gamma_D} dn_D \quad (24.232)$$

-
$$n_C = \frac{\gamma_C}{\gamma_D} n_D \rightarrow dn_C = \frac{\gamma_C}{\gamma_D} dn_D \quad (24.233)$$

Thus, it follows

$$\left(-\mu_A \frac{\gamma_A}{\gamma_D} - \mu_B \frac{\gamma_B}{\gamma_D} + \mu_C \frac{\gamma_C}{\gamma_D} + \mu_D \frac{\gamma_D}{\gamma_D}\right) dn_D = 0 \quad (24.234)$$

respectively

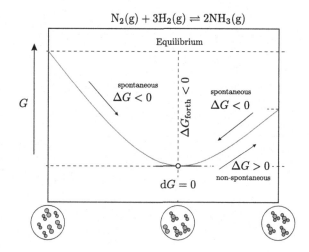

Fig. 24.13 Dynamic chemical reaction

Fig. 24.14 Chemical balancing driven by μ_i

$$N_2(g) + 3H_2(g) \rightleftharpoons 2NH_3(g)$$

$$-\gamma_A \mu_A - \gamma_B \mu_B + \gamma_C \mu_C + \gamma_D \mu_D = 0. \qquad (24.235)$$

In other words it is

$$\sum_{i,\text{products}} \gamma_i \mu_i - \sum_{j,\text{educts}} \gamma_j \mu_j = 0. \qquad (24.236)$$

Example 24.7 This example shows an equilibrium reaction

$$N_2(g) + 3H_2(g) \rightleftharpoons 2NH_3(g) \qquad (24.237)$$

An equilibrium reaction can run back and forth, see Fig. 24.14: Educts on the left hand side are chemically reactive and form the product on the right hand side. However, the mechanism can run vice versa, so that the product is split into the educts again. The driver for that reaction is the change of Gibbs' enthalpy from products to educts. In equilibrium state this $dG = 0$. Thus, it obeys

$$-\mu_{H_2} - 3\mu_{H_2} + 2\mu_{NH_3} = 0. \qquad (24.238)$$

For an isothermal/isobar reaction the cause for the chemical reaction and its equilibrium state is the chemical potential μ of the involved components.

Appendix A
Steam Table (Water) According to IAPWS

(See Tables A.1, A.2, A.3, A.4, A.5, A.6, A.7, A.8, A.9).[1]

Table A.1 Saturated liquid and saturated steam

ϑ °C	p_s bar	h' $\frac{kJ}{kg}$	h''	s' $\frac{kJ}{kg\,K}$	s''	v' $\frac{m^3}{kg}$	v''	c_p' $\frac{kJ}{kg\,K}$	c_p''
0	0.0061122	−0.041588	2500.9	−0.000155	9.1558	0.0010	206.1431	4.22	1.888
0.01	0.0061166	0.00061178	2500.911	0	9.1555	0.0010002	205.9975	4.2199	1.8882
1	0.0065709	4.1767	2502.7299	0.01526	9.1291	0.0010001	192.4447	4.2165	1.8889
2	0.0070599	8.3916	2504.5666	0.030606	9.1027	0.0010001	179.7636	4.2134	1.8895
3	0.0075808	12.6035	2506.4024	0.045886	9.0765	0.0010001	168.0141	4.2105	1.8902
4	0.0081355	16.8127	2508.2375	0.061101	9.0506	0.0010001	157.1213	4.2078	1.8909
5	0.0087257	21.0194	2510.0717	0.076252	9.0249	0.0010001	147.0169	4.2054	1.8917
6	0.0093535	25.2237	2511.9051	0.09134	8.9994	0.0010001	137.6382	4.2031	1.8924
7	0.010021	29.4258	2513.7377	0.10637	8.9742	0.0010001	128.9281	4.2011	1.8932
8	0.01073	33.626	2515.5693	0.12133	8.9492	0.0010002	120.8344	4.1992	1.894
9	0.011483	37.8244	2517.4001	0.13624	8.9244	0.0010003	113.3092	4.1974	1.8949
10	0.012282	42.0211	2519.2298	0.15109	8.8998	0.0010003	106.3087	4.1958	1.8957
11	0.013129	46.2162	2521.0586	0.16587	8.8755	0.0010004	99.7927	4.1943	1.8966
12	0.014028	50.41	2522.8864	0.18061	8.8514	0.0010005	93.7243	4.193	1.8975
13	0.014981	54.6024	2524.7132	0.19528	8.8275	0.0010007	88.0698	4.1917	1.8985
14	0.015989	58.7936	2526.5389	0.2099	8.8038	0.0010008	82.7981	4.1905	1.8994

(continued)

[1] International Association for the Properties of Water and Steam: The listed data has been generated with XSteam, see [20].

© Springer Nature Switzerland AG 2019
A. Schmidt, *Technical Thermodynamics for Engineers*,
https://doi.org/10.1007/978-3-030-20397-9

Table A.1 (continued)

ϑ °C	p_s bar	h' $\frac{kJ}{kg}$	h''	s' $\frac{kJ}{kg\,K}$	s''	v' $\frac{m^3}{kg}$	v''	c_p' $\frac{kJ}{kg\,K}$	c_p''
15	0.017057	62.9837	2528.3636	0.22447	8.7804	0.0010009	77.8807	4.1894	1.9004
16	0.018188	67.1727	2530.1871	0.23898	8.7571	0.0010011	73.2915	4.1884	1.9014
17	0.019383	71.3608	2532.0094	0.25344	8.7341	0.0010013	69.0063	4.1875	1.9025
18	0.020647	75.5479	2533.8307	0.26785	8.7112	0.0010015	65.0029	4.1866	1.9035
19	0.021982	79.7343	2535.6507	0.2822	8.6886	0.0010016	61.2609	4.1858	1.9046
20	0.023392	83.9199	2537.4695	0.2965	8.6661	0.0010018	57.7615	4.1851	1.9057
21	0.024881	88.1048	2539.287	0.31075	8.6439	0.0010021	54.4873	4.1844	1.9069
22	0.026452	92.289	2541.1033	0.32495	8.6218	0.0010023	51.4225	4.1838	1.908
23	0.028109	96.4727	2542.9182	0.3391	8.6	0.0010025	48.5521	4.1832	1.9092
24	0.029856	100.6558	2544.7318	0.3532	8.5783	0.0010028	45.8626	4.1827	1.9104
25	0.031697	104.8384	2546.5441	0.36726	8.5568	0.001003	43.3414	4.1822	1.9116
26	0.033637	109.0205	2548.3549	0.38126	8.5355	0.0010033	40.9768	4.1817	1.9129
27	0.035679	113.2022	2550.1643	0.39521	8.5144	0.0010035	38.7582	4.1813	1.9141
28	0.037828	117.3835	2551.9723	0.40912	8.4934	0.0010038	36.6754	4.1809	1.9154
29	0.040089	121.5645	2553.7788	0.42298	8.4727	0.0010041	34.7194	4.1806	1.9167
30	0.042467	125.7452	2555.5837	0.43679	8.4521	0.0010044	32.8816	4.1803	1.918
31	0.044966	129.9255	2557.3871	0.45056	8.4317	0.0010047	31.154	4.18	1.9194
32	0.047592	134.1057	2559.189	0.46428	8.4115	0.001005	29.5295	4.1798	1.9207
33	0.050351	138.2855	2560.9892	0.47795	8.3914	0.0010054	28.001	4.1795	1.9221
34	0.053247	142.4653	2562.7877	0.49158	8.3715	0.0010057	26.5624	4.1794	1.9235
35	0.056286	146.6448	2564.5846	0.50517	8.3518	0.001006	25.2078	4.1792	1.9249
36	0.059475	150.8242	2566.3797	0.51871	8.3323	0.0010064	23.9318	4.1791	1.9263
37	0.062818	155.0035	2568.1731	0.5322	8.3129	0.0010068	22.7292	4.179	1.9278
38	0.066324	159.1827	2569.9647	0.54566	8.2936	0.0010071	21.5954	4.1789	1.9292
39	0.069997	163.3619	2571.7545	0.55906	8.2746	0.0010075	20.5261	4.1788	1.9307
40	0.073844	167.541	2573.5424	0.57243	8.2557	0.0010079	19.517	4.1788	1.9322
41	0.077873	171.7202	2575.3284	0.58575	8.2369	0.0010083	18.5646	4.1788	1.9337
42	0.08209	175.8993	2577.1125	0.59903	8.2183	0.0010087	17.6652	4.1788	1.9353
43	0.086503	180.0785	2578.8946	0.61227	8.1999	0.0010091	16.8155	4.1788	1.9368
44	0.091118	184.2578	2580.6746	0.62547	8.1816	0.0010095	16.0126	4.1789	1.9384
45	0.095944	188.4372	2582.4526	0.63862	8.1634	0.0010099	15.2534	4.179	1.94
46	0.10099	192.6167	2584.2285	0.65174	8.1454	0.0010103	14.5355	4.1791	1.9416
47	0.10626	196.7963	2586.0023	0.66481	8.1276	0.0010108	13.8562	4.1792	1.9432
48	0.11176	200.9761	2587.7739	0.67785	8.1099	0.0010112	13.2132	4.1794	1.9449
49	0.11751	205.156	2589.5432	0.69084	8.0923	0.0010117	12.6045	4.1796	1.9466
50	0.12351	209.3362	2591.3103	0.70379	8.0749	0.0010121	12.0279	4.1798	1.9482
51	0.12977	213.5166	2593.075	0.71671	8.0576	0.0010126	11.4815	4.18	1.95
52	0.13631	217.6973	2594.8374	0.72958	8.0405	0.0010131	10.9637	4.1802	1.9517
53	0.14312	221.8782	2596.5973	0.74242	8.0235	0.0010136	10.4726	4.1805	1.9534
54	0.15022	226.0594	2598.3548	0.75522	8.0066	0.001014	10.0069	4.1808	1.9552
55	0.15761	230.241	2600.1098	0.76798	7.9899	0.0010145	9.5649	4.1811	1.957
56	0.16532	234.4229	2601.8622	0.7807	7.9733	0.001015	9.1454	4.1814	1.9588
57	0.17335	238.6052	2603.6121	0.79339	7.9568	0.0010155	8.7471	4.1818	1.9607
58	0.18171	242.7878	2605.3592	0.80603	7.9405	0.0010161	8.3688	4.1821	1.9625
59	0.19041	246.9709	2607.1037	0.81864	7.9243	0.0010166	8.0093	4.1825	1.9644
60	0.19946	251.1544	2608.8454	0.83122	7.9082	0.0010171	7.6677	4.1829	1.9664
61	0.20887	255.3383	2610.5843	0.84375	7.8922	0.0010176	7.3428	4.1834	1.9683

(continued)

Table A.1 (continued)

ϑ °C	p_s bar	h' $\frac{kJ}{kg}$	h''	s' $\frac{kJ}{kg\,K}$	s''	v' $\frac{m^3}{kg}$	v''	c_p' $\frac{kJ}{kg\,K}$	c_p''
62	0.21866	259.5228	2612.3203	0.85625	7.8764	0.0010182	7.0338	4.1838	1.9703
63	0.22884	263.7077	2614.0534	0.86872	7.8607	0.0010187	6.7399	4.1843	1.9723
64	0.23942	267.8931	2615.7836	0.88115	7.8451	0.0010193	6.4601	4.1848	1.9743
65	0.25041	272.0791	2617.5107	0.89354	7.8296	0.0010199	6.1938	4.1853	1.9764
66	0.26183	276.2657	2619.2347	0.9059	7.8142	0.0010204	5.9402	4.1859	1.9785
67	0.27368	280.4528	2620.9556	0.91823	7.799	0.001021	5.6986	4.1864	1.9806
68	0.28599	284.6405	2622.6733	0.93052	7.7839	0.0010216	5.4684	4.187	1.9828
69	0.29876	288.8289	2624.3877	0.94277	7.7689	0.0010222	5.249	4.1876	1.985
70	0.31201	293.0179	2626.0988	0.95499	7.754	0.0010228	5.0397	4.1882	1.9873
71	0.32575	297.2076	2627.8066	0.96718	7.7392	0.0010234	4.8402	4.1889	1.9895
72	0.34	301.398	2629.5109	0.97933	7.7245	0.001024	4.6498	4.1896	1.9919
73	0.35478	305.5891	2631.2117	0.99146	7.71	0.0010246	4.4681	4.1902	1.9942
74	0.37009	309.781	2632.909	1.0035	7.6955	0.0010252	4.2947	4.191	1.9966
75	0.38595	313.9736	2634.6026	1.0156	7.6812	0.0010258	4.1291	4.1917	1.999
76	0.40239	318.1669	2636.2926	1.0276	7.6669	0.0010265	3.9709	4.1924	2.0015
77	0.41941	322.3611	2637.9788	1.0396	7.6528	0.0010271	3.8198	4.1932	2.0041
78	0.43703	326.5561	2639.6612	1.0516	7.6388	0.0010277	3.6754	4.194	2.0066
79	0.45527	330.752	2641.3397	1.0635	7.6248	0.0010284	3.5373	4.1948	2.0092
80	0.47415	334.9487	2643.0143	1.0754	7.611	0.001029	3.4053	4.1956	2.0119
81	0.49368	339.1463	2644.6849	1.0873	7.5973	0.0010297	3.279	4.1965	2.0146
82	0.51387	343.3448	2646.3515	1.0991	7.5837	0.0010304	3.1582	4.1974	2.0174
83	0.53476	347.5443	2648.0139	1.1109	7.5701	0.001031	3.0426	4.1983	2.0202
84	0.55636	351.7447	2649.672	1.1227	7.5567	0.0010317	2.9319	4.1992	2.0231
85	0.57867	355.9461	2651.3259	1.1344	7.5434	0.0010324	2.8259	4.2001	2.026
86	0.60174	360.1485	2652.9755	1.1461	7.5301	0.0010331	2.7244	4.2011	2.029
87	0.62556	364.3519	2654.6206	1.1578	7.517	0.0010338	2.6272	4.202	2.0321
88	0.65017	368.5563	2656.2613	1.1694	7.5039	0.0010345	2.5341	4.203	2.0352
89	0.67559	372.7618	2657.8973	1.1811	7.4909	0.0010352	2.4448	4.2041	2.0383
90	0.70182	376.9684	2659.5288	1.1927	7.4781	0.0010359	2.3591	4.2051	2.0415
91	0.7289	381.1762	2661.1555	1.2042	7.4653	0.0010367	2.2771	4.2062	2.0448
92	0.75685	385.385	2662.7775	1.2158	7.4526	0.0010374	2.1983	4.2072	2.0482
93	0.78568	389.595	2664.3946	1.2273	7.44	0.0010381	2.1228	4.2083	2.0516
94	0.81542	393.8062	2666.0068	1.2387	7.4275	0.0010389	2.0502	4.2095	2.0551
95	0.84609	398.0185	2667.6139	1.2502	7.415	0.0010396	1.9806	4.2106	2.0586
96	0.87771	402.2321	2669.216	1.2616	7.4027	0.0010404	1.9138	4.2118	2.0623
97	0.91031	406.447	2670.813	1.273	7.3904	0.0010411	1.8497	4.213	2.066
98	0.9439	410.6631	2672.4047	1.2844	7.3782	0.0010419	1.788	4.2142	2.0697
99	0.97852	414.8805	2673.991	1.2957	7.3661	0.0010427	1.7288	4.2154	2.0736
100	1.0142	419.0992	2675.572	1.307	7.3541	0.0010435	1.6719	4.2166	2.0775
105	1.209	440.2131	2683.3933	1.3632	7.2951	0.0010474	1.4185	4.2232	2.0983
110	1.4338	461.3634	2691.0676	1.4187	7.238	0.0010516	1.2094	4.2304	2.1212
115	1.6918	482.5528	2698.5848	1.4735	7.1827	0.0010559	1.0359	4.2381	2.1464
120	1.9867	503.7846	2705.9342	1.5278	7.1291	0.0010603	0.8913	4.2464	2.174
125	2.3222	525.0618	2713.1055	1.5815	7.077	0.0010649	0.77011	4.2553	2.2042
130	2.7026	546.3878	2720.0878	1.6346	7.0264	0.0010697	0.66808	4.2648	2.237
135	3.132	567.7661	2726.8708	1.6872	6.9772	0.0010747	0.5818	4.2751	2.2726

(continued)

Table A.1 (continued)

ϑ °C	p_s bar	h' $\frac{kJ}{kg}$	h''	s' $\frac{kJ}{kg\,K}$	s''	v' $\frac{m^3}{kg}$	v''	c_p' $\frac{kJ}{kg\,K}$	c_p''
140	3.615	589.2003	2733.4439	1.7393	6.9293	0.0010798	0.50852	4.286	2.3109
145	4.1563	610.6941	2739.7968	1.7909	6.8826	0.001085	0.44602	4.2978	2.352
150	4.761	632.2516	2745.9191	1.842	6.837	0.0010905	0.3925	4.3103	2.3959
155	5.4342	653.8769	2751.8005	1.8926	6.7926	0.0010962	0.3465	4.3236	2.4425
160	6.1814	675.5747	2757.4305	1.9428	6.7491	0.001102	0.30682	4.3379	2.4918
165	7.0082	697.3495	2762.7985	1.9926	6.7066	0.001108	0.27246	4.3532	2.5438
170	7.9205	719.2064	2767.8937	2.0419	6.6649	0.0011143	0.24262	4.3695	2.5985
175	8.9245	741.1507	2772.7045	2.0909	6.6241	0.0011207	0.2166	4.3869	2.656
180	10.0263	763.188	2777.2194	2.1395	6.5841	0.0011274	0.19386	4.4056	2.7164
185	11.2327	785.3243	2781.4259	2.1878	6.5447	0.0011343	0.17392	4.4255	2.7798
190	12.5502	807.566	2785.311	2.2358	6.506	0.0011414	0.15638	4.4468	2.8464
195	13.9858	829.9199	2788.861	2.2834	6.4679	0.0011488	0.14091	4.4696	2.9163
200	15.5467	852.3931	2792.0616	2.3308	6.4303	0.0011565	0.12722	4.494	2.99
205	17.2402	874.9933	2794.8974	2.3779	6.3932	0.0011645	0.11509	4.5202	3.0677
210	19.0739	897.7289	2797.3523	2.4248	6.3565	0.0011727	0.1043	4.5482	3.1496
215	21.0555	920.6086	2799.4096	2.4714	6.3202	0.0011813	0.094689	4.5784	3.2363
220	23.1929	943.6417	2801.051	2.5178	6.2842	0.0011902	0.086101	4.6107	3.328
225	25.4942	966.8384	2802.2577	2.5641	6.2485	0.0011994	0.078411	4.6455	3.4252
230	27.9679	990.2095	2803.0093	2.6102	6.2131	0.001209	0.07151	4.6829	3.5283
235	30.6224	1013.7668	2803.2844	2.6561	6.1777	0.001219	0.065304	4.7233	3.6379
240	33.4665	1037.5228	2803.06	2.7019	6.1425	0.0012295	0.05971	4.7668	3.7545
245	36.5091	1061.4911	2802.3114	2.7477	6.1074	0.0012404	0.054658	4.8139	3.8789
250	39.7594	1085.6868	2801.0121	2.7934	6.0722	0.0012517	0.050087	4.865	4.0119
255	43.2267	1110.126	2799.1333	2.8391	6.037	0.0012636	0.045941	4.9204	4.1545
260	46.9207	1134.8266	2796.6436	2.8847	6.0017	0.0012761	0.042175	4.9807	4.308
265	50.8512	1159.8083	2793.5086	2.9304	5.9662	0.0012892	0.038748	5.0466	4.4741
270	55.0284	1185.0928	2789.6899	2.9762	5.9304	0.001303	0.035622	5.1188	4.6547
275	59.4626	1210.7046	2785.1446	3.0221	5.8943	0.0013175	0.032767	5.1982	4.8524
280	64.1646	1236.671	2779.8245	3.0681	5.8578	0.0013328	0.030154	5.2859	5.0701
285	69.1454	1263.023	2773.6747	3.1143	5.8208	0.0013491	0.027758	5.3832	5.3114
290	74.4164	1289.7957	2766.6326	3.1608	5.7832	0.0013663	0.025557	5.4918	5.5806
295	79.9895	1317.0297	2758.6266	3.2076	5.7449	0.0013846	0.023531	5.6137	5.8825
300	85.8771	1344.7713	2749.5737	3.2547	5.7058	0.0014042	0.021663	5.7515	6.2231
305	92.0919	1373.0748	2739.3776	3.3024	5.6656	0.0014252	0.019937	5.9083	6.6096
310	98.6475	1402.0034	2727.9243	3.3506	5.6243	0.0014479	0.018339	6.0883	7.0513
315	105.558	1431.6321	2715.0772	3.3994	5.5816	0.0014724	0.016856	6.2968	7.561
320	112.8386	1462.051	2700.6677	3.4491	5.5373	0.0014991	0.015476	6.5414	8.1575
325	120.5052	1493.3719	2684.483	3.4997	5.4911	0.0015283	0.014189	6.8331	8.8689
330	128.5752	1525.738	2666.2483	3.5516	5.4425	0.0015606	0.012984	7.1888	9.7381
335	137.0673	1559.3407	2645.6023	3.6048	5.391	0.0015967	0.011852	7.6354	10.83
340	146.0018	1594.4466	2622.0667	3.6599	5.3359	0.0016375	0.010784	8.2166	12.2412
345	155.4015	1631.4365	2595.0082	3.7175	5.2763	0.0016846	0.0097698	9.0023	14.112
350	165.2916	1670.8892	2563.6305	3.7783	5.2109	0.0017401	0.0088009	10.102	16.6415
355	175.7012	1713.7096	2526.4499	3.8438	5.1377	0.0018078	0.007866	11.8584	20.7136
360	186.664	1761.4911	2480.9862	3.9164	5.0527	0.0018945	0.006945	14.8744	27.5691
365	198.2216	1817.5893	2422.0051	4.001	4.9482	0.0020156	0.0060043	21.4744	42.0135
370	210.4337	1892.6429	2333.5016	4.1142	4.7996	0.0022221	0.0049462	47.1001	93.4065
373.946	220.6397	2087.5		4.412		0.0031	∞		

Table A.2 Specific volume v of water in $\frac{m^3}{kg}$ 1/2

p in bar	ϑ in °C										
	0	25	50	75	100	125	150	200	250	300	350
0.01	0.0010002	137.5362	149.0961	160.6471	172.1934	183.737	195.2789	218.3599	241.4391	264.5173	287.5949
0.1	0.0010002	0.001003	14.8674	16.0347	17.1967	18.356	19.5136	21.826	24.1365	26.446	28.755
1	0.0010002	0.001003	0.0010121	0.0010258	1.696	1.8173	1.9367	2.1725	2.4062	2.6389	2.871
5	0.00099995	0.0010028	0.0010119	0.0010256	0.0010433	0.0010648	0.0010905	0.42503	0.47443	0.5226	0.57014
10	0.0009997	0.0010026	0.0010117	0.0010254	0.001043	0.0010645	0.0010902	0.206	0.23274	0.25798	0.28249
20	0.00099919	0.0010021	0.0010113	0.0010249	0.0010425	0.0010639	0.0010895	0.0011561	0.11148	0.1255	0.13859
30	0.00099869	0.0010017	0.0010108	0.0010244	0.001042	0.0010633	0.0010888	0.001155	0.070622	0.081175	0.090555
40	0.00099818	0.0010012	0.0010104	0.001024	0.0010415	0.0010628	0.0010881	0.001154	0.0012517	0.058868	0.066474
50	0.00099768	0.0010008	0.0010099	0.0010235	0.001041	0.0010622	0.0010875	0.001153	0.0012499	0.045347	0.051971
60	0.00099718	0.0010003	0.0010095	0.0010231	0.0010405	0.0010616	0.0010868	0.0011521	0.0012481	0.036191	0.042253
70	0.00099668	0.00099987	0.0010091	0.0010226	0.00104	0.0010611	0.0010862	0.0011511	0.0012463	0.029494	0.035265
80	0.00099619	0.00099942	0.0010086	0.0010222	0.0010395	0.0010605	0.0010855	0.0011501	0.0012446	0.02428	0.029978
90	0.00099569	0.00099898	0.0010082	0.0010217	0.001039	0.00106	0.0010849	0.0011491	0.0012429	0.0014024	0.025818
100	0.0009952	0.00099854	0.0010078	0.0010212	0.0010385	0.0010594	0.0010842	0.0011482	0.0012412	0.001398	0.022442
150	0.00099276	0.00099636	0.0010056	0.001019	0.0010361	0.0010567	0.001081	0.0011435	0.001233	0.0013783	0.011481
200	0.00099035	0.0009942	0.0010035	0.0010168	0.0010337	0.001054	0.0010779	0.001139	0.0012254	0.0013611	0.0016649
250	0.00098799	0.00099208	0.0010014	0.0010146	0.0010313	0.0010514	0.0010749	0.0011346	0.0012181	0.0013459	0.0015988
300	0.00098567	0.00098998	0.00099934	0.0010125	0.001029	0.0010488	0.001072	0.0011304	0.0012113	0.0013322	0.0015529
350	0.00098338	0.00098792	0.00099731	0.0010104	0.0010267	0.0010462	0.0010691	0.0011264	0.0012048	0.0013197	0.0015175
400	0.00098113	0.00098588	0.00099531	0.0010083	0.0010245	0.0010438	0.0010663	0.0011224	0.0011986	0.0013083	0.0014884
450	0.00097891	0.00098388	0.00099334	0.0010063	0.0010223	0.0010413	0.0010635	0.0011186	0.0011927	0.0012977	0.0014638
500	0.00097673	0.0009819	0.00099139	0.0010043	0.0010201	0.0010389	0.0010608	0.0011149	0.0011871	0.0012879	0.0014424

(continued)

Table A.2 (continued)

p in bar	ϑ in °C										
	0	25	50	75	100	125	150	200	250	300	350
600	0.00097247	0.00097802	0.00098758	0.0010003	0.0010159	0.0010343	0.0010555	0.0011077	0.0011764	0.00127	0.0014067
700	0.00096834	0.00097425	0.00098386	0.00099648	0.0010118	0.0010298	0.0010505	0.001101	0.0011666	0.0012541	0.0013773
800	0.00096434	0.00097057	0.00098025	0.00099276	0.0010078	0.0010255	0.0010456	0.0010945	0.0011574	0.0012398	0.0013525
900	0.00096046	0.00096699	0.00097673	0.00098913	0.001004	0.0010212	0.001041	0.0010884	0.0011488	0.0012268	0.0013309
1000	0.00095669	0.00096351	0.00097329	0.0009856	0.0010002	0.0010172	0.0010364	0.0010826	0.0011407	0.0012148	0.0013118

Table A.3 Specific volume v of water in $\frac{m^3}{kg}$ 2/2

p in bar	ϑ in °C								
	400	450	500	550	600	650	700	750	800
0.01	310.6722	333.7492	356.8261	379.9029	402.9796	426.0562	449.1327	472.2092	495.2857
0.1	31.0636	33.372	35.6802	37.9883	40.2963	42.6042	44.9121	47.2199	49.5278
1	3.1027	3.3342	3.5656	3.7968	4.0279	4.259	4.49	4.721	4.952
5	0.61729	0.66421	0.71095	0.75757	0.8041	0.85056	0.89696	0.94333	0.98967
10	0.30659	0.33044	0.35411	0.37766	0.40111	0.4245	0.44783	0.47112	0.49438
20	0.15121	0.16354	0.17568	0.18769	0.19961	0.21146	0.22326	0.23501	0.24674
30	0.099377	0.10788	0.11619	0.12437	0.13244	0.14045	0.1484	0.15631	0.16419
40	0.073432	0.080042	0.086441	0.092699	0.098857	0.10494	0.11097	0.11696	0.12292
50	0.05784	0.063325	0.068583	0.073694	0.078703	0.083637	0.088515	0.09335	0.098151
60	0.047423	0.052168	0.056672	0.061021	0.065264	0.069432	0.073542	0.077609	0.081642
70	0.039962	0.04419	0.048159	0.051966	0.055664	0.059284	0.062847	0.066365	0.069849
80	0.034348	0.038197	0.041769	0.045172	0.048463	0.051673	0.054825	0.057933	0.061005
90	0.029963	0.033528	0.036795	0.039886	0.04286	0.045753	0.048586	0.051374	0.054127
100	0.026439	0.029785	0.032813	0.035655	0.038377	0.041016	0.043594	0.046127	0.048624
150	0.015671	0.018478	0.020828	0.022945	0.024921	0.026803	0.028619	0.030387	0.032118
200	0.0099496	0.01272	0.014793	0.016571	0.018184	0.019694	0.021133	0.02252	0.023869
250	0.0060061	0.0091752	0.011142	0.012735	0.01414	0.01543	0.016643	0.017803	0.018922
300	0.0027982	0.0067381	0.0086903	0.010175	0.011444	0.01259	0.013654	0.014662	0.015629
350	0.0021053	0.0049589	0.0069334	0.0083477	0.0095231	0.010566	0.011524	0.012423	0.01328
400	0.0019109	0.0036921	0.0056249	0.0069853	0.0080891	0.0090538	0.009931	0.010748	0.011523
450	0.0018038	0.002915	0.0046344	0.0059384	0.0069828	0.0078848	0.0086979	0.0094507	0.01016
500	0.001731	0.0024873	0.0038894	0.0051185	0.0061087	0.0069575	0.0077176	0.0084175	0.0090741

(continued)

Table A.3 (continued)

p in bar	ϑ in °C								
	400	450	500	550	600	650	700	750	800
600	0.001633	0.0020847	0.0029518	0.0039548	0.0048336	0.0055908	0.0062651	0.0068818	0.0074568
700	0.0015663	0.0018921	0.0024632	0.0032232	0.0039749	0.0046483	0.0052519	0.0058036	0.0063167
800	0.0015162	0.0017739	0.0021881	0.0027601	0.0033837	0.003975	0.0045161	0.0050133	0.0054762
900	0.0014763	0.0016911	0.0020143	0.0024578	0.0029695	0.0034823	0.0039663	0.0044159	0.004836
1000	0.0014431	0.0016282	0.0018934	0.0022498	0.0026723	0.0031145	0.0035462	0.0039532	0.0043355

Table A.4 Specific enthalpy h of water in $\frac{kJ}{kg}$ 1/2

p in bar	ϑ in °C										
	0	25	50	75	100	125	150	200	250	300	350
0.01	-0.041192	2547.5538	2594.4004	2641.3698	2688.536	2735.9492	2783.6463	2879.9959	2977.7344	3076.9529	3177.7157
0.1	-0.032023	104.8447	2591.9936	2639.8034	2687.4309	2735.1334	2783.0201	2879.5902	2977.4457	3076.7343	3177.5435
1	0.059662	104.9281	209.4118	314.0231	2675.7674	2726.6805	2776.5918	2875.4751	2974.5371	3074.5404	3175.8174
5	0.467	105.2985	209.7568	314.3458	419.3985	525.2465	632.2663	2855.8962	2961.1298	3064.5962	3168.0612
10	0.97582	105.7613	210.1879	314.7492	419.7742	525.5915	632.5749	2828.2675	2943.2222	3051.7032	3158.1633
20	1.9923	106.6864	211.0499	315.556	420.5256	526.2821	633.1931	852.5725	2903.2314	3024.2519	3137.6412
30	3.0072	107.6107	211.9116	316.3627	421.2774	526.9734	633.8126	852.9781	2856.5485	2994.3493	3116.0622
40	4.0206	108.5342	212.773	317.1695	422.0295	527.6654	634.4334	853.3874	1085.6861	2961.6515	3093.3182
50	5.0325	109.457	213.634	317.9762	422.7819	528.3581	635.0554	853.8004	1085.662	2925.644	3069.2942
60	6.0429	110.3791	214.4947	318.7829	423.5345	529.0515	635.6786	854.217	1085.6501	2885.4905	3043.8584
70	7.0517	111.3004	215.355	319.5896	424.2875	529.7455	636.3031	854.6371	1085.65	2839.8277	3016.8497
80	8.0591	112.2209	216.215	320.3962	425.0407	530.4402	636.9287	855.0607	1085.6614	2786.3785	2988.061
90	9.0649	113.1408	217.0747	321.2028	425.7942	531.1355	637.5556	855.4876	1085.6839	1344.2693	2957.2186
100	10.0693	114.0599	217.934	322.0094	426.548	531.8315	638.1836	855.9179	1085.7172	1343.0966	2923.9578
150	15.0694	118.6443	222.2256	326.0417	430.3208	535.3207	641.3403	858.1171	1086.0356	1338.0633	2692.9998
200	20.0338	123.2107	226.5087	330.0728	434.0996	538.8246	644.5238	860.3911	1086.5836	1334.1395	1645.9511
250	24.9636	127.7595	230.7833	334.1026	437.8839	542.3422	647.7322	862.734	1087.3336	1331.0633	1623.8646
300	29.8599	132.2909	235.0496	338.1307	441.6732	545.8726	650.9638	865.1402	1088.2629	1328.6604	1608.7975
350	34.7235	136.8054	239.3074	342.1569	445.4671	549.415	654.2174	867.605	1089.3525	1326.8077	1597.5402
400	39.5556	141.3033	243.5569	346.1812	449.265	552.9686	657.4915	870.1243	1090.5863	1325.414	1588.7405
450	44.357	145.7849	247.798	350.2032	453.0667	556.5327	660.7849	872.6941	1091.9506	1324.4102	1581.7011
500	49.1286	150.2505	252.0309	354.2229	456.8716	560.1065	664.0964	875.3112	1093.4335	1323.7424	1575.9832
600	58.5861	159.1351	260.472	362.2545	464.4903	567.2813	670.7697	880.6748	1096.7157	1323.2514	1567.412

(continued)

Table A.4 (continued)

p in bar	ϑ in °C												
	0	25	50	75	100	125	150	200	250	300	350
700	67.9346	167.9596	268.8806	370.2752	472.1184	574.4884	677.5039	886.1936	1100.3648	1323.6819	1561.5683
800	77.1804	176.7265	277.2572	378.284	479.7538	581.7238	684.2923	891.8495	1104.3275	1324.8526	1557.6673
900	86.3292	185.438	285.6023	386.2804	487.3947	588.9842	691.1293	897.627	1108.5612	1326.6315	1555.2253
1000	95.386	194.0963	293.9166	394.2638	495.0395	596.2666	698.0098	903.5132	1113.031	1328.9193	1553.9225

Table A.5 Specific enthalpy h of water in $\frac{kJ}{kg}$ 2/2

p in bar	ϑ in °C								
	400	450	500	550	600	650	700	750	800
0.01	3280.0756	3384.0797	3489.7698	3597.1812	3706.3406	3817.2647	3929.9603	4044.4266	4160.6592
0.1	3279.9362	3383.9646	3489.6734	3597.0995	3706.2707	3817.2044	3929.9078	4044.3805	4160.6185
1	3278.5396	3382.8123	3488.7086	3596.2822	3705.5715	3816.6009	3929.3827	4043.9201	4160.2118
5	3272.292	3377.6655	3484.4082	3592.6421	3702.4586	3813.9149	3927.046	4041.8714	4158.4023
10	3264.3855	3371.19	3479.0037	3588.0739	3698.5556	3810.549	3924.1188	4039.3058	4156.1368
20	3248.2271	3358.0519	3468.0932	3578.8756	3690.7089	3803.7889	3918.2436	4034.1585	4151.5935
30	3231.571	3344.6585	3457.0405	3569.5916	3682.8068	3796.9905	3912.3407	4028.9905	4147.0344
40	3214.3735	3330.9912	3445.8374	3560.2183	3674.8479	3790.1538	3906.4104	4023.8023	4142.4601
50	3196.5917	3317.032	3434.4761	3550.7526	3666.8311	3783.2785	3900.4531	4018.5944	4137.8714
60	3178.183	3302.7635	3422.9493	3541.1913	3658.7552	3776.3644	3894.469	4013.3674	4133.2689
70	3159.1044	3288.1695	3411.2503	3531.5318	3650.6193	3769.4116	3888.4587	4008.1219	4128.6531
80	3139.3106	3273.234	3399.3726	3521.7715	3642.4227	3762.4199	3882.4224	4002.8585	4124.0248
90	3118.7526	3257.9419	3387.3105	3511.9082	3634.1646	3755.3896	3876.3607	3997.5778	4119.3844
100	3097.3753	3242.2779	3375.0584	3501.9399	3625.8446	3748.3207	3870.2739	3992.2803	4114.7328
150	2975.5477	3157.8415	3310.7911	3450.474	3583.3076	3712.4081	3839.4829	3965.5633	4091.3257
200	2816.8362	3061.5343	3241.1865	3396.2412	3539.2259	3675.5941	3808.1522	3938.5205	4067.7254
250	2578.7494	2950.3799	3165.9152	3339.2842	3493.6905	3637.973	3776.3704	3911.233	4044.0049
300	2152.7882	2820.9123	3084.7924	3279.7892	3446.8724	3599.6773	3744.2417	3883.7843	4020.2341
350	1988.2517	2670.967	2998.0171	3218.0815	3399.0162	3560.8734	3711.8836	3856.2609	3996.4806
400	1931.1915	2511.6302	2906.6872	3154.6483	3350.4327	3521.7571	3679.4249	3828.7509	3972.8094
450	1897.6375	2377.2586	2813.3506	3090.1926	3301.4916	3482.5474	3647.0029	3801.3439	3949.2828
500	1874.3676	2284.3776	2722.5198	3025.7028	3252.6142	3443.4808	3614.7605	3774.1295	3925.9604
600	1843.1916	2179.8228	2570.4044	2902.0646	3156.9527	3366.7648	3551.3945	3720.6353	3880.1539

(continued)

Table A.5 (continued)

p in bar	ϑ in °C									
	400	450	500	550	600	650	700	750	800
700	1822.9002	2123.3881	2466.3096	2795.013	3067.5056	3293.5699	3490.4519	3668.9601	3835.8142
800	1808.7051	2087.5894	2397.6482	2709.9985	2988.0897	3225.6656	3432.9206	3619.7377	3793.3225
900	1798.473	2062.7176	2350.381	2645.2374	2920.7605	3164.4114	3379.5448	3573.5076	3753.0163
1000	1791.0568	2044.579	2316.2634	2596.0109	2865.0706	3110.6026	3330.7556	3530.6802	3715.1889

Table A.6 Specific entropy s of water in $\frac{kJ}{kg\,K}$ 1/2

p in bar	ϑ in °C										
	0	25	50	75	100	125	150	200	250	300	350
0.01	−0.00015452	9.0921	9.243	9.383	9.5138	9.6368	9.753	9.9682	10.1645	10.3456	10.5142
0.1	−0.00015391	0.36725	3.1741	8.3167	8.4488	8.5725	8.6892	8.9048	9.1014	9.2827	9.4513
1	−0.0001478	0.36723	0.70375	1.0156	7.361	7.4931	7.6147	7.8356	8.0346	8.2171	8.3865
5	−0.00012103	0.36713	0.70357	1.0153	1.3067	1.5813	1.8419	7.0611	7.2726	7.4614	7.6345
10	−8.8423e−05	0.367	0.70334	1.015	1.3063	1.5808	1.8414	6.6955	6.9266	7.1247	7.3028
20	−2.6077e−05	0.36674	0.70287	1.0144	1.3055	1.5798	1.8403	2.3301	6.5474	6.7685	6.9582
30	3.2474e−05	0.36648	0.70241	1.0137	1.3048	1.5789	1.8391	2.3285	6.2893	6.5412	6.7449
40	8.7256e−05	0.36622	0.70195	1.0131	1.304	1.578	1.838	2.3269	2.7933	6.3638	6.5843
50	0.0001383	0.36596	0.70149	1.0125	1.3032	1.577	1.8369	2.3254	2.7909	6.2109	6.4515
60	0.00018563	0.36569	0.70103	1.0119	1.3024	1.5761	1.8358	2.3238	2.7885	6.0702	6.3356
70	0.00022927	0.36543	0.70057	1.0112	1.3017	1.5752	1.8347	2.3223	2.7861	5.9335	6.2303
80	0.00026926	0.36516	0.70011	1.0106	1.3009	1.5743	1.8337	2.3207	2.7837	5.7935	6.1319
90	0.00030561	0.3649	0.69965	1.01	1.3001	1.5734	1.8326	2.3192	2.7814	3.2529	6.0378
100	0.00033836	0.36463	0.69919	1.0094	1.2994	1.5724	1.8315	2.3177	2.7791	3.2484	5.9458
150	0.00044893	0.36328	0.69689	1.0063	1.2956	1.5679	1.8262	2.3102	2.7679	3.2275	5.4435
200	0.00047328	0.3619	0.6946	1.0033	1.2918	1.5635	1.8209	2.303	2.7572	3.2087	3.7288
250	0.00041456	0.36051	0.69232	1.0003	1.2881	1.5591	1.8158	2.2959	2.7469	3.1915	3.6803
300	0.00027584	0.35908	0.69004	0.99729	1.2845	1.5548	1.8107	2.289	2.7371	3.1756	3.6435
350	6.0091e−05	0.35764	0.68777	0.99433	1.2809	1.5505	1.8058	2.2823	2.7276	3.1608	3.6131
400	−0.00022982	0.35618	0.68551	0.99139	1.2773	1.5463	1.8009	2.2758	2.7185	3.1469	3.587
450	−0.00059112	0.35469	0.68325	0.98848	1.2738	1.5422	1.7961	2.2693	2.7097	3.1338	3.5638
500	−0.0010211	0.35319	0.68099	0.98558	1.2703	1.5381	1.7914	2.2631	2.7012	3.1214	3.543
600	−0.0020771	0.35012	0.67649	0.97987	1.2634	1.5301	1.7822	2.2509	2.6848	3.0982	3.5064

(continued)

Table A.6 (continued)

p in bar	ϑ in °C										
	0	25	50	75	100	125	150	200	250	300	350
700	−0.0033782	0.34698	0.67201	0.97423	1.2567	1.5223	1.7732	2.2392	2.6694	3.0769	3.4747
800	−0.0049067	0.34377	0.66754	0.96866	1.2501	1.5146	1.7645	2.228	2.6548	3.0572	3.4465
900	−0.0066463	0.34049	0.66309	0.96317	1.2436	1.5071	1.756	2.2171	2.6408	3.0388	3.4211
1000	−0.0085823	0.33716	0.65864	0.95774	1.2373	1.4998	1.7477	2.2066	2.6275	3.0215	3.3978

Table A.7 Specific entropy s of water in $\frac{kJ}{kg\,K}$ 2/2

p in bar	ϑ in °C								
	400	450	500	550	600	650	700	750	800
0.01	10.6722	10.8212	10.9625	11.0971	11.2258	11.3494	11.4682	11.5829	11.6938
0.1	9.6093	9.7584	9.8997	10.0343	10.1631	10.2866	10.4055	10.5202	10.6311
1	8.5451	8.6945	8.8361	8.9709	9.0998	9.2234	9.3424	9.4571	9.5681
5	7.7954	7.9464	8.0891	8.2247	8.3543	8.4784	8.5977	8.7128	8.824
10	7.4668	7.6198	7.764	7.9007	8.0309	8.1557	8.2755	8.3909	8.5024
20	7.129	7.2863	7.4335	7.5723	7.7042	7.8301	7.9509	8.067	8.1791
30	6.9233	7.0853	7.2356	7.3767	7.5102	7.6373	7.759	7.8759	7.9885
40	6.7712	6.9383	7.0919	7.2353	7.3704	7.4989	7.6215	7.7391	7.8523
50	6.6481	6.8208	6.9778	7.1235	7.2604	7.3901	7.5137	7.6321	7.7459
60	6.5431	6.7216	6.8824	7.0306	7.1692	7.3002	7.4248	7.5439	7.6583
70	6.4501	6.6351	6.7997	6.9505	7.0909	7.2232	7.3488	7.4687	7.5837
80	6.3657	6.5577	6.7264	6.8798	7.0221	7.1557	7.2823	7.403	7.5186
90	6.2875	6.4871	6.6601	6.8163	6.9605	7.0955	7.2231	7.3446	7.4608
100	6.2139	6.4217	6.5993	6.7584	6.9045	7.0409	7.1696	7.2918	7.4087
150	5.8817	6.1433	6.3479	6.523	6.6797	6.8235	6.9576	7.0839	7.2039
200	5.5525	5.9041	6.1445	6.339	6.5077	6.6596	6.7994	6.9301	7.0534
250	5.1401	5.6755	5.9642	6.1816	6.3638	6.5246	6.6706	6.8057	6.9324
300	4.4756	5.4419	5.7956	6.0403	6.2374	6.4077	6.5602	6.7	6.8303
350	4.2139	5.1945	5.6331	5.9093	6.1229	6.3032	6.4625	6.6072	6.7411
400	4.1142	4.9446	5.4746	5.7859	6.017	6.2079	6.3743	6.5239	6.6614
450	4.0506	4.7361	5.3209	5.6685	5.9179	6.1197	6.2932	6.4479	6.5891
500	4.0029	4.5892	5.1759	5.5566	5.8245	6.0372	6.218	6.3777	6.5226
600	3.9316	4.4134	4.9357	5.3519	5.6528	5.8867	6.0815	6.2512	6.4034
700	3.8777	4.308	4.7663	5.1786	5.5003	5.7522	5.96	6.139	6.2982
800	3.8338	4.2331	4.6475	5.0391	5.3674	5.6321	5.8509	6.0382	6.2039
900	3.7965	4.1748	4.5593	4.9289	5.254	5.5255	5.7526	5.947	6.1184
1000	3.7638	4.1267	4.49	4.8406	5.158	5.4316	5.664	5.8644	6.0405

Table A.8 Specific heat capacity c_p of water in $\frac{kJ}{kg\,K}$ 1/2

p in bar	ϑ in °C										
	0	25	50	75	100	125	150	200	250	300	350
0.01	4.2199	1.8732	1.8757	1.8823	1.8913	1.902	1.914	1.9405	1.9693	1.9996	2.0311
0.1	4.2199	4.1822	1.9272	1.9058	1.9057	1.9112	1.9201	1.9436	1.9711	2.0007	2.0318
1	4.2194	4.1819	4.1796	4.1915	2.0741	2.0107	1.9857	1.9757	1.9891	2.0121	2.0396
5	4.2174	4.1807	4.1786	4.1907	4.2157	4.2546	4.3102	2.1448	2.0783	2.0657	2.0753
10	4.215	4.1793	4.1775	4.1896	4.2146	4.2533	4.3086	2.4288	2.2116	2.1408	2.1231
20	4.21	4.1764	4.1752	4.1874	4.2123	4.2506	4.3053	4.4914	2.5602	2.3201	2.2301
30	4.2052	4.1736	4.1729	4.1853	4.21	4.248	4.3022	4.4856	3.0772	2.5431	2.3539
40	4.2003	4.1708	4.1706	4.1831	4.2078	4.2455	4.299	4.4799	4.8646	2.8199	2.4967
50	4.1955	4.168	4.1684	4.181	4.2055	4.2429	4.2959	4.4743	4.8511	3.1714	2.661
60	4.1908	4.1652	4.1661	4.1788	4.2033	4.2403	4.2927	4.4687	4.8379	3.6378	2.8504
70	4.1861	4.1625	4.1639	4.1767	4.2011	4.2378	4.2897	4.4632	4.825	4.2919	3.0704
80	4.1814	4.1597	4.1617	4.1746	4.1989	4.2353	4.2866	4.4578	4.8125	5.287	3.3288

(continued)

Table A.8 (continued)

p in bar	ϑ in °C										
	0	25	50	75	100	125	150	200	250	300	350
90	4.1768	4.157	4.1595	4.1726	4.1967	4.2328	4.2836	4.4525	4.8002	5.7305	3.637
100	4.1723	4.1543	4.1573	4.1705	4.1945	4.2303	4.2806	4.4472	4.7883	5.6816	4.0118
150	4.1501	4.1412	4.1466	4.1603	4.1838	4.2182	4.2659	4.4219	4.7325	5.476	8.7885
200	4.129	4.1285	4.1361	4.1503	4.1734	4.2064	4.2518	4.3982	4.6824	5.3168	8.1062
250	4.109	4.1162	4.126	4.1407	4.1633	4.1951	4.2383	4.3758	4.6371	5.1883	6.98
300	4.0899	4.1044	4.1162	4.1313	4.1534	4.1841	4.2252	4.3547	4.5959	5.0814	6.3935
350	4.0717	4.0931	4.1067	4.1221	4.1439	4.1734	4.2126	4.3347	4.5582	4.9906	6.0154
400	4.0544	4.0821	4.0974	4.1132	4.1346	4.163	4.2005	4.3158	4.5234	4.912	5.7424
450	4.038	4.0716	4.0884	4.1046	4.1255	4.153	4.1888	4.2977	4.4912	4.8431	5.5342
500	4.0225	4.0614	4.0797	4.0961	4.1167	4.1432	4.1774	4.2806	4.4613	4.7819	5.37
600	3.9937	4.0422	4.063	4.0798	4.0997	4.1245	4.1558	4.2485	4.4073	4.6775	5.1244
700	3.9678	4.0242	4.0472	4.0644	4.0836	4.1068	4.1355	4.2191	4.3598	4.5912	4.9456
800	3.9446	4.0076	4.0322	4.0497	4.0683	4.09	4.1164	4.1921	4.3174	4.5181	4.8076
900	3.924	3.9921	4.0181	4.0358	4.0537	4.074	4.0984	4.167	4.2794	4.4553	4.6966
1000	3.9057	3.9777	4.0048	4.0225	4.0397	4.0589	4.0813	4.1437	4.2449	4.4003	4.6048

Table A.9 Specific heat capacity c_p of water in $\frac{kJ}{kg\,K}$ 2/2

p in bar	ϑ in °C								
	400	450	500	550	600	650	700	750	800
0.01	2.0635	2.0968	2.1309	2.1656	2.2008	2.2362	2.2716	2.307	2.3423
0.1	2.0641	2.0972	2.1312	2.1659	2.201	2.2364	2.2718	2.3071	2.3424
1	2.0697	2.1015	2.1345	2.1685	2.2031	2.2381	2.2732	2.3083	2.3434
5	2.0952	2.1206	2.1494	2.1803	2.2126	2.2458	2.2795	2.3135	2.3477
10	2.1284	2.1451	2.1682	2.1951	2.2245	2.2555	2.2875	2.3201	2.3532
20	2.1997	2.1964	2.2069	2.2254	2.2486	2.275	2.3035	2.3333	2.3642
30	2.278	2.2508	2.2472	2.2564	2.2732	2.2948	2.3196	2.3467	2.3753
40	2.3642	2.3088	2.2892	2.2883	2.2982	2.3149	2.336	2.3601	2.3865
50	2.459	2.3705	2.333	2.3212	2.3238	2.3353	2.3525	2.3737	2.3978
60	2.5632	2.4364	2.3787	2.355	2.3499	2.3559	2.3692	2.3874	2.4092
70	2.6778	2.5067	2.4265	2.3899	2.3765	2.3769	2.3861	2.4012	2.4206
80	2.8037	2.5817	2.4765	2.4258	2.4038	2.3983	2.4032	2.4152	2.4322
90	2.9425	2.6617	2.5287	2.4629	2.4316	2.4199	2.4205	2.4293	2.4438
100	3.0958	2.747	2.5833	2.5011	2.46	2.442	2.438	2.4435	2.4555
150	4.1778	3.2687	2.896	2.7112	2.612	2.5575	2.5288	2.5166	2.5155
200	6.3601	4.0074	3.2845	2.9552	2.7812	2.6824	2.6251	2.593	2.5775
250	12.9977	5.086	3.7661	3.2354	2.9679	2.8168	2.7267	2.6725	2.6415
300	25.8263	6.6908	4.3597	3.5529	3.1713	2.9598	2.8331	2.7549	2.7072
350	11.6455	8.9762	5.0715	3.9073	3.3894	3.1103	2.9435	2.8396	2.7744
400	8.7042	10.9505	5.8745	4.2938	3.6194	3.2666	3.057	2.9259	2.8428
450	7.4742	10.8642	6.6881	4.7004	3.8571	3.4264	3.1723	3.0133	2.9118
500	6.7794	9.5666	7.309	5.1031	4.0973	3.5873	3.2881	3.1008	2.9813
600	5.9976	7.5398	7.5221	5.7534	4.5557	3.9007	3.5151	3.273	3.1193
700	5.5542	6.5095	6.9698	6.0369	4.9233	4.1825	3.7267	3.4357	3.2526
800	5.2618	5.9182	6.3751	5.9817	5.1372	4.4081	3.9135	3.5834	3.3765
900	5.0515	5.5326	5.9164	5.7779	5.2063	4.5583	4.0692	3.7127	3.4861
1000	4.8915	5.2584	5.576	5.5489	5.1706	4.6275	4.191	3.8235	3.5762

Appendix B
Selected Absolute Molar Enthalpies/Entropies

(See Tables B.1, B.2, B.3, B.4, B.5, B.6, B.7, B.8, B.9, B.10, B.11, B.12, B.13, B.14, B.15, B.16, B.17, B.18, B.19, B.20, B.21, B.22, B.23, B.24, B.25, B.26).[2]

Table B.1 Molar enthalpy and entropy of H_2 at $p_0 = 1$ bar

T in K	C_p in $\frac{J}{mol\,K}$	S_m in $\frac{J}{mol\,K}$	H_m in $\frac{kJ}{mol}$
0	0	0	−8.468
298.15	28.836	130.681	0.000
300	28.849	130.859	0.053
325	28.991	133.175	0.776
350	29.086	135.327	1.503
375	29.149	137.336	2.231
400	29.189	139.218	2.960
425	29.214	140.989	3.690
450	29.231	142.659	4.420
475	29.243	144.240	5.151
500	29.254	145.740	5.883
600	29.318	151.079	8.811
700	29.444	155.607	11.748
800	29.629	159.550	14.701
900	29.873	163.053	17.676
1000	30.206	166.217	20.679
1200	30.983	171.789	26.794
1400	31.866	176.631	33.079
1600	32.732	180.943	39.539
1800	33.539	184.846	46.167
2000	34.276	188.418	52.950
3000	37.078	202.888	88.731
4000	39.087	213.840	126.848
5000	40.793	222.750	166.812

---

[2]The listed data has been taken from NASA Thermo Build, see [30].

© Springer Nature Switzerland AG 2019
A. Schmidt, *Technical Thermodynamics for Engineers*,
https://doi.org/10.1007/978-3-030-20397-9

Table B.2 Molar enthalpy and entropy of H at $p_0 = 1$ bar

T in K	C_p in $\frac{J}{mol\,K}$	S_m in $\frac{J}{mol\,K}$	H_m in $\frac{kJ}{mol}$
0	0	0	211.801
298.15	20.786	114.718	217.999
300	20.786	114.846	218.037
325	20.786	116.510	218.557
350	20.786	118.051	219.077
375	20.786	119.485	219.596
400	20.786	120.826	220.116
425	20.786	122.086	220.636
450	20.786	123.275	221.155
475	20.786	124.398	221.675
500	20.786	125.465	222.195
600	20.786	129.254	224.273
700	20.786	132.459	226.352
800	20.786	135.234	228.430
900	20.786	137.682	230.509
1000	20.786	139.873	232.588
1200	20.786	143.662	236.745
1400	20.786	146.867	240.902
1600	20.786	149.642	245.059
1800	20.786	152.090	249.217
2000	20.786	154.280	253.374
3000	20.786	162.709	274.160
4000	20.786	168.688	294.947
5000	20.786	173.327	315.733

Table B.3 Molar enthalpy and entropy of O_2 at $p_0 = 1$ bar

T in K	C_p in $\frac{J}{mol\,K}$	S_m in $\frac{J}{mol\,K}$	H_m in $\frac{kJ}{mol}$
0	0	0	−8.680
298.15	29.378	205.149	0.000
300	29.388	205.331	0.054
325	29.529	207.689	0.791
350	29.701	209.883	1.531
375	29.898	211.939	2.276
400	30.115	213.875	3.026
425	30.347	215.708	3.782
450	30.589	217.449	4.544
475	30.839	219.110	5.311
500	31.092	220.698	6.086
600	32.090	226.456	9.245
700	32.990	231.472	12.500
800	33.745	235.928	15.838
900	34.361	239.939	19.244
1000	34.883	243.587	22.707
1200	35.695	250.024	29.771
1400	36.288	255.573	36.971
1600	36.808	260.453	44.281
1800	37.302	264.817	51.693
2000	37.784	268.772	59.202
3000	39.980	284.521	98.117
4000	41.707	296.271	139.001
5000	42.997	305.723	181.385

Table B.4 Molar enthalpy and entropy of O at $p_0 = 1$ bar

T in K	C_p in $\frac{J}{mol\,K}$	S_m in $\frac{J}{mol\,K}$	H_m in $\frac{kJ}{mol}$
0	0	0	242.450
298.15	21.912	161.060	249.175
300	21.901	161.196	249.216
325	21.769	162.944	249.761
350	21.658	164.553	250.304
375	21.563	166.044	250.844
400	21.483	167.433	251.382
425	21.414	168.733	251.919
450	21.354	169.955	252.453
475	21.302	171.108	252.986
500	21.257	172.200	253.518
600	21.125	176.063	255.637
700	21.040	179.312	257.745
800	20.984	182.118	259.846
900	20.944	184.587	261.942
1000	20.915	186.792	264.035
1200	20.874	190.601	268.213
1400	20.855	193.817	272.386
1600	20.842	196.601	276.556
1800	20.831	199.055	280.723
2000	20.826	201.250	284.888
3000	20.939	209.706	305.749
4000	21.302	215.775	326.851
5000	21.799	220.581	348.397

Table B.5 Molar enthalpy and entropy of OH at $p_0 = 1$ bar

T in K	C_p in $\frac{J}{mol\,K}$	S_m in $\frac{J}{mol\,K}$	H_m in $\frac{kJ}{mol}$
0	0	0	28.465
298.15	29.886	183.740	37.278
300	29.879	183.924	37.333
325	29.789	186.312	38.079
350	29.714	188.517	38.823
375	29.653	190.565	39.565
400	29.603	192.477	40.306
425	29.563	194.271	41.045
450	29.532	195.960	41.784
475	29.510	197.556	42.522
500	29.495	199.069	43.260
600	29.514	204.446	46.209
700	29.656	209.005	49.167
800	29.913	212.981	52.144
900	30.267	216.523	55.152
1000	30.682	219.733	58.199
1200	31.597	225.406	64.425
1400	32.528	230.348	70.839
1600	33.374	234.748	77.431
1800	34.117	238.723	84.182
2000	34.765	242.351	91.071
3000	37.038	256.919	127.077
4000	38.536	267.790	164.899
5000	39.675	276.518	204.036

Table B.6 Molar enthalpy and entropy of $H_2O(liq)$. The dependency of the enthalpy on the pressure is supposed to be insignificant. Liquid water is treated as an incompressible fluid, i.e. the entropy does not need to be corrected!

T in K	C_p in $\frac{J}{mol\,K}$	S_m in $\frac{J}{mol\,K}$	H_m in $\frac{kJ}{mol}$
0	0	0	−299.108
298.15	75.351	69.942	−285.830
300	75.355	70.408	−285.691
325	75.316	76.438	−283.807
350	75.534	82.026	−281.922
375	75.963	87.251	−280.029
400	76.799	92.174	−278.121
425	77.835	96.862	−276.188
450	79.033	101.343	−274.228
475	80.982	105.664	−272.230
500	83.901	109.888	−270.171
600	125.409	127.516	−260.447

Table B.7 Molar enthalpy and entropy of $H_2O(g)$ at $p_0 = 1$ bar. Water vapour is treated as an ideal gas

T in K	C_p in $\frac{J}{mol\,K}$	S_m in $\frac{J}{mol\,K}$	H_m in $\frac{kJ}{mol}$
0	0	0	−251.730
298.15	33.588	188.829	−241.826
300	33.596	189.037	−241.764
325	33.724	191.731	−240.922
350	33.881	194.236	−240.077
375	34.062	196.579	−239.228
400	34.265	198.784	−238.374
425	34.486	200.868	−237.515
450	34.721	202.846	−236.650
475	34.968	204.729	−235.779
500	35.225	206.529	−234.901
600	36.324	213.047	−231.325
700	37.499	218.734	−227.634
800	38.728	223.821	−223.823
900	39.998	228.456	−219.887
1000	41.291	232.737	−215.823
1200	43.843	240.491	−207.308
1400	46.226	247.432	−198.297
1600	48.340	253.746	−188.836
1800	50.176	259.548	−178.980
2000	51.756	264.918	−168.783
3000	56.823	286.994	−114.168
4000	59.325	303.718	−55.974
5000	61.045	317.146	4.235

Table B.8 Molar enthalpy and entropy of N_2 at $p_0 = 1$ bar

T in K	C_p in $\frac{J}{mol\,K}$	S_m in $\frac{J}{mol\,K}$	H_m in $\frac{kJ}{mol}$
0	0	0	−8.670
298.15	29.124	191.610	0.000
300	29.125	191.790	0.054
325	29.140	194.122	0.782
350	29.165	196.282	1.511
375	29.200	198.295	2.241
400	29.249	200.181	2.971
425	29.311	201.956	3.703
450	29.388	203.634	4.437
475	29.478	205.225	5.173
500	29.582	206.740	5.911
600	30.109	212.177	8.894
700	30.754	216.866	11.937
800	31.434	221.017	15.046
900	32.090	224.758	18.223
1000	32.696	228.171	21.462
1200	33.724	234.227	28.109
1400	34.518	239.488	34.936
1600	35.127	244.138	41.903
1800	35.599	248.304	48.978
2000	35.970	252.075	56.136
3000	37.027	266.891	92.713
4000	37.548	277.621	130.022
5000	37.932	286.041	167.764

Table B.9 Molar enthalpy and entropy of N at $p_0 = 1$ bar

T in K	C_p in $\frac{J}{mol\,K}$	S_m in $\frac{J}{mol\,K}$	H_m in $\frac{kJ}{mol}$
0	0	0	466.483
298.15	20.786	153.302	472.680
300	20.786	153.431	472.718
325	20.786	155.094	473.238
350	20.786	156.635	473.758
375	20.786	158.069	474.277
400	20.786	159.410	474.797
425	20.786	160.671	475.317
450	20.786	161.859	475.836
475	20.786	162.983	476.356
500	20.786	164.049	476.876
600	20.786	167.839	478.954
700	20.786	171.043	481.033
800	20.786	173.818	483.112
900	20.786	176.267	485.190
1000	20.786	178.457	487.269
1200	20.780	182.246	491.425
1400	20.787	185.449	495.582
1600	20.792	188.226	499.740
1800	20.792	190.674	503.898
2000	20.791	192.865	508.056
3000	20.964	201.313	528.896
4000	21.810	207.440	550.214
5000	23.459	212.471	572.790

Table B.10 Molar enthalpy and entropy of NO at $p_0 = 1$ bar

T in K	C_p in $\frac{J}{mol\,K}$	S_m in $\frac{J}{mol\,K}$	H_m in $\frac{kJ}{mol}$
0	0	0	82.092
298.15	29.862	210.748	91.271
300	29.858	210.933	91.327
325	29.827	213.321	92.073
350	29.835	215.532	92.818
375	29.880	217.591	93.565
400	29.957	219.522	94.312
425	30.060	221.341	95.063
450	30.187	223.063	95.816
475	30.333	224.699	96.572
500	30.494	226.259	97.332
600	31.240	231.883	100.418
700	32.032	236.758	103.582
800	32.774	241.084	106.823
900	33.423	244.983	110.133
1000	33.991	248.534	113.505
1200	34.886	254.815	120.397
1400	35.533	260.244	127.443
1600	36.014	265.022	134.599
1800	36.383	269.286	141.841
2000	36.674	273.135	149.148
3000	37.540	288.191	186.310
4000	38.063	299.065	224.121
5000	38.607	307.614	262.444

Table B.11 Molar enthalpy and entropy of NO_2 at $p_0 = 1$ bar

T in K	C_p in $\frac{J}{mol\,K}$	S_m in $\frac{J}{mol\,K}$	H_m in $\frac{kJ}{mol}$
0	0	0	23.985
298.15	37.177	240.171	34.193
300	37.235	240.401	34.262
325	38.036	243.413	35.203
350	38.858	246.262	36.164
375	39.687	248.971	37.146
400	40.514	251.559	38.148
425	41.330	254.039	39.171
450	42.130	256.425	40.215
475	42.907	258.723	41.278
500	43.659	260.943	42.360
600	46.374	269.151	46.865
700	48.603	276.473	51.618
800	50.392	283.084	56.571
900	51.823	289.105	61.685
1000	52.982	294.627	66.927
1200	54.703	304.449	77.706
1400	55.907	312.977	88.774
1600	56.800	320.503	100.049
1800	57.512	327.236	111.482
2000	58.126	333.328	123.047
3000	60.991	357.431	182.589
4000	64.283	375.418	245.192
5000	67.738	390.135	311.209

Table B.12 Molar enthalpy and entropy of CO at $p_0 = 1$ bar

T in K	C_p in $\frac{J}{mol\,K}$	S_m in $\frac{J}{mol\,K}$	H_m in $\frac{kJ}{mol}$
0	0	0	−119.206
298.15	29.141	197.660	−110.535
300	29.143	197.840	−110.481
325	29.170	200.174	−109.752
350	29.211	202.337	−109.023
375	29.268	204.354	−108.292
400	29.341	206.245	−107.559
425	29.431	208.027	−106.825
450	29.538	209.712	−106.087
475	29.659	211.312	−105.348
500	29.794	212.837	−104.604
600	30.438	218.324	−101.594
700	31.171	223.070	−98.514
800	31.900	227.280	−95.360
900	32.572	231.077	−92.136
1000	33.179	234.541	−88.848
1200	34.170	240.682	−82.108
1400	34.914	246.008	−75.196
1600	35.475	250.708	−68.155
1800	35.905	254.913	−61.015
2000	36.242	258.714	−53.799
3000	37.205	273.619	−17.006
4000	37.698	284.395	20.464
5000	38.063	292.847	58.348

Table B.13 Molar enthalpy and entropy of CO_2 at $p_0 = 1$ bar

T in K	C_p in $\frac{J}{mol\,K}$	S_m in $\frac{J}{mol\,K}$	H_m in $\frac{kJ}{mol}$
0	0	0	−402.875
298.15	37.135	213.787	−393.510
300	37.220	214.017	−393.441
325	38.333	217.041	−392.497
350	39.387	219.921	−391.525
375	40.383	222.672	−390.528
400	41.325	225.309	−389.506
425	42.217	227.841	−388.462
450	43.062	230.278	−387.396
475	43.863	232.628	−386.309
500	44.624	234.898	−385.203
600	47.323	243.280	−380.601
700	49.561	250.748	−375.754
800	51.432	257.492	−370.701
900	52.999	263.643	−365.478
1000	54.309	269.297	−360.110
1200	56.347	279.390	−349.032
1400	57.798	288.191	−337.610
1600	58.873	295.982	−325.938
1800	59.693	302.966	−314.078
2000	60.335	309.290	−302.072
3000	62.156	334.152	−240.694
4000	63.204	352.180	−178.002
5000	64.505	366.412	−114.200

Table B.14 Molar enthalpy and entropy of $C_{Graphite}$ at $p_0 = 1$ bar

T in K	C_p in $\frac{J}{mol\,K}$	S_m in $\frac{J}{mol\,K}$	H_m in $\frac{kJ}{mol}$
0	0	0	−1.054
298.15	8.528	5.734	0.000
300	8.592	5.787	0.016
325	9.436	6.508	0.241
350	10.256	7.238	0.487
375	11.052	7.972	0.754
400	11.824	8.711	1.040
425	12.570	9.450	1.345
450	13.286	10.189	1.668
475	13.969	10.926	2.009
500	14.617	11.659	2.366
600	16.835	14.529	3.944
700	18.534	17.257	5.716
800	19.828	19.820	7.637
900	20.826	22.216	9.672
1000	21.612	24.452	11.795
1200	22.762	28.501	16.240
1400	23.572	32.074	20.878
1600	24.185	35.263	25.656
1800	24.676	38.141	30.544
2000	25.089	40.763	35.521
3000	26.609	51.244	61.421
4000	27.794	59.065	88.635
5000	28.890	65.385	116.980

Table B.15 Molar enthalpy and entropy of S at $p_0 = 1$ bar

T in K	C_p in $\frac{J}{mol\,K}$	S_m in $\frac{J}{mol\,K}$	H_m in $\frac{kJ}{mol}$
0	0	0	270.513
298.15	23.674	167.832	277.170
300	23.669	167.978	277.214
325	23.584	169.870	277.805
350	23.477	171.614	278.393
375	23.358	173.229	278.978
400	23.232	174.733	279.561
425	23.105	176.138	280.140
450	22.980	177.455	280.716
475	22.858	178.694	281.289
500	22.742	179.863	281.859
600	22.338	183.972	284.112
700	22.031	187.392	286.330
800	21.801	190.318	288.521
900	21.624	192.875	290.692
1000	21.490	195.146	292.847
1200	21.316	199.048	297.127
1400	21.205	202.324	301.378
1600	21.170	205.153	305.614
1800	21.200	207.647	309.850
2000	21.278	209.885	314.097
3000	21.997	218.639	335.712
4000	22.682	225.067	358.072
5000	23.080	230.176	380.976

Table B.16 Molar enthalpy and entropy of S_2 at $p_0 = 1$ bar

T in K	C_p in $\frac{J}{mol\,K}$	S_m in $\frac{J}{mol\,K}$	H_m in $\frac{kJ}{mol}$
0	0	0	119.468
298.15	32.505	228.166	128.600
300	32.541	228.368	128.660
325	32.993	230.990	129.479
350	33.404	233.451	130.309
375	33.774	235.768	131.149
400	34.107	237.959	131.998
425	34.406	240.036	132.854
450	34.675	242.010	133.718
475	34.916	243.891	134.588
500	35.133	245.688	135.464
600	35.816	252.158	139.013
700	36.305	257.717	142.620
800	36.697	262.591	146.271
900	37.046	266.934	149.958
1000	37.377	270.855	153.680
1200	38.022	277.725	161.219
1400	38.675	283.636	168.889
1600	39.279	288.840	176.685
1800	39.813	293.498	184.596
2000	40.271	297.717	192.605
3000	41.691	314.353	233.682
4000	42.603	326.470	275.823
5000	43.849	336.102	319.010

Table B.17 Molar enthalpy and entropy of SO at $p_0 = 1$ bar

T in K	C_p in $\frac{J}{mol\,K}$	S_m in $\frac{J}{mol\,K}$	H_m in $\frac{kJ}{mol}$
0	0	0	−4.038
298.15	30.176	221.941	4.760
300	30.200	222.128	4.816
325	30.528	224.558	5.575
350	30.870	226.833	6.343
375	31.216	228.975	7.119
400	31.560	231.000	7.903
425	31.896	232.924	8.697
450	32.220	234.756	9.498
475	32.531	236.507	10.308
500	32.826	238.183	11.125
600	33.842	244.261	14.460
700	34.618	249.539	17.885
800	35.213	254.202	21.378
900	35.685	258.378	24.923
1000	36.066	262.158	28.512
1200	36.695	268.793	35.791
1400	37.222	274.489	43.183
1600	37.741	279.493	50.680
1800	38.260	283.969	58.280
2000	38.769	288.026	65.983
3000	40.848	304.169	105.866
4000	42.139	316.111	147.408
5000	43.123	325.621	190.044

Table B.18 Molar enthalpy and entropy of SO_2 at $p_0 = 1$ bar

T in K	C_p in $\frac{J}{mol\,K}$	S_m in $\frac{J}{mol\,K}$	H_m in $\frac{kJ}{mol}$
0	0	0	−307.358
298.15	39.842	248.222	−296.810
300	39.909	248.468	−296.736
325	40.813	251.699	−295.727
350	41.707	254.756	−294.696
375	42.583	257.664	−293.642
400	43.432	260.439	−292.567
425	44.251	263.097	−291.471
450	45.035	265.649	−290.354
475	45.781	268.104	−289.219
500	46.489	270.470	−288.066
600	48.938	279.172	−283.289
700	50.831	286.864	−278.297
800	52.282	293.751	−273.138
900	53.407	299.976	−267.851
1000	54.291	305.651	−262.464
1200	55.565	315.670	−251.470
1400	56.425	324.304	−240.266
1600	57.044	331.881	−228.916
1800	57.516	338.628	−217.458
2000	57.892	344.708	−205.916
3000	59.151	368.441	−147.341
4000	60.144	385.594	−87.692
5000	61.207	399.127	−27.028

Table B.19 Molar enthalpy and entropy of CH_4 at $p_0 = 1$ bar

T in K	C_p in $\frac{J}{mol\,K}$	S_m in $\frac{J}{mol\,K}$	H_m in $\frac{kJ}{mol}$
0	0	0	−84.616
298.15	35.691	186.371	−74.600
300	35.760	186.592	−74.534
325	36.785	189.494	−73.627
350	37.956	192.262	−72.693
375	39.243	194.924	−71.729
400	40.618	197.500	−70.731
425	42.057	200.005	−69.697
450	43.542	202.451	−68.627
475	45.055	204.846	−67.520
500	46.585	207.196	−66.374
600	52.691	216.231	−61.410
700	58.543	224.797	−55.845
800	64.013	232.976	−49.714
900	69.066	240.811	−43.056
1000	73.676	248.331	−35.915
1200	81.628	262.490	−20.359
1400	88.098	275.577	−3.364
1600	93.373	287.696	14.801
1800	97.744	298.954	33.926
2000	101.443	309.449	53.854
3000	114.634	353.281	162.398
4000	124.805	387.673	282.189
5000	134.990	416.604	412.030

Table B.20 Molar enthalpy and entropy of C_2H_6 at $p_0 = 1$ bar

T in K	C_p in $\frac{J}{mol\,K}$	S_m in $\frac{J}{mol\,K}$	H_m in $\frac{kJ}{mol}$
0	0	0	−95.743
298.15	52.501	229.221	−83.852
300	52.725	229.547	−83.754
325	55.825	233.889	−82.398
350	59.010	238.142	−80.962
375	62.233	242.323	−79.447
400	65.458	246.442	−77.851
425	68.657	250.507	−76.174
450	71.808	254.521	−74.418
475	74.898	258.486	−72.584
500	77.916	262.405	−70.674
600	89.173	277.626	−62.308
700	99.121	292.135	−52.883
800	107.914	305.957	−42.523
900	115.707	319.126	−31.334
1000	122.540	331.680	−19.413
1200	133.797	355.061	6.270
1400	142.398	376.362	33.928
1600	148.971	395.825	63.094
1800	154.040	413.677	93.417
2000	157.995	430.120	124.636
3000	168.650	496.566	288.869
4000	172.877	545.742	459.909
5000	174.937	584.564	633.926

Table B.21 Molar enthalpy and entropy of C_3H_8 at $p_0 = 1$ bar

T in K	C_p in $\frac{J}{mol\,K}$	S_m in $\frac{J}{mol\,K}$	H_m in $\frac{kJ}{mol}$
0	0	0	−119.421
298.15	73.589	270.315	−104.680
300	73.956	270.771	−104.544
325	78.963	276.888	−102.632
350	84.010	282.925	−100.595
375	89.034	288.892	−98.432
400	93.989	294.797	−96.144
425	98.838	300.641	−93.733
450	103.561	306.425	−91.203
475	108.140	312.148	−88.556
500	112.568	317.808	−85.797
600	128.738	339.797	−73.712
700	142.656	360.714	−60.125
800	154.758	380.571	−45.241
900	165.385	399.426	−29.222
1000	174.614	417.341	−12.210
1200	189.732	450.576	24.293
1400	201.220	480.726	63.440
1600	209.971	508.193	104.598
1800	216.708	533.329	147.295
2000	221.956	556.445	191.183
3000	236.070	649.600	421.404
4000	241.660	718.385	660.636
5000	244.383	772.635	903.804

Table B.22 Molar Enthalpy and Entropy of C_2H_5OH at $p_0 = 1$ bar

T in K	C_p in $\frac{J}{mol\,K}$	S_m in $\frac{J}{mol\,K}$	H_m in $\frac{kJ}{mol}$
0	0	0	−249.492
298.15	65.309	280.593	−234.950
300	65.593	280.998	−234.829
325	69.490	286.401	−233.141
350	73.441	291.695	−231.354
375	77.386	296.897	−229.469
400	81.279	302.016	−227.485
425	85.088	307.058	−225.405
450	88.790	312.027	−223.232
475	92.370	316.924	−220.967
500	95.820	321.751	−218.614
600	108.294	340.355	−208.391
700	118.855	357.863	−197.019
800	127.925	374.341	−184.670
900	135.841	389.875	−171.473
1000	142.690	404.552	−157.537
1200	153.923	431.605	−127.826
1400	162.515	456.008	−96.144
1600	169.103	478.158	−62.953
1800	174.204	498.382	−28.601
2000	178.202	516.952	6.655
3000	189.077	591.635	191.210
4000	193.434	646.708	382.750
5000	195.567	690.125	577.363

Table B.23 Molar enthalpy and entropy of $C_2H_5OH(liq)$. The dependency of the enthalpy on the pressure is supposed to be insignificant. Liquid methanol is treated as an incompressible fluid, i.e. the entropy does not need to be corrected!

T in K	C_p in $\frac{J}{mol\,K}$	S_m in $\frac{J}{mol\,K}$	H_m in $\frac{kJ}{mol}$
0	0	0	−301.592
298.15	112.250	160.100	−277.510
300	112.978	160.797	−277.302
325	124.074	170.262	−274.343
350	137.506	179.934	−271.078
375	153.212	189.944	−267.449
400	171.054	200.391	−263.400
425	190.834	211.345	−258.880
450	212.304	222.853	−253.844

Table B.24 Molar enthalpy and entropy of CH_3OH at $p_0 = 1$ bar

T in K	C_p in $\frac{J}{mol\,K}$	S_m in $\frac{J}{mol\,K}$	H_m in $\frac{kJ}{mol}$
0	0	0	−212.375
298.15	44.039	239.810	−200.940
300	44.157	240.083	−200.858
325	45.849	243.683	−199.734
350	47.687	247.147	−198.565
375	49.624	250.503	−197.349
400	51.621	253.769	−196.083
425	53.647	256.959	−194.767
450	55.679	260.083	−193.401
475	57.695	263.148	−191.983
500	59.683	266.158	−190.516
600	67.192	277.715	−184.166
700	73.864	288.585	−177.106
800	79.750	298.840	−169.419
900	84.960	308.540	−161.178
1000	89.539	317.734	−152.448
1200	97.122	334.758	−133.751
1400	102.991	350.190	−113.714
1600	107.525	364.251	−92.643
1800	111.057	377.128	−70.771
2000	113.837	388.979	−48.270
3000	121.459	436.835	70.012
4000	124.534	472.255	193.208
5000	126.046	500.224	318.578

Table B.25 Molar enthalpy and entropy of CH_3OH(liq). The dependency of the enthalpy on the pressure is supposed to be insignificant. Liquid methanol is treated as an incompressible fluid, i.e. the entropy does not need to be corrected!

T in K	C_p in $\frac{J}{mol\,K}$	S_m in $\frac{J}{mol\,K}$	H_m in $\frac{kJ}{mol}$
0	0	0	−257.905
298.15	81.080	127.270	−238.910
300	81.431	127.773	−238.760
325	86.840	134.496	−236.659
350	93.563	141.169	−234.407
375	101.637	147.892	−231.969
400	110.969	154.743	−229.314
425	121.290	161.776	−226.413
450	132.122	169.015	−223.246

Table B.26 Molar enthalpy and entropy of air at $p_0 = 1$ bar

T in K	C_p in $\frac{J}{mol\,K}$	S_m in $\frac{J}{mol\,K}$	H_m in $\frac{kJ}{mol}$
0	0	0	−8.775
298.15	29.102	198.822	−0.126
300	29.105	199.002	−0.072
325	29.146	201.334	0.656
350	29.202	203.496	1.386
375	29.271	205.513	2.117
400	29.355	207.404	2.849
425	29.452	209.187	3.584
450	29.563	210.873	4.322
475	29.686	212.475	5.063
500	29.821	214.001	5.807
600	30.442	219.491	8.819
700	31.135	224.235	11.897
800	31.825	228.438	15.046
900	32.467	232.224	18.261
1000	33.050	235.676	21.537
1200	34.023	241.792	28.249
1400	34.767	247.095	35.131
1600	35.352	251.777	42.145
1800	35.825	255.969	49.264
2000	36.216	259.764	56.470
3000	37.502	274.719	93.396
4000	38.270	285.620	131.308
5000	38.841	294.223	169.872

Appendix C
Caloric State Diagrams of Water

(See Figs. C.1, C.2, C.3).

© Springer Nature Switzerland AG 2019
A. Schmidt, *Technical Thermodynamics for Engineers*,
https://doi.org/10.1007/978-3-030-20397-9

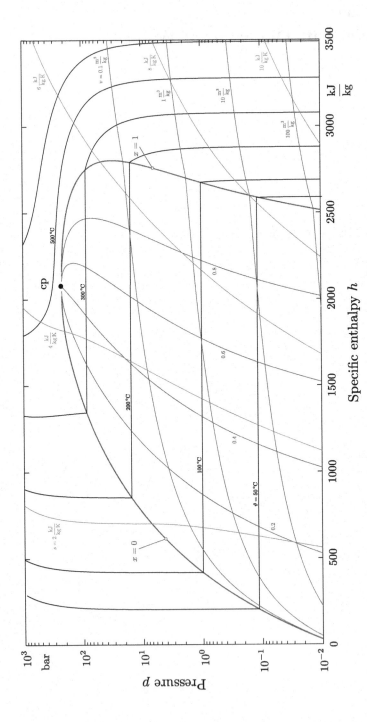

Fig. C.1 log p, h-diagram of water

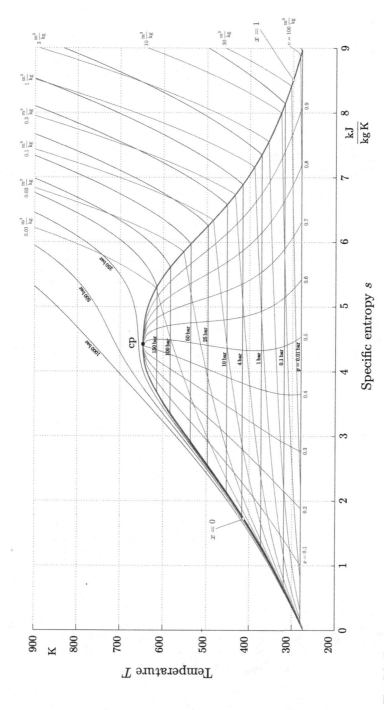

Fig. C.2 T, s-diagram of water

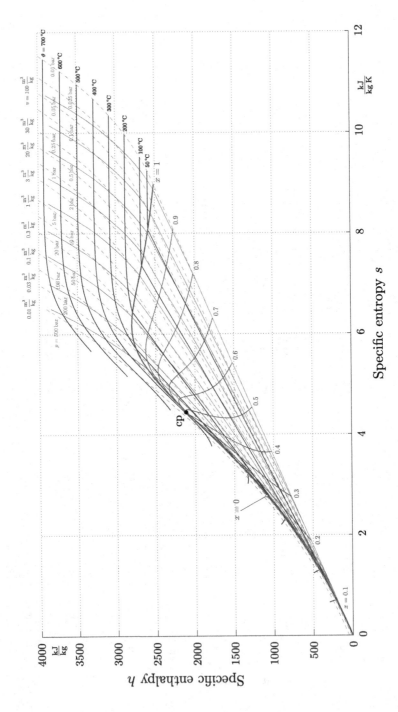

Fig. C.3 h, s-diagram of water

Appendix D
The h_{1+x}, x-Diagram

(See Fig. D.1).

© Springer Nature Switzerland AG 2019
A. Schmidt, *Technical Thermodynamics for Engineers*,
https://doi.org/10.1007/978-3-030-20397-9

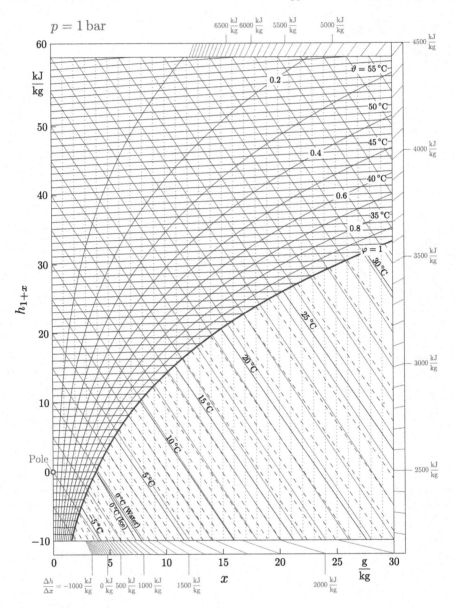

Fig. D.1 h_{1+x}, x-diagram (atmospheric air + water)

References

1. Hahne, E.: Technische Thermodynamik: Einführung und Anwendung. Oldenbourg Verlag, München (2010)
2. Cerbe, G., Wilhelms, G.: Technische Thermodynamik: Theoretische Grundlagen und praktische Anwendungen. Hanser Verlag, Berlin (2013)
3. Doering, E., Schedwill, H., Dehli, M.: Grundlagen der Technischen Thermodynamik - Lehrbuch für Studierende der Ingenieurwissenschaften. Vieweg+Teubner Verlag, Berlin (2012)
4. Baehr, H.D., Kabelac, S.: Thermodynamik: Grundlagen und technische Anwendungen. Springer Verlag, Wiesbaden (2012)
5. Brandt, F.: Brennstoffe und Verbrennungsrechnung. Vulkan Verlag GmbH, Essen (2000)
6. Renz, U.: Feuerungstechnik: Vorlesungsskript. Lehrstuhl für Wärmeübertragung und Klimatechnik, RWTH Aachen (2000)
7. Grothe, K.H., Feldhusen, J.: Dubbel. Taschenbuch für den Maschinenbau. Springer Verlag, Berlin (2014)
8. Bošnjaković, F., Knoche, K.F.: Technische Thermodynamik Teil 1. Steinkopff Verlag, Darmstadt (1988)
9. Watter, H.: Nachhaltige Energiesysteme: Grundlagen. Systemtechnik und Anwendungsbeispiele aus der Praxis. Vieweg+Teubner Verlag, Wiesbaden (2009)
10. Freymann, R., Strobl, W., Obieglo, A.: The turbosteamer: a system introducing the principle of cogeneration in automotive applications. MTZ Worldwide **69**, 20–27 (2008)
11. Kuchling, H.: Taschenbuch der Physik. Carl Hanser Verlag GmbH & Co. KG, München (2010)
12. Schmidt, A.: Untersuchungen zum Wärmeübergang in blasenbildenden Wirbelschichten. RWTH Aachen: Dissertation (2002)
13. Schugger, C.: Experimentelle Untersuchung des primären Strahlzerfalls bei der motorischen Hochdruckeinspritzung. RWTH Aachen: Dissertation (2007)
14. Geller, W.: Thermodynamik für Maschinenbauer: Grundlagen für die Praxis. Springer Verlag, Berlin (2006)
15. Stephan, P., Schaber, K., Stephan, K., Mayinger, F.: Thermodynamik - Grundlagen und technische Anwendungen (Band 2: Mehrstoffsysteme und chemische Reaktionen). Springer Verlag, Berlin (2010)
16. Rauschnabel, K.: Entropie und 2. Hauptsatz der Thermodynamik. http://mitarbeiter.hs-heilbronn.de/~rauschn/5_Thermodynamik/Physik_5_7_Entropie.pdf/ (2011). Accessed 08 Feb 2011
17. Heintz, A.: Gleichgewichtsthermodynamik - Grundlagen und einfache Anwendungen. Springer Verlag, Berlin (2011)

© Springer Nature Switzerland AG 2019
A. Schmidt, *Technical Thermodynamics for Engineers*,
https://doi.org/10.1007/978-3-030-20397-9

18. Sommerfeld, A.: Thermodynamik und Statistik, Nachdruck der 2. Auflage 1962/1977. Frankfurt am Main: Verlag Harri Deutsch (2011)
19. Chemieingenieurwesen, VDI-Gesellschaft V.: VDI-Wärmeatlas. Springer Verlag, Berlin (2006)
20. Holmgren, M.: X Steam, Thermodynamic properties of water and steam. https://de.mathworks.com/matlabcentral/fileexchange/9817-x-steam--thermodynamic-properties-of-water-and-steam (2018). Accessed 01 June 2018
21. Kraus, H.: Die Atmosphäre der Erde. Vieweg Verlag, Braunschweig/Wiesbaden (2000)
22. Stierstadt, K., Fischer, G.: Thermodynamik: Von der Mikrophysik zur Makrophysik. Springer Verlag, Berlin (2010)
23. Schröder, W.: Fluidmechanik. Verlag Mainz, Aachen (2018)
24. Ganzer, U.: Gasdynamik. Springer Verlag, Berlin (1988)
25. Müller, R.: Luftstrahltriebwerke. Grundlagen. Charakteristiken Arbeitsverhalten. Springer Verlag, Berlin (1997)
26. Knoche, K.F.: Technische Thermodynamik. Vieweg, Braunschweig/Wiesbaden (1992)
27. Warnatz, J., Maas, U., Dibble, R.W.: Verbrennung - Physikalisch-Chemische Grundlagen, Modellierung und Simulation. Experimente. Schadstoffentstehung. Springer Verlag, Berlin (2001)
28. Joos, F.: Technische Verbrennung: Verbrennungstechnik. Verbrennungsmodellierung. Emissionen. Springer Verlag, Berlin (2006)
29. Lucas, K.: Thermodynamik: Die Grundgesetze der Energie- und Stoffumwandlungen. Springer Verlag, Berlin (2008)
30. McBride, B.J., Zehe, M.J., Gordon, S.: NASA glenn coecients for calculating thermodynamic properties of individual species. In: Glenn Research Center, NASA/TP-2002-211556 (2002)

Index

© Springer Nature Switzerland AG 2019
A. Schmidt, *Technical Thermodynamics for Engineers*,
https://doi.org/10.1007/978-3-030-20397-9

Printed in the United States
by Baker & Taylor

Printed in the United States
By Bookmasters